91

Helicopter Performance, Stability, and Control

Helicopter Performance, Stability, and Control

RAYMOND W. PROUTY

Robert E. Krieger Publishing Company
Malabar, Florida
1990

Original Edition 1986
Reprint Edition 1990 w/corr.

Printed and Published by
ROBERT E. KRIEGER PUBLISHING COMPANY, INC.
KRIEGER DRIVE
MALABAR, FLORIDA 32950

Library of Congress Cataloging-in-Publication Data
Prouty, Raymond W.
    Helicopter performance, stability, and control / Raymond W.
Prouty.
        p.   cm.
    Reprint, with corrections. Originally published: Boston : PWS
Engineering, c1986.
    ISBN 0-89464-457-2 (alk. paper)
    1. Helicopters--Aerodynamics. 2. Stability of helicopters.
I. Title.
TL716.P725 1990
629.133'352--dc20   90-4119
                                                 CIP

10   9   8   7   6   5   4   3   2

To my wife, Joyce, who has been wondering what I have been up to; and to my boss, Ken Amer, who always knew.

Preface

The purpose of this book is to provide both the student and the practicing helicopter aerodynamicist with the information they need to analyze the performance of an existing helicopter or to participate in the design of a new helicopter. This text should also be helpful for those in the helicopter industry who work with aerodynamicists and may have been baffled by what seemed to be an art rather than a science. The information presented here includes the derivation of the theory behind the various methods of analysis, appropriate experimental data to correlate and supplement the theory, and charts that permit rapid analysis. A special attempt is made to relate helicopter aerodynamics to airplane aerodynamics for those who are making the transition.

To focus on the practical aspects of the subject, an "Example Helicopter" is defined in Appendix A and is consistently used throughout the book to illustrate, by numerical calculations, the application of the analysis. These calculations are listed at the end of each chapter, along with a listing of "How To's," which can be used to shorten the search for a specific subject. Although no problems are presented at the ends of chapters, they can easily be generated using the example helicopter calculations as examples and the helicopter parameters selected by the instructor or the student. The generation of problems involving extension of the theory, wherein the proof is left to the ambitious student, is left to the ambitious instructor.

The first six chapters are devoted to the various aspects of helicopter performance. Chapters 7, 8, and 9 cover stability and control. The final chapter presents the tradeoff considerations that the engineer must face during the preliminary design phase to ensure both good performance and good flying qualities.

Aeroelastic rotor effects, including vibration, oscillating loads, and the stability of nonconventional rotors, are not addressed. For an understanding of these mysterious effects, other sources must be consulted.

The material includes almost all of the information and methods that I have found to be significant as a working aerodynamicist for Hughes, Sikorsky, Bell, and Lockheed since 1952. The outline of the book was first developed as lecture notes used in teaching engineering extension courses at the University of California at Los Angeles.

Contents

Helicopter Performance, Stability, and Control

CHAPTER 1

Aerodynamics of Hovering Flight

MOMENTUM METHOD

Basis of the Theory

Like any physical system, the hovering helicopter must obey the basic laws of physics. One of these laws was stated by Newton: "For every action, there is an equal and opposite reaction." In the case of the helicopter in hovering flight, the action is the development of a rotor thrust equal to the gross weight. The reaction

---

In the subsequent text, there are a number of times when some light can be shed on the theory by using specific numerical examples. For this purpose, a typical helicopter has been defined and designated *the example helicopter*. The characteristics of this aircraft are listed in Appendix A.

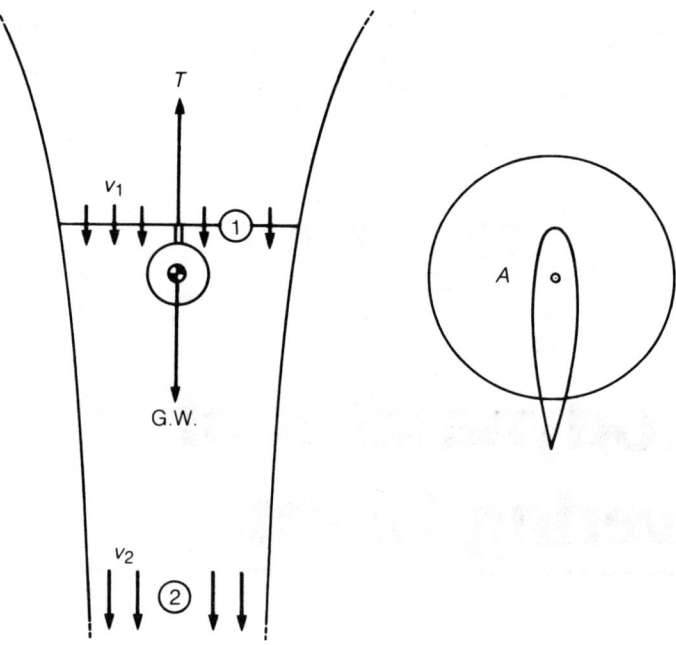

FIGURE 1.1   Induced Velocities in the Vicinity of a Hovering Rotor

is represented by the acceleration of a mass of air from a stagnant condition far above the rotor to a condition with a finite velocity in the wake below the rotor, as shown in Figure 1.1. The conditions at the rotor are governed by the familiar relationship:

$$\text{Force} = (\text{Mass})\,(\text{acceleration})$$

For systems such as rotors that are accelerating a mass of air on a continuous basis, the equation may be written as:

$$\text{Rotor Thrust} = (\text{Mass flow per second})\,(\text{total change in flow velocity})$$

or:

$$T = (m/\text{sec})\ (\Delta v)$$

Induced Velocities

The elements of the thrust equation can be evaluated from a study of the flow field in Figure 1.1. In that figure, the three regions of interest have been numbered 0 for the region high above the rotor, 1 for the plane of the rotor, and 2 for the region far below the rotor in the fully developed rotor wake. The mass flow per second in the plane of the rotor is:

$$m_1/\sec = \rho\, v_1\, A \text{ slugs/sec}$$

where ρ is the density of the air in slugs/ft^3, v_1 is the induced velocity at the rotor plane, and A is the area of the rotor disc. The total change in velocity, Δv, between region 0 and region 2 is:

$$\Delta v = v_2 - v_0 \text{ ft/sec}$$

Since far above the rotor the air has no velocity, the total change in velocity is:

$$\Delta v = v_2 \text{ ft/sec}$$

Thus the expression for rotor thrust may be written:

$$T = \rho\, v_1 A v_2 \text{ lb}$$

The equation for thrust in this form is not very useful. Fortunately, a relationship between v_1 and v_2 can be derived by equating the rate of energy dissipated at the rotor to the rate of energy imparted to the wake. Since the rotor and its wake make up a closed system, these two rates of energy must be equal. The energy per second dissipated by the rotor, $E_R/\sec$, is:

$$E_R/\sec = \text{Force} \times \text{Velocity}$$

or

$$E_R/\sec = T v_1 \text{ ft lb/sec}$$

or

$$E_R/\sec = \rho v_1^2 A\, v_2 \text{ ft lb/sec}$$

The energy per second imparted to the wake, $E_W/\sec$, is the total change in kinetic energy. Since there is no kinetic energy far above the rotor, the total change is the value found in the remote wake:

$$E_W/\sec = \frac{1}{2}(m_2/\sec)\, v_2^2$$

The mass flow per second in the remote wake, m_2/sec, is the same as the mass flow at the rotor, m_1/sec. This is a consequence of the Law of Continuity, which also applies to such things as syrup pouring from a pitcher. Once the flow is established, the amount of syrup reaching the pancake per second is the same as that leaving the lip of the pitcher even though the cross-section of the flow is larger at the top than at the bottom. The energy in the wake thus becomes:

$$E_W/\sec = \frac{1}{2}(m_1/\sec)\ v_2^2 \text{ ft lb/sec}$$

or

$$E_W/\sec = \frac{1}{2}\rho v_1 A v_2^2 \text{ ft lb/sec}$$

Now equating the two rates of energy gives:

$$\rho v_1^2 A v_2 = \frac{1}{2}\rho v_1 A v_2^2$$

or

$$v_2 = 2v_1$$

That is, the induced velocity in the remote wake is twice the induced velocity at the rotor disc.

The equation for thrust becomes:

$$T = 2\rho\ v_1^2\ A \text{ lb}$$

and consequently the induced velocity in the plane of the rotor, v_1, is:

$$v_1 = \sqrt{\frac{T}{2\rho A}} \text{ ft/sec}$$

The rotor thrust, T, divided by the disc area, A, is the disc loading, D.L., in pounds per square foot. The equation for the induced velocity may be written:

$$v_1 = \sqrt{\frac{1}{2\rho}}\ \sqrt{\text{D.L.}}$$

For sea-level standard conditions, the density, ρ, is 0.002378 slugs per cubic foot, so that:

$$v_1 = 14.5\ \sqrt{\text{D.L.}} \text{ ft/sec}$$

and

$$v_2 = 29\ \sqrt{\text{D.L.}} \text{ ft/sec}$$

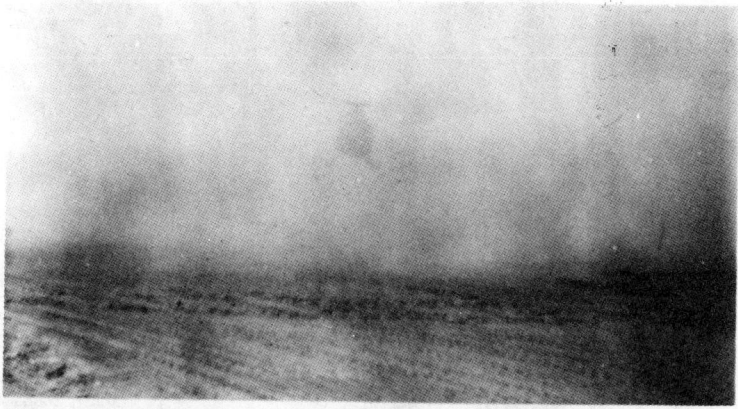

FIGURE 1.2   OH-6 Hovering in Dust

Source: Hill, "A Promising Concept for the Joint Development of Military Intra-Theater and Commercial Inter-Urban VTOL Transports," 24th AHS Forum, 1968.

The example helicopter described in Appendix A has a design gross weight of 20,000 lb and a rotor radius of 30 ft. Assuming that the rotor thrust is equal to the gross weight, the disc loading is 7.1 lb/ft², the velocity at the disc is 39 ft/sec, and the velocity in the remote wake is 78 ft/sec or 46 knots. For computations involving induced velocities produced by tandem-rotor helicopters in which one rotor overlaps the other, only the net projected area should be used in computing the disc loading. Similarly, with a coaxial helicopter, only the disc area of a single rotor should be used.

Effect of Induced Velocities

The induced velocity in the wake of a hovering helicopter can produce operational problems if the hovering is done close to dust, sand, snow, or other loose surfaces. The higher the disc loading, the stronger is the *placer mining* capability of the wake. Helicopter disc loadings range from 4 to 12 lb/ft², and the corresponding downwash velocities in the remote wake range from 35 to 58 knots. Even the lower velocities can lift enough dust or snow to effectively cut off the pilot's view of the ground, as is shown in Figure 1.2. At the higher disc loadings, gravel can be entrained in the wake and forced to circulate through the rotor and the engine intake.

A high rotor downwash velocity also makes it difficult to work under a hovering helicopter while hooking up a sling load or guiding the pilot to a precision landing.

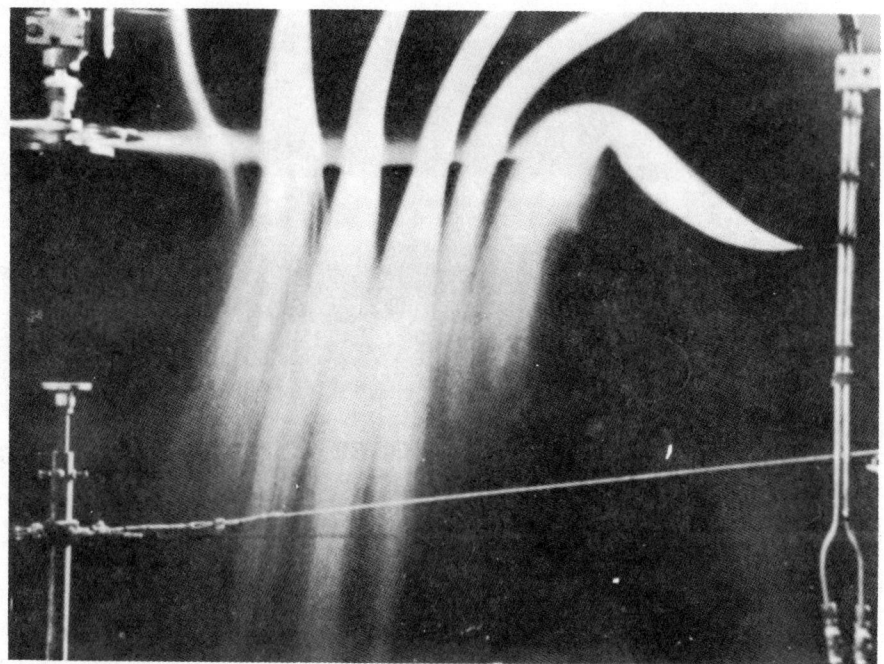

FIGURE 1.3   Smoke Study of Model Rotor in Hover

Source: Landgrebe, "An Analytical and Experimental Investigation of Helicopter Rotor
Hover Performance and Wake Geometry Characteristics," USAAMRDL TR 71-24, 1971.

Thus it may be seen that the higher the disc loading, the more severe are the
operational problems. For nonhelicopter hovering aircraft that use very high disc-
loading devices such as propellers or jet engines for lift, these problems become
severe enough to limit landings and takeoffs to hard-surfaced, prepared areas. In
light of these problems, one might ask: "Why use high disc loadings?" The answer
is that high disc loadings permit design of compact helicopters with low empty
weight, which for many applications are the most efficient aircraft.

Another effect of the rotor-induced velocities in hover is to produce a
download on the fuselage and any other aircraft components that are located under
the rotor. The rotor wake contracts from the diameter of the rotor to its remote
wake size in about a quarter of a rotor radius, as shown in Figure 1.3, which was
obtained using smoke for flow visualization. For most helicopters, the fuselage can
be considered to be immersed in the remote wake and to receive the full effect of
the downwash. The download, or vertical drag, on the fuselage can be computed
from the standard equation for drag:

$$D = C_D q S$$

The dynamic pressure in the remote wake, q_2, is equal to the disc loading of the rotor. This statement may be proved by using the relationships:

$$v_2 = 2v_1 = 2\sqrt{\frac{\text{D.L.}}{2\rho}}$$

and

$$q_2 = \frac{1}{2}\rho v_2^2$$

Thus

$$q_2 = \frac{1}{2}\rho\left(2\sqrt{\frac{\text{D.L.}}{2\rho}}\right)^2$$

or

$$q_2 = \text{D.L. lb/ft}^2$$

A first estimate of the download may be obtained by assuming an effective drag coefficient of 0.3 for all the aircraft components in the remote wake. (A more elegant procedure will be found in Chapter 4.) For this estimate, the vertical drag, D_V, is:

$$D_V = (0.3)(\text{D.L.}) \ (\text{Projected area of all affected components})$$

It is often convenient to express the vertical drag as a fraction of the gross weight:

$$\frac{D_V}{\text{G.W.}} = \frac{(0.3)(\text{D.L.})(\text{Projected area})}{(\text{D.L.})(\text{Disc area})}$$

or:

$$\frac{D_V}{\text{G.W.}} = \frac{0.3(\text{Projected area})}{\text{Disc area}}$$

The example helicopter has a projected area of 380 ft^2 and a disc area of 2,827 ft^2. It thus has a download of 4% of its gross weight. The rotor thrust required to support a hovering helicopter and its vertical drag is:

$$T = \left(1 + \frac{D_V}{\text{G.W.}}\right)\text{G.W.}$$

Ideal Power

The minimum—or ideal—power required to produce rotor thrust may be determined from momentum considerations as the energy per second dissipated at the rotor:

$$E_R/\text{sec} = (\text{Force})(\text{Velocity}) = T\,v_1 \text{ ft lb/sec}$$

Since 550 ft lb/sec is the equivalent of one horsepower, the ideal power is:

$$\text{h.p.}_i = \frac{Tv_1}{550}$$

or

$$\text{h.p.}_i = \frac{T\sqrt{\dfrac{\text{D.L.}}{2\rho}}}{550}$$

At sea level, the ideal power is:

$$\text{h.p.}_i = \frac{T\sqrt{\text{D.L.}}}{38}$$

From this equation, it may be seen that for a given rotor thrust, the higher the disc loading, the higher the power required. During the early days of helicopter development, when power plants were relatively heavy, designers minimized the power required by using low disc loadings of 2 to 3 lb/ft². The introduction of lightweight turbine engines has made it possible to put less emphasis on low power requirements and more emphasis on designing compact helicopters with minimum structural weight. For this reason, over the years disc loadings have been increasing steadily until values of 10 to 12 lb/ft² are considered feasible for helicopters that do not have to operate close to loose surfaces.

Figure of Merit

The ideal power derived from momentum considerations is the power required by a rotor consisting only of an idealized *actuator disc* without regard for the drag of the actual blades. In this sense, the ideal power is analogous to the induced drag of an airplane wing and, as a matter of fact, is generally called *induced power*. The actual power, of course, is higher than the induced, or ideal, power. The ratio of induced power to actual power is known as the *Figure of Merit* (F.M.):

$$\text{F.M.} = \frac{\text{Induced power}}{\text{Actual power}}$$

It will later be shown that the value of the Figure of Merit is a function of the rotor geometry and of the rotor operating conditions. Whirl tower tests of conventional helicopter rotors have shown that the maximum Figure of Merit that can be expected in practice is 0.75 to 0.80.

First Approximation to the Power Required to Hover

The actual power is the induced power divided by the Figure of Merit:

$$\text{h.p.}_{\text{act}} = \frac{\text{h.p.}_i}{\text{F.M.}} = \frac{T \sqrt{\dfrac{\text{D.L.}}{2\rho}}}{550\,(\text{F.M.})}$$

At sea level, this becomes:

$$\text{h.p.}_{\text{act}} = \frac{T \sqrt{\text{D.L.}}}{38\,(\text{F.M.})}$$

The power loading in pounds per horsepower can be obtained by rearranging the equation:

$$\frac{T}{\text{h.p.}_{\text{act}}} = \frac{38\,(\text{F.M.})}{\sqrt{\text{D.L.}}}$$

A plot of this equation is presented in Figure 1.4 for Figure of Merit values of 1.0, 0.8, and 0.6—corresponding to the ideal rotor, to a very good practical rotor, and to an average rotor respectively. (*Note*: A low Figure of Merit in hover does not necessarily represent a poorly designed rotor. More commonly, it represents a rotor that has been designed for high-speed flight and is not being operated at its optimum hovering conditions.) It may be seen from Figure 1.4 that high power loadings go with low disc loadings and high Figures of Merit. The use of this chart allows a first approximation to be made of the power required to hover. The example helicopter has a disc loading of 7.1 lb/ft². If a Figure of Merit of 0.8 is assumed for its rotor, then the power loading is 11.4 lb/h.p. and, since the helicopter weighs 20,000 lb, the power required by the rotor is 1,760 hp. If the rotor radius had been 40 feet instead of 30, the disc loading would have been 4.0 lb/ft², and the power required would have been only 1,320; but the tail boom would have had to be 10 feet longer to achieve clearance between main and tail rotors, and the nose would have had to be longer to balance the tail boom. As a

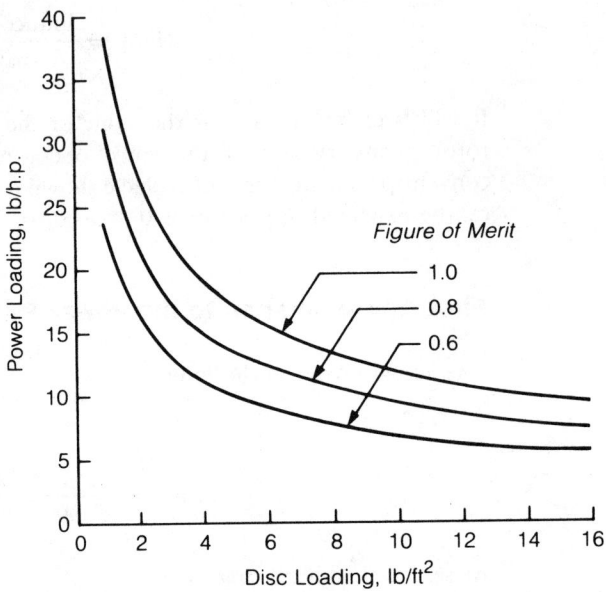

FIGURE 1.4   Effect of Disc Loading on Power Loading

consequence, the reduction in power required would have been obtained at the expense of added structural weight and a larger overall size with, perhaps, little increase in payload capability. Making decisions on this type of trade-off represents much of the engineering effort during the early stages of the development of a new helicopter design.

BLADE ELEMENT METHOD

The Lift on a Blade Element

The momentum method gives much insight into conditions at the rotor and in the wake, but it does not deal with the actual development of thrust—that is, the lift on the individual blade elements. The mechanism by which lift is developed by an airfoil is explained in textbooks on airplane aerodynamics, and that explanation applies just as well to an element of a rotor blade as to a section of a wing.

The geometry of a blade element is shown in Figure 1.5. A blade element is one small portion of the blade a distance, r, from the center of rotation with a spanwise dimension, Δr. The increment of lift, ΔL, on this blade element is:

$$\Delta L = q\, c_i c \Delta r$$

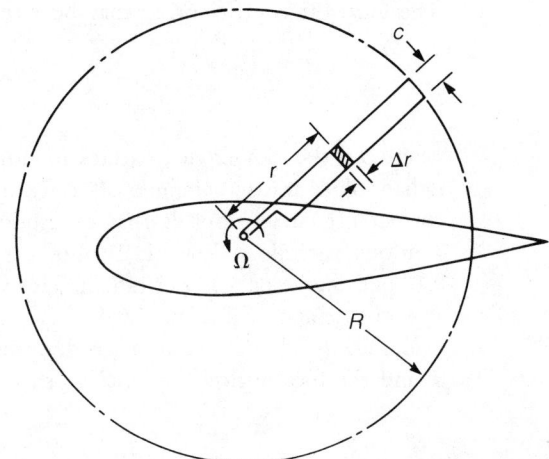

FIGURE 1.5   Geometry of a Blade Element

The local dynamic pressure, q, is a function of the velocity due to rotation at the blade element. This velocity is zero at the center of rotation and increases linearly to the tip. To express this velocity, it is convenient to use the rotational speed in terms of radians per second, Ω. The local velocity at the blade element, V_l, is:

$$V_l = \Omega r \text{ ft/sec}$$

At the tip, the tip speed, V_T, is:

$$V_T = \Omega R \text{ ft/sec}$$

Tip speeds on modern helicopters are selected to be as high as they can be without encountering compressibility effects on the advancing blade in forward flight. These effects will be discussed in more detail in Chapter 3. At this point, it is enough to state that design tip speeds on both the main and tail rotors fall in the region between 500 and 800 ft/sec. Since tip speeds fall into this rather narrow band, it is usually more convenient in studying rotor aerodynamics to use this as a parameter to measure rotor speed rather than rpm, which will vary widely with rotor size. Note that

$$\Omega R = \frac{2\pi R}{60}(\text{rpm}) = (\text{Rotor circumference})(\text{rps})$$

The local dynamic pressure, q, at the blade element is:

$$q = \tfrac{1}{2}\rho(\Omega r)^2$$

The local lift coefficient, c_l, may be written as:

$$c_l = a\alpha$$

where α is the local angle of attack in radians, and a is the slope of the lift curve per radian. Since the actual angle of attack including induced effects will be used, the correct lift curve slope is that coresponding to a two-dimensional airfoil. Conventional airfoils at low Mach numbers have lift curve slopes of approximately 0.10 per degree or 5.73 per radian. This value increases slightly with Mach number and will be assumed to be 6.0 for use in this type of simple rotor analysis.

The local angle of attack is determined by the geometric pitch of the blade, θ, and the local inflow angle, ϕ, as shown in Figure 1.6.

$$\alpha = \theta - \phi$$

The inflow angle, ϕ, is defined by the two mutually perpendicular velocities, Ωr and v_1, such that:

$$\phi = \tan^{-1}\frac{v_1}{\Omega r}$$

If ϕ is less than 10°—and it is for most of the rotor—then the small-angle assumption may be used:

$$\phi = \frac{v_1}{\Omega r} \text{ radians}$$

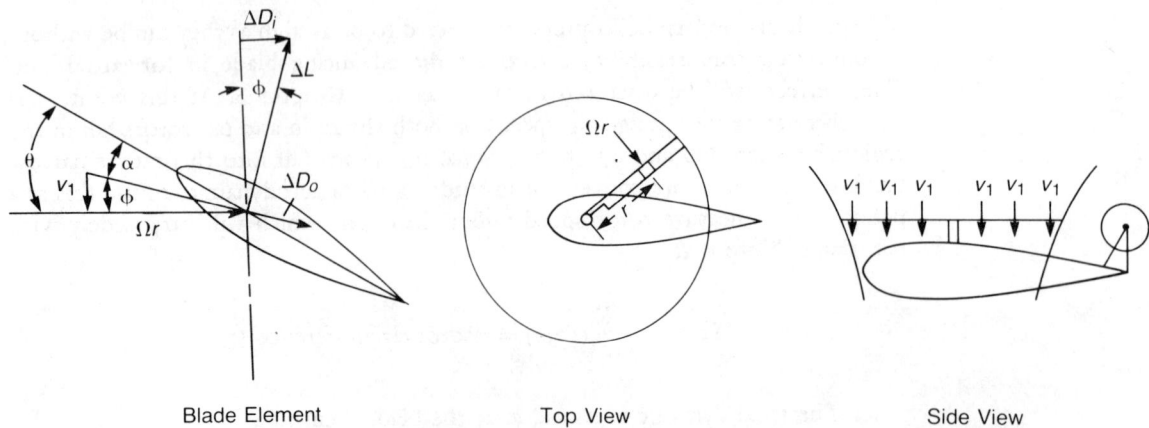

Blade Element   Top View   Side View

FIGURE 1.6 Orientation of Blade Element and Local Velocities

Thus the local angle of attack is:

$$\alpha = \theta - \frac{v_1}{\Omega r} \text{ rad}$$

and

$$c_l = a\left(\theta - \frac{v_1}{\Omega r}\right)$$

The increment of lift on the blade element is:

$$\Delta L = \frac{\rho}{2}(\Omega r)^2 a\left(\theta - \frac{v_1}{\Omega r}\right) c \Delta r$$

Integration for Thrust

The lift on the entire blade is the integration of the lift on all the blade elements from the center of the rotor to the tip. In order to perform the integration in a tidy manner, the blade will be assumed to have *ideal twist*. Most helicopter blades are twisted such that the pitch at the tip is less than the pitch at the root. The most common twist is linear such that:

$$\theta = \theta_0 + \frac{r}{R}\theta_1$$

where θ_0 is the pitch that the blade would have if it extended into the center of rotation and θ_1 is the angle of twist or *washout* between the center of rotation and the tip. The value of linear twist in current use is from $-5°$ to $-16°$. For blades with ideal twist instead of linear twist, the local value of pitch is:

$$\theta = \frac{\theta_t}{r/R}$$

where θ_t is the pitch at the blade tip. Plots of blade pitch for a blade with -10 degrees of linear twist and for a blade with ideal twist are shown in Figure 1.7 for approximately the same rotor lift. It will later be shown that the ideal twist produces better rotor performance than any other type of twist, but that the margin of increased performance over linear twist is relatively small. Actual helicopter blades use linear twist instead of ideal twist because of the ease of manufacture. Because of the ease of analysis, however, ideal twist is assumed in this derivation with the consequences of linear twist shown later.

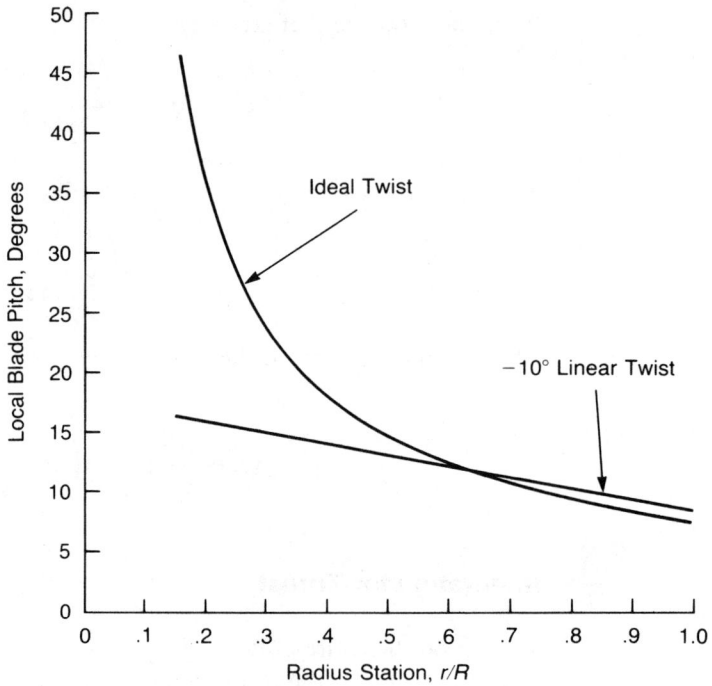

FIGURE 1.7   Ideal and Linear Blade Twist

The inflow angle was earlier shown to be:

$$\phi = \frac{v_1}{\Omega r}$$

This equation can be rewritten as:

$$\phi = \frac{v_1}{\Omega R} \frac{R}{r} = \frac{\phi_t}{r/R}$$

where ϕ_t is the inflow angle at the tip. Substituting the equations just developed for θ and ϕ into the equation for the lift on the blade element gives:

$$\Delta L = \frac{\rho}{2} (\Omega r)^2 a \left(\frac{\theta_t}{r/R} - \frac{\phi_t}{r/R} \right) c \Delta r$$

or

$$\frac{\Delta L}{\Delta r} = \frac{\rho}{2} \Omega^2 Rac(\theta_t - \phi_t)r$$

where $\Delta L/\Delta r$ is the lift per running foot along the blade and is simply a linear—or triangular—lift distribution with respect to blade station. The total lift on the blade is equal to the area of the triangle:

$$L = \frac{\rho}{2}\,\Omega^2 Rac(\theta_t - \phi_t) \int_0^R r\,dr$$

Thus the lift per blade is:

$$L = \frac{\rho}{2}\,\Omega^2 R^2 acR\left(\frac{\theta_t - \phi_t}{2}\right)$$

and the total rotor thrust is the lift per blade times the number of blades, b:

$$T = bL = \frac{\rho}{2}\,(\Omega R)^2 bcRa\left(\frac{\theta_t - \phi_t}{2}\right)$$

Note the similarity of this equation with the equation for the lift of an airplane wing:

$$L_W = \frac{\rho}{2}\,V^2 Sa\alpha$$

In the case of the rotor, $(\rho/2)(\Omega R)^2$ is the dynamic pressure based on tip speed, bcR is the total blade area, a is the slope of the airfoil lift curve, and $(\theta_t - \phi_t)/2$ is an effective angle of attack.

Nondimensional Coefficients

Before developing the analysis further, it is timely to introduce the concept of nondimensional coefficients. Just as does the airplane aerodynamicist, the helicopter aerodynamicist finds it convenient to work with nondimensional coefficients to define rotor characteristics in a form that is independent of rotor size. For this purpose, the system used by NASA will be adopted:

$$\text{Thrust coefficient, } C_T = \frac{T}{\rho A(\Omega R)^2}$$

$$\text{Torque coefficient, } C_Q = \frac{Q}{\rho A(\Omega R)^2 R}$$

$$\text{Power coefficient, } C_P = \frac{P}{\rho A(\Omega R)^3}$$

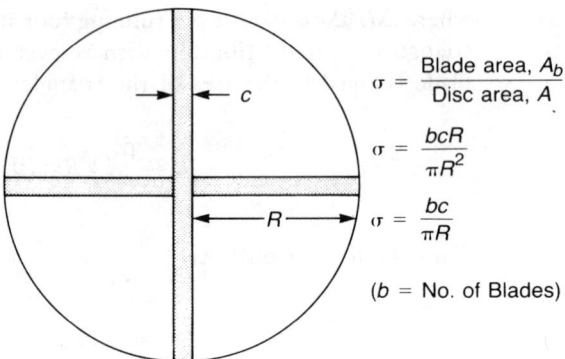

$$\sigma = \frac{\text{Blade area, } A_b}{\text{Disc area, } A}$$

$$\sigma = \frac{bcR}{\pi R^2}$$

$$\sigma = \frac{bc}{\pi R}$$

(b = No. of Blades)

FIGURE 1.8   Illustration of Solidity Ratio

One of the convenient features of this set of coefficients is that the torque and power coefficients are numerically equal. This may be shown by writing the equation for power, P, in ft lb/sec as:

$$P = Q\Omega$$

Substituting this into the equation for C_P gives:

$$C_P = \frac{Q\Omega}{\rho A (\Omega R)^3} = \frac{Q}{\rho A (\Omega R)^2 R} = C_Q$$

Note that the reference area for each of these rotor coefficients is the rotor disc area, A. It is often desirable to use coefficients based on blade area rather than on disc area. In order to do this, the rotor solidity ratio, σ, is defined:

$$\sigma = \frac{\text{Total blade area}}{\text{Disc area}}$$

As shown in Figure 1.8, the solidity is the amount of the rotor disc made solid by the blades:

$$\sigma = \frac{A_b}{A} = \frac{bcR}{\pi R^2} = \frac{bc}{\pi R}$$

For the example helicopter:

$$\sigma = \frac{120}{2,827} = \frac{(4)(2)}{\pi(30)} = 0.085$$

Note that in using this definition, the blades are assumed to extend into the center of rotation where some overlapping of blade area occurs. This assumption is used in the definition of solidity even though, in practice, the actual blade sections do not start until some distance from the center of rotation. This is a satisfactory assumption since the inboard portions of the blades contribute little to the aerodynamic characteristics of the rotor. However, the same cannot be said about the outboard portions of the rotor. If the blade is tapered, the aerodynamic biasing toward the tips should be accounted for by rewriting the equation for solidity to define a *thrust-weighted solidity* as:

$$\sigma_T = \frac{bc_e}{\pi R}$$

where

$$c_e = \frac{\displaystyle\int_{(r/R)=0}^{(r/R)=1} c\left(\frac{r}{R}\right)^2 d\,\frac{r}{R}}{\displaystyle\int_{(r/R)=0}^{(r/R)=1} \left(\frac{r}{R}\right)^2 d\,\frac{r}{R}}$$

and c is the local chord.

For example, a blade planform might consist of a portion with a constant chord, c_R, from the root to a point, r_1, and then taper linearly to a tip chord of c_T. The equation for the effective thrust weighted chord in this case is:

$$c_e = (c_R - c_T)\left(\frac{r_1}{R}\right)^3 - \left(\frac{c_R - c_T}{1 - \frac{r_1}{R}}\right)\left[-\frac{1}{4} + \left(\frac{r_1}{R}\right)^3 - \frac{3}{4}\left(\frac{r_1}{R}\right)^4\right] + c_T$$

The rotor coefficients based on effective blade area are:

$$C_T/\sigma = \frac{T}{\rho A_b(\Omega R)^2} = \frac{T}{\rho\sigma A(\Omega R)^2}$$

$$C_Q/\sigma = \frac{Q}{\rho A_b(\Omega R)^2 R} = \frac{Q}{\rho\sigma A(\Omega R)^2 R}$$

$$C_P/\sigma = \frac{P}{\rho A_b(\Omega R)^3} = \frac{P}{\rho\sigma A(\Omega R)^3} = C_Q/\sigma$$

An analogy with airplane aerodynamics is that C_T corresponds to span loading and governs induced power. On the other hand, C_T/σ corresponds to lift coefficient, and its value is used to determine the margin to blade stall just as C_L is compared to

$C_{L_{max}}$. The numerical value of C_T/σ may be related to an average two-dimensional lift coefficient, $\bar{c}_l$, by defining the lift per running foot as a function of the average lift coefficient, which for this purpose will be assumed to be constant along the blade:

$$\frac{\Delta L}{\Delta r} = \frac{\rho}{2}(\Omega r)^2 \bar{c}_l c$$

Integrating out the blade and multiplying by the number of blades gives an expression for rotor thrust:

$$T = b\frac{\rho}{2}\Omega^2 \bar{c}_l c \int_0^R r^2 dr = \frac{\rho}{6}(\Omega R)^2 bcR\bar{c}_l$$

but by definition:

$$T = \rho bcR(\Omega R)^2 C_T/\sigma$$

Equating these two equations gives:

$$\bar{c}_l = 6C_T/\sigma$$

Most airfoils used for rotor blades at operational Reynolds and Mach numbers have maximum lift coefficients of between 1.0 and 1.4. Thus the maximum value of C_T/σ that may be developed before reaching blade stall is between 0.17 and 0.23. For the example helicopter at sea level with a vertical drag penalty of 4 percent, the value of C_T/σ is:

$$C_T/\sigma = \frac{(20{,}000)(1.04)}{(.002378)(4)(2)(30)(650)^2} = 0.086$$

and the corresponding average lift coefficient, $\bar{c}_l$ is:

$$\bar{c}_l = 6(0.086) = 0.52$$

If the airfoil has a maximum lift coefficient of 1.2, the rotor can lift more than twice the design gross weight of the helicopter before stalling. At 25,000 ft, however, the average lift coefficient is 1.14, and the rotor is on the verge of stall.

Relationship between Thrust and Pitch

The equation derived earlier for the thrust of a rotor with ideal twist can be written in nondimensional form:

$$T = \frac{\rho}{2}(\Omega R)^2 bcRa\left(\frac{\theta_t - \phi_t}{2}\right) = C_T/\sigma\rho bcR(\Omega R)^2$$

or

$$C_T/\sigma = \frac{a}{4}(\theta_t - \phi_t)$$

The induced angle at the tip is:

$$\phi_t = \frac{v_1}{\Omega R}$$

where

$$v_1 = \sqrt{\frac{T}{2\rho A}}$$

thus

$$\phi_t = \frac{1}{\Omega R}\sqrt{\frac{C_T\rho(\Omega R)^2 A}{2\rho A}} = \sqrt{\frac{C_T}{2}} = \sqrt{\frac{\sigma C_T/\sigma}{2}}$$

Substituting this into the equation for C_T/σ gives:

$$C_T/\sigma = \frac{a}{4}\left(\theta_t - \sqrt{\frac{\sigma C_T/\sigma}{2}}\right)$$

or, solving for θ_t:

$$\theta_t = 57.3\left[\frac{4}{a}C_T/\sigma + \sqrt{\frac{\sigma C_T/\sigma}{2}}\right]\text{degrees}$$

A plot of this equation for a solidity of 0.085 corresponding to the example helicopter and for solidities of half and twice this value are shown in Figure 1.9. For the example helicopter operating at a C_T/σ of .086, the required pitch at the tip is 6.7°. Note the relatively low slope of the curve near zero thrust. In this region, the induced velocity is important and a large amount of the blade pitch is used to compensate for it, with only a small proportion available to produce thrust. At higher values of thrust, the induced velocity is of less importance and more of the blade pitch is available to produce thrust.

The collective pitch of a blade with linear twist can be related to the tip pitch of a blade with ideal twist through the approximate relationship:

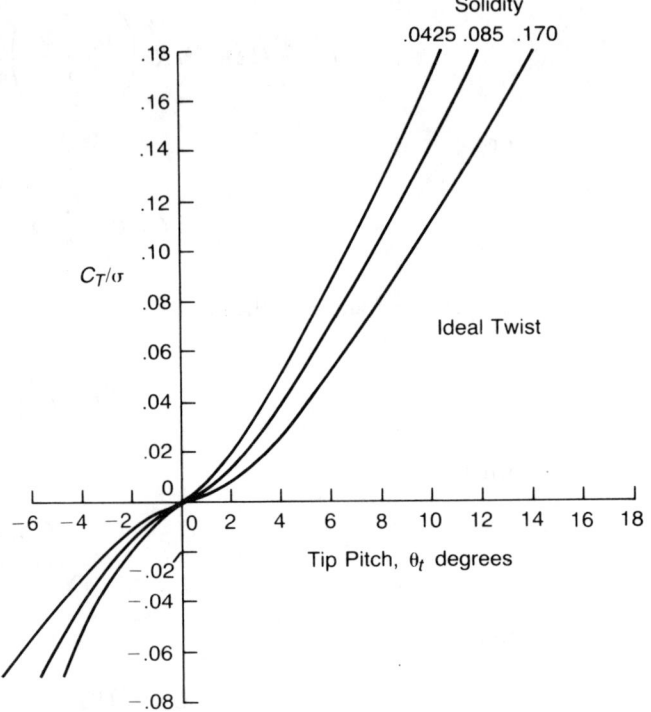

FIGURE 1.9   Thrust/Solidity Coefficient as a Function of Tip Pitch

$$\theta_0 = \frac{3}{2}\,\theta_t - \frac{3}{4}\,\theta_1$$

Thus for the example helicopter with a linear twist of $-10°$, the collective pitch is approximately:

$$\theta_0 = \frac{3}{2}\,(6.7) - \frac{3}{4}\,(-10) = 17.6°$$

The Torque on a Blade Element

The determination of rotor thrust as a function of blade angle is of some importance, but more important is the answer to the question, "How much power is required to produce the thrust?" The first approximation of the power required is obtained using the Figure of Merit method described earlier. The second approximation may be made with the blade element theory.

The increment of power, ΔP, produced by a blade element, is:

$$\Delta P = \Delta Q \Omega$$

where the incremental torque, ΔQ, is equal to the drag on the blade element times the radius of the element. Just as for airplanes, the drag is made up of two parts: induced and profile. From Figure 1.6, it may be seen that since the lift is perpendicular to the local velocity, it is tilted back by the induced angle, ϕ. Thus the induced drag due to lift, ΔD_i, is:

$$\Delta D_i = \Delta L \phi$$

The profile drag, ΔD_0, is the force parallel to the local flow; and, since ϕ is considered to be a small angle, the component of profile drag contributing to torque is taken as being equal to the entire profile drag. Thus the increment of torque, ΔQ, is:

$$\Delta Q = r(\Delta L \phi + \Delta D_0)$$

The drag can be treated in the same way that lift was:

$$\Delta D_0 = \frac{\rho}{2}(\Omega r)^2 c_d c \Delta r$$

where the drag coefficient, c_d, may be considered a constant at this stage of the analysis.

The equation for incremental torque can now be written as:

$$\Delta Q = r\left[\frac{\rho}{2}(\Omega r)^2 c_l c \Delta r \phi + \frac{\rho}{2}(\Omega r)^2 c_d c \Delta r\right]$$

Expressions for c_l and ϕ for an ideally twisted blade have already been derived. Using these, the torque per running foot is:

$$\frac{\Delta Q}{\Delta r} = r\left[\frac{\rho}{2}\,\Omega^2 Rac(\theta_t - \phi_t)r\left(\phi_t \frac{R}{r}\right) + \frac{\rho}{2}(\Omega r)^2 c_d c\right]$$

Integration for Power

The total torque can be obtained by integration in the same way that the total lift was:

$$Q = b\frac{\rho}{2}\Omega^2 c\left[R^2 a \int_0^R (\theta_t - \phi_t)\phi_t r\,dr + c_d \int_0^R r^3 dr\right]$$

or

$$Q = \frac{\rho}{2} \ (\Omega R)^2 bcR \left[a \ \frac{(\theta_t - \phi_t)}{2} \ \phi_t + \frac{c_d}{4} \right] R$$

In nondimensional form:

$$C_Q = \frac{\sigma}{2} \left[\frac{a}{2} (\theta_t - \phi_t)\phi_t + \frac{c_d}{4} \right]$$

Using expressions already derived, this becomes:

$$C_Q = C_T \sqrt{\frac{C_T}{2}} + \frac{c_d \sigma}{8}$$

The first term in this equation is due to the combined effect of the rearward tilt of the incremental lift vectors in Figure 1.6 and is known as the *induced torque coefficient*. The induced torque coefficient could have been derived from the momentum equation by noting that the induced power is:

$$P_i = Tv_1 = T \sqrt{\frac{T}{2\rho A}}$$

Thus:

$$C_{P_i} = \frac{C_T \rho A (\Omega R)^3}{\rho A (\Omega R)^3} \sqrt{\frac{C_T \rho A (\Omega R)^2}{2\rho A}} = C_T \sqrt{\frac{C_T}{2}}$$

and since it was earlier shown that C_P was numerically equal to C_Q:

$$C_{Qi} = C_{P_i} = C_T \sqrt{\frac{C_T}{2}}$$

This shows that the momentum theory and the blade element theory—at least for ideally twisted blades—produce the same expression for induced power. The second term in the equation for C_Q is the profile torque coefficient and is due only to the drag of the blade elements. The Figure of Merit written in nondimensional terms is:

$$\text{F.M.} = \frac{C_{P_i}}{C_P} = \frac{C_T \sqrt{C_T/2}}{C_P} = \sqrt{\frac{\sigma}{2}} \ \frac{(C_T/\sigma)^{3/2}}{C_Q/\sigma}$$

Determination of the Blade Drag Coefficient

The average value of the drag coefficient, $\bar{c}_d$, for use in the analysis may be obtained by using an average value of angle of attack and the corresponding drag coefficient as determined by wind tunnel tests of two-dimensional airfoil sections. Figure 1.10 shows the characteristics of the NACA 0012 airfoil section, which is typical of the airfoils actually used on blades. These characteristics have been taken from reference 1.1 and were synthesized from whirl tower tests of a complete rotor over a wide range of collective pitch and tip speed by choosing the lift and drag coefficients that brought the test results into the best agreement with the theory. The results of two-dimensional wind tunnel tests of the same airfoil were used as a guide. Figure 1.10 shows the lift and drag coefficients as functions of

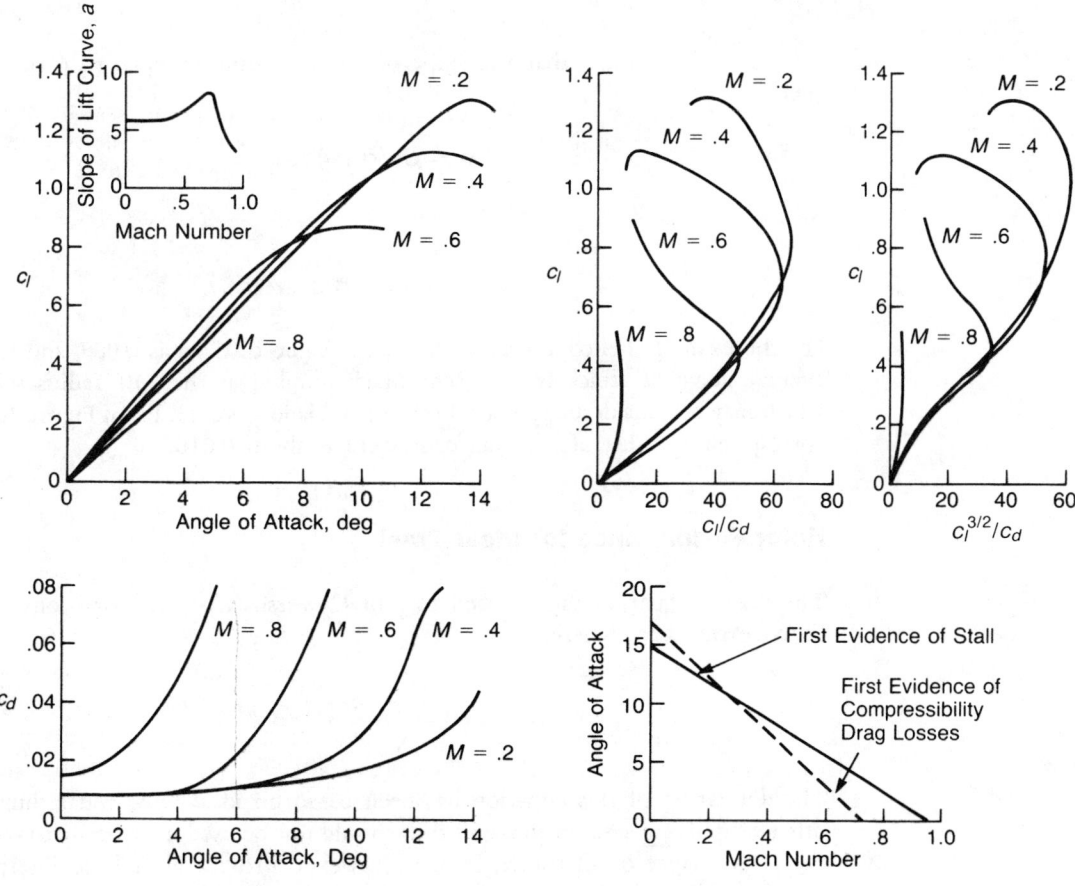

FIGURE 1.10   Characteristics of NACA 0012 Airfoil

both the angle of attack and the Mach number. Also shown are other characteristics of importance to the helicopter aerodynamicist, which will be discussed later. The average lift coefficient for the rotor, $\bar{c}_l$, is:

$$\bar{c}_l = a\bar{\alpha}$$

But it was shown earlier that

$$\bar{c}_l = 6C_T/\sigma$$

therefore,

$$\bar{\alpha} = \frac{6C_T/\sigma}{a}$$

If it is assumed that the slope of the lift curve is equal to 6 per radian, then:

$$\bar{\alpha} = C_T/\bar{\sigma} \text{ radians}$$

or

$$\bar{\alpha} = 57.3 \, C_T/\bar{\sigma} \text{ degrees}$$

For the example helicopter at sea level, the value of C_T/σ is 0.086, and thus the average angle of attack is 4.9°. The Mach number at the 75% radius station—which may be considered typical of the entire blade—is 0.43. From Figure 1.10, the corresponding value of the drag coefficient is about 0.010.

Rotor Performance for Ideal Twist

The torque equation can be used to plot C_Q versus C_T—or, more conveniently, C_Q/σ versus C_T/σ, where:

$$C_Q/\sigma = C_P/\sigma = \sqrt{\frac{\sigma}{2}} \, (C_T/\sigma)^{3/2} + \frac{\bar{c}_d}{8}$$

The derivation of this equation has been based on ideal twist and a number of other simplifying assumptions and thus should not be used uncritically to compute the performance of all rotors. It does, however, provide a guide to interpreting actual performance data such as Figure 1.11, where C_Q/σ has been plotted against $(C_T/\sigma)^{3/2}$ for whirl tower data from Reference 1.1. The theoretical line is based on the value of $\bar{c}_d$ determined from Figure 1.10 as a function of the average angle of attack and the Mach number at the 75% radius. It may be seen that although there

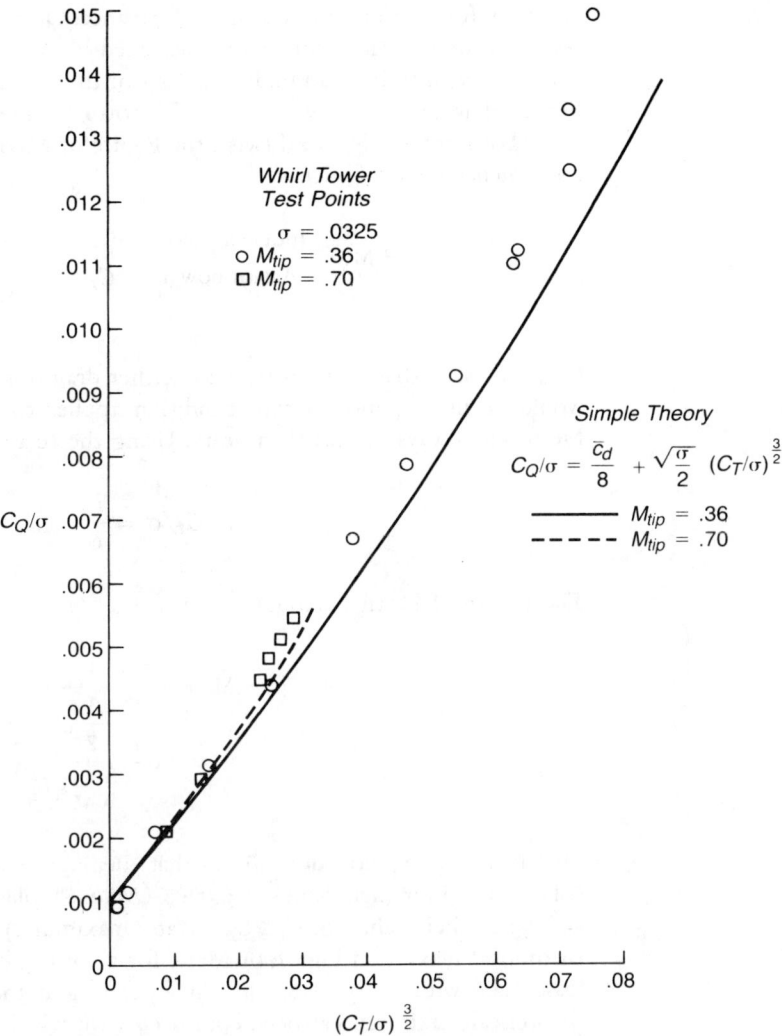

FIGURE 1.11   Comparison of Test Results with Simple Theory

is some discrepancy between the theory and the test points, the general characteristics of the theoretical line are the same as the test results. The reasons for the discrepancies that do exist will be discussed in a later section.

Maximum Figure of Merit

On most helicopters, the critical design condition is high-speed flight. To satisfy this condition, the twist is usually less, and the blade area usually more than is optimum for maximum hovering performance. On some helicopters, such as flying cranes, however, the rotor may be designed by the hover requirement. For these cases, it is well to have an understanding of the maximum theoretical value of the Figure of Merit and how it varies with rotor parameters.

For a rotor with ideal twist, the Figure of Merit may be written in terms of nondimensional coefficients:

$$\text{F.M.} = \frac{\text{Induced power}}{\text{Actual power}} = \frac{C_{P_i}}{C_P} = \frac{C_T\sqrt{C_T/2}}{C_T\sqrt{C_T/2} + \dfrac{\bar{c}_d \sigma}{8}}$$

It may be seen that if the rotor had neither drag nor solidity, the Figure of Merit would be unity. Since neither condition applies to actual rotors, the Figure of Merit will always be less than unity. Using the relationship:

$$C_T/\sigma = \frac{\bar{c}_l}{6}$$

The Figure of Merit becomes:

$$\text{F.M.} = \frac{1}{1 + \dfrac{\dfrac{3}{2}\sqrt{3}}{\sqrt{\sigma}\,\dfrac{\bar{c}_l^{3/2}}{\bar{c}_d}}}$$

This form of the equation shows that the Figure of Merit is highest for high solidities and for high values of $\bar{c}_l^{3/2}/c_d$. (*Note*: Sailplanes have their lowest rates of sink when their values of $C_L^{3/2}/C_D$ are at a maximum.) The equation represents the theoretical maximum Figure of Merit for a rotor with ideal twist, no tip or root losses, and with every blade element operating at the same angle of attack. This theoretical maximum has been compared with whirl tower results in Figure 1.12. The whirl tower Figure of Merit is from reference 1.1 and the theoretical curve is based on the NACA 0012 airfoil data of Figure 1.10, which are from the same set of tests. The difference between the two curves is due to tip and root losses, nonideal twist, nonuniform blade element angles of attack, and rotation of the wake (all of which will be discussed before we finish).

The equation for Figure of Merit can also be written as a function of disc loading and tip speed:

$$\text{F.M.} = \frac{1}{1 + \dfrac{\dfrac{3}{2}(\Omega R)}{\bar{c}_l/\bar{c}_d \sqrt{\dfrac{2\text{D.L.}}{\rho}}}}$$

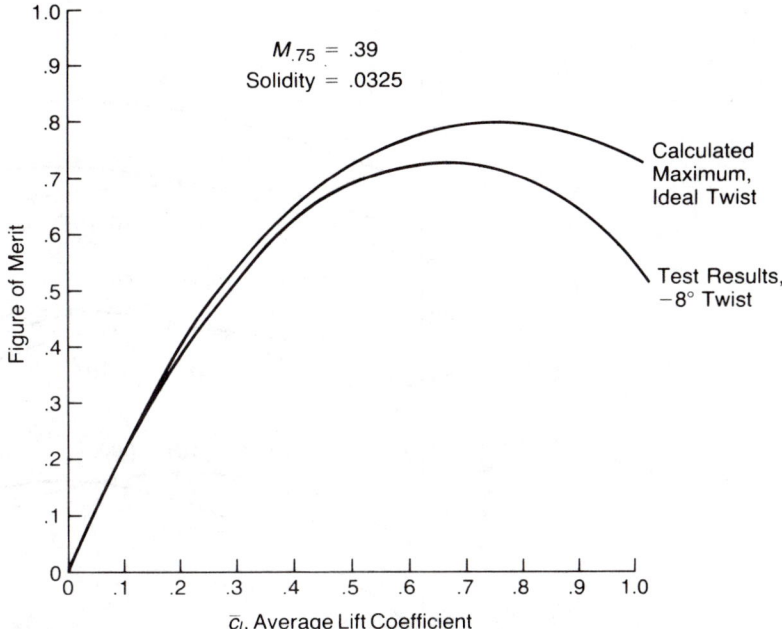

$M_{.75} = .39$
Solidity $= .0325$

FIGURE 1.12   Comparison of Measured Figure of Merit with Calculated Maximum

Source: Carpenter, "Lift and Profile-Drag Characteristics of an NACA 0012 Airfoil Section as Derived from Measured Helicopter-Rotor Hovering Performance," NACA TN 4357, 1958.

This form shows that for a given disc loading and tip speed, the Figure of Merit is maximum when c_l/c_d is a maximum. The apparent conflict with the previous discussion is sometimes called a paradox, but the truth of both statements is shown in Figure 1.13 on which the theoretical Figure of Merit based on the NACA 0012 airfoil data of Figure 1.10 has been plotted as a function of $\bar{c}_l$ for both constant solidities and constant disc loadings. It may be seen that the curves for constant solidity peak at a $\bar{c}_l$ of 0.8, which is the lift coefficient for maximum $c_l^{3/2}/c_d$, and that the curves for constant disc loading peak at a $\bar{c}_l$ of 0.7, which is the lift coefficient for maximum c_l/c_d. It should be noted that the constant disc loading lines are a function of tip speed; the lower the tip speed, the higher the potential Figure of Merit. The benefit, of course, is decreased if the higher solidity required by the lower tip speed requires higher blade weight. The equation leads us to conclusions that can also be obtained graphically, as presented in Figure 1.14. This shows the velocities and forces acting at a typical blade element of three rotors developing the same thrust. The first rotor is the baseline. The second has half the diameter resulting in four times the disc loading, which doubles the induced velocity and thus the induced drag. Since the lift-to-drag ratio is assumed to be the same, the profile drag is the same. The Figure of Merit is proportional to the induced drag

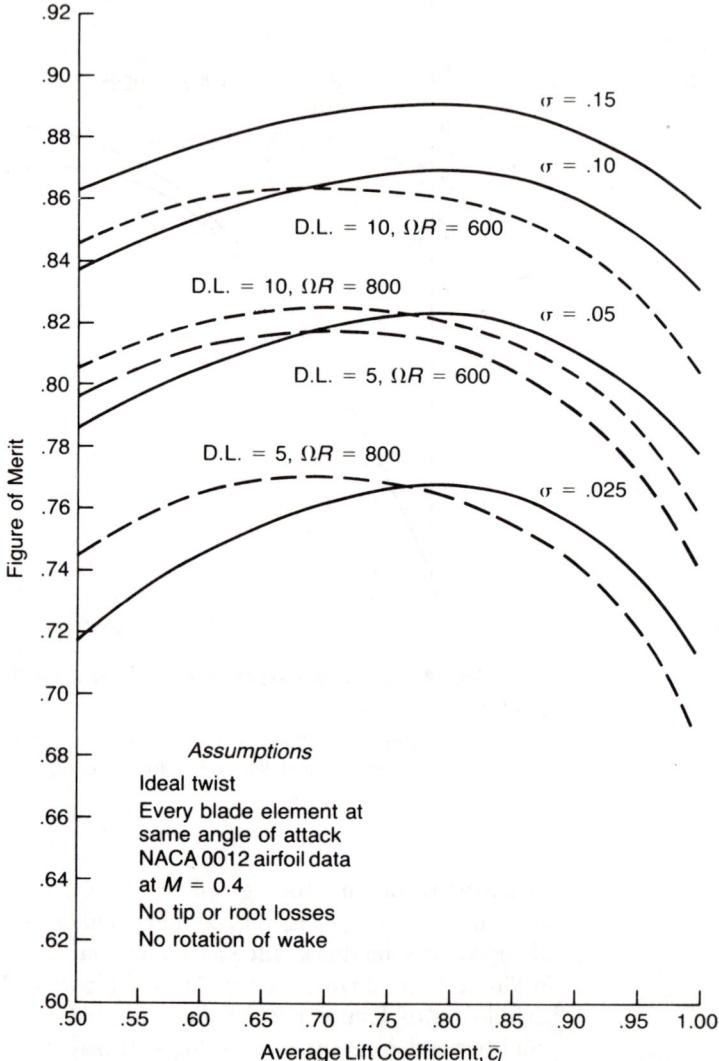

FIGURE 1.13   Calculated Figure of Merit

divided by the sum of induced drag and profile drag. At the higher disc loading, this ratio is closer to 1.0 than for the baseline rotor. This explains why designers of high disc loading aircraft can claim a higher Figure of Merit than can designers of low disc loading aircraft. The third rotor has the same diameter as the baseline but only half the tip speed. The lift vector is tilted further back, leading to twice the induced drag and to a similar increase in Figure of Merit as for the second rotor. It should be noted, however, that whereas the high disc loading rotor takes almost

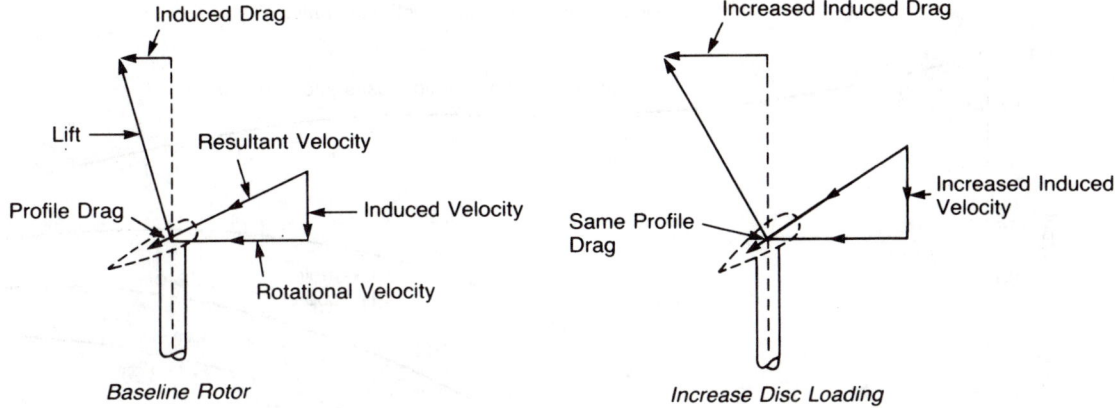

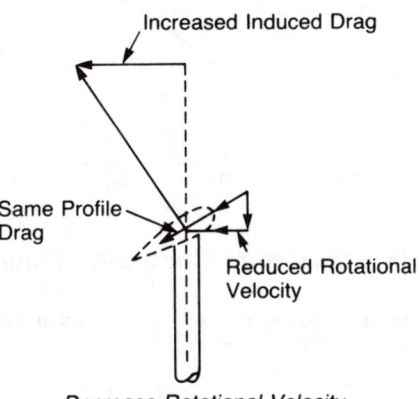

Decrease Rotational Velocity

FIGURE 1.14   Effect of Changing Conditions

twice the power of the baseline, the low tip speed rotor requires about the same since the power is proportional to the product of drag and speed.

Another study of the maximum figure of Merit is given in reference 1.2. Figure 1.15 presents the results for a four-bladed rotor with constant induced velocity as the maximum attainable Figure of Merit versus thrust coefficient. The top line is the Figure of Merit assuming no drag. It is unity only for zero thrust. At higher values, wake swirl and tip vortex interactions (to be discussed later) combine to produce some unavoidable losses. The lower lines show how drag affects the results. The drag characteristics assumed for this study were simply based on skin friction as a function of Reynolds number but independent of angle of attack and Mach number as expressed by the equation:

$$c_d = \frac{0.144}{(\mathrm{RN})^{1/5}}$$

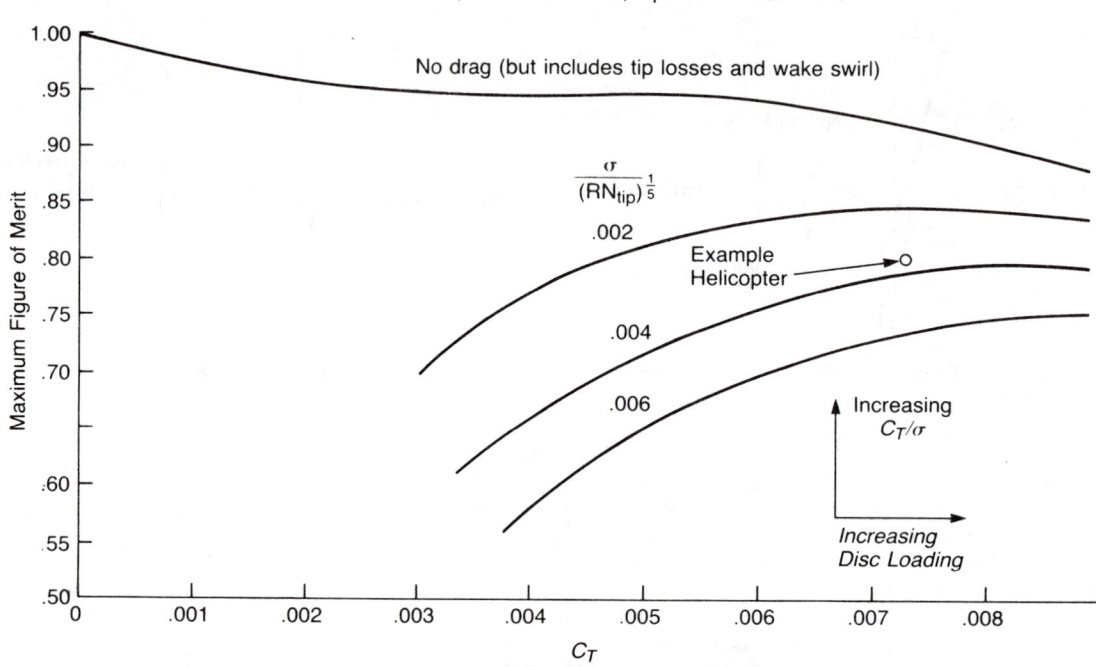

FIGURE 1.15   Maximum Computed Figure of Merit

Source: Harris & McVeigh, "Uniform Downwash with Rotors Having a Finite Number of Blades," *JAHS* 21-1, 1975.

where at sea level:

$$\text{R.N.} = 6{,}400\ cV$$

The family of lines essentially represents constant values of rotor solidity. Going down in solidity (up in C_T/σ) increases the maximum Figure of Merit. For constant solidity, increasing disc loading or decreasing tip speed increases C_T and thus the Figure of Merit. In any case, the study indicates the maximum optimistic hover performance. The calculated value for the Figure of Merit for the example helicopter is shown as a point at about 0.80.

Coning of a Rotor in Hover

A blade—whether hinged or cantilevered from the hub—seeks an equilibrium coning angle that is a function of the lift, centrifugal forces, and blade weight. The magnitude of the coning, a_0, may be found by setting the moments at the hinge—or at the effective hinge—to zero:

$$M_{\text{hinge}} = M_{\text{lift}} + M_{\text{c.f.}} + M_w = 0$$

For most engineering calculations of coning, it is sufficiently accurate to assume that the hinge is at the center of rotation. In this case, the moment due to lift is:

$$M_{\text{lift}} = \int_0^R \frac{dL}{dr} r\, dr$$

or for a blade with ideal twist:

$$M_{\text{lift}} = \int_0^R \frac{\rho}{2}\ \Omega^2 Rac(\theta_t - \phi_t) r^2\, dr = \frac{2}{3}\frac{C_T/\sigma}{a}\ \Omega^2 \rho a c R^4$$

The moment due to centrifugal force is:

$$M_{\text{c.f.}} = -\int_0^R (\Omega^2 m r) a_0 r\, dr = a_0 \Omega^2 I_b$$

where m is the mass per running foot and I_b is the blade's moment of inertia about its flapping hinge.

Setting the sum of the three moments equal to zero and solving for a_0 gives:

$$a_0 = \frac{\rho a c R^4}{I_b} \frac{2}{3} \frac{C_T/\sigma}{a} - \frac{M_w}{I_b \Omega^2} \text{ radians}$$

The combination of terms, $\rho a c R^4/I_b$, is a nondimensional parameter, γ, known as the Lock number after a pioneer British autogiro aerodynamicist, C. N. H. Lock. It represents the ratio of aerodynamic to centrifugal forces. Two blades that have the same airfoil and aspect ratio, and are constructed of material with the same density, will have the same Lock number no matter what their radius. The heavier the blade, the lower the Lock number. For operational blades, it varies from 10 for lightly built blades to 2 for blades of tip-driven helicopters with concentrated masses at the tips. The rotor parameters for the example helicopter are given in Appendix A. From these values:

$$\gamma_{\text{ex.hel.}} = \frac{\rho a c R^4}{I_b} = 8.1$$

The last term in the equation for coning is the contribution of the dead weight of the blade where:

$$M_w = -\int_0^R gmr\, dr$$

If the blade has a uniform mass distribution, then:

$$M_w = -\frac{3}{2}\frac{g}{R}I_b$$

and the coning equation may be written:

$$a_0 = \frac{2}{3}\gamma\,\frac{C_T/\sigma}{a} - \frac{\frac{3}{2}gR}{(\Omega R)^2}\ \text{radians}$$

As a general rule, the weight contribution is negligible except for very large rotors. For the example helicopter at its hover conditions, the calculated coning is 4.4° if the last term is ignored and 4.3° if it is included.

Although the derivation of the equation for coning was based on the assumption of a blade with a flapping hinge, it is also valid for flexible blades cantilevered from the shaft without hinges since the aerodynamic and centrifugal forces overpower whatever structural stiffness the blades might have.

Some past analyses have mistakenly led to the conclusion that there is an upper limit to how big a rotor can be built because of excessive coning. That this is an erroneous conclusion is evident from the foregoing equation provided the Lock number, γ, and the blade loading coefficient, C_T/σ, are constants. As a matter of fact, the coning will actually decrease because of the weight term as radius is increased while holding the same tip speed.

The equation provides a simple method of estimating the maximum coning that a rotor can develop. Earlier it was shown that C_T/σ is approximately 1/6 of the average lift coefficient, $\bar{c}_l$. Assuming a maximum value of 1.2 for this parameter gives a maximum value of 0.2 for C_T/σ. Later discussions will show that this is a reasonable upper limit. For the example helicopter with a blade Lock number of 8.1, this results in a maximum coning of just over 10°.

FACTORS AFFECTING HOVER PERFORMANCE

Review of Assumptions

The discussion of hover performance up to this point has been based on a number of simplifying assumptions. For normal rotors in normal flight conditions, the use of these assumptions gives good results consistent with a first approximation method; but when more accuracy is required, the assumptions must be challenged and their effects evaluated. The most important of the assumptions are:

- *Assumption*: The lifting portion of the blade extends from the center of rotation to the extreme tip.
  Challenge: The lifting portion of the blade actually starts some distance outboard from the center, and the tips are only partially effective.
- *Assumption*: Induced velocities are uniform over the disc.
  Challenge: Induced velocities are nonuniform.
- *Assumption*: The blades have ideal twist.
  Challenge: No actual blades have ideal twist.
- *Assumption*: The blades are rigid torsionally so that no structural twisting takes place.
  Challenge: No blade is infinitely rigid, and a rotating blade is subjected to twisting moments from several sources.
- *Assumption*: Blades have constant chord.
  Challenge: Some blades are tapered.
- *Assumption*: The wake does not rotate.
  Challenge: The wake does rotate.
- *Assumption*: There is no effect of tip vortices on the angle of attack of a following blade.
  Challenge: Tip vortex interference has been found to be significant.
- *Assumption*: The airfoil lift and drag characteristics are the same as the NACA 0012 characteristics of Figure 1.10.
  Challenge: Many rotors use airfoils different than the NACA 0012.
- *Assumption*: The airfoil characteristics are not a function of local stall or compressibility effects.
  Challenge: Stall and compressibility effects impose significant power penalties in some flight conditions.
- *Assumption*: There are no effects due to radial flow.
  Challenge: There are some effects due to radial flow.
- *Assumption*: The rotor is far from the ground.
  Challenge: In some cases the rotor is close to the ground.

Some of these assumptions result in errors in the calculation of the hover performance; others are good assumptions in that their consequences are negligible. In order to differentiate between the two types, the assumptions will be discussed one at a time.

Tip Loss and Root Loss

The lift of a rotor blade—or of a wing—goes to zero at the extreme tip, but it starts falling off some distance inboard. Thus the integration of the equation for rotor lift to the extreme tip is somewhat optimistic. A similar optimism, though of somewhat lesser importance, is introduced by starting the integration at the center of the rotor even though the blade "cutout" may be 10% to 25% of the radius. Figure 1.16 shows the theoretical and realistic lift distributions for a blade with

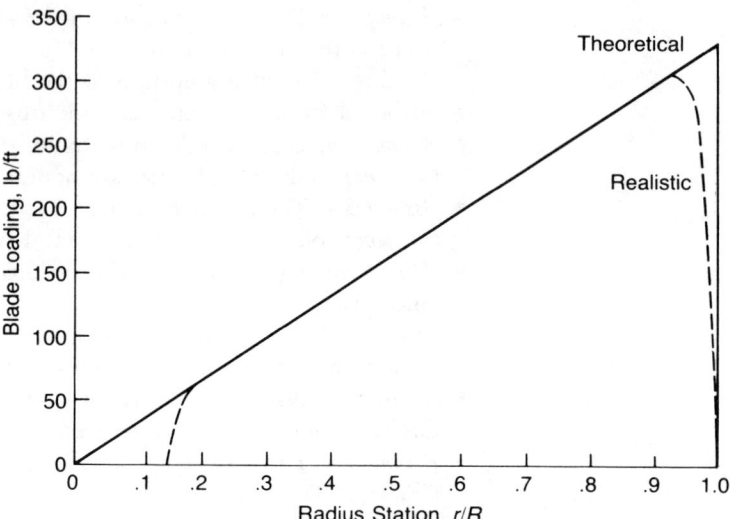

FIGURE 1.16   Theoretical and Realistic Lift Distributions

ideal twist. A numerical method for computing the shape of the lift distribution at the tip has been developed for propeller analysis and adapted for rotors in Reference 1.3. This method, known as the Goldstein-Lock method, is sometimes incorporated in digital computer hover performance programs. A more convenient scheme, however, is to use modified limits of integration such that:

$$T = b \int_{x_0 R}^{BR} \frac{\Delta L}{\Delta r} dr$$

where x_0 is the fraction of root cutout and BR is the effective outer radius, which is picked such that the area under the theoretical lift distribution out to BR is the same as the area under the actual curve out to R. The amount of the tip loss is dependent on the total lift of the blade and its geometry. The higher the lift, and the wider the chord with respect to the radius, the further inboard the lift starts falling off. Both these effects are included in an empirical equation for B that was first derived by Prandtl and gives satisfactory correlation with the Goldstein-Lock calculations for lightly loaded rotors.

$$B = 1 - \frac{\sqrt{2C_T}}{b}$$

For the main rotor of the example helicopter hovering at sea level, the equation gives an effective radius factor, B, of 0.97.

When root and tip losses are accounted for, the momentum equations previously derived should be modified. This can be done by defining an effective disc loading, $D.L._{eff}$, based on the effective disc area, which is smaller than the true area:

$$D.L._{eff} = \frac{T}{A(B^2 - x_0^2)}$$

Thus:

$$v_{1_{eff}} = \sqrt{\frac{D.L._{eff}}{2\rho}}$$

For the example helicopter,

$$D.L. = 7.1 \ \mathrm{lb/ft^2}$$

But

$$D.L._{eff} = \frac{7.1}{.97^2 - .15^2} = 7.7 \ \mathrm{lb/ft^2}$$

Thus at sea level

$$v_{1_{eff}} = 14.5\sqrt{7.7} = 40.2 \ \mathrm{ft/sec}$$

instead of the 39 ft/sec previously calculated.

With root and tip losses accounted for, the equation for C_T/σ becomes:

$$C_T/\sigma = (B^2 - x_0^2)\,\frac{a}{4}\,(\theta_t - \phi_t)$$

The value of the induced angle, ϕ_t, now must be based on only the effective disc area, although the definitions of the nondimensional coefficients are still based on the geometric rotor radius. Thus ϕ_t becomes:

$$\phi_t = \sqrt{\frac{C_T}{2(B^2 - x_0^2)}} = \sqrt{\frac{\sigma C_T/\sigma}{2(B^2 - x_0^2)}} \ \mathrm{radians}$$

Solving the equation for C_T/σ gives for pitch at the blade tip, θ_t:

$$\theta_t = 57.3\left[\frac{4}{a}\frac{C_T/\sigma}{(B^2 - x_0^2)} + \sqrt{\frac{\sigma C_T/\sigma}{2(B^2 - x_0^2)}}\right] \ \mathrm{degrees}$$

For the example helicopter with an ideally twisted rotor, the tip pitch required at the design hover condition is 7.1° instead of the 6.7° calculated without the losses. Since the lift falls off toward the tip, the induced drag also falls off, but the profile drag does not. At the root, there may not be any lifting surface, but there is always a spar or a section of the hub that has drag even if it has no lift. Thus the profile drag must be integrated from root to tip. Considering the increased inflow angle due to the decrease in effective disc area, the expression for C_P/σ of a rotor with ideal twist is:

$$C_P/\sigma \doteq C_T/\sigma \sqrt{\frac{\sigma C_T/\sigma}{2(B^2 - x_0^2)}} + \frac{c_d}{8}$$

The drag coefficient for use in this equation should be found from airfoil data such as in Figure 1.10 and an average angle of attack, which is:

$$\bar{\alpha} = \frac{57.3 C_T/\sigma}{B^2 - x_0^2} \text{ degrees}$$

For the example helicopter with ideal twist, this method gives the rotor power required for hover at the design gross weight as 1,840 horsepower if root and tip losses are not considered and 1,900 horsepower if they are, a difference of about 3.2%.

There is no easy way to verify experimentally the simple hover tip loss equation, but it may at least be partially justified by comparing the results of hover calculations made using it with those made using the more sophisticated prescribed wake vortex method. Figure 1.17 shows the calculated ideal Figure of Merit, assuming no profile drag, as it is affected by number of blades. The solid line is from reference 1.2 and was calculated for uniform inflow using a prescribed wake vortex method. The dashed line is for a rotor with ideal twist and is based on the reduction of disc area due to root and tip losses and also includes the effect of wake rotation, as discussed in a later section. It may be seen that the shape of the two lines is essentially the same, indicating that the simple tip loss equation is in good company.

Calculation of Nonuniform Induced Velocity Distribution

The assumption of a uniform induced velocity distribution simplified the previous analysis, but in order to reflect more accurately actual conditions, it must be replaced by a nonuniform distribution. This distribution can be considered to consist of two effects, a local tip effect due to vortex interference, which will be discussed later, and an overall effect, which can be analyzed by combining the momentum and the blade element systems of analysis at an annulus of the disc, as in Figure 1.18. The increment of thrust on this annulus, ΔT, is:

$$\Delta T = \rho v_1 2\pi r \Delta r v_2$$

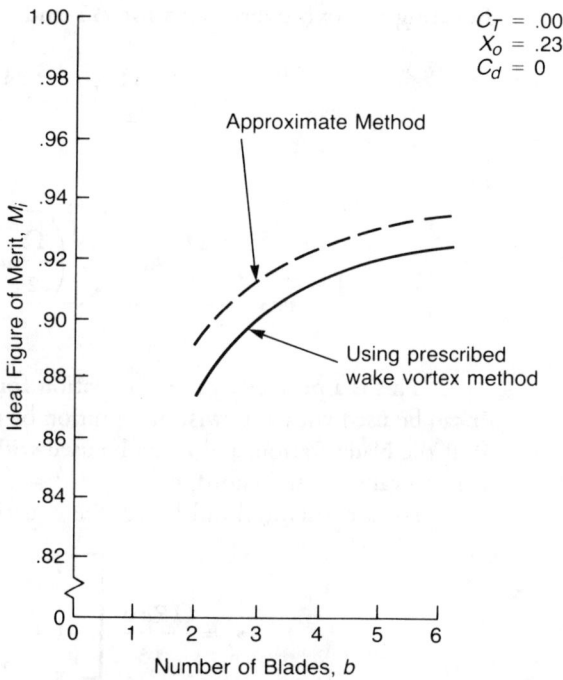

FIGURE 1.17 Effect of Number of Blades on Ideal Figure of Merit as Calculated by Two Methods

Source: Harris & McVeigh, "Uniform Downwash with Rotors Having a Finite Number of Blades," *JAHS* 21-1, 1975.

where $2\pi r \Delta r$ is the area of the annulus and v_1 and v_2 are the induced velocities at the rotor disc and in the remote wake, respectively. Just as in the original derivation of the momentum equation, it may be shown that:

$$v_2 = 2v_1$$

so that the equation for ΔT becomes:

$$\Delta T = 4\rho\pi v_1^2 r \Delta r$$

From the blade element theory, the increment of thrust can also be written:

$$\Delta T = b\frac{\rho}{2}\ (\Omega r)^2 a \left(\theta - \frac{v_1}{\Omega r}\right)c\Delta r$$

Equating the two expressions for ΔT and arranging the result gives:

$$4\pi v_1^2 + \frac{\Omega}{2}\, bacv_1 - \frac{\Omega^2}{2}\, rba\theta c = 0$$

or

$$v_1 = \frac{-\dfrac{\Omega}{2}\, acb + \sqrt{\left(\dfrac{\Omega}{2}\, acb\right)^2 + 8\pi b\Omega^2 ra\theta c}}{8\pi}$$

This is a perfectly general equation for the induced velocity at any radius, r. It can be used with any twist distribution by using the correct value of blade pitch, θ, at the blade station, and it can be used with any blade taper scheme by using the correct value of the chord, c.

For a constant chord blade, the equation can be manipulated to give:

$$v_1 = \frac{\Omega Ra\sigma}{16}\left[-1 + \sqrt{1 + \frac{32\theta\, \dfrac{r}{R}}{a\sigma}}\right]$$

Note: If the analysis is being done for a rotor with cambered airfoils, θ should be replaced by $(\theta - \alpha_{oL})$ where α_{oL} is the angle of attack for zero lift.

For this case, it may be seen that the induced velocity depends on the radius station only as the parameter, $\theta(r/R)$, varies with the radius. If the rotor has ideal twist, then this parameter is a constant being equal to the pitch at the tip:

$$\theta\, \frac{r}{R} = \frac{\theta_t}{r/R}\, \frac{r}{R} = \theta_t$$

For blades with constant chord and ideal twist, the induced velocity is a constant across the disc as was originally assumed in the momentum theory. Thus one definition of *ideal twist* is the twist required for constant chord blades to produce a uniform induced velocity.

The lift distribution corresponding to ideal twist and uniform induced velocity is triangular. This is in contrast to the ideal wing, which produces uniform induced velocity with an elliptical lift distribution. The equation for lift distribution written in terms of the circulation, Γ, is:

$$\frac{dL}{dr} = c\Gamma V = c\Gamma\Omega r$$

For a triangular lift distribution on an ideal blade, the circulation is a constant; and only two trailing vortices are generated: one at the root and one at the tip.

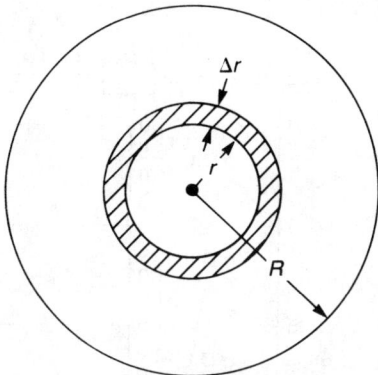

FIGURE 1.18   Geometry of Rotor Annulus

The nonuniform distribution of induced velocity of a rotor that does not have ideal twist manifests itself in the remote wake as a nonuniform dynamic pressure distribution, which is significant in making estimates of fuselage downloads or ground erosion. Figure 1.19 shows the measured distribution from reference 1.4 for several locations downstream of a full-scale rotor with a linear twist of −4°. It may be seen that the wake contracts very rapidly, accomplishing most of the contraction within the first 10% of radius. The tests were conducted in winds of less than 3 knots, but the test results show that even winds this low can deflect the wake significantly.

Nonideal Twist

The primary effect of nonideal twist is to require more induced power than ideal twist. Figure 1.20 shows calculated distributions of pitch, induced velocity, angle of attack, and drag loading for several values of twist for the rotor of the example helicopter at its design gross weight. Several observations may be made about this series of plots:

- Ideal twist gives constant induced velocity and constant induced drag loading.
- A linear twist of −20° comes closest to simulating ideal twist.
- All the linear twist curves have approximately the same pitch at the 75% blade station. (This is a good rule of thumb for all thrust levels.)
- The profile drag loading with ideal twist is high near the root because of high angles of attack.

Figure 1.21 shows the effect of twist on hover performance of the example helicopter. From this series of plots, it may be observed that:

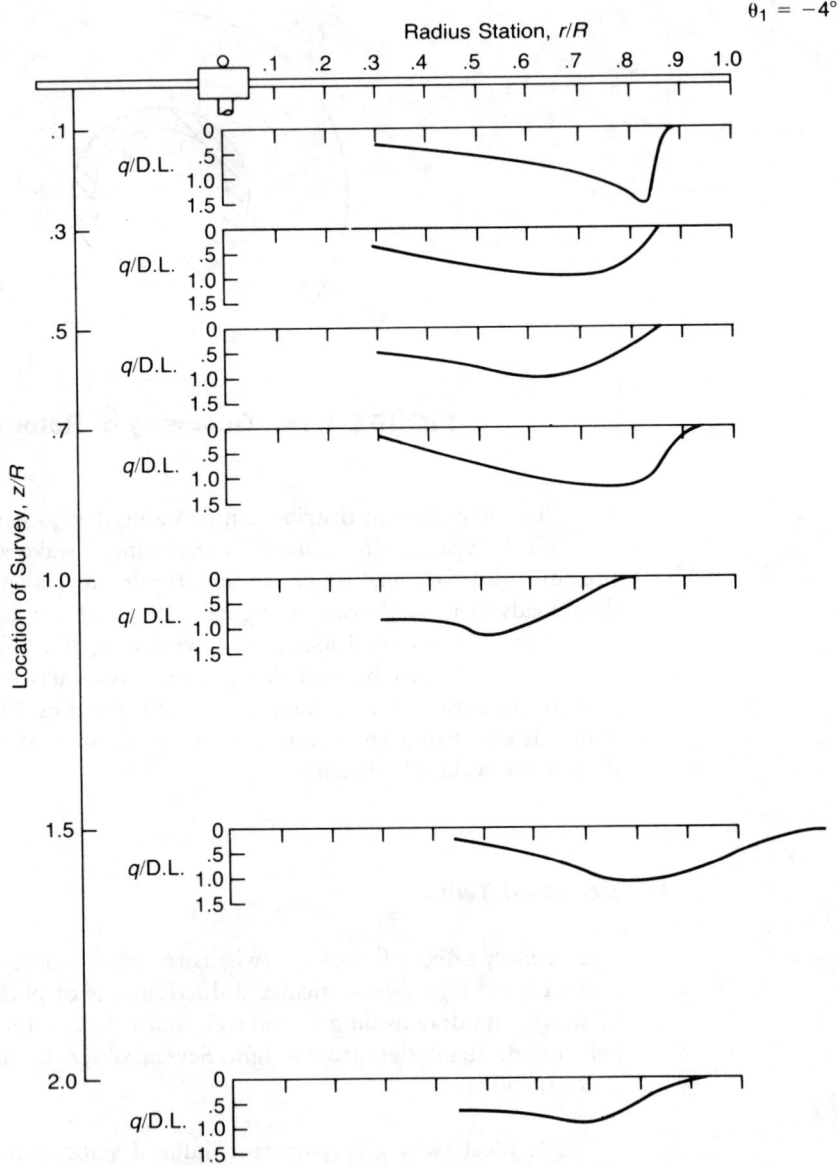

FIGURE 1.19 Measured Wake Characteristics

Source: Boatwright, "Measurements of Velocity Components in the Wake of a Full-Scale Helicopter Rotor in Hover," USAAMRDL TR 72-33, 1972.

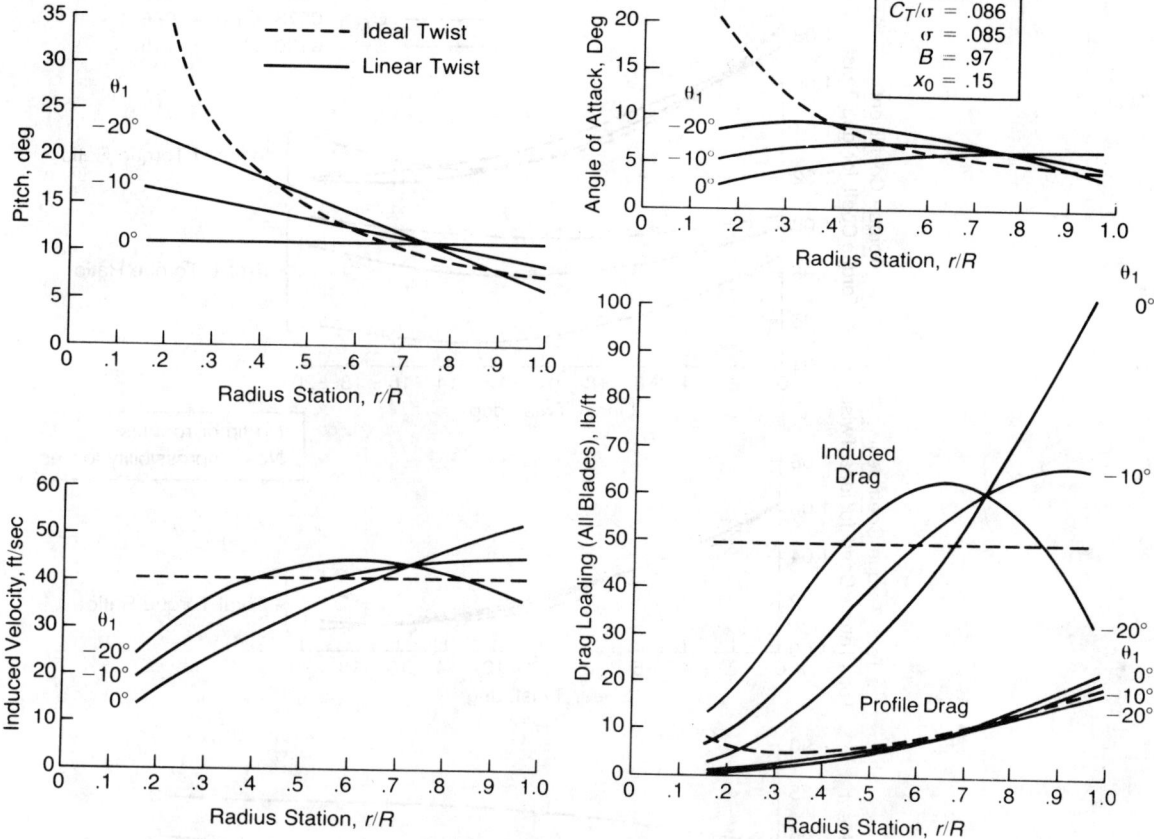

FIGURE 1.20   Effect of Twist on Hover Conditions

- Increasing the twist decreases the induced torque from 8% more than ideal twist to about 2% more.
- It is possible to have too much twist, especially at the light disc loadings.
- A rotor with linear twist in general has less profile torque than a rotor with ideal twist because of the high inboard angles of attack noted on the previous figure.
- Going from no twist to ideal twist can raise the Figure of Merit about 5%.
- Most of the potential benefit of twist is realized in the first 10° of twisting.

Figure 1.22, based on references 1.5 and 1.6, shows that the theoretical effects of twist are verified by model tests.

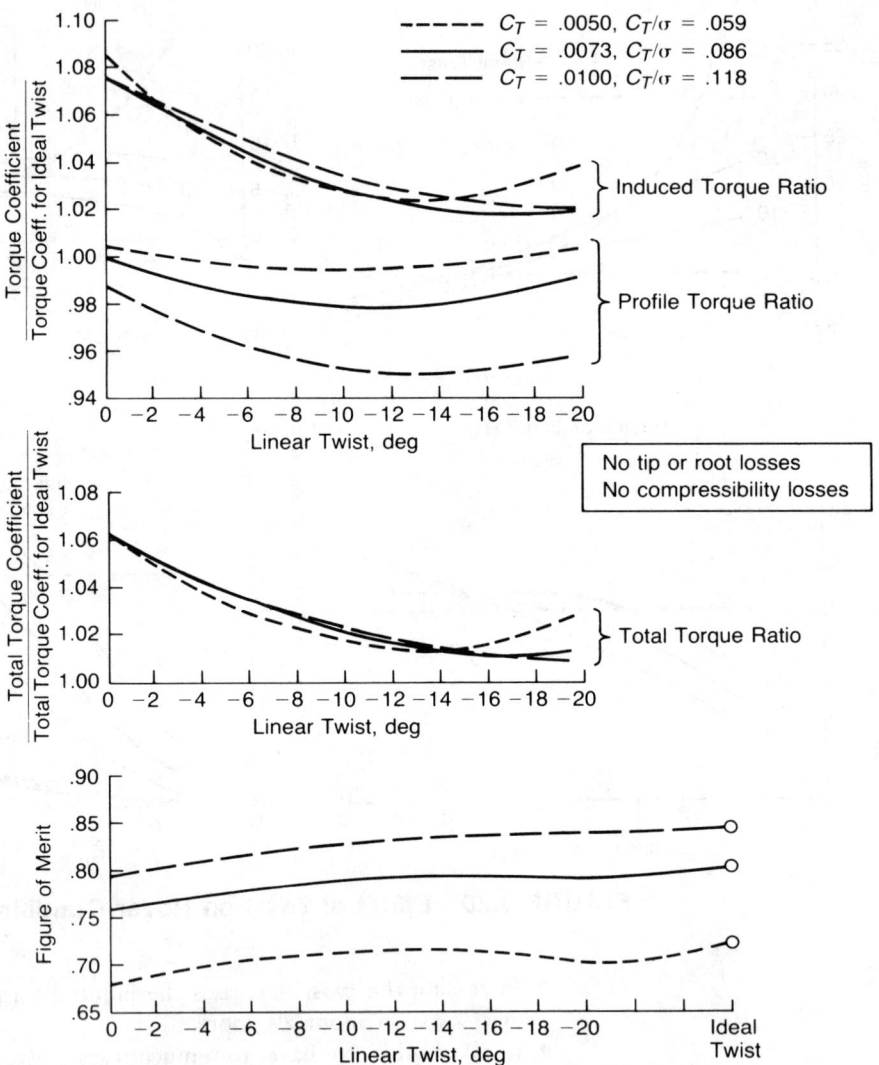

FIGURE 1.21   Effect of Twist on Hover Performance

It should be pointed out that whereas high twist is beneficial in hover, it produces high vibratory loads in high-speed forward flight and thus is usually limited to some compromise value. Currently this compromise is in the neighborhood of −5° to −16°.

One secondary twist consideration is that the negative values that are beneficial in reducing the angles of attack in powered flight are detrimental in autorotation. Reference 1.7 also presents test data showing that the optimum twist for hovering in ground effect is significantly less than is optimum for hovering out of ground effect.

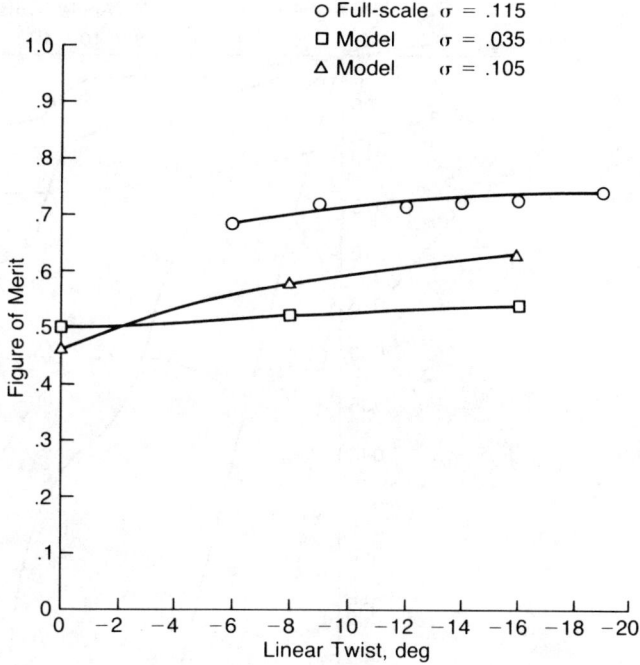

FIGURE 1.22   Effect of Twist on Measured Rotor Performance

Sources: Clark, "Can Helicopter Rotors Be Designed for Low Noise and High Performance?" AHS 30th Forum, 1974; Landgrebe, "An Analytical and Experimental Investigation of Helicopter Rotor Hover Performance and Wake Geometry Characteristics," USAAMRDL TR 71-24, 1971.

Dynamic Twist

A rotating blade may be subjected to torsional moments that can modify the twist distribution significantly from the nonrotating, built-in twist. Reference 1.8 reports measured "dynamic twist" of up to 5° on a full-scale whirl tower rotor. Several sources of torsional moments can be identified. One is the airfoil's aerodynamic pitching moment about the quarter chord, which is a function of blade camber and the local combination of angle of attack and Mach number. Figure 1.23 shows the measured pitching moment characteristics of a symmetrical and of a cambered airfoil from reference 1.9. The plotted parameter is the product of the pitching moment coefficient and the Mach number squared, which is proportional to the actual pitching moment. Even for the symmetrical airfoil, the pitching moments are not small except at combinations of low angles of attack and low Mach numbers. Another significant source of a torsional moment is the position of the airfoil aerodynamic center with respect to the blade flexual axis. An

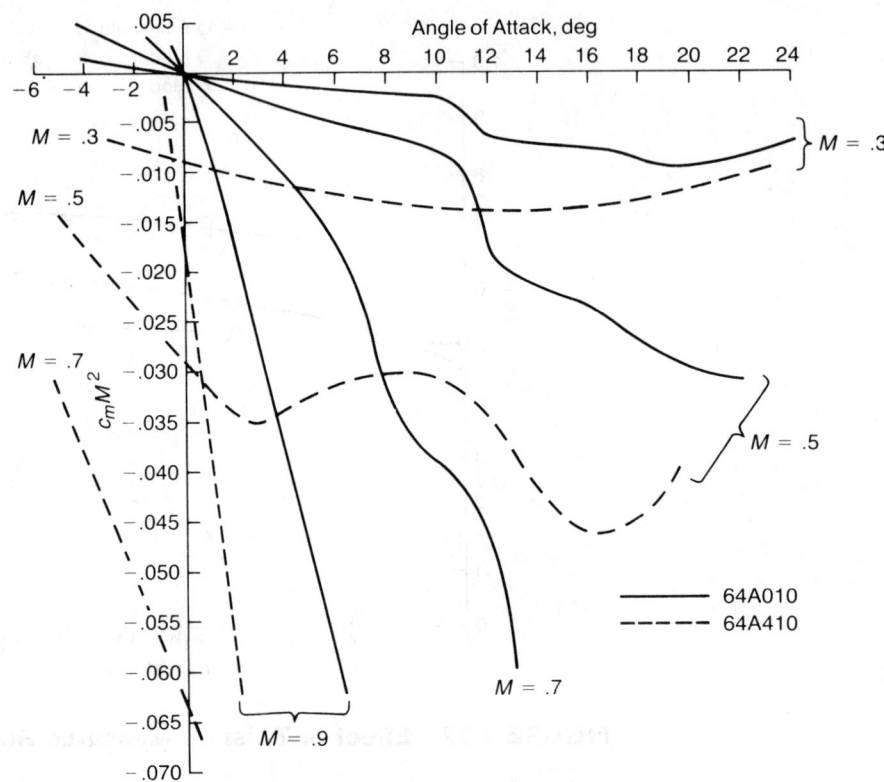

FIGURE 1.23 Pitching Moment Characteristics for a Symmetrical and a Cambered Airfoil

Source: Stivers, "Effects of Subsonic Mach Number on the Forces and Pressure Distributions on Four NACA 64A-Series Airfoil Sections at Angles of Attack as High as 28°," NACA TN 3162, 1954.

aerodynamic center position forward of the flexual axis will result in nose-up twisting moments as lift is increased. Some designers use swept-back tips to counteract this effect.

The last effect is a centrifugal flattening moment sometimes known as the *tennis racket effect*, named for the tendency of a tennis racket to try to align its plane with the plane of rotation as it is swung in an arc. The forces and moments due to centrifugal forces acting on a blade with positive pitch are shown in Figure 1.24. It may be seen that the forces acting on the mass elements at the leading and trailing edges produce a torsional moment that tends to twist the blade toward flat pitch. This moment not only twists the blade but also produces a control moment that must be counterbalanced to hold the blades at a positive pitch. This moment may be reduced by the use of balancing weights located perpendicular to the blade

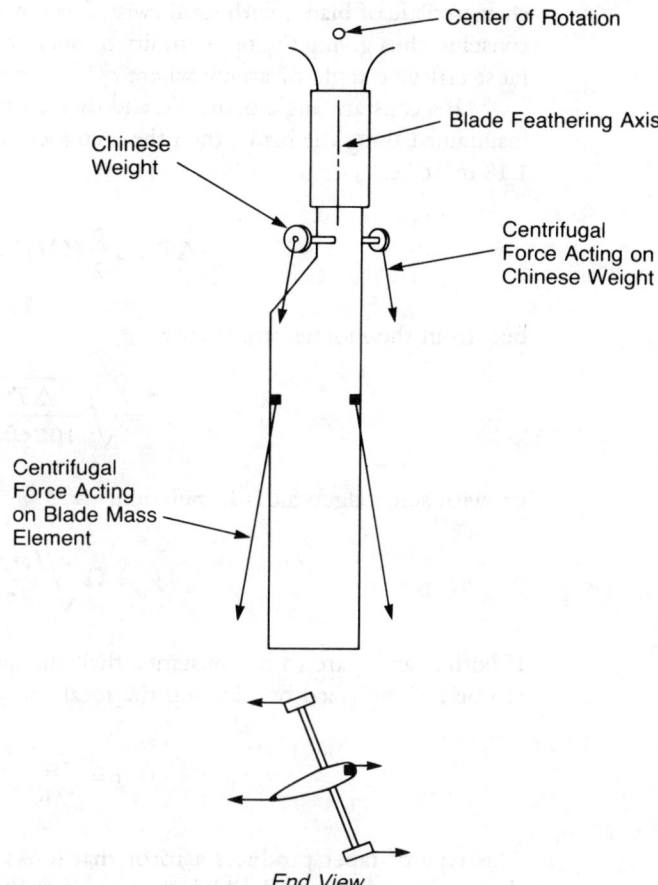

FIGURE 1.24   Twisting Moments Due to Centrifugal Forces

chord, as shown in Figure 1.24. These are usually called Chinese weights, for reasons perhaps better left unexplained.

Dynamic twist has proved to be a problem in correlating measured thrust with measured collective pitch, but since the thrust and the power are both affected to about the same degree, small amounts of dynamic twist have little effect on the power-to-thrust relationships.

Effect of Taper

Constant chord blades are easy to design and to manufacture, but tapered blades can be made to be more efficient aerodynamically. A special combination of taper and twist can produce not only the uniform induced velocity that is the special characteristic of blades with ideal twist, but can also make the local angle of attack constant, thus giving the opportunity to operate each blade element at the airfoil's most efficient angle of attack where $c_l^{3/2}/c_d$ is a maximum.

If a constant angle of attack, and therefore a constant lift coefficient, is to be maintained along the blade, then the increment of thrust on the annulus of Figure 1.18 must be:

$$\Delta T = b\frac{\rho}{2}(\Omega r)^2 c_l c \Delta r$$

but, from the momentum theory,

$$v_1 = \sqrt{\frac{\Delta T}{4\rho\pi r\Delta r}}$$

or, with some algebraic manipulation,

$$v_1 = \Omega\sqrt{\frac{bcrc_l}{8\pi}}$$

If both v_1 and c_l are to be constants, then the quantity cr must be a constant. This can be accomplished by defining the local chord, c, such that:

$$c = \frac{c_{\text{tip}}}{r/R}$$

This type of taper produces a rotor that looks like the one in Figure 1.25—one that is impractical to build but is interesting in being theoretically the most optimal hovering rotor that can be designed. If the lift coefficient is to be kept constant, then the local angle of attack must be constant where:

$$\alpha = \theta - \frac{v_1}{\Omega r}$$

In order that both α and v_1 are constants, then the blade must be twisted so that:

$$\theta = \alpha + \frac{v_1}{\Omega r}$$

This special twist distribution is shown in Figure 1.26, where it is compared with the ideal twist for a constant-chord blade. A series of calculations for the ideally twisted, constant-chord blade discussed in the previous paragraph shows that this rotor has a maximum Figure of Merit of 0.85 at an average lift coefficient of 0.94.

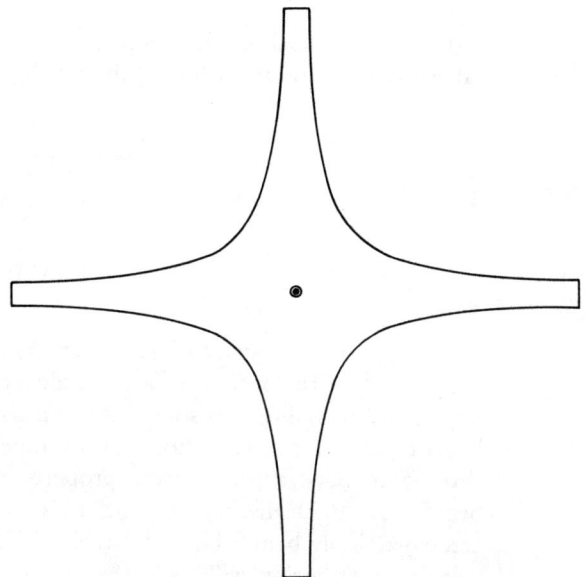

FIGURE 1.25   Rotor with Ideal Taper

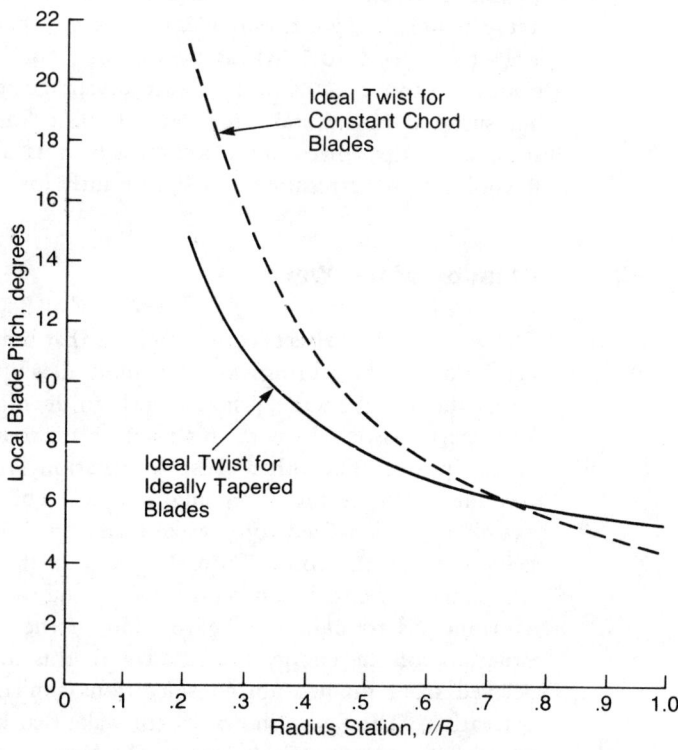

FIGURE 1.26   Ideal Twist for Blades with Constant Chord and Blades with Ideal Taper

If the angle of attack had been a constant, the Figure of Merit would have been the value from the equation for maximum Figure of Merit:

$$\text{F.M.} = \cfrac{1}{1 + \cfrac{\frac{3}{2}\sqrt{3}}{\sqrt{\sigma}\,\dfrac{c_l^{3/2}}{c_d}}} = .89$$

Thus ideal taper can increase the maximum hover performance about 4% over a blade with constant chord and ideal twist. Since the ideally tapered blade is impractical to build, it is sometimes approximated with a blade that has either a linear taper or a constant chord out to some radius station and a linear taper from that point. Such a rotor, when properly twisted, can achieve a portion of the benefit of an ideally tapered and twisted blade. Even without twist, taper is aerodynamically beneficial. A study of the equation for the nonuniform induced velocity distribution will show that a constant induced velocity can be achieved with a constant pitch if the chord is inversely proportional to the blade station, as well as for a constant chord with the pitch inversely proportional to the blade station. Reference 1.10 reports a comparison of two model rotors, each with untwisted blades, which were identical except that one was untapered and one had a taper ratio of 2 to 1. As can be seen in Figure 1.27, the rotor with the tapered blades had about a 10% performance advantage at low and moderate thrust values but suffered earlier stall. Reference 1.10 attributes the earlier stall to a larger amount of tip vortex interference, but it is also possible that the lower tip Reynolds number resulted in a significantly lower maximum lift coefficient.

Rotation of the Wake

The wake has two effects operating on it that tend to produce rotation. One is the profile drag, which brings some air molecules up to the speed of rotation before losing them to the wake, just as a truck on the highway produces a following wake. The energy associated with this rotation is accounted for in the computation of profile power. The other cause of rotation is an induced effect that was not accounted for in the foregoing analysis. This rotation may be visualized by examining an idealized rotor wake made up of tip and root vortices which form helixes under the rotor. From Figure 1.28 it may be seen that the horizontal components of circulation in both the tip and the root vortices are oriented in such a manner as to induce wake rotation in the direction of rotor rotation. The equation for the energy associated with this induced rotation, or swirl, can be derived from momentum considerations similar to those used in the previous derivations. The figure shows a rotor wake that has no rotation above the disc but which has a rotation, ω, below it. The change in total pressure is:

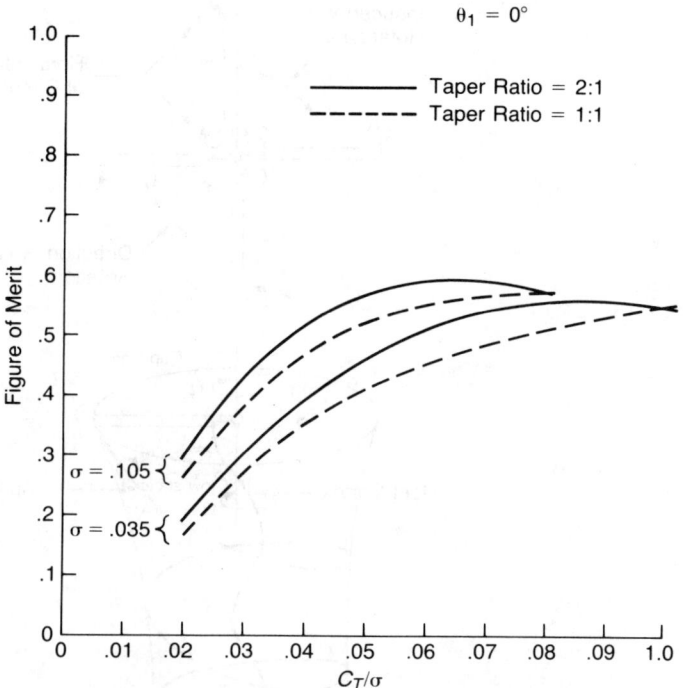

FIGURE 1.27   Effect of Blade Platform Taper on Measured Rotor Hover Performance

Source: Bellinger, "Experimental Investigation of Effects of Blade Section Camber and Planform Taper on Rotor Performance," USAAMRDL TR 72-4, 1972.

$$\Delta p = [p_{\text{low static}} + \tfrac{1}{2}\rho v_1^2 + \tfrac{1}{2}\rho(\omega r)^2] - [p_{\text{up static}} + \tfrac{1}{2}\rho v_1^2]$$

or:

$$\Delta p = \Delta p_{\text{static}} + \tfrac{1}{2}\rho(\omega r)^2$$

The change in static pressure can be related to a change in induced power through the familiar induced power expression:

$$\Delta P_i = \Delta T v_1$$

or

$$\Delta P_i = \Delta p_{\text{static}} \; \Delta A v_1$$

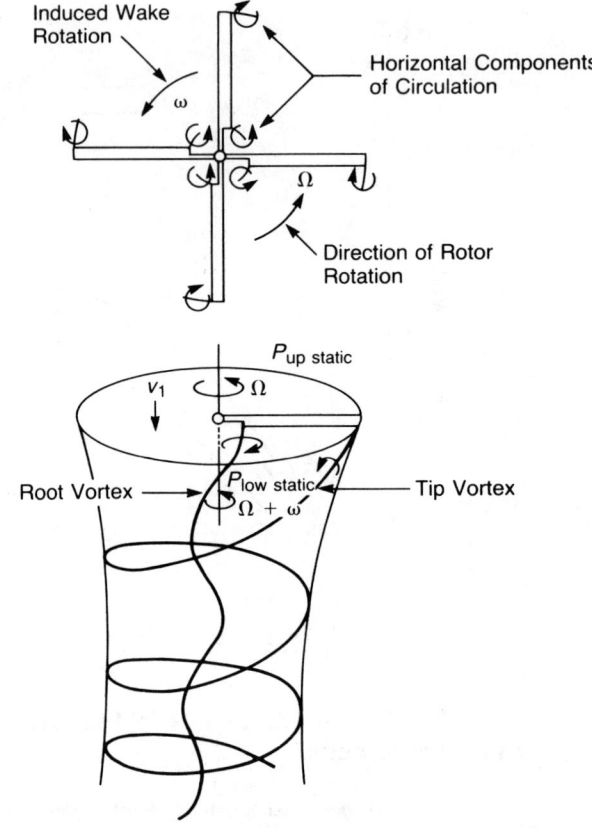

FIGURE 1.28   Components of Vorticity Producing Wake Rotation

thus:

$$P_{i_{\text{thrust}}} = \left(\frac{T}{A}\right) A \sqrt{\frac{T}{2\rho A}}$$

or

$$P_{i_{\text{thrust}}} = T \sqrt{\frac{T}{2\rho A}}$$

The induced power associated with rotation is:

$$\Delta P_{i_{\text{rotation}}} = \frac{1}{2}\rho(\omega r)^2 \Delta A v_1$$

Assuming a uniform induced velocity, v_1, for this analysis:

$$P_{i_{\text{rotation}}} = \rho \pi R^4 v_1 \int_0^1 \left(\frac{r}{R}\right)^3 \omega^2 d\left(\frac{r}{R}\right)$$

which can be rewritten:

$$\frac{P_{i_{\text{rotation}}}}{P_{i_{\text{thrust}}}} = \frac{1}{C_T} \int_0^1 \left(\frac{r}{R}\right)^3 \left(\frac{\omega}{\Omega}\right)^2 d\left(\frac{r}{R}\right)$$

An expression for ω/Ω as a function of r/R can be derived by writing Bernoulli's equation for air flow relative to the blade just above and just below the rotor disc:

$$p_{\text{up static}} + \tfrac{1}{2}\rho(\Omega r)^2 = p_{\text{low static}} + \tfrac{1}{2}\rho(\Omega - \omega)^2 r^2$$

or

$$\Delta p_{\text{static}} = \tfrac{1}{2}\rho r^2 [2\Omega\omega - \omega^2]$$

but

$$\Delta p_{\text{static}} = \text{D.L.} = \rho C_T (\Omega R)^2$$

thus

$$\frac{\omega}{\Omega} = 1 - \sqrt{1 - \frac{2C_T}{\left(\dfrac{r}{R}\right)^2}}$$

Note that ω is imaginary if $(r/R)^2 < 2C_T$. To avoid this in the integration, the lower limit can be set to $\sqrt{2C_T}$ with little loss of validity:

$$\frac{P_{i_{\text{rotation}}}}{P_{i_{\text{thrust}}}} = \frac{1}{C_T} \int_{\sqrt{2C_T}}^1 \left(\frac{r}{R}\right)^3 \left[1 - \sqrt{1 - \frac{2C_T}{\left(\dfrac{r}{R}\right)^2}}\right]^2 d\frac{r}{R}$$

The integral has been evaluated as a function of C_T and is plotted in Figure 1.29. Also shown are the results of two more rigorous analyses from references 1.11 and 1.12, which were made with different assumptions but which resulted in nearly the same values as the approximate method. For the example helicopter in

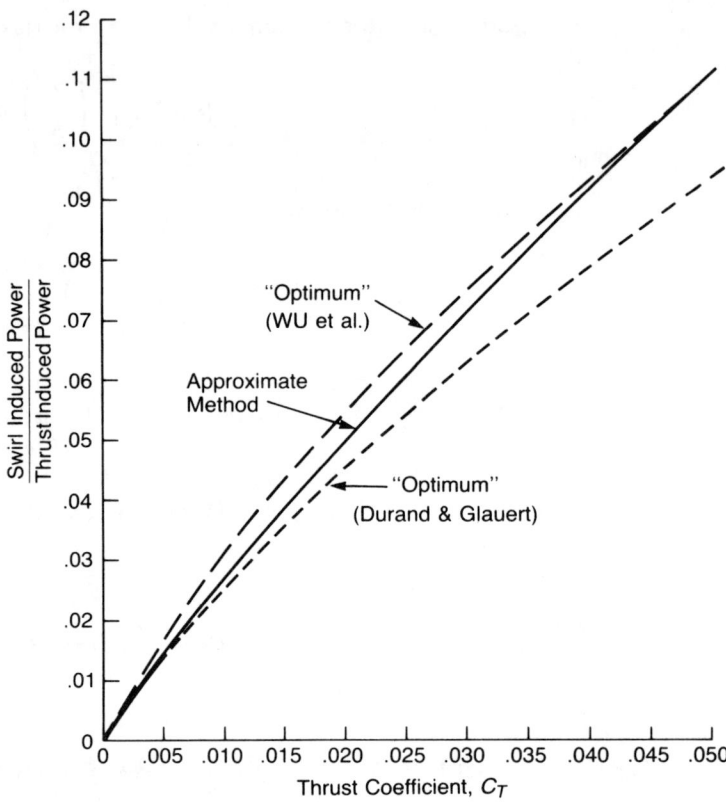

FIGURE 1.29 Power Losses Due to Rotation of Wake

Source: Durand & Glauert, *Aerodynamic Theory*, Division L, "Airplane Propellers," Julius Springer, Berlin, 1935; Wu, Sigman, & Goorjian, "Optimum Performance of Hovering Rotors," NASA TMX 62138, 1972.

hover, the value of C_T is 0.0073 and the corresponding induced power due to wake rotation is about 2% of that corresponding to thrust. This correction is marginally significant for this rotor but would be significant for more heavily loaded rotors or propellers used for static thrust.

The aerodynamicist will occasionally be asked to calculate the rotational velocity in the wake to define conditions at the engine inlet and exhaust or in front of rocket pods. The equation for the induced rotation, $(r\omega)_i$, is

$$(r\omega)_i = \Omega R \ \frac{r}{R} \left[1 - \sqrt{1 - \frac{2C_T}{(r/R)^2}} \right]$$

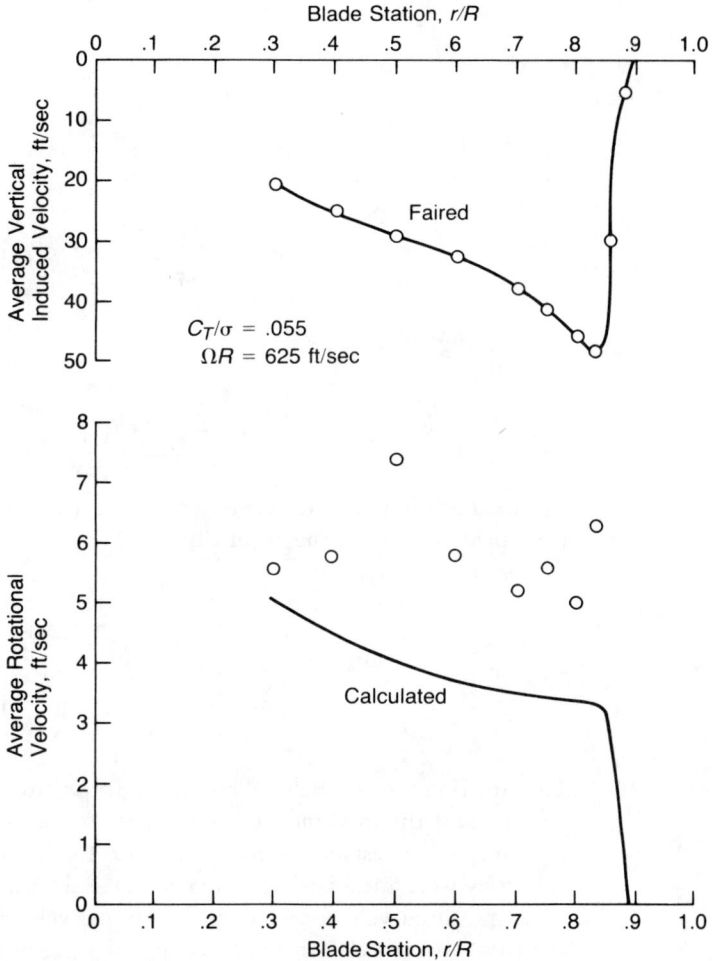

FIGURE 1.30   Vertical and Rotational Velocities at a Location .1 R below Rotor

Source: Test points are from Boatwright, "Measurements of Velocity Components in the Wake of a Full-Scale Helicopter Rotor in Hover," USAAMRDL TR 72-33, 1972.

The corresponding equation for the rotational velocity due to profile drag can be derived from the momentum equation in the annulus of Figure 1.18:

$$F = (m/sec)(\Delta v)$$

or

$$d\Delta D_0 = (\rho \Delta r 2\pi r v_1)(\omega r)_0$$

but

$$\Delta D_0 = \frac{\rho}{2} (\Omega r)^2 c c_d \Delta r$$

so that:

$$(\omega r)_0 = \frac{b \Omega^2 c c_d r}{4 \pi v_1} = \frac{\sigma c_d}{4 v_1} (\Omega R)^2 \frac{r}{R}$$

Using the fact that

$$v_1 = \Omega R \sqrt{\frac{C_T}{2}}$$

the induced and profile terms may be combined to give an equation for the rotational velocity at the rotor disc:

$$\omega r = \frac{\frac{r}{R} v_1}{\sqrt{\frac{C_T}{2}}} \left[1 - \sqrt{1 - \frac{2 C_T}{\left(\frac{r}{R}\right)^2} + \frac{\sigma c_d}{4 \frac{v_1}{\Omega R}}} \right]$$

The equation can also be used below the rotor disc and for rotors that do not have ideal twist if the local induced velocity is used in place of v_1. For example, during the whirl tower tests reported in reference 1.4, both the vertical and the rotational velocities were measured at a location .1 R below the rotor. Figure 1.30 shows these measured velocities and the rotational velocity calculated from the vertical velocity. The correlation indicates that the method is adequate, at least for the region in which engines or rocket pods would be located.

Tip Vortex Interference

The induced velocity previously derived from the combined momentum and blade element theory can also be obtained using a vortex method as shown in reference 1.13. The method is similar to that used for wing analysis, in which trailing vortices are assumed to leave the wing between adjacent wing elements and to have a strength proportional to the change in lift between the elements. In the simplest rotor vortex theory, the trailing vortices from an infinite number of blades are assumed to form concentric cylindrical vortex sheets with no wake contraction. In practice, the wake does contract producing a local distortion of the induced velocity near the blade tips. Figure 1.31 shows the location of the tip vortices from a finite number of blades with and without wake contraction. It may be seen that the contraction of the wake is such as to make the older vortices force the youngest vortex up toward the rotor plane. As a matter of fact, photographs of the

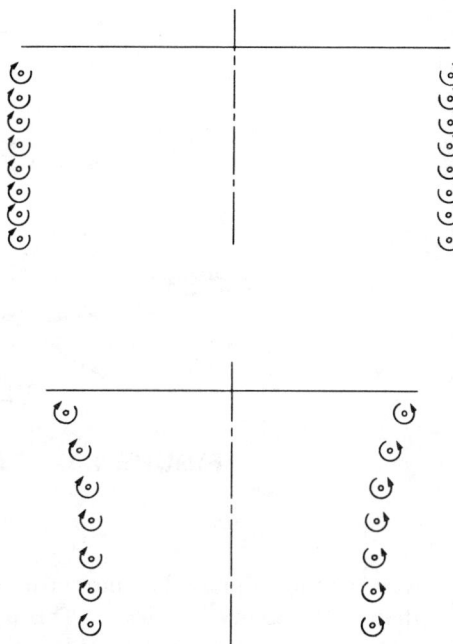

FIGURE 1.31   Tip Vortex Locations with and without Wake Contraction

tip vortex made on a humid day at Sikorsky and reproduced in reference 1.14 have shown that in some cases the tip vortex remains in the tip path plane until the next blade actually strikes it, as sketched in Figure 1.32. The changes in the local induced velocity due to the proximity of the vortex cause large discontinuities in the angle of attack distribution, as shown in the top portion of Figure 1.33, which is based on calculations made by a sophisticated computer program as reported in reference 1.5. The lift vectors near the tip have more rearward tilt, and in some cases the tip is stalled compared with the simpler predictions. The distortions affect the distribution of power along the blade, as shown in the bottom portion of Figure 1.33. An attempt to compensate for the high angles of attack near the tip is reported in reference 1.15. It has resulted in the design of a local region of high and nonlinear blade twist on the Sikorsky Blackhawk.

It is to be hoped that a simple and accurate analytical computing method accounting for vortex interference will eventually be developed. In the meantime it is suggested that the momentum method be used with an empirical correction factor. The suggested factor is based on a study of data presented in reference 1.5, which reports on tests of a family of model rotors that had variations in number of blades, twist, blade aspect ratio, and tip speed. The results of this study are summarized in Figure 1.34 as the ratio of the measured power to the power calculated by the momentum method—accounting for variable tip loss factor and

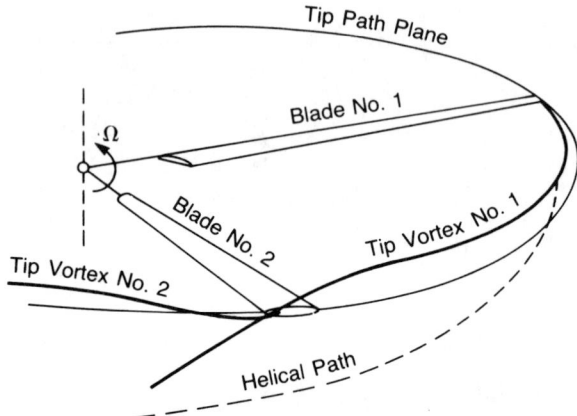

FIGURE 1.32   Tip Vortex Interference

wake rotation—plotted against a parameter that is the product of disc loading and the thrust/solidity coefficient. This parameter has little justification in logic but gives better correlation than either disc loading or C_T/σ alone. It does reflect the idea that the wake contraction effect starts as an induced phenomenon that eventually results in stall of local blade elements. Similar results are obtained using the full-scale rotor performance data of reference 1.14. The use of the suggested empirical correction in Figure 1.34 eliminates the number of blades, twist, blade aspect ratio, and tip speed as significant parameters in the vortex interference problem—an assumption that appears to be justified but will warrant continuous review in light of further experience.

The tip vortices are also responsible for another characteristic of hover performance—the relative unsteadiness of the condition. Smoke studies reported in reference 1.5 show that distinct tip vortices can be traced down into the wake only about one radius. At that point, they tend to couple together in a random way, which sometimes reinforces and sometimes cancels the vortex effect. A similar observation reported in reference 1.16 reveals that usually only four well-defined tip vortices can be identified under the reference blade, regardless of the number of blades in the rotor. Figure 1.35 shows Schlieren photographs of the wake of a model rotor in a quiet room. The single exposure shows the third vortex beginning to twist about itself, which soon resulted in self-destruction. Further observation shows that when the vortex does begin to dissipate, the wake ceases to contract and begins to expand instead, as shown in the third photograph of Figure 1.35. The multiple-exposure photograph shows that the vortex paths do not repeat themselves and that at least once during the 15 revolutions a very significant transient excursion occurred. This type of randomness manifests itself as unsteadiness in the inflow at the rotor disc and in both measured thrust and power.

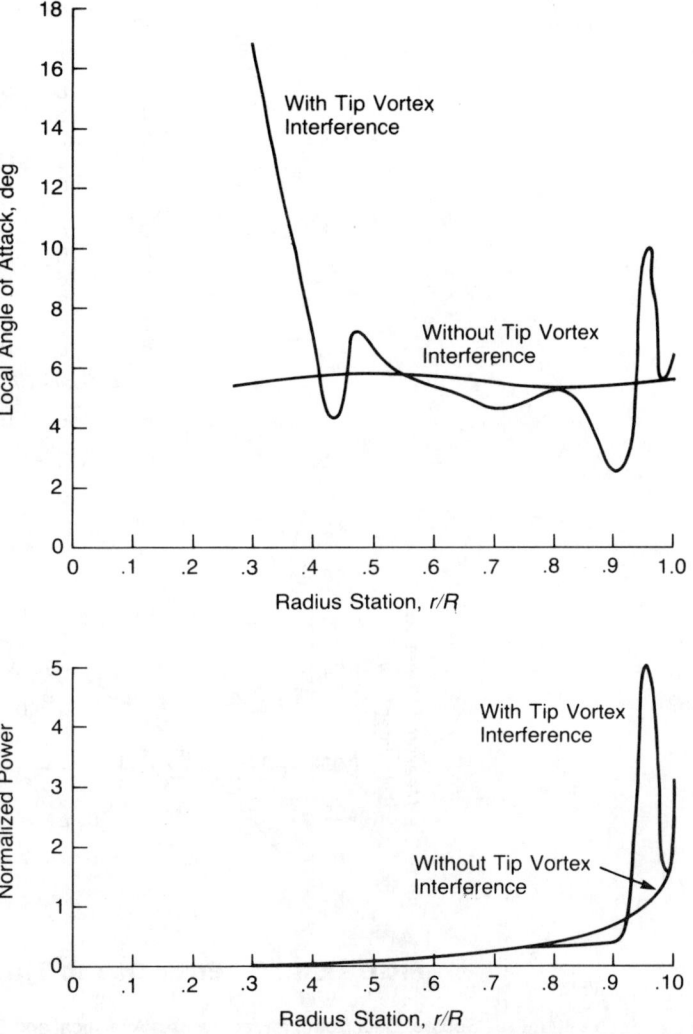

FIGURE 1.33   Calculated Angle of Attack and Power Distributions

Source: Clark, "Can Helicopter Rotors Be Designed for Low Noise and High Performance?" AHS 30th Forum, 1974.

Figure 1.36 presents a 14-second record of the thrust and power variations of the AH-56A rotor on the Lockheed whirl tower. The variations amount to approximately ±3% for thrust and ±6% for power. This type of variation makes hover performance difficult to measure accurately and probably accounts for much of the test scatter in Figure 1.34. Yet another illustration of the nonsteadiness of hover is given by reference 1.14, where it is shown that even on a whirl tower in

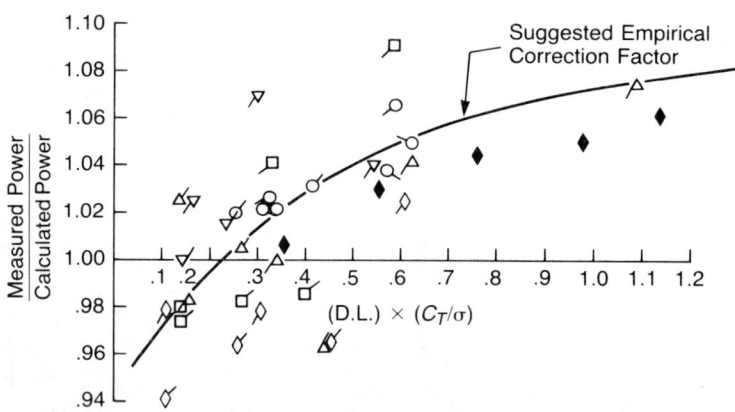

| | | 4.5-foot Dia. Model | | |
Symbol	b	θ_1	AR	M_{tip}
	2	−8	18.2	.625
	4	−8	18.2	.625
	6	−8	18.2	.625
	8	−8	18.2	.625
	2	0	18.2	.625
	6	0	18.2	.625
	2	−16	18.2	.625
	6	−16	18.2	.625
	2	−8	18.2	.47
	6	−8	18.2	.47
	2	−8	13.2	.625
	6	−8	13.2	.625
		Full-Scale Ch-53 Rotor		
◆	6	−6	16.7	.624

FIGURE 1.34   Error Due to Tip Vortex Interference

Source: Data from Landgrebe, "An Analytical and Experimental Investigation of Helicopter Rotor Hover Performance and Wake Geometry Characteristics," USAAMRDL TR 71-24, 1971; and Jenney, Olson, & Landgrebe, "A Reassessment of Rotor Hovering Performance Prediction Methods," *JAHS* 13-2, 1968.

winds of less than 5 knots, the vortex interaction occurs over only one-quarter to one-half of each revolution.

Airfoil Data

The use of the NACA 0012 airfoil data in the hovering analysis is a valid assumption for normal hover conditions, since most airfoils used for rotors have

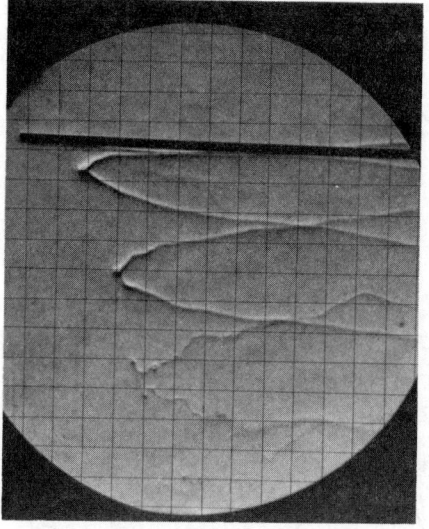

Single Exposure

Fifteen Exposures

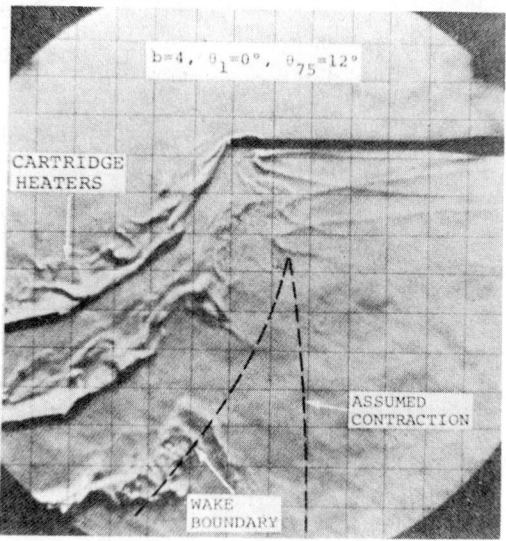

FIGURE 1.35 Schlieren Photographs of the Wake of a Model Rotor

Source: Kocurek & Tangler, "A Prescribed Wake Lifting Surface Hover Performance Analysis," *JAHS* 22-1, 1977.

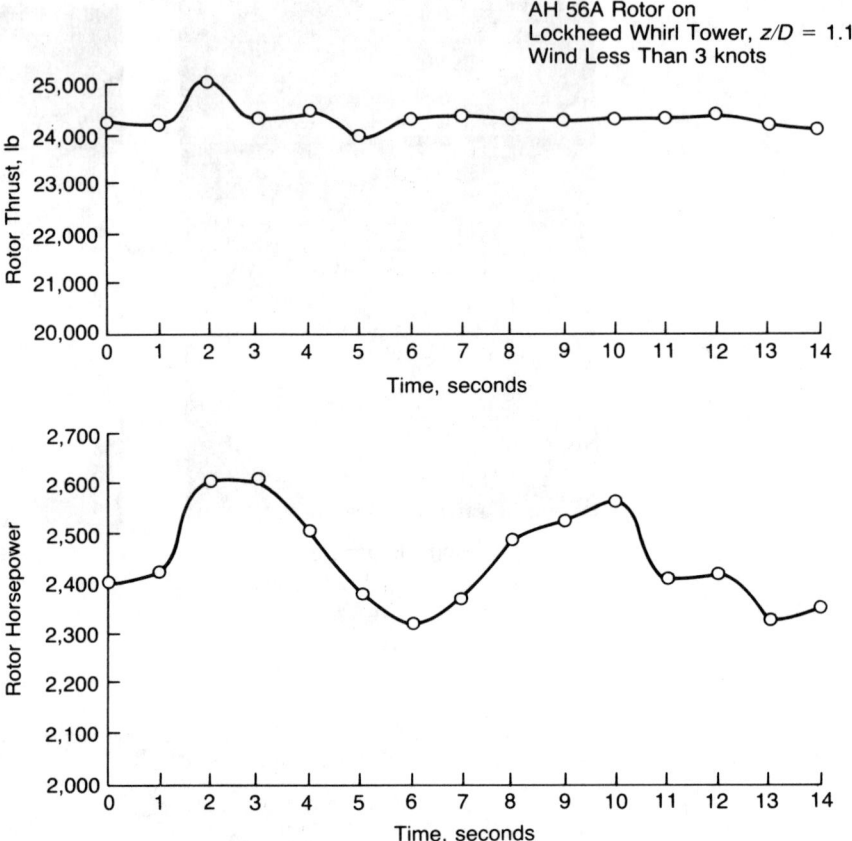

FIGURE 1.36   Fluctuations in Whirl Tower Data

about the same lift and drag characteristics below stall and drag divergence. When an airfoil is used that has significantly different characteristics than the NACA 0012, a more sophisticated method of defining the drag coefficient than simply basing it on the average lift coefficient should be used. One method is to define the drag coefficient as a power series:

$$c_d = c_{d_0} + c_{d_1}\alpha + c_{d_2}\alpha^2$$

where the coefficients are chosen to give the best fit to the experimental airfoil data. Figure 1.37 shows the NACA 0012 drag curve at 0.45 Mach number and three curves that each fit the data at 1°, 5°, and one other arbitrarily chosen angle of attack. The choice of which curve to use depends on the highest blade angle of attack for a given hover condition. Chapter 6 discusses airfoils in detail and lists data sources for a number of specific airfoils.

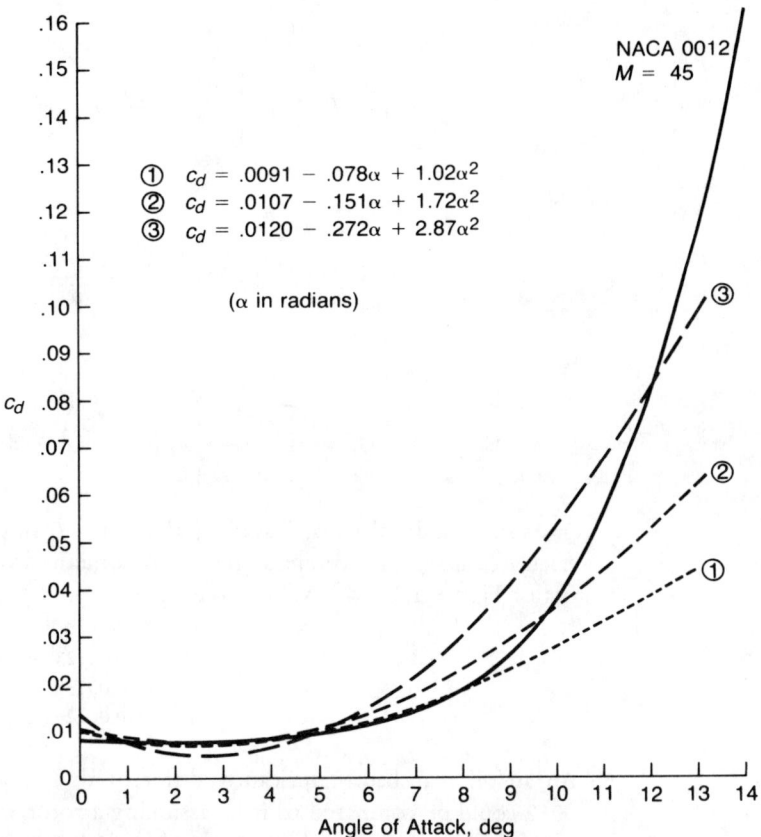

FIGURE 1.37   Three-Term Drag Polars

For a rotor with ideal twist, the local angle of attack is:

$$\alpha = \frac{\theta_t}{r/R} - \frac{\phi_t}{r/R}$$

The profile torque loading is:

$$\frac{\Delta Q_0}{\Delta r} = r\frac{\rho}{2}(\Omega r)^2 c\left[c_{d_0} + \frac{c_{d_1}}{r/R}(\theta_t - \phi_t) + \frac{c_{d_2}}{(r/R)^2}(\theta_t - \phi_t)^2 \right]$$

or:

$$\frac{\Delta Q_0}{\Delta r} = \frac{\rho}{2}\Omega^2 c R^3\left[\left(\frac{r}{R}\right)^3 c_{d_0} + \left(\frac{r}{R}\right)^2 c_{d_1}(\theta_t - \phi_f) + \frac{r}{R}c_{d_2}(\theta_t - \phi_t)^2 \right]$$

and

$$Q_0 = b \frac{\rho}{2} (\Omega R)^2 cR \left[\frac{c_{d_0}}{4} + \frac{c_{d_1}}{3} (\theta_t - \phi_t) + \frac{c_{d_2}}{2} (\theta_t - \phi_t)^2 \right]$$

but

$$\theta_t - \phi_t = \frac{4C_T/\sigma}{a}$$

so

$$C_Q/\sigma_0 = \frac{1}{2} \left[\frac{c_{d_0}}{4} + \frac{4}{3} c_{d_1} \left(\frac{C_T/\sigma}{a} \right) + 8 c_{d_2} \left(\frac{C_T/\sigma}{a} \right)^2 \right]$$

For the ideal rotor hovering at $C_T/\sigma = .086$, the profile torque coefficient calculated using the average angle of attack method was 0.00125. Using the various fits of Figure 1.37, this value would be:

Fit	C_Q/σ
1	0.00123
2	0.00130
3	0.00124

An airfoil that had significantly different drag characteristics than the NACA 0012 could be compared to it by assuming a rotor with ideal twist as above or its data could be used directly in the combined blade element and momentum method described later in this chapter.

Stall and Drag Divergence

Stall and drag divergence are of primary interest in forward flight, where they may limit the maximum speed and the maximum maneuvering capability. There can be cases, however, when either stall or drag divergence affects hover performance. The airfoil data of the NACA 0012 airfoil of Figure 1.10 shows the envelopes of the start of stall as a function of Mach number and also the envelope of the beginning of drag divergence as a function of angle of attack. For a rotor with this airfoil section, any local combination of angle of attack or Mach number above these boundaries produces stall and/or compressibility losses. If a computing procedure is used in which the airfoil lift and drag coefficients are evaluated at each blade element, these effects will automatically be accounted for; but if a simpler method using an average lift coefficient and a corresponding average drag coefficient is used, the effects will be neglected. In this case, however, empirical corrections can be made using the following equation which has been based on a comparison of the measured whirl tower results reported in reference 1.1 with calculations made with simple theory.

$$\Delta C_Q/\sigma = 0.001\Delta\alpha + 0.05\Delta M$$

where $\Delta\alpha$ is the amount the average angle of attack (in degrees) exceeds the stall boundary at a Mach number corresponding to the 75% radius station, and ΔM is the amount the tip Mach number exceeds the drag divergence Mach number at the average angle of attack.

Radial Flow

It is reasonable to expect that the boundary layer on the blade—being composed of molecules directly affected by the surface—would have a centrifugal force that would tend to produce a flow outboard toward the tip and that this radial flow would represent a power loss. Measurements of flow in the boundary layer of a hovering rotor reported in reference 1.17, however, indicate that the flow is slightly inboard instead of outboard. The direction of flow appears to be a function of four effects: centrifugal pumping, wake contraction, spanwise pressure gradient, and undeveloped tip vortex. In different flight conditions, one or more of these predominate to produce either inboard or outboard flow. The conclusion based on what is now known of the problem is that neglecting radial flow is an acceptable assumption.

Ground Effect

Just as with an airplane, the helicopter flying close to the ground requires less power than when it is flying far from the ground. The source of this *ground effect* for a hovering helicopter may be visualized by picturing an *image rotor* flying upside down at the same distance below the ground as the actual rotor is above it, as shown in Figure 1.38. The image wake is considered to be formed of a series of spiral vortex filaments generated at the blade tips and carried up by the image rotor induced velocity. The upward velocity induced in the plane of the actual rotor by the image set of vortex filaments can be calculated and used as a correction to the normal induced velocity term in the power equation. From a blade element standpoint, the reduction in power corresponds to the reduction of rearward tilt of the lift vector, as shown in Figure 1.39. From that figure, it may be seen that in order to maintain the same angle of attack, and thus thrust, the blade pitch must be reduced when flying in ground effect. (That the presence of the ground can influence the flow conditions at the rotor can be demonstrated at the breakfast table. The characteristics of the stream of syrup at the lip of the pitcher can be changed depending on how high it is above the pancake.)

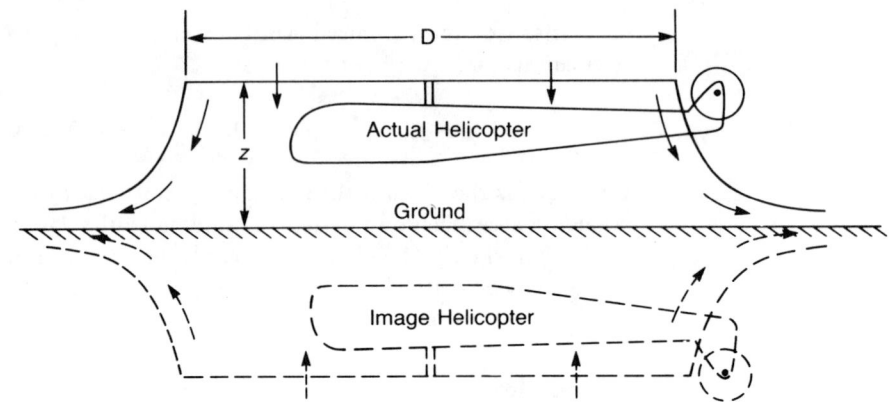

FIGURE 1.38   Image Helicopter Concept for Ground Effect Studies

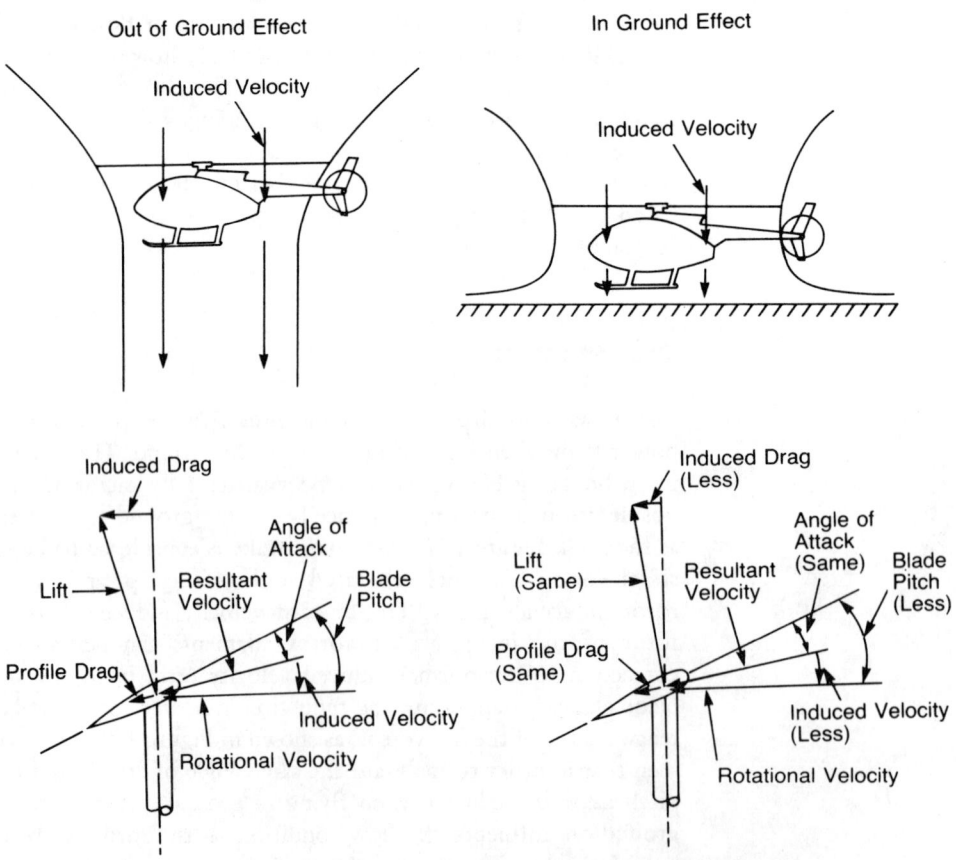

FIGURE 1.39   How Ground Effect Affects Conditions at Blade Element

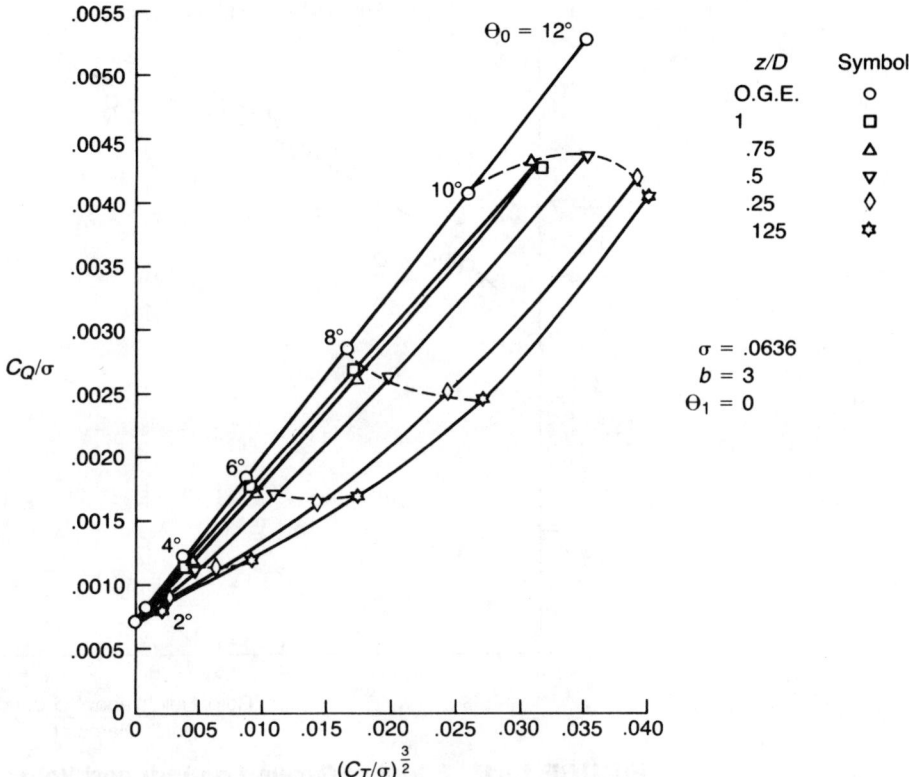

FIGURE 1.40 Effect of Ground on Performance of Model Rotor

Source: Knight & Hefner, "Analysis of Ground Effect on the Lifting Airscrew," NACA TN 835, 1941.

The key to the ground-effect analysis is how much the induced velocity at the rotor disc is reduced compared to what it would be out of ground effect. Figure 1.40 shows the measured effect of the ground on one of several model rotors reported in reference 1.18. These and similar model test results from reference 1.10 have been used to determine the induced velocity ratio as a function of rotor height using the difference in the torque/solidity coefficient at a constant thrust/solidity coefficient. The results of this analysis are shown in Figure 1.41. An extensive study of the hover performance of helicopters tested by the Army is reported in reference 1.19. The results are shown as a dashed line in Figure 1.41.

The classical studies of ground effect on hovering rotors are found in references 1.18 and 1.20. In these studies, primary emphasis was placed on the ratio of thrust in ground effect to thrust out of ground effect at constant power. For many calculations, such as for hover ceiling in ground effect, however, it is more convenient to be able to compute the power required at constant rotor thrust. The

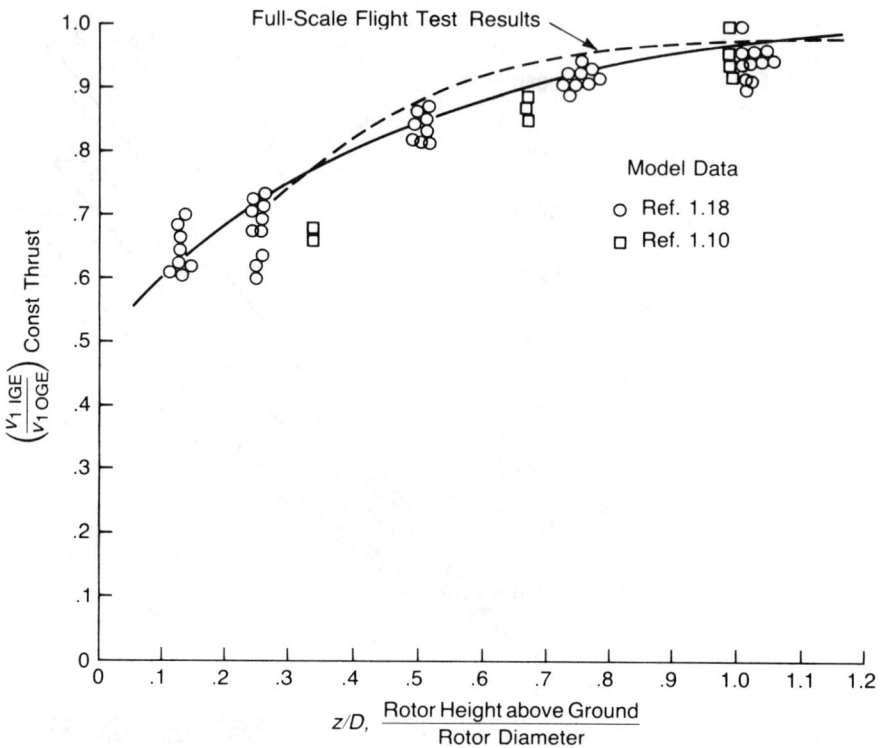

FIGURE 1.41   Effect of Ground on Induced Velocities as Determined by Model and Full-Scale Tests

Sources: Circles: Knight & Hefner, "Analysis of Ground Effect on the Lifting Airscrew," NACA TN 835, 1941; squares: Bellinger, "Experimental Investigation of Effects of Blade Section Camber and Planform Taper on Rotor Performance," USAAMRDL TR 72-4, 1972; dashed line: Hayden, "The Effect of the Ground on Helicopter Hovering Power Required," AHS 32nd Forum, 1976.

out-of-ground-effect main rotor power is the sum of the profile and induced power:

$$\text{h.p.}_{\cdot \text{OGE}} = \text{h.p.}_{\cdot 0_{\text{OGE}}} + \frac{T v_{1_{\text{OGE}}}}{550}$$

If the rotor thrust is held constant while approaching the ground, the angle of attack of each blade element and the corresponding profile power can be considered to be a constant. The main rotor power in ground effect is thus:

$$\text{h.p.}_{\cdot \text{IGE}} = \text{h.p.}_{\cdot 0_{\text{OGE}}} + \frac{T v_{1_{\text{OGE}}}}{550} \left(\frac{v_{1_{\text{IGE}}}}{v_{1_{\text{OGE}}}} \right)_{\text{const. thrust}}$$

And the difference in power is:

$$\Delta \text{h.p.} = \frac{T v_{1_{\text{OGE}}}}{550} \left[1 - \left(\frac{v_{1_{\text{IGE}}}}{v_{1_{\text{OGE}}}} \right)_{\text{const. thrust}} \right]$$

or in nondimensional form:

$$\Delta C_Q / \sigma_{\text{const. thrust}} = - (C_T / \sigma)^{\frac{3}{2}} \sqrt{\frac{\sigma}{2}} \left[1 - \left(\frac{v_{1_{\text{IGE}}}}{v_{1_{\text{OGE}}}} \right)_{\text{const. thrust}} \right]$$

This applies to an isolated rotor. An actual helicopter will generally exhibit somewhat more benefit from ground effect than that measured on an isolated rotor because of the accompanying decrease in fuselage download. Surveys made below a rotor hovering at $Z/D = 0.5$ and reported in reference 1.21 showed an upwash inboard of the 40% radius station. Also shown in the same reference is the download on a disc with a radius of 40% of the rotor radius and located 0.16 R below the rotor. For values of $Z/D < 0.75$, the disc actually produced an upload rather than a download. Further experimental evidence of the effect of the ground on download is presented in reference 1.22, where it is shown that the installation of a wing on a Boelkow BO 105 caused a significant increase in the power required to hover out-of-ground effect but no increase to hover in-ground effect. This loss of vertical drag in ground effect results in a slight modification of the equation for the change in torque:

$$\Delta C_Q / \sigma_{\text{const. weight}} = - (C_T / \sigma_{\text{OGE}})^{3/2} \sqrt{\frac{\sigma}{2}} \left\{ 1 - \left(1 - \frac{Dv}{\text{G.W.}} \right)^{3/2} \left(\frac{v_{1_{\text{IGE}}}}{v_{1_{\text{OGE}}}} \right)_{\text{const. thrust}} \right\}$$

For the example helicopter:

$$C_T / \sigma_{\text{OGE}} = 0.085$$

and

$$\frac{Dv}{\text{G.W.}} = 0.04$$

If it is hovering at 30% of its rotor diameter, from Figure 1.41:

$$\left(\frac{v_{1_{\text{IGE}}}}{v_{1_{\text{OGE}}}} \right)_{\text{const. thrust}} = 0.75$$

The resulting decrease is:

$$\Delta C_Q/\sigma_{\text{const. weight}} = 0.000143$$

which corresponds to a decrease of 407 h.p. out of a total of approximately 2,000 h.p.

For studies in which the thrust ratio at constant power is required, the following equation can be used:

$$\left(\frac{T_{\text{IGE}}}{T_{\text{OGE}}}\right)_{\text{const. power}} = \left\{ \frac{1}{\left(\dfrac{v_{1_{\text{IGE}}}}{v_{1_{\text{OGE}}}}\right)_{\text{const. thrust}}} \left[1 - \frac{(C_Q/\sigma_{0_{\text{IGE}}} - C_Q/\sigma_{0_{\text{OGE}}})}{(C_T/\sigma)^{3/2}_{\text{OGE}} \sqrt{\dfrac{\sigma}{2}}} \right] \right\}^{2/3}$$

If the difference in profile power is considered to be negligible, this equation reduces to:

$$\left(\frac{T_{\text{IGE}}}{T_{\text{OGE}}}\right)_{\text{const. power}} \doteq \frac{1}{\left(\dfrac{v_{1_{\text{IGE}}}}{v_{1_{\text{OGE}}}}\right)^{2/3}_{\text{const. thrust}}}$$

The wake in ground effect is not really a steady flow, as is assumed theory, but a flow with large-scale fluctuations that can be felt as gus observer standing near the helicopter. Reference 1.21 speculates tha fluctuations are associated with the vortex that is made up of the indivi vortices from the blade roots. This vortex apparently writhes like a pinned sna along the ground, causing the entire wake to shift and wobble.

Pilots occasionally report that when hovering in ground effect, experience random yaw disturbances. This is probably due to the effect of the vortex as it writhes near the tail rotor. Although we cannot see the unsteadiness in the rotor's wake, a good analogy is the local unsteadi swiftly moving river. An experimental observation of this phenomenon is repo in reference 1.23, in which an instrumented helicopter model hovering with a main rotor height of one-fourth diameter experienced random vertical stabilizer force variations of 20–30% of the mean value. The variations disappeared when the rotor height was raised to half a diameter.

Many pilots claim that the ground effect over tall grass or water is l over a solid surface. At this time there are no test data either supp refuting the claim.

METHODS FOR ESTIMATING HOVER PERFORMANCE

The sophisticated prescribed wake digital computer program describ reference 1.5 is considered to be an investigative tool for studying hover performance, but not a practical way for making quick engineering estimates. For this purpose, three methods are suggested, one of which should satisfy the needs of the moment:

- Combined momentum and blade element method
- Hover charts
- Adjustment of existing whirl tower data

Each method has its special advantages. The combined momentum and blade element method is flexible but tedious unless a computer is available. The chart method is fast but is restricted by the assumptions that were used in preparing the charts, which may not be strictly applicable to the rotor being analyzed. The adjustment of existing whirl tower data has its greatest usefulness when the rotor being analyzed shares a special feature, such as airfoil section, with a rotor that has already been tested. Each method is described in the following paragraphs.

COMBINED MOMENTUM AND BLADE ELEMENT THEORY WITH EMPIRICAL CORRECTIONS

1. *Given:* Rotor geometry—number of blades, radius, chord, twist, cutout, airfoil data; and test conditions—tip speed, atmospheric density, speed of sound.
2. Select a finite number of blade elements (at least five but not more than fifteen).
3. At the boundary between each blade element, tabulate: nondimensional blade station r/R; nondimensional chord, c/R; local Mach number, $M = (r/R)[(\Omega R)/V_{sound}]$; slope of lift curve, $a = f(M)$, per radian, from airfoil data such as Figure 1.10; local twist, $\Delta\theta$.
4. Choose collective pitch, θ_0; tabulate pitch at boundary between each blade element:

$$\theta = \theta_0 + \Delta\theta - \alpha_{L0}$$

Note: Choose minimum value of θ_0 so that θ is always positive.
5. Calculate the local inflow angle

$$\frac{v_1}{\Omega r} = \frac{ab\dfrac{c}{R}}{16\pi\dfrac{r}{R}}\left[-1 + \sqrt{1 + \frac{32\pi\theta\dfrac{r}{R}}{ab\dfrac{c}{R}}}\;\right]$$

or, for a constant chord blade:

$$\frac{v_1}{\Omega r} = \frac{a\sigma}{16\dfrac{r}{R}}\left[-1 + \sqrt{1 + \frac{32\theta\dfrac{r}{R}}{a\sigma}}\right]$$

Note: θ and a are in radian units in this equation.

6. Calculate the local angle of attack:

$$\alpha = \theta - \tan^{-1}\frac{v_1}{\Omega r} \text{ degrees}$$

7. Using curves of airfoil data such as Figure 1.10 or equations such as those developed in Chapter 6, tabulate c_l and c_d for the local angle of attack and Mach number. *Note*: If airfoil data synthesized from whirl tower or model rig test results are being used, they already include the compressibility tip relief effect. If, on the other hand, airfoil data from a two-dimensional wind tunnel are being used, the tip relief may be accounted for by reducing the local Mach number of the outer 10% of the blade by the increment corresponding to the thickness ratio of the tip airfoil from Figure 3.38 of Chapter 3.

8. Compute the running thrust loading:

$$\frac{dC_T}{d\dfrac{r}{R}} = \frac{b\left(\dfrac{r}{R}\right)^2\left(\dfrac{c}{R}\right)c_l}{2\pi} \text{ or } \frac{\left(\dfrac{r}{R}\right)^2\sigma c_l}{2}$$

9. Integrate either graphically or numerically to obtain the thrust coefficient without tip loss:

$$C_{T_{\text{no tip loss}}} = \int_{x_0}^{1}\frac{dC_T}{d\dfrac{r}{R}}\,d\frac{r}{R}$$

10. Calculate tip loss factor:

$$B = 1 - \frac{\sqrt{2C_{T_{\text{no tip loss}}}}}{b}$$

Note: All the airfoil data tabulated in Chapter 6 as originating from whirl tower or model rig test results were synthesized from two-bladed rotors with the assumption that the effective radius was constant at 0.97 R. If these airfoil data are used, the same assumption should be made, at least in the range of thrust coefficients used in the tests. A study using the airfoil data from reference 1.1 indicates that satisfactory correlation is obtained for a wide range of rotor parameters if the following effective radius equations are used:

$$\text{for } C_T < 0.006 \qquad B = 1 - \frac{0.06}{b}$$

$$\text{for } C_T > 0.006 \qquad B = 1 - \frac{\sqrt{2.27C_T - .01}}{b}$$

11. Calculate the corrected thrust coefficient using either graphical or numerical integration:

$$C_T = C_{T_{\text{no tip loss}}} - \int_B^1 \frac{dC_T}{d\frac{r}{R}} \, d\frac{r}{R}$$

12. Compute the running profile torque loading:

$$\frac{dC_{Q_0}}{d\frac{r}{R}} = \frac{b\left(\frac{r}{R}\right)^3 \left(\frac{c}{R}\right) c_d}{2\pi}$$

13. Integrate for the profile torque coefficient:

$$C_{Q_0} = \int_0^1 \frac{dC_{Q_0}}{d\frac{r}{R}} \, d\frac{r}{R}$$

14. Compute the running induced torque loading:

$$\frac{dC_{Q_i}}{d\frac{r}{R}} = \frac{b\left(\frac{r}{R}\right)^3 \frac{c}{R} c_l \frac{v_1}{\Omega r}}{2\pi}$$

15. Integrate for the induced torque coefficient:

$$C_{Q_i} = \int_{x_0}^B \frac{dC_{Q_i}}{d\frac{r}{R}} \, d\frac{r}{R}$$

16. Compute ΔC_{Q_i} due to the rotation of the wake using Figure 1.29 to obtain the correction ratio as a function of C_T:

$$\Delta C_{Q_i} = \left(\frac{\text{Swirl induced power}}{\text{Thrust induced power}} \right) C_{Q_i}$$

17. Find the disc loading as:

$$\text{D.L.} = C_T \rho (\Omega R)^2$$

18. Find C_T/σ as:

$$C_T/\sigma = \frac{C_T}{\dfrac{bc}{\pi R}}$$

19. From Figure 1.34 obtain the empirical correction factor for the contraction of the wake as a function of $(\text{D.L.}) \times (C_T/\sigma)$:

20. Calculate the total torque coefficient:

$$C_Q = (C_{Q_0} + C_{Q_i} + \Delta C_{Q_i}) \left(\frac{\text{Measured power}}{\text{Calculated power}} \right)$$

21. Calculate the rotor thrust and power:

$$T = \rho A (\Omega R)^2 C_T$$

$$\text{h.p.} = \frac{\rho A (\Omega R)^3 C_Q}{550}$$

The method has been used with the synthesized 0012 airfoil characteristics on several rotors for which whirl tower data have been published. Figure 1.42 shows correlation with the NACA two-bladed, low-solidity rotor of reference 1.1; the moderate solidity CH-53A rotor of reference 1.5; and the high solidity AH-56A tail rotor of reference 1.24. It may be seen that the correlation is satisfactory both in power and in collective pitch. Figure 1.43 presents test results from two Sikorsky whirl tower tests, one for a main rotor and the other for a tail rotor. The test points are compared with the results of two calculation schemes: the momentum–blade element method of this book; and the Circulation Coupled Hover Analysis Program (CCHAP) described in reference 1.25. It may be seen that the momentum-blade element method gives satisfactory correlation except at very high thrust levels and is therefore adequate for most hover calculations of a practical, engineering nature. Figure 1.44 shows the performance of the main and tail rotors of the example helicopter calculated by the simple method, and

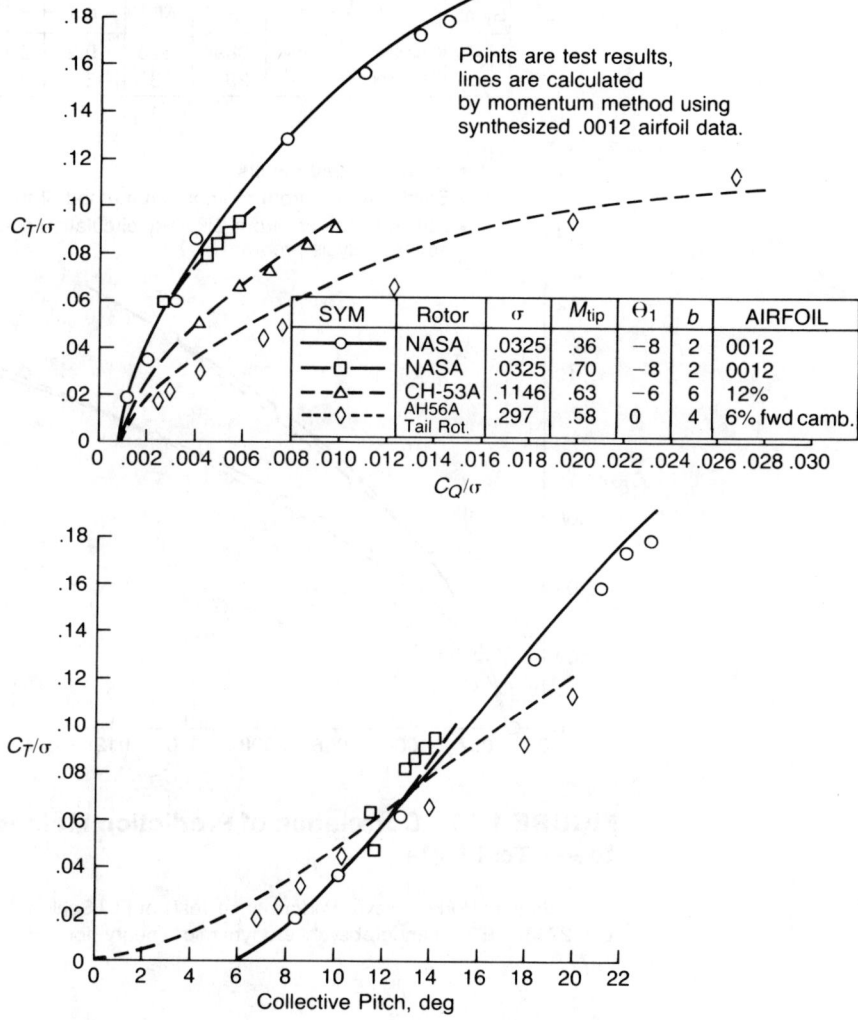

Points are test results,
lines are calculated
by momentum method using
synthesized .0012 airfoil data.

SYM	Rotor	σ	M_{tip}	Θ_1	b	AIRFOIL
—○—	NASA	.0325	.36	−8	2	0012
—□—	NASA	.0325	.70	−8	2	0012
—△—	CH-53A	.1146	.63	−6	6	12%
—◇—	AH56A Tail Rot.	.297	.58	0	4	6% fwd camb.

FIGURE 1.42   Correlation of Test Results with Calculated Results

Source: Carpenter, "Lift and Profile-Drag Characteristics of a NACA 0012 Airfoil Section as Derived from Measured Helicopter-Rotor Hovering Performance," NACA TN 4357, 1958; Landgrebe, "An Analytical and Experimental Investigation of Helicopter Rotor Hover Performance and Wake Geometry Characteristics," USAAMRDL TR 71-24, 1971; Johnston & Cook, "AH-56A Vehicle Development," AHS 27th Forum, 1971.

Symb.	Rotor	σ	M_{tip}	Θ_1	b	Airfoil	Ref.
○	"Baseline" main rotor	.0894	.523	−9.25	6	0012	1.40
△	Tail rotor	.20	.63	−8	4	0012	1.25

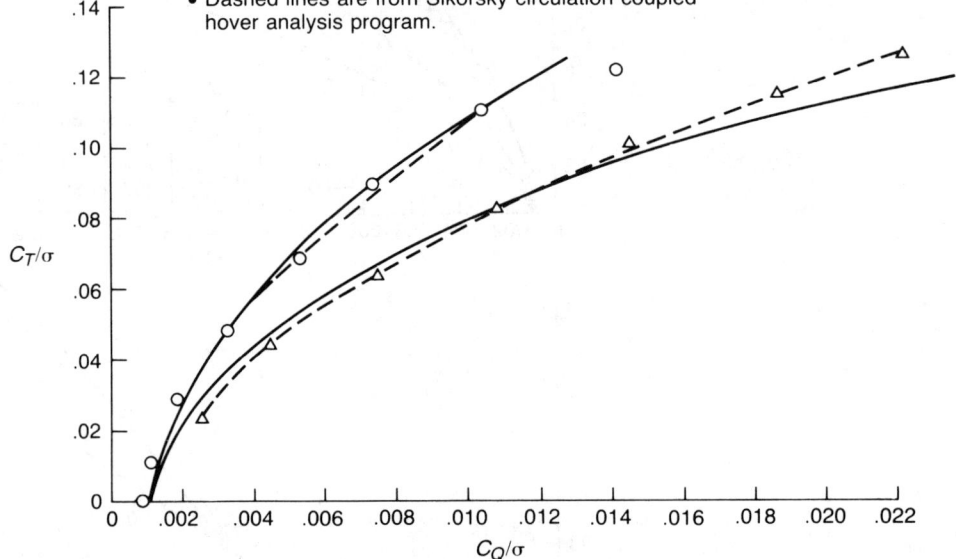

- Points are test results.
- Solid lines are from momentum method of this book.
- Dashed lines are from Sikorsky circulation coupled hover analysis program.

FIGURE 1.43   Correlation of Prediction Methods with Sikorsky Whirl Tower Test Data

Source: Rorke, "Hover Performance Tests of Full Scale Variable Geometry Rotors," NASA CR 2713, 1976; Landgrebe, "Aerodynamic Theory for Advanced Rotorcraft" *JAHS* 22-2, 1977.

Figure 1.45 shows the various elements of the calculation for the main rotor at a collective pitch of 17.5°.

Hover Charts

The hover charts starting on page 81 were prepared using the combined momentum and blade element method just described. The charts are based on the following assumptions:

- Constant chord
- Linear twist

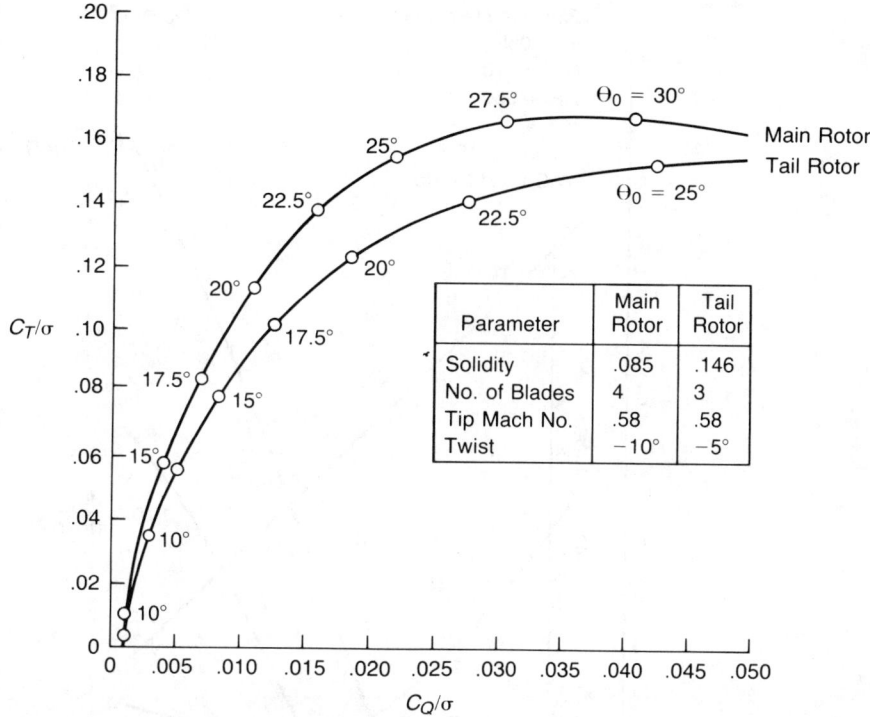

FIGURE 1.44 Calculated Performance of Isolated Main and Tail Rotors of Example Helicopter

- NACA 0012 airfoil characteristics based on the whirl tower tests of reference 1.1. Thus the Reynolds number characteristics correspond to a 16-inch chord.
- Blade cutout = 0.15 R

Adjustment of Existing Whirl Tower Data

The hover performance of a new rotor design may be estimated using test data from a previously tested rotor, even if they are not identical. The effect of differences in solidity, twist, number of blades, and tip Mach number can be estimated from the hover charts and used as correction factors.

Table 1.1 lists a number of whirl tower tests and the characteristics of the rotors tested.

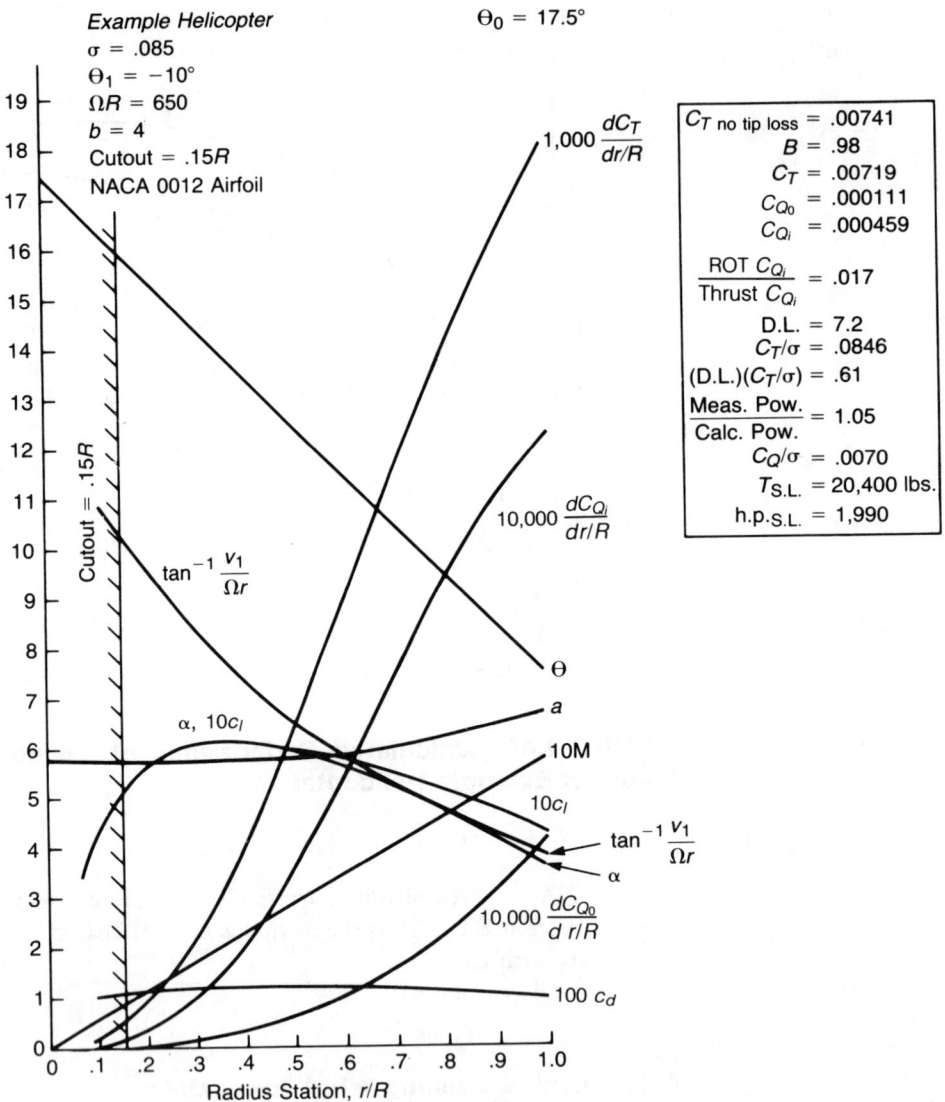

FIGURE 1.45   Results of Sample Calculation

TABLE 1.1
Tabulation of Published Whirl Tower Tests

Radius (ft)	Chord (in) @ .75 R	No. of Blades	Twist (deg)	Taper c_T/c_R	Solidity @ .75 R	Airfoil	Cutout (% R)	Max. Test M_{tip}	Max. Test D.L.	Max. Test C_T/σ	Ref.	Date
2.5	2	2 to 5	0	1	.042 to .106	0015	14	.22	3	.117	1.26	1937
20	9.6	3	−8	.5	.038	23015	12	.69	6.6	.177	1.27	1952
20	9.6	3	0	.5	.038	23015	12	.69	5.7	.164	1.27	1952
18.6	14.4	2	0	1	.041	8H12	10	.58	5.9	.146	1.28	1954
18.4	13	2	−6.5	1	.037	63_2−015	15	.71	6.1	.192	1.29	1956
7.5	14	2	0	1	.097	0012	20	.45	2.5	.066	1.30	1956
18.7	11.3	2	−5.5	.77	.032	0017 to 0009	14	.72	4.9	.158	1.31	1958
18.8	11.6	2	−5.5	.77	.033	0015	14	.81	4.7	.189	1.32	1958
26.8	16.4	2	−8	1	.033	0012	15	.70	4.4	.191	1.1	1958
5	9.4	2	−12	.5	.10	0012	18	.98	22	.118	1.33	1960
18.8	11.6	2	−5.5	.91	.033	63_{215}A018	14	.76	7.0	.177	1.34	1960
4.5	2.7	5	−14	1	.079	0012	25	.95	14	.136	1.35	1961
18.7	10.6	2	−6	.63	.030	63_2A015(230)	14	.67	3.8	.195	1.8	1961
18	10.9	2	−6	.67	.032	63_2A012(130)	15	.75	5.1	.179	1.36	1962
2	2	4	−8	1	.11	0012	10 to 50	.67	13.7	.108	1.37	1970
5	14	4	0	1	.297	13006	10	.58	33	.111	1.24	1971

TABLE 1.1 (continued)

Radius (ft)	Chord (in) @ .75 R	No. of Blades	Twist (deg)	Taper c_T/c_R	Solidity @ .75 R	Airfoil	Cutout (% R)	Max. Test M_{tip}	Max. Test D.L.	Max. Test C_T/σ	Ref.	Date
36	26	6	−6	1	.115	0012	24	.63	12.5	.095	1.5	1971
2.2	1.5 & 2	2 to 8	0,−8,−16	1	.034 to .14	0012	15	.63	11.7	.11	1.5	1971
2.2	1.5	2 to 8	0	1 & .5	.034 to .14	0012 & 23112	15	.63	3.0	.112	1.10	1972
17.5	13	2	−4.2	.71	.039	0015	9	.55	1.9	.103	1.4	1972
3	2.3	3	−7 to −9	1	.061	0012,23012,VR7/8	15	.72	9.8	.15	1.38	1973
2.3	1.5	6	0	1	.102	0012	17	.58	8.1	.09	1.39	1974
26 & 29	16.4	3 & 6	−8 & −9	1	.049 & .089	0012	22	.64	12.4	.125	1.40	1976
2	5	1	0	1	.066	0012	20	.25	2.8	.076	1.41	1976
8	15.4	4	−8	1	.20	0012	18	.60	27.8	.13	1.25	1977

EXAMPLE HELICOPTER CALCULATIONS

HOW TO'S

The following items can be evaluated by the methods in this chapter.

TWIST = 0° 2 BLADES

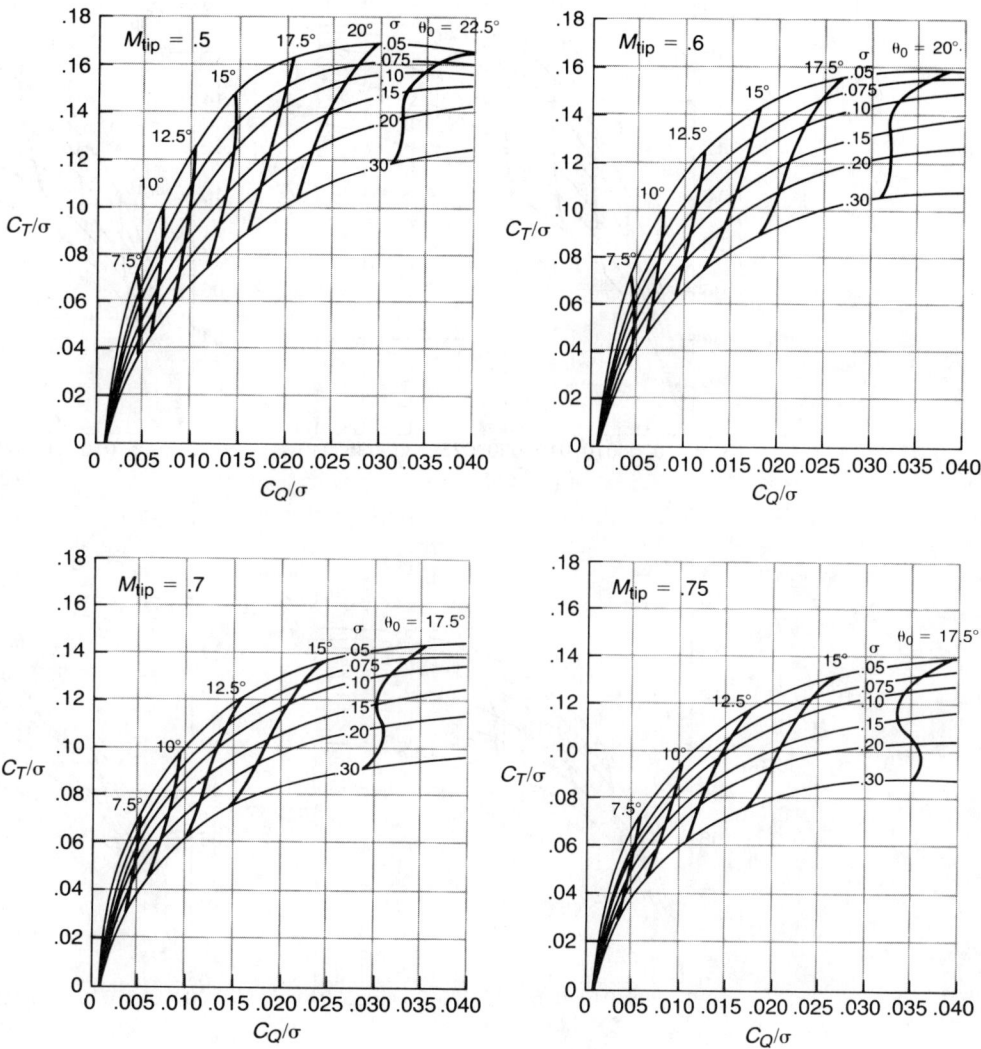

TWIST = 0°   4 BLADES

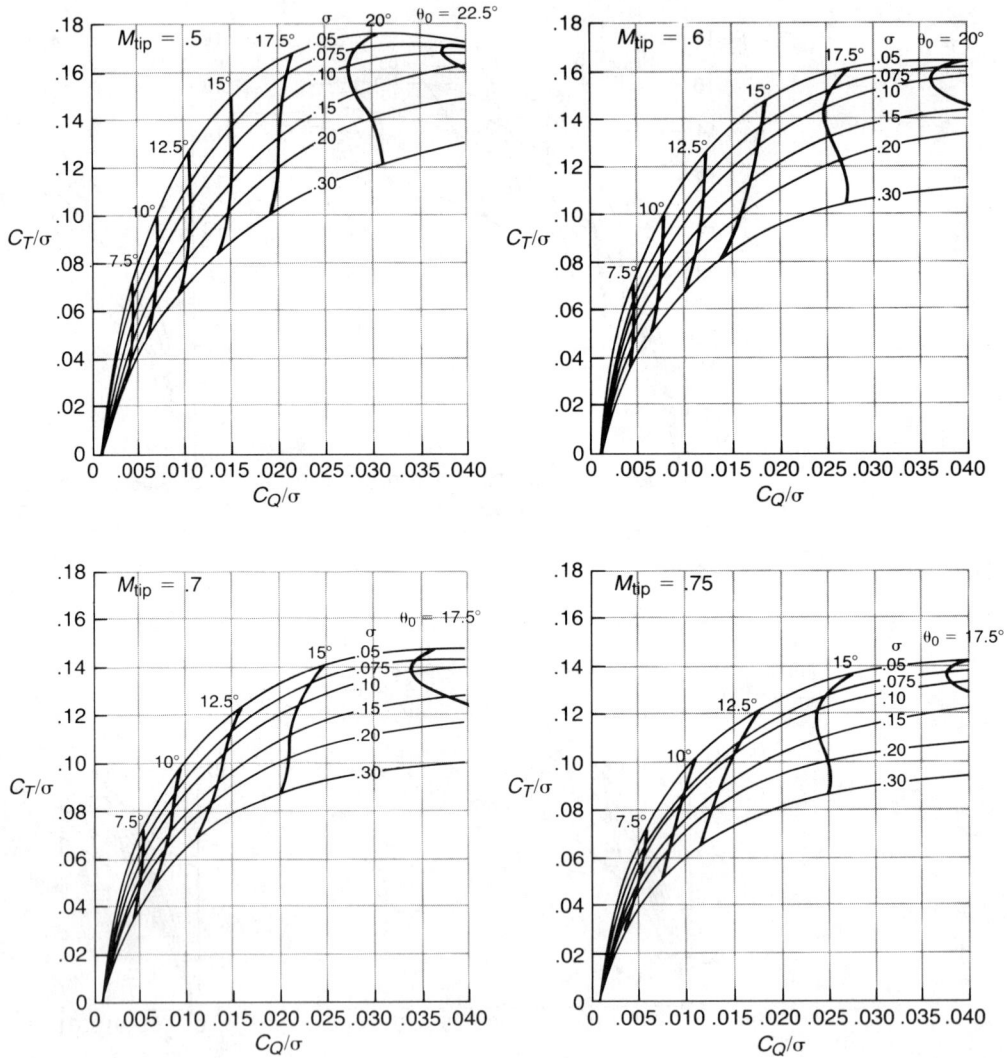

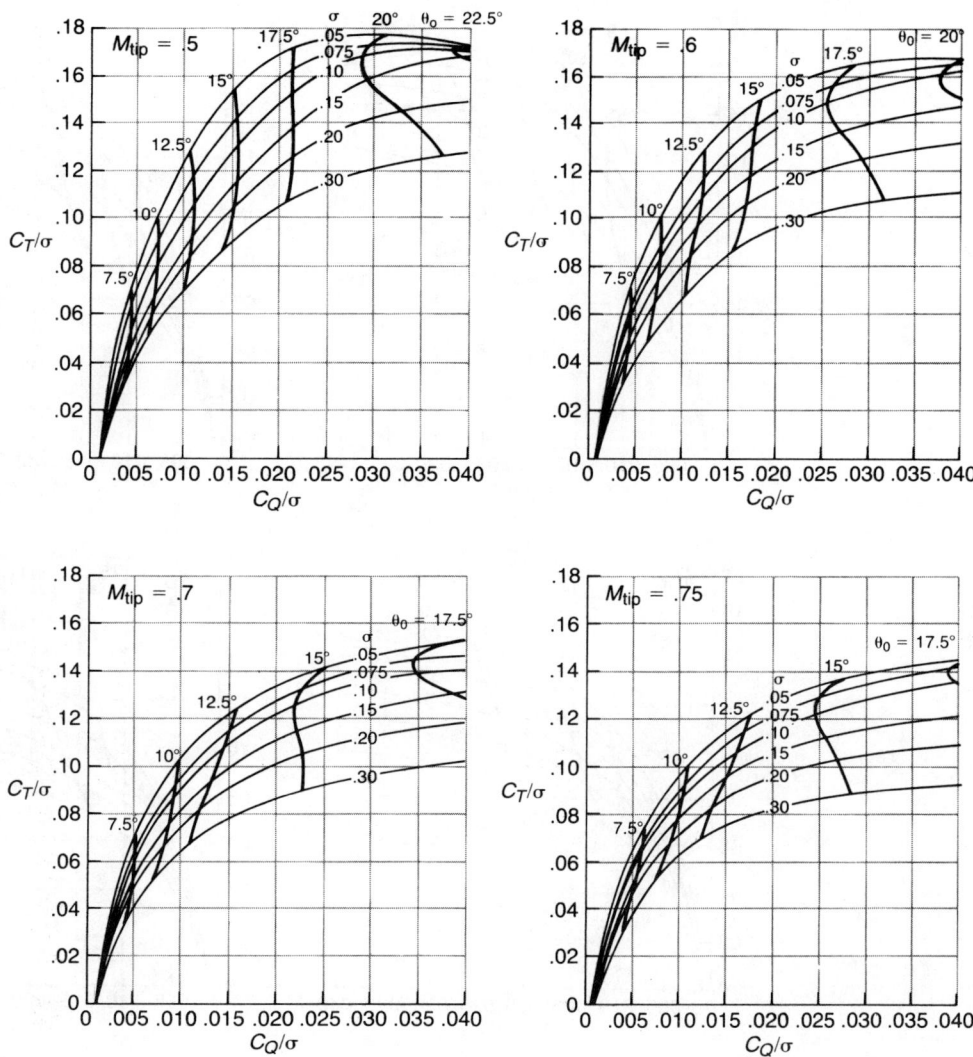

TWIST = −5°   2 BLADES

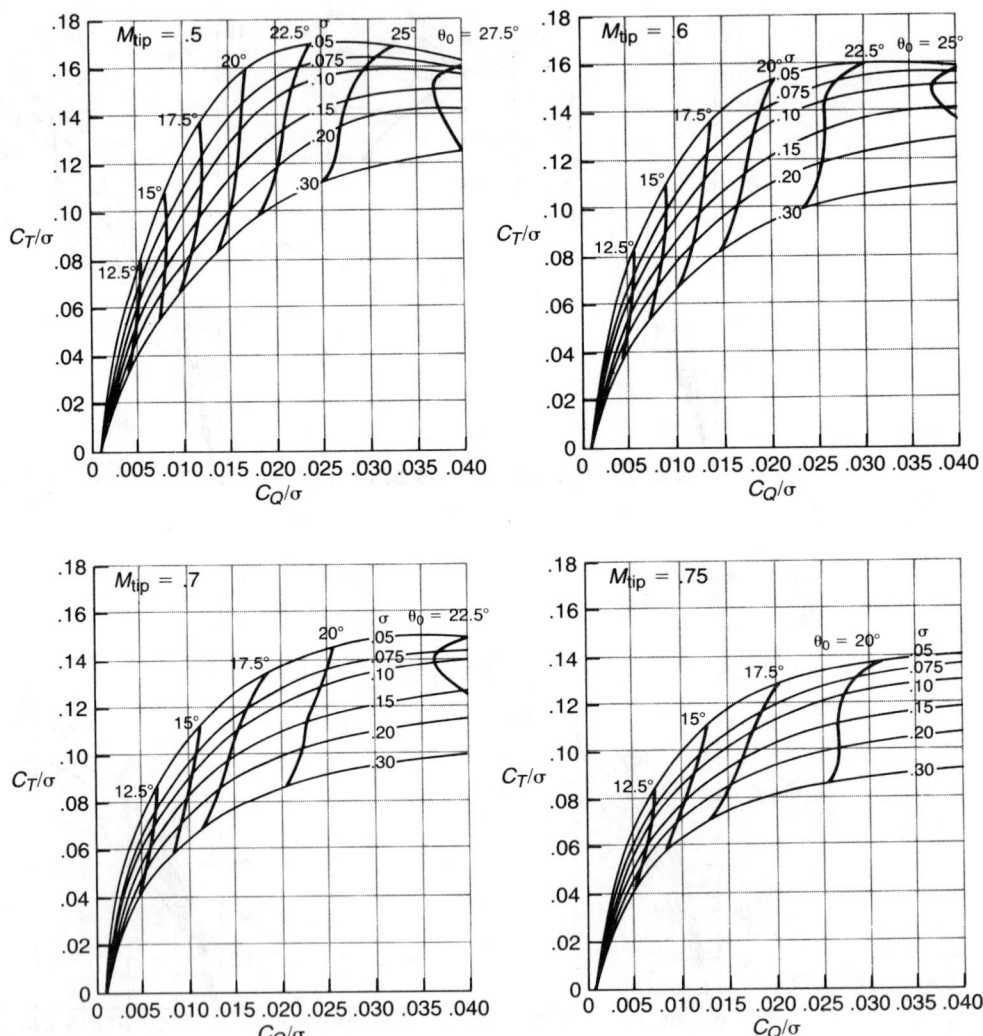

84

TWIST = −5°   4 BLADES

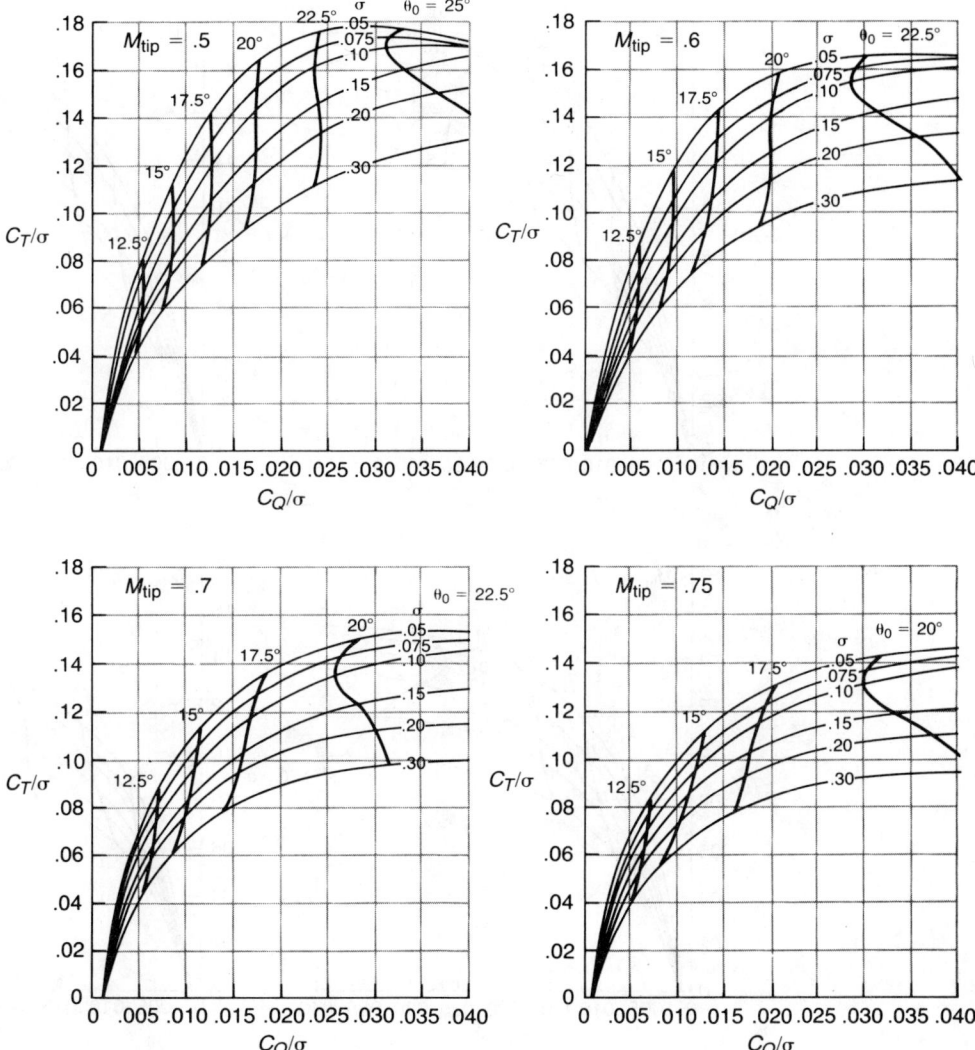

TWIST = −5°   6 BLADES

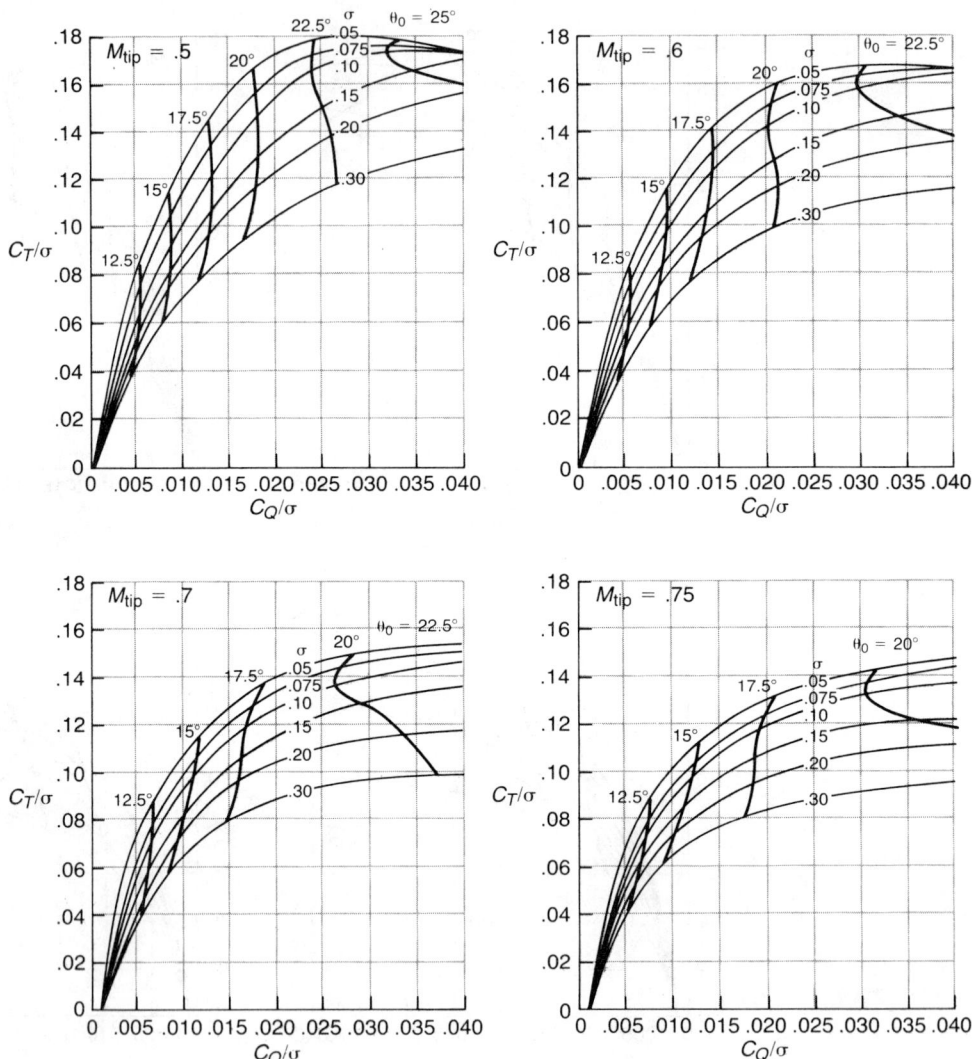

TWIST BLADES = −10°   2 BLADES

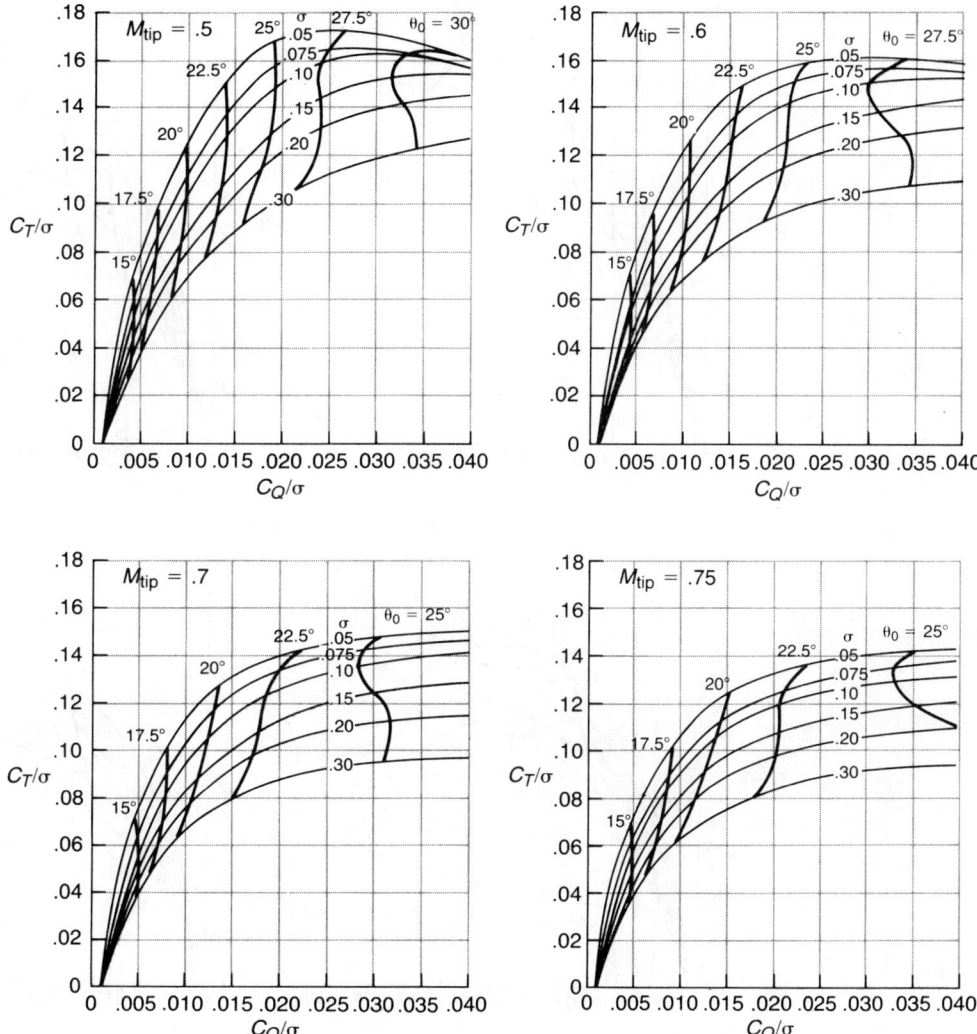

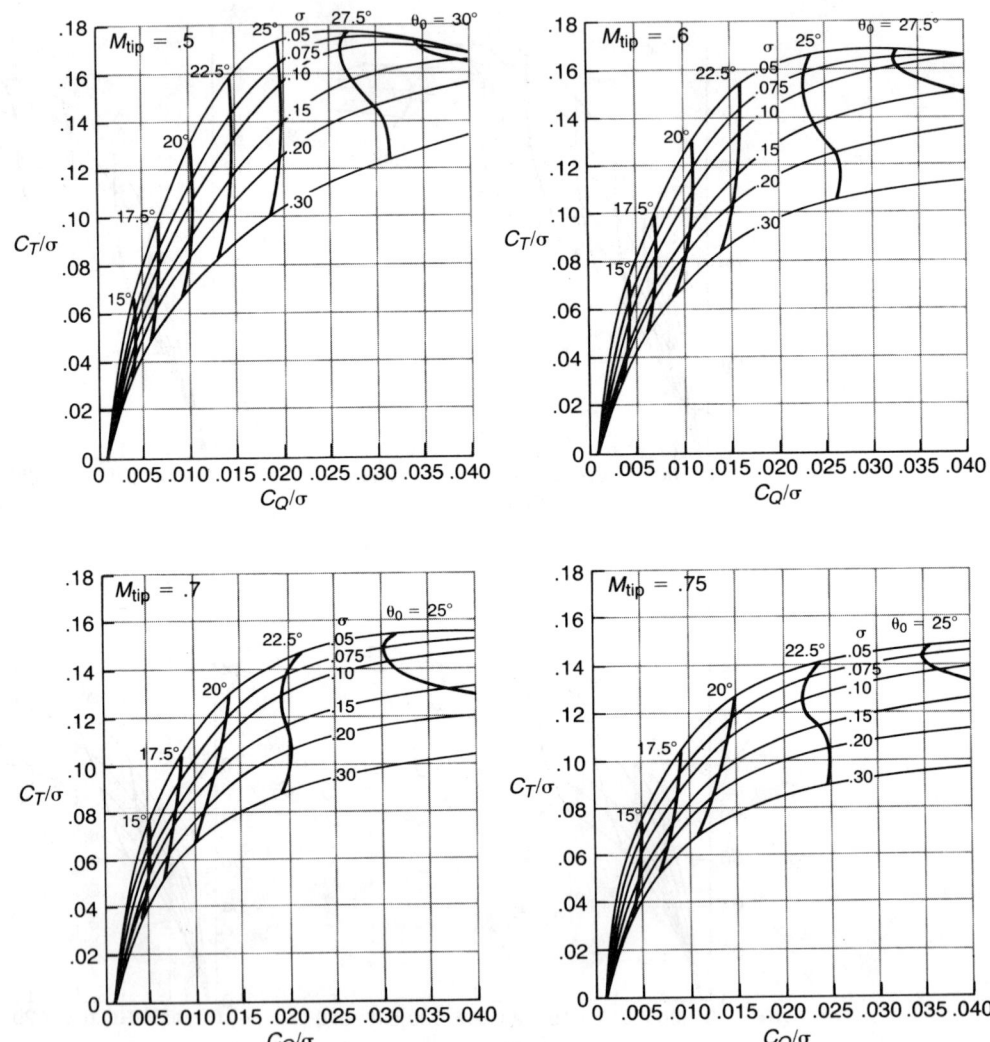

TWIST = −10° 6 BLADES

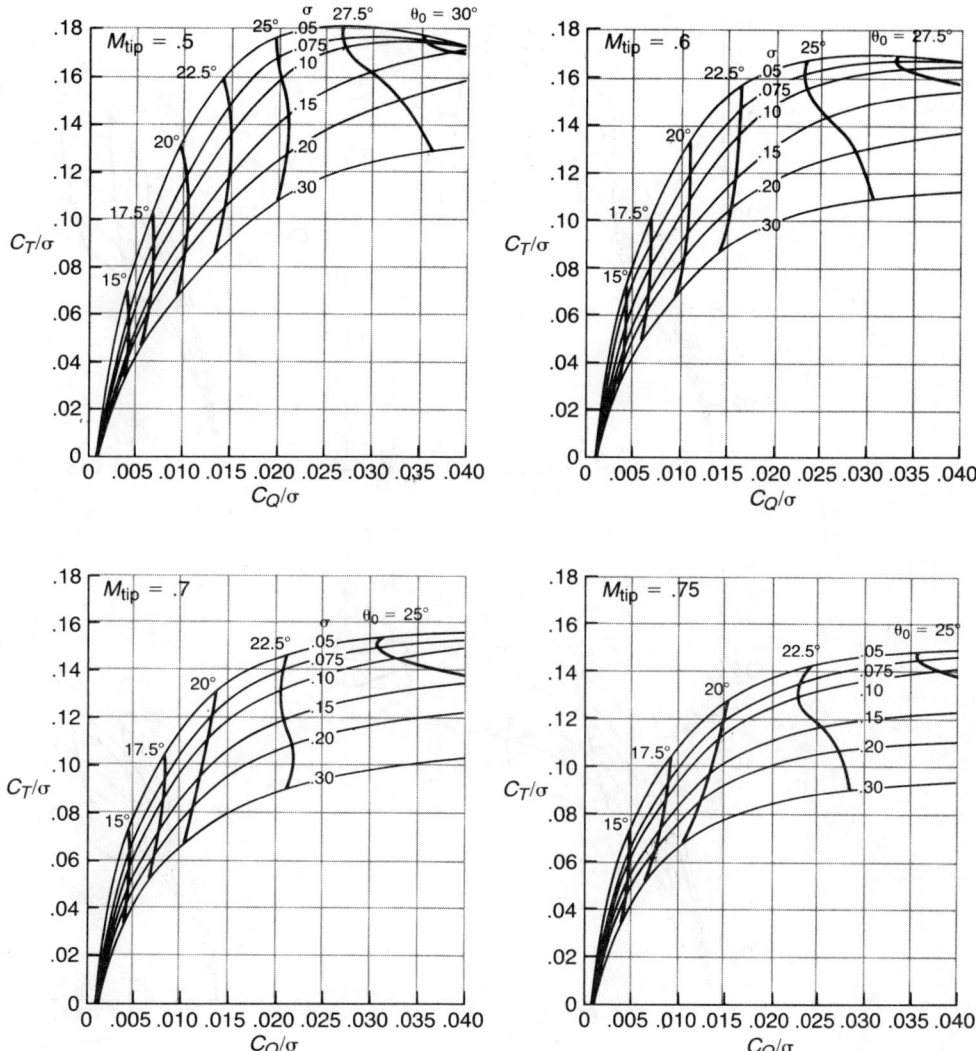

TWIST = −15°   2 BLADES

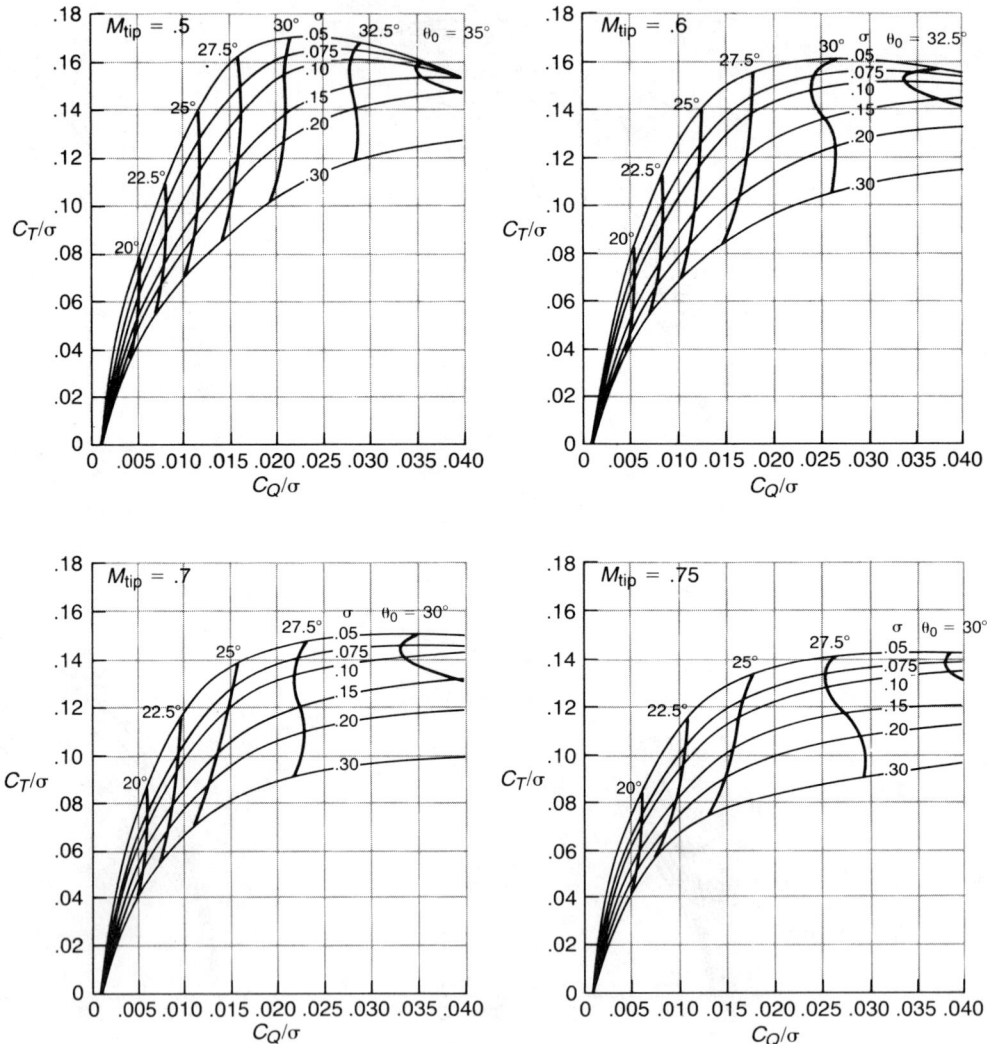

TWIST = −15°   4 BLADES

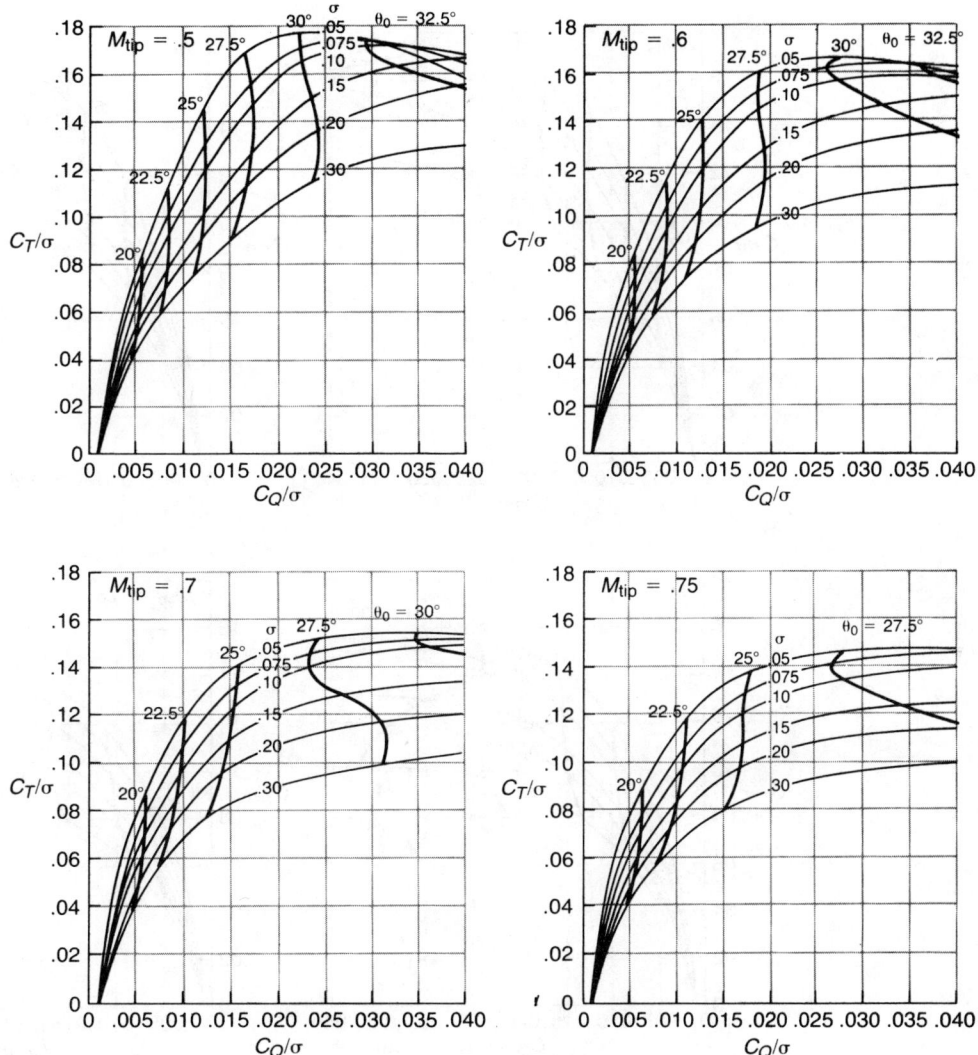

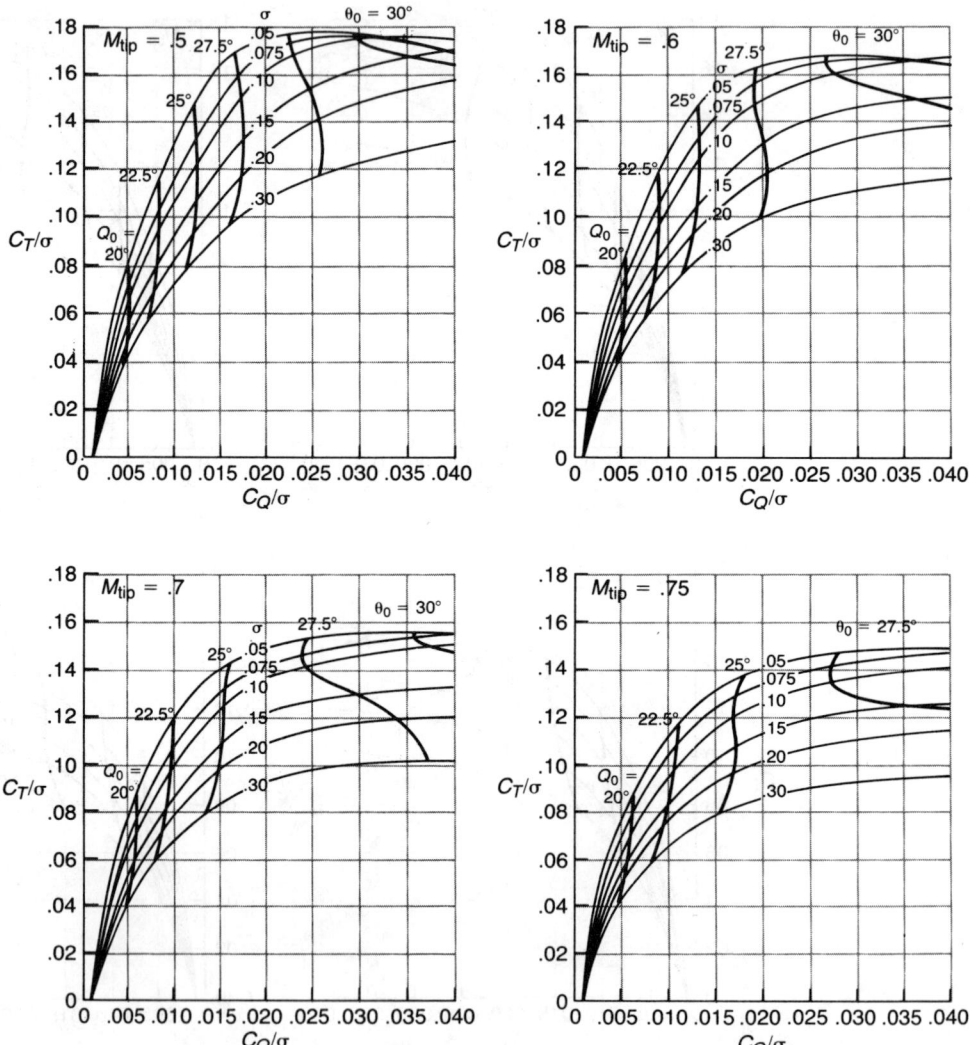

CHAPTER 2

Aerodynamics of Vertical Flight

STATES OF FLOW

Just as in hover, there are two ways to analyze the rotor in vertical flight, the momentum method and the blade element method. In some vertical flight conditions, however, the relationships do not exist that were used to develop the hover methods. To obtain a physical understanding of the problem, let us examine the flow around a rotor mounted in an open-ended vertical wind tunnel as in Figure 2.1. Several different types of flight conditions can be simulated:

Hover

The tunnel fan is stopped and the rotor produces flow down through the tunnel. The governing momentum equation is:

$$v_{1_{\text{hov}}} = \sqrt{\frac{T}{2\rho A}}$$

Climb

The tunnel fan is used to induce a downflow in addition to that of the rotor. Both the local flow at the rotor and the remote flow are down. The induced velocity at the rotor is:

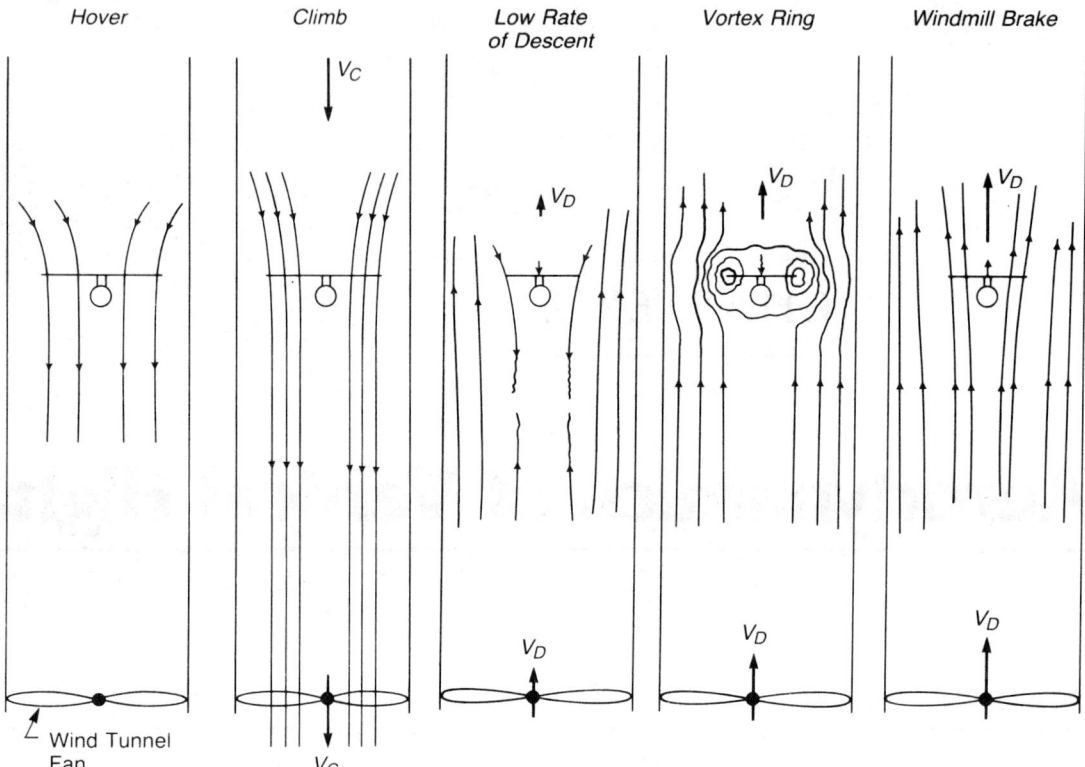

FIGURE 2.1 Vertical Flight Flow States as Illustrated by Wind Tunnel Conditions

$$v_{1_C} = -\frac{V_C}{2} + \sqrt{\left(\frac{V_C}{2}\right)^2 + \frac{T}{2\rho A}}$$

or assuming constant thrust conditions:

$$v_{1_C} = -\frac{V_C}{2} + \sqrt{\left(\frac{V_C}{2}\right)^2 + v_{1_{hov}}^2}$$

where V_C is the climb velocity in feet per second.

Low Rate of Descent

The tunnel fan is used to induce a small upflow in the tunnel. For this case the flow is mixed. The local flow near the rotor is dominated by the rotor-induced velocity and is down, but the rest of the flow field is up.

From a momentum standpoint, the equation for the induced velocity is:

$$v_{1_D} = \frac{V_D}{2} + \sqrt{\left(\frac{V_D}{2}\right)^2 + v_{1_{hov}}^2}$$

This equation is semivalid for very low rates of descent; but, because of the mixed conditions, the continuous-flow assumption on which the momentum theory is based breaks down when the rate of descent is approximately a quarter of the hover-induced velocity.

Vortex Ring State

The tunnel fan induces an upward velocity of about the same magnitude as the rotor-induced velocity. The rotor operates in a doughnut-shaped mass of revolving air, which goes down through the rotor and up around the outside of the tips. This is the *vortex ring state*. Since no definite and continuous wake exists, the momentum concepts cannot be used. In the extreme case, when the rate of descent is equal to the induced velocity, there is no net mass flow through the rotor, and the momentum equations would indicate that no thrust could be developed. In reality, however, the rotor does have a significant thrust capability even in this condition. The classical boundaries of the vortex ring state are from hover to a rate of descent equal to twice the hover-induced velocity.

Windmill Brake State

If the tunnel upflow is increased to the point where it is higher than the rotor-induced velocity, both the local flow at the rotor and the flow in the remote field are up; there is a well-behaved wake above the rotor, and the momentum concepts can again be used. This condition is known as the *windmill brake state*, and its governing equation is:

$$v_{1_{W.B.}} = \frac{V_D}{2} - \sqrt{\left(\frac{V_D}{2}\right)^2 - v_{1_{hov}}^2}$$

CONDITIONS AT THE BLADE ELEMENT

The conditions at the blade element for hover, climb, and descent are shown in Figure 2.2. In climb, the rearward tilt of the lift vector produces extra *inflow drag*, which is represented by extra power required to climb. In a steady climb, the helicopter is not being accelerated, so the rotor thrust has the same value as in

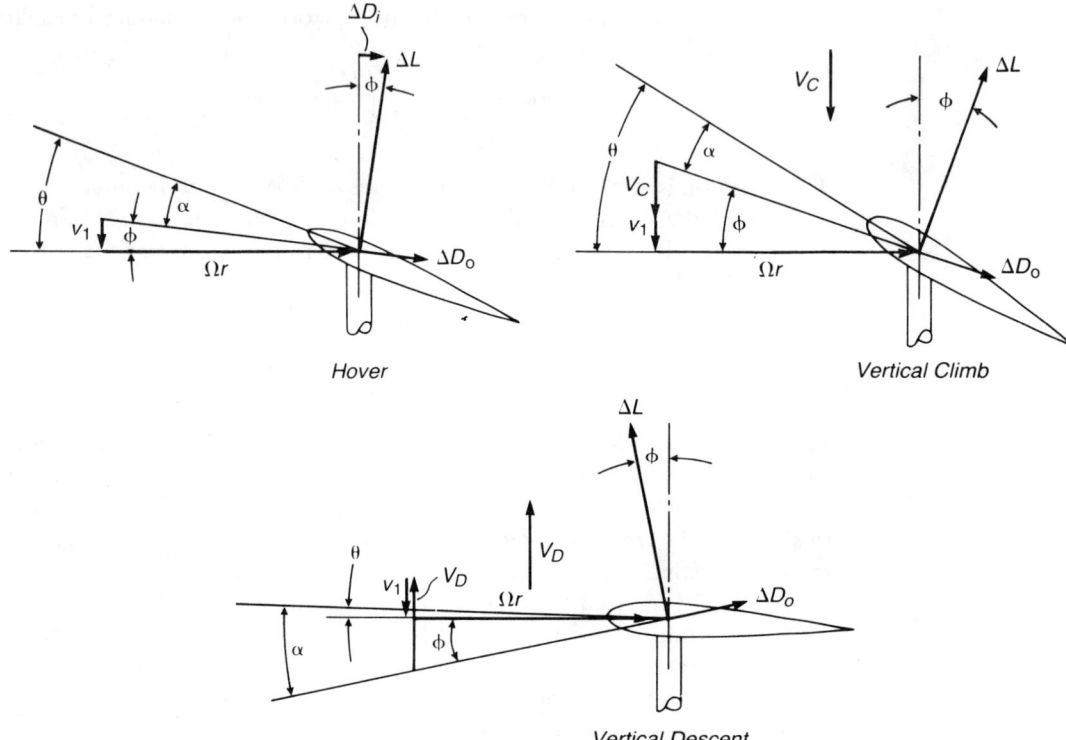

FIGURE 2.2   Blade Element Environments in Vertical Flight

hover except for minor increases due to the increased vertical drag of the fuselage. This means that the average angle of attack is essentially the same as in hover; but, because of the extra inflow through the rotor, the blade pitch must be increased over the hover value. For calculations of the vertical climb, the combined momentum and blade element method developed for the hover analysis can be modified to include the effect of climb by using the general equation for induced velocity:

$$
v_1 = \frac{-\left(\dfrac{\Omega}{2} acb + 4\pi V_C\right) + \sqrt{\left(\dfrac{\Omega}{2} acb + 4\pi V_C\right)^2 + 8\pi b \Omega^2 acr \left(\theta - \dfrac{V_C}{\Omega r}\right)}}{8\pi}
$$

In descent, the blade element is subjected to air coming up at it from below. This reduces the rearward tilt of the thrust vector, thus reducing the power required. The upward velocity also means that the blade pitch must be reduced to hold the average angle of attack and the same rotor thrust.

At some rate of descent, the forward tilt of the lift vector is equal to the drag component. This is the condition of *autorotation*, since no torque need be applied to maintain rotor speed. In an actual rotor, some blade elements will have more drag than the forward component of lift; but on other elements the situation will be reversed. Autorotation occurs when the integration of torque along the blade is zero—or, for an actual helicopter, is sufficiently negative to make up for losses in the tail rotor, transmissions, and accessories. All helicopters are equipped with an overrunning clutch between the transmission and the engine, so that the rotor does not have to drive a dead engine in autorotation.

Vertical autorotation is a balanced, steady flight condition in which the rotor thrust is equal to the gross weight and the rotor speed remains constant. For a given helicopter at a given gross weight, there is a unique combination of rate of descent, rotor speed, and blade pitch that defines the condition. Autorotation, once established, is stable: if the rotor speed decreases, the horizontal velocity vector at the blade element, Ωr, will shorten, and the lift vector will be tilted further forward, thus tending to increase rotor speed. The opposite effect occurs if the rotor speed increases from its original value. The pilot can control rotor speed by adjusting the blade pitch. A reduction of pitch initially decreases the rotor lift so that the rate of descent increases, thus tilting the lift vector forward and accelerating the rotor. As the rotor speeds up, the lift again reaches a value equal to the weight of the helicopter, and a new set of equilibrium conditions is reached with a higher rotor speed and a slightly different rate of descent than before the blade pitch was decreased.

For each helicopter, there is an upper and a lower limit of rotor speed that the pilot must observe. The upper limit is the speed at which the centrifugal forces in the blades and hub reach the structural design limit. The lower limit is the speed at which, in order to maintain rotor thrust equal to the gross weight, each blade element is being operated at or near its stall angle of attack. Once the blade elements stall, the drag increases rapidly, as shown in Figure 1.10 of Chapter 1, and becomes greater than any forward tilt of the lift vector can compensate. At this point the rotor slows down to a stop, and the flight becomes an accident. The upper limit is generally 10–20% above the normal speed for hover, and the lower limit may be as low as 20–30% below the normal speed.

POWER REQUIRED IN A VERTICAL CLIMB

The power required in a vertical climb is higher than that required in hover primarily because of the change in potential energy. Secondary effects are the increased vertical drag, increased tail rotor power, and decreased induced power since the rotor is handling more air. For simple calculations, the profile power of the main and tail rotors can be considered to be the same in climb as in hover.

For this analysis, the equation for vertical drag will be written:

$$D_V = 4\left(\frac{D_V}{\text{G.W.}}\right)_{\text{hov}} \frac{\rho}{2}(v_{1_C} + V_C)^2 A_M + (\Delta A_Z C_D)\frac{\rho}{2}V_C^2$$

where ΔA_z is the area of the portion of the airframe not in the wake and C_D is the drag coefficient of the additional area.

From the momentum equation:

$$v_{1_C} + V_C = \frac{V_C}{2} + \sqrt{\left(\frac{V_C}{2}\right)^2 + v_{1_{hov}}^2}$$

The tail rotor thrust required to balance main rotor torque is:

$$T_T = \frac{P_M}{(\Omega R)_M} \frac{R_M}{l_T}$$

and the tail rotor power is:

$$P_T = P_{0_T} + \frac{P_M}{(\Omega R)_M} \frac{R_M}{l_T} v_{1_T}$$

For this analysis, it will be assumed that the tail rotor-induced velocity remains a constant at its hover value and that the tail rotor power is directly proportional to tail rotor thrust alone. The rationale for this assumption is that even though the thrust of the tail rotor increases in climb, it goes into a forward-flight condition where it handles more air and thus produces less induced velocity for the same thrust. The total power is:

$$P_{tot} = P_M \left[1 + \frac{v_{1_{hov_T}}}{(\Omega R)_M} \frac{R_M}{l_T} \right] + P_{0_T}$$

and the difference in power between climb and hover is:

$$\Delta h.p. = \frac{1}{550} \left\{ \left[G.W. \cdot (v_{1_C} + V_C) + 4 \left(\frac{D_V}{G.W.}\right)_{hov} \frac{\rho}{2} (v_{1_c} + V_c)^3 A_M + (\Delta A_z C_D) \frac{\rho}{2} V_C^3 \right] \right.$$

$$\left. - \left[G.W. \cdot (v_{1_{hov}}) + 4 \left(\frac{D_V}{G.W.}\right)_{hov} \frac{\rho}{2} v_{1_{hov}}^3 A_M \right] \right\} \left\{ 1 + \frac{v_{1_{hov_T}}}{(\Omega R)_M} \frac{R_M}{l_T} \right\}$$

An opportunity to check the validity of this equation is provided by the flight test data of reference 2.1 for an AH-1G. Figure 2.3 shows the correlation and also the difficulty in obtaining this type of test data. For example, at a power increment of 150 h.p., the measured rate of climb varied from 570 to 1,180 ft/min.

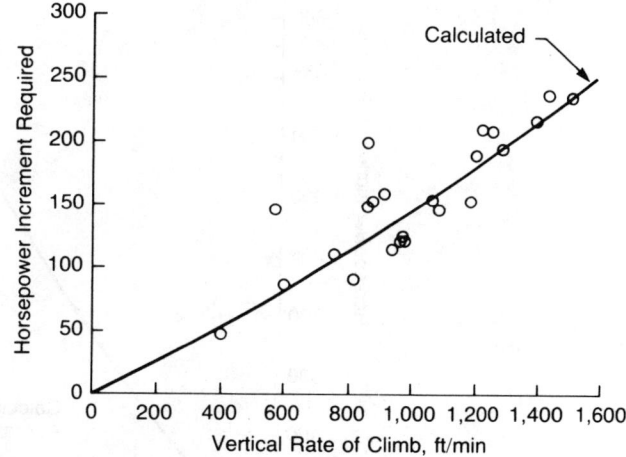

AH-IG Physical Parameters

R_M = 22 ft
R_T = 4.67 ft
ℓ_T = 27.5 ft
$\dfrac{D_V}{\text{G.W.}}$ = .025 (estimated)
$\Delta A_z C_d$ = .5 (tail rotor pylon)

Test Conditions

G.W.= 7,600 lb
ΩR_M = 746 ft/sec
ρ/ρ_0 = .91
h.p.$_{\text{hov}}$ = 870

FIGURE 2.3   Power Required in Vertical Climb, AH-1G Helicopter

Source: Ferrell & Frederickson, "Flight Evaluation Compliance Test Techniques for Army Hot Day Hover Criteria," USAASTA Project 68-55, 1974.

Figure 2.4 presents the results of the same type of analysis for the example helicopter.

A simple method for making a quick estimate of the incremental power for low rates of climb can be obtained by ignoring the vertical drag and tail rotor effects in the above equation. Then:

$$\Delta\text{h.p.} = \frac{\text{G.W.}}{550}\left(v_{1_C} + V_C - v_{1_{\text{hov}}}\right)$$

or

$$\Delta\text{h.p.} = \frac{\text{G.W.}}{550}\left[\frac{V_C}{2} + \sqrt{\left(\frac{V_C}{2}\right)^2 + v_{1_{\text{hov}}}^2} - v_{1_{\text{hov}}}\right]$$

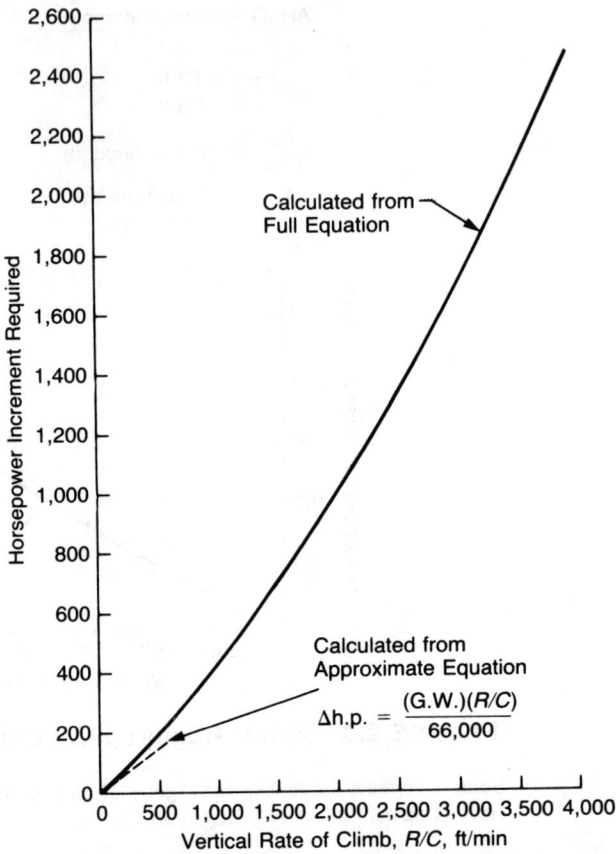

FIGURE 2.4 Power Required in Vertical Climb, Example Helicopter

It may be seen that for low rates of climb such that

$$\left(\frac{V_C}{2}\right)^2 \ll (v_{1_{hov}})^2$$

the change in power is simply:

$$\Delta h.p. = \frac{G.W.}{550}\frac{V_c}{2}$$

which is just half of the value which would be computed based on the rate of change of potential energy. For the example helicopter at a vertical rate of climb of 400 ft/min—or 8.33 ft/sec—the two velocity terms are:

$$\left(\frac{V_C}{2}\right)^2 = \left(\frac{8.33}{2}\right)^2 = 17$$

and

$$(v_{1_{hov}})^2 = (40.4)^2 = 1{,}530$$

Thus the criterion can be considered to be satisfied for this case. The first approximation to the extra power required to climb at this rate for the example helicopter is 150 h.p., as shown in Figure 2.4. A more detailed analysis of vertical climb is given in reference 2.2.

COLLECTIVE PITCH REQUIRED

The increment in collective pitch above that required to hover can be estimated from the same principles used for the power increment. The change in pitch can be based on the change in inflow velocity and the tangential flow at the three-quarter radius station.

$$\Delta\theta = \frac{(v_{1_C} + V_C - v_{1_{hov}})}{.75(\Omega R)}$$

For the example helicopter with a vertical rate of climb of 400 ft/min, the additional collective pitch is 1.0°.

ROTOR THRUST DAMPING

The momentum relationships can be used to derive an approximate equation for the change in rotor thrust coefficient as a result of a change in axial velocity at constant collective pitch. This allows the damping of main rotors and tail rotor to be calculated for use in studies of such maneuvers as jump takeoffs, power failures, and response to pedal inputs.

A hovering rotor with ideal twist was discussed in Chapter 1. If this rotor is in vertical flight, the equation for C_T/σ is:

$$C_T/\sigma = \frac{a}{4}\left[\theta_T - \frac{(v_{1_C} + V_C)}{\Omega R}\right]$$

For small rates of climb:

$$C_T/\sigma = \frac{a}{4}\left[\theta_T - \frac{\left(\dfrac{V_C}{2} + \sqrt{\dfrac{T}{2\rho A}}\right)}{\Omega R}\right]$$

or

$$C_T/\sigma = \frac{a}{4}\left[\theta_T - \frac{V_C}{2\Omega R} - \sqrt{\frac{C_T}{2}}\right]$$

The partial derivative evaluated at the initial value of C_T/σ is:

$$\left[\frac{\partial C_T/\sigma}{\partial V_C}\right]_{\theta=\text{const.}} = \frac{-1}{\Omega R\left(\dfrac{8}{a} + \dfrac{\sqrt{\dfrac{\sigma}{2}}}{\sqrt{C_T/\sigma}}\right)}$$

This can be used for small rates of descent as well as climb and for inflow through the tail rotor due to turns or sideward flight where the rotor is not in the vortex ring state.

CHARACTERISTICS OF THE VORTEX RING STATE

Flight in the vortex ring state is characterized by very unstable flow conditions, which produce vibration and erratic thrust variations. Flight test experience on a small tandem rotor helicopter reported in reference 2.3 showed that the characteristic vibration began at a rate of descent equal to about 23% of the hover-induced velocity and persisted until the rate of descent exceeded 125% of the hover-induced velocity.

For the example helicopter, these boundaries would cover the region between 400 and 2,900 ft/min rate of descent. The results also showed that for the test helicopter, forward speeds above about 10 knots were sufficient to avoid vortex ring vibration at all rates of descent. Similar tests on a larger single-rotor helicopter reported in reference 2.4 showed that a forward speed of 25 knots was enough to avoid the vibration. Thus, in flight, it is relatively easy to get out of the vortex ring state by flying with a small amount of forward speed. As a matter of fact, it is a fairly difficult piloting task to stay in the vortex ring state for any length of time.

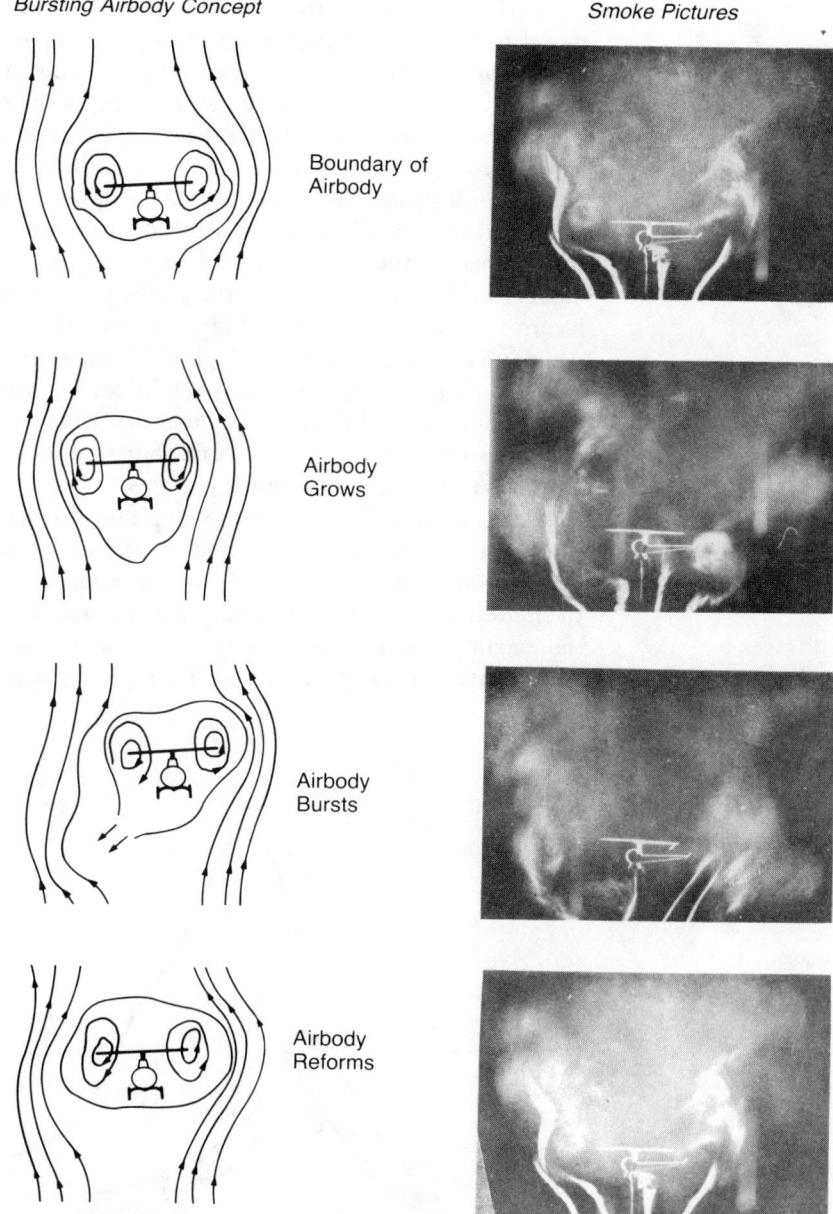

FIGURE 2.5   Vortex Ring Conditions

Source: Drees & Hendal, "Airflow Patterns in the Neighborhood of Helicopter Rotors," *Aircraft Engineering*, Vol. 23, April, 1951. Photos courtesy of NLR.

The instability of the vortex ring flow studied with smoke going through a model rotor in a wind tunnel was reported in reference 2.5. Two observations in that report are of interest in visualizing the flow. The first speculates about an effective *airbody* within which air is circulated by the rotor, as shown in Figure 2.5. The airbody is originally roughly spherical; but, as the rotor pumps energy into it, it lengthens downward until it bursts like a bubble, permitting the accumulated energy to dissipate into the surrounding upward-moving air. The airbody then returns to its original shape and starts the cycle again. The fluctuating airflow, of course, makes the thrust and/or rate of descent also fluctuate. The other observation based on a study of the smoke pictures (especially vivid in the motion pictures made during this study) is that the airflow fluctuations are not symmetrical around the rotor. Figure 2.5 shows several frames from the smoke movies during a steady vortex ring condition. The vortex can be seen to be formed first on one side and then on the other in a completely random fashion. Thus not only does the rotor thrust fluctuate, but so also does the flapping as the rotor responds to the changing inflow patterns.

The magnitude of the vortex ring flow fluctuations as measured on a model rotor in a wind tunnel are reported in reference 2.6 and are shown in Figure 2.6. The region of roughness is about the same as that found in flight test. A theoretical study of the vortex ring state presented in reference 2.7 concludes that the maximum wake instability should occur for the condition in which the tip vortices stay in the plane of the rotor. Reference 2.7 shows by a momentum

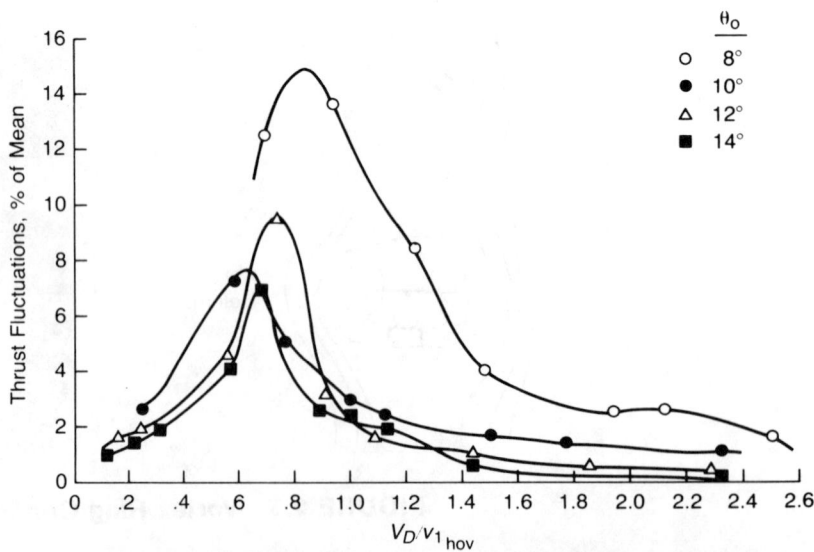

FIGURE 2.6   Thrust Fluctuations in Vertical Descent

Source: Azuma & Obata, "Induced Flow Variation of the Helicopter Rotor Operating in the Vortex Ring State," *Jour. of Aircraft*, July–August, 1968.

analysis that this condition exists when the rate of descent is equal to 0.707 times the induced velocity in hover. This conclusion is verified by Figure 2.6. A further study of vortex ring theory is presented in reference 2.8.

The experimental data of reference 2.6 also illustrates the phenomenon known to pilots as *power settling*, where more power is required to descend than to hover. Figure 2.7 shows normalized thrust and torque values as a function of the vertical descent velocity ratio. A reference thrust level has been selected—corresponding to hover at 12° of collective pitch—and the required collective

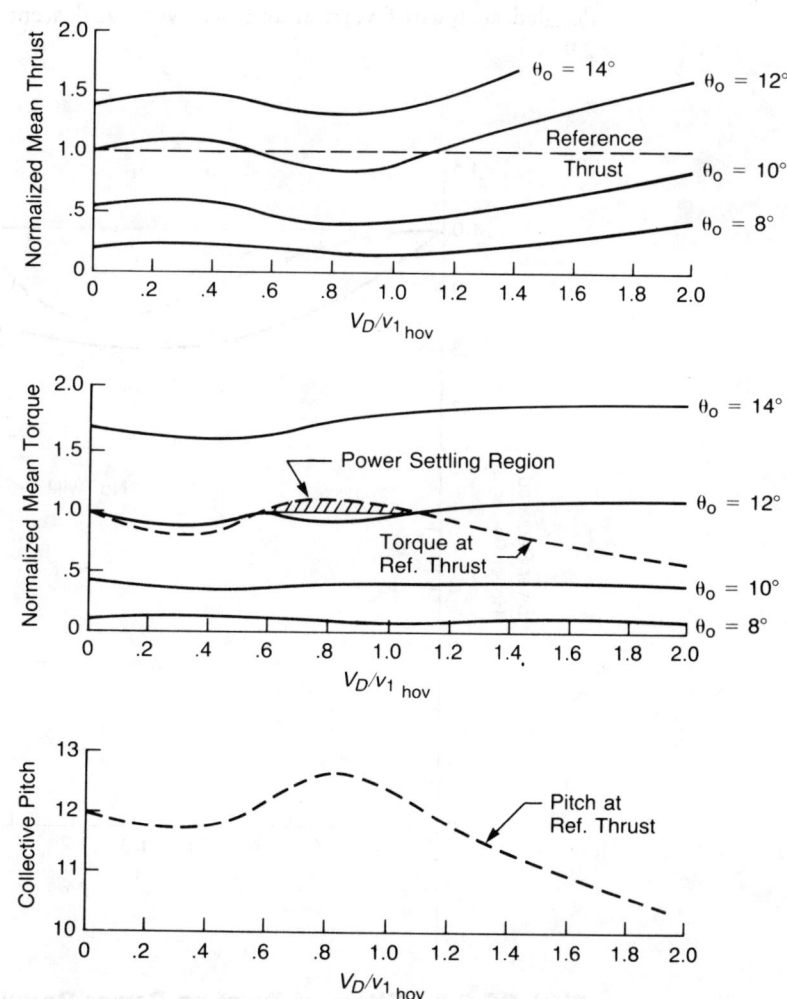

FIGURE 2.7   Effect of Rate of Descent on Rotor Conditions

Source: Azuma & Obata, "Induced Flow Variation of the Helicopter Rotor Operating in the Vortex Ring State," *Jour. of Aircraft*, July–August, 1968.

pitch and torque as a function of rate of descent have been found by cross-plotting. It may be seen that both the required pitch and torque first decrease, as would be expected, and then increase in the region of maximum flow fluctuation before decreasing again as the rotor approaches autorotation. The power-settling condition is of practical importance if during a vertical takeoff of a multiengined helicopter, one engine fails and the pilot attempts to return vertically to the takeoff point. If during this return the pilot enters the power-settling region, the power required from the remaining engines—or the final impact velocity—will be higher than would be estimated from simple transfer-of-energy equations. This region also roughly corresponds to the region of negative thrust damping. A detailed analysis of vertical and near-vertical descent will be found in reference 2.9.

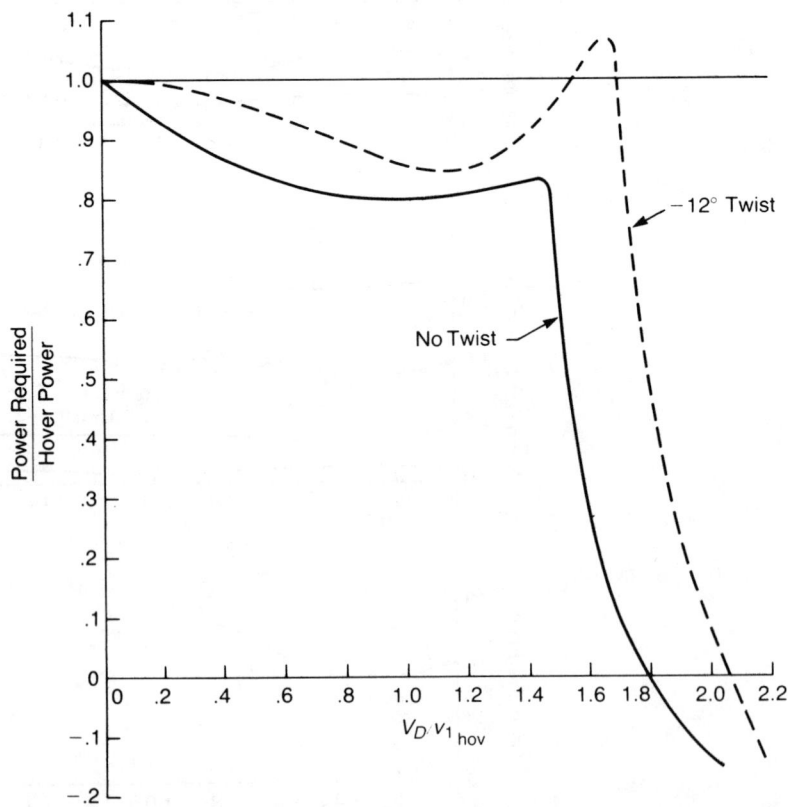

FIGURE 2.8   Effect of Twist on Power Required in Vertical Descen

Source: Castles & Gray, "Empirical Relation between Induced Velocity, Thrust, and Rate of Descent of a Helicopter Rotor as Determined by Wind-Tunnel Tests of Four Model Rotors," NAC TN 2474, 1951.

An as yet unexplained mystery is indicated by Figure 2.8, which presents an analysis of wind tunnel data reported in reference 2.10 for two model rotors, which were identical except for blade twist. The rotor with the twisted blades demonstrated a definite power-settling regime, whereas the rotor with untwisted blades did not. In addition, the authors comment: "Also, the fluctuations in the forces and moments on the rotor with twisted blades were very much larger at the higher rates of descent than for the rotors with untwisted blades."

THE TAIL ROTOR AND THE VORTEX RING STATE

Vertical descent is not the only flight condition in which the vortex ring state is of importance. A tail rotor may be operated in the vortex ring state during sideward flight or during a hover turn over a spot. Figure 2.9 shows the tail rotor collective pitch as a function of sideward velocity for the UH-1 as reported in reference 2.11. The "lump" is attributed to the vortex ring state. In the region of neutral and negative slope to the left of hover, pilots find it very difficult to hold a steady

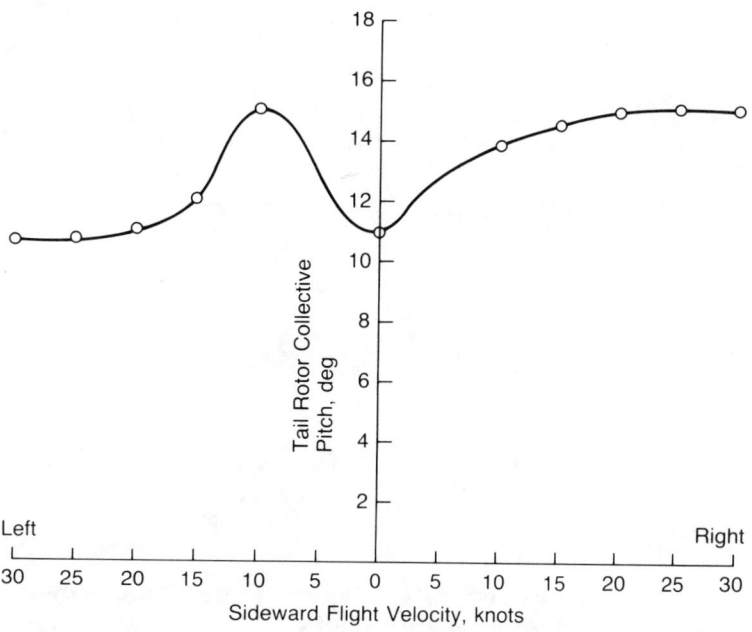

FIGURE 2.9   Tail Rotor Pitch Required in Sideward Flight

Source: Lehman, "Model Studies of Helicopter Tail Rotor Flow Patterns In and Out of Ground Effect," USAAVLABS TR 71-12, 1971.

heading. Hover turns that put the tail rotor in this condition, create a situation that pilots call "falling into a hole." The basic cause of this is best illustrated by the curves in the top portion of Figure 2.7. The curves are unstable between velocity ratios of 0.4 and 0.8, and any increase in the turn rate will lower the tail rotor thrust, thus increasing the rate even further. The hover-induced velocity for the UH-1 tail rotor is about 48 ft/sec. The maximum vortex ring effect would be expected to occur at 70% of this value or at about 20 knots. The fact that it occurs at a lower speed appears to be a result of the main rotor wake interacting with the tail rotor. Figure 2.10, taken from reference 2.12, shows a dramatic difference in pedal position in sideward flight resulting from a change in the direction of tail rotor rotation for the Lockheed AH-56A. Rotation with the bottom blade going aft apparently entrains the tip vortex of the main rotor, causing a premature vortex ring condition at the tail rotor. Rotation in the opposite direction seemed

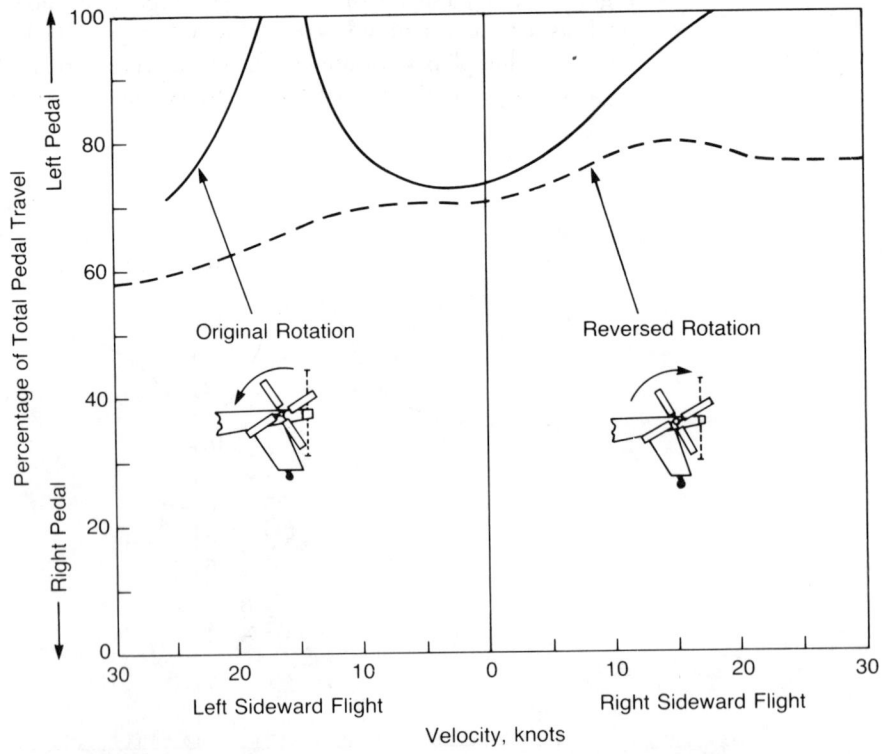

FIGURE 2.10 Typical Pedal Requirements for AH-56A in Right and Left Sideward Flight

Source: Wiesner and Kohler, "Tail Rotor Design Guide," USAAMRDL TR 73-99, 1973.

to cause an indefinite delay in the vortex ring condition resulting from a favorable interference of the main rotor tip vortex.

Experimental evidence of this interference is given in reference 2.13. Figure 2.11, based on that report, shows the pitch of the tail rotor required to balance main rotor torque as a function of sideward speed for four different model configurations. The top portion shows the effect of the presence of the main rotor and the lower portion shows the effect of tail rotor direction of rotation in the presence of the main rotor. The evidence of premature vortex ring conditions is not as dramatic as in the preceding two figures, but it is there.

Another study of left sideward flight motivated by some less-than-satisfactory sideward-flight characteristics discovered during the initial development flying on the Hughes AH-64 is reported in reference 2.14. Figure 2.12 shows the pedal activity required to hold heading in left sideward flight with two different tail configurations: a T-tail with a 100-inch diameter tail rotor and a stabilator with a 110-inch diameter tail rotor. The improvement observed with the second configuration was also present even when flown with the smaller tail rotor. Thus the beneficial change appears to be due to raising the tail rotor, which apparently made it less susceptible to effects from the main rotor tip vortices. Two further observations from this program are worth mentioning. The same test pilots who were investigating the unsatisfactory left-sideward-flight characteristics of the AH-64 with the T-tail found that a Huey Cobra that was available as a chase aircraft had no trouble in the same flight condition even with its stability augmentation system turned off! The Cobra tail rotor blades are untwisted, whereas those on the AH-64 have −8° of twist. The differences seem to correlate with the wind tunnel experience quoted during the discussion of Figure 2.8.

The other observation is that on the AH-64, pure left sideward flight was not the most critical condition. Instead, the unsteadiness was worse, with the aircraft flying about 20° off of pure left with the tail rotor *leading* the main rotor. Why this should be more critical than with the tail rotor following the main rotor is not known.

RATE OF DESCENT IN VERTICAL AUTOROTATION

Even though vertical autorotation occurs in the vortex ring state, a first approximation procedure for calculating the rate of descent may be derived from a combination of blade element and momentum concepts. Setting the torque equation for the ideally twisted rotor to zero gives:

$$C_Q = \frac{\sigma}{2} \left[\frac{a}{2} (\theta_T - \phi_T) \phi_T + \frac{c_d}{4} \right] = 0$$

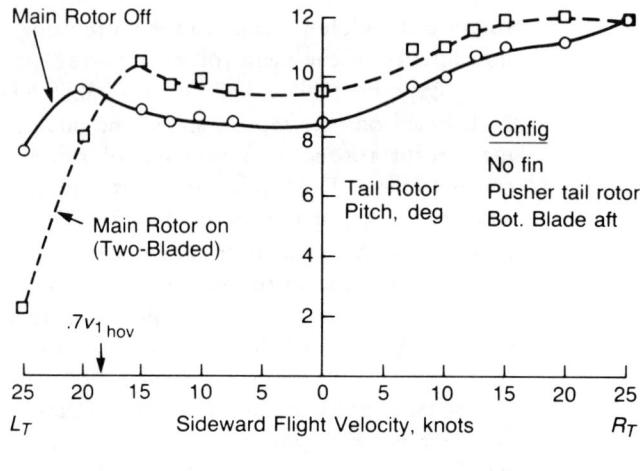

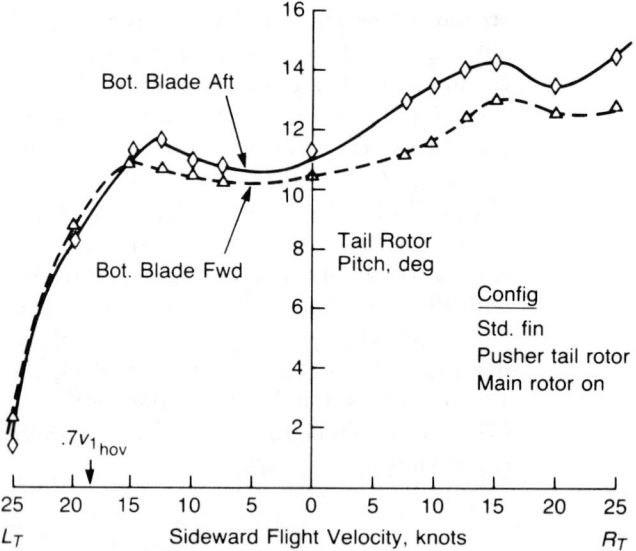

FIGURE 2.11   Effect of Main Rotor Wake and Direction of Tail Rotor Rotation at Yaw Trim Conditions

Source: Yeager, Young, & Mantay, "A Wind-Tunnel Investigation of Parameters Affecting Helicopter Directional Control at Low Speeds in Ground Effect," NACA TND-7694, 1974.

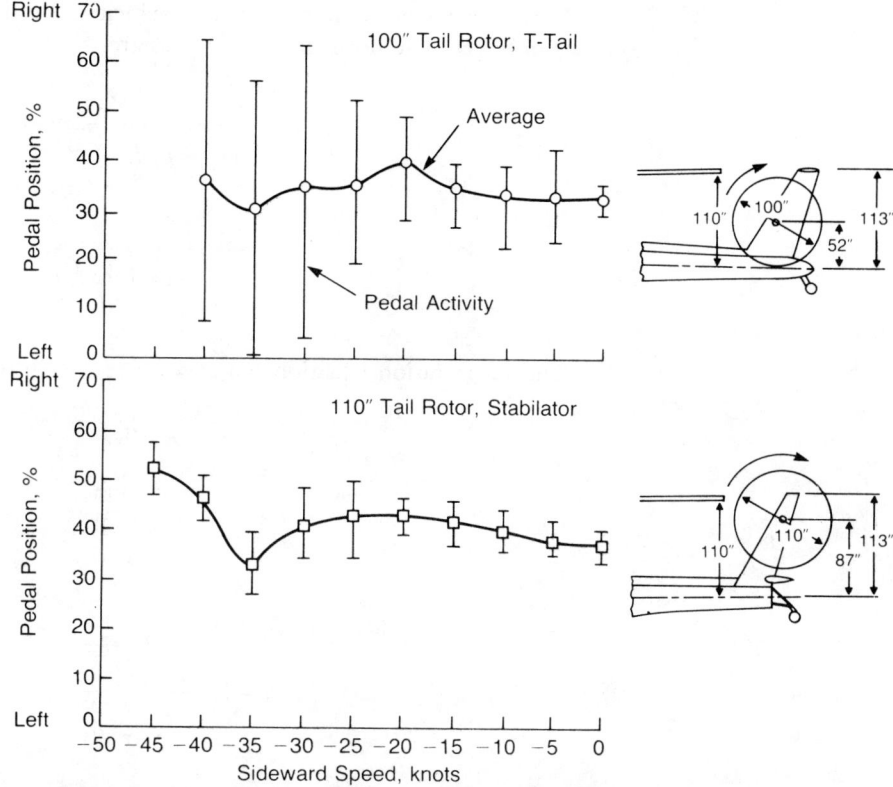

FIGURE 2.12   Pedal Activity in Left Sideward Flight of Hughes AH-64

Source: Prouty, "Development of the Empennage Configuration of the YAH-64 Advanced Attack Helicopter," USAAVRADCOM TR-82-D-22, 1983.

where from the hovering derivation:

$$(\theta_T - \phi_T) = C_T \frac{4}{\sigma a}$$

The inflow angle at the tip, ϕ_T, is:

$$\phi_T = -\left[\frac{V_D - v_1}{\Omega R}\right]$$

Thus:

$$-C_T \frac{(V_D - v_1)}{\Omega R} + \frac{\sigma c_d}{8} = 0$$

It is convenient to normalize the velocities by dividing by the induced velocity in hover. The normalized velocities are:

$$\overline{V}_D = \frac{V_D}{v_{1_{\text{hov}}}} = \frac{V_D}{\Omega R \sqrt{\dfrac{C_T}{2}}}$$

$$\bar{v}_1 = \frac{v_1}{v_{1_{\text{hov}}}} = \frac{v_1}{\Omega R \sqrt{\dfrac{C_T}{2}}}$$

The autorotation equation can now be rewritten to give:

$$\overline{V}_D - \bar{v}_1 = \frac{\dfrac{\sigma c_d}{8}}{\dfrac{C_T^{\frac{3}{2}}}{\sqrt{2}}}$$

or

$$\overline{V}_D - \bar{v}_1 = \frac{\dfrac{3}{2}\sqrt{3}}{\sqrt{\sigma}\dfrac{\bar{c}_l^{\frac{3}{2}}}{c_d}}$$

(Note the similarity to the Figure of Merit equation of Chapter 1.)

Although the momentum equations cannot be strictly applied in the vortex ring state where vertical autorotation occurs, enough experimental data was obtained from the wind tunnel tests of model rotors reported in reference 2.10 to produce the empirical relationships between $\overline{V}_D$ and $(\overline{V}_D - \bar{v}_1)$ that are plotted in Figure 2.13. For the example helicopter at normal gross weight and rotor speed:

$$\frac{\dfrac{3}{2}\sqrt{3}}{\sqrt{\sigma}\dfrac{\bar{c}_l^{\frac{3}{2}}}{c_d}} = 0.22 = \overline{V}_D - \bar{v}_1$$

From Figure 2.13 at a twist of $-10°$:

$$\overline{V}_D = 1.97$$

and

$$V_D = 1.97 v_{1_{\text{hov}}} = 1.97(40.4) = 79.6 \text{ ft/sec} = 4{,}780 \text{ ft/min}$$

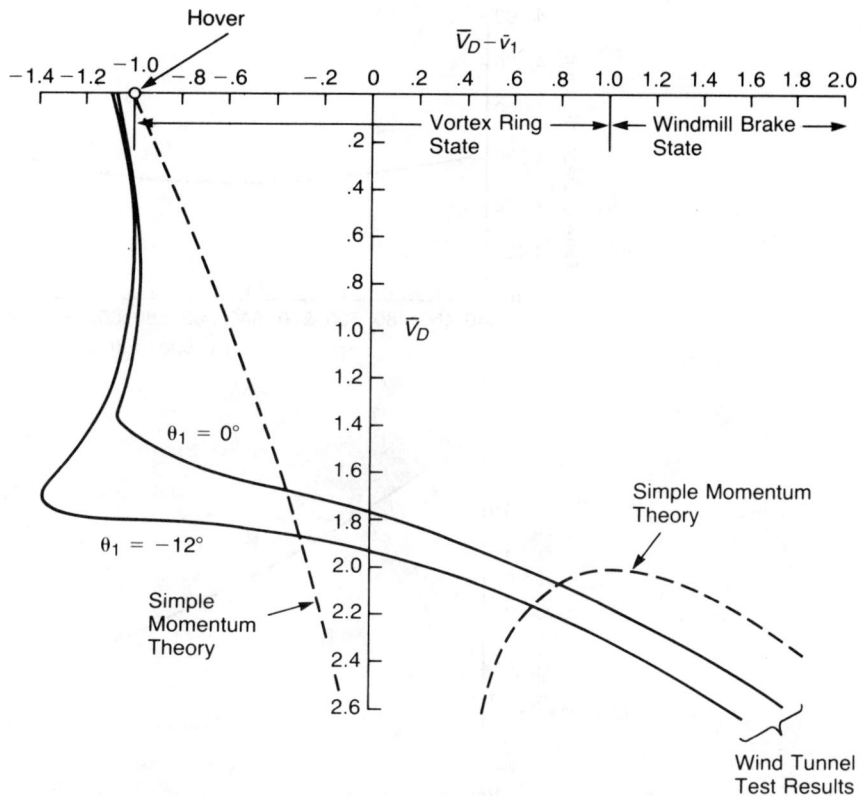

FIGURE 2.13   Nondimensional Velocities in Vertical Descent

Source: Castles & Gray, "Empirical Relation between Induced Velocity, Thrust, and Rate of Descent of a Helicopter Rotor as Determined by Wind-Tunnel Tests of Four Model Rotors," NACA TN 2474, 1951.

This result is so typical that a rule of thumb is that the rate of descent in vertical autorotation is twice the hover-induced velocity. This is also borne out by an examination of Figure 2.8. Figure 2.13 shows that an untwisted rotor is better than a twisted one for vertical autorotation in that a lower rate of descent is required.

The rate of descent is a function of rotor speed; the minimum occurring at the rotor speed that corresponds to the maximum $\bar{c}_l^{3/2}/c_d$. Figure 2.14 shows the calculated rate of descent of the example helicopter as a function of rotor speed. The minimum rate of descent occurs at a tip speed of about 550 ft/sec. Although this represents a theoretical optimum, it is not a practical condition since the rotor is on the verge of blade stall, which could be triggered by small changes in flight conditions, and because of the relatively low value of kinetic energy stored in the rotor for use in the landing flare. It is more likely that the pilot would try to hold

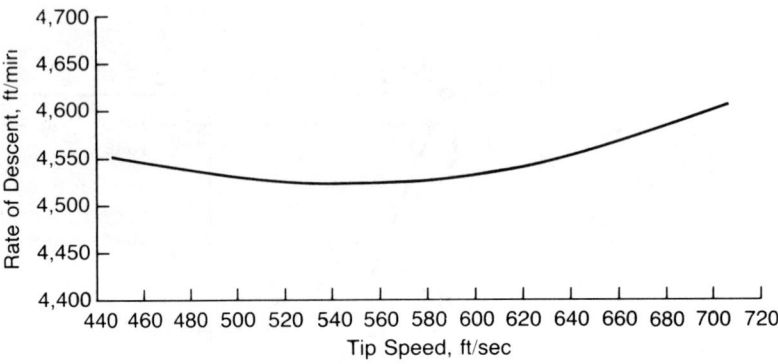

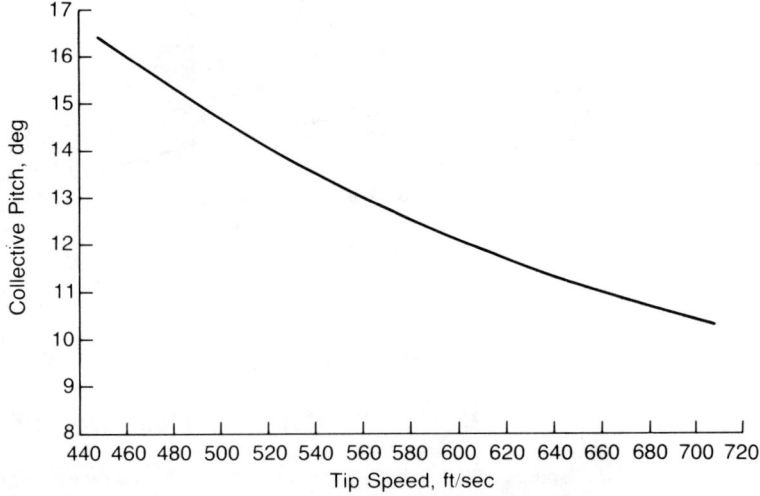

FIGURE 2.14   Effect of Tip Speed on Rate of Descent in Vertical Autorotation

the normal power-on tip speed or even a value slightly higher. The collective pitch required as a function of tip speed can be estimated by combining equations already derived.

$$\theta_0 = \frac{3}{2}\,\theta_{T_{\text{Ideal twist}}} - \frac{3}{4}\,\theta_1$$

$$\theta_0 = \frac{3}{2}\left\{57.3\left[\frac{4}{a}\,C_T/\sigma - \frac{v_{1_{\text{hov}}}}{\Omega R}\,(\overline{V}_D - \overline{v}_1)\right]\right\} - \frac{3}{4}\,\theta_1 \quad \text{degrees}$$

For these calculations, no power losses due to the transmission, accessories, or tail rotor have been considered. For actual cases where the rotor must supply the extra power, Δh.p., the equation for the velocities becomes:

$$\overline{V}_D - \dot{v}_1 = \frac{\frac{3}{2}\sqrt{3}}{\frac{\sqrt{\sigma}}{\bar{c}_l^{3/2}/c_d}}\left[1 + \frac{4{,}400\Delta\text{h.p.}}{\rho A_b(\Omega R)^3 c_d}\right]$$

For the example helicopter, it is estimated that the rotor must develop 40 hp during vertical autorotation. This will increase the value of $(\overline{V}_D - \bar{v}_1)$ by 11%, but the rate of descent will increase by only 1.5%. It is of some interest to note that the rate of descent of a parachute is given by the equation:

$$R/D = 60\sqrt{\frac{\text{D.L.}}{\frac{\rho}{2}C_D}} \quad \text{ft/min}$$

For a parachute with the same disc loading as the example helicopter and a drag coefficient of 1.2, the rate of descent is 4,260 ft/min—approximately the same as the helicopter. This comparison adds validity to the observation that the rate of descent of the helicopter in autorotation is approximately twice the hover-induced velocity. (A parachute with a drag coefficient of 1.0 would be an exact analogy of the helicopter in this condition.)

EFFECT OF RAPID PITCH CHANGES

The instantaneous thrust and power associated with a vertical flight condition is affected by the rate of change of collective pitch. This is of some significance in its effect on performance during jump take-offs and in landing flares from vertical autorotation. It is of even more significance in its effect on the design of a tail rotor drive system.

The phenomenon may be illustrated by visualizing a rotor with untwisted blades on a whirl tower at full rotor speed and flat pitch. If the collective pitch is suddenly increased to 10°, the angle of attack of each blade element will instantaneously be 10° and then will decrease as the equilibrium-induced flow pattern is established. Figure 2.15 taken from reference 2.15, shows the measured thrust coefficient of a 33-foot-diameter rotor on a whirl tower for several rates of collective pitch inputs. It may be seen that for very rapid increases, a transient thrust overshoot to twice the final steady value is possible. Not only does the thrust have an overshoot but the torque may also; especially if the transient angles of attack produce stall.

For the main rotor, the transient increase in torque required may exceed that available from the engine and the rotor will slow down, the extra power required being produced by the loss in rotational kinetic energy. Such a decrease is usually safe unless the rotor slows to a point where it can no longer produce the required

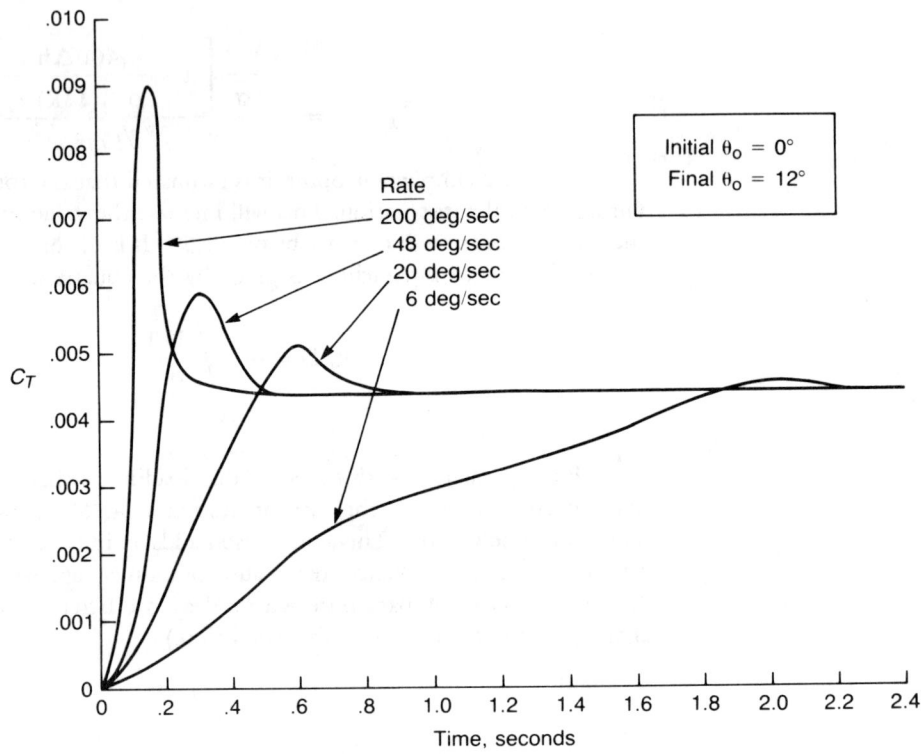

FIGURE 2.15   Effect of Rapid Pitch Change on Rotor Thrust

Source: Carpenter & Fridovich, "Effect of a Rapid Blade-Pitch Increase on the Thrust and Induced-Velocity Response of a Full-Scale Helicopter Rotor," NACA TN 3044, 1953.

thrust. The situation is different for a tail rotor, however. The transient increase in torque is small compared with the total inertia of the rotating system so that little or no slowing takes place and the tail rotor simply demands whatever torque is required from its drive system. This torque may be several times the maximum steady torque which might reasonably be used in the design of the tail rotor drive system. The most critical flight maneuver is stopping a turn over a spot in the torque direction (analogous to vertical descent of the main rotor) with full opposite pedal. For this case, the instantaneous change in angle of attack may be as much as 30° which will insure fully stalled blades and a correspondingly high torque requirement compared with normal flight. In lieu of designing the tail rotor drive system to take this abnormally high torque, the current practice is to design for some more moderate loading—for example, twice the maximum hover torque—and then to ask the pilots to refrain from extreme turn-reversal maneuvers. In some cases, helicopters have been equipped with dampers on the rudder pedals to make it difficult for the pilot to apply sudden pitch changes.

EXAMPLE HELICOPTER CALCULATIONS

HOW TO's

The following items can be evaluated by the methods in this chapter.

CHAPTER 3

Aerodynamics of Forward Flight

MOMENTUM AND ENERGY CONSIDERATIONS

Just as in hover and vertical flight, there are two basic methods for analyzing the characteristics of a rotor in forward flight: the momentum, or energy, method; and the blade element method. The blade element method is necessary for accurate performance estimation and for establishing the limits of rotor performance, but the momentum method provides a rapid means of obtaining a first estimate of the performance as well as a valuable insight into the physics of the system. The balance of forces that governs the helicopter in forward flight is shown in Figure 3.1. The rotor thrust vector must balance not only the gross weight, as it does in hover, but also the horizontal drag of the rotor blades and the lift and drag of the fuselage, hub, landing gear, and other necessary—but draggy—items by which helicopters achieve their role as useful vehicles. The drag of the rotor blades is known as the rotor horizontal force, or H-force, and all other drag items are classified as parasite drag.

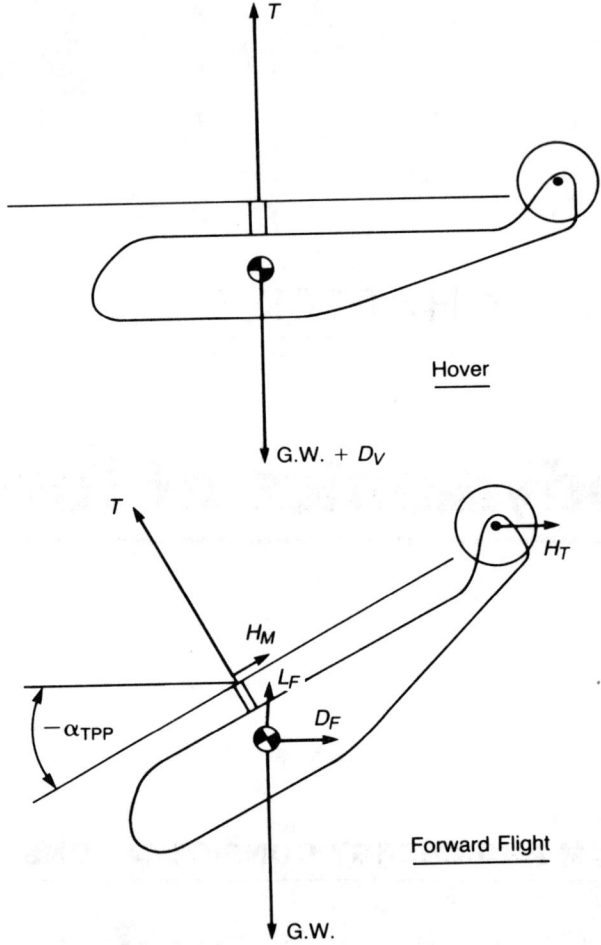

FIGURE 3.1 Balance of Forces on Helicopter

Induced Velocities in Forward Flight

Equations for the induced velocities in hover were derived using the momentum method. A variation of the same method applies to forward flight and starts with the same equation:

$$T = \text{m/sec} \times \Delta v$$

In order to define the mass flow and the change in velocity, it is helpful to go back to the aerodynamics of a wing for an analogy. When a wing generates lift, at least in theory, it affects all the air, both near and far. Of course, the air adjacent to the wing is affected the most, whereas at large distances from the wing the effect is negligible. It may be shown that for an ideal wing with an elliptical lift distribution, the downward acceleration of the total mass of air is mathematically equivalent to uniformly accelerating only the mass of air in a round stream tube whose diameter is equal to the wing span. This purely mathematical coincidence should not be given a physical connotation. For the wing shown in Figure 3.2, the momentum equation may be written:

$$L = \rho \sqrt{V^2 + v_1^2} \ \pi \left(\frac{b}{2}\right)^2 v_2, \ \text{lb}$$

For most forward flight conditions, it may be assumed that v_1 is small compared with V, so that:

$$L = \rho V \pi \left(\frac{b}{2}\right)^2 v_2$$

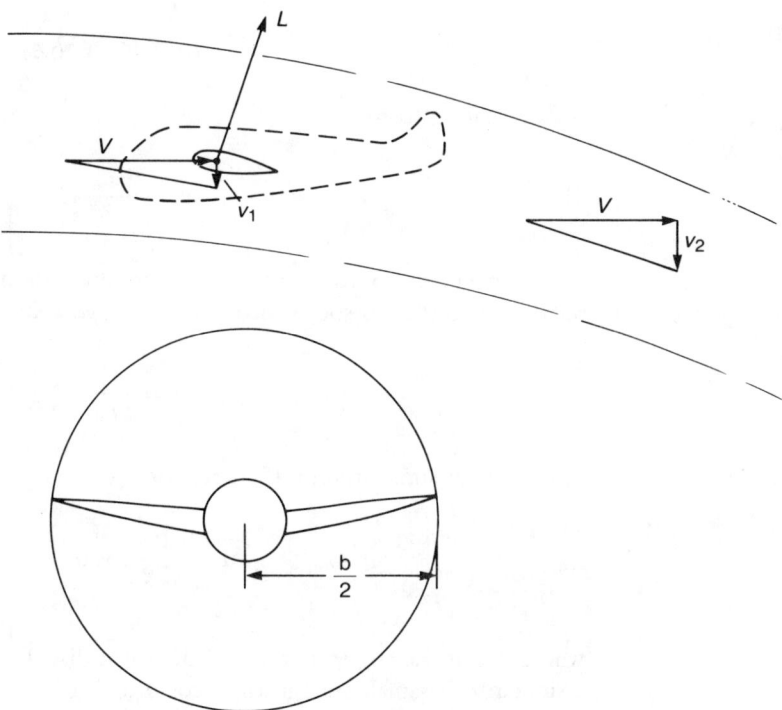

FIGURE 3.2 Induced Velocity at an Airplane Wing

and, by the methods used for the rotor in hover, it may again be shown that:

$$v_2 = 2v_1$$

so that:

$$v_1 = \frac{L}{2\rho\pi \left(\dfrac{b}{2}\right)^2 V} \ , \text{ft/sec}$$

The induced angle, α_{ind}, can be written in the form familiar to the airplane aerodynamicist:

$$\alpha_{\text{ind}} = \frac{v_1}{V} = \frac{C_L S}{\pi b^2} = \frac{C_L}{\pi \text{A.R.}} \ , \text{rad}$$

The rotor obeys the same momentum laws as the wing. Once the validity of the stream tube concept is accepted, the equation for the induced velocity of a rotor in forward flight can be written by analogy to the wing equation:

$$v_1 = \frac{T}{2\rho\pi R^2 V} = \frac{T}{2\rho A V}$$

An alternative form is:

$$v_1 = \frac{v_{1_{\text{hov}}}^2}{V}$$

It is convenient in rotor analysis to nondimensionalize the forward speed by relating it to the tip speed through the *tip speed ratio*, μ, where:

$$\mu = \frac{V}{\Omega R}$$

Note that in some studies, the definition of μ is given as:

$$\mu = \frac{V}{\Omega R} \cos \alpha_{\text{TPP}}$$

where α_{TPP} is the angle of attack of the rotor disc. In this book, α_{TPP} will always be assumed to be small enough that: $\cos \alpha_{\text{TPP}} = 1$.

Using this definition, another alternative form for the induced velocity equation is:

$$v_1 = \frac{\Omega R}{\mu} \frac{C_T}{2}$$

These equations apply to conditions where the forward velocity is relatively large with respect to the induced velocity. It is sometimes necessary to make calculations at low speeds, where this assumption is not valid. For these cases, the momentum equation is:

$$T = \rho A \sqrt{V^2 + v_1^2}\, 2v_1$$

Solving this without making any assumptions gives:

$$v_1 = \sqrt{-\frac{V^2}{2} + \sqrt{\left(\frac{V^2}{2}\right)^2 + \left(\frac{T}{2\rho A}\right)^2}}$$

or:

$$v_1 = \sqrt{-\frac{V^2}{2} + \sqrt{\left(\frac{V^2}{2}\right)^2 + v_{1_{hov}}^4}}$$

A comparison of the results of using this equation and the simpler one derived earlier is shown in Figure 3.3, where v_1 in ft/sec is plotted against forward speed in knots for the example helicopter. It may be seen that for the example helicopter, the two equations give essentially the same results above about 30 knots.

The simple equation for induced velocity derived here is known as the *constant momentum induced velocity*. A more realistic view sees the induced velocity as the result of a very complex vorticity pattern, consisting of trailing, shed, and bound vortex elements associated with the lift and the change of lift on each blade element. This complexity is of great importance when studying blade loads and helicopter vibration, but it has been found that for most performance calculations the use of the constant momentum value—which represents the average of the complex velocity field—gives reasonably accurate results. There are, however, two helicopter trim problems that require modifications to the simple theory. The first of these is the calculation of the effects of the rotor wake impinging on a horizontal stabilizer in low-speed forward flight. In hover, the average dynamic pressure in the fully developed wake is equal to the disc loading. Experience has shown, however, that local dynamic pressures in the wake in hover and forward flight may be significantly higher than this. Some feeling for this may be obtained by studying the plots of the wake dynamic pressure below a hovering rotor, which were shown in Figure 1.19 of Chapter 1. Test results reported in reference 3.1 show that the download pressure to which a horizontal stabilizer may be subjected in low-speed flight can be as much as three times the main rotor disc loading. Assuming a drag coefficient of 2 for the stabilizer in this condition gives a dynamic

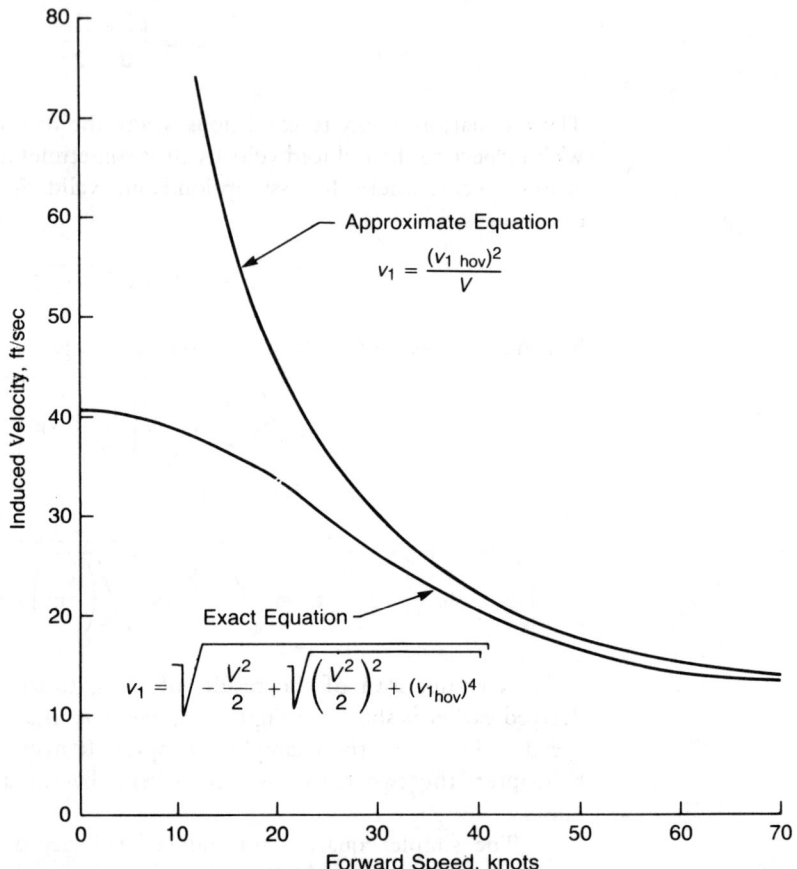

Induced Velocity, ft/sec

Approximate Equation

$$v_1 = \frac{(v_{1\,hov})^2}{V}$$

Exact Equation

$$v_1 = \sqrt{ -\frac{V^2}{2} + \sqrt{\left(\frac{V^2}{2}\right)^2 + (v_{1hov})^4} }$$

Forward Speed, knots

FIGURE 3.3   Induced Velocities in Low-Speed Flight for Example Helicopter

pressure in the wake of 1.5 times the disc loading. This can be the source of disconcerting trim shifts in transition flight.

Another analysis problem that will be examined in detail later is that of calculating the lateral blade flapping in forward flight. For this, it has been found necessary to represent the local induced velocity at the disc as:

$$v_L = v_1 \left(1 + K \frac{r}{R} \cos \psi \right)$$

where ψ is the blade azimuth position, being zero over the tail. This equation defines an induced velocity distribution that is small at the leading edge of the disc and large at the trailing edge, as shown in Figure 3.4. Smoke studies of the induced velocities around wind tunnel models of rotors in forward flight reported in references 3.2 and 3.3 show that the average velocity distribution does indeed have

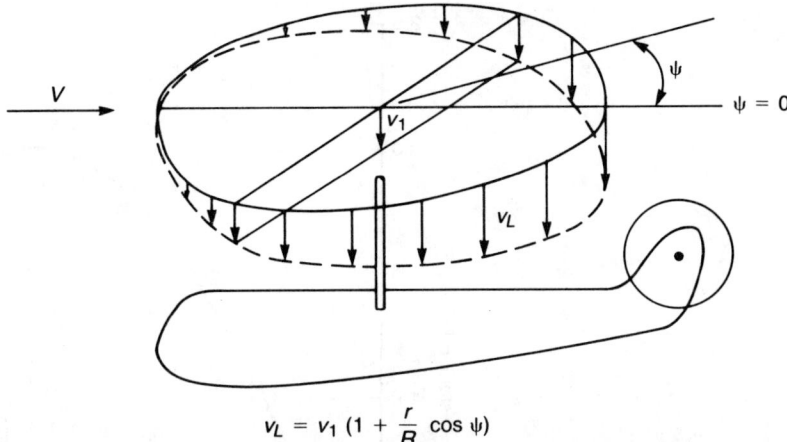

$$v_L = v_1 \left(1 + \frac{r}{R} \cos \psi \right)$$

FIGURE 3.4   Assumed Induced Velocity Distribution

this general pattern and, furthermore, that the induced flow at the leading edge of the disc is essentially zero. This observation leads to assigning the distortion factor, K, a value of unity, so that:

$$v_L = v_1 \left(1 + \frac{r}{R} \cos \psi \right)$$

This applies to relatively high speeds—say above 100 knots—but at low speeds, the distortion may be somewhat higher. This is shown in Figure 3.5, taken from reference 3.4, which reports on flight tests and wind tunnel experience on the very rigid rotor Sikorsky Advancing Blade Concept helicopter. From the cyclic pitch required to trim, the distortion factor in the induced velocity equation could be determined. The figure indicates that while a value of unity may be good for high speeds, in the transition region it may be as high as 2.

Several alternative forms of this equation are discussed in reference 3.5, a study of the surprisingly large lateral flapping that occurs at low speeds. A more rigorous derivation of the induced velocity distribution using vortex theory is given in reference 3.6.

Induced Power

The rotor obeys the same momentum equations as the wing. The equation for the induced drag of the ideal rotor in forward flight can be written by analogy to the wing:

$$D_{\text{ind}} = T\alpha_{\text{ind}} = \frac{Tv_1}{V} = \frac{T^2}{2\rho AV^2}, \text{ lb}$$

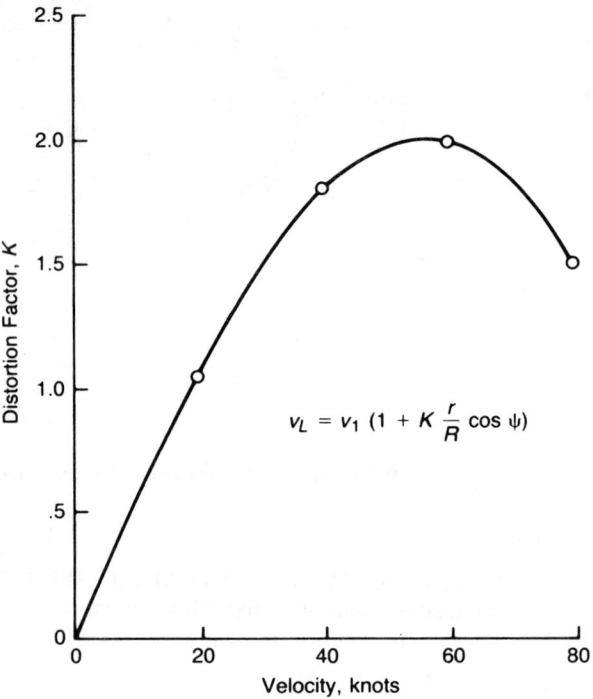

FIGURE 3.5   Induced Velocity Distortion Factor at Low Airspeeds

Source: Ruddell, "Advancing Blade Concept (ABC) Development," *JAHS* 22-1, 1977

The corresponding induced power is:

$$\text{h.p.}_{\text{ind}} = \frac{D_{\text{ind}}V}{550} = \frac{T^2}{1,100\rho AV}$$

Note that, for a constant rotor thrust, the induced power decreases as forward speed increases. This is a consequence of the rotor handling more air per second and having to accelerate it downward less to achieve the same thrust.

The derivation so far has been based on an ideal rotor, analogous to an ideal wing with an elliptical lift distribution. The airplane aerodynamicist accounts for the fact that a wing seldom has an elliptical lift distribution by using an *Oswald efficiency factor*, *e*, in the denominator of the induced drag equation:

$$C_{D_{\text{ind}}} = \frac{C_L^2}{\pi \text{A.R.} e}$$

where *e* is less than unity. For a rotor, a corresponding *rotor efficiency factor* can be evaluated by comparing the distribution of circulation in the wake with an ideal elliptical distribution. This was done in reference 3.7, in which the circulation distribution in the wake of an infinite-bladed rotor was computed from a blade element method assuming an uniform induced velocity distribution at the rotor. Figure 3.6 shows the computed distribution of circulation for a specific flight condition and the value of e that corresponds to this distribution. Figure 3.7 shows the results of the evaluation in terms of e as a function of the angle of attack of the tip path plane for several values of tip speed ratio, C_T/σ, and twist. The angle of attack of the tip path plane for use in this type of analysis can be approximated by:

$$\alpha_{\text{TPP}} = -57.3 \frac{D}{\text{G.W.}} , \text{deg}$$

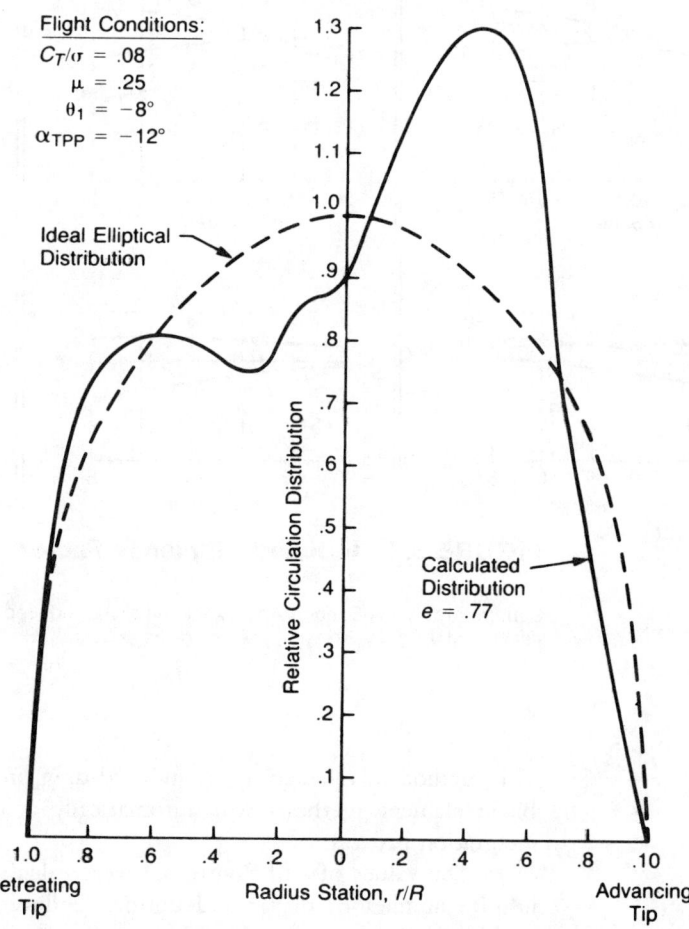

FIGURE 3.6   Comparison of Calculated and Ideal Circulation Distribution in Rotor Wake

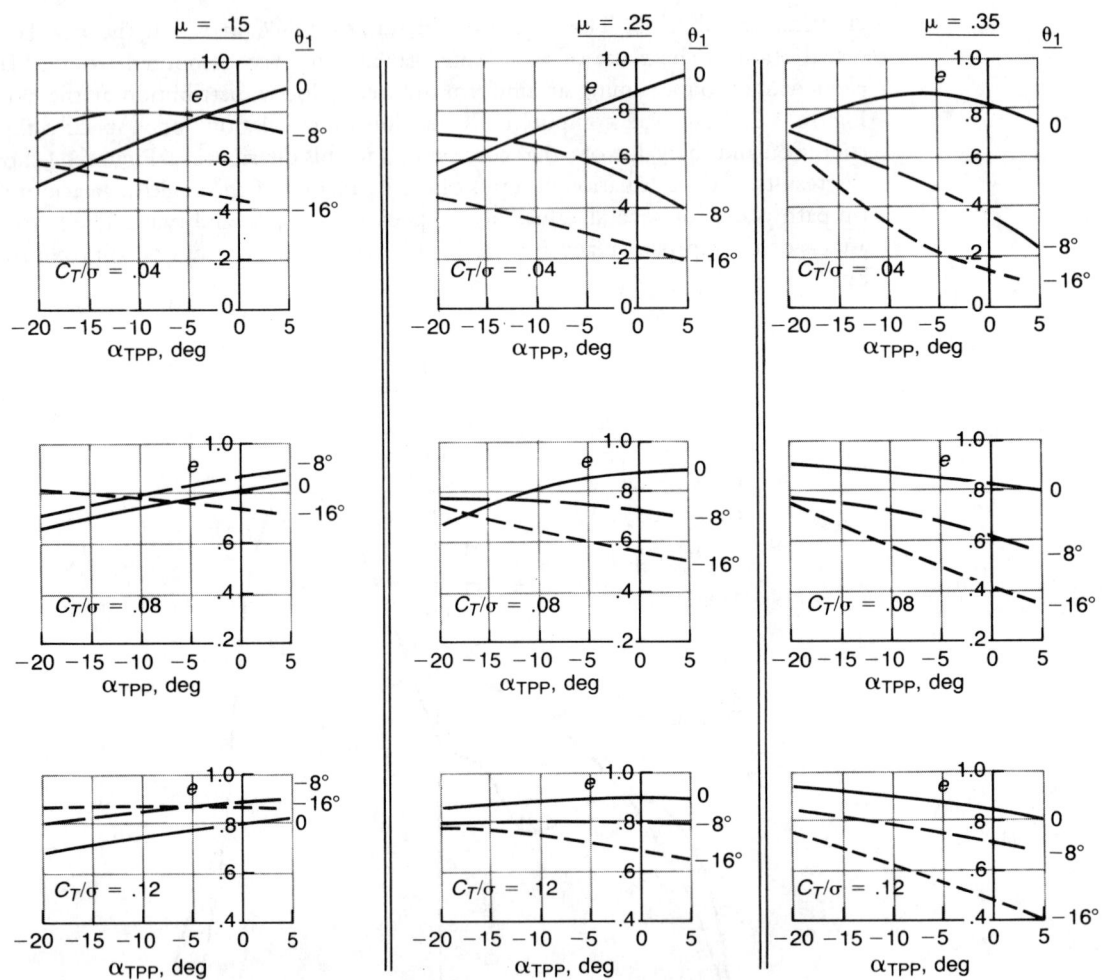

FIGURE 3.7   Induced Efficiency Factor

Source: Prouty, "A Second Approximation to the Induced Drag of a Helicopter Rotor in Forward Flight," *JAHS* 21-3, 1976.

This method of calculating the induced drag only applies to the energy method. Blade element methods will automatically account for the effect during the integration process.

The values of e of Figure 3.7 were calculated using the assumptions of an infinite number of blades and uniform induced velocity at the rotor. A more sophisticated analysis of this problem is reported in reference 3.8. In this study, the actual induced velocities at the blades produced by shed and trailing vortices are computed, and the increments of induced drag due to these velocities are integrated to give the total rotor-induced drag. This type of analysis, of course, is more accurate but also more work. Reference 3.8 gives the results for only one point corresponding to a two-bladed UH-1 helicopter with −12° of twist at a tip speed ratio of .26, a C_T/σ of 0.09, and a rotor angle of attack of about −9°. The value of e calculated by this method is 0.26 compared to 0.75 interpolated from Figure 3.7. Some, if not all, of the difference may be due to the difference in the number of blades used in the two analyses. Further calculations of this type should lead to a better understanding of the effects of different numbers of blades.

The equation for the induced power is:

$$\text{h.p.}_{\text{ind}} = \frac{T^2}{1,100 \rho A V e}$$

or, alternatively:

$$\text{h.p.}_{\text{ind}} = \frac{TV}{550e} \left(\frac{v_{1_{\text{hov}}}}{V} \right)^2$$

or

$$C_P/\sigma_{\text{ind}} = C_Q/\sigma_{\text{ind}} = (C_T/\sigma)^2 \frac{\sigma}{2\mu e}$$

Since most performance calculations are done with a blade element approach, the use of the rotor efficiency factor is limited. It does, however, provide some insight into what portion of the total power is induced power and how this portion is affected by flight conditions. It can also be used in the analysis of flight and wind tunnel test data, where it is desired to divide the total power into its various components.

The calculated induced power for the example helicopter using the efficiency factor of Figure 3.7 is shown in Figure 3.8 as a function of forward speed.

The rotors of a tandem rotor helicopter cannot be treated as two independent rotors since they aerodynamically interfere with each other. The effect is primarily one of the rear rotor operating in a "climb" condition, due to the downwash of the front rotor. The induced velocity at the rear rotor due to the front rotor can be estimated using theoretical charts such as those presented in reference 3.9. Figure 3.9 summarizes the results for several relative rotor positions as the ratio of induced velocity at the rear rotor to the induced velocity at the front rotor, $\Delta v/v_1$ as a function of the wake skew angle, x, where:

$$x = \tan^{-1} \frac{V}{v_1}$$

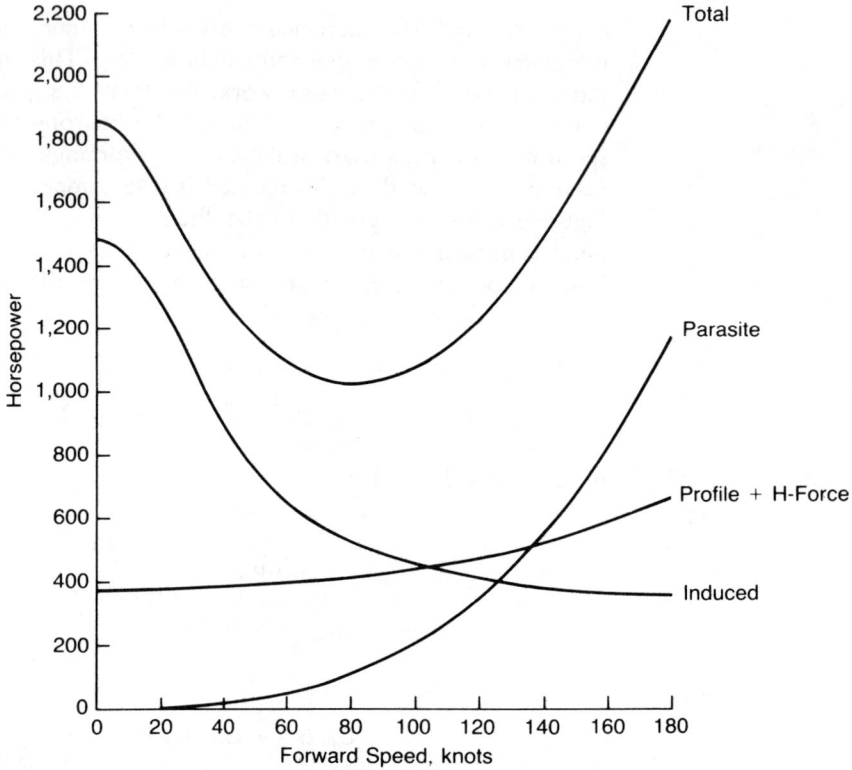

FIGURE 3.8 Main Rotor Power Calculated Using Energy Method for Example Helicopter

($x = 0$ in hover and approaches 90° in high-speed flight.)

The increase in power is:

$$\Delta h.p._{ind} = \frac{T\left(\dfrac{\Delta v}{v_1}\right)v_1}{550}$$

where T is the thrust of one rotor.

For side-by-side configurations, the streamtube has the same diameter as the total span of the rotors, and thus the induced velocity corresponding to a given thrust is less than on a single rotor. The same consideration applies to a tandem rotor helicopter in sidewards flight, and, as a matter of fact, tandems are flown sideways at low speeds when maximum takeoff performance is desired.

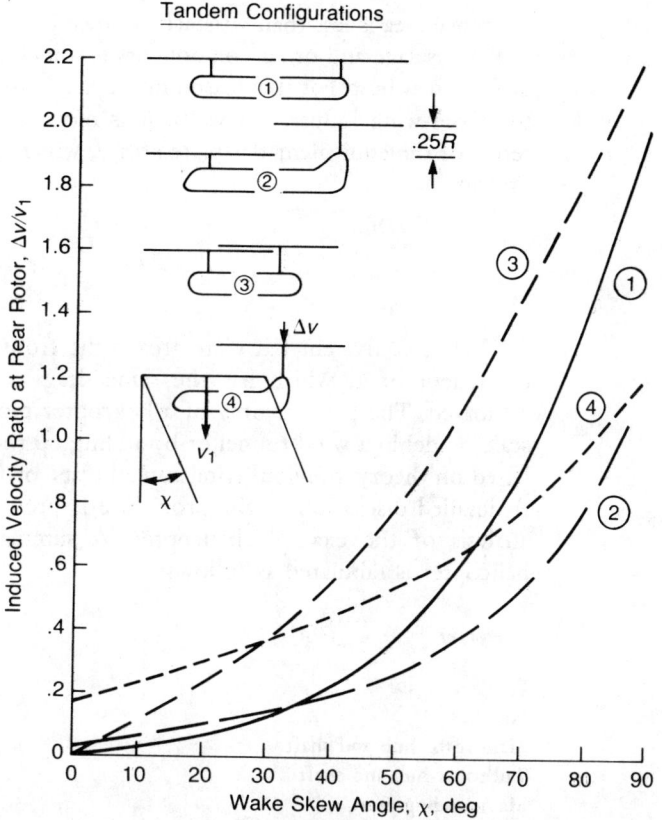

FIGURE 3.9   Interference Effect for Tandem Rotor Helicopters

Parasite Power

The power required to overcome the drag of all the nonrotor components is known as the parasite power, h.p._p:

$$\text{h.p.}_p = \frac{D_p V}{550}$$

The parasite drag could be expressed as a function of a drag coefficient as it is in airplanes:

$$D_p = C_D q S$$

but in the case of a helicopter without a wing, the assignment of a reference area, S, is not a straightforward procedure. For instance, using the rotor disc area as a reference area is less than satisfactory since it tends to imply that the parasite drag of the fuselage and other components is somehow a function of the rotor size—a half truth at best. For this reason in the helicopter industry (and to some extent in the fixed-wing industry as well), it is more common to define parasite drag in terms of the equivalent flat plate area, f, which is simply drag divided by dynamic pressure:

$$f = \frac{D}{q}, \text{ ft}^2$$

The equivalent flat plate area is the frontal area of a flat plate with a drag coefficient of 1, which has the same drag as the object whose drag is being estimated. The parasite drag of a helicopter can be estimated either by testing a scale model in a wind tunnel or by adding up the drag of the various components based on theory, previous wind tunnel tests, or flight tests of similar components. A detailed discussion of the procedure is presented in Chapter 4, along with an analysis of the example helicopter. A summary of the drag of the example helicopter is tabulated as follows:

Component	Equivalent Flat Plate Area, f, ft^2
Fuselage	5.8
Nacelles	1.1
Main rotor hub and shaft	7.0
Tail rotor hub and shaft	.7
Main landing gear	1.2
Tail landing gear	0.8
Horizontal stabilizer	0.2
Vertical stabilizer	0.2
Rotor-fuselage interference	1.3
Exhaust system	0.5
Miscellaneous	0.5
Total	19.3

Typical values of f range from 5 ft² for small, clean helicopters to 60 ft² for large flying cranes. The parasite drag is:

$$D_p = fq = f\frac{\rho}{2}V^2, \text{ lb}$$

and the parasite power is:

$$\text{h.p.}_p = \frac{D_p V}{550} = \frac{f\rho V^3}{1,100}$$

or, nondimensionally:

$$C_P/\sigma_p = C_Q/\sigma_p = \frac{f}{A_b}\frac{\mu^3}{2}$$

Figure 3.8 shows how the parasite power varies with speed for the example helicopter.

An approximation for the angle of attack of the tip path plane can be written in nondimensional terms using the equivalent flat plate area:

$$\alpha_{\text{TPP}} \doteq -57.3\,\frac{D}{\text{G.W.}}\,,\text{ deg}$$

$$\alpha_{\text{TPP}} \doteq -57.3\,\frac{f}{A_b}\frac{\mu^2/2}{C_T/\sigma}\,,\text{ deg}$$

Profile Power

Profile power in forward flight has the same source as it did in hover: the air friction drag of the individual blade elements. In the derivation of the blade element equations in the following section, it will be shown that the profile torque/solidity coefficient has the form:

$$C_Q/\sigma_0 = \frac{c_d}{8}\,(1 + \mu^2)$$

The H-force coefficient due to the drag of the blade elements will be shown to be:

$$C_H/\sigma_0 = \frac{c_d\mu}{4}$$

Thus the total power dissipated by blade drag as both torque and H-force is:

$$\text{h.p.}_{\text{O+H}} = \frac{\rho A_b(\Omega R)^3}{550}\left[\frac{c_d}{8}\,(1 + \mu^2)\right] + \rho A_b(\Omega R)^2\left[\frac{c_d}{4}\mu\right]\frac{V}{550}$$

or

$$\text{h.p.}_{\text{O+H}} = \frac{\rho A_b(\Omega R)^3}{550}\frac{c_d}{8}\,(1 + 3\mu^2)$$

or, in nondimensional form:

$$C_P/\sigma_{O+H} = \frac{c_d}{8}(1 + 3\mu^2)$$

where c_d is the average value, which may be assumed to be the same as that found in hover for this analysis. Figure 3.8 shows this component of power for the example helicopter as a function of forward speed.

Characteristics of the Total Main Rotor Power Required Curve

The total power required for the main rotor out of ground effect is:

$$\text{h.p.}_M = \frac{T^2}{1{,}100\rho AVe} + \frac{\rho f V^3}{1{,}100} + \frac{\rho A_b(\Omega R)^3 \frac{c_d}{8}(1 + 3\mu^2)}{550}$$

or, in nondimensional form:

$$C_P/\sigma = C_Q/\sigma = (C_T/\sigma)^2 \frac{\sigma}{2\mu e} + \frac{f}{A_b}\frac{\rho}{2}\mu^3 + \frac{c_d}{8}(1 + 3\mu^2)$$

The total main rotor power required for the example helicopter is plotted in Figure 3.8. At hover it has zero slope since the rotor cannot distinguish between forward flight and rearward flight. The power required at moderate speeds is less than at hover because of the rapid decrease in induced power. This leads to the observation that it is possible to fly a helicopter in forward flight that does not have enough power to hover provided that some means is available to achieve the takeoff. These means might include a strong wind or the use of ground effect. At some forward speed, the power required rises rapidly as a result of the cubic relationship of parasite power with speed. The location of the minimum power "bucket" depends on the disc loading and the cleanliness of the design. It varies from about 40 knots for "dirty," low-disc-loading helicopters to 100 knots for clean, high-disc-loading aircraft.

The energy method that produced the power required curve of Figure 3.8 is only a first approximation. It tends to underestimate power required at high speeds. For the results of the more sophisticated blade element method, see Figure 4.38 of Chapter 4.

Tail Rotor Power

The tail rotor power can be analyzed with the same approximations used for the main rotor. The tail rotor thrust required to balance the main rotor thrust is:

$$T_T = 550 \left(\frac{\text{h.p.}_M}{\Omega R_M} \right) \left(\frac{R_M}{l_T} \right)$$

where the subscript, M, applies to the main rotor and l_T is the tail rotor moment arm with respect to the main rotor shaft. The tail rotor is not used to overcome drag, so its equation has no parasite power term as the main rotor does. To a first approximation, the tail rotor power in forward flight is:

$$\text{h.p.}_T = \left[\frac{T^2}{1{,}100 \rho A V} + \frac{\rho A_b (\Omega R)^3 \frac{C_d}{8}}{550} (1 + 3\mu^2) \right]_T$$

where the subscript, T, indicates that all physical parameters are those of the tail rotor.

Effect of Density Altitude

The characteristics of the power required equations make it possible to easily calculate the power required at altitude if the power required at sea level at several gross weights has already been determined. Assuming that the main rotor thrust is equal to the gross weight, the main rotor power equation may be rewritten:

$$\left(\frac{\text{h.p.}}{\rho/\rho_0} \right)_M = \frac{\left(\frac{\text{G.W.}}{\rho/\rho_0} \right)^2}{1{,}100 \rho_0 A V e} + \frac{\rho_0 f V^3}{1{,}100} + \frac{\rho_0 A_b (\Omega R)^3 \frac{C_d}{8}(1 + 3\mu^2)}{550}$$

and the equation for the tail rotor power in the same form is:

$$\left(\frac{\text{h.p.}}{\rho/\rho_0} \right)_T = \frac{\left[\frac{550}{(\Omega R)_M} \frac{R_M}{l_T} \right]^2 \left(\frac{\text{h.p.}}{\rho/\rho_0} \right)_M^2}{1{,}100 \rho_0 A_T V e} + \left[\frac{\rho_0 A_b (\Omega R)^3 \frac{C_d}{8}(1 + 3\mu^2)}{550} \right]_T$$

The form of these equations shows that for calculating purposes, weight and density do not have to be treated as separate variables, but that one set of calculations made at sea level at several gross weights can also be used for altitude calculations. For example, Figure 4.38 of Chapter 4 shows the power required by the example helicopter at sea level at several gross weights. To obtain the power required at altitude, the gross weight is divided by the density ratio and the gross weight curve corresponding to this equivalent weight is used. The corresponding values of equivalent power read from the curves are multiplied by the density ratio to obtain the actual power.

This method is valid except in those cases where compressibility effects on profile power are significant. Since these effects are a function of temperature rather than of density ratio, they will have to be estimated separately, by methods discussed in a later section.

Ground Effect in Forward Flight

In forward flight the rotor obtains a beneficial effect from flying near the ground, just as it does in hover. The ground effect may be expressed in terms of the ratio of the induced velocity in and out of ground effect. This ratio has been determined for a rotor from the work of reference 3.10 and is plotted in Figure 3.10. Also shown is the same ratio for a wing from reference 3.11 and for a hovering rotor from Figure 1.41 of Chapter 1. Note the similarity of all three curves. The induced power required is proportional to the induced velocity ratio and thus is reduced for flight near the ground. If the ratio applied at all flight conditions, it could be concluded that the induced power would always decrease as the helicopter went from hover to forward flight near the ground as well as away from the ground.

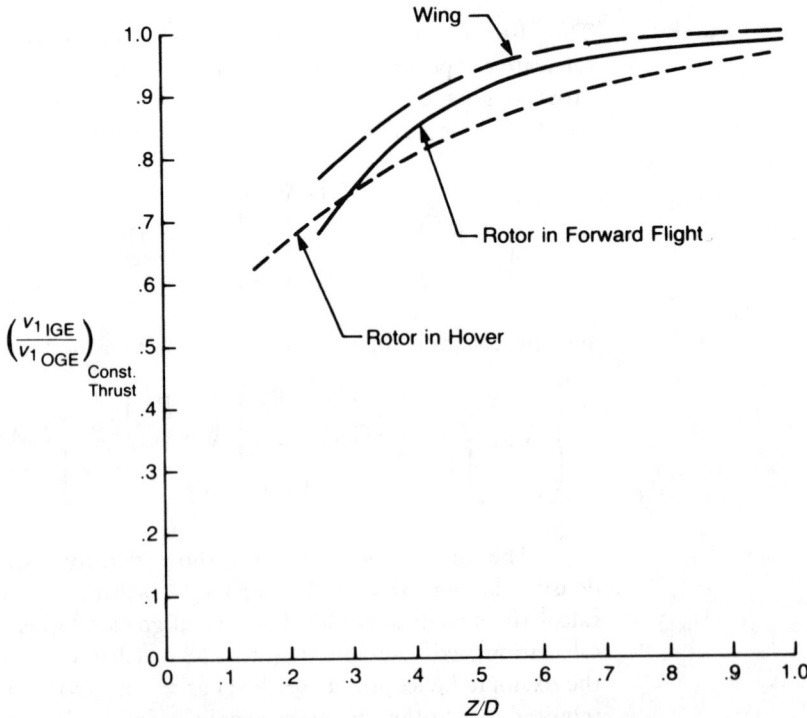

FIGURE 3.10 Effect of Ground on Induced Velocity Ratios in Forward Flight and in Hover

Source for rotor in forward flight: Heyson, "Ground Effect for Lifting Rotors in Forward Flight," NASA TND-234, 1960; for rotor in hover: Figure 1.41; for wing: Hoak, "USAF Stability and Control DATCOM," 1960.

Test experience, however, shows that during transition from hover to forward flight at heights less than about half the rotor diameter, the power may actually increase rather than decrease. Pilots speak of this as "running off the ground cushion." Figure 3.11 shows the effect as measured in flight and reported in reference 3.12 and in a wind tunnel from reference 3.13. The reversal of ground effect is due to the helicopter overrunning the ground vortex, as illustrated in Figure 3.12, which is based on the wind tunnel observations of reference 3.13. As the leading edge of rotor approaches the ground vortex, the inflow is increased just as if part of the rotor were in a climb, thus increasing the power required. The recovery to a more normal inflow pattern occurs suddenly as the vortex passes under the rotor. The effect of the ground vortex on tail rotor performance in sideward and rearward flight is discussed in references 3.1 and 3.14.

Analysis of Climbs and Autorotation by the Energy Method

Energy methods based on the power required in level flight are useful for making first approximations of the extra power required to climb or the rate of descent in

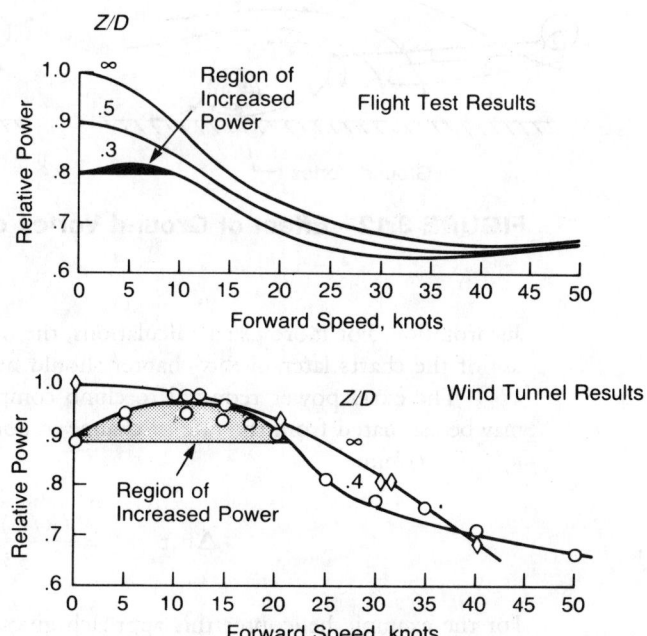

FIGURE 3.11   Experimental Evidence of Ground Effect

Sources: Cheesman & Bennett, "The Effect of the Ground on a Helicopter Rotor in Forward Flight," British R&M 3021, 1957; Sheridan & Wiesner, "Aerodynamics of Helicopter Flight Near the Ground," AHS 33rd Forum, 1977.

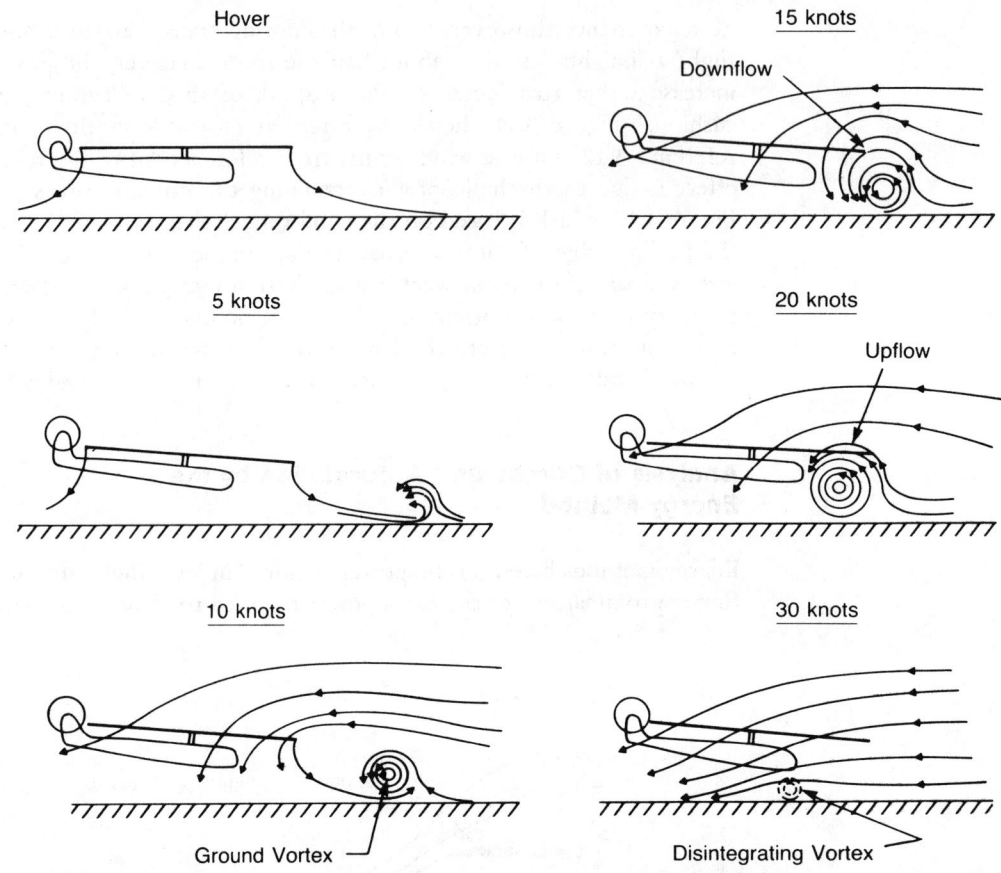

FIGURE 3.12   Effect of Ground Vortex on Inflow Patterns

autorotation. For more exact calculations, the methods outlined in the section on use of the charts later in this chapter should be used.

The extra power required to climb compared to that required to fly level may be estimated from the rate of change in potential energy. For a rate of climb, R/C, in ft/min:

$$\Delta \text{h.p.}_{\text{climb}} \doteq \frac{(R/C)(\text{G.W.})}{33,000}$$

For the example helicopter this approach gives a first approximation:

$$\Delta \text{h.p.}_{\text{climb}} \doteq 0.61(R/C)$$

The best rate of climb will be achieved at the speed for which the power required for level flight is a minimum, since at this speed the Δ h.p. available from the engine will be a maximum. Figure 3.8 shows that for the example helicopter this speed is about 75 knots.

In autorotation, the same approach may be used; that is, the rate of change of potential energy must be large enough to supply power equal to that required for level flight:

$$R/D_{auto} = \frac{33,000 \text{ h.p.}}{\text{G.W.}}$$

The minimum rate of descent also is achieved at the speed for minimum power required in level flight. The effect of several significant parameters on the autorotative rate of descent can be seen by writing the power-required equation as:

$$\text{h.p.} = \frac{(\text{G.W.})^2}{1,100 \rho A V e} + \frac{\rho V^3 f}{1,100} + \text{h.p.}_0$$

Thus:

$$R/D_{auto} = \frac{30 \text{ D.L.}}{\rho V e} + \frac{30 \rho V^3 f}{\text{G.W.}} + \frac{33,000 \text{ h.p.}_0}{\text{G.W.}}, \text{ ft/min}$$

The effect of varying disc loading, equivalent flat plate area, and gross weight on the autorotative rate of descent of the example helicopter at 75 knots is shown below. (The profile power is assumed to remain constant).

Parameter \ Case	One	Two	Three	Four	Five
Disc loading, D.L.	7.1	3.55	7.1	3.55	7.1
Equiv. flat plate area, f, ft^2	19.3	19.3	19.3	19.3	9.65
Gross weight, G.W.	20,000	20,000	10,000	10,000	20,000
h.p., level flight	1,070	780	780	635	1,025
R/D_{Auto}, ft/min	1,765	1,295	2,580	2,100	1,690

Comparing cases 1 and 4 leads to the somewhat surprising conclusion that the lighter a helicopter, the faster its autorotative rate of descent. This paradox can be explained by comparing the total potential energy at the two weights and the power required to fly in level flight. The potential energy is only half as much in case 4 as in case 1, but the power required is more than half as high. To obtain this power from the loss in potential energy, the lighter helicopter must come down faster.

Figure 3.13 shows results obtained during flight tests of the Lockheed Model 286 helicopter. The decrease in rate of descent with increasing gross weight

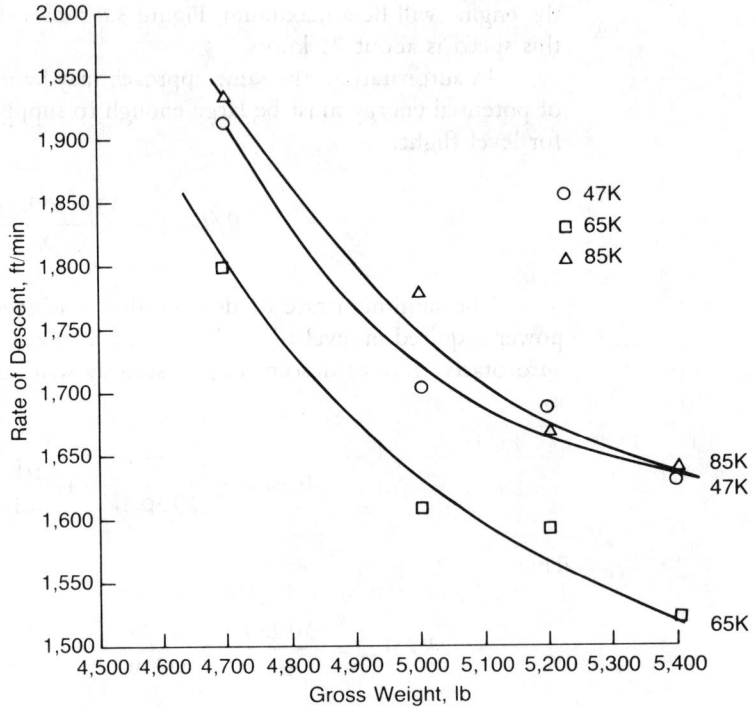

FIGURE 3.13   Measured Autorotative Rates of Descent for Lockheed Model 286 Helicopter

Source: Internal Lockheed report.

is evident. The trend would be expected to hold only until the rotor was so heavily loaded that blade stall would cause the profile power to increase rapidly.

BLADE ELEMENT METHODS

Tangential Velocity

Just as in hovering, the momentum, or energy, method is useful in understanding the physics of forward flight and for making rough calculations; but the blade element theory must be used to define flight limitations and to do more accurate calculations. In forward flight, the velocity acting on the blade element is a function of both the radial station and the blade azimuth position.

The azimuth angle, ψ, is defined as shown in Figure 3.14 with $\psi = 0$ over the tail. The velocity acting on the blade element is the vector sum of the velocity due to rotation, Ωr, and the forward speed of the helicopter, V. The study of swept-wing aerodynamics has shown that the component of velocity perpendicular to the leading edge is the only velocity that is important in establishing aerodynamic forces. The velocity perpendicular to the leading edge—or tangential to the chord of the element—U_T, is:

$$U_T = \Omega r + V\sin\psi$$

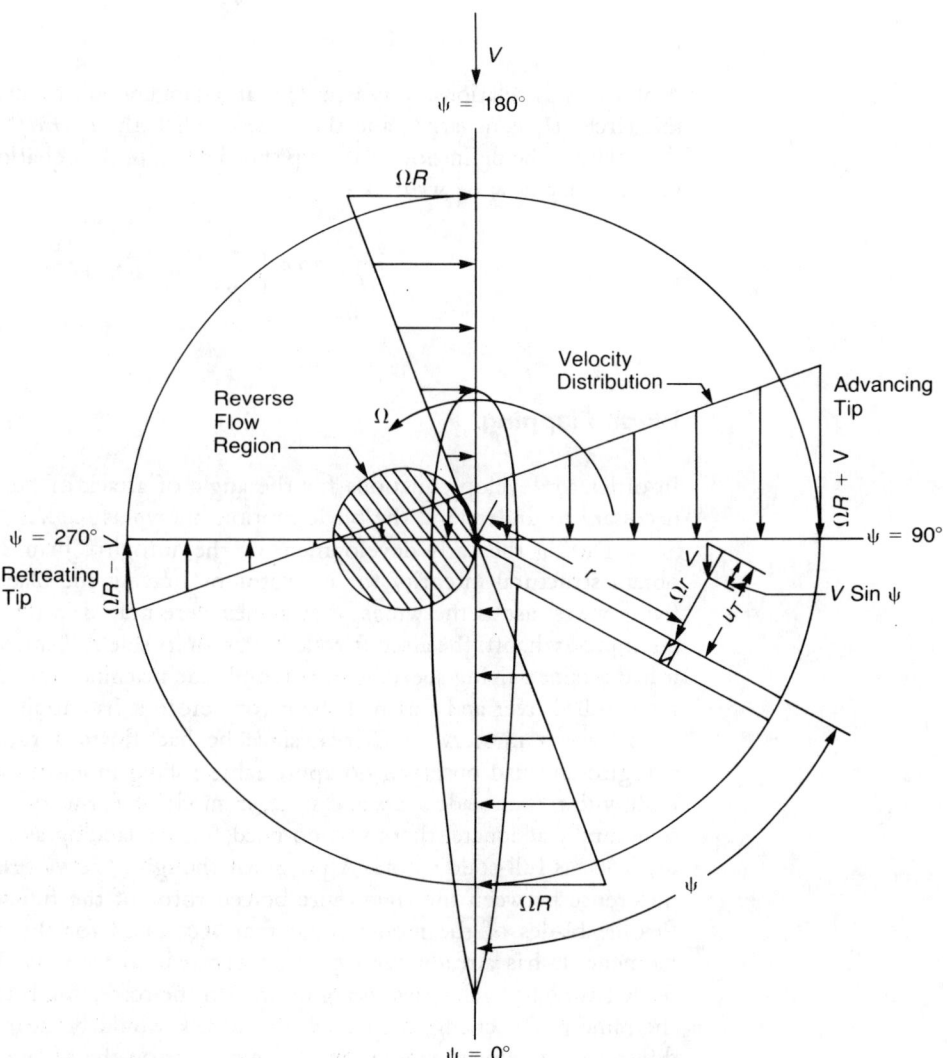

FIGURE 3.14   Tangential Velocities in Forward Flight

Over the tail and over the nose, the blade element sees the same velocity as it would in hover, but on the advancing blade it sees a higher velocity, and on the retreating blade a lower. As a matter of fact, on the retreating blade there are elements where the velocity perpendicular to the leading edge is actually negative—that is, air strikes the trailing edge rather than the leading edge of the blade. Figure 3.14 shows the vector addition of the velocities for the blades at the cardinal azimuth positions. If the equation for U_T is set to zero, the radius station at which the velocity vanishes is:

$$r = -\frac{V}{\Omega} \sin \psi$$

A plot of this relationship is a circle that is tangent to the rotor centerline. Inside this circle, U_T is negative, and the zone is called the *reverse flow region*.

Using the definition of the tip speed ratio, μ, the equation for the tangential velocity may now be written:

$$U_T = \Omega R \left(\frac{r}{R} + \mu \sin \psi \right)$$

Blade Flapping

In order to develop equations for the angle of attack of the blade element, it is necessary to understand the blade motion known as *flapping*.

During the early development of the autogiro, Juan de la Cierva tried to obtain structural integrity for his rotor by bracing the blades with landing and flying wires just as the wings of airplanes were braced in those days. On the first attempted takeoff, the aircraft rolled over on its side and smashed the rotor before it had attained flying speed. Cierva rebuilt the machine, but on the next takeoff it again rolled over and smashed the rotor before it had attained flying speed. This action was a mystery to Cierva, since he had flown a rubber-powered model autogiro and had observed no appreciable rolling moment. The model had been built with rattan blade spars and since, in model size, this type of construction was structurally adequate, there was no need for the landing and flying wires that he used on his full-scale rotors. After much thought, Cierva realized that it was the difference between the rigid, wire-braced rotor of the full-scale aircraft and the flexible blades of the model rotor that accounted for the difference in rolling moment. It has already been pointed out that in forward flight, the advancing blade has higher velocities acting on it than the retreating blade. If each blade had the same pitch setting, their angles of attack would be nearly the same, but the difference in velocity would produce more lift on the advancing side than on the retreating side. This would produce an unbalanced rolling moment, as shown in

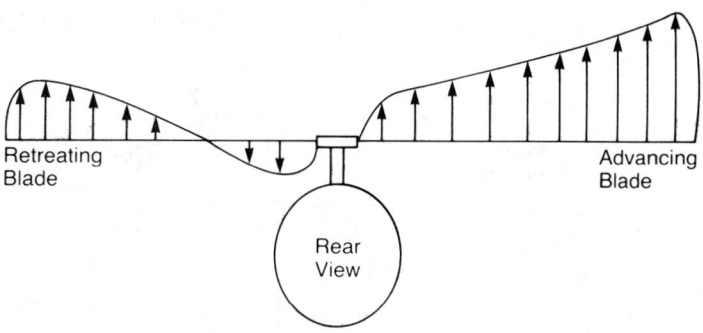

(a) Lift Distribution without Flapping

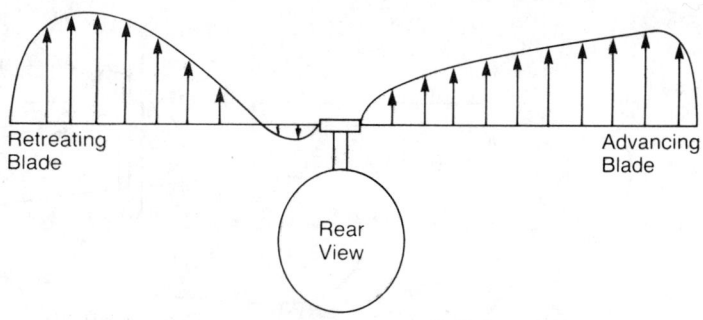

(b) Lift Distribution with Flapping

FIGURE 3.15   Effect of Blade Flapping on Lateral Lift Distribution

Figure 3.15a, which would be expected to roll the aircraft over. On the model, however, because of the flexibility of the rattan blade spars, the blades could bend up and down. Thus the advancing blade, which initially had high lift, began to accelerate upward. As it was accelerating upward, it was also being rotated toward the nose, where the local velocity was reduced to its mean value, so that no unbalanced lift existed and the blade stopped accelerating. The retreating blade was undergoing a similar experience except that it was accelerating downward as it rotated to a position over the tail. The flapping produced a climbing condition on the advancing blade as shown in Figure 3.16 and thus decreased its angle of attack. The retreating blade, on the other hand, was descending and thus experiencing an increased angle of attack. The rotor came to a flapping equilibrium when the local changes in angle of attack were just sufficient to compensate for the local changes in dynamic pressure. In this equilibrium condition, the rotor was not tilted sideways but was tilted fore and aft as shown in Figure 3.17, and the lift distribution was balanced as shown in Figure 3.15b rather than unbalanced as in Figure 3.15a.

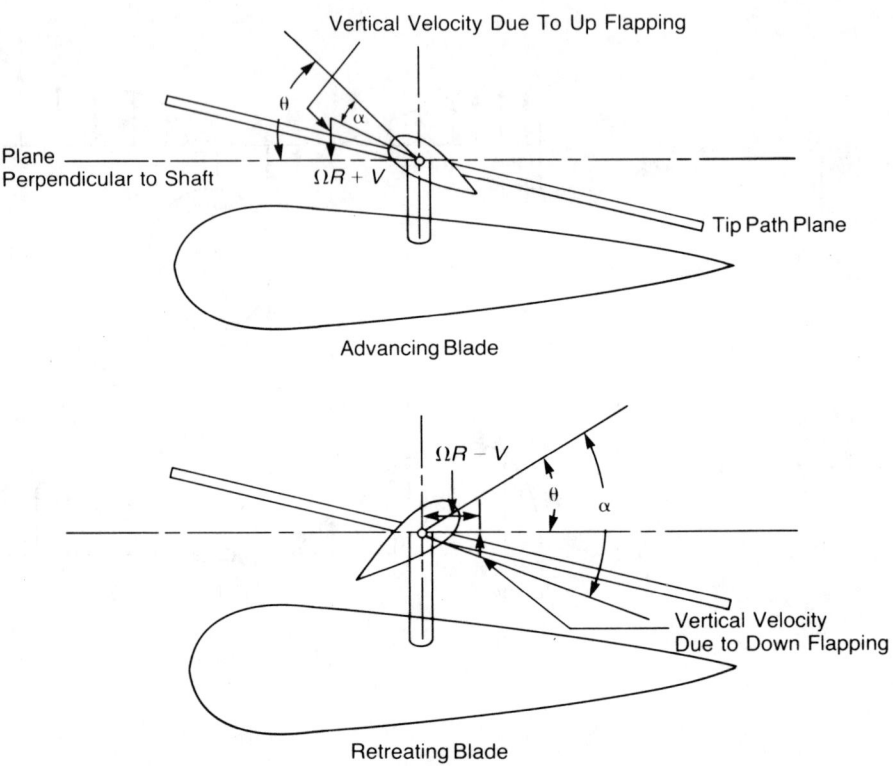

FIGURE 3.16   Change in Angle of Attack Due to Flapping

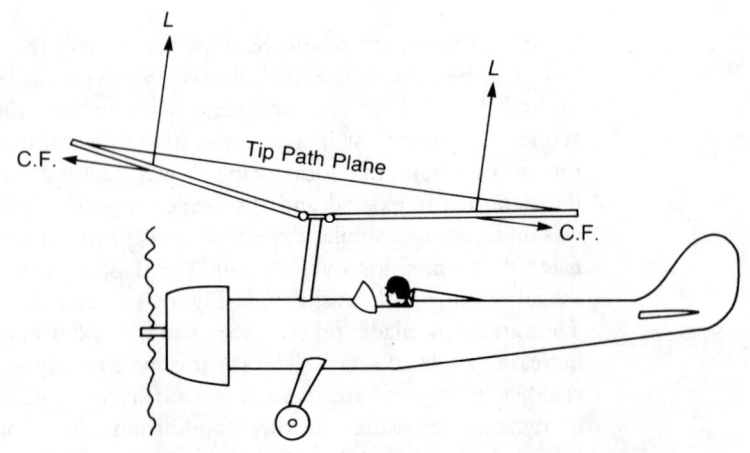

FIGURE 3.17   Autogiro in Level Flight

When this phenomenon became clear to Cierva, he knew what he had to do: add flexibility to his full-scale rotor. He decided that the simplest solution would be to use a mechanical hinge that would allow the blades to flap and automatically eliminate the rolling moment that had given him such serious problems. The only wires retained were those used to prevent the blades from flapping down too far as the rotor was stopped. In flight, the blades were kept extended by a combination of lift and centrifugal forces, as in Figure 3.17. When such a rotor is established in its stable flapping position, there are no accelerating forces on it causing it to flap with respect to an axis system fixed in the tip path plane. Thus in order that the moments at the hinges are zero, the aerodynamic moment must be a constant around the azimuth and numerically equal to the centrifugal force moment produced by the coning, which is also a constant since the tip path is assumed to lie in a plane. Any change in the distribution of the aerodynamic moments due to changes in flight conditions will force the rotor to seek a new stable flapping position where the aerodynamic and centrifugal moments will again be constant, equal, and opposite around the azimuth.

With this technological breakthrough, Cierva was able to fly his autogiro and began a long line of development that has resulted in most present-day helicopters having rotors with flapping hinges. The invention of the flapping hinges did not cure all of Cierva's rotor problems, however. He was plagued with high stresses and structural failures in his early blades resulting from moments in the plane of the rotor disc, the so-called *inplane moments*. These were due to a combination of cyclically varying drag and inertia (or Coriolis) loads. The drag variation was caused by the previously discussed nonuniform aerodynamic environment in which even though the blade lift was made uniform around the azimuth by flapping, the drag was not. The nonuniform inertial loads were the result of the blades obeying a physical law called the *law of conservation of momentum*, whose effects are familiar to those old enough to remember the swiveling piano stool and the trick of holding a pair of heavy books at arm's length while being set spinning by a friend. As the books were pulled in toward the body, the spin became faster and faster. The same principle is used by a figure skater to produce a high-speed spin. The law of conservation of momentum states that the product of the moment of inertia and the rate of spin is a constant. As the blade flaps up, its center of gravity moves in toward the spin axis, as shown in Figure 3.18. The same thing happens for the down-flapping blade. This reduces the moment of inertia so each blade tries to speed up. If they are resisted by the inertia of other blades, they will try to bend forward, thus producing inplane bending moments and corresponding stresses in the blade roots. Note that each blade of flapping rotor goes through this cycle twice in a revolution, thus producing two-per-rev—or 2 P—stresses. Had Cierva's rotor had only two blades, these stresses would have been minimized since the blades would have acted in unison with nothing to restrain them. As it was, Cierva's rotor had four blades, and it had problems. He decided that if one hinge in a blade was good, two must be better; so he incorporated vertical hinges such that the blades could move back and forth in the plane of the rotor without generating stresses in the roots. These hinges are now known as *lead-lag hinges* and, in conjunction with the flapping hinges, produce the

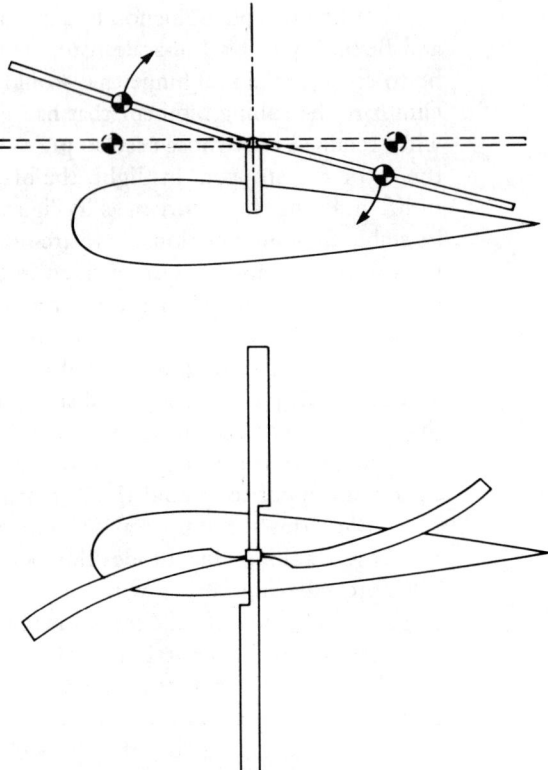

FIGURE 3.18   Inplane Blade Bending Due to Flapping Motion

fully articulated rotor systems used on many helicopters today. The lead-lag hinges are important in the study of rotor loads, vibration, and such specialized problems as ground resonance; but they have no effect on performance, stability, or control. For this reason, they will be neglected in the remainder of this book.

A fully articulated rotor is shown in Figure 3.19. Some designers have elected to put enough structure into the blades and hub that the inplane stresses can be kept to a low and safe level without lead-lag hinges as in the rotor of Figure 3.20 and some designers have even eliminated the flapping hinge by substituting a flexible section of the hub. This latter system is the so-called *rigid rotor* or *hingeless rotor* of Figure 3.21, which will later be shown to be equivalent to a hinged rotor with offset flapping hinges.

The 90° lag between the maximum aerodynamic input and the maximum flapping is typical of systems in resonance, and it may be shown that the rotor is a resonating system by the analogy of a mass supported on a spring, as shown in Figure 3.22. The mass will have a characteristic natural frequency at which it will oscillate if it is plucked and then released. The natural frequency will be equal to $\sqrt{k/m}$ rad/sec. If, instead of plucking the mass, a variable-speed electric motor

FIGURE 3.19   Fully Articulated Rotor

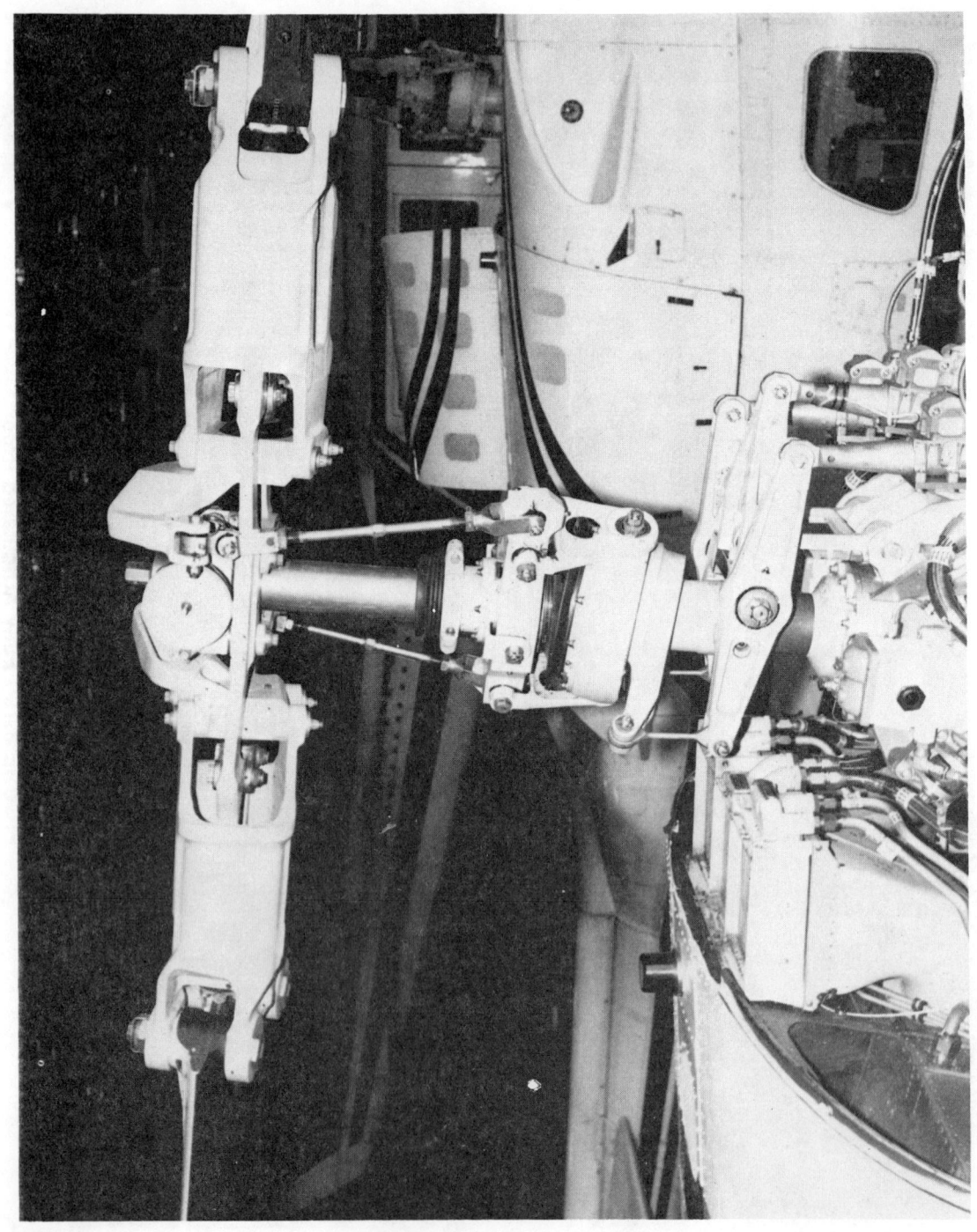

FIGURE 3.20 Teetering Rotor

FIGURE 3.21 Hingeless Rotor

149

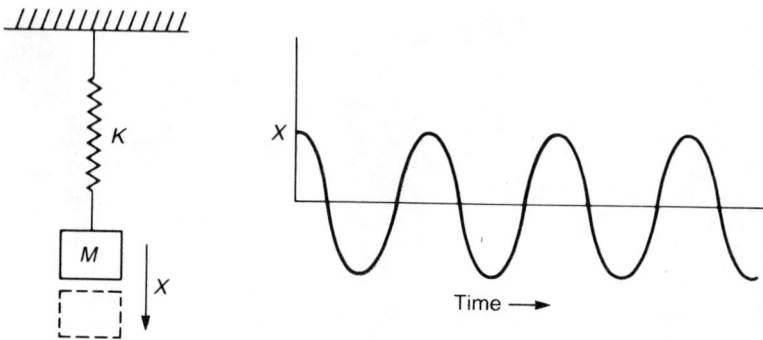

FIGURE 3.22 Characteristic Natural Frequency of Mass-Spring System

with an unbalanced flywheel is used to excite it at various motor speeds, it will be found that the speed at which the motion of the mass is the greatest will be exactly the same as the natural frequency found by plucking the system. At this speed, it may be said that the mass is in resonance with the exciting frequency.

For a rotating system such as a rotor blade, the natural frequency is given by an equation similar to that for the spring and mass:

$$\omega_n = \sqrt{\frac{K}{I}}$$

where K is the rate of the restoring moment about the flapping hinge in foot pounds per radian of flapping angle, and I is the moment of inertia of the blade about the flapping hinge. The restoring moment is due to the centrifugal force acting though a moment arm which is a function of the flapping angle, β, as shown in Figure 3.23. The centrifugal force acting on a blade element a distance r from the center of rotation is:

$$\Delta C.F. = \Omega^2 rm\,\Delta r$$

where m is the mass per running foot. The restoring moment due to this centrifugal force is thus:

$$\Delta M = \Delta C.F.\, r\beta$$

or

$$\Delta M = \Omega^2 r^2 m \beta \Delta r$$

and the total restoring moment obtained by integration is:

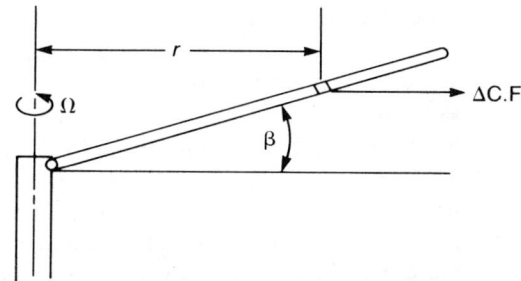

FIGURE 3.23   Centrifugal Force Acting on a Blade Element

$$M = \int_0^R \Omega^2 r^2 m \beta \, dr = \Omega^2 \beta \int_0^R m r^2 dr \ \ \text{ft-lb}$$

The spring rate to use in the equation is:

$$K = \frac{M}{\beta} = \Omega^2 \int_0^R m r^2 dr \ \ \text{ft-lb/rad}$$

The moment of inertia is defined as:

$$I = \int_0^R m r^2 dr \ \ \text{slug ft}^2$$

so that:

$$K = \Omega^2 I \ \ \text{ft-lb/rad}$$

Substituting this into the equation for the natural frequency gives:

$$\omega_n = \sqrt{\frac{K}{I}} = \sqrt{\frac{\Omega^2 I}{I}} = \Omega \ \ \text{rad/sec}$$

This says that the natural frequency of the flapping motion of a blade is equal to its rotational speed no matter what that speed is, and that the blade will always be in resonance with the exciting forces caused by the once-per-revolution unbalance of the aerodynamic forces. (Strictly speaking, this is true only for a blade without hinge offset, but it is nearly true for all blades. See Chapter 7 for the effects of offset.) The blade in resonance will exhibit the characteristic 90° lag between input and response that is inherent in all resonating systems. The result is that the rotor flaps fore and aft rather than laterally in response to unequal dynamic pressure on the advancing and retreating blades. In addition, a small amount of lateral flapping will be generated due to another aerodynamic effect. As the rotor produces lift, it is *coned* by the combination of lift and centrifugal forces. In forward flight the blade over the nose experiences air coming toward its lower surface, whereas the blade over the tail experiences air approaching it from on top, as shown in Figure 3.24. The result is that the angle of attack on the blade at $\psi = 180°$ is increased and the angle of attack of the blade at $\psi = 0°$ is decreased. The rotor compensates for this inequality in the same manner as it did for the nonsymmetric velocity patterns on the advancing and retreating blades: it responds 90° later by tilting up on the retreating side and down on the advancing side. This causes flapping velocities over the nose and over the tail that are exactly enough to compensate for the difference in angle of attack caused by the coning.

The flapping motion may be represented by an infinite Fourier series:

$$\beta = a_0 - a_{1_s} \cos\psi - b_{1_s} \sin \psi - a_{2_s} \cos 2\psi - b_{2_s} \sin 2\psi \cdots - a_{n_s} \cos n\psi - b_{n_s} \sin n\psi$$

(where ψ is as defined in Figure 3.14.)

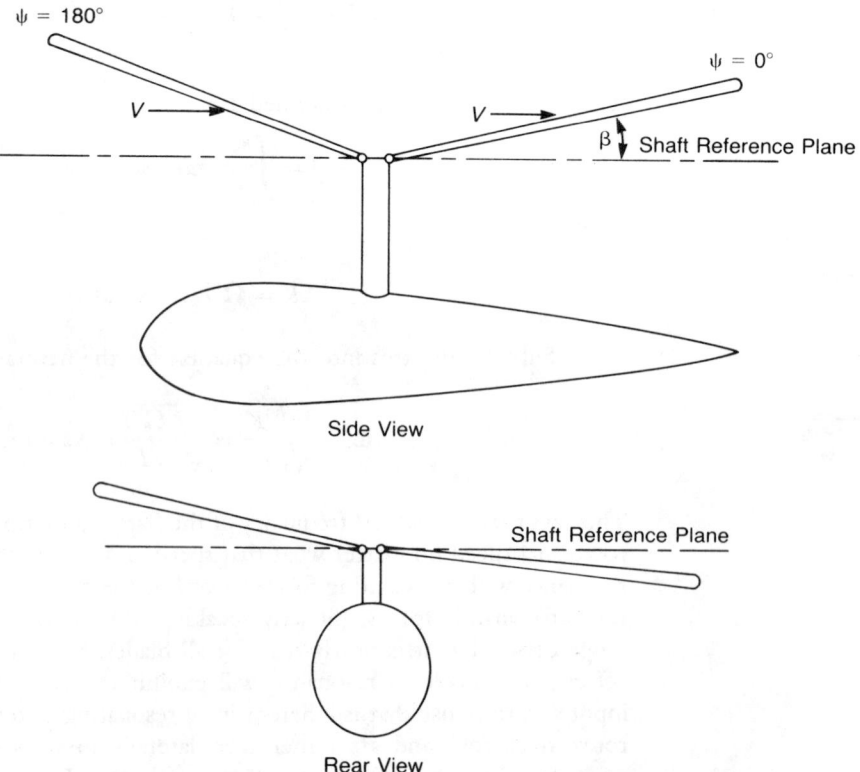

FIGURE 3.24   Velocity Orientation Causing Lateral Rotor Tilt

Only the first three terms will be used in the subsequent analysis since it may be shown using the equations of reference 3.15 that the second and higher harmonics represented by the remaining terms are relatively small and have very little effect on rotor thrust and torque. Thus it will be assumed that:

$$\beta = a_0 - a_{1_s} \cos \psi - b_{1_s} \sin \psi$$

where a_0 represents the average value, or coning; a_{1_s} is the longitudinal flapping with respect to a plane perpendicular to the shaft defined as positive when the blade flaps down at the tail and up at the nose, and b_{1_s} is the lateral flapping defined as positive when the blade flaps down on the advancing side and up on the retreating side—the normal condition. For the flapping shown in Figure 3.25, the coefficients are:

$$a_0 = 4.0$$

$$a_{1_s} = 1.0$$

$$b_{1_s} = 0.5$$

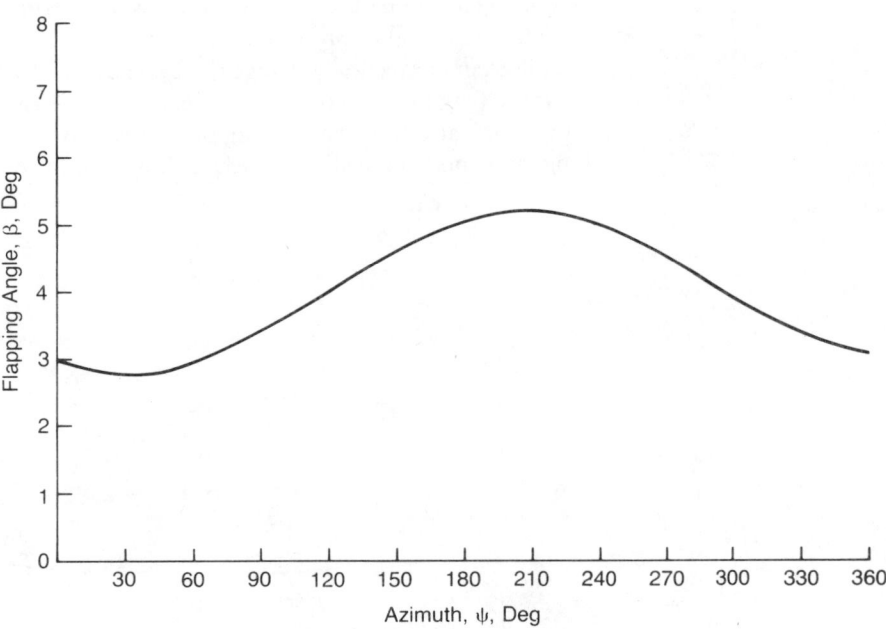

FIGURE 3.25   Typical Flapping Angle Response

Cyclic Pitch

In the early Cierva autogiros, the rotor was simply a lifting device and roll and pitch control were obtained by ailerons on stub wings and by a conventional elevator, neither of which was very effective at low speed. Following his early successes, Cierva developed a means of obtaining direct control by tilting the rotor on a gimbal with respect to the shaft. With this scheme, pitch and roll control were generated by tilting the rotor thrust vector to give it a moment arm with respect to the center of gravity, as shown in Figure 3.26. This allowed Cierva to do away with the airplane control surfaces, but as autogiros became larger the control forces required to tilt the rotor became so high that flight was difficult. At this point, a means of rotor control called *cyclic pitch* was developed. In this system— which is almost universally used at present—the pilot cyclically changes the pitch of the blades about feathering bearings by tilting a mechanism known as a *swashplate*. A schematic of this system is shown in Figure 3.27. It may be seen that if the swashplate is perpendicular to the rotor shaft, the blade angle is constant around the azimuth, but that if the swashplate is tilted, the blade pitch will go through one complete feathering cycle each revolution. If the pilot pushes the stick forward, the swashplate is tilted forward. Since the pitch horn from the blade is attached to the swashplate 90° ahead, the blade has its pitch reduced when it is on the right side and has its pitch increased when it is on the left side. When the blade is over the nose or the tail, the forward tilt of the swashplate has no effect on the blade pitch.

Cyclic pitch can be used for two purposes: to trim the tip path plane with respect to the mast, and to produce control moments for maneuvering. In the first case, the pilot can mechanically change the angle of attack of the blades by the same amount as the flapping motion would have, thus eliminating the flapping. This can be used to eliminate all of the flapping or to leave just enough to balance pitching or rolling moments on the aircraft such as those due to an offset center of gravity.

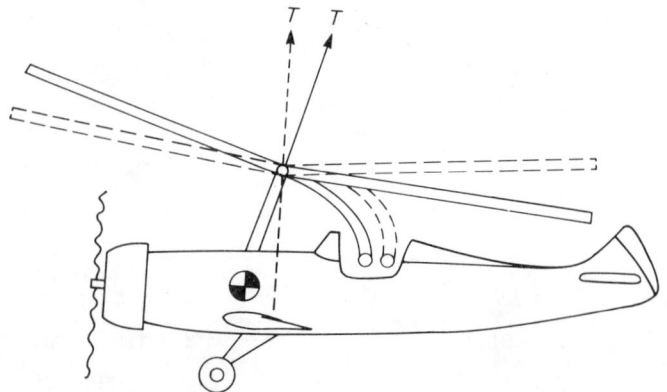

FIGURE 3.26   Pitch Control by Direct Rotor Tilt

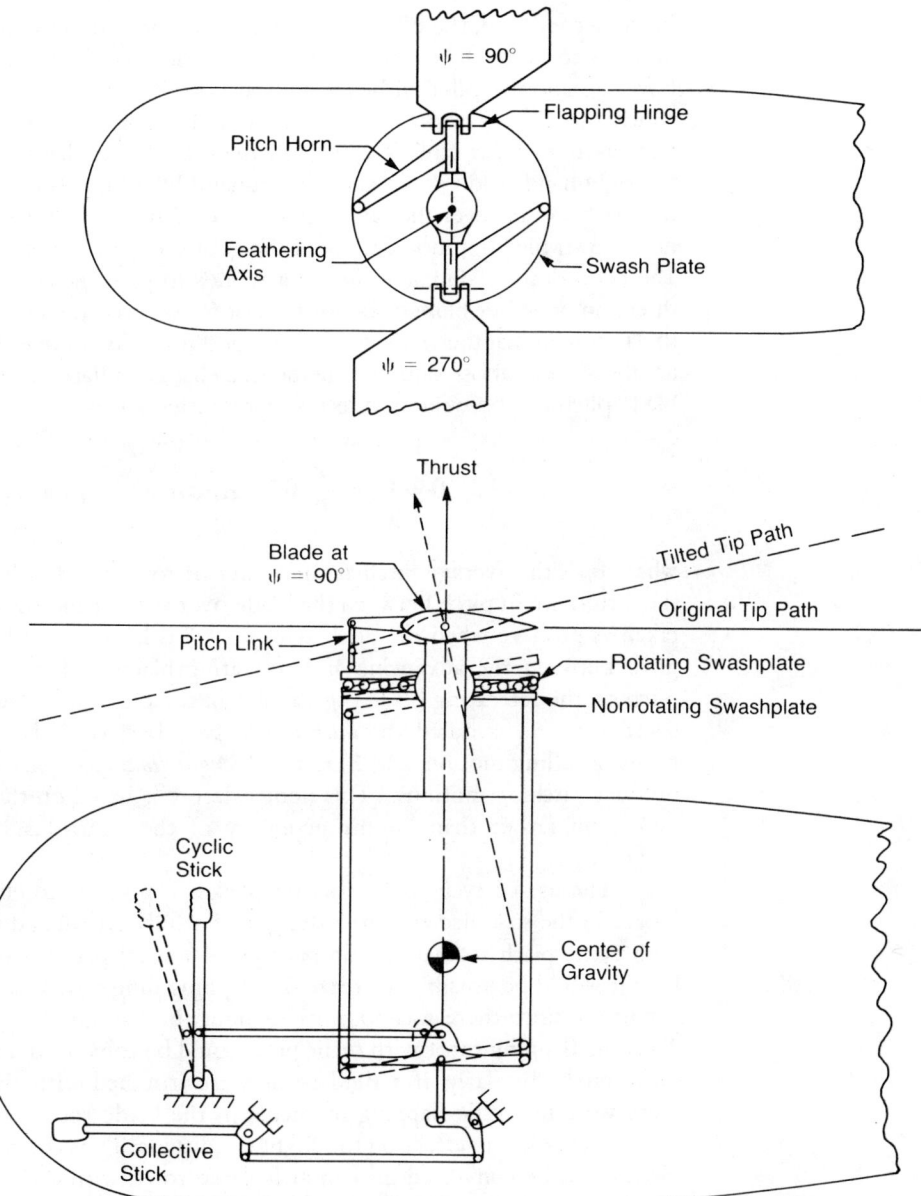

FIGURE 3.27 Schematic of Swashplate Control System

In the second case, the pilot deliberately introduces an unbalanced lift distribution in order to make the rotor tilt for maneuvering. For example, if the helicopter is hovering and the pilot wishes to tilt the nose down, he pushes the stick forward, which tilts the swashplate down in front. The pitch of the blade at $\psi = 90°$ is decreased and that at $\psi = 270°$ is increased. The resultant imbalance accelerates the right-hand blade down and the left-hand blade up. The rotor flaps down over the nose and up over the tail, tilting the rotor thrust vector forward to produce a nose down pitching moment about the center of gravity, as shown in Figure 3.27. The procedure is similar if the pilot wishes to pitch nose up or to roll in either direction. Whether being used for trim or for control, the cyclic pitch is equivalent to flapping in that the changes in rotor conditions due to one degree of cyclic pitch are the same as those due to a one-degree change in flapping. Like the flapping, the blade pitch can be written in terms of a Fourier series:

$$\theta = \theta_0 + \frac{r}{R}\,\theta_1 - A_1 \cos\psi - B_1 \sin\psi$$

where θ_0 is the average pitch at the center of rotation, θ_1 is linear twist, A_1 is half the difference in pitch between the blade over the tail and the blade over the nose, taken as positive when the pitch at $\psi = 180°$ is larger. B_1 is half the difference in pitch between the advancing and retreating blades, taken as positive when the pitch on the retreating blade is greater than the pitch on the advancing blade. The coefficient, A_1, is called the *lateral cyclic pitch* because it is used by the pilot to produce rolling motion, and B_1 is called *longitudinal cyclic pitch* because it is used to produce pitching motions. (This nomenclature is based on the results as the pilot sees them rather than on the geometry of the control system as seen by the designer.)

The use of cyclic pitch for trim makes it possible to eliminate the flapping hinges in the so-called *rigid rotor* designs. If a fully articulated rotor were trimmed with cyclic pitch so that the tip path plane was perpendicular to the rotor shaft, then it would be possible to freeze the flapping hinges with no change in the rotor condition, since there was no motion about the flapping hinges in the first place. Thus the flapping rotor with cyclic pitch could be converted into a rigid rotor with cyclic pitch. Similarly, if a rigid rotor were trimmed with cyclic pitch such that there were no cyclic flapping moments in the blade roots, then hinges could be introduced with no effect on the flight conditions. Thus the rigid rotor with cyclic pitch could be converted into an articulated rotor with cyclic pitch. From this, it may be seen that for the rotor trimmed perpendicular to the mast, or nearly so, there is no essential difference between an articulated rotor and a rigid rotor. Where differences do exist will be discussed in Chapter 7.

Besides cyclic control, which the pilot obtains by tilting the swashplate with the cyclic stick held in his right hand, he has also the collective control that changes the pitch of all blades simultaneously by raising and lowering the entire swashplate. This is done with the collective stick, as shown in Figure 3.27.

Angle of Attack of the Blade Element

In order to define increments of lift and drag acting on a blade element in forward flight, it is necessary to write the equation for the angle of attack at the element as a function of the radial station and the azimuth position. For this analysis, it will be assumed that the rotor has no hinge offset and that flapping harmonics above the first may be neglected. The local angle of attack, shown in Figure 3.28, is made up of two angles just as it is in hover—the blade pitch and the inflow angle:

$$\alpha = \theta + \tan^{-1}\left(\frac{U_P}{U_T}\right)$$

The use of U_T in this equation is based on the concept originally proved for swept wing airplanes—that only the velocity normal to the leading edge counts. The blade pitch has already been defined as a Fourier series:

$$\theta = \theta_0 + \theta_1 \frac{r}{R} - A_1 \cos\psi - B_1 \sin\psi$$

and the equation for the tangential velocity, U_T, has been derived:

$$U_T = \Omega R\left(\frac{r}{R} + \mu \sin\psi\right)$$

The perpendicular velocity, U_P, is a vector that is perpendicular to the blade quarter-chord line and lies in a plane that contains the rotor shaft. It is positive going up and consists of several components, as shown in Figure 3.29.

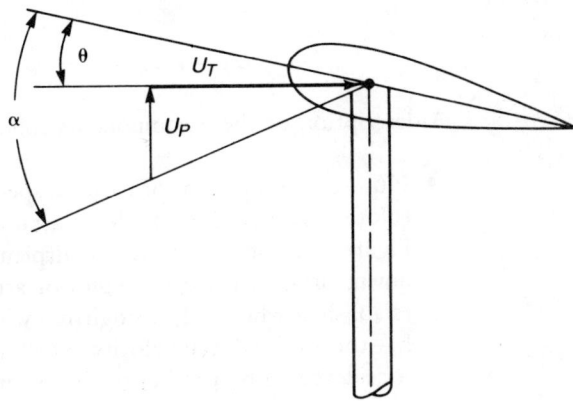

FIGURE 3.28   Angle of Attack at Blade Element

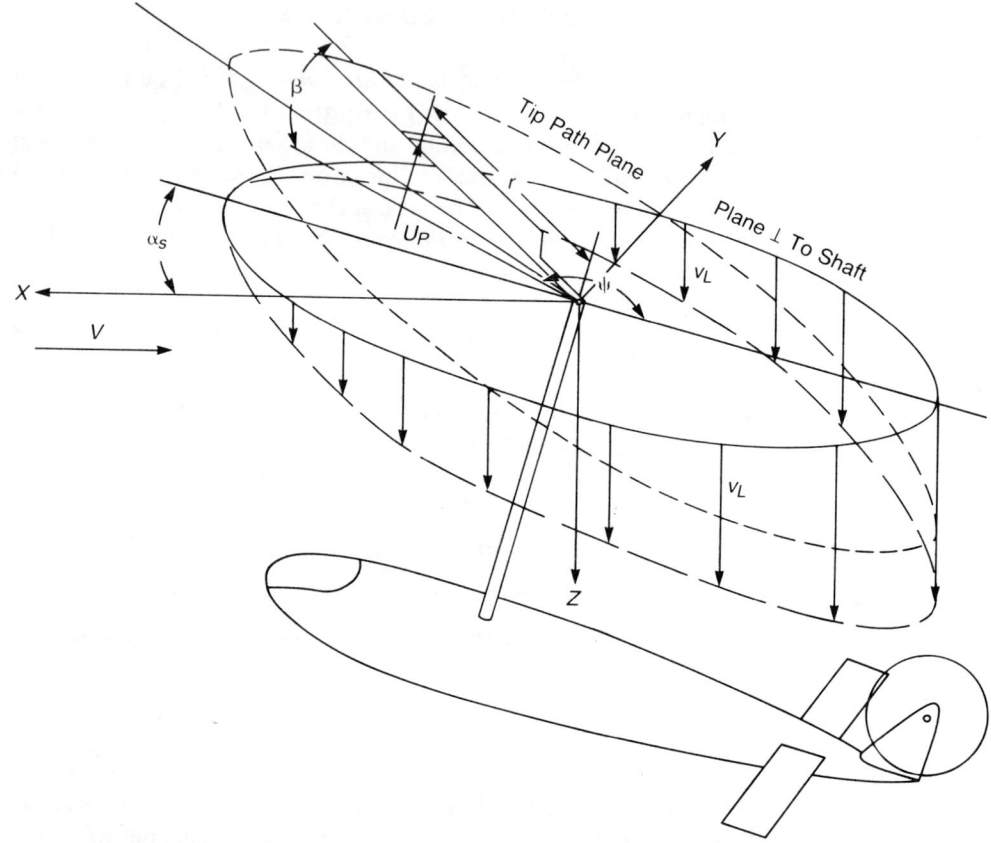

FIGURE 3.29   Components of U_p

$$U_P = V\alpha_s - v_L - r\dot\beta - V\beta \cos \psi$$

A description of these components follows:

- $V\alpha_s$—the component of forward speed that is parallel to the rotor shaft. (Note that a positive angle of attack is with the nose up, as shown in Figure 3.29, just as it is for airplanes. Helicopters generally fly nose down, or with negative angles of attack, but the sign convention was established when only autogiros—which do fly nose up—had rotors.)
- v_L—the local induced velocity, which for normal flight conditions may be considered to be parallel to the rotor shaft:

$$v_L = v_1 \left(1 + \frac{r}{R} \cos \psi \right)$$

where

$$v_1 = \frac{\Omega R}{\mu} \frac{C_T}{2}$$

- $r\dot{\beta}$—the contribution of blade flapping. Since the flapping angle is:

$$\beta = a_0 - a_{1_s} \cos \psi - b_{1_s} \sin \psi$$

The flapping velocity is:

$$\dot{\beta} = a_{1_s} \sin \psi \frac{d\psi}{dt} - b_{1_s} \cos \psi \frac{d\psi}{dt}$$

or

$$\dot{\beta} = \Omega (a_{1_s} \sin \psi - b_{1_s} \cos \psi)$$

and the contribution to U_p is:

$$r\dot{\beta} = \Omega R \frac{r}{R} (a_{1_s} \sin \psi - b_{1_s} \cos \psi)$$

- $V\beta \cos \psi$—the flapping effect that was illustrated with Figure 3.24:

$$V\beta \cos \psi = V(a_0 - a_{1_s} \cos \psi - b_{1_s} \sin \psi) \cos \psi$$

Putting all the contributions together gives:

$$U_P = V\alpha_s - v_1 \left(1 + \frac{r}{R} \cos \psi \right) - \Omega R \frac{r}{R} (a_{1_s} \sin \psi - b_{1_s} \cos \psi)$$
$$- V(a_0 - a_{1_s} \cos \psi - b_{1_s} \sin \psi) \cos \psi$$

For subsequent analysis, it will be convenient to use the angle of attack of the tip path plane as a reference angle rather than the angle of attack of the plane perpendicular to the shaft. Figure 3.30 shows the relationships between the various planes which are used in rotor analyses. Much of the early work, such as in references 3.16 and 3.17, used the control plane as a basic reference system. This is sometimes called the "plane of no feathering." Later investigators for convenience have switched to the shaft plane or to the tip path plane—which is the "plane of no flapping"—as their basic reference planes. From Figure 3.30, it may be seen that:

$$\alpha_s = \alpha_{\text{TPP}} - a_{1_s}$$

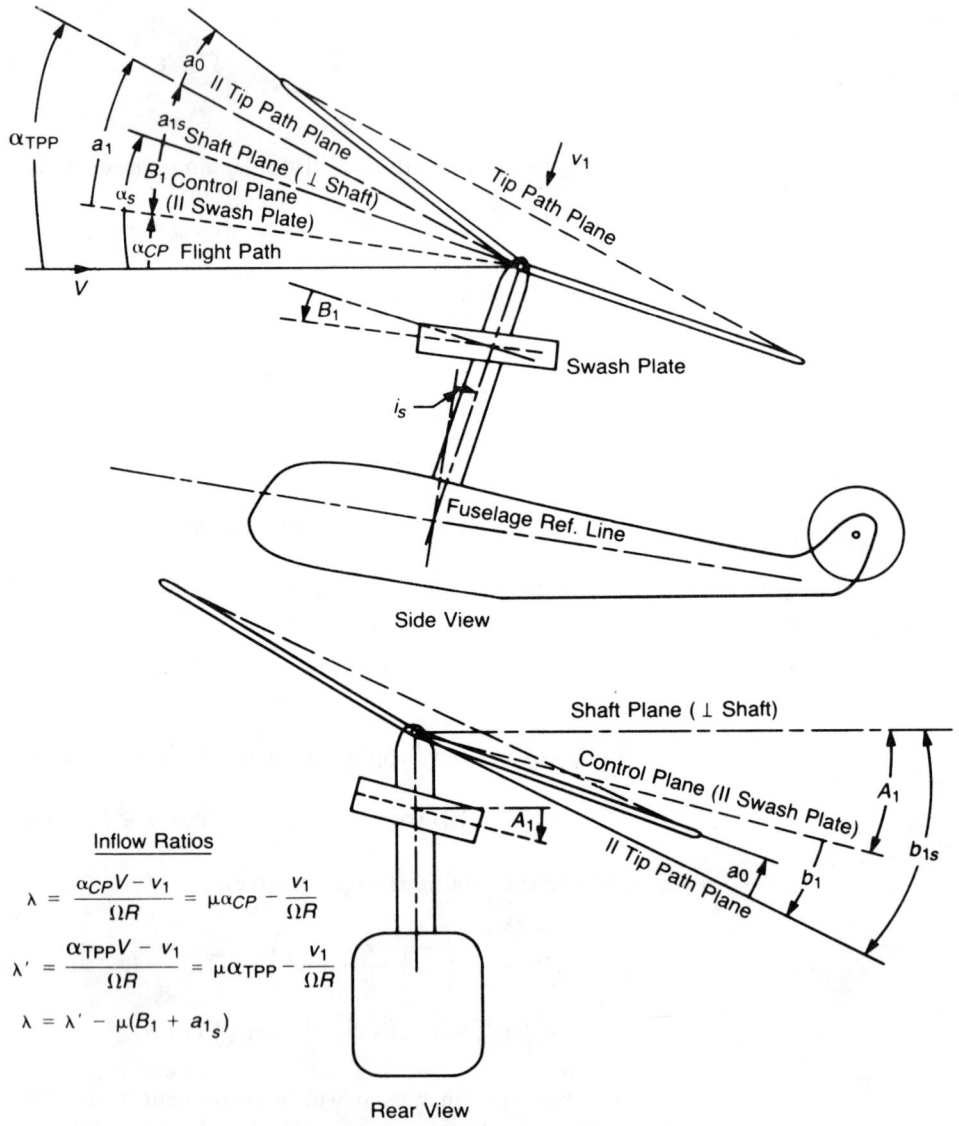

FIGURE 3.30   Rotor Angle Relationships

and thus:

$$U_P = \Omega R \left[\mu(\alpha_{TPP} - a_{1_s}) - \frac{v_1}{\Omega R} \left(1 + \frac{r}{R} \cos \psi \right) - \frac{r}{R} (a_{1_s} \sin \psi - b_{1_s} \cos \psi) \right.$$

$$\left. - \mu(a_0 - a_{1_s} \cos \psi - b_{1_s} \sin \psi) \cos \psi \right]$$

The local angle of attack can now be written:

$$
\alpha = \frac{1}{\dfrac{r}{R} + \mu \sin \psi} \left\{ \frac{r}{R} \left[\theta_0 + \theta_1 \frac{r}{R} - (A_1 - b_{1_s}) \cos \psi - (B_1 + a_{1_s}) \sin \psi \right] \right.
$$

$$
- \frac{v_1}{\Omega R} \left(1 + \frac{r}{R} \cos \psi \right) + \mu \left[\alpha_{\text{TPP}} + \left(\theta_0 + \theta_1 \frac{r}{R} \right) \sin \psi \right.
$$

$$
\left. \left. - a_0 \cos \psi - (A_1 - b_{1_s}) \sin \psi \cos \psi - (B_1 + a_{1_s}) \sin^2 \psi \right] \right\}
$$

If all angles—including $\tan^{-1}(v_1/\Omega R)$—are expressed in degrees, then α is also in degrees. Note that when α is expressed in this form, the combinations of cyclic pitch and flapping, $(A_1 - b_{1_s})$ and $(B_1 + a_{1_s})$, occur as primary variables. This is a consequence of the equivalence of flapping and feathering, which says that as far as rotor aerodynamics are concerned, one degree of cyclic pitch produces the same effect as one degree of flapping. (This equivalence strictly applies only to rotors with zero flapping hinge offset, but it is a good assumption for the analysis of performance of rotors that have moderate flapping hinge offsets or even for hingeless rotors which flap through structural bending. For analysis of stability and control, the hinge offset is significant and is discussed in Chapter 7.) In some of the early rotor analyses, such as in reference 3.17, these combinations of cyclic pitch and flapping are referred to as flapping with respect to the "plane of no feathering" and are designated a_1 and b_1:

$$
a_1 = B_1 + a_{1_s}
$$

$$
b_1 = -(A_1 - b_{1_s})
$$

The equation for the local angle of attack contains several types of quantities that are either known or which may be computed:

- θ_0, collective pitch that is required to produce enough rotor thrust to balance the weight and to compensate for the inflow.
- θ_1, blade twist.
- a_{1_s} and b_{1_s}, tilt of the tip path plane with respect to the shaft that is required to produce enough moment about the center of gravity to balance or trim the helicopter.
- A_1 and B_1, cyclic pitch required to compensate for the unsymmetrical velocity pattern and to produce the amount of a_{1_s} and b_{1_s} required to trim.

- μ, tip speed ratio $= V/\Omega R$.
- α_{TPP}, angle of attack of the tip path plane required to put rotor thrust, helicopter weight, and helicopter drag into equilibrium.
- a_0, coning of blades, which is established by equilibrium between lift and centrifugal forces.
- $v_1/\Omega R$, induced velocity ratio, where v_1 is from momentum equation.

The flight conditions for the example helicopter have been determined at a tip speed ratio of 0.3 (115 knots) by methods to be outlined later, and the resultant angle of attack distribution has been plotted in Figure 3.31. Note that the angle of attack of the advancing tip is in the neighborhood of 0°, while the retreating tip is nearly 10°. The angle of attack on each side of the boundary of the reverse flow region goes to ±90°, but since the local velocity is zero along this boundary, these high angles have little practical significance.

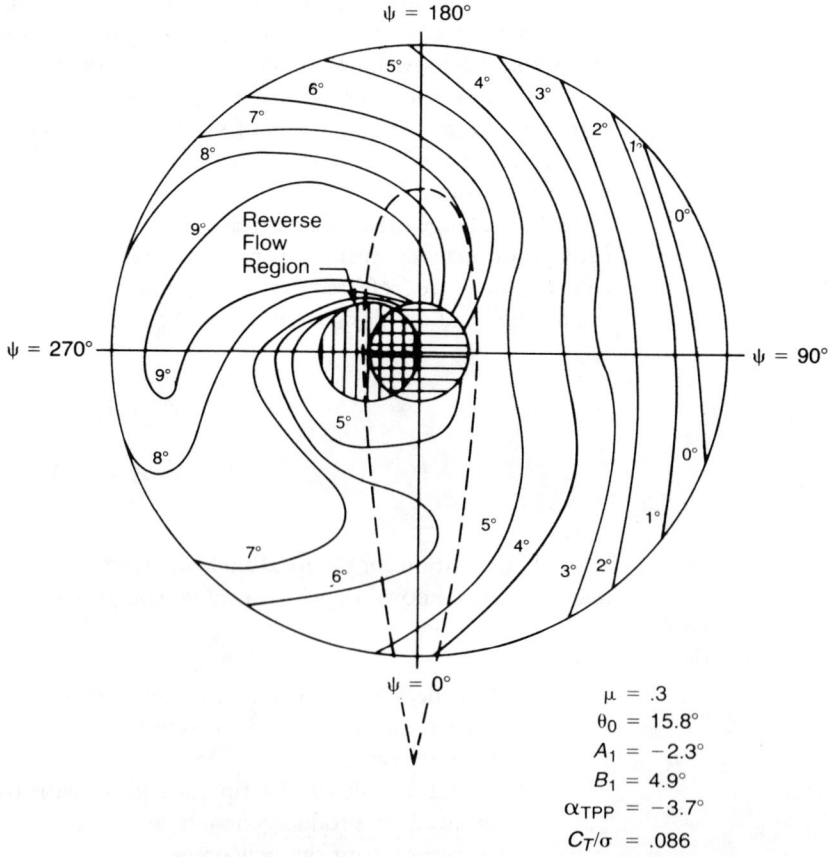

FIGURE 3.31 Angle of Attack Distribution for Example Helicopter at 115 Knots

CLOSED-FORM EQUATIONS

Again, just as in hover, there are two schemes for integrating the forces on the blade elements to give rotor performance. The first method involves performing the integration of the equations mathematically to produce closed-form equations as a function of rotor geometry and flight conditions. The second method, suitable for a computer, performs the integrations by numerical methods. The closed-form integration method produces equations that are useful for rough calculations and for giving an understanding of the important factors; but since it must use simplifying assumptions in order to make the integration feasible, accuracy is sacrificed for convenience. The numerical integration method, which will be outlined later, uses a minimum of assumptions so that the resulting accuracy is high, but it is inconvenient without investing considerable time and effort in programming and checking out the computer.

Closed-Form Integration for Thrust

Just as in hover, the rotor thrust in forward flight can be computed from the integration of the lift on each blade element along the blade and around the azimuth. The same equation for the incremental lift applies:

$$\Delta L = q c_i c \Delta r$$

or in terms of blade element conditions:

$$\Delta L = \frac{\rho}{2} U_T^2 a \alpha c \Delta r$$

where both the velocity, U_T, and the angle of attack, α, have been derived as functions of r/R and of ψ. The integration procedure is simple in principle but tedious in application because of the large number of terms involved. For this reason, the method will be outlined in general, but detailed steps will be left to the ambitious student. The integration starts with the equation of lift per running foot:

$$\frac{\Delta L}{\Delta r} = \frac{\rho}{2} a c U_T^2 \alpha$$

For a given value of ψ, the lift per running foot is as shown in Figure 3.32. The lift on the blade is designated as L_{b_ψ} and is the area under the curve.

$$L_{b_\psi} = \int_0^R \frac{\Delta L}{\Delta r} dr$$

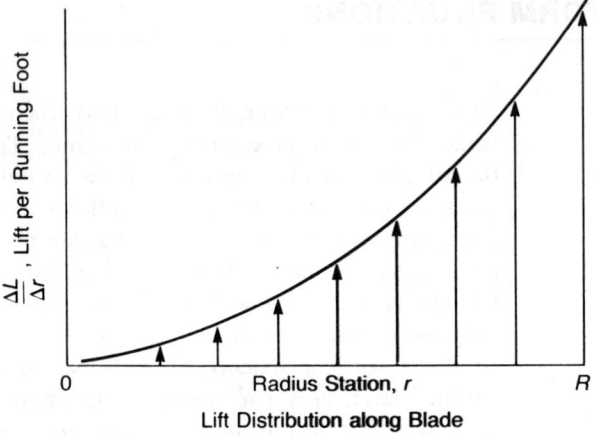

Lift Distribution **along Blade**

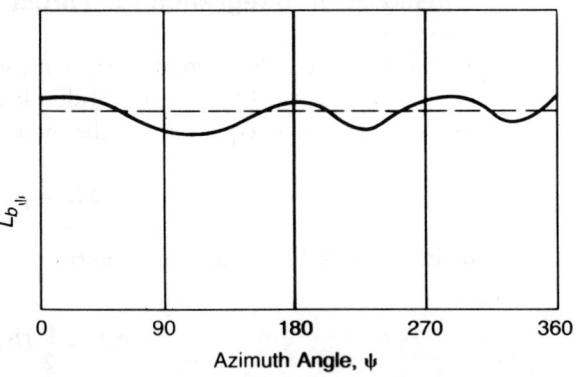

Lift Distribution **Around Azimuth**

FIGURE 3.32   Illustration of Integration for Thrust

For each value of ψ, the integral can be evaluated and the result plotted as a function of ψ, as in Figure 3.32. The average lift per blade, $\bar{L}_b$, around the azimuth is the area under the curve divided by the length of the horizontal axis which is $360°$ or 2π radians:

$$\bar{L}_b = \frac{1}{2\pi} \int_0^{2\pi} L_{b_\psi} d\psi$$

Substituting for L_{b_ψ}, the average lift per blade becomes a double integral on r and ψ. The total thrust is equal to the number of blades times the average lift per blade:

$$T = \frac{b}{2\pi} \int_0^{2\pi} \int_0^R \frac{\Delta L}{\Delta r} \, dr d\psi = \frac{b\rho ac}{4\pi} \int_0^{2\pi} \int_0^R U_T^2 \alpha \, dr d\psi$$

In nondimensional form, the equation is:

$$C_T/\sigma = \frac{a}{4\pi} \int_0^{2\pi} \int_0^1 (\bar{U}_T^2 \theta + \bar{U}_P \bar{U}_T) d \frac{r}{R} \, d\psi$$

where:

$$\theta = \theta_0 + \frac{r}{R} \theta_1 - A_1 \cos \psi - B_1 \sin \psi$$

$$\bar{U}_T = U_T/\Omega R = r/R + \mu \sin \psi$$

$$\bar{U}_P = U_P/\Omega R = \mu\alpha_{TPP} - v_1/\Omega R - \frac{r}{R} a_{1_s} \sin \psi - \mu a_{1_s} \sin^2 \psi$$

$$+ \left[(b_{1_s} - v_1/\Omega R) \frac{r}{R} - \mu a_0 \right] \cos \psi + \mu b_{1_s} \sin \psi \cos \psi$$

(Note: Just as in the discussion of hovering, the tip and root losses will be neglected at this stage of the analysis and discussed later along with the effect of the reversed flow region.)

The integration is straightforward and may be done using tables of integrals. The result is:

$$C_T/\sigma = \frac{a}{4} \left\{ \left[\frac{2}{3} + \mu^2 \right] \theta_0 + \left[\frac{1}{2} + \frac{\mu^2}{2} \right] \theta_1 + \mu \left[\alpha_{TPP} - (B_1 + a_{1_s}) \right] - v_1/\Omega R \right\}$$

The term, $\alpha_{TPP} - (B_1 + a_{1_s})$, is the angle of attack of the swashplate or *control plane* with respect to the flight path, as shown in Figure 3.30. In classical treatments of rotor aerodynamics, such as in references 3.16 and 3.17, the flow perpendicular to this control plane is called the inflow:

$$\text{Inflow} = V[\alpha_{TPP} - (B_1 + a_{1_s})] - v_1$$

and normalizing by dividing by tip speed gives the inflow ratio, λ:

$$\lambda = \mu[\alpha_{TPP} - (B_1 + a_{1_s})] - v_1/\Omega R$$

or in terms of shaft angle:

$$\lambda = \mu[\alpha_s - B_1] - v_1/\Omega R$$

As a function of θ, μ, and λ, the thrust coefficient is:

$$C_T/\sigma = \frac{a}{4}\left\{\left(\frac{2}{3} + \mu^2\right)\theta_0 + \left(\frac{1}{2} + \frac{\mu^2}{2}\right)\theta_1 + \lambda\right\}$$

This form is convenient when working with flapping rotors without cyclic pitch such as tail rotors or the rotors of simple autogiros. It is, however, somewhat awkward to use for the analysis of helicopters with cyclic pitch, since the value of the inflow ratio cannot be evaluated directly from the flight conditions. For these rotors, a more convenient equation results from using the angle of attack of the tip path plane as a basic parameter, rather than the angle of attack of the control plane. This is obtained by using the equation for the sum of longitudinal cyclic pitch and longitudinal rotor flapping, $(B_1 + a_{1_s})$, which is derived by writing the equation for the aerodynamic rolling moment on the disc and then equating it to zero since the rotor must be trimmed aerodynamically with respect to rolling moments. The increment of rolling moment is:

$$\Delta R = -\Delta L r \sin \psi$$

and the total rolling moment is:

$$R = -\frac{1}{2\pi}\int_0^{2\pi}\int_0^R \frac{\rho}{2}acU_T^2\alpha r \sin \psi\, dr\, d\psi$$

When this equation is expanded in terms of U_T and α, equated to zero, and then solved for $(B_1 + a_{1_s})$, the result is:

$$B_1 + a_{1_s} = \frac{\mu}{1 + \frac{3}{2}\mu^2}\left[\frac{8}{3}\theta_0 + 2\theta_1 + 2(\mu\alpha_{TPP} - v_1/\Omega R)\right]$$

In this equation, the inflow term is with respect to the tip path plane and will be designated λ':

$$\lambda' = \mu\alpha_{TPP} - v_1/\Omega R$$

[Note that $\lambda' = \lambda + \mu(B_1 + a_{1_s})$.]

Thus:

$$B_1 + a_{1_s} = \frac{\mu}{1 + \frac{3}{2}\mu^2}\left[\frac{8}{3}\theta_0 + 2\theta_1 + 2\lambda'\right]$$

Using this in the equation for λ and performing the algebra gives a new equation for C_T/σ as a function of θ, μ, and λ'.

$$C_T/\sigma = \frac{\dfrac{a}{4}}{1 + \dfrac{3}{2}\mu^2}\left[\left(\frac{2}{3} - \frac{2}{3}\mu^2 + \frac{3}{2}\mu^4\right)\theta_0 + \frac{1}{2}\left(1 - \frac{3}{2}\mu^2 + \frac{3}{2}\mu^4\right)\theta_1 + \left(1 - \frac{\mu^2}{2}\right)\lambda'\right]$$

The new inflow ratio, λ', can be evaluated easily from known flight conditions. The angle of attack of the tip path plane is:

$$\alpha_{\text{TPP}} = -\tan^{-1}\frac{(D_F + H_M + H_T)}{(\text{G.W.} - L_F)}$$

Methods for evaluating the main and tail rotor H-forces and the fuselage lift in forward flight will be discussed later, but since they are generally small, it may be assumed that for a first approximation:

$$\alpha_{\text{TPP}} = -\tan^{-1}\frac{D_F}{\text{G.W.}}$$

or

$$\alpha_{\text{TPP}} = -\frac{f}{A_b}\frac{\mu^2/2}{C_T/\sigma}\ \text{rad}$$

From the momentum equation previously derived:

$$v_1/\Omega R = C_T/\sigma\ \frac{\sigma}{2\mu}$$

Thus λ' can be evaluated for a first approximation from the flight conditions and the helicopter configuration:

$$\lambda' = -\left[\frac{f}{A_b}\frac{\mu^3/2}{C_T/\sigma} + C_T/\sigma\ \frac{\sigma}{2\mu}\right]$$

Average Blade Element Lift Coefficient

In Chapter 1 it was shown that if each blade element operated at the same lift coefficient, $\bar{c}_l$, a relationship with C_T/σ could be established that helped give a feeling for how close the rotor was to stall:

$$C_T/\sigma = \frac{\bar{c}_l}{6}$$

A similar calculation can be made in forward flight, although it must be admitted that the usefulness is somewhat limited considering that in forward flight the stall is localized on the retreating side. Let

$$T = \frac{b}{2\pi} \int_0^{2\pi} \int_0^R \frac{\rho}{2} U_T^2 \bar{c}_l \, c \, dr \, d\psi$$

or

$$C_T/\sigma = \frac{\bar{c}_l}{4\pi} \int_0^{2\pi} \int_0^1 \bar{U}_T^2 \, d\frac{r}{R} \, d\psi$$

then

$$C_T/\sigma = \frac{\bar{c}_l}{6} \left(1 + \frac{3}{2} \mu^2 \right)$$

Trim Control Angles

Generally in a helicopter analysis, the value of C_T/σ is known from its definition:

$$C_T/\sigma = \frac{T}{\rho A_b (\Omega R)^2} \doteq \frac{G.W.}{\rho A_b (\Omega R)^2}$$

and thus the primary use of the equation for C_T/σ is to determine the collective pitch, θ_0, which will be used in subsequent analysis:

$$\theta_0 = \frac{\dfrac{4}{a} \left(1 + \dfrac{3}{2} \mu^2 \right) C_T/\sigma - \dfrac{1}{2} \left(1 - \dfrac{3}{2} \mu^2 + \dfrac{3}{2} \mu^4 \right) \theta_1 - \left(1 - \dfrac{\mu^2}{2} \right) \lambda'}{\dfrac{2}{3} - \dfrac{2}{3} \mu^2 + \dfrac{3}{2} \mu^4}$$

The longitudinal cyclic pitch B_1 has already been derived:

$$B_1 + a_{1_s} = \frac{\mu}{1 + \dfrac{3}{2} \mu^2} \left[\frac{8}{3} \theta_0 + 2\theta_1 + 2\lambda' \right]$$

The magnitude of the flapping, a_{1_s}, is determined by the pitching moment that the rotor must produce to balance such things as an offset center-of-gravity position, a lift force on the horizontal stabilizer, and/or an aerodynamic pitching moment on the fuselage. If the net of these effects is zero, then a_{1_s} will also be zero. If the net effect is not zero, a_{1_s} can be evaluated by the methods derived in Chapter 8, "The Helicopter in Trim."

For some types of problems, such as determining the stability and control characteristics of the rotor or for calculating the flapping of a tail rotor or of a rotor in a wind tunnel with fixed cyclic pitch, it is convenient to rewrite the equation in terms of the angle of attack of the plane perpendicular to the shaft, α_s, rather than the angle of attack of the tip path plane. In this case:

$$a_{1_s} = \frac{\mu}{1 - \dfrac{\mu^2}{2}} \left[\frac{8}{3}\theta_0 + 2\theta_1 + 2\left(\mu\alpha_s - \frac{v_1}{\Omega R}\right)\right] - \left(\frac{1 + \dfrac{3}{2}\mu^2}{1 - \dfrac{\mu^2}{2}}\right)B_1$$

The equation for the lateral cyclic pitch, A_1, may be derived by setting the pitching moment on the rotor to zero. The result is:

$$A_1 - b_{1_s} = -\frac{\left(\dfrac{4}{3}\mu a_0 + \dfrac{v_1}{\Omega R}\right)}{1 + \mu^2/2}$$

where the lateral flapping, b_{1_s}, is that required to trim the helicopter for external rolling moments such as those produced by tail rotor thrust or a lateral center of gravity offset. For most steady flight conditions, b_{1_s} can be considered to be negligible.

Coning in Forward Flight

An expression for the steady portion of blade flapping, or coning, a_0, may be derived in forward flight using the same procedure as was used in hover—that is, setting all of the blade bending moments to zero at the hinge. Figure 3.33 shows a rotor blade with aerodynamic, weight, inertial, and centrifugal forces acting on a blade element. At the flapping hinge—which for this analysis will be assumed to be at the center of rotation—the net moment must be zero since a hinge cannot support a moment. The equations for the four moments are:

$$\text{Aerodynamic: } M_A = \int_0^R r\,\frac{\Delta L}{\Delta r}\,dr = \frac{\rho}{2}\,ac \int_0^R r\alpha U_T^2 dr$$

$$\text{Weight: } M_w = -\int_0^R mg\,r\,dr = -mg\,\frac{R^2}{2}$$

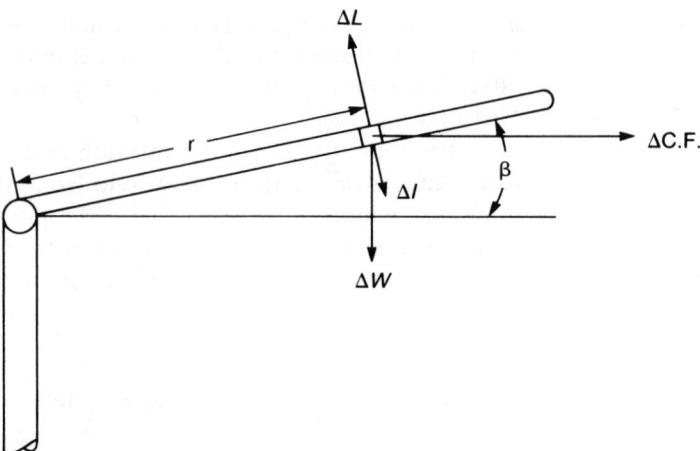

FIGURE 3.33   Forces Acting on a Blade Element

Inertia: $M_I = - \int_0^R (rm\ddot{\beta}) r \, dr = - \dfrac{mR^3}{3} \ddot{\beta}$

Centrifugal: $M_{C.F.} = - \int_0^R m\Omega^2 r(r\beta) \, dr = - \dfrac{m\Omega^2 R^3 \beta}{3} = -I_b\Omega^2\beta$

Since only the steady portion of the flapping is being derived, only those terms that do not contain $\sin \psi$, $\cos \psi$, $\sin 2\psi$, or $\cos 2\psi$ need be retained. When M_A is expanded and the trignometric identity

$$\sin^2\psi = \frac{1}{2} - \frac{\cos 2\psi}{2}$$

is used, the constant terms remaining are:

$$M_{A_{const}} = \frac{\rho}{2} ac(\Omega R)^2 R^2 \times \left\{ \theta_0 \left[\frac{1}{4} + \frac{\mu^2}{4} \right] + \theta_1 \left[\frac{1}{5} + \frac{\mu^2}{6} \right] + \frac{\mu}{3} \left[\alpha_{TPP} - (B_1 + a_{1_s}) \right] - \frac{v_1/\Omega R}{3} \right\}$$

The constant portion of the weight equation is simply:

$$M_{w_{const}} = - \frac{mgR^2}{2}$$

The inertia moment is:

$$M_I = -\frac{mR^3}{3}\Omega^2(a_{1_s}\cos\psi + b_{1_s}\sin\psi)$$

and as such contains no constant terms.

The centrifugal force moment has one constant term:

$$M_{C.F._{const}} = -I_b\Omega^2 a_0$$

Summing the moments to zero at the flapping hinge gives:

$$M_A + M_w + M_I + M_{C.F.} = 0$$

which results in:

$$a_0 = \frac{\rho acR^4}{2I_b} \times \left\{\theta_0\left[\frac{1}{4}+\frac{\mu^2}{4}\right] + \theta_1\left[\frac{1}{5}+\frac{\mu^2}{6}\right] + \frac{\mu}{3}[\alpha_{TPP} - (B_1 + a_{1_s})] - \frac{v_1/\Omega R}{3}\right\} - \frac{mgR^2}{2I_b\Omega^2}$$

using the substitutions which were used in the hovering derivation:

$$a_0 = \frac{\gamma}{6}\left\{\theta_0\left[\frac{3}{4}+\frac{3}{4}\mu^2\right] + \theta_1\left[\frac{3}{5}+\frac{\mu^2}{2}\right] + \mu[\alpha_{TPP} - (B_1 + a_{1_s})] - v_1/\Omega R\right\} - \frac{\frac{3}{2}gR}{(\Omega R)^2}$$

If the bracketed term is compared with the equation originally derived for C_T/σ in forward flight, it will be seen that for all practical purposes:

$$a_0 \doteq \frac{2}{3}\gamma\,\frac{C_T/\sigma}{a} - \frac{\frac{3}{2}gR}{(\Omega R)^2}\ \text{radians}$$

Note that this is identical to the equation for coning derived for the hovering rotor with ideal twist. Although this equation has been derived for a rotor with flapping hinges, it is a good assumption even for two-bladed teetering rotors and for rigid rotors, since the aerodynamic and centrifugal forces overpower whatever structural stiffness the blades may have.

Correlation of Flapping with Test Results

Reference 3.5 presents flapping obtained during a wind tunnel test of a model rotor at low tip speed ratios. The measured values of coning, longitudinal flapping, and lateral flapping are shown in Figure 3.34 along with values calculated from the

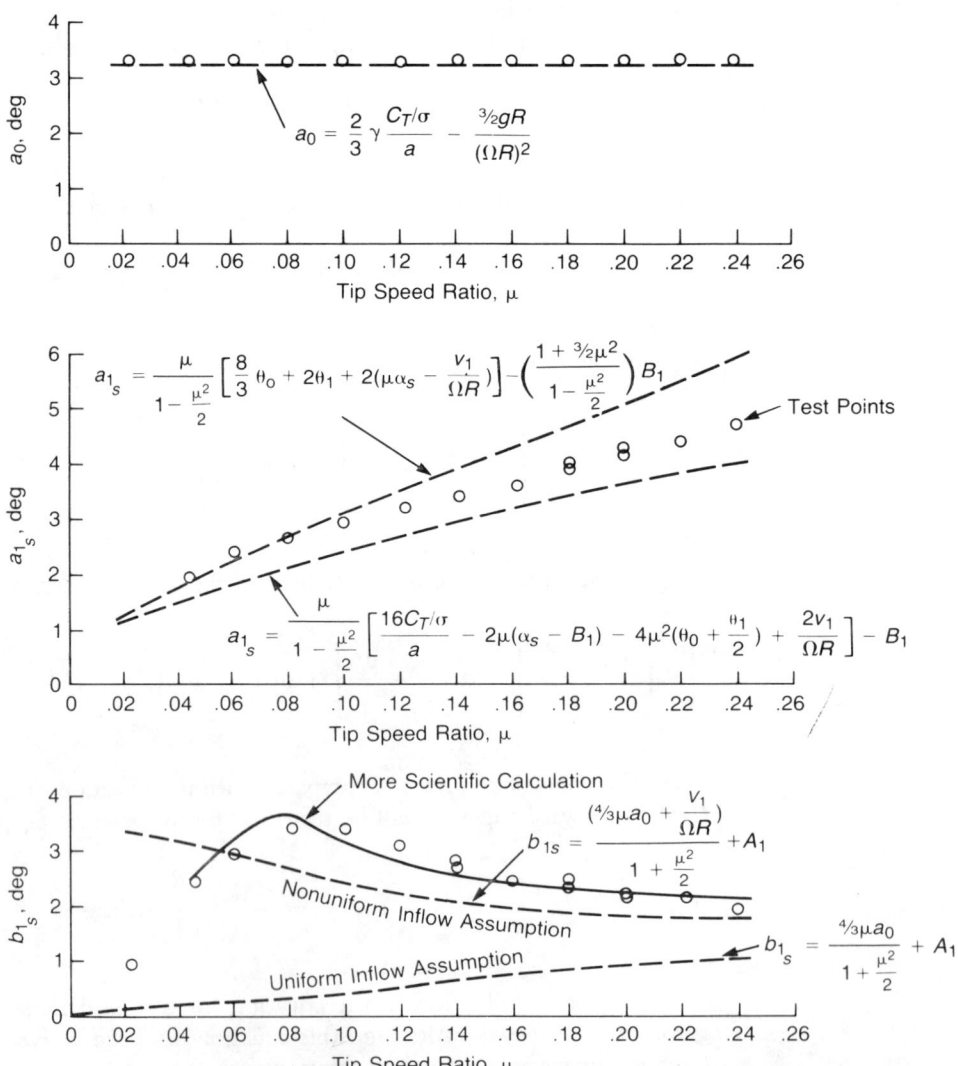

FIGURE 3.34   Measured and Calculated Flapping for Wind Tunnel Model

Sources: Harris, "Articulated Rotor Blade Flapping Motion at Low Advance Ratio," *JAHS* 17-1, 1972; Johnson, "Comparison of Calculated and Measured Helicopter Rotor Lateral Flapping Angles," AVRADCOM TR 80-A-11 NASA TM 81213, 1980.

equations derived here. Two calculated lines are shown for longitudinal flapping. The first is based on the equation derived earlier and the second on an alternative form generated by combining the equations for a_{1_s} and C_T/σ:

$$a_{1_s} + B_1 = \frac{\mu}{1 - \dfrac{\mu^2}{2}} \cdot \left[\frac{16 C_T/\sigma}{a} - 2\mu(\alpha_s - B_1) - 4\mu^2(\theta_0 + \tfrac{1}{2}\theta_1) + \frac{2v_1}{\Omega R} \right]$$

This second equation gives lower calculated flapping because the wind tunnel model produced a somewhat lower value of C_T/σ than would be predicted from the test conditions. (This correlation of longitudinal flapping is slightly different from that given in reference 3.5 because of a term equivalent to $4\mu^2(\theta_0 + \tfrac{1}{2}\theta_1)$ was omitted from that study.)

The presence of the term $v_1/\Omega R$ in the lateral flapping equation is the result of assuming that the induced velocity distribution is nonuniform—specifically, that it has the form:

$$v_L = v_1 \left(1 + K \frac{r}{R} \cos \psi \right)$$

where K is assumed to equal to unity. The assumption is obviously not valid near hover, as shown in Figure 3.34, but above a tip speed ratio of about 0.05 it gives better correlation than the calculations based on uniform inflow. More detailed analysis of the role of the induced velocity distribution will be found in references 3.5 and 3.6. The large lateral flapping at low forward speeds requires left cyclic stick to trim the helicopter (for rotors turning counterclockwise). (Pilots refer to this as the "transverse flow" effect.) The magnitude of this cross-coupling effect is shown by the lateral flapping equation to be a function of v_1, which in turn is a function of the disc loading. In sideward flight, forward stick must be held in flight to the right and aft stick in flight to the left. This effect is discussed in reference 3.18.

A much more scientific approach can be used—one involving a free wake analysis—as is reported in reference 3.19. This results in an almost perfect correlation with the wind tunnel data for lateral flapping.

Closed-Form Integration for Torque

The solution of the closed-form blade element method for the torque required in forward flight follows the procedures developed for torque in hover and for thrust in forward flight. The increment of torque as shown in Figure 3.35, is:

$$\Delta Q = r(\Delta D - \Delta L \phi) = r \left(\Delta D - \Delta L \frac{U_P}{U_T} \right)$$

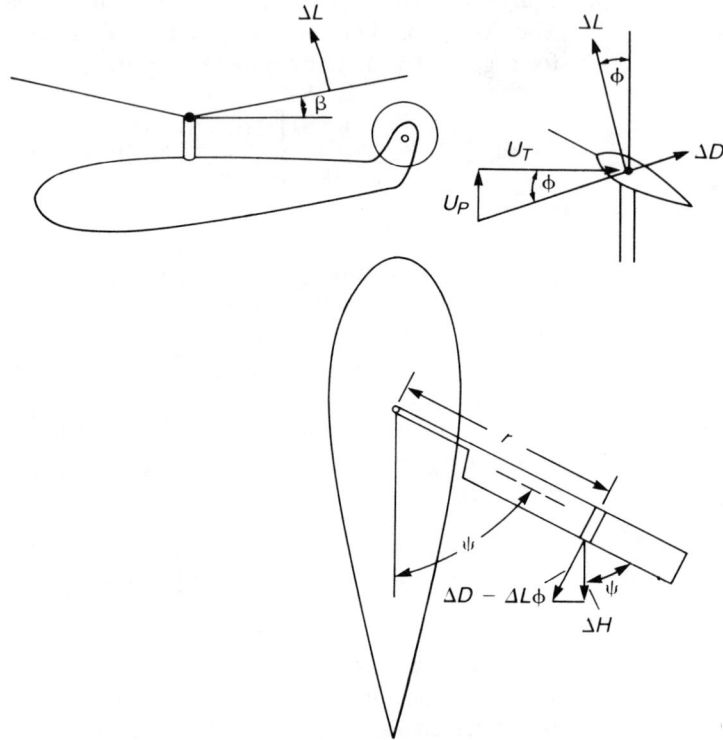

FIGURE 3.35 Force Increments Acting on Blade Element Which Contribute to Rotor Torque and H-Force

The torque per running foot is:

$$\frac{\Delta Q}{\Delta r} = r \frac{\rho}{2} U_T^2 c \left[c_d - a\alpha \frac{U_P}{U_T} \right]$$

and the total torque is:

$$Q = \frac{b}{2\pi} \int_0^{2\pi} \int_0^R r \frac{\rho}{2} U_T^2 c c_d dr d\psi - \frac{b}{2\pi} \int_0^{2\pi} \int_0^R r \frac{\rho}{2} U_T^2 c a\alpha \frac{U_P}{U_T} dr d\psi$$

Carrying through the integration and nondimensionalizing gives:

$$C_Q/\sigma = \frac{c_d}{8}(1 + \mu^2) - \frac{a}{4}\left\{ (\lambda' - \mu a_{1_s})^2 + \frac{3}{2}\mu a_{1_s}(\lambda' - \mu a_{1_s}) \right.$$

$$\left. + \left(\frac{2}{3}\theta_0 + \frac{\theta_1}{2}\right)(\lambda' - \mu a_{1_s}) + \frac{\mu^2 a_{1_s}^2}{2} + (B_1 + a_{1_s})\left[\frac{a_{1_s}}{4} - \frac{\mu}{2}(\lambda' - \mu a_{1_s}) - \mu^2 \frac{a_{1_s}}{8}\right] \right.$$

$$- (A_1 - b_{1_s}) \times \left[\frac{1}{4} \left(b_{1_s} - \frac{v_1}{\Omega R} \right) - \frac{\mu a_0}{3} + \frac{\mu b_{1_s}}{8} \right] - \frac{\mu a_0 b_{1_s}}{3} + \frac{\mu^2 a_0^2}{2}$$

$$- \frac{v_1}{\Omega R} \left(b_{1_s} - \frac{v_1}{\Omega R} \right) + \frac{2}{3} \frac{v_1}{\Omega R} \mu a_0 \Bigg\}$$

(The equation has been written in this form to show the equivalence with the similar equation in reference 3.17).

    The first part is the profile torque coefficient due only to the drag coefficient, and the second part is the inflow torque coefficient due only to the tilt of the lift vector. The inflow term includes both the induced and the parasite contributions as defined by the momentum equations. It can be simplified by using the equations for $(B_1 + a_{1_s})$ and $(A_1 - b_{1_s})$ previously derived. With this substitution, all of the cyclic pitch and flapping terms drop out:

$$C_Q/\sigma = \frac{c_d}{8} (1 + \mu^2) - \frac{a}{4} \left(\frac{\lambda'}{1 + \frac{3}{2}\mu^2} \right) \left[\frac{\theta_0}{3} (2 - \mu^2) + \frac{\theta_1}{2} \left(1 - \frac{\mu^2}{2} \right) + \lambda' \left(1 + \frac{\mu^2}{2} \right) \right]$$

$$- \frac{a}{4} \left(\frac{\mu^2}{1 + \frac{1}{2}\mu^2} \right) \left[\frac{a_0^2}{2} \left(\frac{1}{9} + \frac{\mu^2}{2} \right) + \frac{1}{3} \mu a_0 \frac{v_1}{\Omega R} + \frac{1}{8} \left(\frac{v_1}{\Omega R} \right)^2 \right]$$

    The last term, which is a function of a_0 and $v_1/\Omega R$, is generally negligible when compared with the other terms. For this type of analysis, the value of c_d to use is the average obtained as a function of C_T/σ as it was in hover—that is, as a function of angle of attack equal to 57.3 (C_T/σ) degrees and at a Mach number corresponding to 75% of the tip speed. A more sophisticated way of handling the drag coefficient will be discussed later.

Closed-Form Integration for *H*-Force

The *H*-force is the horizontal force perpendicular to the rotor shaft and is derived in the same manner as that of the torque. Figure 3.35 shows the arrangement of the forces on the blades that contribute to the H-forces. At any blade element the incremental H-force is:

$$\Delta H = \left(\Delta D - \Delta L \frac{U_P}{U_T} \right) \sin \psi - \Delta L \beta \cos \psi$$

In level flight, ΔH is positive on the advancing side and negative on the retreating side, although in autorotation the reverse can be true. Thus the total H-force can be either positive or negative, depending on which side of the rotor is dominant.

Because of this, the magnitude of the **calculated H-force** can vary drastically with relatively small changes in flight conditions—or, as will be shown later, with changes in the assumptions made in **the integration**. When the integration is carried out with the same assumptions **as for the thrust and torque equations, the** result is:

$$C_H/\sigma = \frac{c_d\mu}{4} - \frac{a}{4}\left\{ \lambda'\left[\mu\left(\theta_0 + \frac{\theta_1}{2}\right) - \frac{1}{2}(B_1 + a_{1_s}) \right] \right.$$

$$-a_{1_s}\left[\left(\frac{2}{3} + \mu^2\right)\theta_0 + \left(\frac{1}{2} + \frac{\mu^2}{2}\right)\theta_1 + \lambda' - \mu(B_1 + a_{1_s}) \right]$$

$$\left. + (A_1 - b_{1_s})\left[\frac{1}{8}\mu\frac{v_1}{\Omega R} - \frac{a_0}{3} \right] - a_0\left[\frac{1}{3}\left(\frac{v_1}{\Omega R}\right) + \frac{\mu a_0}{2} \right] \right\}$$

Again, using the equations for $(B_1 + a_{1_s})$ and $(A_1 - b_{1_s})$ and also the equation for C_T/σ, the equation simplifies to:

$$C_H/\sigma = \frac{c_d\mu}{4} - \frac{a}{4}\left(\frac{\mu\lambda'}{1 + \frac{3}{2}\mu^2}\right)\left[\theta_0\left(-\frac{1}{3} + \frac{3}{2}\mu^2\right) + \frac{\theta_1}{2}\left(-1 + \frac{3}{2}\mu^2\right) - \lambda' \right]$$

$$+ a_{1_s}C_T/\sigma + \frac{a}{4}\left(\frac{\mu}{1 + \frac{1}{2}\mu^2}\right)\left[\frac{a_0^2}{2}\left(\frac{1}{9} + \frac{\mu^2}{2}\right) + \frac{1}{3}\mu a_0\frac{v_1}{\Omega R} + \frac{1}{8}\left(\frac{v_1}{\Omega R}\right)^2 \right]$$

In this case the last term is generally **small but perhaps** not negligible.

If the inflow portions of the rotor **equations** are combined, it may be **seen** that:

$$-C_Q/\sigma_{\text{inflow}} = \lambda'C_T/\sigma + \mu(C_H/\sigma_{\text{inflow}} - a_{1_s}C_T/\sigma)$$

This identity is illustrated graphically by **examining the rotor mounted in a wind** tunnel at a positive angle of attack, **as shown in Figure 3.36. The momentum** power that the rotor extracts from the **airflow is negative rotor power and is equal**

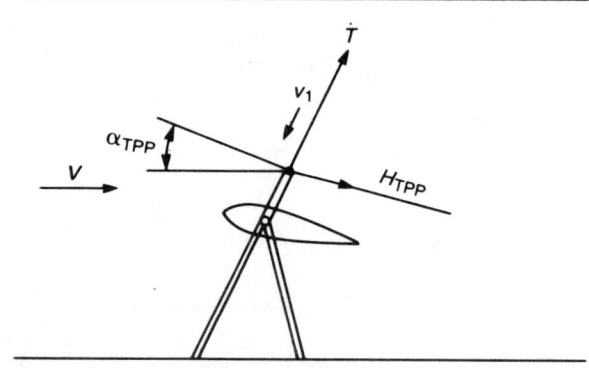

FIGURE 3.36   Forces Acting on Rotor in Wind Tunnel

to the two forces, thrust and H-force, multiplied by the corresponding velocities parallel to them:

$$-P_{\text{inflow}} = (V\alpha_{\text{TPP}} - v_1)T + VH_{\text{TPP}}$$

This reduces to the nondimensional identity expressed earlier.

The relationship between the total coefficients is:

$$C_Q/\sigma = \frac{c_d}{8}(1 + 3\mu^2) - (\lambda' - \mu a_{1_s})C_T/\sigma - \mu C_H/\sigma$$

which can be used as an aid in reducing the computation of either C_Q/σ or C_H/σ if the other is known.

Compressibility Correction and Compressibility Tip Relief

If any blade element exceeds the drag divergence Mach number for its airfoil, the power required will be higher than that calculated from the closed-form equations. Before developing the method for estimating this power loss, it is appropriate to discuss the three-dimensional phenomenon known as *compressibility tip relief*.

The relief effect is seen clearly by comparing the streamlines on a two-dimensional airfoil and on a three-dimensional body of revolution with the same thickness ratio shown in Figure 3.37. On the two-dimensional airfoil, the streamlines are constrained to remain parallel to each other but on the body of revolution, they spread out as they approach the maximum thickness thus reducing the local velocity and increasing the free-stream Mach number at which drag

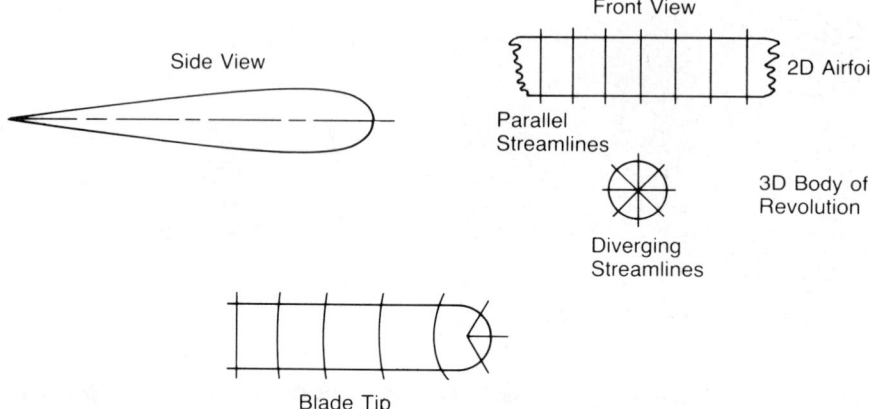

FIGURE 3.37   Comparison of Two- and Three-Dimensional Bodies

divergence occurs. The blade tip may be thought of as a combination of a two- and three-dimensional body and, as such, would be expected to have a drag divergence Mach number that lies between those of the other two bodies. Wind tunnel tests of wings, reported in reference 3.20, verify this expectation by showing that the drag divergence Mach number increases as the aspect ratio is decreased.

Tip relief consists of two parts: a large one due to the reduction of the effective Mach number, which delays the formation of the shock wave, and a smaller one due to the reduction in the local dynamic pressure, which reduces the actual drag and, thus, the drag coefficient based on the free-stream dynamic pressure.

A method for estimating the tip relief for rotor blades is given in reference 3.21 based on the similar analysis for wings of reference 3.22. With this method, both the reduction of Mach number and that of drag coefficient may be calculated as a function of the blade thickness ratio and the distance from the tip. The method is suitable for detailed computer studies, but for a quick estimate of tip relief, a simpler method is available. The simpler method is based on the fact that the tip of a blade with a thick airfoil will act more like a three-dimensional body and thus will have a greater tip relief than the tip of a blade with a thin airfoil. A convenient velocity parameter that is a function of the airfoil thickness ratio is the incompressible two-dimensional velocity ratio, (v/V), and the average of this ratio along the chord is used to define the effective Mach number at the tip.

$$M_{\text{eff}} = \frac{M}{(\bar{v}/V)}$$

The local velocity ratio, v/V, as a function of chord can be calculated using airfoil theory or can be derived from wind tunnel measurements of the pressure coefficient distribution since:

$$\left(\frac{v}{V}\right)^2 = 1 - C_P$$

For many airfoils, the local value of $(v/V)^2$ is plotted against chord in Appendix I of reference 3.23. These plots have been used to find the average velocity ratio at zero angle of attack and the corresponding tip relief for several families of airfoils. The correction has been applied to the test data shown in Figure 6.28 of Chapter 6 to produce Figure 3.38, which shows the two- and three-dimensional drag divergence Mach numbers as a function of the effective thickness ratio. The method has been used to correlate with flight test data given in references 3.24 and 3.25. The results shown in the following table demonstrate a satisfying degree of agreement.

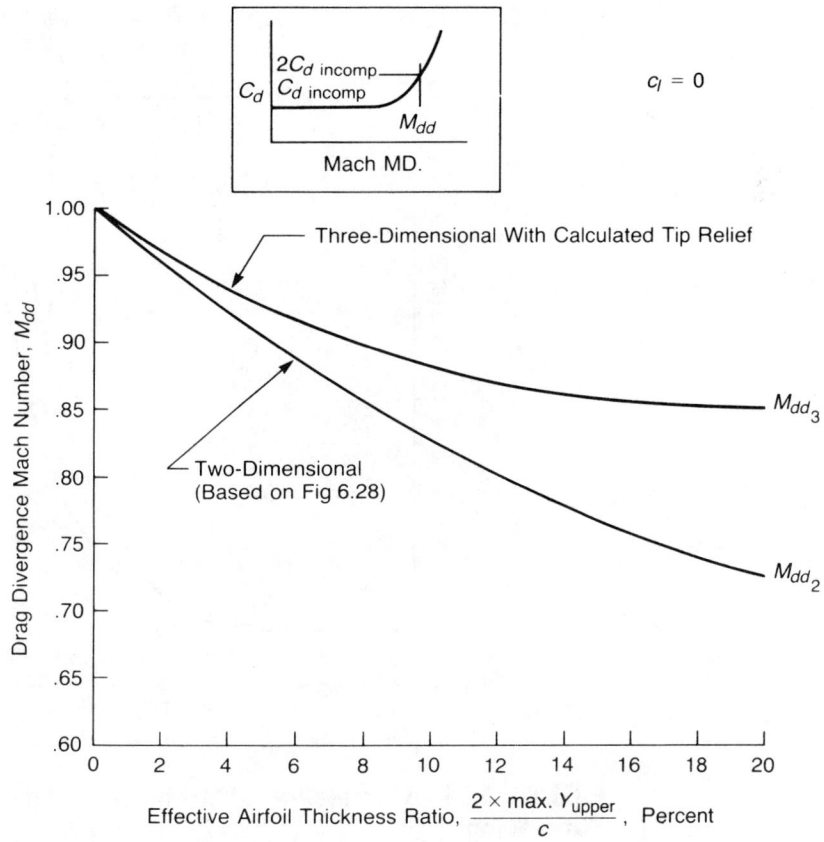

FIGURE 3.38 Effect of Tip Relief on Drag Divergence Mach Number

Ref.	Tip Airfoil	$M_{dd_{2D}}$	$M_{dd_{3D}}$	$M_{dd_{observed}}$
3.24	0012	.80	.87	.875
3.24	0006	.88	.92	.925
3.25	23012	.80	.87	.852

A further check has been made applying the method to the fixed-wing example of reference 3.22. For this case, the wing had a NACA 0012 airfoil and an aspect ratio of 3.25. The method of this section was adapted to the wing by assuming that the tip relief linearly decreased to zero one chord length in from each tip. The average Mach number relief for the wing, therefore, was less than the extreme tip value. Figure 3.39 shows the experimental drag measurements for the

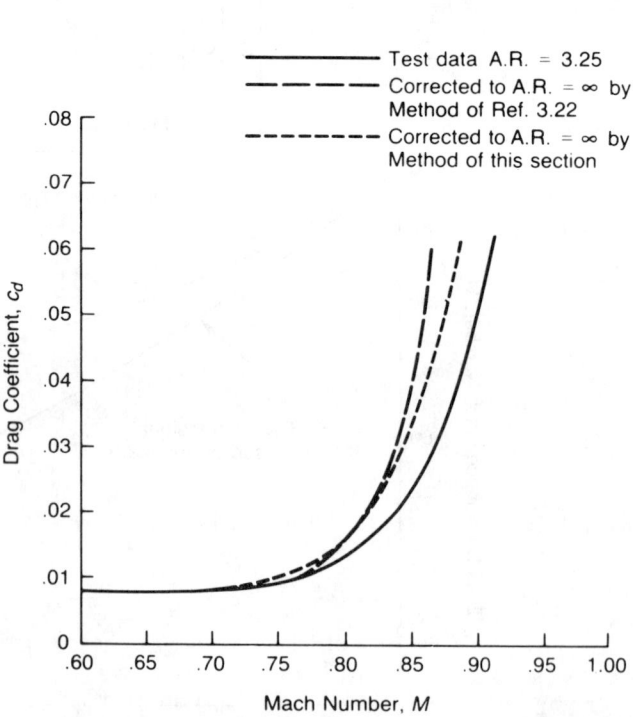

FIGURE 3.39   Correlation of Methods for Tip Relief on a Low Aspect Ratio Wing

Source of test data: Anderson, "Aspect Ratio Influence at High Subsonic Speeds," *Jour. of Aero. Sci.* 23-9, 1956.

low-aspect-ratio wing and estimates for a wing of infinite aspect ratio using both the sophisticated method of reference 3.22 and the simple method of this section. Again, the degree of correlation is satisfactory at least near the drag rise area, which determines the drag divergence Mach number.

Additional justification of the procedure is developed by comparison with calculations of the spanwise distribution of the maximum local Mach number for blades of aspect ratios of 10, 15, and 20, shown in Figure 3.40 taken from reference 3.26. The airfoil is the NACA 0012. The calculated curves are compared with straight lines, which follow the assumptions just discussed. The assumptions appear to be very good for the aspect ratios of 15 and 20, though somewhat extreme for the aspect ratio of 10.

An approximation to the additional torque due to compressibility may be obtained by integrating the extra drag from the radius station at which the drag first rises above the incompressible value to the tip. In forward flight, the advancing tip normally operates at very low angles of attack so that to a first approximation, the compressibility losses can be related to the airfoil drag characteristics at zero lift. Test results show, however, that the losses are also a function of C_T/σ. The procedure adopted for this analysis has been to derive the

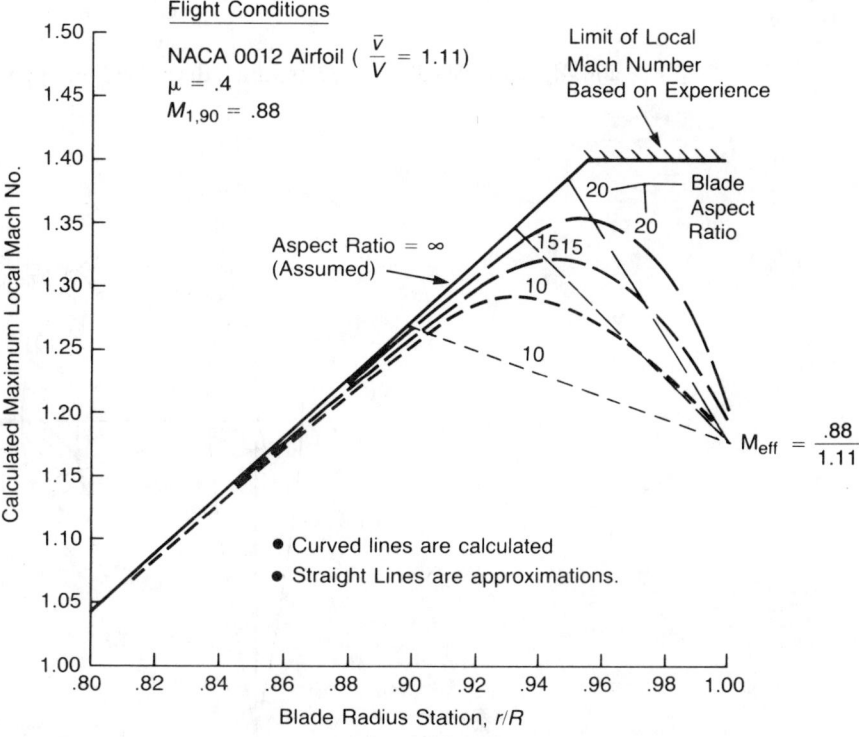

FIGURE 3.40 Comparison of Calculated and Approximate Tip Relief

basic equations using the zero-lift drag characteristics of the airfoil and then to establish an empirical correction to the drag rise Mach number as guided by wind tunnel and flight tests of actual rotors producing thrust.

The geometric relationships that enter into the analysis are shown in Figure 3.41, where $(r/R)_{dr}$ is the radius station at which drag rise is first experienced and ψ_1 and ψ_2 are the azimuth angles bounding the drag rise region. The equations for these two parameters are:

$$(r/R)_{dr} = \frac{M_{dr}}{M_{\Omega R}} - \mu \sin \psi$$

and

$$\psi_1, \psi_2 = \sin^{-1} \left[\frac{1}{\mu} \left(\frac{M_{dr}}{M_{\Omega R}} - 1 \right) \right]$$

where

$$M_{\Omega R} = \frac{\Omega R}{\text{Speed of sound}}$$

and M_{dr} is the Mach number at which drag rise first appears.

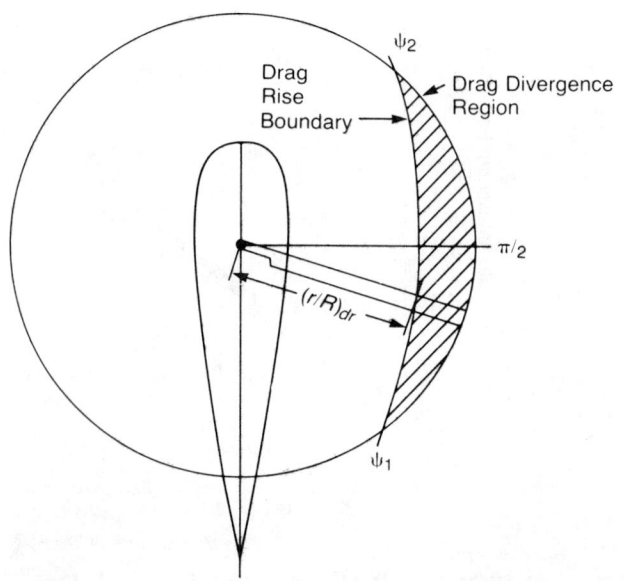

FIGURE 3.41 Geometry of Drag Divergence Region

The excess torque per running foot due to compressibility is:

$$\frac{d\Delta Q_{comp}}{dr} = \frac{\rho}{2} U_T^2 \Delta c_d c r$$

or

$$\Delta C_Q/\sigma_{comp} = \frac{1}{2\pi} \int_{\psi_1}^{\psi_2} \int_{(r/R)dr}^{1} \left(\frac{r}{R} + \mu \sin \psi\right)^2 \frac{r}{R} \Delta c_d d \frac{r}{R} d\psi$$

The drag increment, Δc_d, is a function of Mach number. Figure 3.42 shows the drag for a 0012 airfoil at zero lift. A satisfactory fitting of the rising portion of the curve is a cubic:

$$\Delta c_d = K_1 (M - M_{dr})^3$$

or

$$\Delta c_d = K_1 M_{\Omega R}^3 \left(\frac{r}{R} + \mu \sin \psi - \frac{M_{dr}}{M_{\Omega R}}\right)^3$$

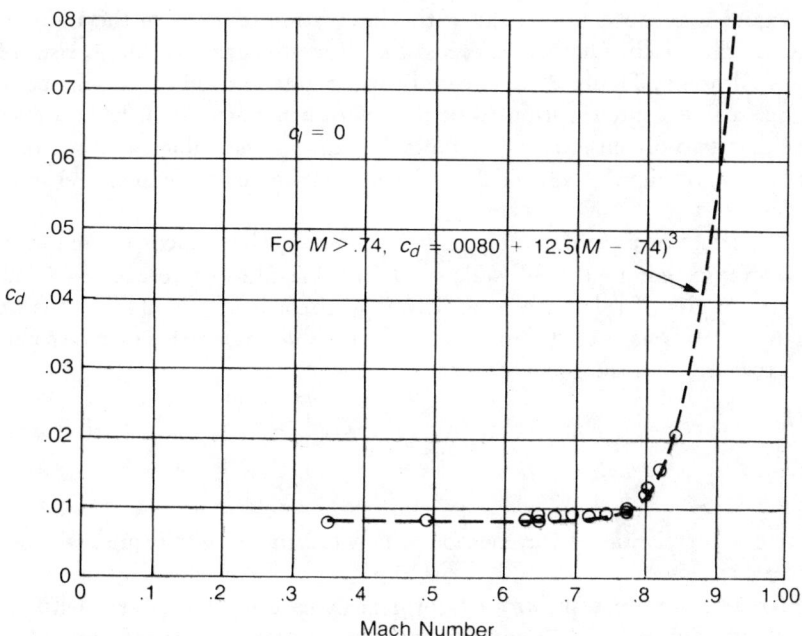

FIGURE 3.42   Zero Lift Drag Measurements for NACA 0012 Airfoil

Source: Hughes, unpublished.

The value of M_{dr} is a function of the spanwise position of the blade element, being the two-dimensional value, M_{dr_2}, inboard of one chord length, and being the three-dimensional value, M_{dr_3}, at the tip. Thus for the inboard section:

$$\frac{r}{R} < \left(1 - \frac{c}{R}\right) \qquad \frac{M_{dr}}{M_{\Omega R}} = \left(\frac{M_{dr_2}}{M_{dr_3}}\right)\left(\frac{M_{dr_3}}{M_{\Omega R}}\right)$$

and for the tip section:

$$\frac{r}{R} > \left(1 - \frac{c}{R}\right) \qquad \frac{M_{dr}}{M_{\Omega R}} = \left[1 - \left(1 - \frac{M_{dr_2}}{M_{dr_3}}\right)\left(\frac{1 - \frac{r}{R}}{c/R}\right)\right]\left[\frac{M_{dr_3}}{M_{\Omega R}}\right]$$

With these relationships, it is possible to evaluate the quantity, $(\Delta C_Q/\sigma_{\text{comp}})/M_{\Omega R}^3$, as a function of the airfoil characteristics, K_1 and M_{dr_2}/M_{dr_3}; the chord to radius ratio, c/R; and the flight conditions, $M_{dr_3}/M_{\Omega R}$, and μ. As an example, the integration has been carried out assuming the 0012 airfoil with:

$$K_1 = 12.5$$

$$M_{dr_2}/M_{dr_3} = .92$$

(Assume $M_{dr_2}/M_{dr_3} = M_{dd_2}/M_{dd_3}$ from Figure 3.38.)

Figure 3.43 shows the results of this integration in terms of the amount the advancing tip Mach number exceeds the three-dimensional drag rise Mach number. Although Figure 3.43 is based on the 0012 airfoil, it should be valid enough as a first approximation to be used with other airfoils suitable for rotors using either two-dimensional wind tunnel results or the value of M_{dr_3} from the upper portion of Figure 3.43, which has been generated using figures 3.38 and 3.42 as guides.

The effect of C_T/σ on compressibility losses has been incorporated into the method after examining the full-scale wind tunnel results of reference 3.27 and the flight test results of reference 3.28. Although there is not complete consistency between the various sets of test data, a reasonable approach appears to be to assume that M_{dr_3} has the form:

$$M_{dr_3} = M_{dr_{3_{c_l=0}}}[1 - 6(C_T/\sigma)^2]$$

Satisfactory correlation of the method with wind tunnel test results is shown in Figure 3.44.

For those more sophisticated rotor analyses using computers with stored tables of two-dimensional lift and drag coefficients as a function of angle of attack

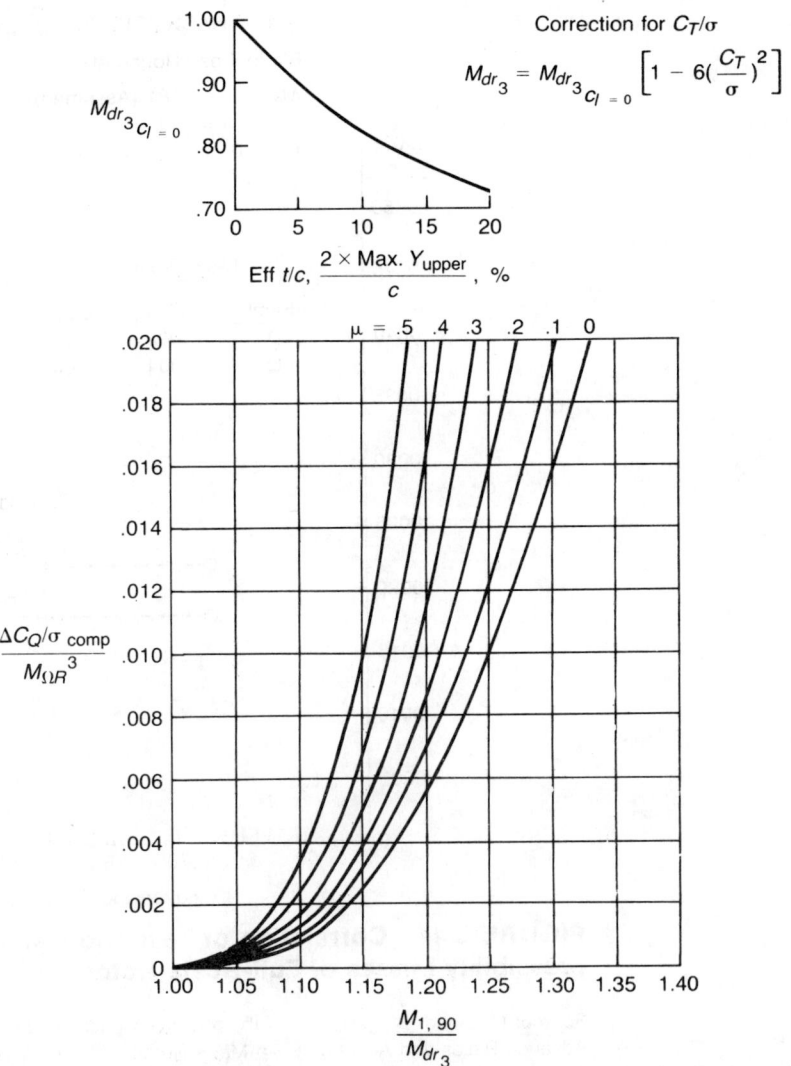

FIGURE 3.43   Torque Coefficient Correction for Compressibility

and Mach number, the tip relief may be accounted for by calculating an effective Mach number for the blade elements within one chord length of the tip such that:

$$M_{\text{eff}} = M_{\text{actual}} \left[\frac{M_{dr_2}}{M_{dr_3}} + \left(1 - \frac{M_{dr_2}}{M_{dr_3}} \right) \left(\frac{1 - \dfrac{r}{R}}{c/R} \right) \right]$$

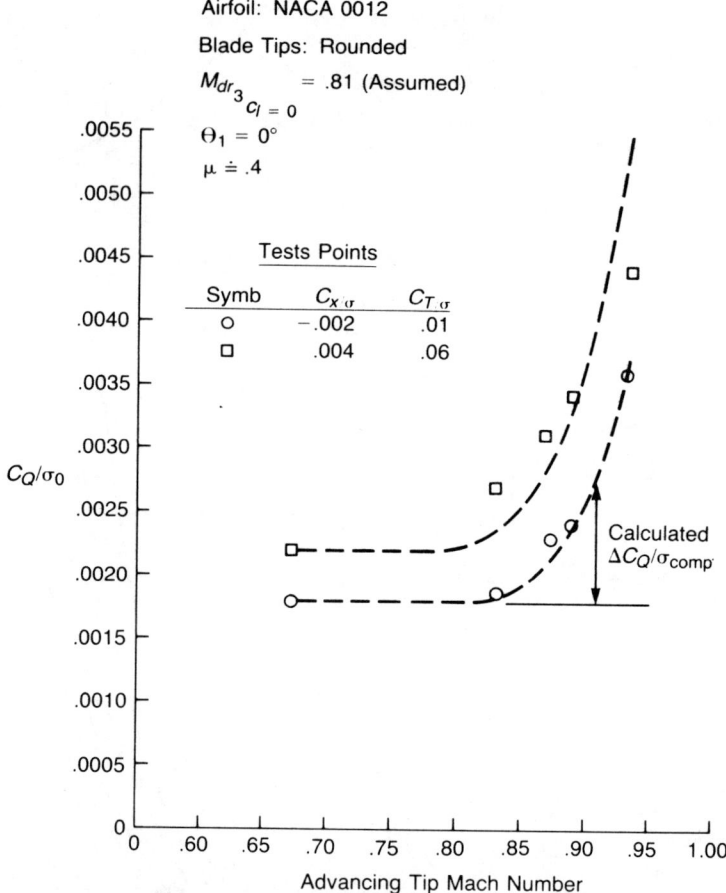

FIGURE 3.44   Correlation of Test and Calculated Results for Compressibility Losses of Full-Scale Rotor

Source: McCloud, Biggers, & Stroub, "An Investigation of Full-Scale Helicopter Rotors at High Advance Ratios and Advancing Tip Mach Numbers," NASA TN D4632, 1968.

This effect can generally be easily incorporated in the program by defining an effective speed of sound for the outboard elements such that:

$$\text{Speed of sound}_{\text{eff.}} = \text{Speed of sound}_{\text{act.}} \left(\frac{M_{\text{act.}}}{M_{\text{eff.}}} \right)$$

A further complication of the whole question of drag divergence on the advancing tip is the evidence that there is a finite delay in compressibility effects due to the rate of change of Mach number with time. Wind tunnel tests of a nonlifting rotor reported in reference 3.29 indicate that the maximum compressibility effects occur somewhat later than $\psi = 90°$, where the local Mach number is the highest. There is not yet any simple method for accounting for this effect.

Closed-Form Equations for the Tail Rotor

The tail rotor is a flapping rotor without cyclic pitch and is thus similar to the early autogiro rotors for which the blade element theory was originally developed. (Note: Even in tail rotors that have no hinges and are thus classified as *rigid rotors*, the aerodynamic and centrifugal forces generally predominate over the structural stiffness, and the rotor will flap by bending the blades almost as much as if it had actual mechanical hinges. The equations developed here can, therefore, be assumed to apply in the first approximation to these rotors.) Most tail rotors—and some main rotors—are designed with a mechanical coupling between flapping and feathering such that a change in flapping produces a change in feathering:

$$\Delta\theta = \Delta\beta \tan\delta_3$$

where δ_3 is the slant angle of the flapping hinge as shown in Figure 3.45. The angle shown there is negative, which is the usual (but not the universal) application on tail rotors. The local blade pitch angle is:

$$\theta = (\theta_0 + a_0 \tan\delta_3) + \theta_1 \frac{r}{R} - a_{1_s} \tan\delta_3 \cos\psi - b_{1_s} \tan\delta_3 \sin\psi$$

The last two terms have the form of cyclic pitch and can be used in the basic rotor equations by treating them as such:

$$A_{1_{\text{eff.}}} = a_{1_s} \tan\delta_3$$

$$B_{1_{\text{eff.}}} = b_{1_s} \tan\delta_3$$

Unlike the case of the main rotor, where the angle of attack of the tip path plane is known from equilibrium conditions, for a tail rotor the angle of attack of the shaft is known. It is generally zero unless the tail rotor shaft is deliberately tilted or the helicopter is flying with sideslip. For this reason, the tail rotor equations are written in terms of the inflow ratio, λ. The tail rotor equations become:

$$C_T/\sigma = \frac{a}{4}\left\{\left(\frac{2}{3} + \mu^2\right)\left(\theta_0 + a_0 \tan\delta_3\right) + \left(\frac{1}{2} + \frac{\mu^2}{2}\right)\theta_1 - \mu b_{1_s} \tan\delta_3 + \lambda\right\}$$

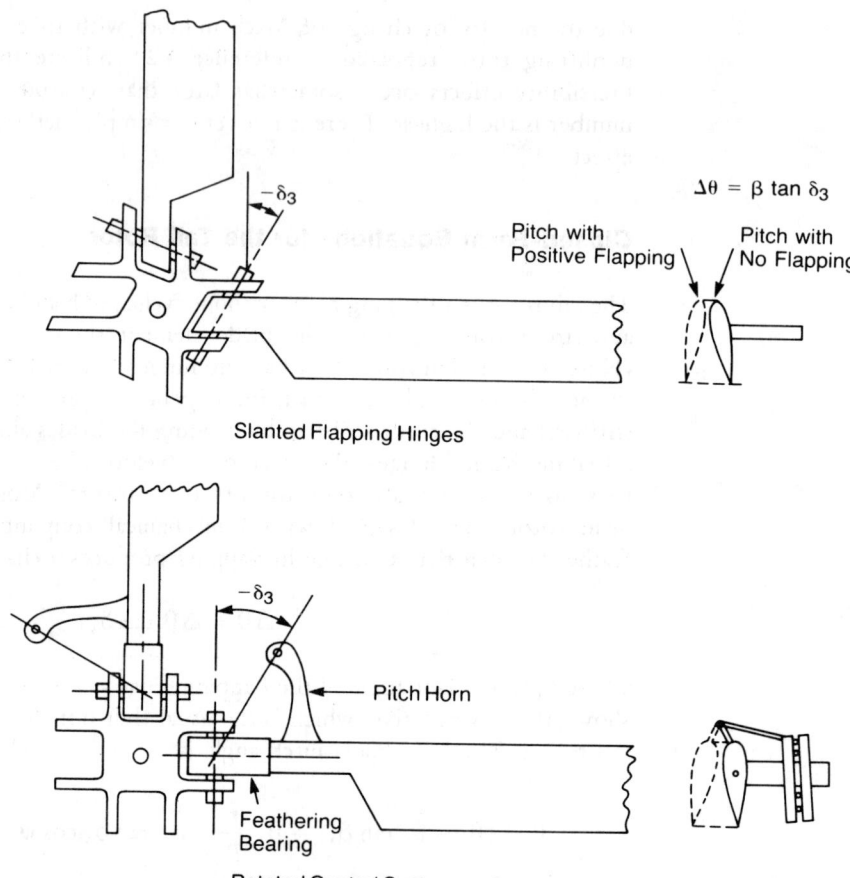

$\Delta\theta = \beta \tan \delta_3$

Pitch with Positive Flapping

Pitch with No Flapping

Slanted Flapping Hinges

Pitch Horn

Feathering Bearing

Rotated Control System

FIGURE 3.45   Two Methods for Obtaining Pitch-Flap Coupling

$$a_{1_s} + b_{1_s} \tan \delta_3 = \frac{\mu}{1 + \frac{3}{2}\mu^2} \left[\frac{8}{3}(\theta_0 + a_0 \tan \delta_3) + 2\theta_1 + 2(\mu a_{1_s} + \lambda) \right]$$

$$b_{1_s} - a_{1_s} \tan \delta_3 = \frac{\frac{4}{3}\mu a_0 + \frac{v_1}{\Omega R}}{1 + \frac{\mu^2}{2}}$$

These equations can be combined to produce closed-form equations for longitudinal and lateral flapping of rotors with δ_3 coupling. For the case where the tail rotor shaft is perpendicular to the flight path:

$$\lambda = -\frac{v_1}{\Omega R} = -C_T/\sigma \frac{\sigma}{2\mu}$$

and the flapping equations become:

$$\frac{4\mu}{2+3\mu^2}\left\{\frac{4}{2+3\mu^2}\left[\left(\frac{4}{a}+\frac{\sigma}{2\mu}\right)C_T/\sigma - (1+\mu^2)\frac{\theta_1}{2}\right] + \theta_1 - C_T/\sigma\frac{\sigma}{2\mu}\right\}$$

$$a_{1_s} = \frac{-2\left[\dfrac{\dfrac{4}{3}\mu a_0 + C_T/\sigma\dfrac{\sigma}{2\mu}}{2+\mu^2}\right]\left[1-\left(\dfrac{4\mu}{2+3\mu^2}\right)^2\right]\tan\delta_3}{\dfrac{2-\mu^2}{2+3\mu^2} + \left[1-\left(\dfrac{4\mu}{2+3\mu^2}\right)^2\right]\tan^2\delta_3}$$

$$b_{1_s} = 2\left[\frac{\dfrac{4}{3}\mu a_0 + C_T/\sigma\dfrac{\sigma}{2\mu}}{2+\mu^2}\right] + a_{1_s}\tan\delta_3$$

Figure 3.46 shows the results of calculating flapping angles for the tail rotor of the example helicopter. It may be seen that the total flapping is reduced by using δ_3 of either sign. (Note: The magnitude of rotor flapping calculated from these simple closed-form equations may be as much as 50% less than that calculated by more elaborate methods, such as the chart methods given later; but the trends are valid.)

The equation for the *H*-force and torque coefficients are:

$$C_H/\sigma_T = \left\{\frac{c_d\mu}{4} - \frac{a}{4}\left[\frac{\mu(\lambda+\mu a_{1_s})}{1+\dfrac{3}{2}\mu^2}\right]\left[(\theta_0 + a_0\tan\delta_3)\left(-\frac{1}{3}+\frac{3}{2}\mu^2\right) + \frac{\theta_1}{2}\left(-1+\frac{3}{2}\mu^2\right) - (\lambda+\mu a_{1_s})\right]\right.$$

$$\left. + \left[\frac{\mu}{1+\dfrac{1}{2}\mu^2}\right]\left[\frac{a_0^2}{2}\left(\frac{1}{9}+\frac{\mu^2}{2}\right) + \frac{1}{3}\mu a_0\frac{v_1}{\Omega R} + \frac{1}{8}\left(\frac{v_1}{\Omega R}\right)^2\right] + a_{1_s}C_T/\sigma\right\}_T$$

$$C_Q/\sigma_T = \left[\frac{c_d}{8}(1+3\mu^2) - \lambda C_T/\sigma - \mu C_H/\sigma_T\right]_T$$

where the subscript T denotes that all parameters refer to the tail rotor. Table 3.1 shows the trim conditions for the tail rotor of the example helicopter at 115 knots for δ_3 values of 0°, 30°, and −30°. It may be seen that the effect of the delta-three

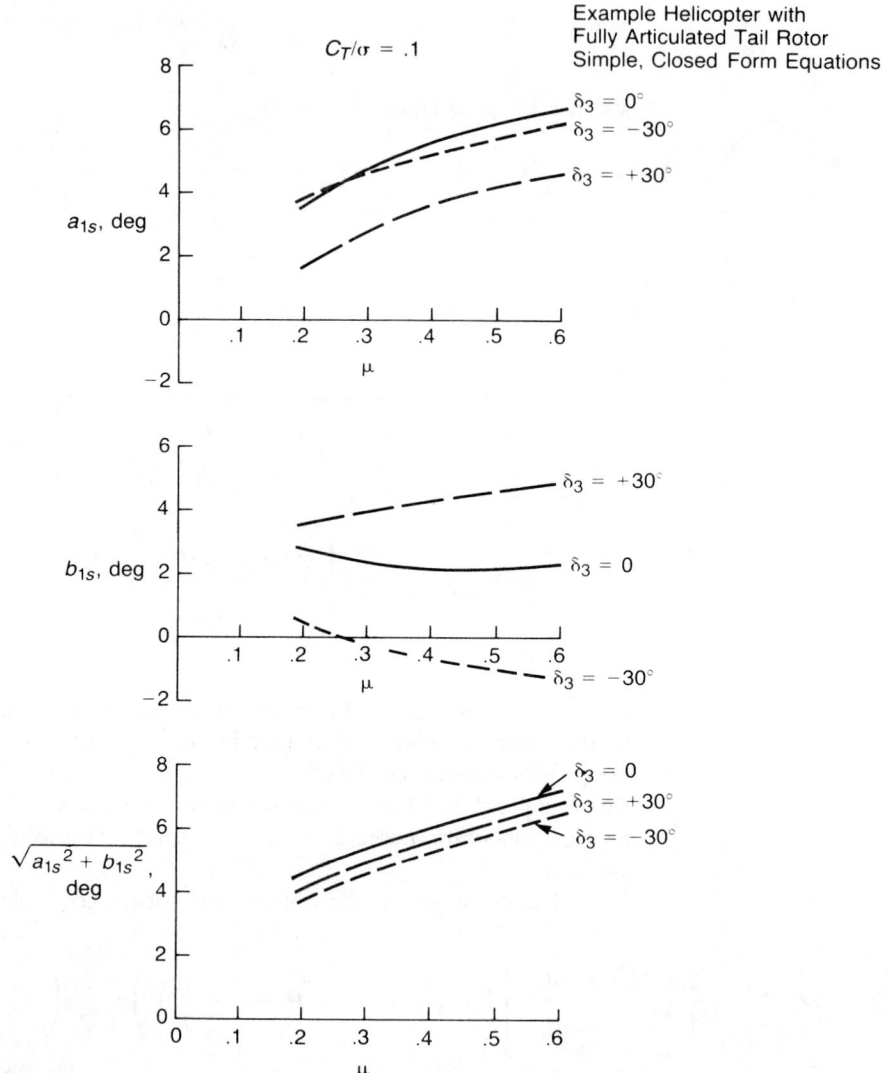

FIGURE 3.46   Effect of Delta-Three Angle on Steady Tail Rotor Flapping

hinge on the trim conditions is small enough to be considered negligible for performance calculations.

Although it has little effect on steady-state conditions, the delta-three angle is important in reducing flapping during maneuvers. Figure 3.47 from reference 3.30 shows that both positive and negative values are effective.

TABLE 3.1
Tail Rotor Trim Conditions

Parameter	$\delta_3 = 0°$	$\delta_3 = 30°$	$\delta_3 = -30°$
bP_M	1,097	1,097	1,097
T_T	755	755	755
a_0	0.95°	0.95°	0.95°
$\theta_0 - a_0 \tan \delta_3$	6.25°	6.82°	5.67°
θ_0	6.25°	6.27°	6.22
$a_{1_s} - b_{1_s} \tan \delta_3$	1.77°	1.89°	1.78°
a_{1_s}	1.77°	1.03°	1.72°
$b_{1_s} + a_{1_s} \tan \delta_3$	0.89°	0.90°	0.89°
b_{1_s}	0.89°	1.49°	-0.10°
bP_T	26.5	29.9	26.7
H_T	36.1	26.9	35.5

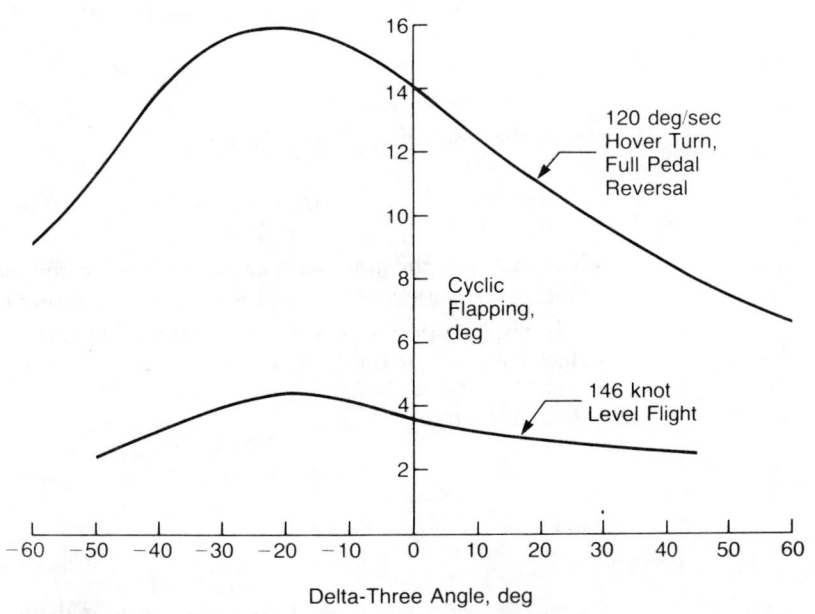

FIGURE 3.47 Effect of Delta-Three Angle on Rotor Flapping During Maneuvers

Source: Huber & Frommlett, "Development of a Bearingless Helicopter Tail Rotor," 6th European Rotorcraft Forum, 1980.

CALCULATION OF TRIM CONDITIONS

Level Flight

Since the trim conditions are dependent on each other, the calculations are done as an iteration of the angle of attack of the tip path plane and the rotor thrust, using the equations:

$$\alpha_{\text{TPP}} = -\tan^{-1}\left(\frac{D_F + H_M + H_T}{\text{G.W.} - L_F}\right)$$

$$T = \sqrt{(\text{G.W.} - L_F)^2 + (D_F + H_M + H_T)^2}$$

The fuselage lift and drag characteristics as a function of angle of attack of the fuselage may be estimated from previous experience or measured in a wind tunnel. Typical fuselage aerodynamics that will be assumed to apply to the example helicopter are shown in Figures A2 through A4 of Appendix A. The lift and drag forces are:

$$L_F = q(L_F/q)$$

$$D_F = qf$$

where the fuselage angle of attack is:

$$\alpha_F = \alpha_{\text{TPP}} - i_s - a_{1_s} - \Delta\alpha_{\text{D.W.}}$$

where $\Delta\alpha_{\text{D.W.}}$ is the downwash angle produced at the fuselage by the rotor. A wind tunnel investigation reported in reference 3.31 shows that the rotor downwash at the fuselage is approximately equal to the value corresponding to the momentum-induced velocity at the plane of the rotor:

$$\Delta\alpha_{\text{D.W.}} = \frac{v_1}{V} \text{ radians}$$

Thus

$$\alpha_F = \frac{\lambda'}{\mu} - i_s - a_{1_s} \text{ radians}$$

Trim conditions for a helicopter can now be evaluated from the equations just developed. A typical case for the example helicopter has been prepared. The assumptions that were used in the iterative process were:

First Iteration	*Subsequent Iterations*
$H = H_0$	$H = (H_0 + H_i)_{\text{prev. iter.}}$
$\alpha_F = 0$	$\alpha_F = (\alpha_F)_{\text{prev. iter.}}$

The results of the calculations are shown in Table 3.2.

Figure 3.48 shows the distribution of the local angle of attack along the advancing and retreating blades for level flight and the two other flight conditions to be discussed next. For the example helicopter with 10° of twist, the highest local angle of attack is not at the retreating tip but at the 70% station.

Climbs and Descents

The closed-form equations can also be used to calculate the power and trim conditions in a climb or in a descent by incorporating the component of gross

TABLE 3.2
Trim Condition Iteration

G.W. = 20,000 lb
$\Omega R = 650$ ft/sec
$\mu = .3$ (115 kts)
$\rho = .002378$ (sea-level standard)
$a_{1_s} = b_{1_s} = 0$ (This is a valid assumption for performance analysis, but not for stability and control analysis. See Chapter 7)

	1st Iteration	*2nd Iteration*	*3rd Iteration*
α_F(start), deg	0	−5.9	−6.1
L_F, lb	−200	−730	−746
D_F, lb	872	904	904
H_M(start), lb	181	364	399
H_T(start), lb	12	30	36
T, lb	20,230	20,770	20,790
α_{TPP}, rad	−.0526	−.0629	−.0647(−3.70°)
λ'	−.0276	−.0311	−.0316
a_0	.0716	.0743	.0744(4.26°)
θ_0, rad	.267	.276	.277(15.85°)
H_M, lb	364	399	401
h.p.$_M$	969	1,089	1,097
T_T, lb	677	750	755
H_T, lb	30	36	36
α_F, deg	−5.9	−6.1	−6.1
A_1, deg			−2.3
B_1, deg			4.9
h.p.$_T$			25
$\alpha_{1,270}$, deg			8.3

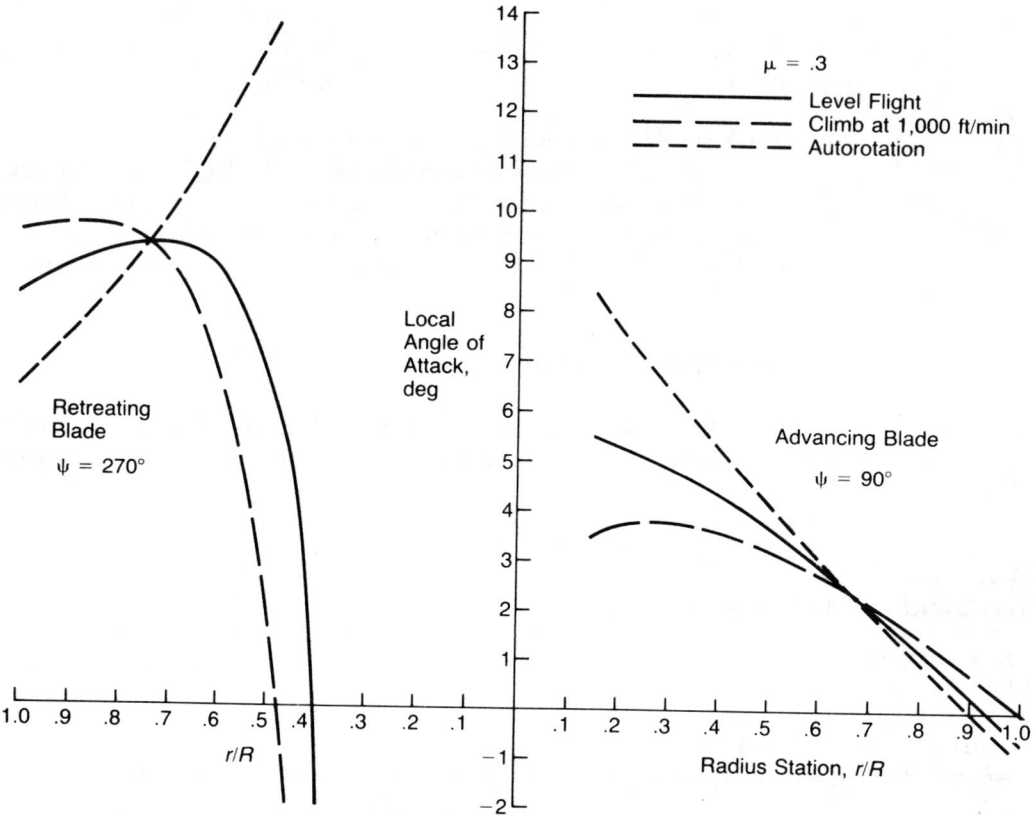

FIGURE 3.48 Angle of Attack Distribution on Advancing and Retreating Blades

weight that acts along the flight path in the equation for the angle of attack of the tip path plane. Figure 3.49 shows the force balance in a descent and in a climb. The equation for the equilibrium tip path plane angle of attack—using the same derivation as in forward flight—is:

$$\alpha_{TPP} = -\tan^{-1}\left[\frac{D_F + H_M + H_T + G.W. \tan \gamma}{G.W. - L_F}\right]$$

where

$$\gamma = \sin^{-1}\left(\frac{(R/C)/60}{V}\right)$$

with the rate of climb, R/C, in feet/minute and the forward speed, V, in feet/second. The trim conditions for the example helicopter at 115 knots ($\mu = .3$)

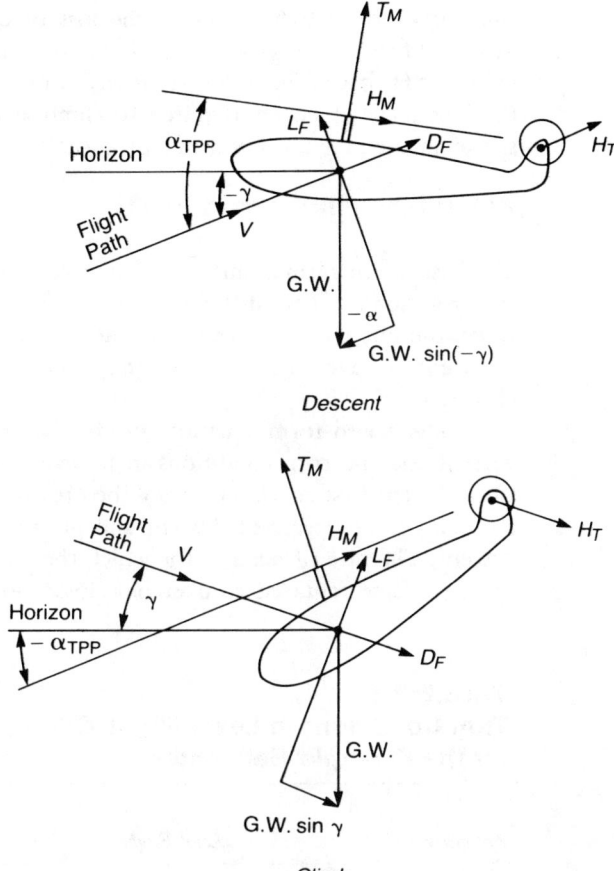

Descent

Climb

FIGURE 3.49   Forces Acting on the Helicopter in Descending and Climbing Flight

and at a rate of climb of 1,000 feet/minute have been calculated by the same iterative method used for level flight. The results are shown in Table 3.3 and the angle of attack distribution in Figure 3.48.

The difference in power required to climb compared to that required for level flight could have also been roughly estimated from the rate of change of potential energy:

$$\Delta h.p._{est} = \frac{R/C(GW)}{33,000}$$

For this case, the estimated difference in power is 606 h.p., whereas the calculated difference (including the tail rotor) is 668 h.p. Thus the example helicopter has a climb *efficiency* of 91%. Most of the loss of efficiency can be attributed to the increased fuselage negative lift while a smaller part is associated with the increased tail rotor H-force. This value of efficiency can be used to make rough estimates of the increment in power required to climb at other weights, rates of climb, and speeds in the neighborhood of 115 knots.

Autorotation in Forward Flight

The basic mechanism of autorotation in vertical flight was described in Chapter 2. Autorotation in forward flight is based on the same concept—that over the entire rotor, the integrated effect of the tilt of the lift vector at each blade element is sufficient to overcome the integrated effect of the drag at all of the blade elements.

The closed-form equations can be used to calculate the autorotative rate of descent and the trim conditions in forward flight. Two calculating methods are available; the first consists of using the previous method for climb and descents by assuming several rates of descent and plotting main rotor power versus rate of descent. The rate of descent for which the main rotor power is zero—or a small negative value required to overcome losses—is the autorotative rate of descent.

TABLE 3.3
Trim Conditions in Level Flight, Climb, and Autorotation at 115 Knots for the Example Helicopter

Parameter	Level Flight	Climb at 1,000'/min	Autorotation
α_F, deg	−6.1	−11.7	2.0
L_F, lb	−746	−1,228	−23
D_F, lb	904	976	882
T_M, lb	20,790	21,290	20,060
α_{TPP}, deg	−3.7	−9.2	4.3
λ'	−.0316	−.0607	.0105
a_0, deg	4.3	4.4	4.1
θ_0, deg	15.8	18.6	12.0
A_1, deg	−2.3	−2.4	−2.2
B_1, deg	4.9	6.0	3.5
H_M, lb	401	660	690
h.p.$_M$	1,097	1,760	−40
T_T, lb	755	1,211	−27
H_T, lb	36	77	12
h.p.$_T$	25	30	25
γ, deg	0	4.9	−8.8
R/C, ft/min	0	1,000	−1,803

This method is convenient if the general procedure has been programmed on a computer. For simpler calculations, a more direct method may be used. This involves using the equation for the torque coefficient:

$$C_Q/\sigma = \frac{c_d}{8}(1 + \mu^2) - \frac{a}{4}\left\{\frac{\lambda'}{1 + \frac{3}{2}\mu^2}\left[\left(1 - \frac{\mu^2}{2}\right)\left(\frac{2}{3}\theta_0 + \frac{\theta_1}{2}\right) + \lambda'\left(1 + \frac{1}{2}\mu^2\right)\right]\right.$$

$$\left. + \left(\frac{\mu^2}{1 + \frac{1}{2}\mu^2}\right)\left[\frac{a_0^2}{2}\left(\frac{1}{9} + \frac{\mu^2}{2}\right) + \frac{1}{3}\mu a_0\left(\frac{v_1}{\Omega R}\right) + \frac{1}{8}\left(\frac{v_1}{\Omega R}\right)^2\right]\right\}$$

and the equation for the collective pitch:

$$\theta_0 = \frac{\frac{4}{a}\left(1 + \frac{3}{2}\mu^2\right)C_T/\sigma - \frac{1}{2}\left(1 - \frac{3}{2}\mu^2 + \frac{3}{2}\mu^4\right)\theta_1 - \left(1 - \frac{\mu^2}{2}\right)\lambda'}{\frac{2}{3} - \frac{2}{3}\mu^2 + \frac{3}{2}\mu^4}$$

The first equation is set equal to the value of torque coefficient corresponding to the power required to drive the tail rotor at flat pitch and to overcome the transmission and accessory losses:

$$C_Q/\sigma = -\frac{[h.p._{\cdot 0_T} + h.p._{\cdot trans} + h.p._{\cdot acc}]550}{\rho A_b(\Omega R)^3}$$

The two equations can be solved simultaneously for the value of the inflow ratio, λ'. Just as in level flight, the rotor thrust is:

$$T = \sqrt{(G.W. - L_F)^2 + (D_F + H_M + H_T)^2}$$

The main rotor H-force is obtained from the thrust and torque coefficients:

$$H_M = \frac{\rho A_b(\Omega R)^2}{\mu}\left[\frac{c_d}{8}(1 + 3\mu^2) - \lambda' C_T/\sigma - C_Q/\sigma\right]$$

The tail rotor is essentially unloaded so that its H-force is simply:

$$H_T = \left[\rho A_b(\Omega R)^2 \frac{c_{d_{min}}}{4}\mu\right]_T$$

The fuselage lift and drag are a function of the angle of attack of the fuselage where—assuming that the rotor downwash at the fuselage is equal to the momentum value at the rotor:

$$\alpha_F = 57.3 \frac{\lambda'}{\mu} - i_s \text{ deg.}$$

The trim values of λ' and θ_0 can now be found with an iterative procedure similar to that used for level flight and climb. For the first step in the iteration, assume:

$$T = \text{G.W.}$$

$$H_M = H_{M_0}$$

Once the trim value of λ' has been found, the angle of descent is calculated from:

$$\sin \gamma = -\frac{1}{C_w/\sigma} \left\{ C_T/\sigma \sin \alpha_{\text{TPP}} + C_H/\sigma_M + C_H/\sigma_T \frac{[A_b(\Omega R)^2]_T}{[A_b(\Omega R)^2]_M} + \frac{\mu^2}{2} \frac{f}{A_b} \right\}$$

where

$$\alpha_{\text{TPP}} = \frac{\lambda'}{\mu} + \frac{\sigma}{2\mu^2} C_T/\sigma$$

and

$$C_w/\sigma = \frac{\text{G.W.}}{\rho A_b(\Omega R)^2}$$

and the rate of descent is:

$$R/D = -60 \, V \sin \gamma$$

The calculation has been made for the example helicopter assuming that transmission and accessory losses amounted to 15 h.p. The results of the calculations are listed in Table 3.3.

The rate of descent in autorotation could have been roughly estimated from the power required in level flight and the rate of change of potential energy required to produce this power:

$$R/D_{\text{est.}} = \frac{33,000 \text{ h.p.}}{\text{G.W.}}$$

The calculated power required for level flight (including 15 h.p. for transmission and accessory losses) is 1,137 h.p. The corresponding estimated rate of descent is

1,880 ft/min. Comparing this with the calculated rate of descent of 1,790 ft/min indicates an "efficiency" of 105%. The apparent gain is primarily due to the reduction in the negative lift on the fuselage and a reduction in the tail rotor H-force. This value can be used for rough estimates of the autorotative rates of descent of the example helicopter at other weights and forward speeds in the neighborhood of 115 knots. The angle of attack distribution in Figure 3.48 shows that the angle of attack is decreased on the retreating tip but increased inboard compared to level flight. For this reason, some helicopter aerodynamicists consider that the critical position on the retreating blade in autorotation is the station at which the tangential velocity is 40% of the tip speed, or:

$$V/R_{\text{crit.}} = 0.4 + \mu$$

If the local angle of attack exceeds the stall angle of the airfoil at this station, high drag can be expected to compromise the ability of the rotor to sustain autorotation.

ELIMINATING THE ASSUMPTIONS

Up to this point, the closed-form equations have been kept simple in order to illustrate the derivation and to produce results that can be readily used for quick calculations. The primary assumptions that have been used are:

- There is no tip loss or root cutout.
- The reverse flow region is ignored.
- A constant blade element drag coefficient is used.

When one or all of these assumptions are eliminated, the accuracy of the method is improved but the work in deriving the equations becomes greater, and the equations themselves become more unwieldly to use. In order to evaluate the effects of the assumptions, they will be examined one at a time.

Tip Loss and Root Cutout

In Chapter 1, a tip loss factor to account for the gradual falling off of the lift at the tip was used in the integration for hover thrust, and an approximate method was given for evaluating the tip loss factor as a function of the thrust coefficient and the number of blades. The same phenomenon exists in forward flight, but there is not yet any comparable method of evaluating the tip loss factor. Nevertheless, the need is recognized, so the closed-form equations are often derived using *BR* as the upper limit of integration, and calculations are based on the value of B obtained with the hover equation or with an arbitrary constant such as 0.97. When the tip loss factor and the root cutout, x_0, are incorporated in the derivation, the rotor equations become:

$$\lambda' = \mu a_{TPP} - \frac{v_1}{\Omega R}$$

$$\frac{v_1}{\Omega R} = C_T/\sigma \, \frac{\sigma}{2\mu B^2}$$

$$C_T/\sigma = \frac{a}{4}\left\{\left[\frac{2}{3}(B^3 - x_0^3) + \mu^2(B - x_0)\right]\theta_0 + \left[\frac{(B^4 - x_0^4)}{2} + \frac{\mu^2}{2}(B^2 - x_0^2)\right]\theta_1 + \lambda'(B^2 - x_0^2)\right.$$

$$\left. + \lambda'(B^2 - x_0^2) - \mu(B_1 + a_{1_1})(B^2 - x_0^2)\right\}$$

$$B_1 + a_{1_1} = \frac{\mu\left[\frac{8}{3}(B^3 - x_0^3)\theta_0 + 2(B^4 - x_0^4)\theta_1 + 2(B^2 - x_0^2)\lambda'\right]}{(B^4 - x_0^4) + \frac{3}{2}(B^2 - x_0^2)\mu^2}$$

$$\theta_0 = \frac{\frac{4}{a}\left[(B^4 - x_0^4) + \frac{3}{2}(B^2 - x_0^2)\mu^2\right]C_T/\sigma - \frac{1}{2}\left[(B^4 - x_0^4)^2 - \frac{3}{2}\mu^2(B^4 - x_0^4)(B^2 - x_0^2) + \frac{3}{2}\mu^4(B^2 - x_0^2)^2\right]\theta_1 - \left[(B^2 - x_0^2)\left(B^4 - x_0^4 - \frac{\mu^2}{2}\right)\right]\lambda'}{\frac{2}{3}(B^3 - x_0^3)(B^4 - x_0^4) - \frac{5}{3}\mu^2(B^3 - x_0^3)(B^2 - x_0^2) + \mu^2(B - x_0)(B^4 - x_0^4) + \frac{3}{2}\mu^4(B - x_0)(B^2 - x_0^2)}$$

$$a_0 = \frac{2}{3}\gamma\,\frac{C_T/\sigma}{a(B^2 - x_0^2)} - \frac{\frac{3}{2}gR}{(\Omega R)^2}$$

$$A_1 - b_{1_1} = -\left[\frac{\frac{4}{3}\mu a_0(B^3 - x_0^3) + \frac{v_1}{\Omega R}(B^4 - x_0^4)}{(B^4 - x_0^4) + \frac{1}{2}\mu^2(B^2 - x_0^2)}\right]$$

$$C_Q/\sigma = \frac{c_d}{8}(1 + \mu^2) - \frac{a}{4}\left\{\frac{\lambda'}{(B^4 - x_0^4) + \frac{3}{2}\mu^2(B^2 - x_0^2)}\left\{\left[\frac{2}{3}\theta_0(B^3 - x_0^3) + \frac{\theta_1}{2}(B^4 - x_0^4)\right]\left[B^4 - x_0^4 - \frac{\mu^2}{2}(B^2 - x_0^2)\right]\right.\right.$$

$$\left. + \lambda'(B^2 - x_0^2)\left[(B^4 - x_0^4) + \frac{1}{2}(B^2 - x_0^2)\mu^2\right]\right\}$$

$$+ \frac{\mu^2}{(B^4 - x_0^4) + \frac{\mu^2}{2}(B^2 - x_0^2)}\left\{a_0^2\left[-\frac{4}{9}(B^3 - x_0^3)^2 + \frac{(B^2 - x_0^2)(B^4 - x_0^4)}{2} + \frac{(B^2 - x_0^2)^2\mu^2}{4}\right]\right.$$

$$\left.\left. + \frac{1}{3}a_0\mu\,\frac{v_1}{\Omega R}(B^3 - x_0^3)(B^2 - x_0^2) + \frac{1}{8}\left(\frac{v_1}{\Omega R}\right)^2(B^4 - x_0^4)(B^4 - x_0^4)\right\}\right\}$$

$$C_H/\sigma = \frac{c_d}{4}\mu - \frac{a}{4}\left[\frac{\lambda'\mu}{(B^4 - x_0^4) + \frac{3}{2}\mu^2(B^2 - x_0^2)}\left\{\theta_0\left[(B - x_0)(B^4 - x_0^4) + \frac{3}{2}\mu^2(B - x_0)(B^2 - x_0^2)\right.\right.\right.$$

$$\left.- \frac{4}{3}(B^2 - x_0^2)(B^3 - x_0^3)\right] + \theta_1\left[-\frac{(B^2 - x_0^2)}{2}(B^4 - x_0^4) + \frac{3}{4}\mu^2(B^2 - x_0^2)^2\right] - \lambda^1(B^2 - x_0^2)^2\Big\}$$

$$- \frac{\mu}{(B^4 - x_0^4) + \frac{\mu^2}{2}(B^2 - x_0^2)}\left\{a_0^2\left[-\frac{4}{9}(B^3 - x_0^3)^2 + \frac{(B^2 - x_0^2)(B^4 - x_0^4)}{2} + \frac{(B^2 - x_0^2)^2\mu^2}{4}\right]\right.$$

$$\left.\left. + \frac{1}{3}a_0\mu\frac{v_1}{\Omega R}(B^3 - x_0^3)(B^2 - x_0^2) + \frac{1}{8}\left(\frac{v_1}{\Omega R}\right)^2(B^2 - x_0^2)(B^4 - x_0^4)\right\}\right] + C_T/\sigma\, a_{1_s}$$

or

$$C_H/\sigma = \frac{1}{\mu}\left[\frac{c_d}{8}(1 + 3\mu^2) - (\lambda' - \mu a_{1_s})\frac{C_T}{\sigma} - C_Q/\sigma\right]$$

These equations are more complicated looking than the corresponding equations without tip and root losses; but since B and x_0 are constants, the resulting numerical equations, which are actually used for calculations, are no more complicated.

Reverse-Flow Region

The integration process used so far has ignored the reverse-flow region, shown in Figure 3.14, by integrating around the azimuth from zero to 2π with no regard for the fact that in the reverse-flow region this method assigns the wrong sign to lift forces, as shown in Figure 3.50, and thus results in a calculated thrust value that is too high. For small tip speed ratios (less than about .25), the error is insignificant since both the size and the dynamic pressures in the reverse-flow region are small. At higher tip speed ratios, however, the error becomes more and more significant. The deficiency can be corrected during the integration by dividing the rotor into three regions, as shown in Figure 3.51, each with its separate limits of integration. Using this method, the thrust equation becomes:

$$T = \frac{bR}{2\pi}\left\{\int_0^\pi\int_0^1\frac{\Delta L}{\Delta r}d\,\frac{r}{R}d\psi + \int_\pi^{2\pi}\int_{-\mu\sin\psi}^1\frac{\Delta L}{\Delta r}d\,\frac{r}{R}d\psi - \int_\pi^{2\pi}\int_0^{-\mu\sin\psi}\frac{\Delta L}{\Delta r}d\,\frac{r}{R}d\psi\right\}$$

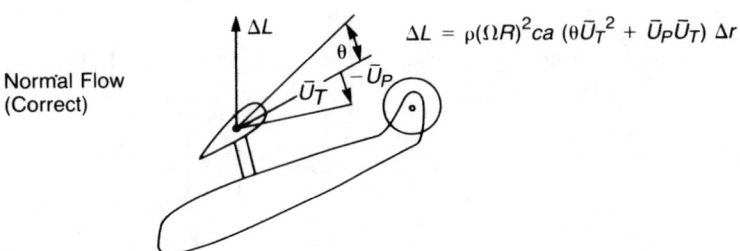

Normal Flow (Correct)

$$\Delta L = \rho(\Omega R)^2 ca\,(\theta\bar{U}_T^2 + \bar{U}_P\bar{U}_T)\,\Delta r$$

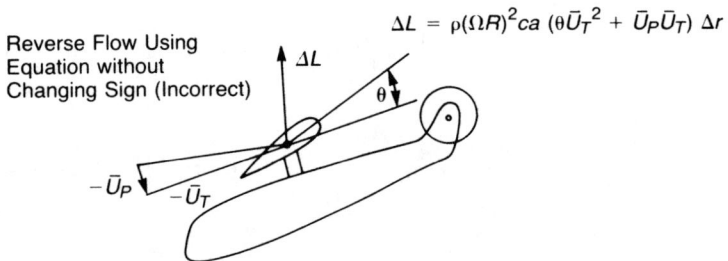

Reverse Flow Using Equation without Changing Sign (Incorrect)

$$\Delta L = \rho(\Omega R)^2 ca\,(\theta\bar{U}_T^2 + \bar{U}_P\bar{U}_T)\,\Delta r$$

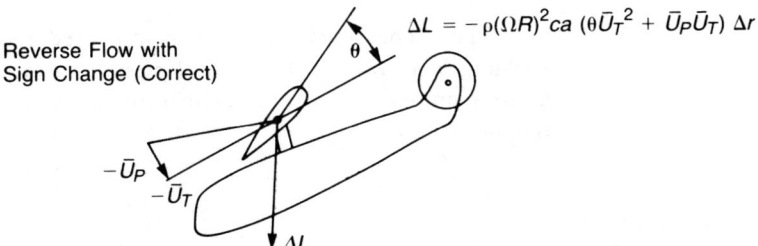

Reverse Flow with Sign Change (Correct)

$$\Delta L = -\rho(\Omega R)^2 ca\,(\theta\bar{U}_T^2 + \bar{U}_P\bar{U}_T)\,\Delta r$$

FIGURE 3.50   Illustration of the Sign Change Required in the Reverse Flow Region

The resulting equation for C_T/σ is:

$$C_T/\sigma = \frac{a}{4\pi}\left[\int_0^{2\pi}\int_0^1 (\bar{U}_T^2\theta + \bar{U}_T\bar{U}_P)d\,\frac{r}{R}\,d\psi - 2\int_\pi^{2\pi}\int_0^{-\mu\sin\psi} (\bar{U}_T^2\theta + \bar{U}_T\bar{U}_P)d\,\frac{r}{R}\,d\psi\right]$$

or

$$C_T/\sigma = \frac{a}{4}\left\{\left(\frac{2}{3} + \mu^2 - \frac{8\mu^3}{9\pi}\right)\theta_0 + \left(1 + \mu^2 - \frac{\mu^4}{8}\right)\frac{\theta_1}{2} + \left(1 + \frac{\mu^2}{2}\right)\lambda' - \left(\mu + \frac{\mu^3}{4}\right)(B_1 + a_{1_s})\right\}$$

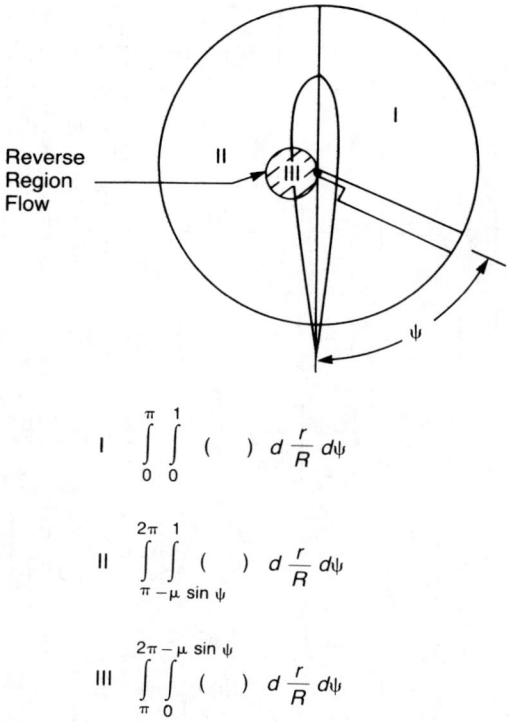

$$\text{I} \quad \int\limits_{0}^{\pi} \int\limits_{0}^{1} \; (\quad) \; d\frac{r}{R} \, d\psi$$

$$\text{II} \quad \int\limits_{\pi}^{2\pi} \int\limits_{-\mu \sin \psi}^{1} \; (\quad) \; d\frac{r}{R} \, d\psi$$

$$\text{III} \quad \int\limits_{\pi}^{2\pi - \mu \sin \psi} \int\limits_{0}^{} \; (\quad) \; d\frac{r}{R} \, d\psi$$

FIGURE 3.51   Regions of Integration

and

$$B_1 + a_{1_s} = \frac{\mu \left[\left(\dfrac{8}{3} + \dfrac{32}{45} \dfrac{\mu^3}{\pi} \right) \theta_0 + \left(2 + \dfrac{\mu^4}{12} \right) \theta_1 + \left(2 - \dfrac{\mu^2}{2} \right) \lambda' \right]}{1 + \dfrac{3}{2} \mu^2 - \dfrac{5\mu^4}{24}}$$

$$\theta_0 = \frac{\dfrac{4}{a} \left(1 + \dfrac{3}{2} \mu^2 - \dfrac{5}{24} \mu^4 \right) C_T/\sigma - \left(1 - \dfrac{3}{2} \mu^2 + \dfrac{\mu^4}{6} - \dfrac{9}{16} \mu^6 - \dfrac{3}{192} \mu^8 \right) \dfrac{\theta_1}{2} - \left(1 + \dfrac{13}{24} \mu^4 + \dfrac{\mu^6}{48} \right) \lambda'}{\dfrac{2}{3} - \dfrac{2}{3} \mu^2 - \dfrac{8}{9\pi} \mu^3 + \dfrac{25}{36} \mu^4 - \dfrac{92}{45\pi} \mu^5 - \dfrac{5}{24} \mu^6 + \dfrac{\mu^7}{135\pi}}$$

$$A_1 - b_{1_s} = -\frac{\left[a_0\left(\frac{4}{3}\mu + \frac{16}{45\pi}\mu^4\right) + \frac{v_1}{\Omega R}\left(1 + \frac{\mu^4}{24}\right)\right]}{1 + \frac{\mu^2}{2} - \frac{\mu^4}{24}}$$

$$\begin{aligned}
C_H/\sigma = {} & \frac{c_d}{4}\left(\mu + \frac{\mu^3}{4}\right) - \frac{a}{4}\left\{\frac{\mu\lambda'}{1 + \frac{3}{2}\mu^2 - \frac{5}{24}\mu^4}\left[\frac{\theta_0}{3}\left(-1 - \frac{4}{\pi}\mu + \frac{9\mu^2}{2} - \frac{106}{15\pi}\mu^3 - \frac{5}{8}\mu^4 + \frac{\mu^5}{30\pi}\right)\right.\right. \\
& \left.+ \frac{\theta_1}{2}\left(-1 - \frac{\mu^2}{4} - \frac{2}{3}\mu^4 - \frac{\mu^6}{96}\right) + \lambda'\left(\mu^2 - \frac{\mu^4}{48}\right)\right] \\
& + \frac{-\mu}{1 + \frac{\mu^2}{2} - \frac{\mu^4}{4}}\left\{a_0^2\left[\frac{1}{2} + \frac{\mu^2}{9} - \frac{64}{135\pi}\mu^3 - \frac{\mu^4}{6} - \left(\frac{128}{45^2\pi^2} - \frac{1}{96}\right)\mu^6\right]\right. \\
& \left.\left.+ a_0\frac{v_1}{\Omega R}\frac{\mu}{3}\left[1 - \frac{8\mu}{5\pi} - \frac{\mu^2}{6} - \frac{8}{15\pi}\mu^3 + \frac{\mu^5}{95\pi}\right] + \frac{1}{8}\left(\frac{v_1}{\Omega R}\right)^2\left[1 - \frac{\mu^2}{2} - \frac{\mu^4}{8} + \frac{\mu^6}{144}\right]\right\}\right\}
\end{aligned}$$

$$\begin{aligned}
C_Q/\sigma = {} & \frac{c_d}{8}\left[1 + \mu^2 - \frac{\mu^4}{8}\right] - \frac{a}{4}\left\{\frac{\lambda'}{1 + \frac{3}{2}\mu^2 - \frac{5}{24}\mu^4}\left[\frac{\theta_0}{3}\left(2 - \mu^2 + \frac{4}{3\pi}\mu^3 + \frac{7}{12}\mu^4 + \frac{14}{15\pi}\mu^5 - \frac{\mu^7}{135\pi}\right)\right.\right. \\
& \left.+ \frac{\theta_1}{2}\left(1 - \frac{\mu^2}{2} + \frac{5}{12}\mu^4 + \frac{5}{48}\mu^6 - \frac{1}{192}\mu^8\right) + \lambda'\left(1 - \frac{11}{24}\mu^4 + \frac{\mu^6}{24}\right)\right] \\
& + \frac{\mu^2}{1 + \frac{\mu^2}{2} - \frac{\mu^4}{24}}\left\{a_0^2\left[\frac{1}{2} + \frac{\mu^2}{9} - \frac{64}{135\pi}\mu^3 - \frac{\mu^4}{6} - \left(\frac{128}{45^2\pi^2} - \frac{1}{96}\right)\mu^6\right]\right. \\
& \left.\left.+ a_0\frac{v_1}{\Omega R}\frac{\mu}{3}\left[1 - \frac{8\mu}{5\pi} - \frac{\mu^2}{6} - \frac{8}{15\pi}\mu^3 + \frac{\mu^5}{45\pi}\right] + \frac{1}{8}\left(\frac{v_1}{\Omega R}\right)^2\left[1 - \frac{\mu^2}{2} - \frac{\mu^4}{8} + \frac{\mu^6}{144}\right]\right\}\right\}
\end{aligned}$$

or

$$C_Q/\sigma = \frac{c_d}{8}\left(1 + 3\mu^2 + \frac{3\mu^4}{8}\right) - (\lambda' - \mu a_{1_s})C_T/\sigma - \mu C_H/\sigma$$

Three-Term Drag Polar

Another refinement to the closed-form equations involves the use of an expression for the coefficient of drag as a function of the local angle of attack rather than of a single average value. The classical form, which was discussed in Chapter 1, is the three-term power series—usually referred to as a *three-term drag polar*:

$$c_d = c_{d_0} + c_{d_1}\alpha + c_{d_2}\alpha^2$$

where α is in radians and the constants are picked to make the best fit with two-dimensional airfoil data. The profile torque and H-force coefficient equations without reverse-flow considerations but with the three-term drag polar are:

$$
C_Q/\sigma_0 = \frac{1}{4}\left\{ \frac{c_{d_0}}{2}(1 + \mu^2) + \frac{c_{d_1}}{1 + \dfrac{3}{2}\mu^2}\left[\theta_0\left(\frac{1}{2} - \frac{19}{36}\mu^2 + \frac{3}{4}\mu^4\right) \right.\right.
$$

$$
\left. + \theta_1\left(\frac{2}{5} - \frac{2}{5}\mu^2 + \frac{\mu^4}{2}\right) + \lambda'\left(\frac{2}{3} - \frac{\mu^2}{3}\right)\right] + \frac{c_{d_2}}{\left(1 + \dfrac{3}{2}\mu^2\right)^2}\left[\theta_0^2\left(\frac{1}{2}\right.\right.
$$

$$
\left. + \frac{2}{9}\mu^2 - \frac{\mu^4}{24} + \frac{9}{8}\mu^6\right) + \theta_1^2\left(\frac{1}{3} + \frac{\mu^2}{4} + \frac{9}{16}\mu^6\right) + \lambda'^2\left(1 + 2\mu^2 + \frac{3}{4}\mu^4\right)
$$

$$
+ \theta_0\theta_1\left(\frac{4}{5} + \frac{2}{5}\mu^2 - \frac{\mu^4}{5} + \frac{3}{2}\mu^6\right) + \theta_0\lambda'\left(\frac{4}{3} + \frac{4}{3}\mu^2 - \mu^4\right)
$$

$$
\left. + \theta_1\lambda'\left(1 + \mu^2 - \frac{3}{4}\mu^4\right)\right] + \frac{c_{d_2}\mu^2}{\left(1 + \dfrac{1}{2}\mu^2\right)^2}\left[a_0^2\left(\frac{1}{18} + \frac{\mu^2}{6} - \frac{\mu^4}{8}\right)\right.
$$

$$
\left.\left. + \left(\frac{v_1}{\Omega R}\right)^2\left(\frac{1}{8} + \frac{\mu^2}{16}\right) + a_0\frac{v_1}{\Omega R}\left(\frac{\mu}{3} + \frac{\mu^3}{6}\right)\right]\right\}
$$

$$
C_H/\sigma_0 = \frac{\mu}{4}\left\{ c_{d_0} + \frac{c_{d_1}}{1 + \dfrac{3}{2}\mu^2}\left[\theta_0\left(\frac{1}{9} - \frac{\mu^2}{2}\right) - \frac{\theta_1}{2}\mu^2 + \frac{\lambda'}{3}\right]\right.
$$

$$
+ \frac{c_{d_2}}{\left(1 + \dfrac{3}{2}\mu^2\right)^2}\left[\theta_0^2\left(-\frac{7}{9} + \frac{5}{3}\mu^2 - \frac{15}{4}\mu^4\right) + \theta_1^2\left(-\frac{1}{2} + \frac{3}{2}\mu^2 - \frac{9}{8}\mu^4\right)\right.
$$

$$
- 2\lambda'^2 + \theta_0\theta_1\left(-\frac{4}{3} + 3\mu^2 - \frac{9}{2}\mu^4\right) + \theta_0\lambda'(-2 + 5\mu^2)
$$

$$
\left. + \theta_1\lambda'(-2 + 3\mu^2)\right] - \frac{c_{d_2}\mu^2}{\left(1 + \dfrac{1}{2}\mu^2\right)^2}\left[a_0^2\left(\frac{2}{9}\mu + \frac{\mu^3}{3}\right) + \frac{\mu}{8}\left(\frac{v_1}{\Omega R}\right)^2\right.
$$

$$
\left.\left. + a_0\frac{v_1}{\Omega R}\left(\frac{1}{6} + \frac{5}{12}\mu^2\right)\right]\right\}
$$

Airfoil data are usually better fit with a power series with more than three terms, but the use of any power above the second leads to difficulties in the integration which do not justify the increased accuracy that might result. The three constants in the drag polar should be chosen to fit the airfoil data at a representative Mach number corresponding to 75% of the tip speed and the highest angle of attack on the retreating side. A discussion of the problem of fitting a three-term polar to test data was given in Chapter 1, and Figure 1.37 shows three polars that approximate the NACA 0012 data.

Trim Conditions Based on Elimination of Assumptions

The trim conditions in level flight have been calculated while eliminating the assumptions one at a time. The results are shown in Table 3.4 for tip speed ratios of 0.30 and 0.45. (In reality, the trim conditions at the tip speed ratio of 0.45 are unobtainable for the example helicopter because of excessive blade stall, as will later be shown; but as a basis of comparing the effects of the various assumptions, it is an illustrative case.) It may be seen from Table 3.4 that the elimination of the assumption that there are no root or tip losses makes relatively little difference in the trim conditions at the low speed but makes a significant difference at high speeds. This is due to the large reduction in the H-force and the consequent reduction in the nose-down attitude of the rotor. The reduction in H-force can be traced to the variation in the calculated induced drag in the tip annulus as a function of azimuth position. Because of the higher dynamic pressure on the advancing tip, the induced drag is higher on the advancing side, where it contributes to positive H-force, than it is on the retreating side, where it contributes to negative H-force. Thus the elimination of the induced drag in this annulus by considering it lost due to tip effects results in a net decrease of the calculated H-force.

The effect of treating the reverse flow region correctly is shown in Table 3.4. At a tip speed ratio of 0.3, the main rotor power is increased approximately 6%, but at a tip speed ratio of 0.45 it is decreased about 10% compared to the corresponding calculation with the reverse-flow region treated incorrectly. At the lower speed, the increase in rotor thrust in the normal-flow region requires higher power. At the higher speed, the reduction in the H-force due to the reversal of the lift vectors in the reverse-flow region, as shown in Figure 3.50, has resulted in a large reduction in the rotor angle and thus has put both the rotor and the fuselage into a more favorable condition with respect to overall power.

The use of a three-term drag polar in place of an average value has an effect similar to those caused by eliminating the first two assumptions—relatively little effect at the low speed, but a dramatic decrease in rotor power at the high speed. Again, this reduction can be traced to the decrease in H-force and the more favorable rotor and fuselage attitude that results. Paradoxically, this is due to the increase in drag on the retreating side, where the contribution to H-force is negative.

The logical next step would be to present all the closed-form rotor equations with the assumptions eliminated. This step, however, has been bypassed in favor of developing the charts at the end of this chapter, which are based on a digital computer program. This program, in using rigorous equations for the

TABLE 3.4
Effect of Eliminating Assumptions on Trim Conditions for Example Helicopter

Assumption	μ = .30 (115K)					μ = .45 (174K)				
No Tip or Root Loss	Held	Eliminated	Held	Held		Held	Eliminated	Held	Held	
No Reverse Flow	Held	Held	Eliminated	Held		Held	Held	Eliminated	Held	
Average C_d	Held	Held	Held	Eliminated	From Rotor Charts	Held	Held	Held	Eliminated	From Rotor Charts
α_F, deg	-6.0	-6.0	-5.7	-5.5	-4.4	-13.0	-9.1	-6.1	-8.5	
L_F, lb	-738	-738	-705	-690	-588	-3,051	-2,288	-1,708	-2,187	
D_F, lb	904	904	899	897	886	2,258	2,105	2,034	2,085	
T_M, lb	20,780	20,780	20,720	20,725	20,606	23,537	22,099	21,838	22,370	
α_{TPP}, deg	-3.7	-3.3	-3.4	-3.1	-2.1	-11.8	-7.9	-6.3	-7.4	
λ'	-.0315	-.0315	-.0298	-.0286	-.023	-.1022	-.0713	-.0580	-.0666	
a_0, deg	4.3	4.8	4.2	4.2	4.2	4.8	4.6	4.6	4.6	
θ_0, deg	15.8	16.8	16.9	15.6	17.2	23.5	22.7	22.5	19.9	
A_1, deg	-2.3	-2.7	-2.3	-2.3	-2.1	-3.1	-3.1	-3.2	-3.0	
B_1, deg	4.9	6.3	5.7	4.8	7.8	10.4	11.6	11.1	8.7	
H_M, lb	396	253	188	200	-145	2,333	812	139	657	
h.p.$_M$	1,080	1,105	1,146	1,086	1,368	2,193	2,046	2,002	1,967	
T_T, lb	745	760	790	748	942	1,511	1,409	1,379	1,355	
H_T, lb	35	42	40	30	68	158	158	129	99	
h.p.$_T$	27	17	27	14	19	-30	-20	-6	36	

Main and tail rotors heavily stalled; iterative procedure does not converge.

forces at the blade elements, automatically eliminates the assumptions while also including the effects of compressibility and unsteady aerodynamics, which are difficult to include in any closed-form analysis.

NUMERICAL INTEGRATION METHODS

Numerical integration methods use the same basic equations for the forces on a blade element as do closed integration methods, but because of the nature of the computer, which doesn't mind boring, repetitive computations, fewer assumptions need be used. A primary advantage of these methods is their ability to use two-dimensional airfoil data as a function of angle of attack and Mach number throughout the complete ranges of these two parameters as they exist on the rotor. In addition, the methods can be made—with ever-increasing complexity—to handle unsteady aerodynamics, yawed flow characteristics, complicated induced velocity patterns, blade flexibility, and lifting surface (rather than lifting line) aerodynamics.

A requirement of any program is that it have the capability to find the trim condition in which the blade inertial, centrifugal, and aerodynamic forces and moments are all in equilibrium, as they must be for a rotor in steady flight. There are two starting approaches for a hinged rotor. In the most common, the computer calculates the flapping by following one blade around the azimuth while continually evaluating its flapping acceleration, velocity, and displacement. For a reasonable set of shaft angles and control settings, the calculations will converge on a condition where the flapping repeats itself from one revolution to another. The other blades in the rotor are assumed to fly in the same path. This method is sometimes referred to as a *flapdoodle*. If it is required to trim the first harmonic flapping to zero with respect to the shaft, the computer changes cyclic pitch in a logical way to do this. Once this condition is achieved, the flapping hinges could be locked. In this situation, the longitudinal and lateral hub moments would be zero, since the tip path plane is perpendicular to the shaft.

The second method—and the one that will be used here—obtains the same results by starting first with locked hinges. On this rotor, changes in cyclic pitch result in aerodynamic pitching and rolling moments. The computer searches for the cyclic pitch that reduces these moments to zero, thus giving the same results as those obtained with the more common method. Because of the equivalence of flapping and feathering demonstrated in the previous analyses, the trim values of cyclic pitch can be reidentified as $(B_1 + a_{1_s})$ and $(A_1 - b_{1_s})$ and thus applicable to flapping rotors in which the tip path plane is not necessarily perpendicular to the shaft.

The second method gives the same results as the first except for the loss of flapping harmonics above the first. Studies have shown that these have little effect on rotor performance, so their loss is not serious for our purposes. Comparison of the two methods indicates that the second is more economical in computer time and will often converge to a trim solution at extreme conditions, where the flapdoodle procedure will diverge because of second-order effects.

The method outlined in the following paragraphs is suitable for a medium-sized computer and is based on the following limitations and assumptions:

- Obtaining performance and trim conditions is the primary objective.
- Two-dimensional airfoil lift and drag characteristics are available.
- The blades do not bend or twist elastically.
- The induced velocity distribution is given by the equation:

$$v_L = v_1 \left(1 + \frac{r}{R} \sin \psi \right)$$

Basic Equations

For this analysis, the procedure used in deriving the closed-form equations will be modified somewhat by resolving both the lift and drag forces into a normal—or thrust—force where before only lift was assumed to contribute to thrust. Figure 3.52 shows the forces on a blade element.

The normal force coefficient is:

$$c_N = c_l \frac{\bar{U}_T}{\bar{U}_B} + c_d \frac{\bar{U}_P}{\bar{U}_B}$$

Where the velocities have been nondimensionalized by dividing by the tip speed:

$$\bar{U}_T = \frac{U_T}{\Omega R}$$

$$\bar{U}_P = \frac{U_P}{\Omega R}$$

and

$$\bar{U}_B = \sqrt{\bar{U}_T^2 + \bar{U}_P^2}$$

The nondimensionalized thrust loading is:

$$\frac{d\,C_T/\sigma}{d\,r/R} = \frac{\bar{U}_B^2}{2}\, c_N$$

The contribution of the entire blade is:

$$\Delta C_T/\sigma = \int_{x_0}^{B} \frac{d\,C_T/\sigma}{d\,r/R}\, dr/R$$

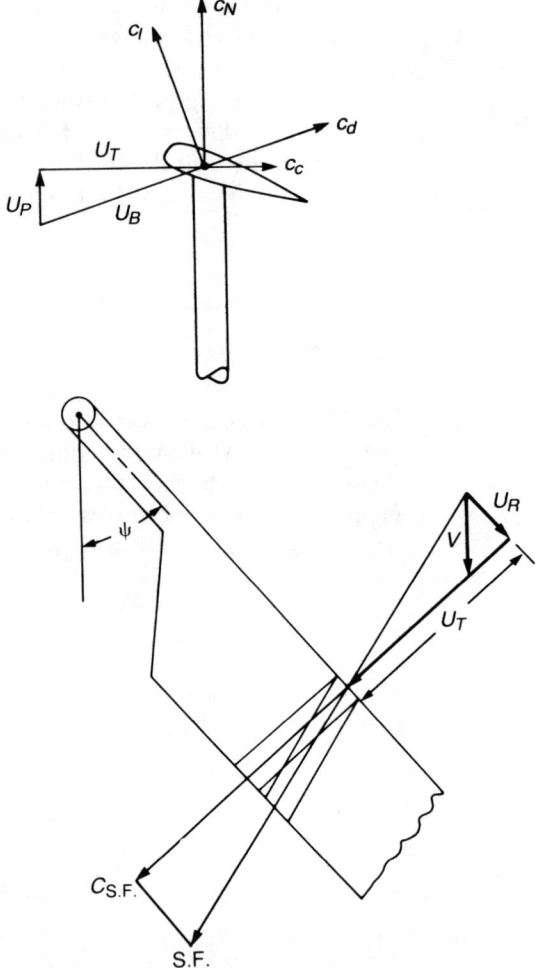

FIGURE 3.52 Resolution of Forces Acting on a Blade Element

The integration can be performed by any of several methods suitable for numerical analysis. One of the simplest is Simpson's rule. If the blade is evenly divided into ten elements from the center of rotation to the tip, the integration uses the eleven points calculated at the margin of each element:

$$I = \frac{1}{30}[y_0 + 4y_1 + 2y_2 + 4y_3 + 2y_4 + 4y_5 + 2y_6 + 4y_7 + 2y_8 + 4y_9 + y_{10}]$$

where in this case y is the calculated value of $dC_T/\sigma/dr/R$. The root and tip losses can be handled as trapezoidal corrections to the total integral:

$$\Delta C_T/\sigma = I - \frac{x_0(y_0 + y_{x_0})}{2} - (1 - B)\frac{(y_B + y_{10})}{2}$$

The total thrust coefficient is the average of C_T/σ evaluated at N equally spaced azimuth positions around the rotor:

$$C_T/\sigma = \frac{1}{N} \sum_{n=1}^{N} \Delta C_T/\sigma_n$$

The number of azimuth stations required is a matter of compromise between accuracy and computer time which will have to be made to satisfy the particular needs of the analyzer. (Our experience indicates that for most rotor conditions in which retreating blade stall is not a factor, eight azimuth stations are sufficient for engineering accuracy.)

The equivalence of flapping and feathering allows performance calculations to be based on a rigid rotor whose tip path plane is perpendicular to the shaft and whose pitching and rolling moments are trimmed out with cyclic pitch. These calculations will then apply to any condition in which the thrust and the angle of attack of the tip path plane are the same. The normal force coefficients are used to produce the pitching and rolling moment coefficient loadings:

$$\frac{d\,C_M/\sigma}{d\,r/R} = -\frac{\bar{U}_B^2}{2} \frac{r}{R} \cos \psi \; c_N$$

and

$$\frac{d\,C_R/\sigma}{d\,r/R} = -\frac{\bar{U}_B^2}{2} \frac{r}{R} \sin \psi \; c_N$$

The integrations for the entire moments are carried out with the same method used for thrust.

The chordwise force that contributes to torque and H-force is made up of three components: those due to pressure drag, skin friction, and tilt of the lift vector. In the closed-form derivation, pressure drag and skin friction were treated as one, being assumed to be governed by the component of velocity perpendicular to the leading edge. This is strictly only true for pressure drag. Skin friction is a function of the magnitude and direction of the total local velocity as shown in Figure 3.52. The skin friction force, S.F., is:

$$\text{S.F.} = \frac{\rho}{2} c \Delta r (U_T^2 + U_R^2) c_f$$

where

$$\bar{U}_R = \mu \cos \psi$$

The chordwise component of S.F. is:

$$C_{\text{S.F.}} = \text{S.F.} \; \frac{U_T}{\sqrt{U_T^2 + U_R^2}} = \frac{\rho}{2} c \Delta r U_T \sqrt{U_T^2 + U_R^2} \; c_f$$

The chordwise coefficient is:

$$c_{c_{\text{S.F.}}} = \frac{C_{\text{S.F.}}}{\dfrac{\rho}{2} c \Delta r U_B^2} = \frac{U_T \sqrt{U_T^2 + U_R^2}}{U_B^2} c_f$$

The total chordwise coefficient due to both pressure drag and skin friction is:

$$c_{c_0} = c_{d_p} \frac{\bar{U}_T}{\bar{U}_B} + c_f \frac{\bar{U}_T \sqrt{\bar{U}_T^2 + \bar{U}_R^2}}{\bar{U}_B^2}$$

For most cases it may be assumed that:

$$c_f = .006$$

so that

$$c_{d_p} = c_d - .006$$

The component of the chordwise force due to the tilt of the lift vector is:

$$c_{c_{\text{ind}}} = - c_l \frac{\bar{U}_P}{\bar{U}_B}$$

and the total is:

$$c_c = c_{c_0} + c_{c_{\text{ind}}}$$

The contribution to the torque loading is:

$$\frac{dC_Q/\sigma}{d\,r/R} = \frac{\bar{U}_B^2}{2} \frac{r}{R} c_c$$

The H-force consists also of pressure drag, skin friction, and tilt of the lift vector. The second two produce only chordwise forces, but skin friction can also produce a spanwise force. The total equation for the increment of H-force is:

$$\frac{dC_H/\sigma}{d\,r/R} = \frac{\bar{U}_B^2}{2} \left[c_c \sin \psi + c_f \frac{\mu}{\bar{U}_T} \cos^2 \psi \right]$$

The integration of these terms is performed as with the thrust coefficient except that no root cutout or tip loss is applied to the portion of the chordwise force produced by pressure drag and skin friction.

The lift and drag coefficients are functions of the local angle of attack perpendicular to the leading edge:

$$\alpha = \theta + \tan^{-1} \frac{\bar{U}_P}{\bar{U}_T}$$

$$\theta = \theta_0 + \frac{r}{R}\,\theta_1 - A_1\cos\psi - B_1\sin\psi$$

$$\bar{U}_T = \frac{r}{R} + \mu\sin\psi$$

$$\bar{U}_P = \lambda' - \frac{v_1}{\Omega R}\,\frac{r}{R}\cos\psi - \mu a_0\cos\psi$$

Where the closed-form equations can be used with sufficient accuracy for a_0 and for $v_1/\Omega R$:

$$a_0 = \frac{2}{3}\,\gamma\,\frac{C_T/\sigma}{a} - \frac{\frac{3}{2}gR}{(\Omega R)^2}$$

$$v_1/\Omega R = C_T/\sigma\,\frac{\sigma}{2\mu}$$

(The value of C_T/σ is initially either known or can be estimated from the closed-form equations. The value will be updated after the first full cycle of calculations.)

For cases in which the rotor has prescribed pitching or rolling velocities, the equation for $\bar{U}_p$ is modified by adding:

$$\Delta\bar{U}_p = \frac{r}{R}\,\frac{\dot{\Theta}}{\Omega}\cos\psi + \frac{r}{R}\,\frac{\dot{\Phi}}{\Omega}\sin\psi$$

In these cases, trim is the condition in which the rotor aerodynamic pitching and rolling moments are equal to the rotor gyroscopic moments instead of to zero as in steady flight conditions. The gyroscopic pitching moment reacted on the shaft of a round flat disc that has a rolling velocity, $\dot{\Phi}$, is:

$$M_{\text{gyro}} = J\dot{\Phi}\,\Omega$$

where J is the polar moment of inertia. For a single blade instead of a disc, the equation becomes:

$$M_{\text{gyro}} = 2I_b\dot{\Phi}\,\Omega$$

and

$$C_M/\sigma_{\text{gyro}} = \frac{2I_b\dot{\Phi}\Omega}{\rho c R^2(\Omega R)^2}$$

The blade flapping moment of inertia can be related to the Lock number:

$$I_b = \frac{c\rho a R^4}{\gamma}$$

so

$$C_M/\sigma_{\text{gyro}} = \frac{2a\,\dot{\Phi}/\Omega}{\gamma}$$

Similarly

$$C_R/\sigma_{\text{gyro}} = \frac{2a\,\dot{\Theta}/\Omega}{\gamma}$$

The rotor is in trim when:

$$C_M/\sigma_{\text{aero}} + C_M/\sigma_{\text{gyro}} = 0$$

$$C_R/\sigma_{\text{aero}} + C_R/\sigma_{\text{gyro}} = 0$$

Since the sign and magnitude of α can vary widely over the rotor disc, it is necessary that the airfoil characteristics be defined throughout the entire range. Figure 3.53 shows the relationships between the angle of attack and the ratio of U_P to U_T. Also shown are the lift and drag coefficients of an NACA 0012 airfoil measured in a wind tunnel and reported in reference 3.32. When the airfoil characteristics are used as a function of α with the signs as shown, the normal and chordwise force equations will automatically resolve the lift and drag components into forces with the correct orientation. (Note: for the cases where U_P and U_T are both negative, the angle is in the third quadrant and computers will usually assign it a minus sign—for example, $-135°$. It is necessary in this procedure to add $360°$ so that the third quadrant angle is positive—for example, $225°$.) The airfoil lift and drag coefficients can be written as equations that are a function of α, and the local Mach number, M, by the method outlined in Chapter 6. The local Mach number is:

$$M = \frac{\bar{U}_B \Omega R}{\text{Speed of sound}}$$

Yawed Flow and Unsteady Aerodynamics

The analysis so far has been based on the assumption that the lift coefficient on the blade element is only a function of the Mach number and of the instantaneous angle of attack corresponding to the component of velocity perpendicular to the leading edge. Correlation of the results of wind tunnel tests with the results of analytical studies based on this assumption give satisfactory agreement at low thrust levels, but at high thrust levels the analysis predicts lower thrust than is measured. One such comparison, taken from reference 3.33, is shown in Figure

$$\alpha = \theta + \tan^{-1}\left(\frac{U_P}{U_T}\right)$$

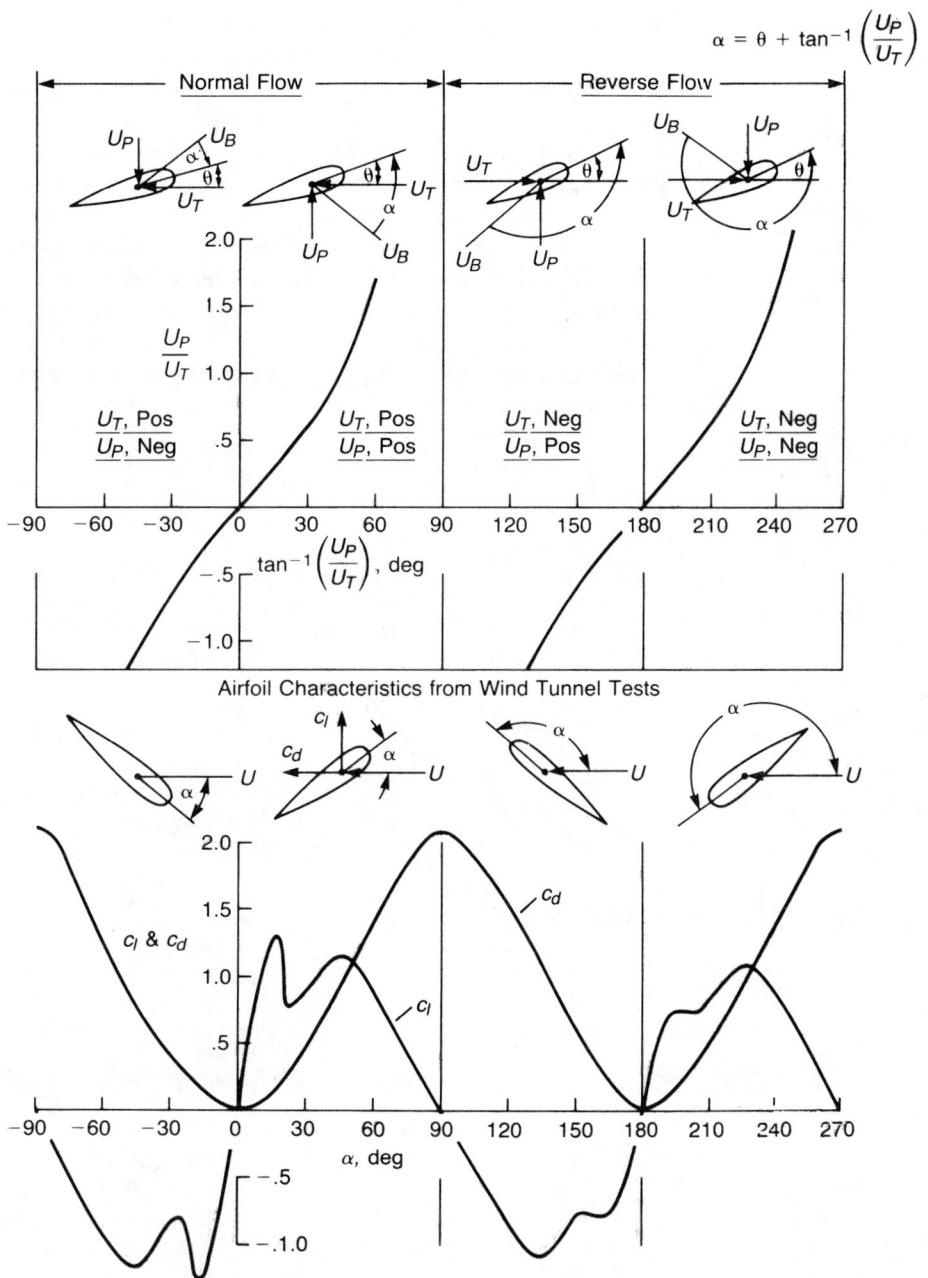

FIGURE 3.53   **Angle of Attack Relationships**

3.54. The discrepancy is explained in reference 3.34 as primarily due to two effects: the beneficial effect of yawed flow on the maximum lift coefficient, and the finite time required for the boundary layer to separate during the stall.

The significance of the yawed flow on maximum lift coefficient was first discussed in reference 3.35. The effect has been measured in wind tunnel tests of yawed wings. Figure 3.55 shows data from reference 3.36—corrected to infinite aspect ratio conditions—first as based on velocities and angles of attack referenced to the tunnel axis system, and then as based on the velocities and angles of attack perpendicular to the leading edge of the wing. In this latter system, which is consistent with that used in the blade element computing method, it may be seen that the measured maximum lift coefficient increases from 0.9 to 2.2 as the sweep is increased. Reference 3.35 suggests that the increase in maximum lift coefficient be represented by the engineering approximation:

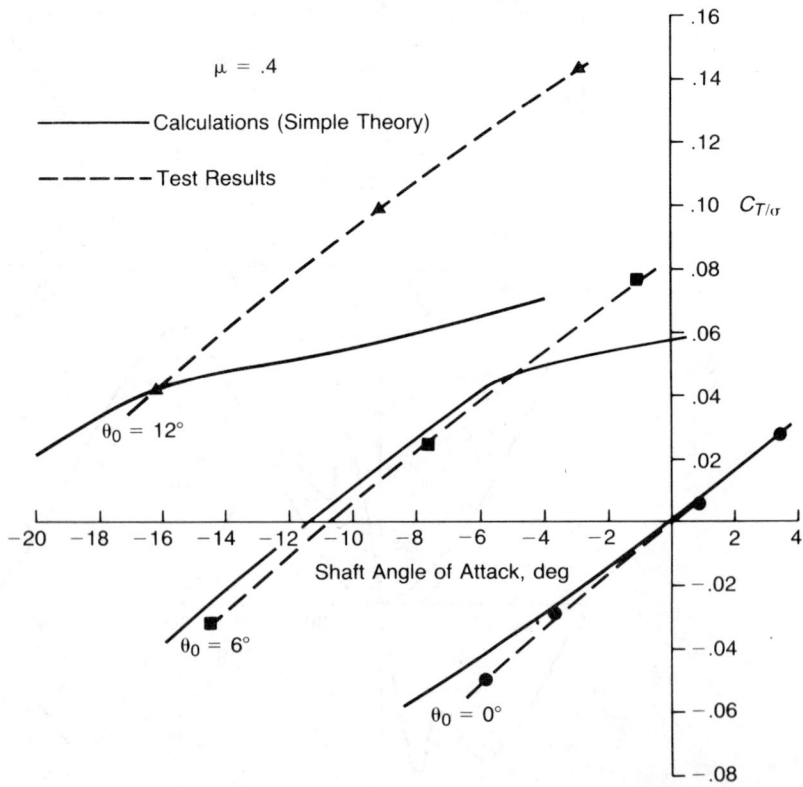

FIGURE 3.54   Failure of Simple Theory to Correlate with Measured Rotor Thrust

Source: Arcidiacono, "Aerodynamic Characteristics of a Model Helicopter Rotor Operating under Nominally Stalled Conditions in Forward Flight," *JAHS* 9-3, 1964.

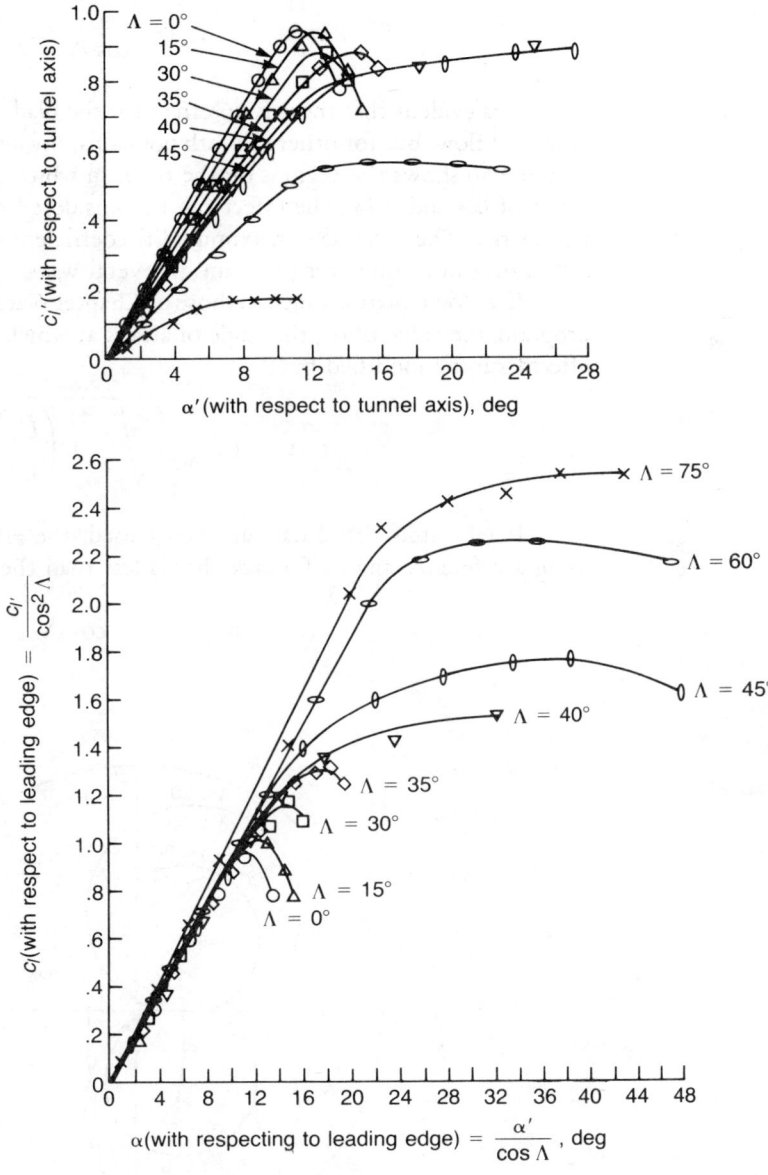

FIGURE 3.55   Effect of Sweep on Two-Dimensional Lift Characteristics

Source: Purser & Spearman, "Wind-Tunnel Tests at Low Speed of Swept and Yawed Wings Having Various Plan Forms," NACA TN 2445, 1951.

$$c_{l_{max}} = \frac{c_{l_{max}\,\Lambda=0}}{\cos\Lambda}$$

It is evident that there is no effect for the blade at $\psi = 270°$, where there is no yawed flow; but for other azimuth positions, the delay in stall can be significant. Figure 3.56 shows the regions on the rotor in which Λ exceeds 30° for tip speed ratios of 0.3 and 0.45. The effect can be considered to be important inside these boundaries. The increase in maximum lift coefficient can be incorporated into the airfoil data in a computer program in several ways.

If airfoil equations such as those in Chapter 6 are being used in the computer program, the value of α_L, the angle of attack at which the airfoil first exhibits stall effects, can be modified by:

$$\alpha_L = (\alpha_{L_{\Lambda=0}})\sqrt{1 + \left(\frac{\bar{U}_R}{\bar{U}_T}\right)^2}$$

If tabulated airfoil data are being used, the effect can be incorporated by using a reference angle of attack that is less than the actual angle of attack:

$$\alpha_{ref.} = \alpha_{actual}\cos\Lambda$$

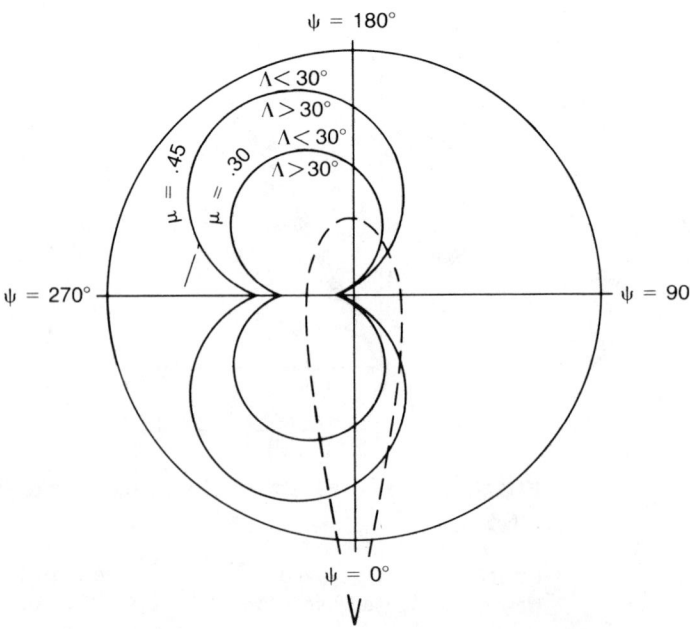

FIGURE 3.56   Boundaries for 30° of Local Sweep

This is used to define a fictitious lift curve slope such that:

$$c_l = \left(\frac{c_{l_{\text{at }\alpha_{\text{ref.}}}}}{\alpha_{\text{ref.}}} \right) \alpha_{\text{actual}}$$

If $\alpha_{\text{ref.}}$ is less than the stall angle, the c_l will be as if stall did not exist. If $\alpha_{\text{ref.}}$ is above the static stall angle, the c_l will be somewhere above the static c_l, as shown in Figure 3.57.

The wind tunnel tests of reference 3.36 have also been used to determine the effect of sweep on the drag and pitching moment characteristics. These studies show that in contrast to lift, sweep angles under 60° have little effect on the drag and moment characteristics.

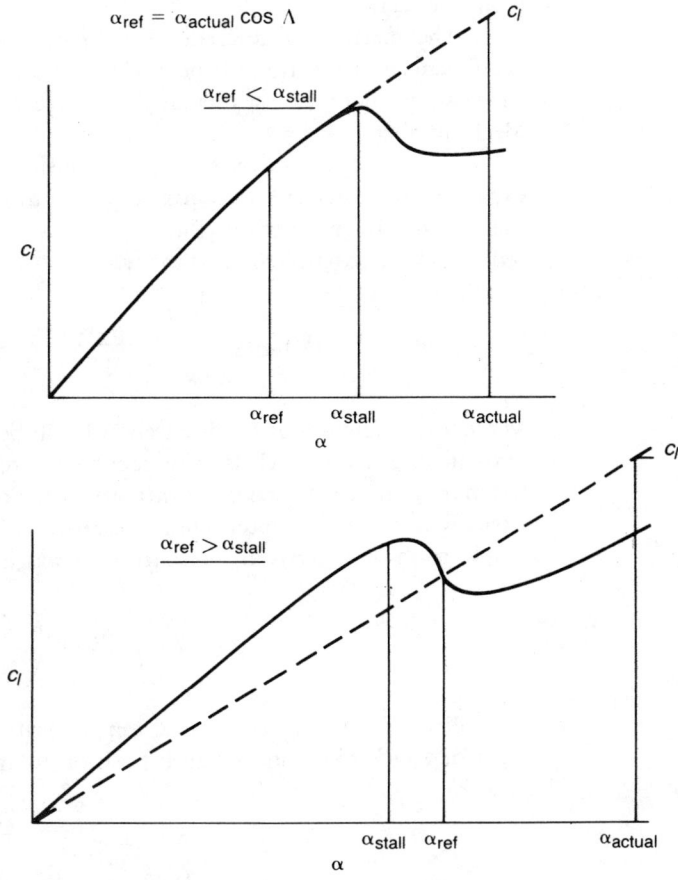

FIGURE 3.57   Use of the Reference Angle Concept

If the angle of attack on a blade element is increasing rapidly, the stall will be delayed until some angle is reached considerably above the normal stall angle. An extensive discussion of this *lift overshoot* phenomenon is given in Chapter 6. Several ways of handling this effect have been developed for use in computer programs. These vary from the sophisticated analytical approaches of references 3.37 and 3.38 to the simpler empirical approaches of references 3.39, 3.40, and 3.41. The first two use potential flow and boundary layer equations to predict the separation angle and are not suited for programs whose primary objective is to calculate rotor performance. The last three depend on characteristics derived from oscillating airfoil wind tunnel tests, and each has been used in rotor performance programs.

The method of reference 3.39 calculates changes in the effective camber and angle of attack due to the pitching velocity and uses these with an empirical time lag to establish the characteristics of the lift overshoot and subsequent hysteresis loop boundaries.

The method of reference 3.40 makes use of tabulated lift and moment coefficients as a function of the angle of attack and its first two time derivatives. These supplement the usual tables, which are a function of angle of attack and the Mach number.

The final method was originally outlined in reference 3.34 and was later expanded in reference 3.41. It has been used to produce the isolated rotor charts in this chapter. In this method, the dynamic overshoot is expressed as an increase in stall angle due to a pitching velocity:

$$\Delta\alpha_{\text{stall}} = \gamma \sqrt{\left| \frac{c/2\dot{\alpha}}{V} \right|} \; x \; (\text{sign of } \dot{\alpha}\,), \text{ deg}$$

where γ is a function of Mach number. It can be evaluated from wind tunnel tests of oscillating airfoils such as those reported in references 3.42 and 3.43 from those test points in which maximum lift was attained before the maximum angle of attack was reached. Chapter 6 has a discussion of this type of test. For the charts, a simple form of γ given in reference 3.44 was used:

$$\gamma = 1.76 \ln \left(\frac{.6}{M} \right)$$

The term within the radical can be written as a function of the change in angle of attack from one azimuth position to another at the same radius station, since:

$$\dot{\alpha} = \frac{\Delta\alpha}{\Delta t}, \text{ deg/sec}$$

where

$$\Delta t = \frac{2\pi R}{\Omega R} \frac{\Delta \psi}{360}, \text{ sec}$$

thus

$$\Delta \alpha_{\text{stall}} = \gamma \sqrt{\frac{c}{R} \left(\frac{1}{4\pi}\right) \frac{1}{\bar{U}_B} \frac{|\Delta \alpha|}{\Delta \psi/360}}, \text{ deg}$$

where both $\Delta \alpha$ and $\Delta \psi$ are in degrees.

This angle increment should be added to the static stall angle in whatever type of airfoil data presentation is used. If the equation form developed in Chapter 6 is used, $\Delta \alpha_{\text{stall}}$ is simply added to α_L after it has been modified for sweep effects. If tabulated airfoil data are being used, the stall delay can be incorporated using the technique that was used for the sweep effect—that is, by defining a reference angle of attack, which is now:

$$\alpha_{\text{ref.}} = \alpha_{\text{actual}} \cos \Lambda - \Delta \alpha_{\text{stall}}$$

Along the boundary of the reverse-flow region, the calculated angle of attack, the sweep, and the rate of change of angle of attack on the blade element, are very high. This combination can result in the equations predicting unreasonably high—or low—lift coefficients. Even though the dynamic pressure is low along the boundary, significant errors can be introduced when an entire blade element is assigned a lift coefficient of 100 or more! For this reason, it has been found desirable to set bounds on the allowable lift coefficient. Figure 3.58 shows the limits on the minimum and maximum values that have been used for the 0012 airfoil. The maximum boundary is purely arbitrary and is based on a lift curve slope of 2π and on the assumption that the sweep and stall delay effects will produce no benefits beyond an angle of attack of 45°. The minimum boundary represents the approximate lift characteristics of a sharp-edged flat plate with extreme thin airfoil stall characteristics.

In Chapter 6 it will be shown that the drag coefficient is affected by dynamic overshoot, which delays the drag rise in the same manner that it delays lift stall. The drag, however, is not affected by sweep as maximum lift is.

Another unsteady aerodynamic effect which may be included in the program is the classical unsteady potential flow. The lift of an airfoil is affected by the rate of change of angle of attack and by the rate of plunge, both of which produce shed vorticity lying behind the trailing edge, which induces velocities at the front of the airfoil. The basic theory was developed in reference 3.45. In that reference, potential theory was used to analyze the effects of unsteady aerodynamics on the

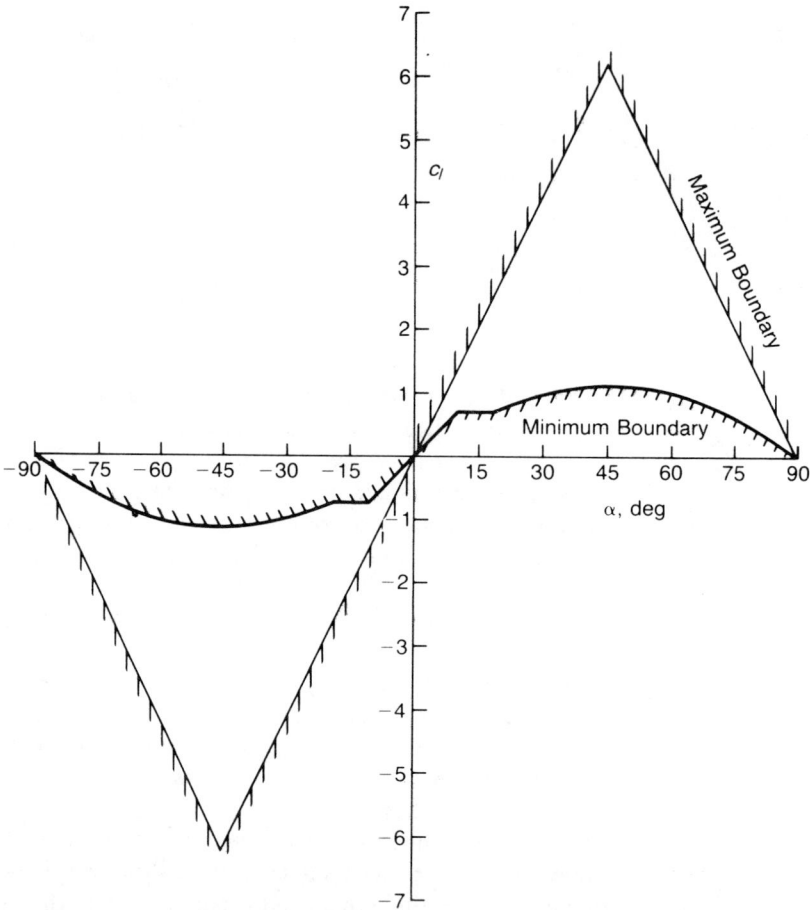

FIGURE 3.58   Boundaries Applied to 0012 Lift Coefficient

lift of an airfoil that is initially at zero angle of attack and is then subjected to oscillating pitch and plunge motions. For an airfoil that is pitching about its quarter chord—a good assumption for most rotors—the equation for the lift coefficient may be written:

$$c_l = a \left\{ C \left[\alpha' + \frac{\dot{h}}{V} + \left(\frac{c}{2V} \right) \dot{\alpha}' \right] + \left[\left(\frac{c}{4V} \right) \dot{\alpha}' + \left(\frac{c}{4V} \right) \frac{\ddot{h}}{V} + \left(\frac{c}{4V} \right)^2 \ddot{\alpha}' \right] \right\}$$

where C is a complex variable that is a function of the frequency of oscillation; α' is the geometric angle of attack; and $\dot{h}$ is the plunge velocity. For a blade element that is oscillating about a mean angle of attack, these parameters may be redefined as:

$$\alpha' = \Delta\theta$$

$$\dot{b} = \Delta U_P$$

$$V = U_T$$

so that

$$\alpha' + \frac{\dot{b}}{V} = \Delta\theta + \frac{\Delta U_P}{U_T} = \Delta\alpha \text{ (from mean value)}$$

and

$$\dot{\alpha}' + \frac{\ddot{b}}{V} = \dot{\alpha}$$

Also

$$\dot{\alpha}' = \dot{\theta}$$

and

$$\ddot{\alpha}' = \ddot{\theta}$$

Further, the parameter c/V can be expressed in terms of the reduced frequency, k, where:

$$k = \frac{c\Omega}{2V}$$

(k is the angular displacement during the time the air moves half a chord). When these substitutions are made and the lift coefficient is related to a mean angle of attack, the equation becomes:

$$c_l = a \left\{ \alpha_{\text{mean}} + C\left[\Delta\alpha + k\frac{\dot{\theta}}{\Omega} \right] + \frac{k}{2}\frac{\dot{\alpha}}{\Omega} + \left(\frac{k}{2}\right)^2 \frac{\ddot{\theta}}{\Omega^2} \right\}$$

The complex variable, C, is:

$$C = F + iG$$

where F represents a lift deficiency and G represents a lag. Both are functions of the reduced frequency and are defined in reference 3.45. Thus:

$$c_l = a \left\{ \alpha_{\text{mean}} + F\left[\Delta\alpha + k\frac{\dot{\theta}}{\Omega} \right] + G\left[i\Delta\alpha + ki\frac{\dot{\theta}}{\Omega} \right] + \frac{k}{2}\frac{\dot{\alpha}}{\Omega} + \left(\frac{k}{2}\right)^2 \frac{\ddot{\theta}}{\Omega^2} \right\}$$

In order to eliminate the imaginary terms, $\Delta\alpha$ and $\dot\theta$ will be assumed to be pure oscillations at the rotor frequency, such that:

$$\Delta\alpha = Ae^{i\Omega t}$$

Thus

$$\Delta\dot\alpha = \dot\alpha = i\Omega Ae^{i\Omega t} = i\Omega\Delta\alpha$$

and

$$i\Delta\alpha = \frac{\dot\alpha}{\Omega}$$

Similarly:

$$\dot\theta = Be^{i\Omega t}$$

Thus

$$\ddot\theta = i\Omega Be^{i\Omega t} = i\Omega\dot\theta$$

and

$$i\dot\theta = \frac{\ddot\theta}{\Omega}$$

The equation may now be written:

$$c_l = a\left\{\alpha_{mean} + F\left[\Delta\alpha + k\frac{\dot\theta}{\Omega}\right] + G\left[\frac{\dot\alpha}{\Omega} + \frac{k\ddot\theta}{\Omega^2}\right] + \frac{k}{2}\frac{\dot\alpha}{\Omega} + \left(\frac{k}{2}\right)^2\frac{\ddot\theta}{\Omega^2}\right\}$$

For use in the computing program, it is convenient to write the equation in terms of a correction due to unsteady effects, which can be simply added to the lift coefficient based on steady conditions. Assume that:

$$c_{l_{steady}} = a\alpha = a(\alpha_{mean} + \Delta\alpha)$$

and

$$c_l = c_{l_{steady}} + \Delta c_{l_{unsteady}}$$

Thus

$$\Delta c_{l_{unsteady}} = a\left\{(F-1)\Delta\alpha + Fk\frac{\dot\theta}{\Omega} + G\left[\frac{\dot\alpha}{\Omega} + \frac{k\ddot\theta}{\Omega^2}\right] + \frac{k}{2}\frac{\dot\alpha}{\Omega} + \left(\frac{k}{2}\right)^2\frac{\ddot\theta}{\Omega^2}\right\}$$

where

$$\Delta\alpha = \alpha - \alpha_{mean},$$

$$\alpha_{mean} = \theta_0 + \theta_1 \frac{r}{R} + \frac{1}{2}\tan^{-1}\left(\frac{\lambda'}{r/R}\right) + \frac{1}{4}\tan^{-1}\left(\frac{\lambda'}{r/R + \mu}\right) + \frac{1}{4}\tan^{-1}\left(\frac{\lambda'}{r/R - \mu}\right)$$

$$\frac{\dot{\theta}}{\Omega} = A_1 \sin\psi - B_1 \cos\psi$$

$$\frac{\ddot{\theta}}{\Omega^2} = A_1 \cos\psi + B_1 \sin\psi$$

$$\frac{\dot{\alpha}}{\Omega} = \frac{\Delta\alpha}{\Delta\psi}$$

$$k = \frac{c\Omega}{2V} = \frac{c\Omega}{2\Omega R \sqrt{\bar{U}_T^2 + \bar{U}_R^2}} = \frac{\frac{1}{2}c/R}{\sqrt{\bar{U}_T^2 + \bar{U}_R^2}}$$

The functions F and G shown on Figure 3.59 can be approximated by:

If $1/R < 15$, then $F = .9 - .00178\,(15 - 1/k)^2$

If $1/R > 15$, then $F = .864 + .0024\,(1/k)$

and $G = -.2 + .0025\,(1/k)$

These equations can be incorporated directly in the computer program. Figure 3.60 shows the correlation of the theory against the test results of an oscillating airfoil, which were reported in reference 3.42. It may be seen that the measured effect is somewhat greater than predicted. The relative importance of the effect on rotor performance is shown in Figure 3.61, where the calculated lift coefficient at the 75% radius for a typical flight condition is shown with and without the unsteady effect. The overall result is to increase the rotor thrust slightly for a given set of rotor conditions. The drag of the blade is assumed to be unaffected by the shed vorticity and, if required, the effect on aerodynamic pitching moment may be calculated using the methods given in reference 3.41.

Using the Computer Program

The equations derived in the previous sections can be used to set up a program to compute nondimensional, isolated rotor performance or the performance and trim conditions for an entire helicopter. An isolated rotor program was used to produce the charts at the end of this chapter. It used λ', θ_0, θ_1, μ, and $M_{1,90}$ (advancing tip Mach number) as inputs. The first estimates of C_T/σ and of the cyclic pitch required to trim the pitching and rolling moments to zero were made using the

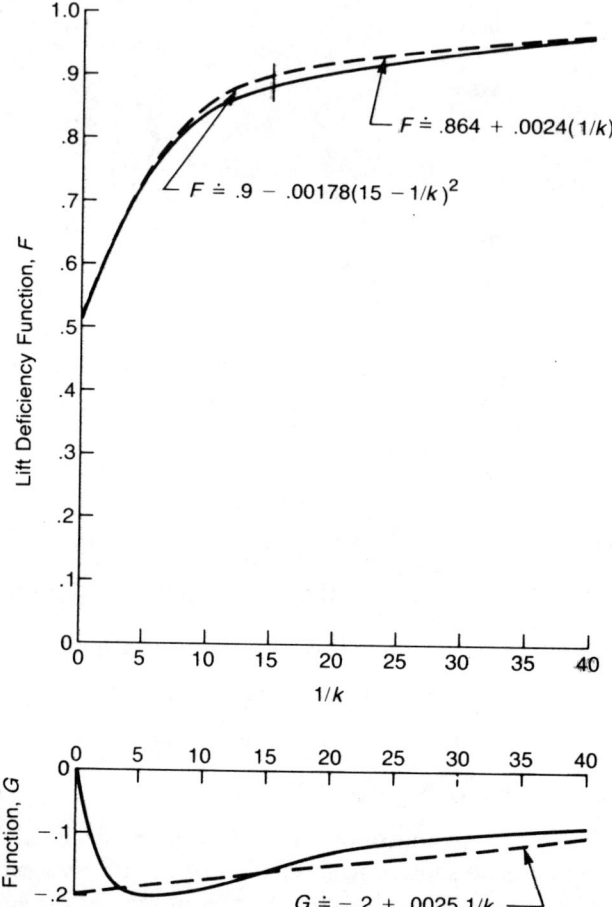

$$F \doteq .864 + .0024(1/k)$$

$$F \doteq .9 - .00178(15 - 1/k)^2$$

$$G \doteq -.2 + .0025\, 1/k$$

FIGURE 3.59   Unsteady Aerodynamic Lift Deficiency and Lag Functions

Source: Theodorsen, "General Theory of Aerodynamic Instability and the Mechanism of Flutter," NACA TR 496, 1935.

closed-form equations. The program was then cycled and the thrust and moment coefficients calculated. If the moment coefficients were outside a predetermined tolerance, the cyclic pitch was modified and another iteration was performed. The actual tolerances used were ± 0.0005 on C_R/σ and ± 0.0020 on C_M/σ. These tolerances correspond to approximately ± 0.1 degree for B_1 and ± 0.5 degree for A_1, which are sufficient to ensure engineering accuracy of the calculations. The equations for the corrections of cyclic pitch that gave satisfactory convergence in most flight conditions were:

$$\Delta B_1 = -190\,(1 - \mu)\, C_R/\sigma \text{ deg}$$

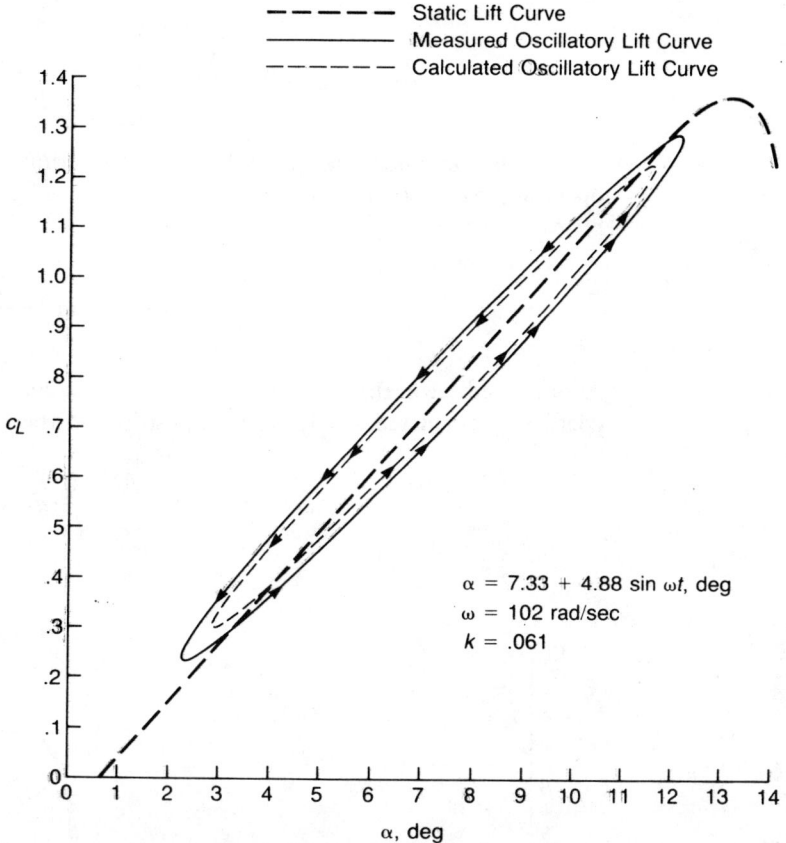

FIGURE 3.60   Correlation of Theory and Test Results for an Oscillating Airfoil

Source: Liiva, Davenport, Gray, & Walton, "Two-Dimensional Tests of Airfoils Oscillating Near Stall," USAAVLABS TR 68-13, 1968.

$$\Delta A_1 = -60 \; (2 + \mu^2) \; C_M/\sigma \; \text{deg}$$

Once the cyclic pitch that satisfied the moment coefficient criteria was established, the program evaluated the thrust, torque, and H-force coefficients.

The ability of the isolated rotor to overcome parasite drag if it were actually installed on a helicopter is given by the equilibrium equation:

$$fq = - T\alpha_{TTP} - H$$

This can be nondimensionalized into a primary parameter, which appears on the rotor charts:

$$f/A_b + (C_T/\sigma)^2 \frac{\sigma}{\mu^4} = -\left[\frac{C_T/\sigma \dfrac{\lambda'}{\mu} + C_H/\sigma}{\mu^2/2}\right]$$

It is sometimes useful to have an indication of how close the rotor is operating to its capability. One such indicator is the angle of attack of the retreating tip:

$$\alpha_{1,270} = \theta_0 + \theta_1 + B_1 + \tan^{-1}\frac{\lambda'}{1-\mu}$$

Another indicator that has been used is the maximum value of profile torque coefficient calculated at any azimuth position on the rotor:

$$\Delta C_Q/\sigma_0 = \int_0^1 \frac{dC_Q/\sigma_0}{dr/R}\, dr/R$$

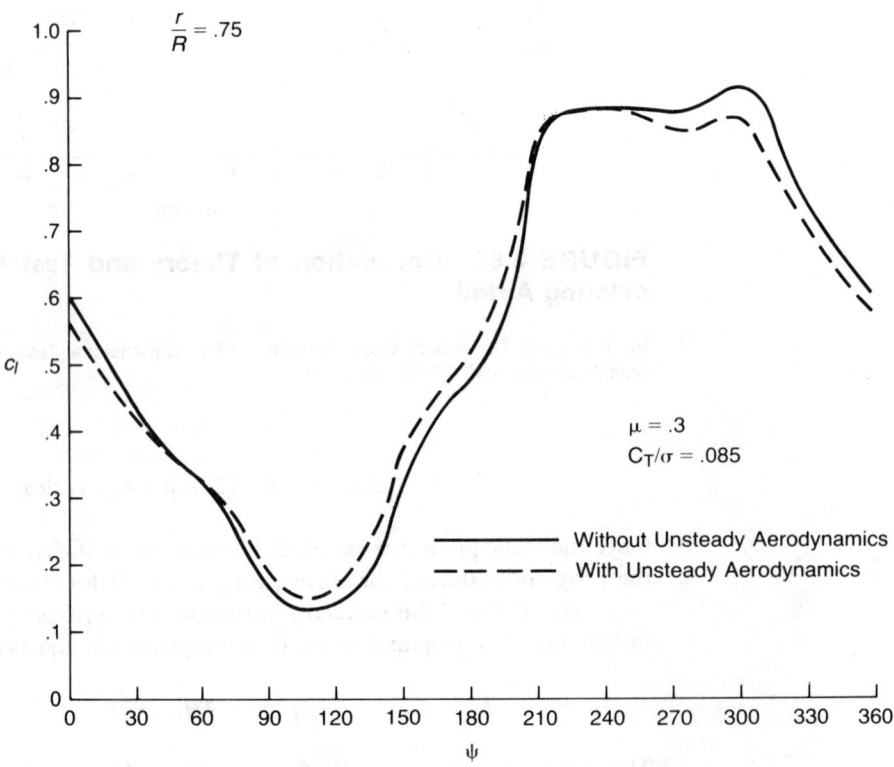

FIGURE 3.61   Effect of Unsteady Aerodynamics below Stall

where

$$\frac{dC_Q/\sigma_0}{dr/R} = \frac{\bar{U}_B^2}{2} \frac{r}{R} c_{c_0}$$

This parameter is computed at each azimuth during the integration for torque, and the computer can be made to remember the highest value obtained and to print it and its azimuth as part of the output.

For an analysis of a complete helicopter, the physical parameters of the main and tail rotors and the lift and drag characteristics of the airframe are used as inputs along with the gross weight, the forward speed, the atmospheric density, and the speed of sound. The program is made to iterate on θ_0, λ', and θ_{0_T}, as well as on A_1 and B_1, to find the final trim conditions in a process identical to that used with the closed-form equations.

ISOLATED ROTOR PERFORMANCE CHARTS

The charts at the end of this chapter have been produced with the methods just described and may be used to solve all of the common forward flight performance problems involving either main or tail rotors. The charts are based on a series of arbitrarily selected rotor parameters, although—as will be discussed—they are flexible enough to be used for rotors with different parameters.

The first chart of each pair can be used to obtain the power required in forward flight at a given tip speed ratio for known values of f/A_b, σ, and C_T/σ. With these values defined, the top chart gives the collective pitch and the bottom chart the corresponding value of C_Q/σ. Two types of stall limits are shown on this chart: the angle of attack of the retreating tip, $\alpha_{1,270}$, and the maximum value of the profile torque at any azimuth station, $\Delta C_Q/\sigma_0$. Reference 3.46 uses values of 0.004 and 0.008 for $\Delta C_Q/\sigma_0$ as "lower and upper stall limits," respectively, to indicate the region where many rotors begin to get into trouble because of stall effects. Most rotors can operate up to the higher set of limits as a helicopter rotor and well beyond it in the autogiro mode. The second set of charts can be used to obtain trim values of C_H/σ, λ', $A_1 - b_{1_s}$, and $B_1 + a_{1_s}$.

The charts have been constructed using the following rotor parameters:

Twist:	−5°, linear
Airfoil:	NACA 0012
Advancing tip Mach no:	.7 (i.e., below the speed for compressibility losses)
Blade chord/radius ratio:	0.079
Tip loss factor:	.97

None of these assumed parameters imposes significant limits on the use of the charts for rotors with different parameters. For rotors with twist other than

−5°, the charts may be used by referencing the collective pitch to the pitch at the 75% radius. The collective pitch to use with the charts is:

$$\theta_{0_{\theta_1=-5°}} = \theta_0 + .75(\theta_1 + 5°)$$

For operation away from retreating blade stall, this is the only correction advised. For operation in the stall regime, twist affects the location of the limit lines of $\Delta C_Q/\sigma_0$ and $\alpha_{1,270}$ on the torque plot. These limits move to higher values of C_T/σ for more than −5° of twist and to lower values for less twist. The displacement of the limit lines for other than −5° is approximately:

$$\Delta C_T/\sigma = -0.003(\theta_1 + 5°)$$

The same displacement applies to the points on the torque curves where sudden changes in curvature occur because of retreating tip stall.

Many modern rotors have airfoils with higher stalling angles than the NACA 0012 on which the charts were based. These airfoils will also shift the limit lines and the points of sudden curvature change to the right. The magnitude of the shift can be estimated by the increase in the airfoil stall angle and by the distance between the two limit lines for $\alpha_{1,270} = 12°$ and $\alpha_{1,270} = 16°$.

Operation of the rotor at advancing tip Mach numbers that cause drag divergence can be accounted for using the method of Figure 3.43. This allows a correction to be made to the torque coefficient based on the amount the three-dimensional drag rise Mach number of the tip airfoil is exceeded. Thus the method can be used with airfoils other than the NACA 0012 if the airfoil drag characteristics are known.

The chord/radius ratio, c/R, enters into the analysis of unsteady aerodynamic effects both above and below stall. For the charts, a ratio of 0.079 was used. It will be assumed that these effects are small enough that the charts can safely be used for blades with other aspect ratios. Similarly, the tip loss factor of 0.97—which is a guess at any rate—can be assumed to apply to any reasonable rotor.

To illustrate the various uses of the charts, a series of numerical examples is given in Table 3.5.

Use of the Charts

The charts* can be used to determine performance and approximate trim conditions for several forward flight conditions by one or more of the procedures provided in Table 3.5. For a more exact computation of trim conditions, the refinements of Chapter 8 should be used.

---

*The chart format was originally developed by M.C. Cheney at Lockheed.

The cases illustrated include:

- Main rotor in level flight.
- Main rotor in level flight with compressibility losses.
- Tail rotor.
- Entire helicopter in level flight.
- Helicopter in autorotation.
- Helicopter in climb.
- Helicopter in dive at constant collective pitch.
- Helicopter with auxiliary propulsion.
- Rotor in a wind tunnel.

(Hint: whenever possible, use even values of tip speed ratios rather than even values of forward speed, so that the charts can be used without interpolation.)

CORRELATION WITH TEST RESULTS

The chart method has been used to produce results that can be compared with a set of wind tunnel test results for a full-scale rotor reported in reference 3.27 and the correlation is shown in Figure 3.62. (The test results have been corrected to free-air conditions for wind tunnel wall effects by the method of reference 3.47.)

During the correlation process, it was found that the input collective pitch had to be adjusted somewhat from that given in the test report. The difficulty appears to be connected with the test conditions rather than with the theory. During the tests of reference 3.27, a rotor with no twist was operated at zero shaft angle of attack. At a collective pitch of zero, it produced positive thrust. The data indicated that at a collective pitch of −1°, the thrust would be zero. The senior author of reference 3.27 believed that the effect was the result of local flow distortions around the body used to fair the rotor support system. Similar discrepancies have been noted in other wind tunnel tests. For this reason, the values of collective pitch selected for correlation in figure 3.62 are 1° higher than those stated in the test report.

A more extensive correlation study of the results of reference 3.27 is reported in reference 3.48. In that study, the effects of including blade flexibility, vortex-caused variable-induced velocity distribution, and unsteady aerodynamics in the calculations were investigated. The conclusions reached were that with respect to performance, the incorporation of unsteady aerodynamics is of primary importance, but that blade flexibility and variable-induced velocity distribution have little significance. It thus appears that the methods outlined in this chapter, which do not account for flexibility or variable inflow but do account for unsteady aerodynamics, are adequate for performance calculations.

TABLE 3.5

| 1. *Main Rotor in Level Flight* | *Example* |

1. *Main Rotor in Level Flight*

Given: σ, θ_1, A_b, f, $M_{dr_{3_{c_{l=0}}}}$ of tip airfoil, γ,

    speed of sound, ΩR, μ, C_T/σ

Find: C_Q/σ, C_H/σ, θ_0, α_{TPP}, $A_1 - b_1$, $B_1 + a_{1_s}$

$\sigma = 0.085$, $\theta_1 = -10°$, $A_b = 240$ ft^2

$f = 19.3$ ft^2, $M_{dr_{3_{c_{l=0}}}} = .80$ (NACA 0012)

$\gamma = 8.1$, speed of sound $= 1,117$ ft/sec

$\Omega R = 650$ ft/sec, $\mu = .3$, $C_T/\sigma = .086$

Procedure:

a. Calculate $f/A_b + \dfrac{\sigma}{\mu^4}(C_T/\sigma)^2$

    (Note: $1/\mu^4$ is already included in the
    chart ordinate.)

$\dfrac{19.3}{240} + 123\,(.085)(.086)^2 = .158$

b. Find $\theta_{0_{\theta_1 = -5°}}$ from top chart of first page.

$\theta_{0_{\theta_1 = -5°}} = 13.2°$

c. Find C_Q/σ from bottom chart.

$C_Q/\sigma = .0046$

d. Is rotor operating beyond limit line
    of $\Delta C_Q/\sigma = 0.004$?

Yes

e. Calculate effective shift of limit lines due to
    twist different than chart value

    $\Delta C_T/\sigma = -.003\,(\theta_1 + 5°)$

$\Delta C_T/\sigma = -.003(-10 + 5) = .015$

f. Estimate new value of C_Q/σ with limit lines and point
    of sudden change of curvature shifted to the right
    by $\Delta C_T/\sigma$.

$C_Q/\sigma = .0044$

g. Calculate advancing tip Mach number, $M_{1,90}$
    and drag rise Mach number, M_{dr_3}.

    $M_{1,90} = \dfrac{(1 + \mu)\,\Omega R}{\text{Speed of sound}}$

$M_{1,90} = \dfrac{1.3(650)}{1,117} = .756$

    $M_{dr_3} = M_{dr_{3_{c_{l=0}}}}[1 - 6(C_T/\sigma)^2]$

$M_{dr_3} = .80\,[1 - 6(.086)^2] = .764$

h. If $M_{1,90} > M_{dr_3}$, use Figure 3.43 to find compressibility
    loss.

No compressibility loss for this example.

i. Find C_H/σ, λ', $A_1 - b_1$, $B_1 + a_{1_s}$ from charts.

For $a_{1_s} = 0$, $C_H/\sigma = -.0008$, $\lambda' = -.023$

$A_1 - b_{1_s} = \dfrac{8}{8.1}(-1.8) = -1.8°$

$B_1 + a_{1_s} = 7.9°$

j. Find α_{TPP} from

$$\alpha_{TPP} = \tan^{-1}\left[\frac{\lambda'}{\mu} + \frac{\sigma C_T/\sigma}{2\mu^2}\right]$$

$$\alpha_{TPP} = \tan^{-1}\,[3.33\,(-.023) + 5.56\,(.085)(.086)]$$
$$= -2.1°$$

k. Find θ_0 from

$$\theta_0 = \theta_{0\theta_1=-5°} - .75(\theta_1 + 5°)$$

$$\theta_0 = 13.2 - .75(-10 + 5) = 16.9°$$

2. Main Rotor in Level Flight with Compressibility Losses

Example

Given: σ, θ_1, A_b, f, $M_{dr_{3C_{l=0}}}$ of tip airfoil, γ,
speed of sound, ΩR, μ, C_T/σ

$\sigma = 0.085$, $\theta_1 = -10°$, $\alpha_{TPP} = 240$ ft^2

$f = 19.3$ ft, $M_{dr_{C_{l=0}}} = .80$ (NACA 0012)

$\gamma = 8.1$, speed of sound $= 1{,}117$ ft/sec
$\Omega R = 750$ ft/sec, $\mu = .3$, $C_T/\sigma = .086$

Find: C_Q/σ, C_H/σ, θ_0, $A_1 - b_{1_s}$, $B_1 + a_{1_s}$

Procedure:
Same as example 1 up to step g.

g. Calculate advancing tip Mach number, $M_{1,90}$
and drag rise Mach number, M_{dr_3}

$$M_{1,90} = \frac{(1 + \mu)\Omega R}{\text{Speed of sound}}$$

$$M_{1,90} = \frac{(1.3)(750)}{1{,}117} = .873$$

$$M_{dr_3} = M_{dr_{C_{l=0}}}[1 - 6(C_T/\sigma)^2]$$

$$M_{dr_3} = .80[1 - 6(.086)^2] = .764$$

h. Calculate $M_{1,90}/M_{dr_3}$.

$$\frac{M_{1,90}}{M_{dr_3}} = \frac{.873}{.764} = 1.143$$

i. Find $\dfrac{\Delta C_Q/\sigma \text{ comp}}{M_{\Omega R}{}^3}$ from Figure 3.43.

$$\frac{\Delta C_Q/\sigma_{\text{comp}}}{M_{\Omega R}{}^3} = .0043$$

j. Find $\Delta C_Q/\sigma_{\text{comp}}$.

$$\Delta C_Q/\sigma_{\text{comp}} = .0043\left(\frac{750}{1{,}117}\right)^3 = .0013$$

k. Find new value of C_Q/σ

$$C_Q/\sigma = .0044 + .0013 = .0057$$

Remainder same as example 1.

3. Tail Rotor

Given: σ, θ_1, α_s, M_{dr_3} of tip airfoil, γ, l_T/R_M,
   speed of sound, ΩR, μ, C_Q/σ_M, ρ/ρ_0, δ_3

$$[\rho_0 A_b (\Omega R)^2]_T, \left[\frac{\rho_0 A_b (\Omega R)^3}{550}\right]_M$$

Find: C_T/σ, θ_0, C_Q/σ, a_{1_s}, C_H/σ

Procedure:

a. Find T_T from:

$$T_T = \frac{550}{\Omega R_M} \left(\frac{1}{l_T/R_M}\right) \left[\frac{\rho_0 A_b (\Omega R)^3}{550}\right]_M C_Q/\sigma_M$$

b. Find $C_T/\sigma_T = \dfrac{T_T}{\rho A_b (\Omega R)^2}$

c. For tail rotor, $\alpha_{TPP} = a_{1_s}$ and $B_1 = 0$.

Plot α_{TPP} and $B_1 + a_{1_s}$ vs. θ_0, find intersection.

$$\alpha_{TPP} = \tan^{-1}\left(\frac{\lambda'}{\mu} + \frac{\sigma C_T/\sigma}{2\mu^2}\right)$$

d. Find C_Q/σ from chart.

e. Calculate advancing tip Mach no. and drag rise Mach no.

$$M_{1,90} = \frac{(1+\mu)\Omega R}{\text{Speed of sound}}$$

$$M_{dr_3} = M_{dr_{3C_{l=0}}}[1 - 6(C_T/\sigma)^2]$$

f. If $M_{1,90} > M_{dr_3}$, use Figure 3.43 to find compressibility correction.

Example

$\sigma = .146$, $\theta_1 = -5°$, $\alpha_s = 0°$,
$M_{dr_3} = .80$ (NACA 0012), $\gamma = 4$

$$l_T/R_M = \frac{37}{30} = 1.23$$

Speed of sound $= 1,117$ ft/sec
$\Omega R = 650$ ft/sec, $\mu = .3$
$C_Q/\sigma_M = .0044$, $\rho/\rho_0 = 1.0$, $\delta_3 = 0$
$[\rho_0 A_b (\Omega R)^2]_T = 19,550$

$$\left[\frac{\rho_0 A_b (\Omega R)^3}{550}\right]_M = 285,000$$

$$T_T = \frac{(550)(285,000)(.0044)}{(650)(1.23)} = 864$$

$$C_T/\sigma_T = \frac{864}{19,550} = .044$$

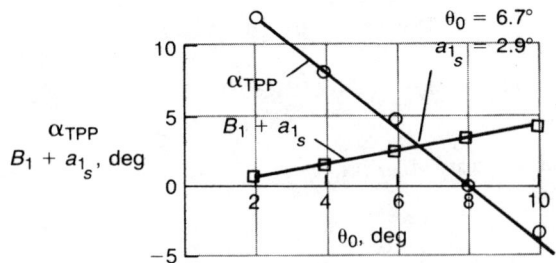

$$M_{1,90} = \frac{(1.3)(650)}{1,117} = .756$$

$$M_{dr_3} = .80[1 - 6(.044)^2] = .791$$

No compressibility loss in this example.

g. Find C_H/σ from

$$C_H/\sigma = (C_H/\sigma - C_T/\sigma \tan a_{1_s}) + C_T/\sigma \tan a_{1_s}$$

$$C_H/\sigma = .0008 + (.044)(\tan 2.9°) = .0031$$

h. Find θ_0 from

$$\theta_{0_T} = \theta_{0_{\theta_1 = -5°}} - .75(\theta_1 + 5)$$

$$\theta_{0_T} = 6.7 - .75(-5 + 5) = 6.7°$$

4. *Entire Helicopter in Level Flight*

Given: Main rotor; σ, θ_1, M_{dr_3} of tip airfoil,
γ, ΩR, A_b, $\rho_0 A_b (\Omega R)^2$

Tail rotor; σ, θ_1, M_{dr_3} of tip airfoil, α_s,
γ, ΩR, l_T/R_M, $\dfrac{\rho_0 A_b (\Omega R)^3}{550}$

Airframe aero; lift and drag versus α_F
Flight conditions; G.W., μ, speed of sound, ρ/ρ_0

Find: Main rotor power, tail rotor power, trim conditions.

Procedure:

a. Calculate q from:

$$q = \rho/\rho_0 (\rho_0/2)(\mu \, \Omega R)^2$$

b. Make first estimate of α_F from:

$$\alpha_F = -\tan^{-1} \frac{qf}{\text{G.W.}}$$

c. From Figure 2, Appendix A, find L_F/q and f.

d. Calculate C_T/σ from:

$$C_T/\sigma = \frac{1}{\rho/\rho_0}\left(\frac{\text{G.W.} - L_F}{241,100}\right)$$

e. Find $f/A_b + \dfrac{\sigma}{\mu^4}(C_T/\sigma)^2$.

Example

$$\sigma = .085, \; \theta_1 = -10°, \; M_{dr_3} = .80$$

$$\gamma = 8.1, \; \Omega R = 650 \text{ ft/sec}, \; A_b = 240 \text{ ft}^2,$$
$$\rho_0 A_b (\Omega R)^2 = 241,100$$

$$\sigma = .146, \; \theta_1 = -5°, \; M_{dr_3} = .80$$

$$\alpha_s = 0°, \; \gamma = 4, \; \Omega R = 650 \text{ ft/sec},$$
$$l_T/R_M = 1.23 \;, \; \frac{\rho_0 A_b (\Omega R)^3}{550} = 23,100$$

See Figure 2, Appendix A.
G.W. = 20,000 lb, $\mu = .3$ (115 kt)
$\rho/\rho_0 = 1$, speed of sound = 1,117 ft/sec

$$q = 1(.001189)[.3(650)]^2 = 45.2 \text{ lb/ft}^2$$

$$\alpha_F = -\tan^{-1} \frac{(45.2)(19.3)}{20,000} = -2.5°$$

$$L_F/q = -9.5, f = 19.4$$

$$C_T/\sigma = 1\left[\frac{20,000 - (-9.5)(45.2)}{241,100}\right] = .085$$

$$\frac{19.4}{240} + 123 \,(.085)(.085)^2 = .157$$

f. From chart, find $\theta_{0_{\theta_1=-5°}}$.

$$\theta_{0_{\theta_1=-5°}} = 13.2°$$

g. From chart, find λ'.

$$\lambda' = -.021$$

$$\alpha_F = \tan^{-1} \frac{\lambda'}{\mu}$$

$$\alpha_F = \tan^{-1}\left(\frac{-.021}{.3}\right) = -4.0°$$

h. From Figure 2, Appendix A, find second values of L_F/q and f.

$$L_F/q = -12.5 \qquad f = 19.5$$

i. Calculate new value of C_T/σ.

$$C_T/\sigma = \frac{20,000 - (-12.5)(45.2)}{241,100} = .085$$

j. Find new value of $f/A_b + \dfrac{\sigma}{\mu^4}(C_T/\sigma)^2$.

$$\frac{19.5}{240} + .076 = .157$$

k. Find new value of $\theta_{0_{\theta_1=-5°}}$.

$$\theta_{0_{\theta_1=-5°}} = 13.2°$$

l. Find C_Q/σ from chart.

$$C_Q/\sigma = .0046$$

m. Is rotor operating beyond limit line of $\Delta C_Q/\sigma = 0.004$?

Yes

n. Calculate effective shift of limit lines due to twist different than chart value

$$\Delta C_T/\sigma = -.003(\theta_1 + 5°)$$

$$\Delta C_T/\sigma = -.003(-10 + 5) = .015$$

o. Estimate new value of C_Q/σ with limit lines and point of sudden change of curvature shifted to the right by $\Delta C_T/\sigma$.

$$C_Q/\sigma = .0044$$

p. Calculate advancing tip Mach no.

$$M_{1,90} = \frac{(1 + \mu)\Omega R}{\text{Speed of sound}}$$

$$M_{1,90} = \frac{(1.3)(650)}{1,117} = .756$$

q. If $M_{1,90} > M_{dr_3}$, use Figure 3.43 to find compressibility loss.

No compressibility loss for this example

r. Find h.p.$_M$ = 285,000 $\rho/\rho_0(C_Q/\sigma)$.

h.p.$_M$ = 285,000 (1)(.0044) = 1,254

s. Find $T_T = \dfrac{550 \text{ h.p.}_M}{\Omega R}\left(\dfrac{1}{l_T/R}\right)$.

$$T_T = \frac{(550)(1,254)}{650}\left(\frac{1}{1.23}\right) = 864$$

t. Find $C_T/\sigma_T = \dfrac{T_T}{\rho/\rho_0(19,550)}$.

$$C_T/\sigma_T = \frac{864}{1(19,550)} = .044$$

u. Find θ_{0_T} and a_{1_s} by cross-plotting as in example 3.

$\theta_{0_T} = 6.7°$, $a_{1_s} = 2.9°$

v. From chart find $(C_H/\sigma - C_T/\sigma \tan a_{1_s})_T$

$(C_H/\sigma - C_T/\sigma \tan a_{1_s})_T = .0008$

w. Calculate C_H/σ_T and H_T.

$C_H/\sigma_T = .0008 + .044 \tan 2.9° = .0030$

$H_T = .0030 \, \rho/\rho_0 \, (19,550) = 59$

x. Find C_H/σ_M from chart (assume $a_{1_s} = 0$).

$C_H/\sigma_M = -.0005$

y. Calculate H_M.

$H_M = \rho/\rho_0(241,100 \, C_H/\sigma)$

$H_M = (1)(241,100)(-.0005) = -121$

z. Find D_F and L_F.

$D_F = 45.2 \, (19.5) = 881$

$L_F = 45.2 \, (-12.5) = -565$

aa. Find T_M from

$$T_M = \sqrt{(G.W. - L_F)^2 + (D_F + H_M + H_T)^2}$$

$T_M = \sqrt{(20,000 + 565)^2 + (881 - 121 + 59)^2}$

$T_M = 20,581$

bb. Find $C_T/\sigma = \dfrac{T_M}{\rho/\rho_0(241,100)}$

$C_T/\sigma = \dfrac{20,581}{(1)(241,100)} = .085$

cc. Calculate new f from

$f = f + H_T/q$.

$f = 19.5 + 59/45.2 = 20.8$

dd. Iterate from step j until values repeat.

ee. When iteration converges, find:

$\theta_0 = \theta_{0_{\theta_1 = -5°}} - .75(\theta_1 + 5°)$

$\theta_0 = 13.5 - .75 \, (-10 + 5) = 17.25°$

ff. From chart, find C_Q/σ_T.

$C_Q/\sigma_T = .0008$

gg. Find $h.p._T = 23,100 \, C_Q/\sigma_T$.

$h.p._T = 19$

hh. List trim values from final iteration step.

$\alpha_F = -4.4°$

$L_F = -588 \text{ lb}$

$D_F = 886 \text{ lb}$

$T_M = 20,606 \text{ lb}$

$$\alpha_{TPP} = -2.1°$$

$$C_T/\sigma = .086$$

$$\lambda' = -.023$$

$$A_1 = -2.1° \qquad (b_{1_s} = 0)$$

$$B_1 = 7.8° \qquad (a_{1_s} = 0)$$

$$H_M = -145$$

$$\text{h.p.}_M = 1,368$$

$$T_T = 942$$

$$H_T = 68$$

$$\text{h.p.}_T = 19$$

$$\theta_{0_T} = 7.1°$$

$$a_{1_{s_T}} = 3.2°$$

5. Helicopter in Autorotation

Example

Given: Main rotor; σ, θ_1, M_{dr_3} of tip airfoil, ΩR, A_b.

$\sigma = .085$, $\theta_1 = -10°$, $M_{dr_3} = .80$

$\Omega R = 650$ ft/sec, $A_b = 240$ ft^2

Tail rotor; power to run at flat pitch.

25 h.p.

Transmission and accessory losses.

15 h.p.

Airframe aerodynamics; Lift and drag versus α_F.

See figure 2, Appendix I.

Flight Conditions; G.W., μ, ρ/ρ_0, speed of sound.

G.W. = 20,000, $\mu = .3$, $\rho/\rho_0 = 1$
Speed of sound = 1,117 ft/sec

Find: Rate of descent and trim conditions.

Procedure:

a. Convert tail rotor and transmission and accessory losses to $-C_Q/\sigma$ for main rotor.

$$C_Q/\sigma = -\frac{(25 + 15)}{285,000} = -.00014$$

b. Find compressibility losses as in example 1.

No compressibility losses.

c. Estimate α_F, find L_F/q and calculate L_F.

$\alpha_F = 0$, $L_F/q = -4.5$, $L_F = -203$

d. Calculate C_T/σ

$$C_T/\sigma = \frac{20,000 - (-203)}{241,100} = .084$$

e. Is rotor operating beyond $\Delta C_Q/\sigma_0 = .004$?

No. No twist correction required.

f. Find $\theta_{0_{\theta_1=-5°}}$ using C_T/σ and C_Q/σ.

$\theta_{0_{\theta_1=-5°}} = 8.5°$

g. Find $\alpha_F = \tan^{-1} \dfrac{\lambda'}{\mu}$

$\lambda' = .021$ from chart

$$\alpha_F = \tan^{-1} \frac{.021}{.3} = 4.0$$

h. Find new values of L_F/q, L_F, C_T/σ.

$L_F/q = 3.1$, $L_F = 140$, $C_T/\sigma = .082$

i. Find new value of $\theta_{0_{\theta_1=-5°}}$.

$\theta_{0_{\theta_1=-5°}} = 8.2°$

j. Find

$$\alpha_{TPP} = \tan^{-1} \left(\frac{\lambda'}{\mu} + \frac{\sigma C_T/\sigma}{2\mu^2} \right)$$

$$\alpha_{TPP} = \tan^{-1} \left[\frac{.022}{.3} + 5.56 \, (.085)(.082) \right] = 6.4°$$

k. Find f from Figure 2, Appendix A.

$f = 19.7$

l. Calculate $C_{D_F}/\sigma = \dfrac{\mu^2 f/A_b}{2}$.

$$C_{D_F}/\sigma = \frac{(.3)^2 \, (19.7)}{2 \, (240)} = .0037$$

m. Find C_H/σ (assume $a_{1_s} = 0$)

$C_H/\sigma = .0004$

n. Find γ_D from:

$$\sin \gamma_D = \frac{C_T/\sigma \dfrac{\alpha_{TPP}}{57.3} + C_{D_F}/\sigma + C_H/\sigma}{C_T/\sigma}.$$

$$\sin \gamma_D = \frac{(.082)\left(\dfrac{6.4}{57.3}\right) + .0037 + .0004}{.082} = .162$$

o. Find R/D from

$R/D = 60 \, \mu(\Omega R) \sin \gamma_D$, ft/min.

$R/D = 60 \, (.3)(650)(.162) = 1,895$ ft/min

p. Find trim conditions.

$\theta_0 = 8.2 - (.75)(-10 + 5) = 12.0°$

$B_1 + a_{1_s} = 5.3°$

$A_1 - b_{1_s} = -1.5°$

6. Helicopter in Climb

Given: Main rotor; σ, θ_1, M_{dr_3} of tip airfoil, γ, ΩR, A_b.

$\sigma = .085$, $\theta_1 = -10°$, $M_{dr_3} = .80$

$\gamma = 8.1$, $\Omega R = 650$ ft/sec, $A_b = 240$ ft^2

Tail Rotor; σ, θ_1, M_{dr_3} of tip airfoil, α_s, ΩR, l_T/R_M.

$\sigma = .146$, $\theta_1 = -5°$, $M_{dr_3} = .80$

$\alpha_s = 0$, $\Omega R = 650$ ft/sec, $l_T/R_M = 1.23$

Airframe lift and drag.

Figure 2, Appendix I.

Flight conditions; G.W., μ, ρ/ρ_0, speed of sound, R/C.

G.W. = 20,000, $\mu = .3$, $\rho/\rho_0 = 1$

Speed of sound = 1,117 ft/sec, R/C = 1,000 ft/min

Find: Main rotor power, tail rotor power, trim conditions.

Procedure:

a. Calculate q.

$q = (1)(.001189)(\mu\,\Omega R)^2 = 45.2$

b. Calculate climb angle, γ_c, from

$$\gamma_c = \tan^{-1}\frac{\dfrac{R/C}{60}}{\mu\,\Omega R}$$

$$\gamma_c = \tan^{-1}\left[\frac{\dfrac{1,000}{60}}{(.3)(650)}\right] = 4.9°$$

c. Estimate angle of attack of fuselage,

$$\alpha_F = -\tan^{-1}\frac{qf}{\text{G.W.}} - \gamma_c.$$

$$\alpha_F = -\tan^{-1}\frac{(45.2)(19.3)}{20,000} - 4.9 = -7.4°$$

d. Find L_F/q and f from Figure 2, Appendix I.

$L_F/q = -19$, $f = 20.1$

e. Calculate C_T/σ from

$$C_T/\sigma = \frac{1}{\rho/\rho_0}\left(\frac{\text{G.W.} - L_F}{241,100}\right).$$

$$C_T/\sigma = (1)\left[\frac{20,000 - (-19)(45.2)}{241,100}\right] = .087$$

f. Find f_{climb} from

$$f_{\text{climb}} = f + \frac{\text{G.W.}\tan\gamma_c}{q}.$$

$$f_{\text{climb}} = 20.1 + \frac{20,000\tan 4.9°}{45.2} = 58.0$$

g. Calculate $\dfrac{f_{\text{climb}}}{A_b} + \dfrac{\sigma}{\mu^4}(C_T/\sigma)^2$

$$\frac{58}{240} + 124\,(.085)(.087)^2 = .321$$

h. From chart, find $\theta_{0_{\theta_1 = -5°}}$.

$\theta_{0_{\theta_1 = -5°}} = 15.8°$

i. From chart, find value of λ' and

$$\alpha_F = \tan \frac{\lambda'}{\mu}.$$

$\lambda' = -.046$

$$\alpha_F = \tan^{-1}\left(\frac{-.046}{.3}\right) = -8.7°$$

j. Find second values of L_F/q and f.

$L_F/q = -21.5, f = 20.5$

k. Calculate new value of C_T/σ.

$$C_T/\sigma = \frac{20,000 - (-21.5)(45.2)}{241,100} = .087$$

l. Find new value of

$$f_{\text{climb}}/A_b + \frac{\sigma}{\mu^4}(C_T/\sigma)^2$$

$$\left[\frac{20.5 + \dfrac{20,000 \tan 4.9°}{45.2}}{240}\right] + 123(.085)(.087)^2 = .323$$

m. Find new value of $\theta_{0_{\theta_1 = -5°}}$.

$\theta_{0_{\theta_1 = -5°}} = 15.9°$

n. Find C_Q/σ from chart.

$C_Q/\sigma = .0076$

o. Correct for twist effect as in example 1.

$C_Q/\sigma = .0074$

p. Check for compressibility as in example 1.

No compressibility losses.

q. Find h.p.$_M$

$$\text{h.p.}_M = 285,000 \, \rho/\rho_0 \, C_Q/\sigma$$

$$\text{h.p.}_M = 285,000(1)(.0074) = 2,109$$

r. Find T_T from

$$T_T = \frac{550 \, \text{h.p.}_M}{\Omega R}\left(\frac{1}{l_T/R_M}\right)$$

$$T_T = \frac{(550)(2,109)}{650}\left(\frac{1}{1.23}\right) = 1,451$$

s. Find C_T/σ_T.

$$C_T/\sigma_T = \frac{1,451}{(1)(19,550)} = .074$$

t. Find θ_{0_T} and $a_{1_{s_T}}$ as in example 3.

$\theta_{0_T} = 9.2°, \ a_{1_{s_T}} = 5.1°$

u. From chart, find

$$[C_H/\sigma - C_T/\sigma \tan a_{1_s}]_T$$

$$[C_H/\sigma - C_T/\sigma \tan a_{1_s}]_T = .0005$$

v. Calculate C_H/σ_T and H_T.

$$C_H/\sigma_T = .0005 + .074 \tan 5.1° = .0071$$

$$H_T = .0071(1)(19,550) = 139$$

w. Find C_H/σ_M from chart (assume $a_{1_{s_M}} = 0$)

$$C_H/\sigma_M = -.0011$$

x. Calculate H_M from

$$H_M = \rho/\rho_0 (241,100) \, C_H/\sigma_M.$$

$$H_M = (1)(241,100)(-.0011) = -265$$

y. Find new value of λ' from chart, calculate α_f

$$\lambda' = -.046$$

$$\alpha_F = \tan^{-1} \frac{-.046}{.3} = -8.70$$

z. Find new values of L_F/q and f.

$$L_F/q = -21.6, f = 20.6$$

aa. Find L_F and D_F.

$$L_F = 45.2 \, (-21.6) = -976$$

$$D_F = 45.2 \, (20.6) = 931$$

bb. Find T_M from

$$T_M = \sqrt{(G.W. - L_F)^2 + (D_F + H_M + H_T)^2}$$

$$T_M = \sqrt{(20,000 + 976)^2 + (931 - 265 + 143)^2} = 20,992$$

cc. Find C_T/σ

$$C_T/\sigma = \frac{20,992}{(1)(241,100)} = .087$$

dd. Calculate new f_{climb}

$$f_{climb} = f + H_T/q + G.W./q \tan \gamma$$

$$f_{climb} = 20.6 + \frac{139}{45.2} + \frac{20,000}{45.2} \tan 4.9° = 61.6$$

ee. Iterate from step 1 until values repeat.

ff. List trim conditions from final step of iteration.

$$\theta_0 = 16.1 - .75 \, (\theta_1 + 5) = 19.9°$$

$\alpha_F = -9.5°$
$L_F = -999$ lb
$D_F = 936$ lb
$T_M = 21,012$ lb
$\alpha_{TPP} = -7.1°$
$\lambda' = -.050$
$A_1 = -2.1°$
$B_1 = 9.4°$
$H_M = -337$ lb
$h.p._M = 2,138$
$T_T = 1,471$
$H_T = 143$
$h.p._T = 26$

7. Helicopter in Dive at Constant Collective Pitch

Example

Given: Main rotor; σ, θ_1, M_{dr_3} of tip airfoil, γ, ΩR, A_b

$\sigma = 0.085$, $\theta_1 = -10°$, $M_{dr_3} = .80$,

$\gamma = 8.1$, $\Omega R = 650$ ft/sec, $A_b = 240$ ft^2

| Tail rotor; σ, θ_1, M_{dr_3} of tip airfoil, α_s, γ, ΩR, l_T/R_M | $\sigma = .146$, $\theta_1 = -5°$, $M_{dr_3} = .80$, $\alpha_s = 0$, $\gamma = 4$, $\Omega R = 650$, $l_T/R_M = 1.23$ |

Airframe aerodynamics — See Figure 2, Appendix A.

Flight conditions; G.W., μ, ρ/ρ_0, speed of sound, θ_0

G.W. $= 20,000$, $\mu = .3$, $\rho/\rho_0 = 1$
Speed of sound $= 1,117$ ft/sec, $\theta_0 = 13°$

Find: Main rotor power, tail rotor power, rate of descent.

Procedure:

a. Calculate q.

$$q = (1)(.001189)[(.3)(650)]^2 = 45.2$$

b. Assume a value of α_F.

$$\alpha_F = 0°$$

c. Find L_F/q from Figure 2, Appendix A.

$$L_F/q = -4.5$$

d. Calculate C_T/σ.

$$C_T/\sigma = \frac{20,000 - (-4.5)(45.2)}{241,100} = .084$$

e. Find $\theta_{0_{\theta_1 = -5°}}$

$$\theta_{0_{\theta_1 = -5°}} = \theta_0 + .75(\theta_1 - 5).$$

$$\theta_{0_{\theta_1 = -5}} = 13 + .75(-10 + 5) = 9.25°$$

f. Find $\alpha_F = \tan^{-1} \dfrac{\lambda'}{\mu}$

$$\alpha_F = \tan^{-1}\left(\frac{.016}{.3}\right) = 3.1°$$

g. Find new L_F/q.

$$L_F/q = 1.3$$

h. Find new C_T/σ.

$$C_T/\sigma = \frac{20,000 - (1.3)(45.2)}{241,100} = .083$$

i. $\alpha_F = \tan^{-1} \dfrac{\lambda'}{\mu}$.

$$\alpha_F = \tan^{-1}\left(\frac{.015}{.3}\right) = 2.9°$$

j. Find C_Q/σ from chart.

$$C_Q/\sigma = .0008$$

k. Calculate h.p.$_M$.

$$\text{h.p.}_M = 1(285,000)(.0008) = 228$$

l. Calculate T_T.

$$T_T = \frac{\text{h.p.}_M(550)}{\Omega R} \frac{1}{l_T/R_M}$$

$$T_T = \frac{(228)(550)}{650} \frac{1}{1.23} = 157$$

m. Calculate C_T/σ_T.

$$C_T/\sigma_T = \frac{157}{19,550} = .008$$

n. Find θ_{0_T} and a_{1_s} by cross-plotting as in example 3.

$$\theta_{0_T} = 4.2° \qquad a_{1_s} = .6°$$

o. From chart, find

$$[C_H/\sigma - C_T/\sigma \tan a_{1_s}]_T$$

$$[C_H/\sigma - C_T/\sigma \tan a_{1_s}]_T = .0007$$

p. Calculate C_H/σ_T.

$$C_H/\sigma_T = .0007 + .008 \tan .6° = .0008$$

q. Calculate H_T.

$$H_T = (1)(.0008)(19,550) = 16 \text{ lb}$$

r. From chart, find

$$f/A_b + \frac{\sigma}{\mu^4}(C_T/\sigma)^2$$

$$f/A_b + \frac{\sigma}{\mu^4}(C_T/\sigma) = -.10$$

s. Find equivalent flat plate drag area of rotor, f_M

$$f_{M_{equiv}} = -A_b \left\{ \left[f/A_b - \frac{\sigma}{\mu^4}(C_T/\sigma)^2 \right] - \frac{\sigma}{\mu^4}(C_T/\sigma)^2 \right\}$$

$$f_M = -240\,[-.10 - 123(.085)(.083)^2] = 41.3$$

t. Find f of fuselage at last α_F.

$$f = 19.5$$

u. Find component of gravity required to overcome drag of rotor and fuselage

$$\text{G.W. } \sin \gamma_D = q(f + f_M) + H_T.$$

$$\text{G.W. } \sin \gamma_D = 45.2\,(19.5 + 41.3) + 16 = 2,764 \text{ lb}$$

Note: For level flight, $f_M = -(f + H_T/q)$

v. Find angle of dive

$$\gamma_D = \sin^{-1}\left(\frac{\text{G.W. } \sin \gamma_D}{\text{G.W.}} \right)$$

$$\gamma_D = \sin^{-1} \frac{2,764}{20,000} = 7.9°$$

w. Find rate of descent

$$R/D = 60\mu\,(\Omega R) \sin \gamma_D, \text{ ft/min}$$

$$R/D = 60(.3)(650) \sin 7.9 = 1,608 \text{ ft/min}$$

x. Find C_Q/σ_T from chart

$$C_Q/\sigma_T = .0010$$

y. Calculate h.p.$_T$

$$\text{h.p.}_T = (1)(23,100)(.0010) = 23 \text{ h.p.}$$

8. *Helicopter with Auxiliary Propulsion*

Given: Main rotor; σ, θ_1, M_{dr_3} of tip airfoil, γ, ΩR, A_b

Tail rotor; σ, θ_1, M_{dr_3} of tip airfoil, α_s, γ, ΩR, l_T/R_M

Airframe aerodynamics

Flight conditions; G.W., μ, ρ/ρ_0, speed of sound, θ_0

Find: Main rotor power, tail rotor power, auxiliary thrust required.

Procedure:

a. Calculate q

b. Assume a value of α_F

c. Find L_F/q from Figure 2, Appendix A

d. Calculate C_T/σ

e. Find $\theta_{0_{\theta_1 = -5^\circ}} = \theta_0 + .75(\theta_1 - 5)$

f. Find α_F from
$$\alpha_F = \tan^{-1} \frac{\lambda'}{\mu}$$

g. Find new L_F/q

h. Find new C_T/σ

i. $\alpha_F = \tan^{-1} \dfrac{\lambda'}{\mu}$

j. Find C_Q/σ from chart

k. Calculate h.p.$_M$

Example

$\sigma = .085$, $\theta_1 = -10^\circ$, $M_{dr_3} = .80$

$\gamma = 8.1$, $\Omega R = 650$ ft/sec, $A_b = 240$ ft^2

$\sigma = .146$, $\theta_1 = -5^\circ$, $M_{dr_3} = .80$

$\alpha_s = 0$, $\gamma = 4$, $\Omega R = 650$, $l_T/R_M = 1.23$

See Figure 2, Appendix A

G.W. $= 20,000$, $\mu = .3$, $\rho/\rho_0 = 1$
Speed of sound $= 1,117$ ft/sec, $\theta_0 = 13^\circ$

a. $q = (1)(.001189)\,[(.3)(650)]^2 = 45.2$

b. $\alpha_F = 0^\circ$

c. $L_F/q = -4.5$

d. $C_T/\sigma = \dfrac{20,000 - (-4.5)(45.2)}{241,100} = .084$

e. $\theta_{0_{\theta_1 = -5^\circ}} = 13 + .75(-10 + 5) = 9.25^\circ$

f. $\alpha_F = \tan^{-1} \dfrac{(.016)}{.3} = 3.1^\circ$

g. $L_F/q = 1.3$

h. $C_T/\sigma = \dfrac{20,000 - (1.3)(45.2)}{241,100} = .083$

i. $\alpha_F = \tan^{-1} \dfrac{(.015)}{.3} = 2.9^\circ$

j. $C_Q/\sigma = .0008$

k. h.p.$_M = 1(285,000)(.0008) = 228$

245

l. Calculate T_T

$$T_T = \frac{\text{h.p.}_M(550)}{\Omega R}\left(\frac{1}{l_T/R_M}\right)$$

$$T_T = \frac{(228)(550)}{650}\left(\frac{1}{1.23}\right) = 157$$

m. Calculate C_T/σ_T

$$C_T/\sigma_T = \frac{157}{19,550} = .008$$

n. Find θ_{0_T} and a_{1_s} by cross-plotting as in example 3.

$\theta_{0_T} = 4.2°$, $a_{1_s} = .6°$

o. From chart, find

$$[C_H/\sigma - C_T/\sigma \tan a_{1_s}]_T$$

$[C_H/\sigma - C_T/\sigma \tan a_{1_s}]_T = .0007$

p. Calculate C_H/σ_T

$C_H/\sigma_T = .0007 + .008 \tan .6° = .0008$

q. Calculate H_T

$H_T = (1)(.0008)(19,550) = 16$ lb

r. From chart, find

$$f/A_b + \frac{\sigma}{\mu^4}(C_T/\sigma)^2$$

$$f/A_b + \frac{\sigma}{\mu^4}(C_T/\sigma)^2 = -.10$$

s. Find equivalent flat plate drag area of rotor, f_M

$$f_{M_{equiv}} = -A_b\left[f/A_b - \frac{\sigma}{\mu^4}(C_T/\sigma)^2\right]$$

$$f_{M_{equiv}} = -240[-.10 - 123(.085)(.083)^2] = 41.3$$

t. Find f of fuselage at last α_F

$f = 19.5$

u. Find T_{aux} from

$$T_{aux} = q(f + f_M) + H_T$$

$T_{aux} = 45.2 (19.5 + 41.3) + 16 = 2,764$ lb

v. Find C_Q/σ_T from chart

$C_Q/\sigma_T = .0011$

w. Calculate h.p.$_T$

h.p.$_T = (1)(23,100)(.0011) = 26$ h.p.

(Note similarity to results of previous case.)

9. Rotor in a Wind Tunnel

Given: Rotor; σ, θ_1, γ, M_{dr_3} of tip airfoil

Test conditions; μ, $M_{1,90}$, $\theta_{.75}$, α_s, a_{1_s}, b_{1_s}

Tunnel proportions; $\dfrac{\text{Width}}{\text{Height}} = \gamma_{\text{tun}}$

$\dfrac{\text{Model height}}{\text{½ tunnel height}} = \zeta$

$\dfrac{\text{Disc area}}{\text{Test section area}} = \dfrac{A_M}{A_{\text{tun}}}$

Find: C_T/σ, C_P/σ, C_{X_R}/σ, A_1, B_1

Procedure:

a. Find $\alpha_{\text{TPP}_{\text{uncorr}}} = \alpha_s + a_{1_s}$

b. Find $\theta_{0_{\theta_1 = -5°}} = \theta_{.75} + .75(5°)$

c. At $\theta_{0_{\theta_1 = -5°}}$ plot λ' versus C_T/σ from chart.
Also plot λ' versus C_T/σ from:

$$\lambda' = \dfrac{\mu\,\alpha_{\text{TPP}_{\text{uncorr}}}}{57.3} - C_T/\sigma\,\dfrac{\sigma}{2\mu}$$

d. Read C_T/σ at intersection

e. Calculate downwash angle,

$$\chi = \cot^{-1}\left(\dfrac{2v_1}{V}\right) = \cot^{-1}\left(C_T/\sigma\,\dfrac{\sigma}{\mu^2}\right)$$

(H-34 Rotor in 40 × 80 foot wind tunnel, test 276, run 3, reference 3.27)

$\sigma = .062$, $\theta_1 = -8°$, $\gamma = 9.3$, $M_{dr_3} = .80$

$\mu = .3$, $M_{1,90} = .74$, $a_{1_s} = .3°$, $b_{1_s} = .2°$

$\theta_{.75} = 9°$ (corrected from 8° to account for identified discrepancy)

$\alpha_s = 5°$ (uncorrected for wall effects)

$\gamma_{\text{tun}} = 2$

$\zeta = 1$

$\dfrac{A_M}{A_{\text{tun}}} = \dfrac{1,820}{3,200} = .57$

$\alpha_{\text{TPP}_{\text{uncorr}}} = 5 + .3 = 5.3°$

$\theta_{0_{\theta_1 = -5°}} = 9 + 3.75 = 12.75°$

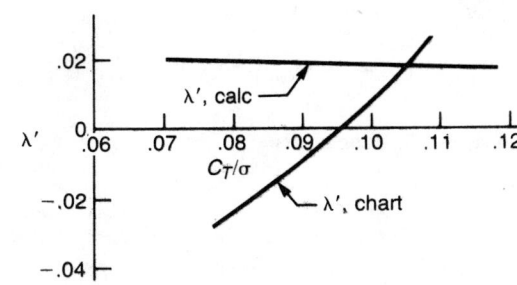

d. Read C_T/σ at intersection

$C_T/\sigma = .105$

$\chi = \cot^{-1}\dfrac{(.105)(.062)}{.3^2} = 86°$

f. From reference 3.47, find $\delta_{W,L}$ at prescribed values of γ, ζ, χ

$\delta_{W,L} = -.54$

g. Find

$$\Delta\alpha_{TPP} = -\delta_{W,L}\left(\frac{A_M}{A_{tun}}\right) 57.3 \; C_T/\sigma \; \frac{\sigma}{2\mu^2}$$

$$\Delta\alpha_{TPP} = -(-.54)(.57)(57.3) \frac{(.105)(.062)}{2(.3)^2} = .6°$$

h. Calculate $\alpha_{TPP_{corr}} = \alpha_{TPP_{uncorr}} + \Delta\alpha_{TPP}$.

$\alpha_{TPP_{corr}} = 5.3 + .6 = 5.9°$

i. Repeat steps c and d to find corrected value of C_T/σ using

$$\lambda' = \frac{\mu\alpha_{TPP_{corr}}}{57.3} - C_T/\sigma \; \frac{\sigma}{2\mu}$$

$C_T/\sigma = .106$

j. Find $C_Q/\sigma = C_P/\sigma$ from chart

$C_Q/\sigma = .0044$

k. Is rotor operating beyond limit line of $\Delta C_Q/\sigma_0 = 0.004$?

Yes

l. Calculate effective shift of limit lines due to different twist than chart value

$$\Delta C_T/\sigma = -.003(\theta_1 + 5°)$$

$$\Delta C_T/\sigma = -.003(-8 + 5) = .009$$

m. Estimate new value of C_Q/σ with limit lines and point of sudden change of curvature shifted by $\Delta C_T/\sigma$

$C_Q/\sigma = .0037$

n. Find C_H/σ from chart and

$C_H/\sigma = -.0033$

$$C_{X_R}/\sigma = -C_T/\sigma \sin \alpha_{TPP_{uncorr}} - C_H/\sigma$$

$$C_{X_R}/\sigma = -(.106)(\sin 5.3°) - (-.0033) = -.0065$$

o. Find A_1 and B_1 using charts

$$\frac{\gamma}{8}(A_1 - b_{1_s}) = -3.5°$$

$$A_1 = \frac{8}{9.2} - 3.5 + .2 = -2.8°$$

$$B_1 + a_{1_s} = 10.2° \qquad B_1 = 10.2 - .3 = 9.9°$$

p. Compare with test results (Table IV-1, Reference 3.27, $\theta_{.75} = 8°$, $\alpha_s = 5°$)

(See Figure 3.62 for complete correlation.)

Parameter	Test	Chart Method
C_T/σ	.110	.106
C_P/σ	.0040	.0037
C_{X_R}/σ	−.0058	−.0065
A_1	−3.3°	−2.8°
B_1	10°	9.9°

WIND TUNNEL TESTS OF ROTORS

Table 3.6 lists the important parameters of several wind tunnel tests of rotors, which have been selected to show the state of the art of rotor testing and to provide candidates for correlation with theory.

TABLE 3.6
Summary of Wind Tunnel Tests of Rotors

Radius (ft)	Chord (in.)	No. of Blades	Twist (deg)	Solidity	Airfoil	Max. Test Values			Ref.	Date
						μ	$M_{1,90}$	C_T/σ		
22	16.5	3	−7	.065	0012	.4	.74	.09	3.49	1958
22	16.5	3	−7	.064	23012	.4	.74	.13	3.49	1958
1	2	2	0	.106	0012	.4	.46	.16	3.33	1964
7.6	14	2	0	.097	0012	.54	.30	.16	3.50	1964
7.6	14	2	0	.097	0012	1.45	.14	.18	3.51	1965
23	21	3	−10.5	.073	0012	.4	.86	.10	3.52	1965
28	16	4	−8	.062	0012	.46	.82	.12	3.27	1968
28	16	4	0	.062	0012	1.05	.93	.12	3.27	1968
24	21	2	−10.9	.046	0012	.4	1.00	.18	3.27	1968
17	21	2	−7.7	.066	0012	.79	.65	.18	3.27	1968
4.6	2.7	4	0	.062	0012	.5	.94	.10	3.53	1974
4.6	2.7	4	−8	.062	0012	.3	.81	.11	3.53	1974
4.6	2.7	4	0	.062	0012 w/ 5° "flap"	.3	.81	.10	3.53	1974
8.8	3.1	4	−10	.075	SC1095	.37	.94	.10	3.54	1983
4.4	15.5	4	−10	.075	SC1095	.37	.94	.10	3.54	1983

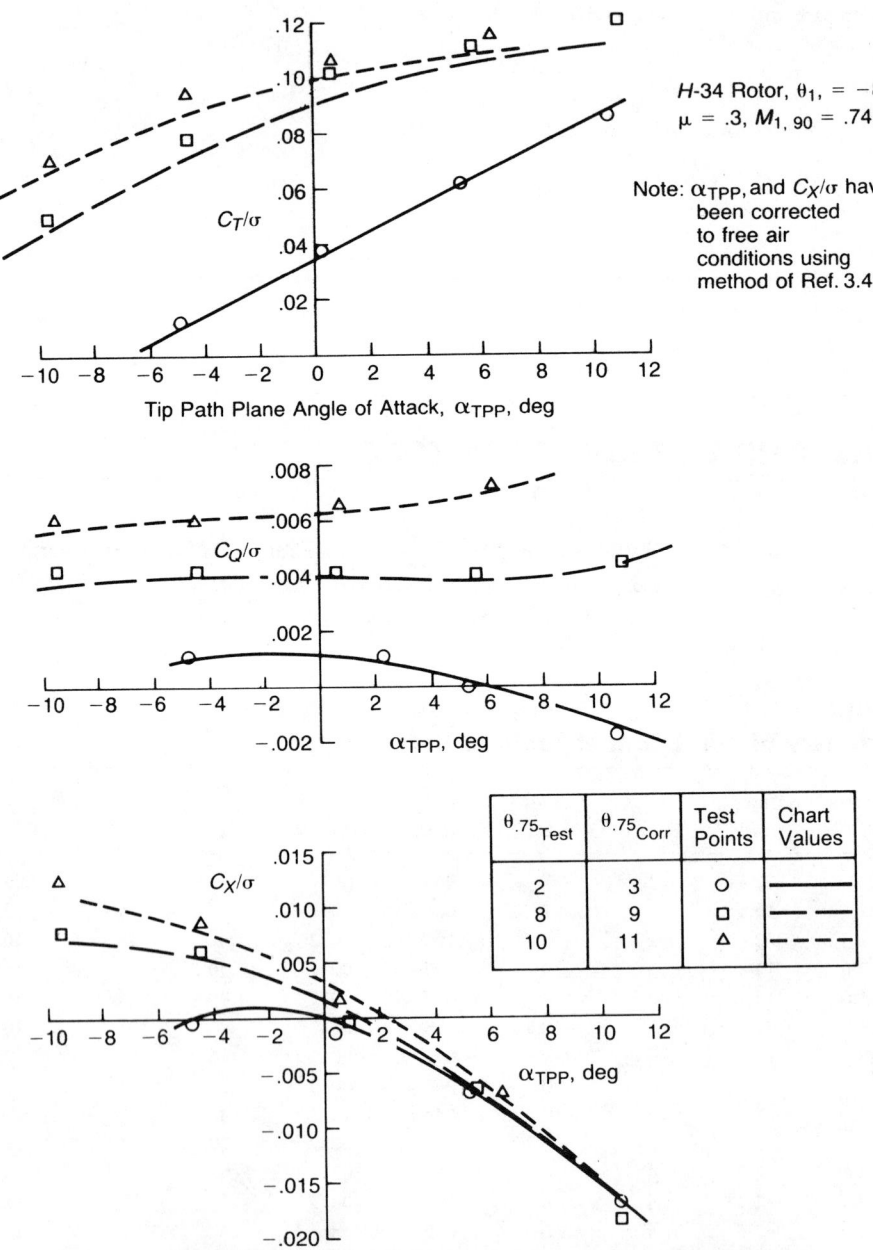

FIGURE 3.62 Correlation of Calculations with Test Data

Source: McCloud, Biggers, & Stroub, "An Investigation of Full-Scale Helicopter Rotors at High Advance Ratios and Advancing Tip Mach Numbers," NASA TN D4632, 1968.

EXAMPLE HELICOPTER CALCULATIONS

HOW TO'S

The following items can be evaluated by the methods of this chapter.

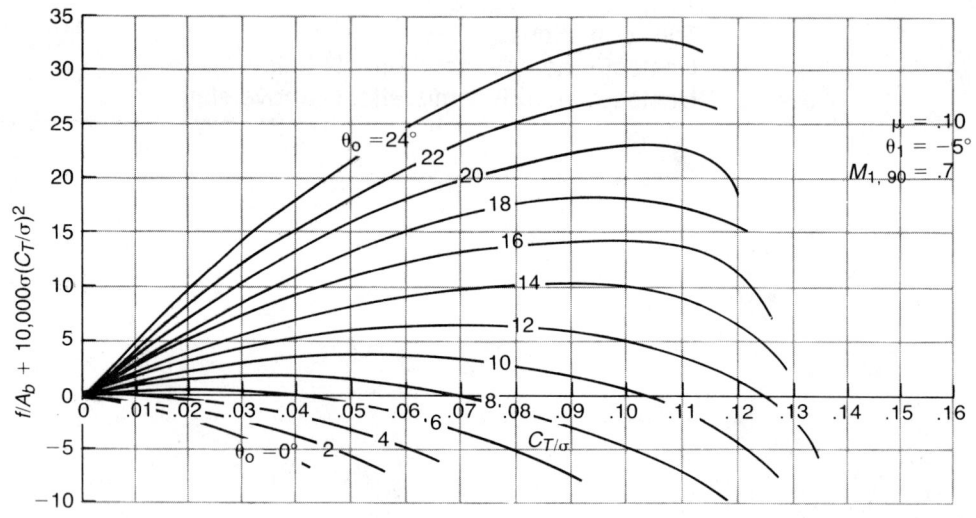

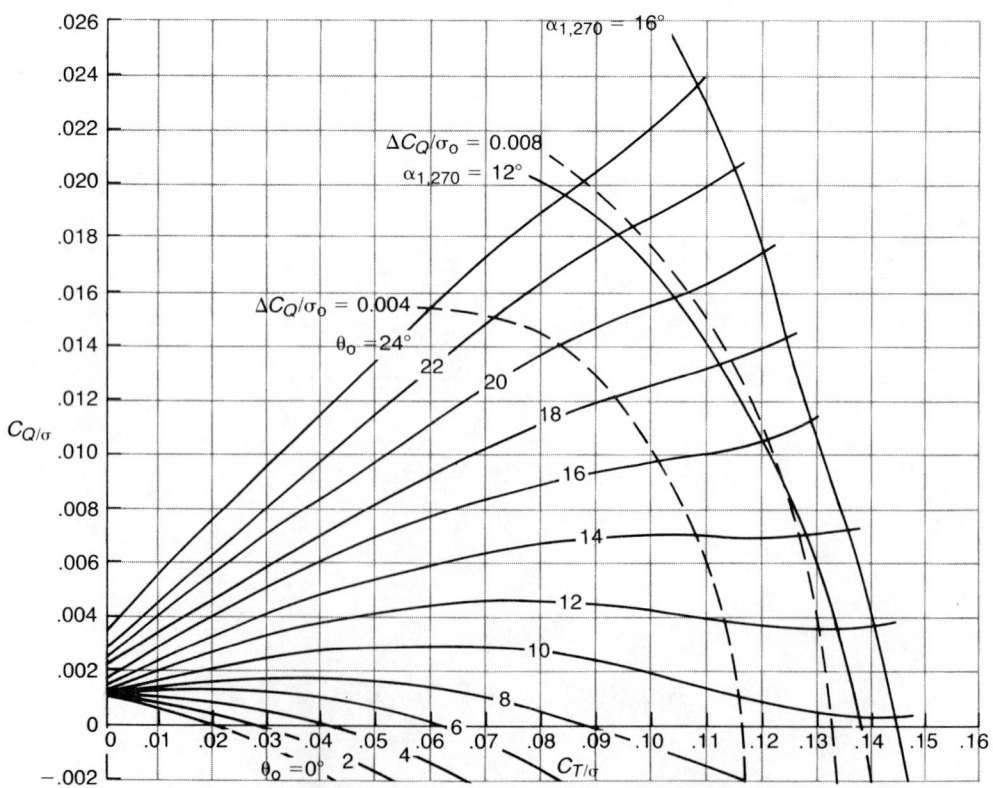

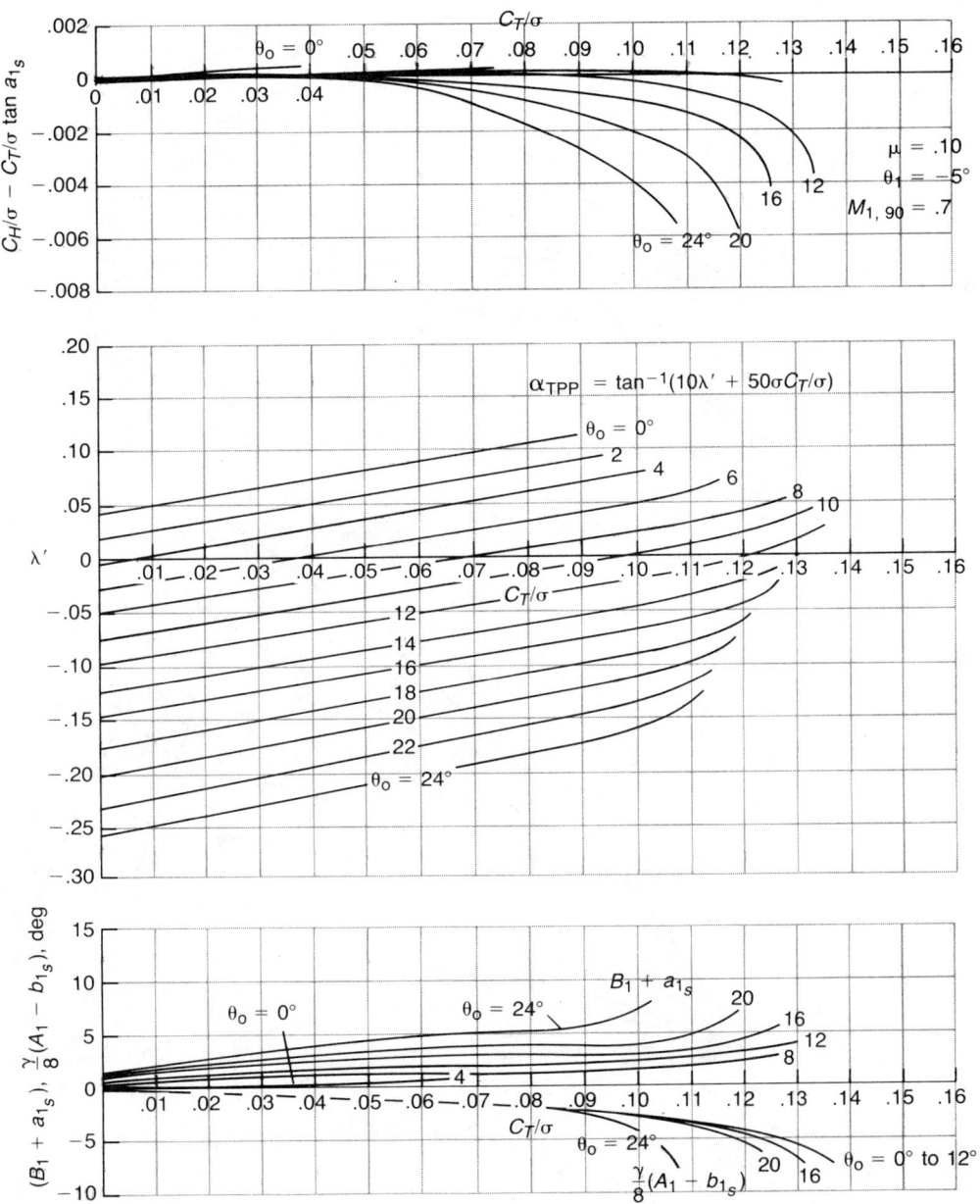

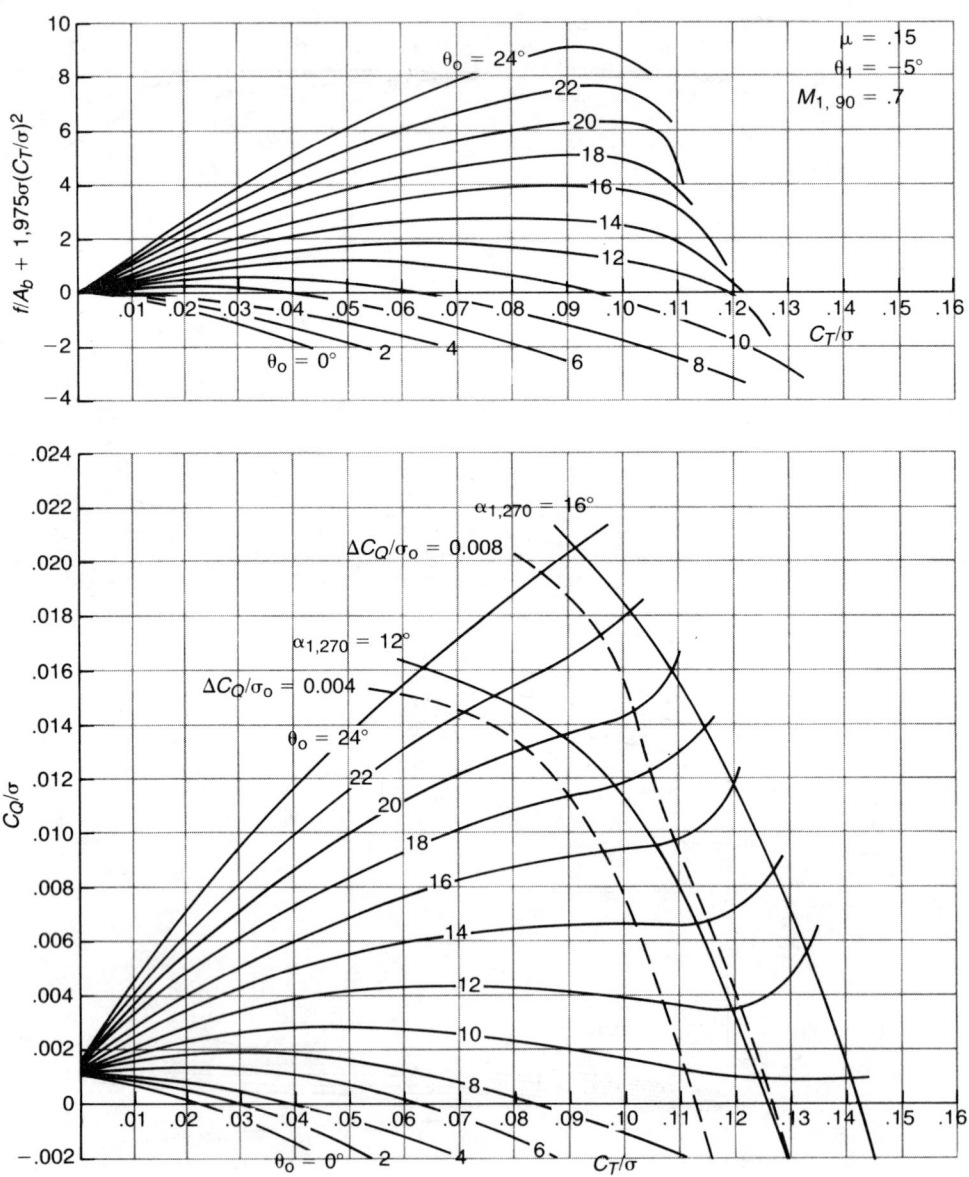

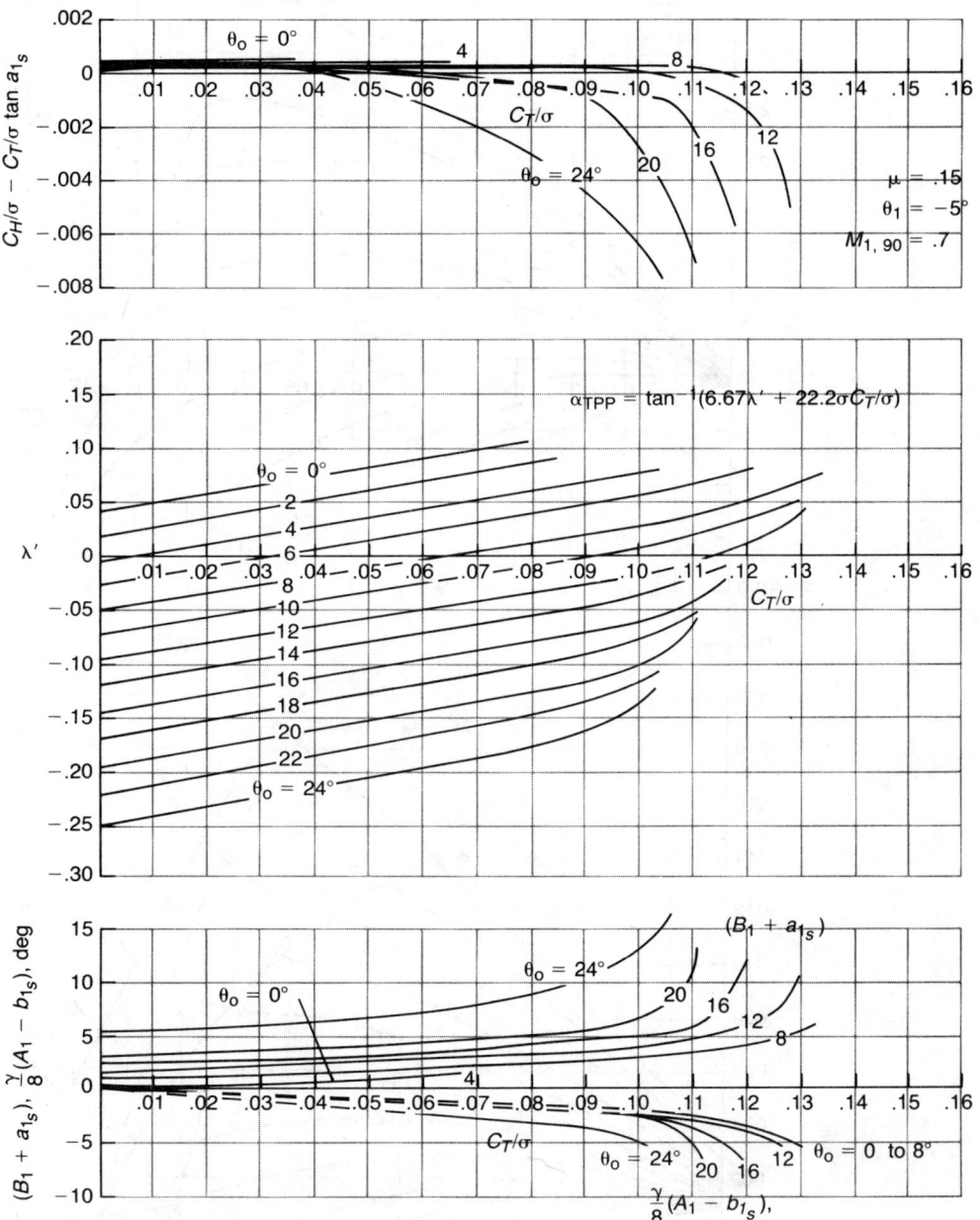

$C_H/\sigma - C_T/\sigma \tan a_{1_s}$

$\theta_o = 0°$

4

8

C_T/σ

$\theta_o = 24°$ 20   16   12

$\mu = .15$
$\theta_1 = -5°$
$M_{1,90} = .7$

$\alpha_{TPP} = \tan^{-1}(6.67\lambda' + 22.2\sigma C_T/\sigma)$

λ'

$\theta_o = 0°$
2
4
6
8
10
12
14
16
18
20
22
$\theta_o = 24°$

C_T/σ

$(B_1 + a_{1_s})$, $\dfrac{\gamma}{8}(A_1 - b_{1_s})$, deg

15

$(B_1 + a_{1_s})$

$\theta_o = 24°$ 20   16   12   8

$\theta_o = 0°$

4

C_T/σ

$\theta_o = 24°$ 20   16   12 $\theta_o = 0$ to $8°$

$\dfrac{\gamma}{8}(A_1 - b_{1_s})$,

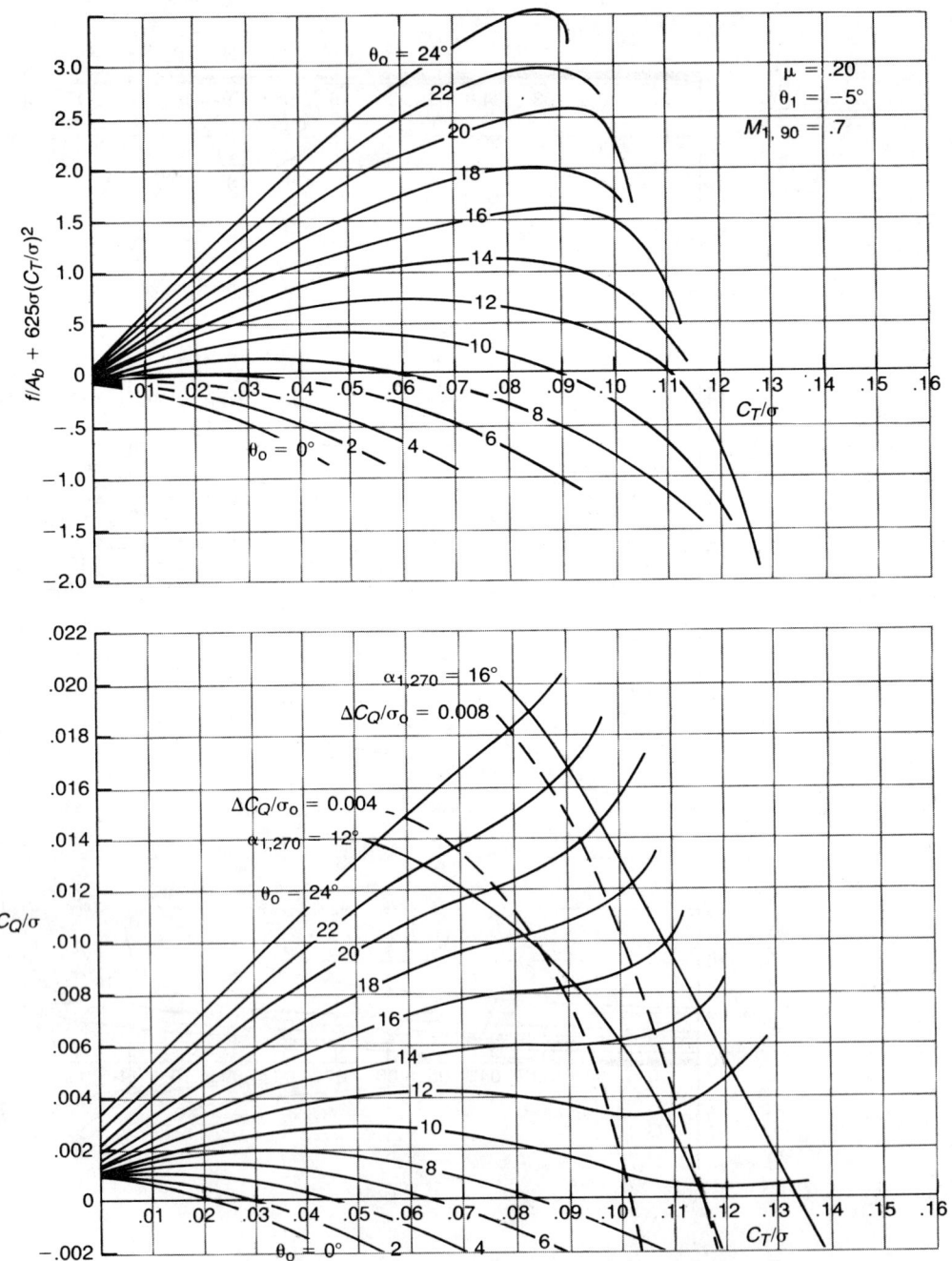

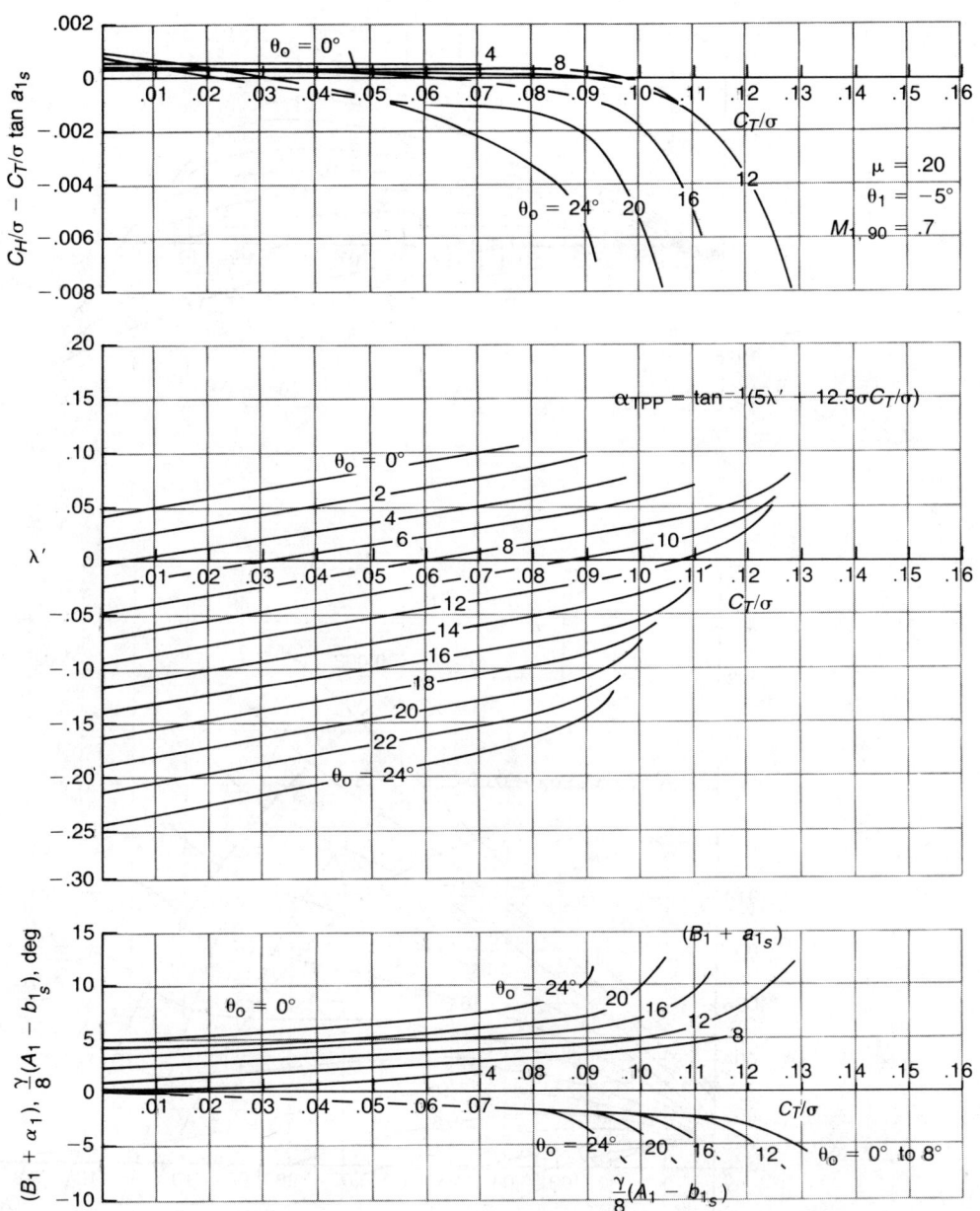

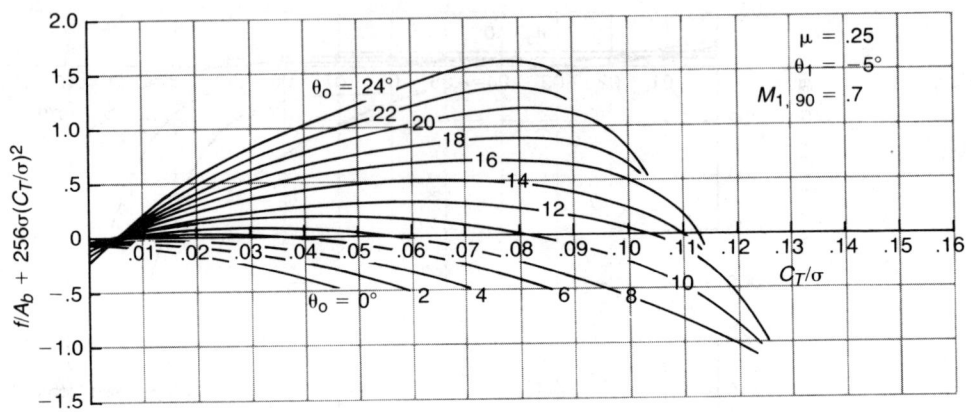

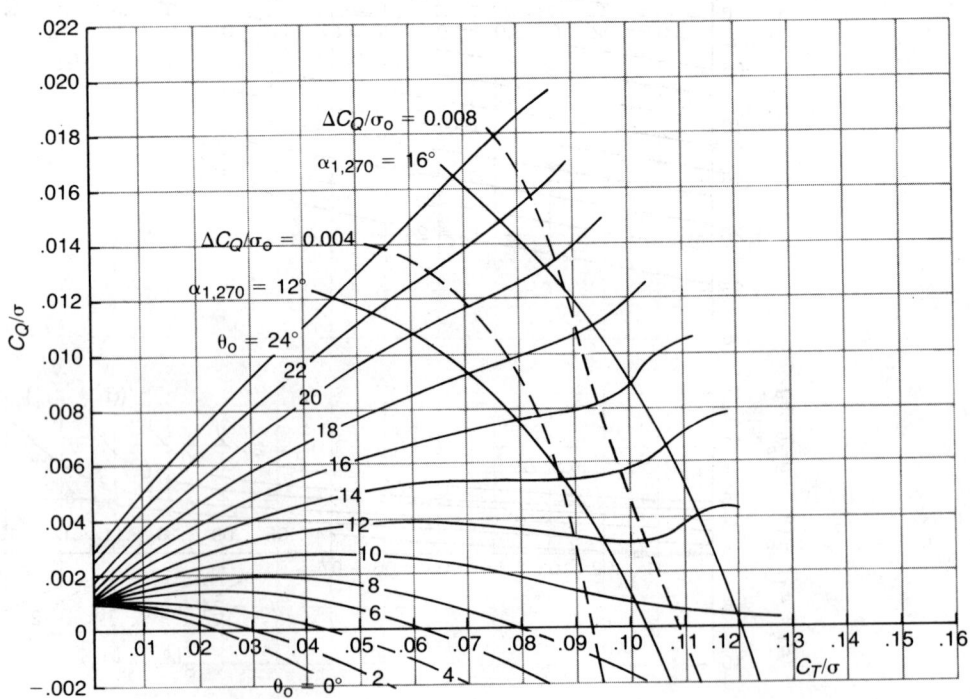

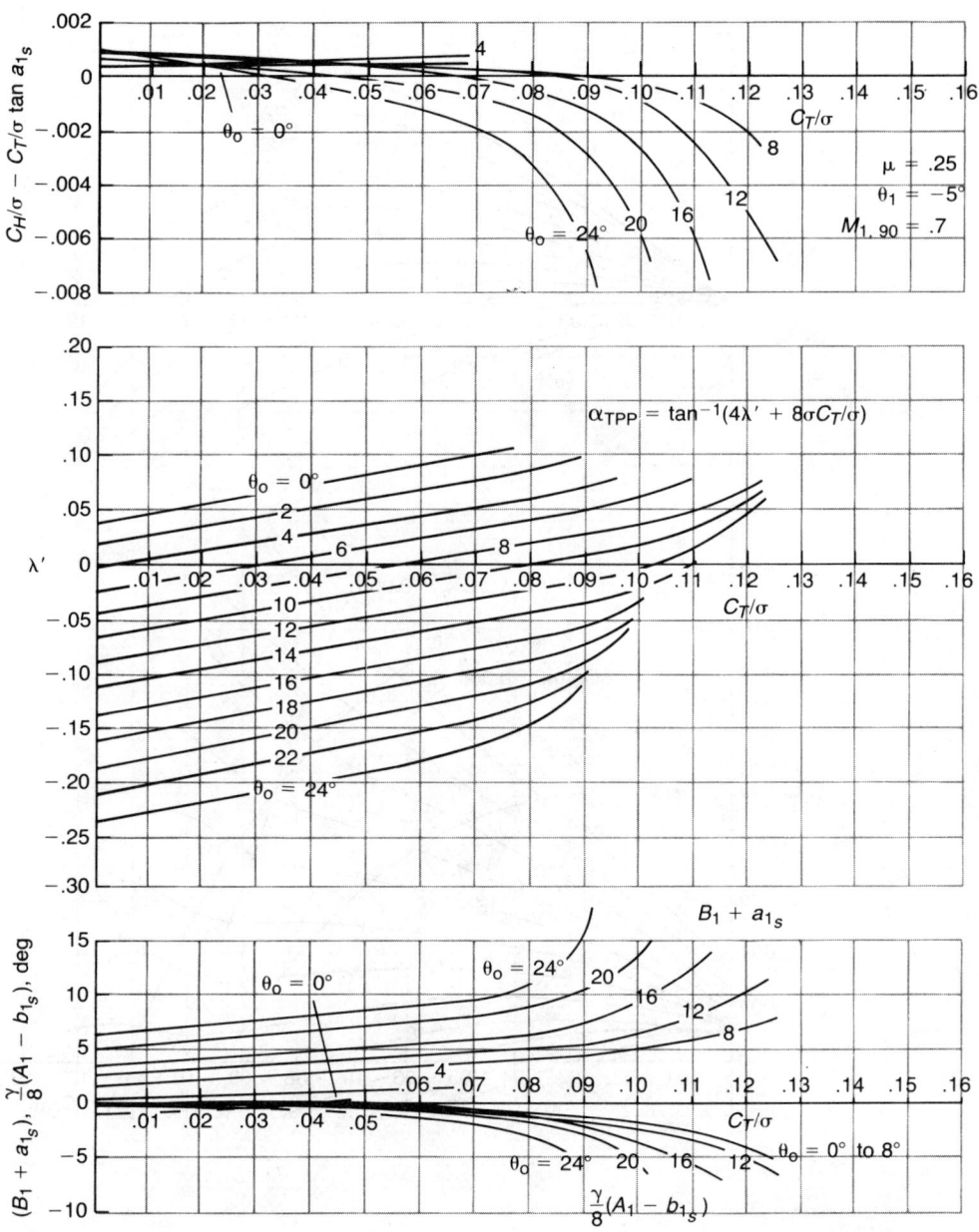

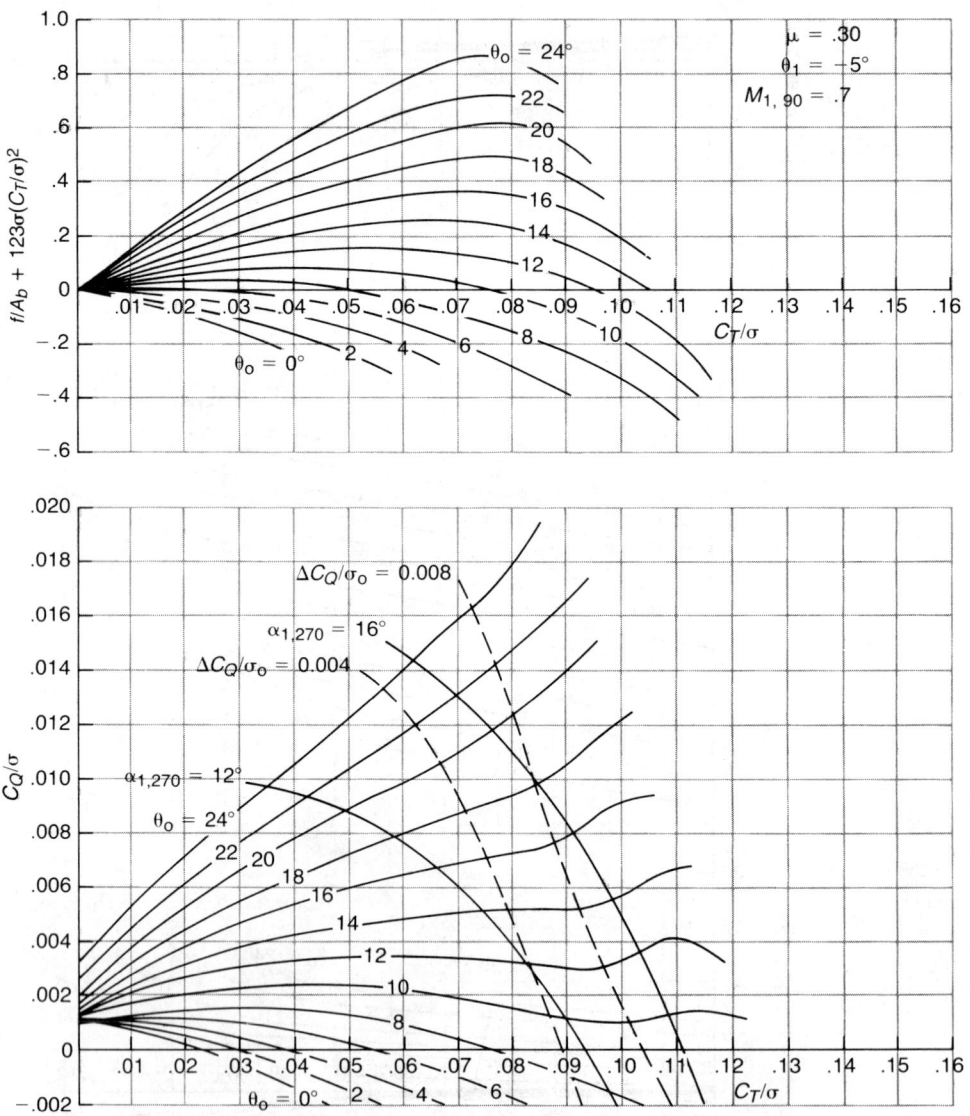

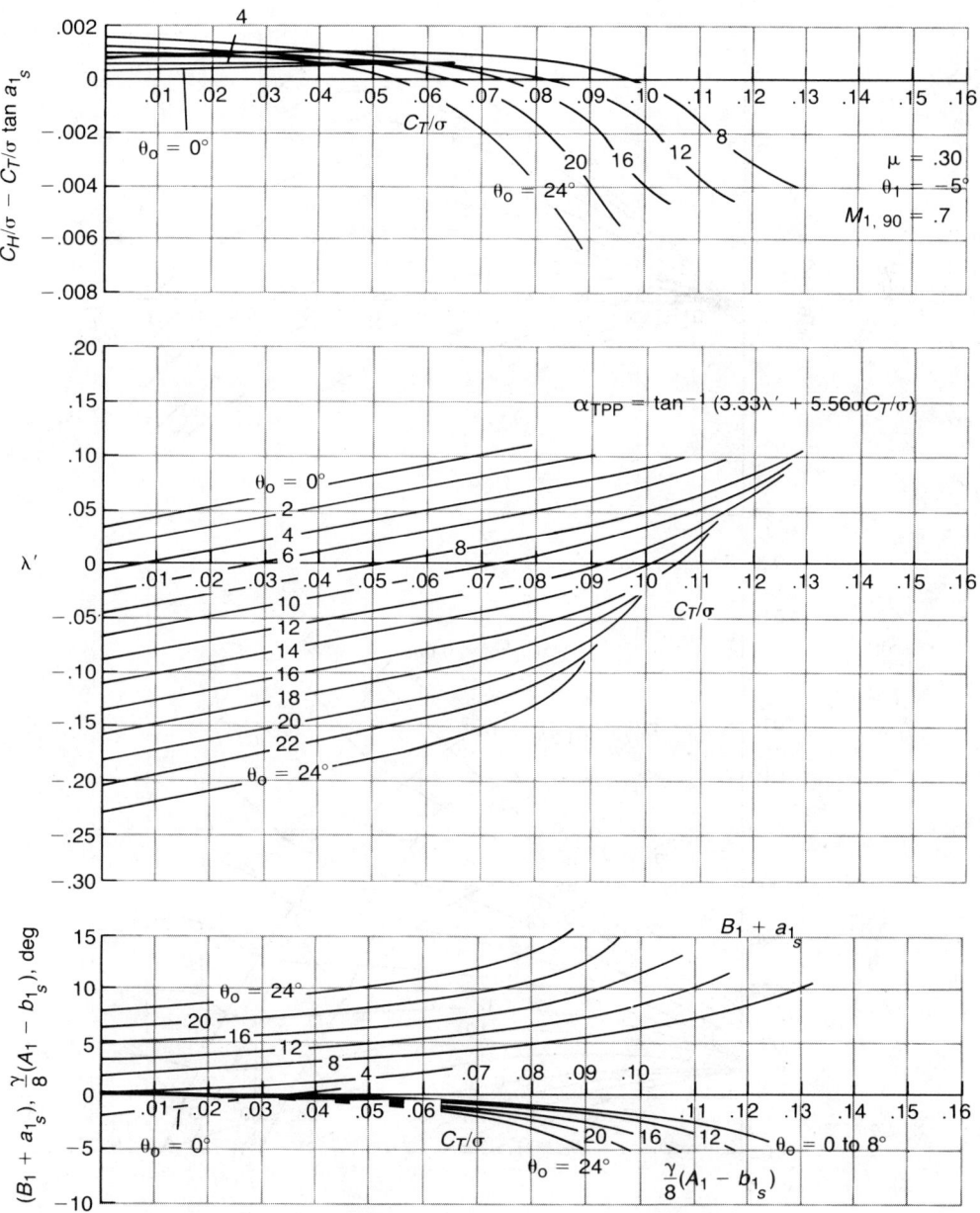

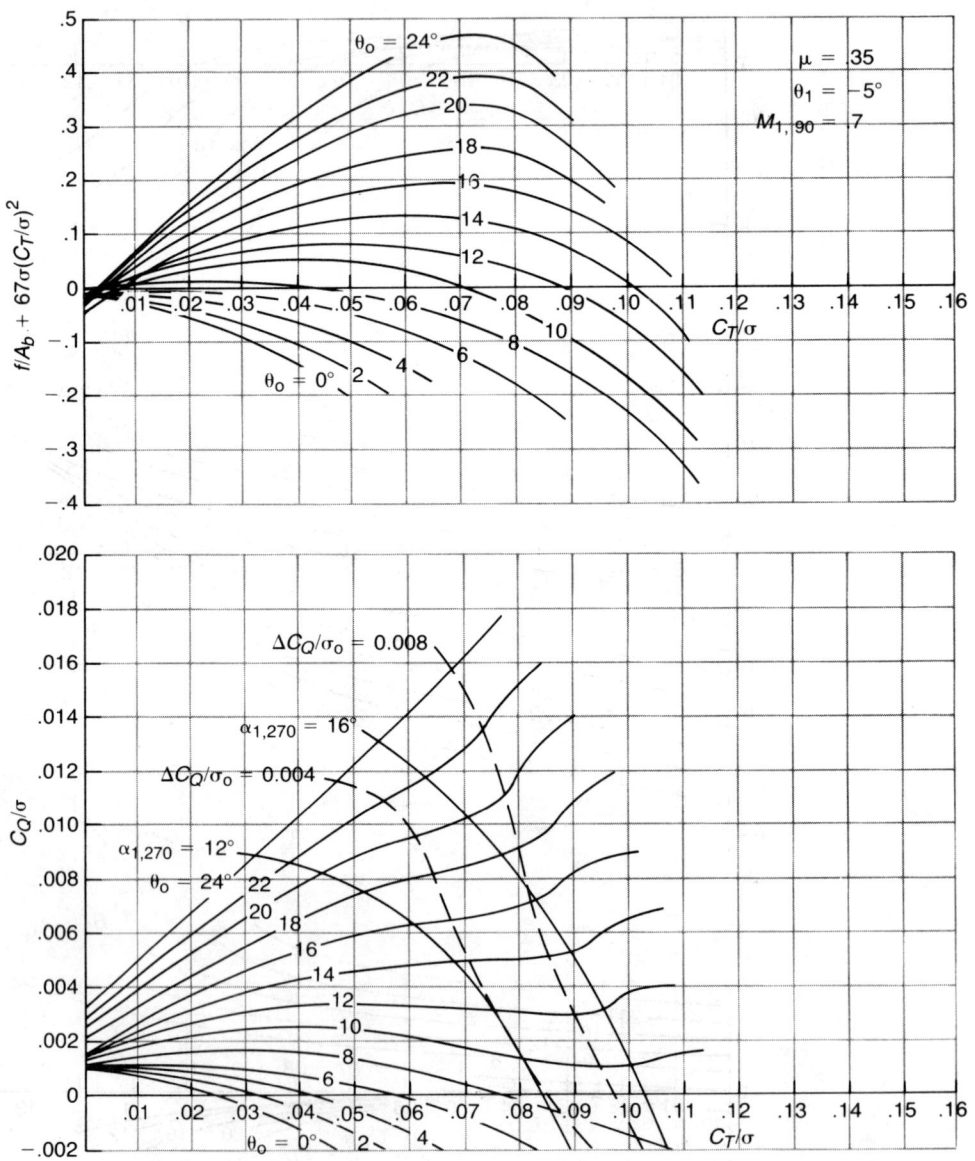

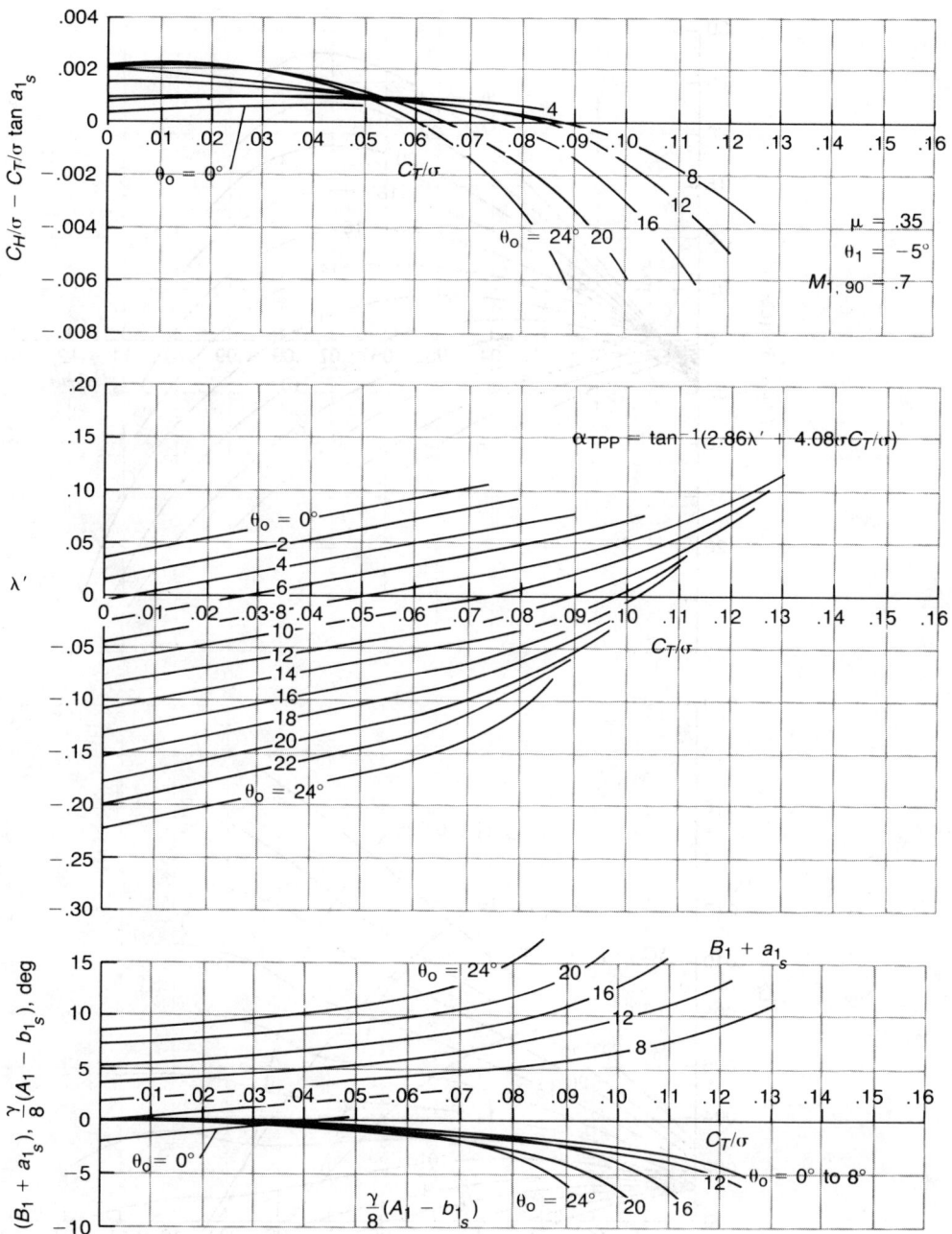

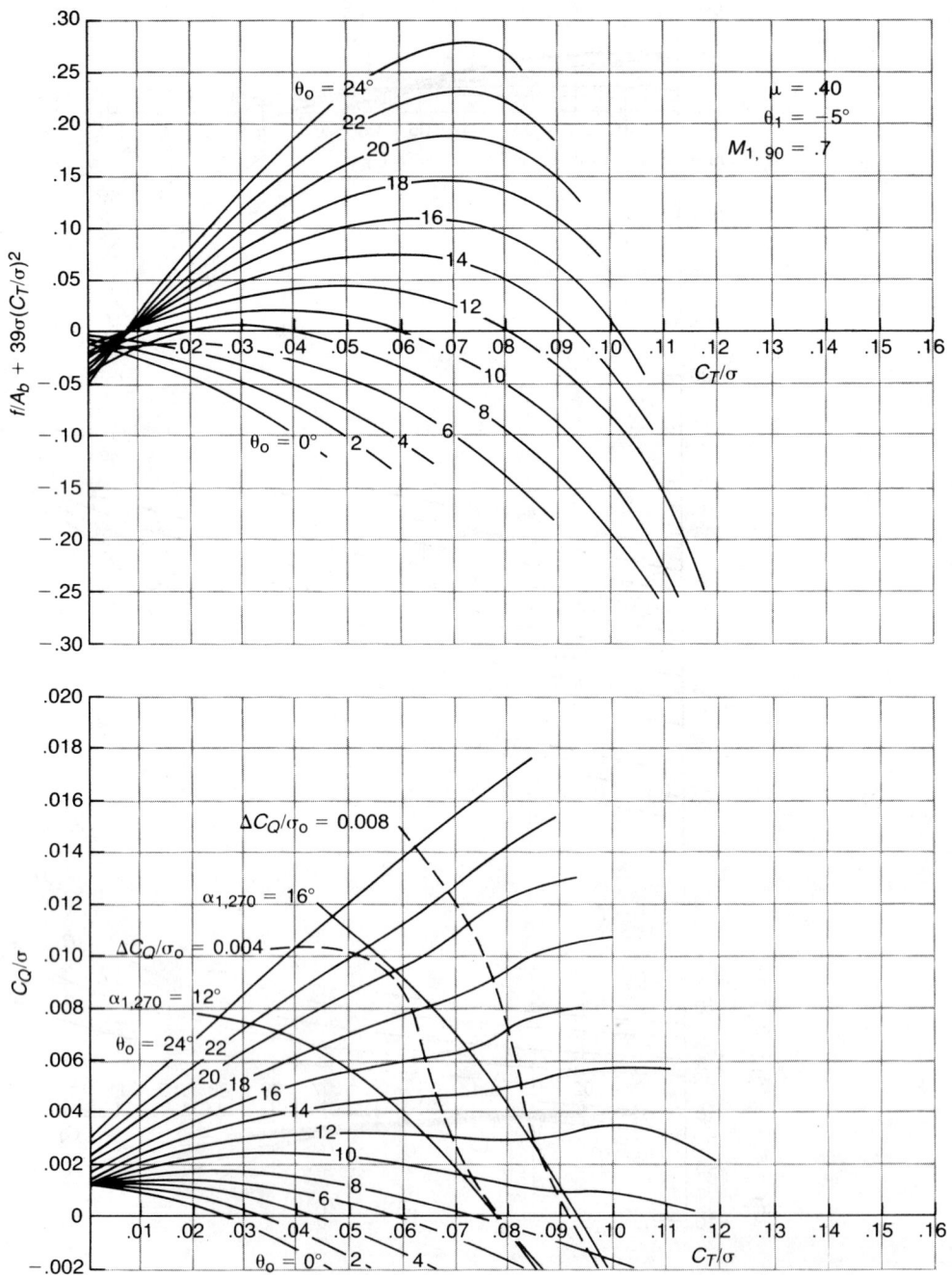

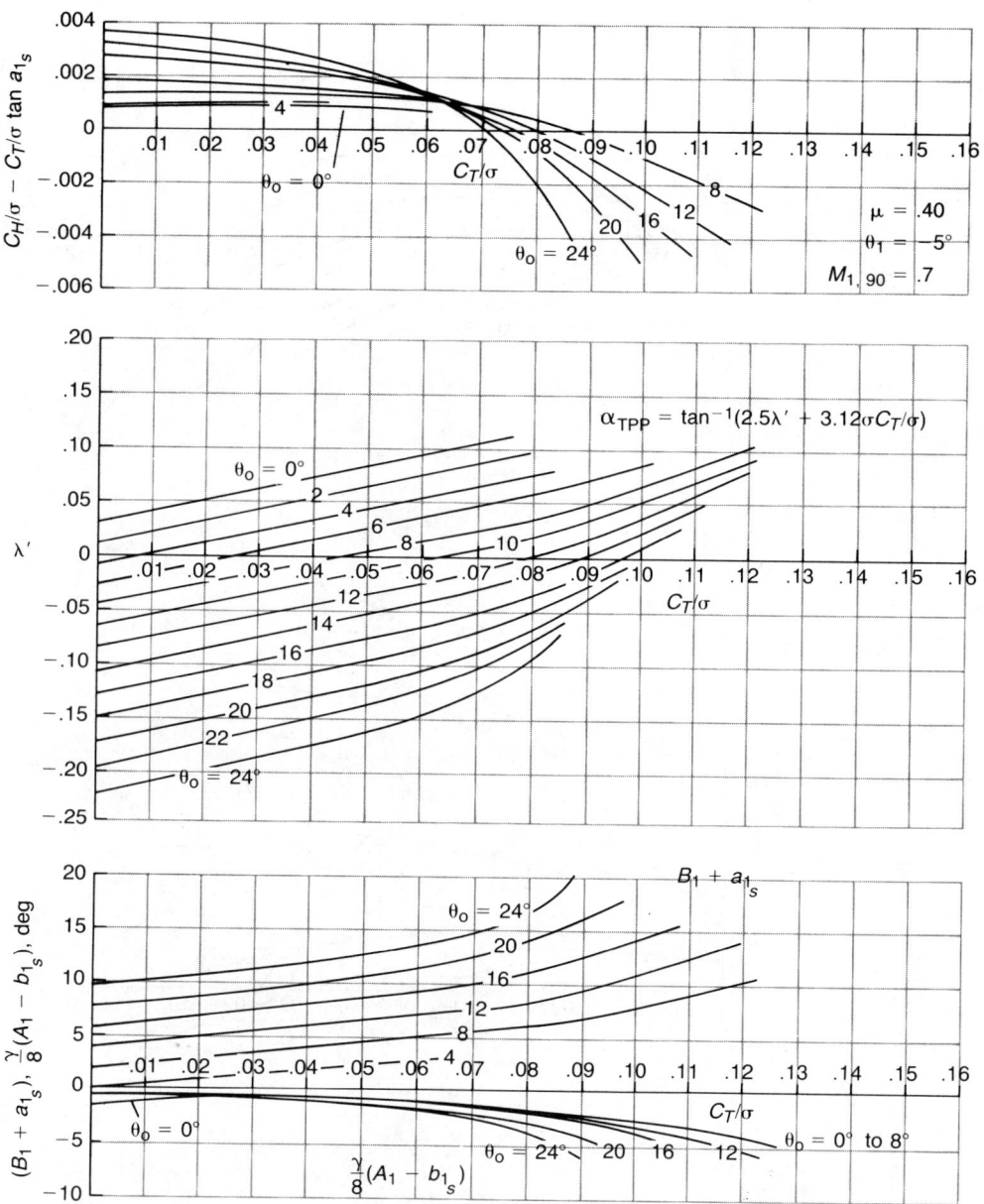

$\alpha_{TPP} = \tan^{-1}(2.5\lambda' + 3.12\sigma C_T/\sigma)$

$\mu = .40$
$\theta_1 = -5°$
$M_{1,90} = .7$

267

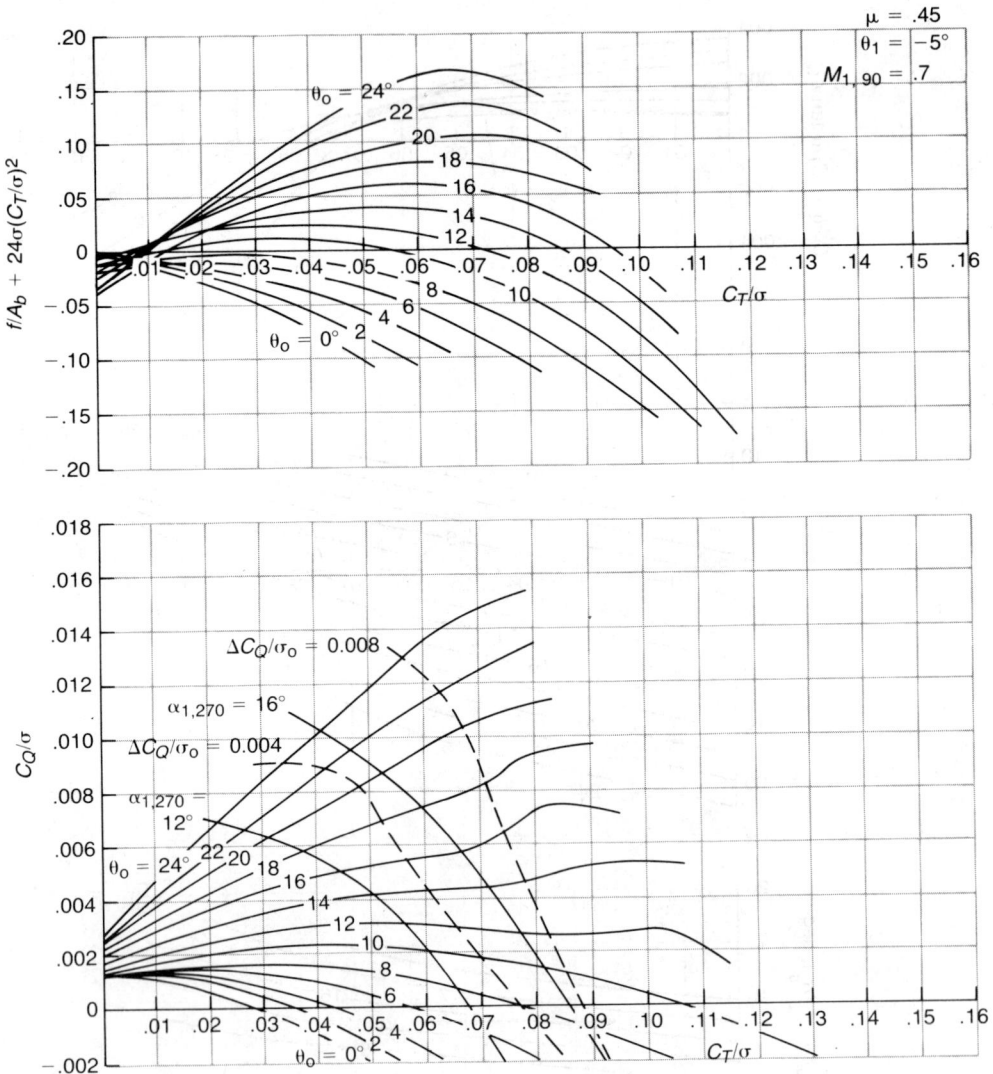

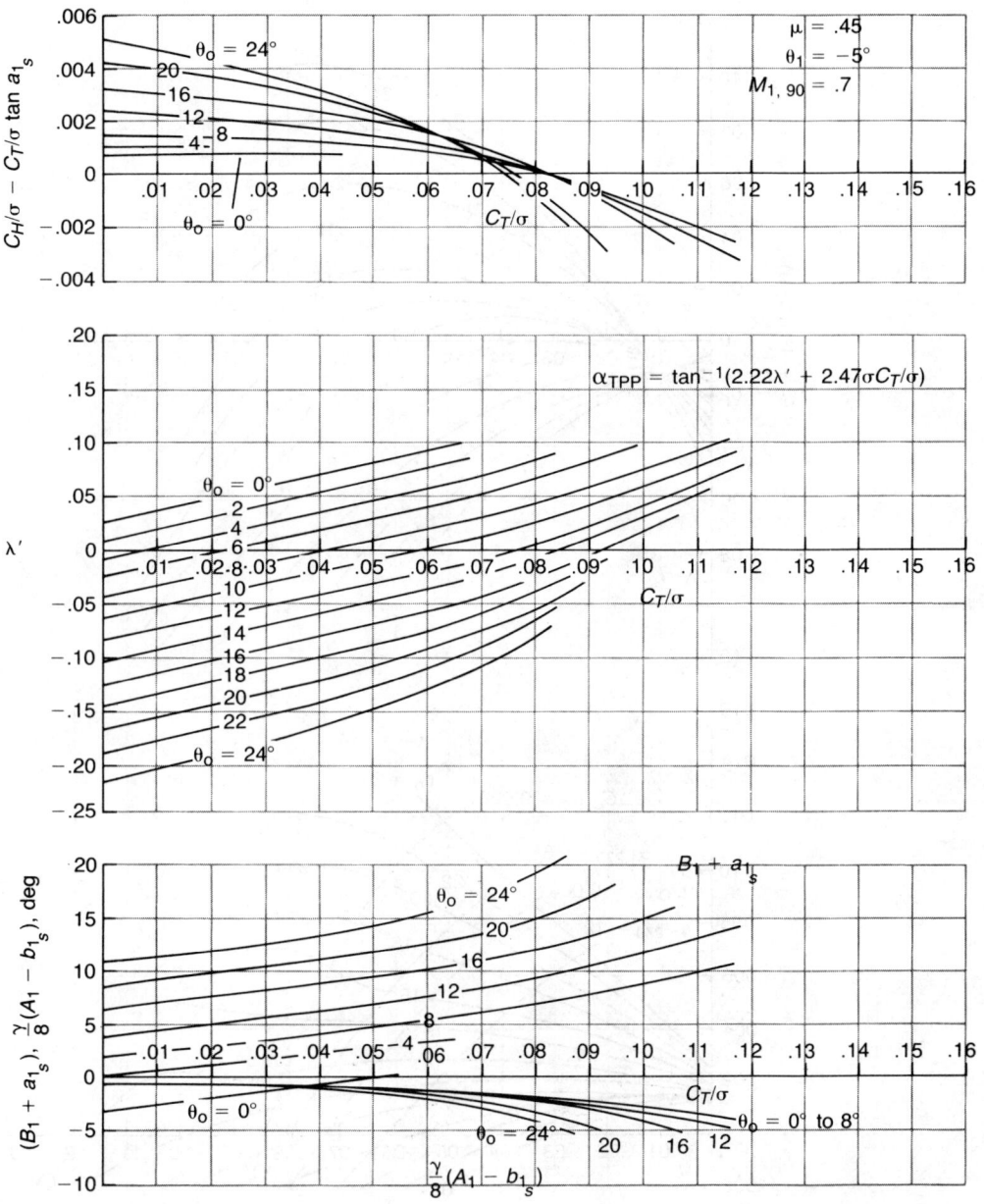

$$\mu = .45$$
$$\theta_1 = -5°$$
$$M_{1,90} = .7$$

$$\alpha_{TPP} = \tan^{-1}(2.22\lambda' + 2.47\sigma C_T/\sigma)$$

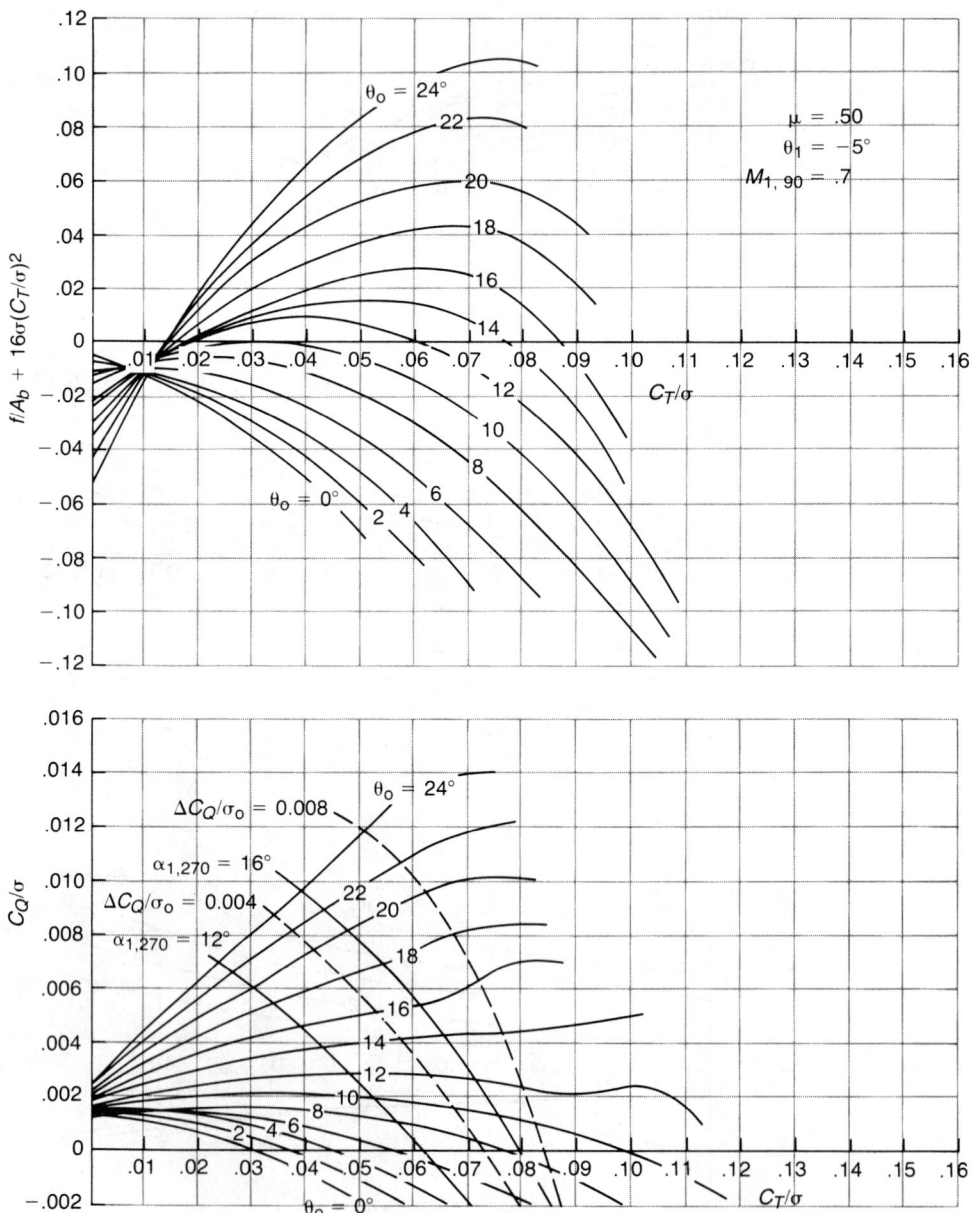

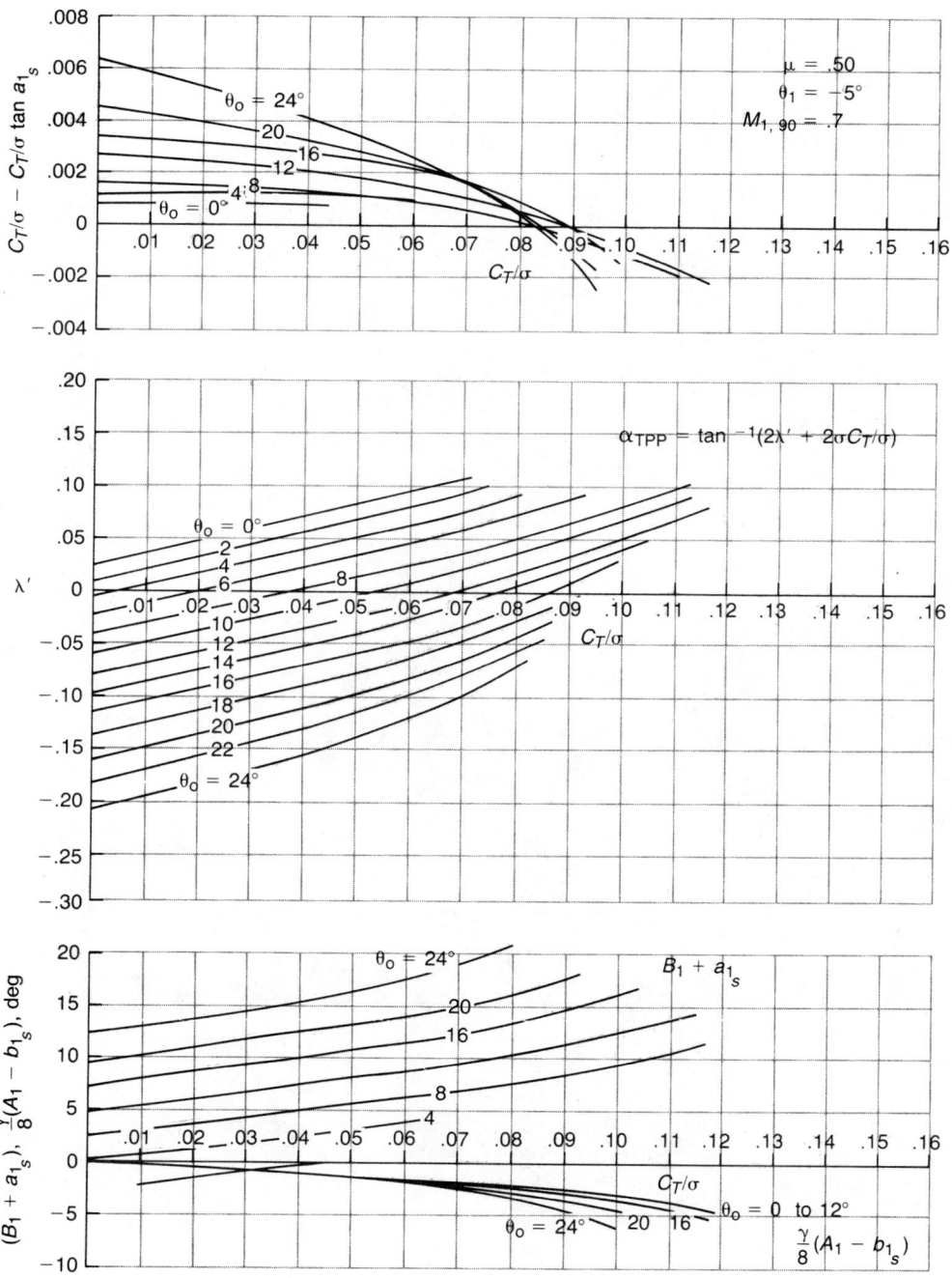

CHAPTER 4

Performance Analysis

INTRODUCTION

A helicopter performance analysis is made to answer the questions:

How high?

How fast?

How far?

How long?

The results of the analysis may be used in design tradeoff studies, in a pilot's handbook, in a set of military Standard Aircraft Characteristic charts (SAC charts), or in a sales brochure. Before the analysis can be done, it is necessary to collect the individual items of information that are required: the performance of the individual rotors, the installed engine performance, the power losses in transmissions and accessories, the vertical drag in hover, the tail rotor-fin interference, and the parasite drag in forward flight. Methods of estimating rotor performance have been described in the preceding chapters. The other items will be discussed in the following paragraphs.

ENGINE PERFORMANCE

Almost all of the performance characteristics of the helicopter depend on engine performance. Engine manufacturers describe the performance of their engines in a specification. Among other things, these documents present the engine ratings and fuel consumptions under various conditions, based on the engine performance as measured in a test cell and then modified by standard and accepted techniques to account for the effects of altitude, temperature, and forward flight that cannot be duplicated in the test cell. In general, there are three types of ratings of interest to the helicopter engineer, which apply to both reciprocating and turboshaft engines:

Rating	*Allowable Time Limit*
Emergency, takeoff, or contingency	2–10 minutes
Military or intermediate	30 minutes
Maximum continuous or normal	No limit

On turboshaft engines, the rating is limited by the maximum allowable turbine inlet temperature, torque, or fuel flow. On reciprocating engines the limits are usually based on maximum intake manifold pressure and rpm. The ratings are a function of altitude, temperature, and forward speed. Figure 4.1 shows the zero

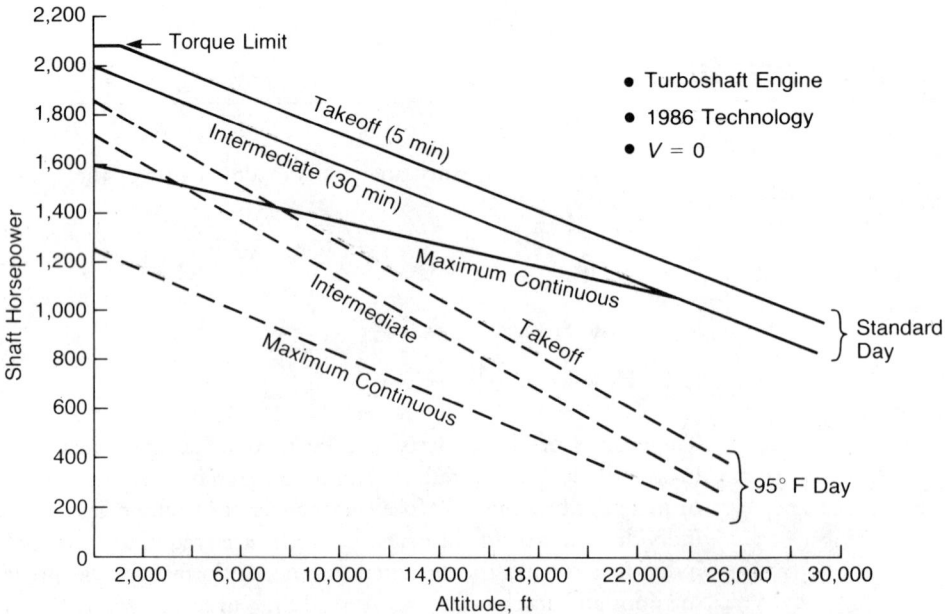

FIGURE 4.1 Typical Uninstalled Engine Ratings as a Function of Altitude and Temperature

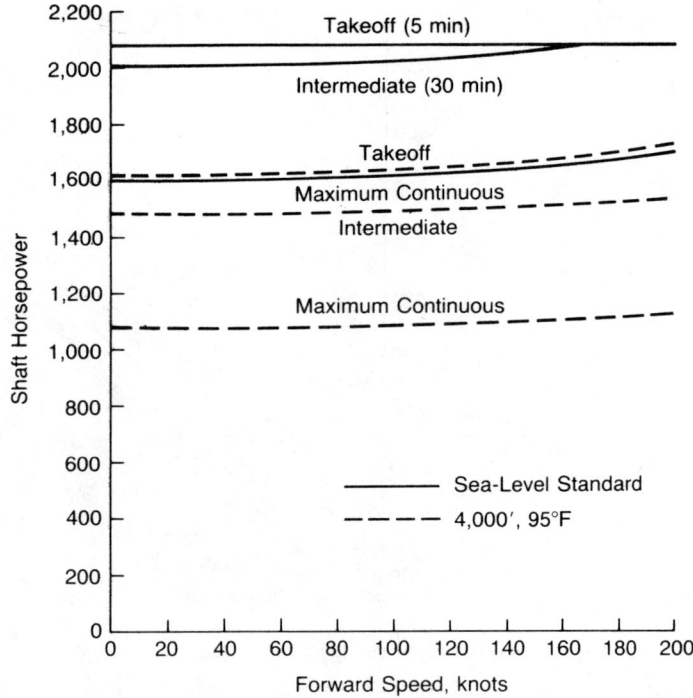

FIGURE 4.2   Effect of Forward Speed on Uninstalled Engine Ratings

forward speed engine ratings for a typical turboshaft engine corresponding to the one installed in the example helicopter. Figure 4.2 shows how the ratings are affected by forward speed due to the ram effect on the performance of the compressor.

The fuel flow at a given power is also a function of altitude and temperature. Fuel flow or specific fuel consumption values in the engine specification apply only to new engines. The military requires that when preparing SAC charts, all fuel flow figures be raised 5 percent to account for possible engine deterioration. Figure 4.3 shows typical fuel flow curves incorporating the 5 percent increase.

POWER LOSSES

There are two types of losses that must be considered in the performance analysis: those that affect the power available from the engine and those that must be added to the power required by the rotors. A convenient dividing line to distinguish between the two types of losses is the engine torquemeter, a standard component

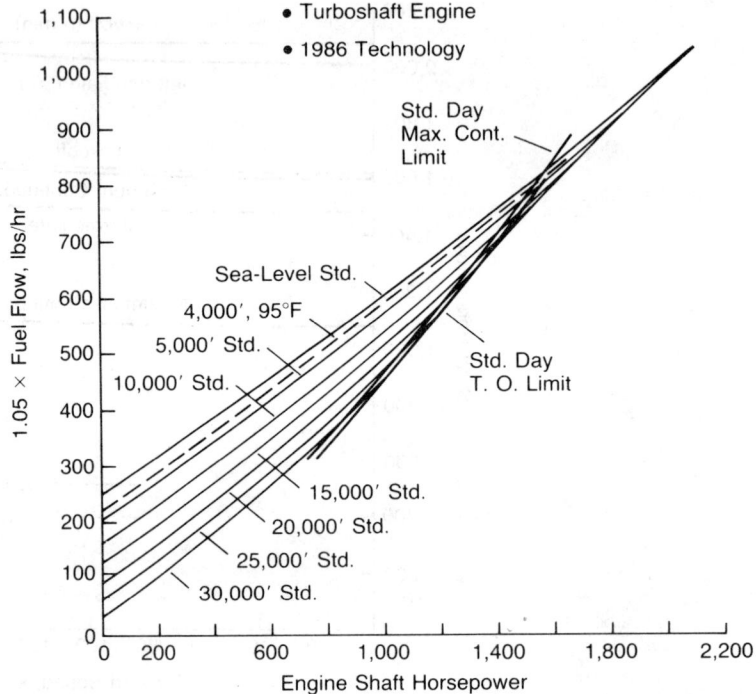

FIGURE 4.3   Typical Engine Fuel Flow Characteristics

of turboshaft engines. (On reciprocating engines, the input shaft to the transmission is the logical division point.)

Engine Installation Losses

The engine as installed in the helicopter will usually not deliver the same power as it does in the engine manufacturer's test cell for a variety of reasons. Some of these reasons and their possible effects are as follows:

	Typical Values
Inlet pressure losses due to duct friction	1–4% of power
Inlet pressure losses due to a particle separator	3–10% of power
Exhaust back pressure due to friction	0.5–2% of power
Exhaust back pressure due to an infrared suppressor	3–15% of power
Rise in inlet temperature due to exhaust reingestion	1–4 degrees F
Compressor bleed	1–20% of power
Engine-mounted accessories	Up to 100 h.p.

Since it is not practical, or even possible, to give methods here for evaluating the magnitude of these various losses, the aerodynamicist must enlist the aid of a specialist to make a study of the specific engine installation. For the example helicopter, it will be assumed that inlet friction decreases the engine power by 2% and that no other losses exist. Thus 98% of the ratings shown in Figures 4.1 and 4.2 will be used for illustrative calculations.

Power Required Losses

Losses which occur between the torquemeter and the rotors must be made up by the engine and are additive to the power required by the rotors. Sources of these losses include:

- Main rotor transmission
- Tail rotor gearboxes
- Engine nose gearboxes
- Transmission-mounted accessories such as generators and hydraulic pumps
- Cooling fans driven from the drive system

Gearbox and transmission losses are produced by friction between the gear teeth and in the bearings and by aerodynamic drag, or "windage." The losses are a function of the size of the gearbox as well as the power being transmitted at any given time. For preliminary design purposes, the losses in gearboxes—including power to run their lubrication systems—can be estimated from the following equation:

Power loss per stage $= K$[Design max. power $+$ Actual power], h.p.

where $K = 0.0025$ for spur or bevel gears and $K = 0.00375$ for planetary gears.

The example helicopter will be assumed to have two engine nose gearboxes with one stage of bevel gears, each designed for 2,000 h.p. The main rotor transmission has two spur gears stages and one planetary gear stage, with a design capacity of 4,000 h.p. The tail rotor drive system contains two gear boxes with one stage of bevel gears each and a design capability of 750 h.p. The total transmission losses are thus:

$$\text{h.p.}_{\text{trans.}} = 0.0025 \ (4,000 + \text{Eng. power}) + 0.00875 \ (4,000 + \text{Main rotor power}) + 0.0050 \ (750 + \text{Tail rotor power})$$

Assuming that the engine power is the sum of the main and tail rotor powers, the losses become:

$$\text{h.p.}_{\text{trans.}} = 49 + .0112 \text{ main rotor power} + .0075 \text{ tail rotor power}$$

Losses due to generators and hydraulic pumps are a function of the load on them during any given flight condition. Assuming typical efficiencies:

$$\text{Generator loss} = \frac{\text{Load in watts}}{(.75)(746)}, \text{h.p.}$$

$$\text{Hydraulic pump loss} = \frac{(\text{Design pressure, psi})(\text{Flow rate, gpm})}{(.80)(1,714)}, \text{h.p.}$$

For performance calculations on the example helicopter, it will be assumed that the generator produces a constant 2,200 watts, which results in a loss of 4 h.p. Hydraulic pumps deliver most of their flow during maneuvers, but even during steady flight they are pumping some fluid. For the example helicopter a minimum flow rate of 1.3 gpm with a 3,000 psi system will be assumed. This gives a hydraulic pump loss of 3 h.p. No separate shaft-driven cooling fans are in the configuration, but, of course, electrically or hydraulically driven fans are already accounted for.

Since the rotor performance is a function of the atmospheric density ratio, the calculations of total engine power will be simplified if it is assumed that both the transmission and accessory losses are also proportional to the density ratio. (This is an assumption more weighted toward convenience than accuracy, but valid enough for power estimating.) Using this assumption for the example helicopter:

$$\text{h.p.}_{\text{trans.+acc.}} = \frac{\rho}{\rho_0} (56 + .0112 \text{ h.p.}_M + .0075 \text{ h.p.}_T)$$

VERTICAL DRAG IN HOVER

A method for making a rough estimate of the vertical drag penalty in hover was given in Chapter 1. This method will now be refined in order to raise the confidence level in the hover performance calculations. The method consists of the following steps:

- Divide the plan view of the airframe into segments.
- Estimate the drag coefficient of each segment as a function of its shape.
- Determine the distribution of dynamic pressure in the rotor wake.
- Sum the effects of each segment.
- Calculate the rotor thrust as the sum of the weight and the vertical drag.
- Correct the rotor power at this thrust for the "ground effect" due to the airframe.

Figure 4.4 shows a half-plan view of the example helicopter divided into segments by concentric circles around the rotor mast. Also shown is a side view with wake stations designated below the rotor. The drag coefficient for each segment depends on the shape of its cross section. Drag measurements of cylinders and flat plates under a rotor are reported in reference 4.1. It is concluded in that report that the turbulence in the rotor wake is always high enough to insure that fully turbulent, or supercritical, boundary-layer conditions exist. Supercritical drag coefficients for a number of two-dimensional shapes with flow from above are shown in Figure 4.5. These drag coefficients are based on data presented in references 4.1 and 4.2.

Tilt-rotor aircraft have a large wing to be accounted for. Deflecting a trailing edge flap can reduce its vertical drag penalty but wind tunnel test results reported in reference 4.3 indicate that the optimum flap angle is about 60° rather than the 90° that would be expected to produce less drag.

The measured distribution of dynamic pressure under a full-scale rotor with −4° of twist was shown on Figure 1.19 of Chapter 1. For rotors with different twist, the measured distribution can be modified by multiplying by the square of the ratio of induced velocities calculated by the method of Chapter 1 for the two values of twist. Figure 4.6 shows the −4° twist distributions modified for −10° of twist for use with the example helicopter.

Using the drag coefficients and the dynamic pressure corresponding to each of the airframe segments, the vertical drag penalty can now be calculated by

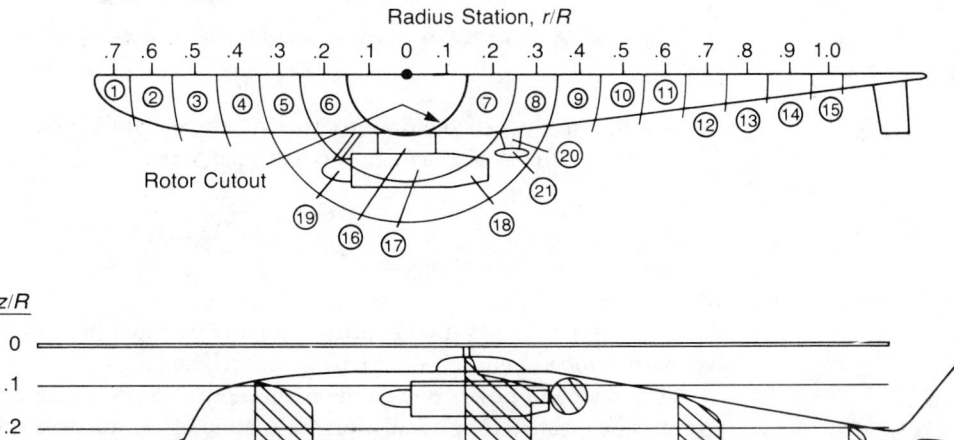

FIGURE 4.4   Aircraft Elements Used in Vertical Drag Analysis

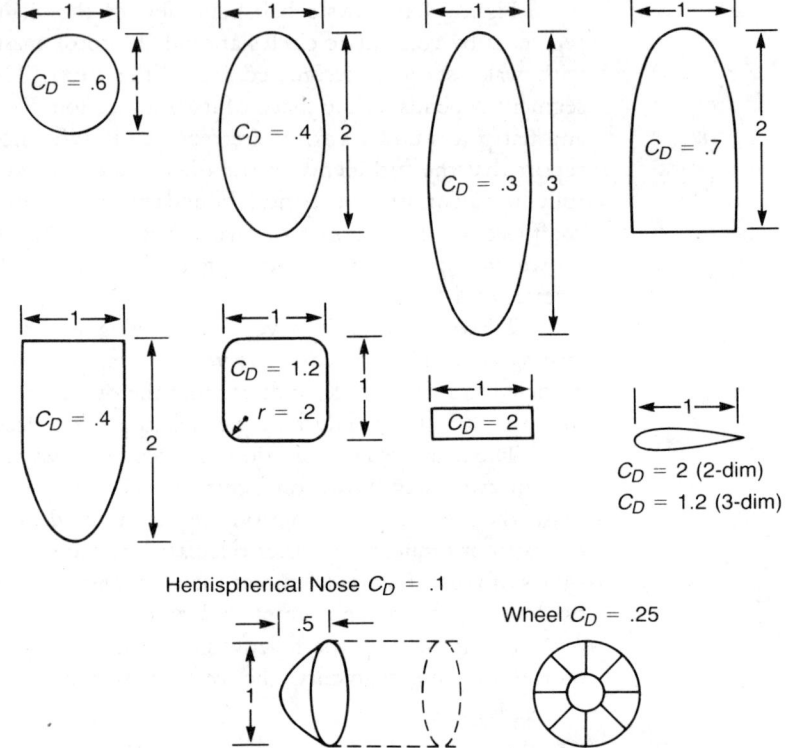

FIGURE 4.5   Drag Coefficients of Typical Component Shapes

Note: Drag coefficients are based on super critical flow and on area projected in plan view.

summing the product of the drag coefficient, the dynamic pressure ratio, and the projected area of each segment in the half-plan view:

$$\frac{D_v}{\text{G.W.}} = \frac{2\sum_{n=1}^{N} C_{d_n}\,(q/\text{D.L.})_n A_n}{A}$$

Table 4-1 presents the calculation for the example helicopter, which shows that the download penalty is 4.2% of gross weight.

The saving in power due to the pseudo ground effect of the fuselage on the rotor may be estimated by calculating the ground effect due to a full ground plane at the mean position of the fuselage by the method of Chapter 1 and then multiplying it by the ratio of projected fuselage planform area in the wake to disc area. Thus:

$$\Delta C_Q/\sigma = \frac{-\Sigma A_n}{A}\left\{ (C_T/\sigma)^{3/2}\sqrt{\frac{\sigma}{2}}\left[1 - \frac{v_{IGE}}{v_{OGE}}\right]\right\}$$

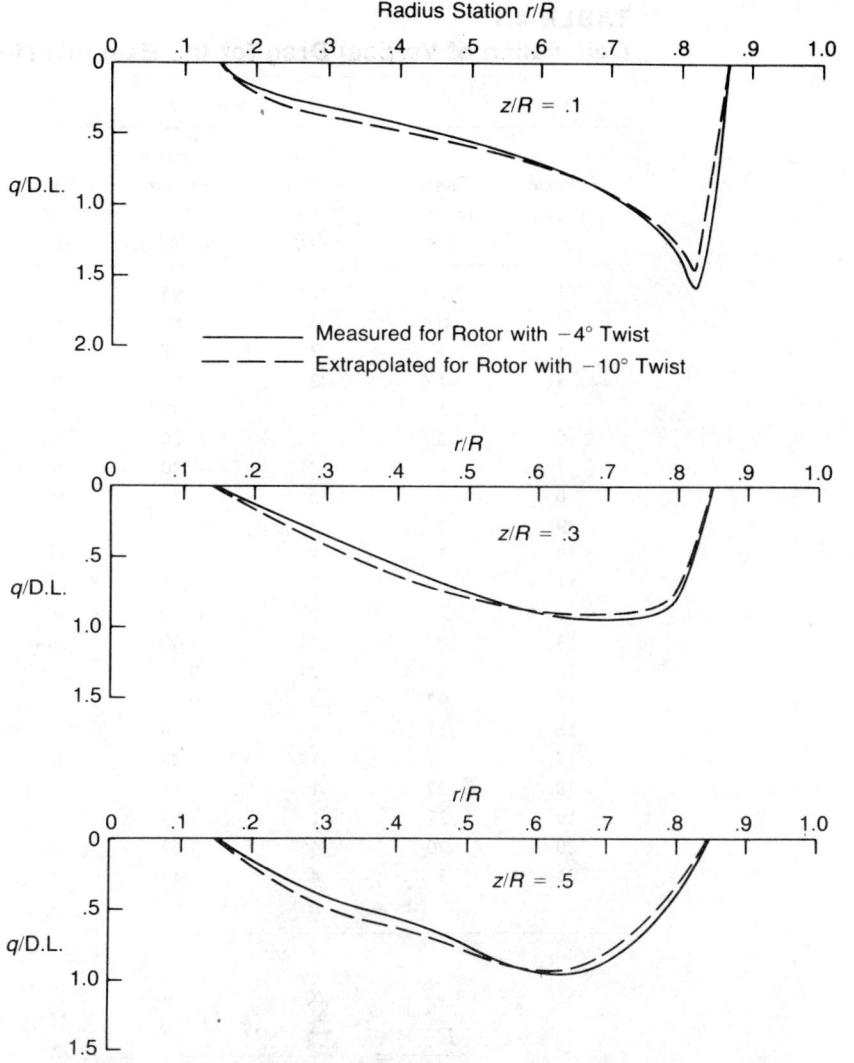

FIGURE 4.6   Distribution of Dynamic Pressure in Wake

Source: Boatwright, "Measurements of Velocity Components in the Wake of a Full-Scale Helicopter Rotor in Hover," USAAMRDL TR 72-33, 1972.

For the example helicopter the mean position of the fuselage is at $Z/D = .12$ and from Figure 1.41 of Chapter 1:

$$\frac{v_{\text{IGE}}}{v_{\text{OGE}}} = .62$$

TABLE 4.1
Calculation of Vertical Drag for the Example Helicopter

1	2	3	4	5	6	7
Airframe Segment, n	Radial Position, r/R	Vertical Position, Z/R	Dynamic Pressure Ratio, $(q/D.L.)_n$	Drag Coeff., C_{D_n}	Segment Area, A_n, ft^2	$4 \times 5 \times 6$
1	.7	.2	.95	.1	6	.57
2	.6	.2	.80	.9	10	7.20
3	.5	.2	.68	.9	12	7.34
4	.4	.2	.55	.9	12	5.94
5	.3	.2	.45	.9	12	4.86
6	.2	.2	.20	.9	12	2.16
7	.2	.2	.20	.9	12	2.16
8	.3	.2	.45	.9	11	4.46
9	.4	.22	.57	.9	10	5.13
10	.5	.22	.70	.6	9	3.78
11	.6	.23	.83	.5	8	3.32
12	.7	.23	.90	.45	7	2.83
13	.8	.24	1.00	.40	6	2.40
14	.9	.24	0	.35	5	0
15	.97	.25	0	.35	2	0
16	.17	.1	.10	2.0	6	1.20
17	.22	.1	.25	.6	14	2.10
18	.27	.1	.35	.6	12	2.52
19	.27	.1	.35	.1	2	.07
20	.20	.35	.42	2.0	1.5	1.26
21	.3	.4	.47	.25	1.5	.18
				Totals	171	59.48

$$\frac{D_v}{G.W.} = \frac{2 \sum_{n=1}^{N} c_{d_n} (q/D.L.)_N A_n}{A} = \frac{2(59.48)}{2,827} = .042$$

$$\frac{\Sigma A_n}{A} = \frac{2(171)}{2,827} = .12$$

The power saving for the sample helicopter is:

$$\Delta C_Q/\sigma = -.12 \left[C_T/\sigma^{3/2} \sqrt{\frac{.085}{2}} (.038) \right] = -.0094 (C_T/\sigma)^{3/2}$$

For hovering at $C_T/\sigma = .086$:

$$\Delta C_Q/\sigma = -.00024$$

which is the equivalent of 68 horsepower.

The method has been used to correlate with the test data reported in reference 4.4. These tests used a model helicopter suspended under a separate rotor. Both elements were mounted on individual balance systems and wings of various sizes and positions could be installed. Figure 4.7 shows the test results in terms of both vertical drag and pseudo ground effect (which corresponds to the "thrust recovery" of reference 4.4). Also shown are calculated values for the two quantities determined by the foregoing method. It may be seen that in this case both the download and the pseudo ground effect calculations are optimistic for the fuselage alone, but for the fuselage with the large wing, the download is optimistic whereas the ground effect is pessimistic. Some of the discrepancy may be due to the relatively small size of the model and the possibility that some components actually experienced high drag, subcritical conditions in spite of the wake turbulence.

Measurements of vertical drag in ground effect are reported in reference 4.5 and are summarized in Figure 4.8, which shows that both the vertical drag and the pseudo ground effect are reduced, or even reversed, when hovering less than a rotor diameter above the ground. Reference 4.6 reports that model tests on the Sikorsky S-76 show that the download ratio changes from +3% out of ground effect to −1% at a wheel height of one foot.

TAIL ROTOR-FIN INTERFERENCE IN HOVER

Another form of interference similar to vertical drag is the mutual interference of the tail rotor and the fin. The interference manifests itself in two ways: as a force on the fin that decreases the effective antitorque force generated by the tail rotor; and as a change in the flow conditions at the rotor that may either increase or decrease the tail rotor power required. Figure 4.9, based on the test data in references 4.7 and 4.8, shows the nondimensionalized interference force on the fin as a function of the blockage area ratio and the separation distance for both pusher and tractor arrangements. The gross thrust required of the tail rotor is:

$$T_{T_{\text{gross}}} = \frac{T_{T_{\text{req.}}}}{1 - F/T}$$

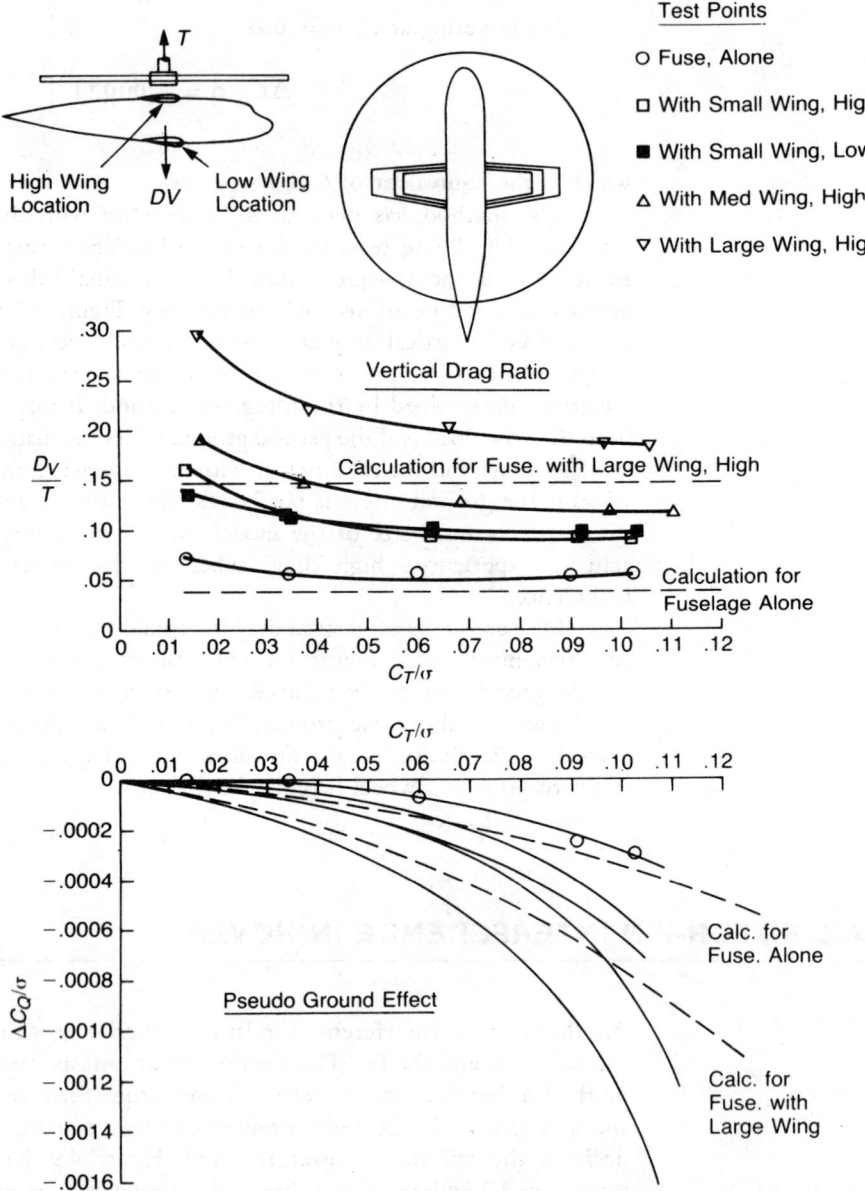

FIGURE 4.7   Vertical Drag Test Results

Source: Cassarino, "Effect of Rotor Blade Root Cutout on Vertical Drag," AAVLABS TR 70-59, 1970.

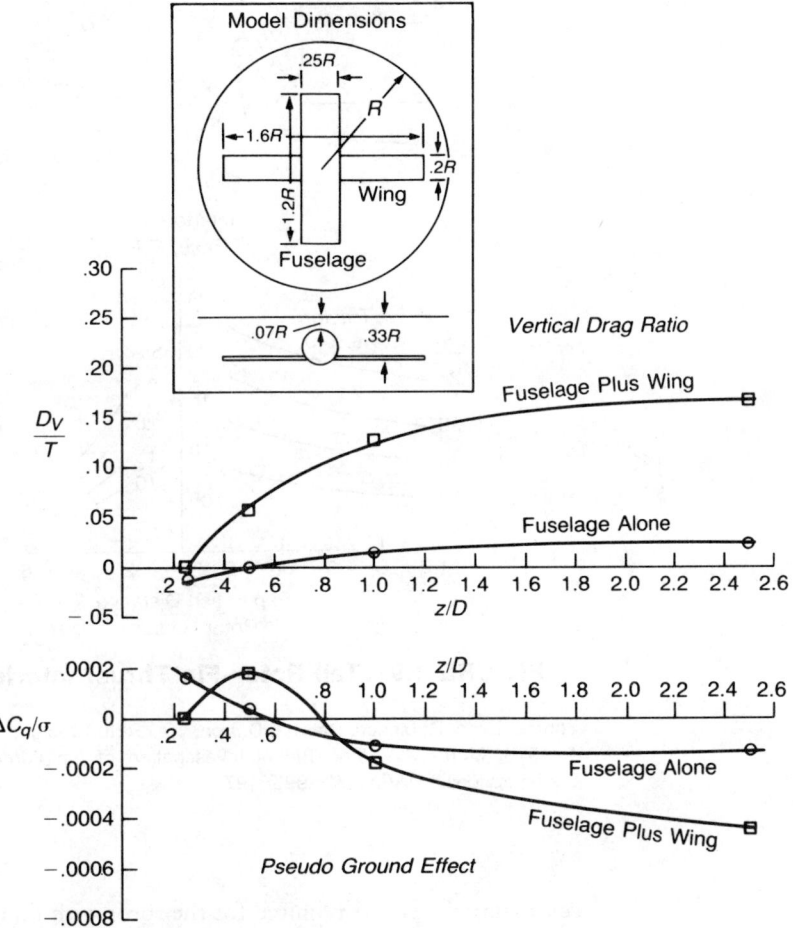

FIGURE 4.8 Effect of Proximity to Ground on Vertical Drag and Pseudo Ground Effect

Source: Fradenburgh, "Aerodynamic Factors Influencing Overall Hover Performance," AGARD CP 1111, 1972.

where $T_{T_{req.}}$ is the net thrust required to balance the main rotor torque.

For a tractor installation (i.e., with the wake blowing on the fin), the tail rotor benefits from a pseudo ground effect just as the main rotor does in the vertical drag situation. For a pusher installation, the fin slows the inflow, thus producing a beneficial *pseudo ceiling* effect. The fin also puts discontinuities into the flow, which might result in local stall. Figure 4.10 shows one set of experimental results for a tail rotor with and without a fin. For this configuration, the value of F/T from Figure 4.9 is 0.13. The effect of the fin on power may be found by

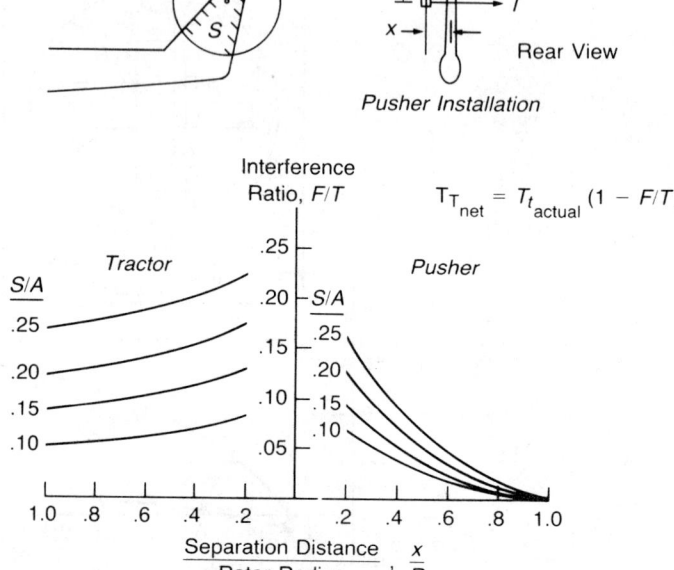

FIGURE 4.9   Tail Rotor-Fin Thrust Interference Ratio in Hover

Source: Lynn, Robinson, Batra, & Duhon, "Tail Rotor Design," Part I: "Aerodynamics," *JAHS* 15-4, 1970; Morris, "A Wind-Tunnel Investigation of Fin Force for Several Tail-Rotor and Fin Configurations," NASA LWP-995, 1971.

comparing the power required for the rotor with fin off at a C_T/σ 13% higher with the measured power with the fin on. Such a comparison at $C_T/\sigma = 0.08$ shows that the measured power is approximately 94% of what would have been predicted from the fin off data. Based on this one set of test data, it is suggested that the tail rotor hover performance be based on the empirical equation:

$$h.p._T = \left(1 - \frac{F/T}{2}\right)(h.p.\ for\ T_{T_{gross}})$$

The example helicopter has an area ratio, S/A, of 0.25, and a separation ratio, x/R, of 0.3. Thus from Figure 4.9 the interference ratio, F/T, is 0.125. The corresponding equations for the gross thrust and the power are:

$$T_{T_{gross}} = 1.125\ T_{net}$$

$$h.p._T = .94\ (h.p.\ for\ T_{T_{gross}})$$

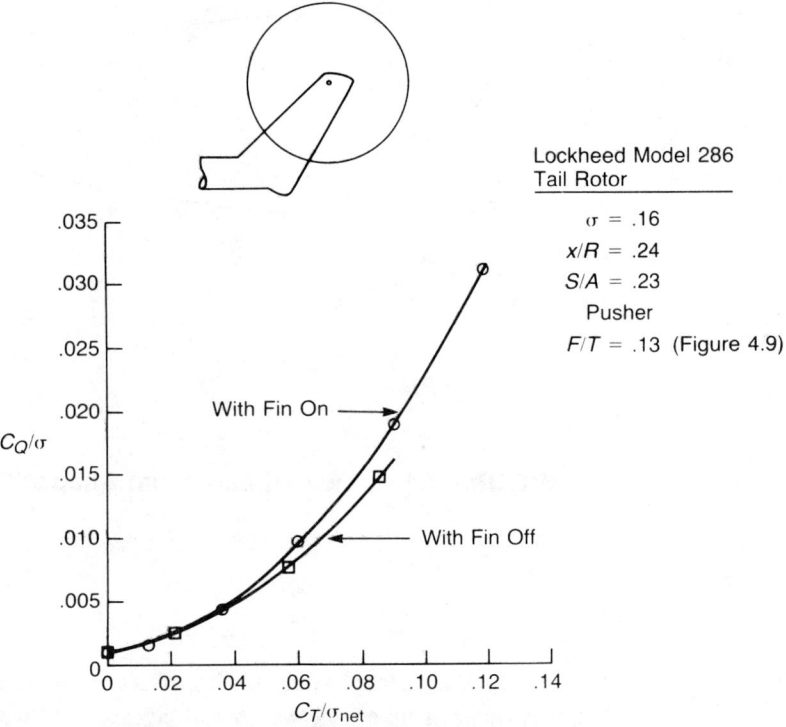

FIGURE 4.10   Tail Rotor-Fin Power Interference

Source: Internal Lockheed document.

PARASITE DRAG IN FORWARD FLIGHT

The parasite drag of a helicopter consists of two types of drag: *streamline drag*, where the flow closes smoothly behind the body; and *bluff body drag*, where the flow separates behind the body. The difference in drag between these two types is dramatically illustrated by Figure 4.11, which shows three two-dimensional bodies with equal drag. The strut has streamline drag, the flat wire has bluff body drag, and the round wire has a combination of both.

Streamline Drag

The primary component of streamline drag is skin friction, which is produced by the surface capturing air molecules and slowing them down—with respect to the aircraft—to zero velocity. At the nose of the body only the layer of molecules immediately adjacent to the surface is slowed. Further downstream, the slowed air

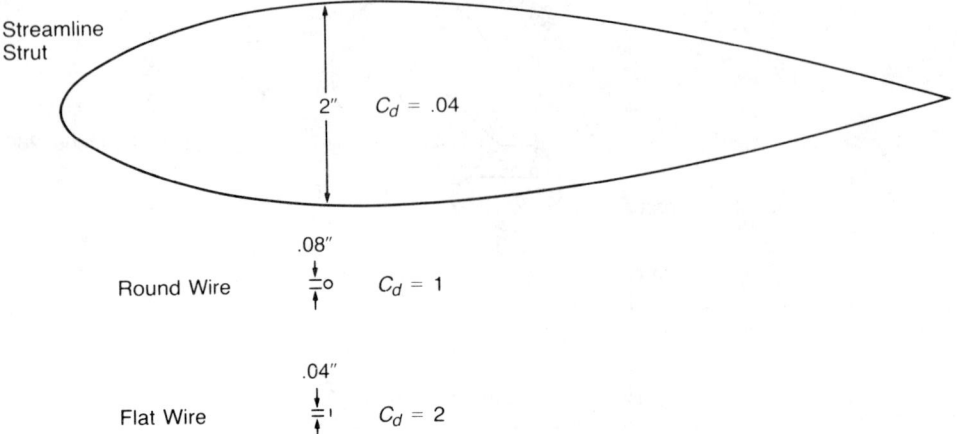

FIGURE 4.11   Two-Dimensional Shapes That Have Same Drag

molecules themselves have a slowing effect on molecules further from the surface, finally building up a *boundary layer* of air with a velocity distribution varying from zero at the skin surface to the free stream velocity at the outer edge. Skin friction is a measure of the total momentum that has been lost by the air in being slowed down. The magnitude of the skin friction is a function of the Reynolds number:

$$\text{R.N.} = \frac{\rho V L}{\mu} \doteq 6{,}400 \; VL \text{ at sea level}$$

where ρ is density in slugs/ft³, μ is dynamic viscosity, V is velocity in ft/sec, and L is length in ft.

The skin friction also depends on whether the boundary layer is laminar or turbulent, as shown in Figure 4.12 taken from reference 4.2. This figure shows the skin friction coefficient for a smooth, flat plate and is based on many measurements made in the last hundred years. Airplane wings and helicopter rotor blades operate in the range of Reynolds numbers in which natural transition will occur somewhere on the surface. The so-called *laminar boundary layer* airfoils are designed to operate as far down the laminar line as possible. Long bodies such as the fuselage of a large jet transport operate at Reynolds numbers in the neighborhood of 10^9 and thus experience low skin friction coefficients even though the boundary layer is almost completely turbulent.

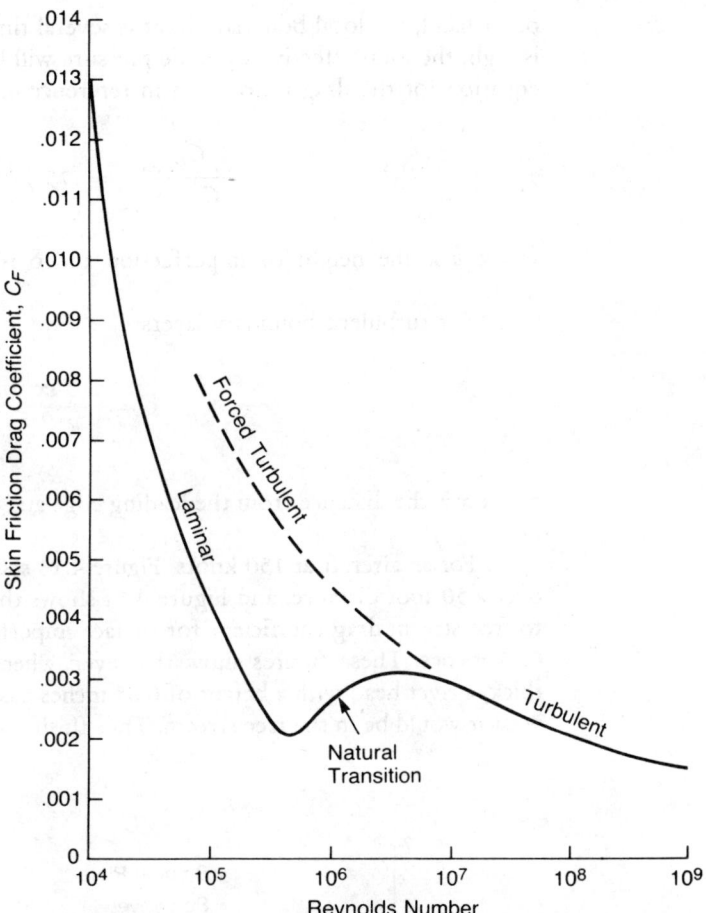

FIGURE 4.12   Skin Friction Drag for Smooth Plates

Source: Hoerner, "Fluid Dynamic Drag," published by author, 1965.

Natural transition from laminar flow to turbulent flow is not limited to the flow across surfaces. It can be readily observed in the smoke rising from a cigarette in a calm room. The smoke rises a few inches as laminar flow until its critical Reynolds number is reached based on velocity, density, viscosity, and distance traveled. At this point it spontaneously and suddenly becomes turbulent.

Surface imperfections such as rivet heads, skin joints, gaps, and so on produce drag according to their effective frontal area and the local dynamic pressure. If the imperfection extends through the boundary layer into clean air flow, the drag will be almost the same as if it were entirely in free air. If, on the other hand, the local boundary layer is several times deeper than the imperfection is high, the local effective dynamic pressure will be low, and so will the drag. The equation for the drag ratio given in reference 4.2 is:

$$\frac{C_{D_{actual}}}{C_{D_{freestream}}} = .75\sqrt[3]{h/\delta}$$

where h is the height of imperfection and δ is the thickness of the boundary layer.

For turbulent boundary layers:

$$\delta = \frac{.154x}{R_x^{1/7}}$$

where x is the distance from the leading edge and R_x is the Reynolds number based on x.

For an aircraft at 150 knots, Figure 4.13 shows the boundary layer thickness over a 50-foot distance, and Figure 4.14 shows the ratio of actual drag coefficient to free stream drag coefficient for surface imperfections with heights of 0.05 and 0.25 inches. These figures show that even when the boundary layer is 5 inches thick, a rivet head with a height of 0.05 inches has a drag coefficient that is 16% of what it would be in the free stream. Thus flush riveting reduces drag even near the

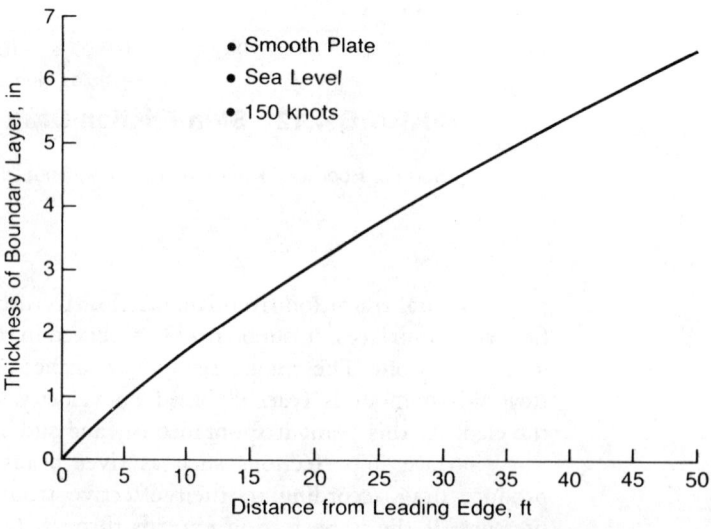

FIGURE 4.13   Boundary-Layer Thickness

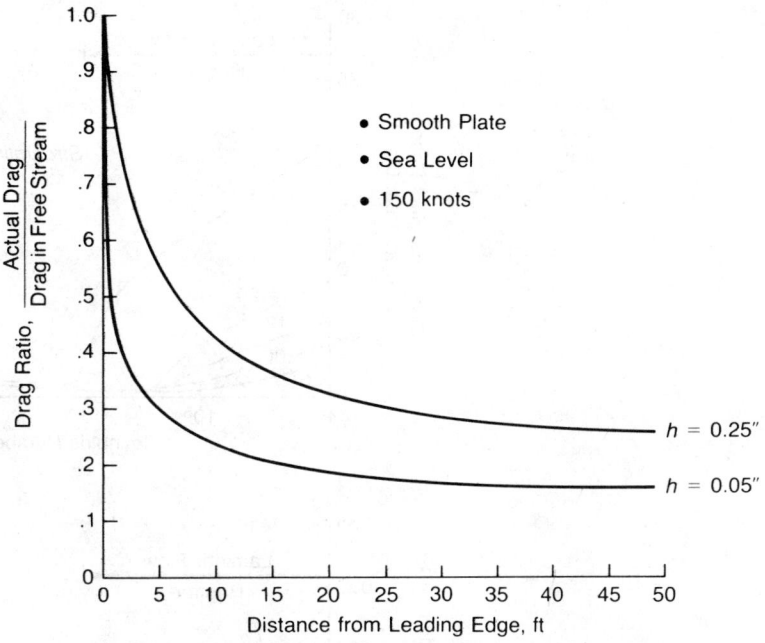

FIGURE 4.14   Drag Ratio of Surface Imperfections

rear of the fuselage. Some surface imperfections will exist even on well-designed aircraft. They may be accounted for individually by the methods outlined earlier or more simply by increasing the computed skin friction coefficient by a factor that is a function of the relative dirtiness of the aircraft. For example, the analysis in reference 4.2 for the ME-109, a propeller-driven World War II fighter, increased the calculated skin friction drag of the fuselage by 12% to account for surface imperfections.

Besides the effect of surface imperfections, streamline aircraft components have more drag than calculated from the drag of a flat plate because of *form drag*. This is caused by the increased velocities over the thick part of the body and the forced thickening of the boundary layer due to the slowing of the air to free stream velocities as the contours are brought together at the rear. This applies to both two- and three-dimensional bodies as shown in Figure 4.15. This figure and Figure 4.16 show the dilemma the aerodynamicist faces in using small-scale models in low-speed wind tunnels for drag measurements. Such testing is necessarily done at lower Reynolds numbers than on the full-scale aircraft, and thus the drag coefficient is higher. In some cases the helicopter aerodynamicist will use the high measured drag of the wind tunnel model on the basis that the model does not have surface imperfections of the actual aircraft—thus two wrongs make a right. A more realistic view of the situation, however, is that absolute full-scale drag values

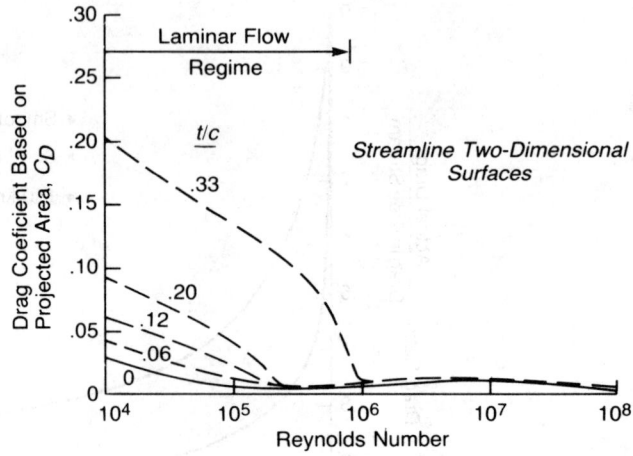

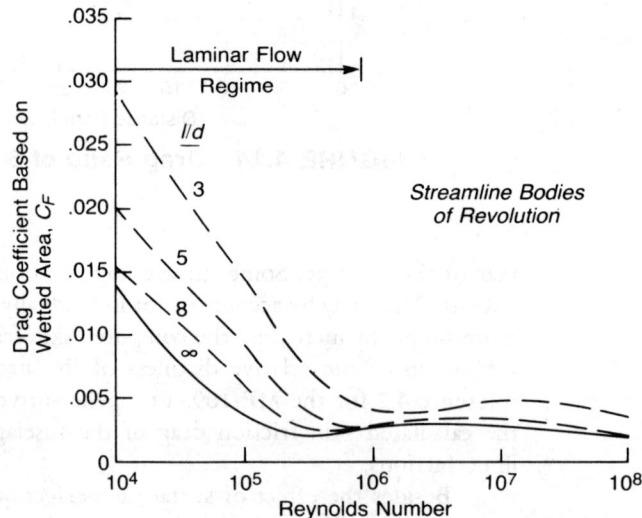

FIGURE 4.15   Effects of Form Drag

Source: Hoerner, "Fluid Dynamic Drag," published by author, 1965.

cannot be obtained from a small-scale wind tunnel test, and that only approximate changes in drag due to changes in configuration can be determined.

For the purposes of preliminary design, the fuselage drag can be estimated from past experience on other fuselages. Figure 4.17 shows drag coefficients measured in wind tunnels for several airplane and helicopter fuselages at zero lift and reported in references 4.9, 4.10, 4.11, and 4.12. As a reference, a minimum drag based on theoretical skin friction at a Reynolds number of 7×10^7,

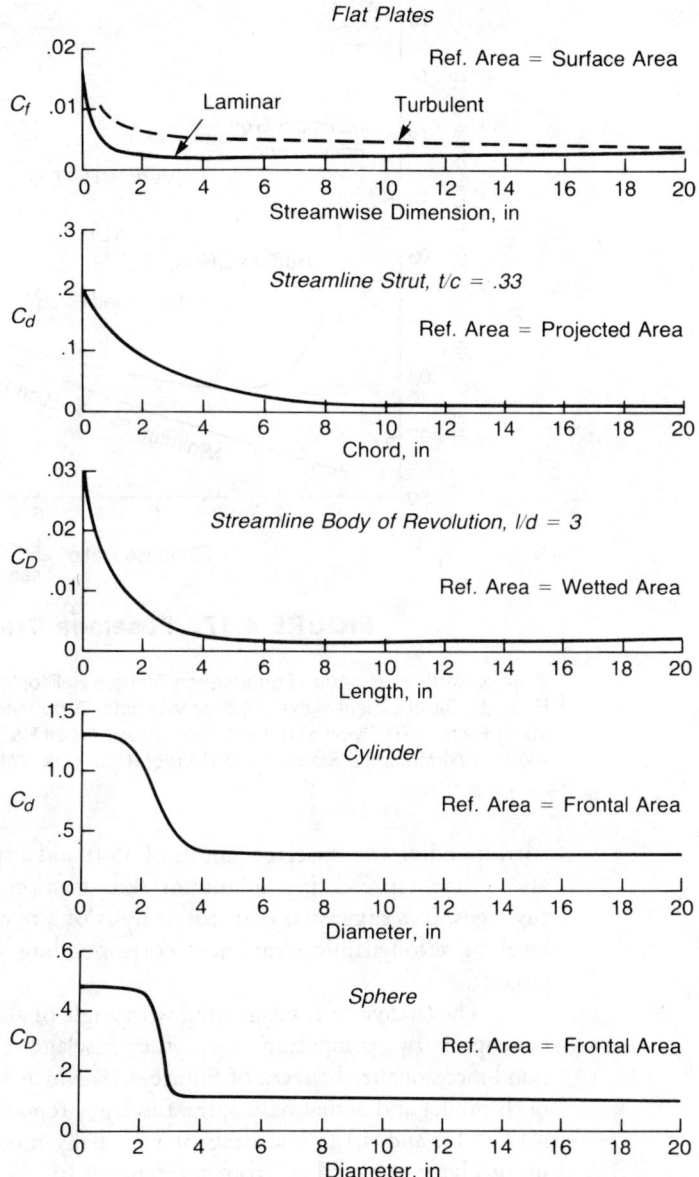

FIGURE 4.16   Drag Coefficients at 150 Knots

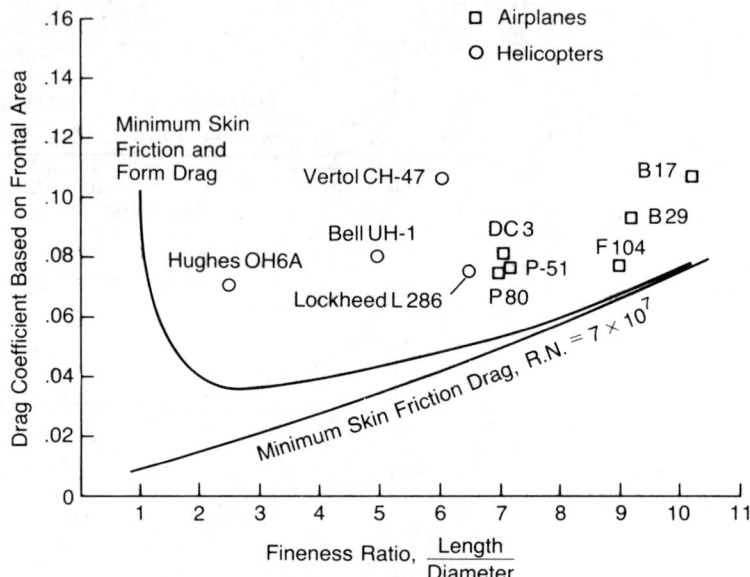

FIGURE 4.17   Fuselage Drag at Zero Lift

Source: Harris et al., "High Performance Tandem Helicopter Study," USATREC TR 61-42, 1961; Harned, "Development of the OH-6 for Maximum Performance and Efficiency," AHS 20th Forum, 1964; Foster, "Tilt-Pylon and Wind Tunnel Tests," Bell R&D Conference, 1961; Perkins & Hage, *Airplane Performance Stability and Control* (New York: Wiley, 1949).

corresponding to a fuselage length of 45 ft and a speed of 150 knots is shown, and also a line representing minimum skin friction and form drag for streamline fuselages. It is suggested that, for analysis of a new design, this figure be used at a level of aerodynamic cleanliness corresponding to that for one of the known aircraft.

The change of fuselage drag with angle of attack can be estimated for a given helicopter by comparing with the fuselage shapes and the corresponding nondimensionalized curves of Figure 4.18, which are based on wind tunnel tests of both model and actual helicopter fuselages reported in references 4.10, 4.13, 4.9, 4.14, 4.15, and 4.12. The drag of externally mounted nacelles can be estimated using Figure 4.19, taken from reference 4.16.

The drag of wings and stabilizer surfaces consists not only of skin friction but of induced drag as well. The total drag equation is:

$$f = \frac{q_L}{q} \left[C_{do}A + \frac{1}{\pi e} \left(\frac{L/b}{q} \right)^2 \right]$$

where C_{do} is the drag coefficient at zero lift, A is the projected area, q_L/q is the ratio of local dynamic pressure at the component to free stream dynamic pressure, L/b is

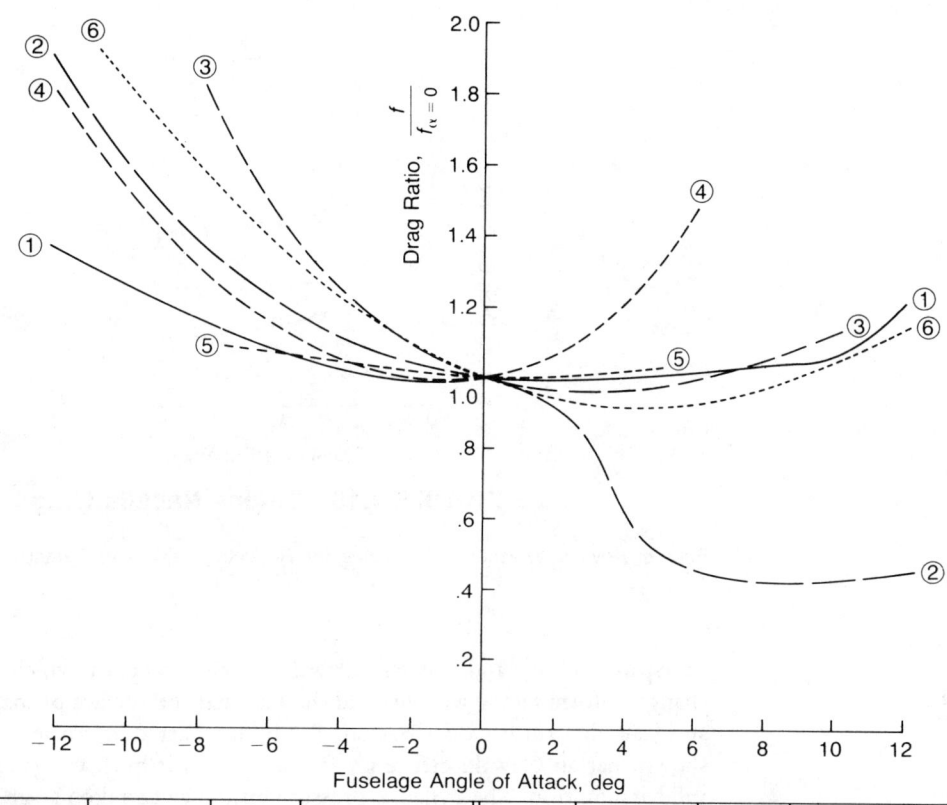

Line	Ref.	Helic.	Sketch	Line	Ref.	Helic.	Sketch
① ————	4.10	Hughes OH-6		④ — — —	4.14	Wind Tunnel Model	
② — · —	4.13	Wind Tunnel Model		⑤ ·····	4.15	Bell UH-1	
③ — — —	4.9	Vertol 107		⑥ ··········	4.12	Sikorsky S-65	

FIGURE 4.18   Effect of Angle of Attack on Fuselage Drag

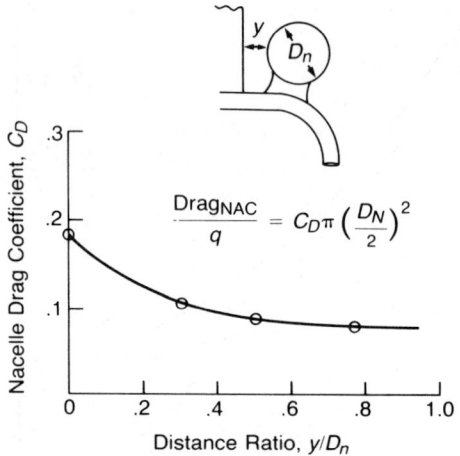

$$\frac{\text{Drag}_{NAC}}{q} = C_D \pi \left(\frac{D_N}{2}\right)^2$$

FIGURE 4.19   Engine Nacelle Drag

Source: Keys & Wiesner, "Guidelines for Reducing Helicopter Parasite Drag," *JAHS* 20-1, 1975.

the span loading, and e is the Ozwald efficiency factor which accounts for the change in form factor with lift and the fact that the surface probably does not have an ideal elliptical lift distribution. For preliminary drag estimates, experience has shown that an Ozwald efficiency factor of 0.8 for both wings and stabilizers is a valid assumption. The dynamic pressure ratio may be taken as unity for a wing but for both vertical and horizontal stabilizers on helicopter fuselages, the dynamic pressure ratio can vary from 0.8 to 0.5, depending on the size of the wake generated by all of the aircraft components ahead of the stabilizers and how much of the area of the stabilizers this wake affects. Some experimental data on this problem will be found in Chapter 8.

Some military helicopters are designed with flat plane canopies which are intended to reduce detectability by limiting the reflection of the sun to a narrow viewing range. Experience with these canopies, both in wind tunnel tests and in flight test, show that they produce a drag penalty primarily due to separation behind the sharp front corners. Wind tunnel tests on a World War II fighter reported in reference 4.2 showed that a flat panel canopy has five times the drag of a rounded canopy. The results of another wind tunnel test, this time of a helicopter, are shown in Figure 4.20 from reference 4.16, where the drag coefficient of the entire fuselage is plotted as a function of the corner radius ratio. The substantial increase in fuselage drag when going to sharp corners shown in Figure 4.20 has been substantiated by the flight test program reported in reference 4.17 in which a H-58A was equipped with a flat panel canopy with sharp forward corners. At its cruise speed of 102 knots, the equivalent flat plate area was increased by 0.7 square feet at its forward center of gravity position and by 2.2 at its aft.

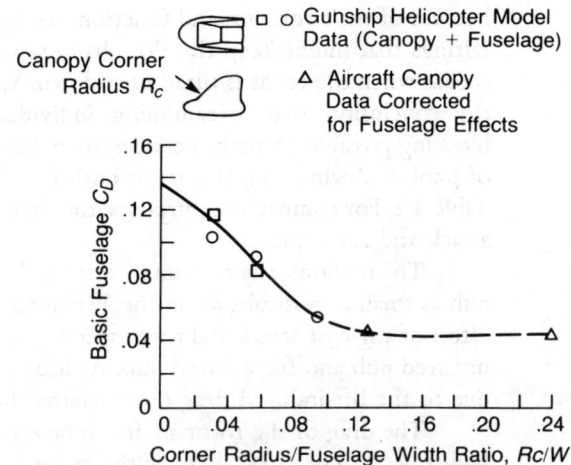

FIGURE 4.20   Effect of Canopy Corner Radius on Fuselage Drag

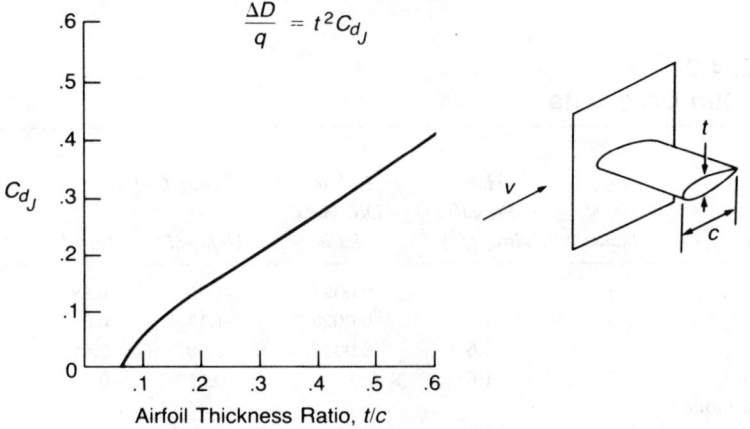

FIGURE 4.21   Drag of Junction between Fuselage and Surface

Source: Hoermer, "Fluid Dynamic Drag," published by author, 1965.

Drag due to the junction of surfaces with the fuselage can be estimated from Figure 4.21, which is based on the test data of reference 4.2.

Bluff Body Drag

A large portion of the drag of a helicopter is due to the bluff body drag of the rotor hubs and landing gear. The main and tail rotor hubs are bluff bodies, which, because of their rotation and function, are impossible to fit with simple streamline fairings that might keep the flow from separating. This is especially true at high speeds when the rotor is tilted nose down. Wind tunnel tests have generally led to the conclusion that streamlining individual bits and pieces without unduly blocking possible air paths between them leads to the lowest hub drag. A summary of published wind tunnel tests of both faired and unfaired rotor hubs is shown in Table 4.2. For comparison purposes, the drag values are presented for zero angle of attack and zero rpm.

The obvious way to minimize the hub drag is to keep the relative size of the hub as small as possible, as on the Vertol CH-47 and on the Hughes OH-6A. The effect of angle of attack and rotor speed on hub drag is shown in Figure 4.22 for an unfaired hub and for a faired hub. At least part of the difference between hubs is due to the lift-induced drag on the faired hub.

The drag of the rotor shaft can be estimated from the cylinder drag data of Figure 4.23. The total drag of the rotor hub and mast in close proximity to a fuselage or mast pylon may be higher than if they were isolated. This is due to separation on the fuselage or pylon triggered by the neighboring bluff body wake

TABLE 4.2
Rotor Hub Drag Data

Helicopter	No. of Blades	Hub Frontal Area, (ft^2)	Hub-to-Disc Area Ratio	Drag Coefficient		Equivalent Flat Plate Area (ft^2)		Reference
				Unfaired	Faired	Unfaired	Faired	
CH-47	3	5.0	0.0027	1.38	0.88	6.9	4.4	4.9
OH-6A	4	1.5	0.0028	1.13	0.80	1.7	1.2	4.10
UH-1B	2	5.6	0.0037	0.98	0.45	5.5	2.5	4.11
LOH wind tunnel model[a]	3	1.65		0.61	0.53	1.0	0.9	4.18
LOH wind tunnel model[a]	2	1.15		0.47	—	0.54	—	4.18
S-58	4	7.5	0.0031	1.53	0.57[b]	11.5	4.3	4.19
S-65	4	16.6	0.0041	1.01	—	16.8	—	4.12
S-65	4	21.7[c]		—	0.59	—	12.9	4.12
S-65	4	33.9[d]		—	0.22	—	7.6	4.12
Wind tunnel model	3	0.062		1.26	—	0.078	—	4.14
	3	0.102		—	0.76	—	0.078	4.14
AS Twinstar	3	2.5	0.0026	1.55		3.9		4.20
AS Puma	4	5.8	0.0030	0.98		5.7		4.20
AS Dauphin	4	4.3	0.0037	1.56		6.7		4.20

[a]Model had no control system.
[b]Model had a boundary-layer control system.
[c]*Rigid* head fairing.
[d]*Floating* head fairing.

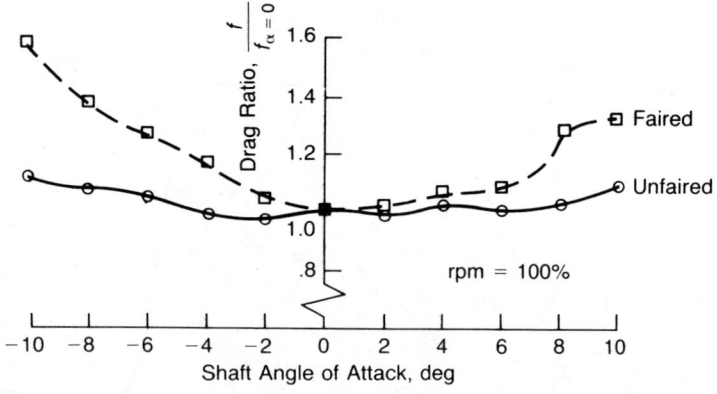

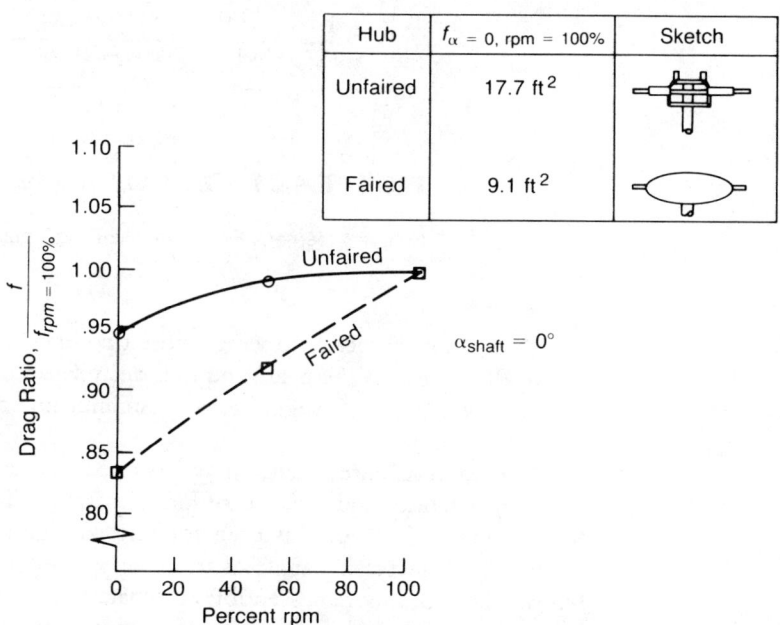

FIGURE 4.22   Effect of Angle of Attack and rpm on Hub Drag

Source: Linville, "An Experimental Investigation of High-Speed Rotorcraft Drag," USAAMRDL TR 71-46, 1971.

on the hub. Obviously, this interference drag is a function of the exact configuration, but one set of tests reported in reference 4.21 and summarized in Figure 4.24 may be taken as typical. One way of decreasing the interference drag is to suck off the low energy boundary layer, as was done on the S-58 wind tunnel model in Table 4.2. Another way of doing the same thing—though perhaps not as effectively—is to use a pylon cap, as on some Sikorsky and Vertol helicopters. This

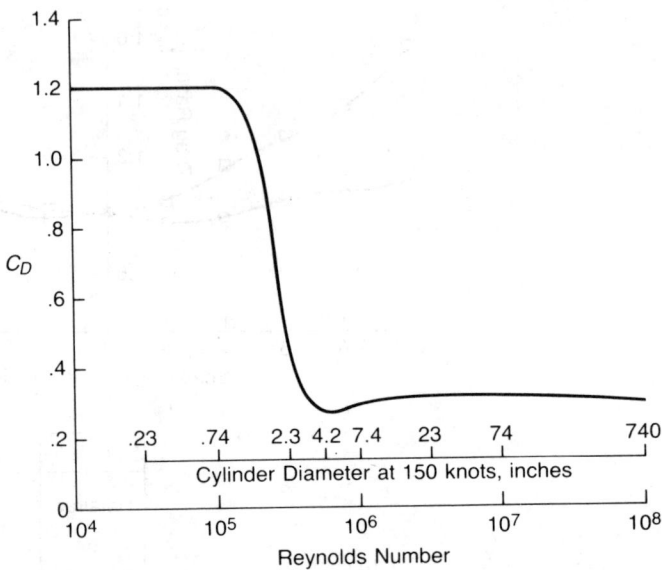

FIGURE 4.23   Drag of Circular Cylinders

Source: Hoerner, "Fluid Dynamic Drag," published by author, 1965.

cap acts as a low aspect ratio wing whose tip vortices energize the boundary layer on the aft portion of the pylon and thus delay separation. A comprehensive survey of hub drag, along with suggestions for minimizing it, is to be found in reference 4.22.

Yet another interference drag is caused by the rotor downwash on the aft fuselage, which can induce areas of local separation. Figure 4.25 shows test results from reference 4.23 of this drag for one configuration. The interference drag increases with increasing angle of attack, apparently because the aft portion of the fuselage becomes more susceptible to separation triggered by wake turbulence as its pressure gradient becomes more and more unfavorable. The wake behind the hub constitutes a low-pressure sink that can draw the flow off of the upper fuselage, thus producing separation. Again it is obvious that this type of interference drag is highly dependent on the configuration; but in order to evaluate it accurately for a specific design, rather elaborate wind tunnel models are required with a rotor that has the correct disc loading mounted separately from the drag model, but in the correct relative position. In lieu of this, it is suggested that Figure 4.25 be used for most applications.

Nonretracting landing gears also generally produce bluff-body drag. Reference 4.2 gives the drag coefficients of several types of wheeled landing gear. Several of its examples are shown in Figure 4.26. Skid gear are combinations of tubes and struts of various shapes, and two examples are also shown in Figure 4.26. Measurements of landing gear drag on small-scale wind tunnel models may be high

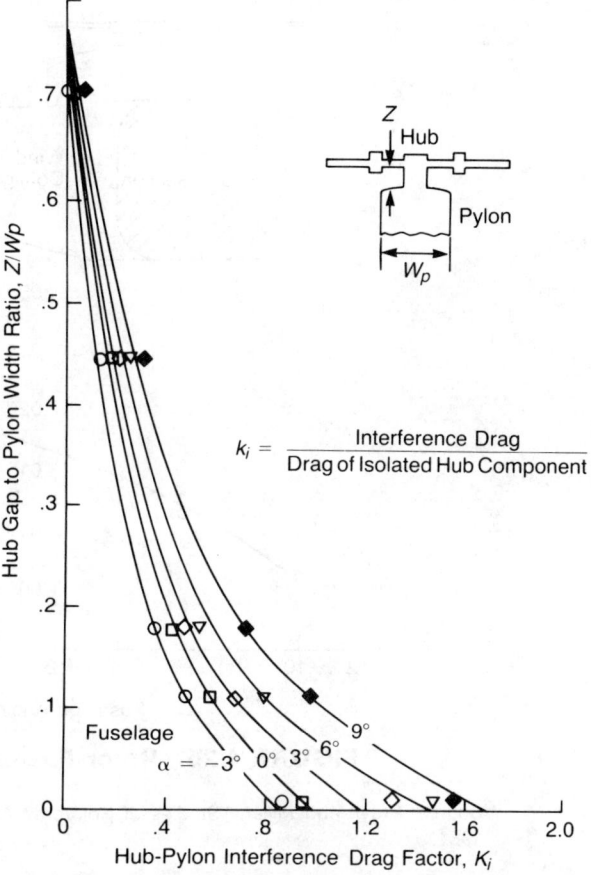

FIGURE 4.24   Hub-Pylon Interference Drag

Source: Keys & Wiesner, "Guidelines for Reducing Helicopter Parasite Drag," *JAHS* 20-1, 1975.

because of the low test Reynolds numbers of the components. Figure 4.23 can be used to estimate this effect.

Miscellaneous Drag

In addition to the major drag items discussed earlier, a modern helicopter has many minor sources of drag, individually small but significant in total. A partial list of these miscellaneous items includes:

Antennas   Windshield wipers
Door handles   External store mounting points

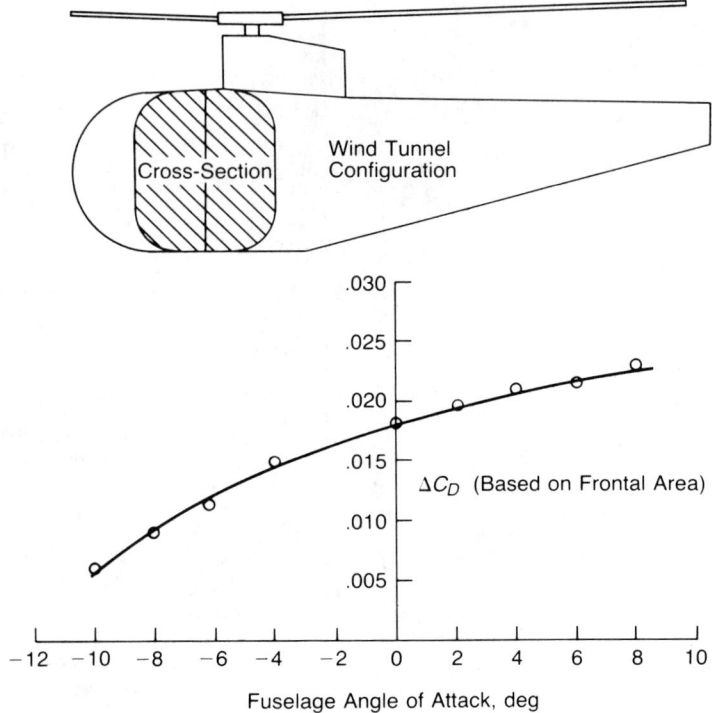

FIGURE 4.25   Rotor-Fuselage Interference Drag

Source: Pruyn and Miller, "Studies of Rotorcraft Aerodynamic Problems," WADD TR 61-124, 1961.

Hinges and latches   Overflow drain tubes
Pitot-static tubes   Abrasive walkways
Temperature probes   Lights: anticollision, formation, landing
Hand holds   Fueling receptacles
Steps   Ground electrical receptacles
External jacking points   Skin gaps, steps, and mismatches

Most of these items have characteristic dimensions of less than 4 inches and thus operate at subcritical Reynolds numbers at normal flight speeds, with correspondingly high drag coefficients. The data and methods of reference 4.2 can be used to evaluate the drag of these items. Such an evaluation for the Lockheed AH 56A "Cheyenne" produced an estimate of slightly more than one square foot of flat plate area.

Less obvious sources of drag are those due to cooling and leakage. When air is taken aboard to cool or to ventilate and then dumped overboard with less velocity than the forward speed of the helicopter, its loss of momentum manifests itself as a drag force. Similar drag will be produced by air that simply leaks into the aircraft at one point and out another without serving any useful purpose.

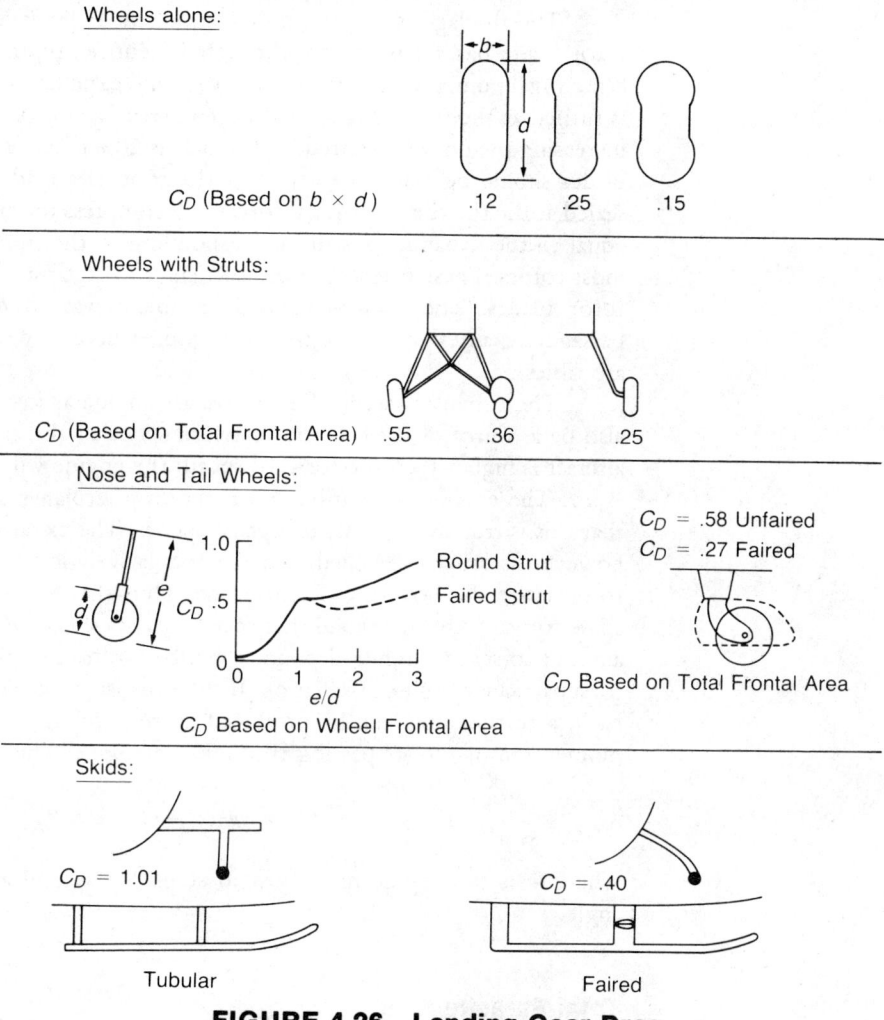

Wheels alone:

C_D (Based on $b \times d$)   .12   .25   .15

Wheels with Struts:

C_D (Based on Total Frontal Area)   .55   .36   .25

Nose and Tail Wheels:

C_D Based on Wheel Frontal Area

$C_D = .58$ Unfaired
$C_D = .27$ Faired

C_D Based on Total Frontal Area

Skids:

$C_D = 1.01$ Tubular

$C_D = .40$ Faired

FIGURE 4.26   Landing Gear Drag

Source: Hoerner, "Fluid Dynamic Drag," published by author, 1965; Sweet & Jenkins, "Wind-Tunnel Investigation of the Drag and Static Stability Characteristics of Four Helicopter Fuselage Models," NASA TN D 1363, 1962.

Compared to a solid wind tunnel model, the actual helicopter in flight has some of the characteristics of a sieve. Some indication of the magnitude of miscellaneous and leakage drag can be obtained from the wind tunnel tests of a Bell UH-1 fuselage, reported in reference 4.15. When all gaps were sealed and all protuberances removed, the drag was reduced by more than 2 square feet of equivalent flat plate area.

Cooling and leakage drag are difficult to estimate without a detailed thermodynamic and internal aerodynamic analysis. During preliminary design, they are usually accounted for by increasing the basic fuselage drag by 10–20%.

The fuselage is not the only component that can suffer from leakage. If a rotor blade has a passageway through its entire length, the rotor will act as a centrifugal pump, taking air in at the root and expelling it at the tip. Unless the air is turned at the tip to align it with the external velocity, losses can be substantial, increasing the power required by as much as 20% in extreme cases. For this reason, blades should be sealed, especially at the root. If a blade is open at the root but sealed at the tip, the centrifugal forces will compress the air inside the tip to a value equal to the dynamic pressure corresponding to the tip speed, about 3–5 psi for most rotors. These pressures were sufficient to collapse solid ribs in early built-up rotor blades. The solution to this problem was to open up a spanwise air passageway, a design feature that is no longer necessary since most modern blades are ribless.

The exhaust system of a turboshaft engine as installed on a helicopter can also be a source of drag. If the rearward speed of the exhaust gas relative to the aircraft is higher than the forward speed, the engine will produce positive *residual thrust*. The engine installations on turboprop airplanes are generally designed so that this is true even for their highest speeds. The exhaust system on a helicopter, however, is usually designed to produce relatively low exhaust velocities in order to optimize the hovering performance. Thus there is residual thrust only up to some forward speed; beyond that speed there is *residual drag*. This drag, which may amount to several hundred pounds, can be estimated by the engine manufacturer for a particular engine installation. If the exhaust stack is canted away from straight back, a further drag can be produced corresponding to the loss of the rearward momentum of the air passing through the engine. Thus the exhaust drag is:

$$D_{ex} = \dot{m} \ (V - V_{ex} \cos \chi)$$

where $\dot{m}$ is the engine mass flow in slugs per second and χ is the exhaust cant angle.

Total Parasite Drag

Figure 4.27, taken from reference 4.24, presents the state of the art of parasite drag of both helicopters and airplanes with several levels of aircraft cleanliness. This plot can be used to make first estimates of the drag of a helicopter in the early stages of the preliminary design. Table 4.3 lists published drag breakdowns for three typical helicopters, which can also be used to guide first estimates. (*Note of caution*: I have never known of an airplane or a helicopter drag estimator who was pleasantly suprised by flight test results showing that he had overestimated the drag of the aircraft. The estimating methods outlined in this chapter must be considered to produce minimum estimates and thus are suitable for the wishful-thinking phases of proposals and sales brochures. For realistic engineering estimates, I recommend that at least another 20% be added to the total to include those items that were not initially included or that will grow during the normal development of the helicopter.)

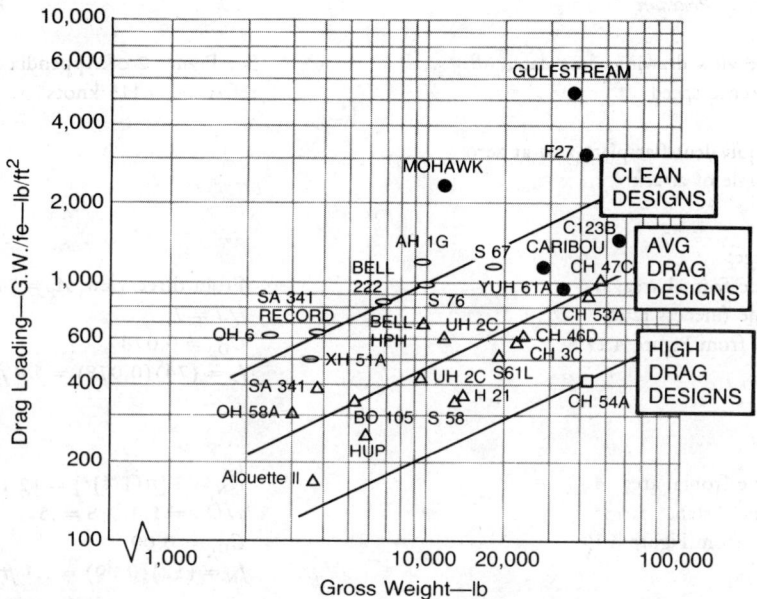

FIGURE 4.27   Helicopter and Airplane Drag State-of-the-Art

Source: Rosenstein & Stanzione, "Computer-Aided Helicopter Design," AHS 37th Forum, 1981.

TABLE 4.3
Total Drag Breakdown, Equivalent Flat Plate Area

Helicopter	OH6A	UH-1B	CH-47
Design gross weight	2,550	9,500	33,000
Main rotor disc area	550	1,520	5,900
Reference	4.10	4.11	4.9
Component			
Fuselage and engine nacelles	1.5	5.0	16.1
Rotor hubs	1.2	5.5	14.1
Landing gear	0.5	3.0	7.9
Empennage	0.1	0.9	—
Miscellaneous	1.7	5.1	5.1
TOTAL	5.0	14.5	43.2

Drag Estimate for Example Helicopter

For purposes of illustration, the parasite drag of the example helicopter has been estimated using the methods and data presented in the preceding section.

Procedure	*Results*

Given: three-view drawing, drag data, reference speed

See Figure 2 of Appendix A, Figures 4.15 to 4.26, Table 4.2
ref speed $= 115$ knots

Estimate: equivalent flat plate area at zero angle of attack

Basic fuselage:
    Determine frontal area, A_F
    Determine fineness ratio
    Find C_{D_F} from Figure 4.17
    $f_F = A_F C_{D_F}$

From three-view, $A_F = 74$ ft^2
$l/d = 7$
$C_{D_F} = 0.078$
$f_F = (74)(0.078) = 5.8$ *ft*^{2}

Nacelles:
    Determine frontal area, A_N
    Determine distance ratio
    Find C_{D_N} from Figure 4.19
    $f_N = A_N C_{D_N}$

$A_N = 2\,[\pi(1.4)^2] = 12$ ft^2
$y/D_N = 1.4/2.8 = .5$
$C_{D_N} = 0.09$
$f_N = (12)(0.09) = 1.1$ *ft*^{2}

Main rotor hub and shaft:
    Determine frontal area of hub
    Determine frontal area of shaft
    Determine diameter of shaft
    From Table 4.2, estimate $C_{D_{MH_{RPM=0}}}$

$A_{MH} = 5$ ft^2
$A_{MS} = 1$ ft^2
$D_{ms} = 0.5$ ft
$C_{D_{MH}} = 1.1$

    From Figure 4.22 find drag ratio
    Calculate $C_{D_{MH_{corr}}}$

$\alpha_s = 0°$, RPM $= 100\%$, D.R. $= 1.00/.95 = 1.05$
$C_{D_{MH_{corr}}} = 1.16$

    $f_{MH} = A_{MH} C_{D_{MH_{corr}}}$

$f_{MH} = 5(1.16)) = 5.8$ ft^2

    Calculate R.N. of shaft (115 k)
    Determine $C_{D_{MS}}$ from Figure 4.23
    $f_{MS} = A_{MS} C_{D_{MS}}$
    Total $f_M = f_{MH} + f_{MS}$
    Determine Z/W_p (see Figure 4.24)
    Find K_i from Figure 4.24
    $f_M = (1 + K_i) f_M$

R.N. $= (6{,}400)(115)(1.69)(0.5) = 0.6 \times 10^6$
$C_{D_{MS}} = 0.3$
$f_{MS} = 1(0.3)$
$f_{M_{uncorr}} = 5.8 + .3 = 6.1$ ft^2
$Z/W_p = 2.8/9 = 0.3$
$K_i = .15\ (\alpha_F = -5°)$
$f_M = 1.15(6.1) = 7.0$ ft^2

Tail rotor hub and shaft:
    Determine frontal area
    From Table 4.2, estimate $C_{D_{TH}}$
    From Figure 4.22, find drag ratio
    Calculate $C_{D_{TH_{corr}}}$

$A_T = .6$ ft^2
$C_{D_{TH}} = 1.1$
For $\alpha_S = 0$, RPM $= 100\%$, D.R. $= 1/.95 = 1.05$
$C_{D_{TH_{corr}}} = (1.1)(1.05) = 1.16$

    $f_T = A_T C_{D_{TH_{corr}}}$
    $f_T = (0.6)(1.16) = 0.7$ ft^2

Main landing gear:
  Determine frontal area
  From Figure 4.26, estimate $C_{D_{MLG}}$.
  $f_{MLG} = A_{MLG}C_{D_{MLG}}$

$A_{MLG} = 4 \text{ ft}^2$
$C_{D_{MLG}} = 0.3$
$f_{MLG} = (4)(0.3) = 1.2 \text{ ft}^2$

Nose landing gear:
  Determine frontal area
  Determine e/d (see Figure 4.26)
  From Figure 4.26, estimate $C_{D_{NLG}}$
  $f_{NLG} = A_{NLG}C_{D_{NLG}}$

$A_{NLG} = 1.5 \text{ ft}^2$
$e/d = 3.4/2 = 1.7$
$C_{D_{NLG}} = 0.56$
$f_{NLG} = 1.5(0.56) = 0.8 \text{ ft}^2$

Horizontal stabilizer:
  Area, A_H (Appendix A)
  Span, b_H
  Aspect Ratio, A.R. $= b^2{}_H/A_H$
  Thickness Ratio, t/c
  Mean aerodynamic chord, MAC
  Reynolds number at 115 knots
  Estimate C_{D_0} from Figure 4.15
  Estimate C_{L_H} from trim conditions
    (see Chapter 8)
  Estimate span efficiency factor, δ

$A = 18 \text{ ft}^2$
$b_H = 9 \text{ ft}$
A.R. $= 9^2/18 = 4.5$
$t/c = .12$
$MAC = 2 \text{ ft}$
R.N. $= 6,400(115)(1.69)(2) = 2 \times 10^6$
$C_{D_0} = 0.010$

$C_{L_H} = -.3$
$\delta = 0.2$

Calculate induced drag coeff.

$$C_{D_i} = \frac{C_{L_H}^2(1 + \delta)}{\pi \text{ A.R.}} = \frac{0.3^2 (1.2)}{\pi (4.5)} = 0.008$$

Calculate root thickness

$t = (0.12)(2) = 0.24 \text{ ft}$

Estimate junction drag coeff., C_{D_j} from Figure 4.21

$C_{D_j} = 0.072$

Compute equiv. junction drag coeff.,

$$C_{D_{j_{equiv.}}} = 2 \left[\frac{C_{D_j} (t^2)}{A_H} \right]$$

$$C_{D_{j_{equiv.}}} = 2 \left[\frac{0.072(0.24)^2}{18} \right] = 0.001$$

Total drag coefficient $= C_{D_0} + C_{D_i} + C_{D_{j_{equiv.}}}$
Estimate q_H/q

$C_{D_H} = 0.010 + 0.008 + 0.001 = 0.019$
$q_H/q = 0.75$

Calculate $f_H = \dfrac{q_H}{q} (C_{D_H}A_H)$

$f_H = 0.75(0.019)(18) = 0.2 \text{ ft}^2$

Vertical stabilizer:
  Area, A_V (Appendix A)
  Thickness ratio, t/c
  Mean aerodynamic chord, MAC
  Reynolds number at 115 knots
  Estimate C_{D_0} from Figure 4.15
  Estimate q_V/q
  Calculate $f_V = q_V/q (C_{D_V}A_V)$

$A_V = 24 \text{ ft}^2$
$t/c = 0.12$
$MAC = 3 \text{ ft}$
R.N. $= 6,400(115)(1.69)(3) = 4 \times 10^6$
$C_{D_0} = 0.010$
$q_V/q = 0.75$
$f_V = 0.75(.010)(24) = 0.2 \text{ ft}^2$

Procedure (cont.)

Results (cont.)

Rotor-fuselage interference drag:
   From Figure 4.25 estimate ΔC_D at $\alpha_F = 0$
   $f_{int.} = \Delta C_D A_F$

$\Delta C_D = 0.018$
$f_{int.} = 0.018(74) = 1.3 \text{ ft}^2$

Exhaust drag:
   Ask engine manufacturer to estimate net exhaust thrust
   for engine installation on example helicopter
   $f_{ex.} = -T_{net}/q$

$T_{net} = 2(-11) = -22 \text{ lb}$

$f_{ex.} = -\dfrac{(-22)}{45} = 0.5 \text{ ft}^2$

Miscellaneous drag items:
   Estimate total drag of antennas, door handles,
   lights, steps, skin gaps,
   cooling leakage, ventilation, etc.

$f_{misc.} = 0.5 \text{ ft}^2$

Total equivalent flat plate area:

$$f_{tot.} = \quad f_F$$
$$+ f_N$$
$$+ f_M$$
$$+ f_T$$
$$+ f_{MLG}$$
$$+ f_{NLG}$$
$$+ f_H$$
$$+ f_V$$
$$+ f_{int.}$$
$$+ f_{ex.}$$
$$+ f_{misc.}$$

$f_{tot.} = $
5.8
1.1
7.0
0.7
1.2
0.8
0.2
0.2
1.3
0.5
0.5
19.3 ft²

HOVER PERFORMANCE

The hover performance of a helicopter is a function of elements discussed in the preceding sections. The analysis is best illustrated by applying it to the example helicopter.

The nondimensional performance of the isolated main and tail rotors in hover was calculated in Chapter 1 and are presented again in Figure 4.28. The main rotor nondimensional performance has been converted into dimensional performance in the form of the horsepower$/\rho/\rho_0$ versus gross weight$/\rho/\rho_0$ in and out of ground effect in Figure 4.29. The performance out of ground effect has been calculated using a vertical drag penalty of 4.2% of the gross weight and a pseudo ground effect on power corresponding to

$$\Delta C_{Q/\sigma} = -0.0094 \, (C_{T/\sigma})^{3/2}$$

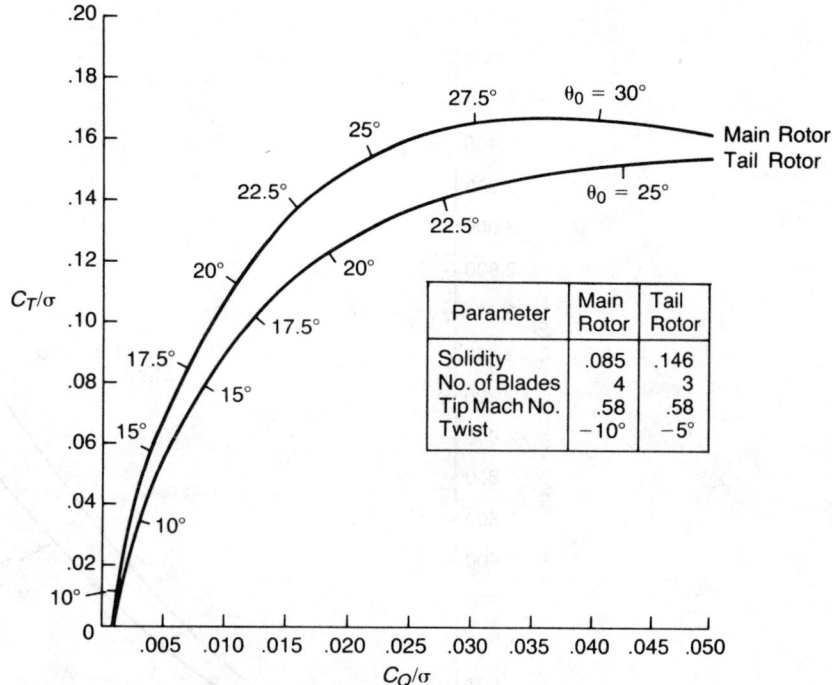

FIGURE 4.28   Calculated Performance of Isolated Main and Tail Rotors of Example Helicopter

For hover in ground effect, no vertical drag or pseudo ground effects have been used. The ground effect has been taken from Figure 1.41 of Chapter 1 for a 5-ft wheel height.

The corresponding tail rotor power required is found by calculating the net tail rotor thrust required to balance the main rotor torque:

$$T_{T_{net}} = \frac{550 \, \text{h.p.}_M}{(\Omega R)_M} \frac{R_M}{l_T}$$

or for the example helicopter:

$$T_{T_{net}} = 0.69 \, \text{h.p.}_M$$

The gross tail rotor thrust due to fin interference is:

$$T_{T_{gross}} = 1.125 \, T_{T_{net}}$$

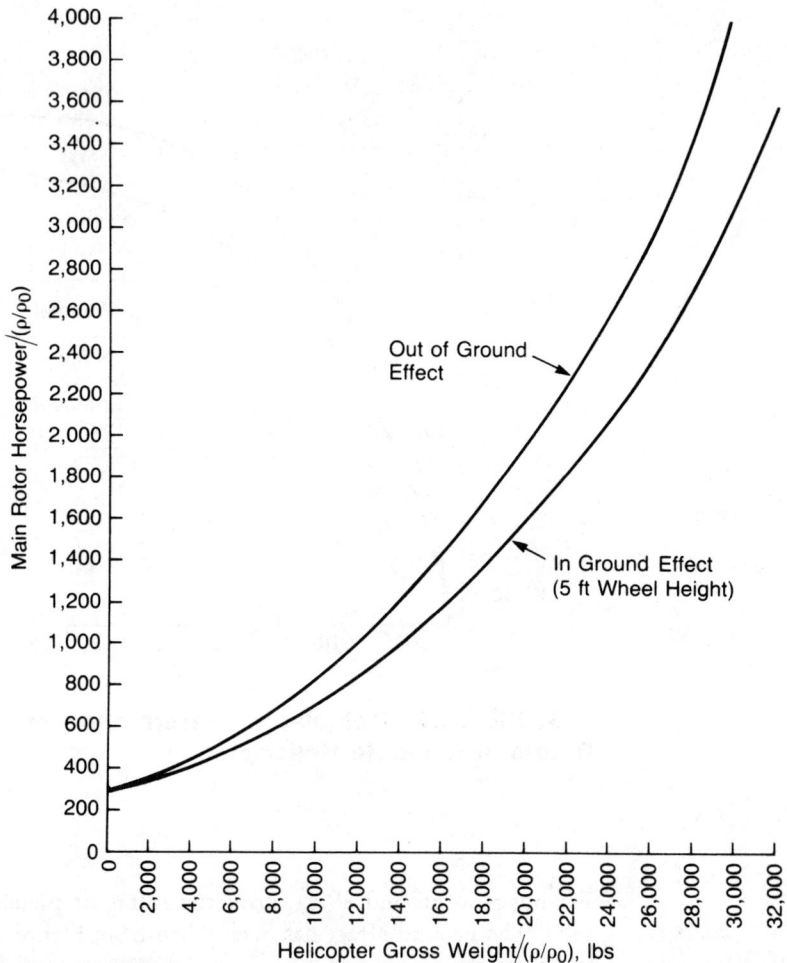

FIGURE 4.29 Main Rotor Performance in Hover as Installed on Example Helicopter

and the tail rotor power is:

$$\text{h.p.}_T = 0.94[\text{h.p. for } T_{T_{\text{gross}}}]$$

The resultant tail rotor power corresponding to the main rotor power of Figure 4.29 is shown in Figure 4.30. A comparison of the hover values of C_T/σ on the main and tail rotors of the example helicopter reveals a mismatch that would probably generate a redesign effort in an actual project. This comparison, shown in

Figure 4.31, indicates that at high gross weights the tail rotor is more heavily loaded than the main rotor—especially when compared to their respective maximum capabilities, which were shown in Figure 4.28. This means that the high gross weight or altitude performance will be limited by the tail rotor rather than the main rotor. Possible redesigns include increasing tip speed, chord, radius, or some combination of these to lower the tail rotor C_T/σ.

The engine power measured at the torquemeters is the sum of the main rotor, tail rotors, transmission, and accessory powers. For the example helicopter:

$$\frac{\text{h.p.}_{\text{eng.}}}{\rho/\rho_0} = [56 + 1.0112 \, \text{h.p.}_M + 1.0075 \, \text{h.p.}_T]_{\rho/\rho_0=1}$$

The engine power required in hover in and out of ground effect is shown in Figure 4.32 as a function of gross weight. The next step is illustrated by figures 4.33 and 4.34, where the power required for various density ratios is plotted. The curves for

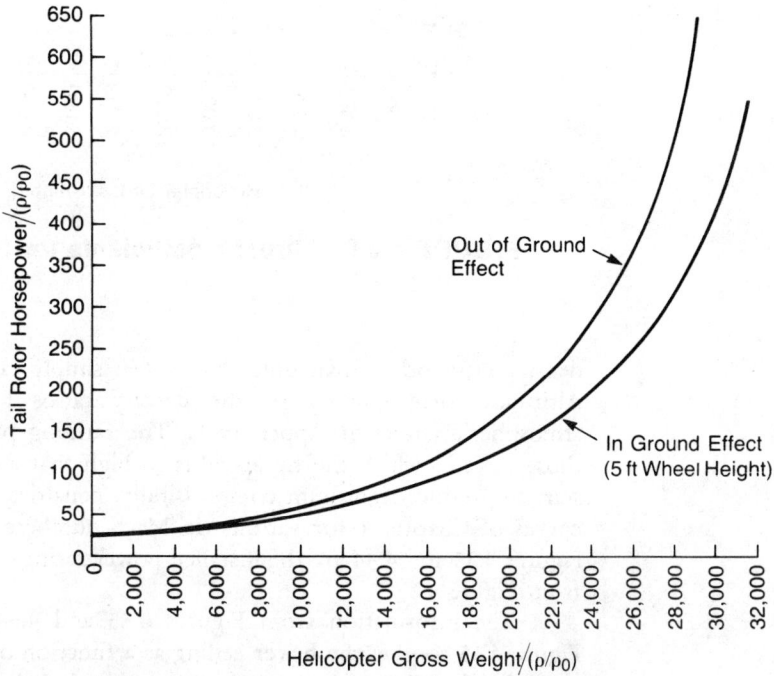

FIGURE 4.30 Tail Rotor Performance in Hover as Installed on Example Helicopter

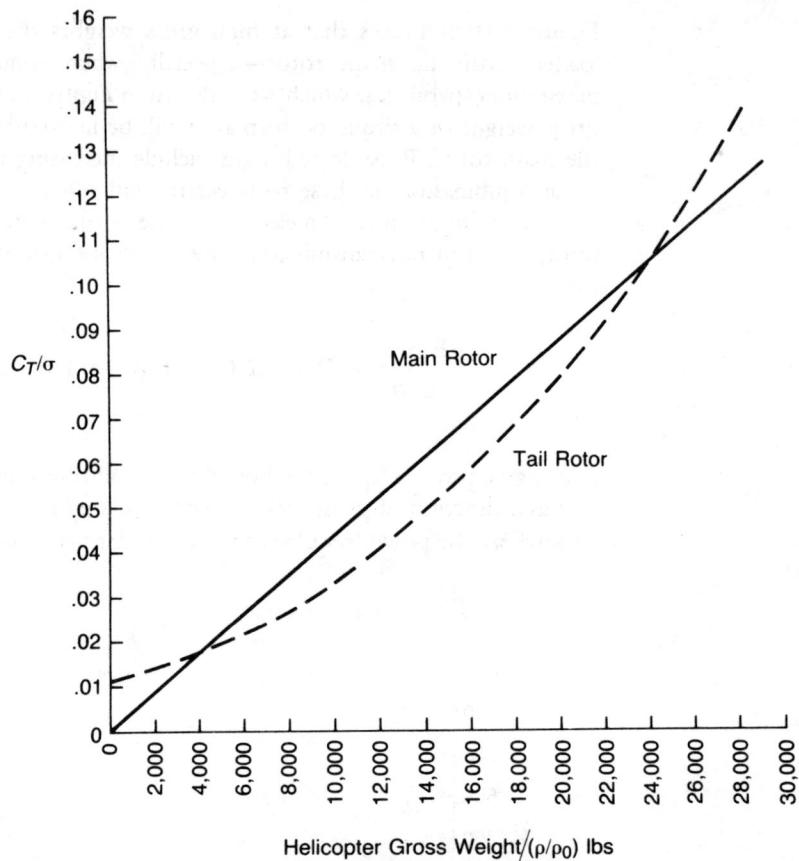

FIGURE 4.31 Thrust Coefficients for Main and Tail Rotors

density ratio other than unity have been simply ratioed from the basic curve. Altitudes corresponding to the density ratios have been taken from the atmospheric charts of Appendix C. The ratioing procedure is valid except for those cases in which the tip speed is so high that a decrease in temperature will start to produce significant compressibility penalties. For this situation, the hover curves of Chapter 1 for various tip Mach numbers can be used. Also shown in Figures 4.33 and 4.34 are the installed power ratings, which are 98% of the ratings from Figure 4.1.

The information from Figures 4.33 and 4.34 has been cross-plotted on Figure 4.35 to give the hover ceiling as a function of gross weight in and out of ground effect. It may be seen that the example helicopter can hover OGE at sea level, standard conditions at a gross weight of 27,800 lb and has a hover ceiling of 7,000 ft. at its design gross weight of 20,000 lb on a 95° day.

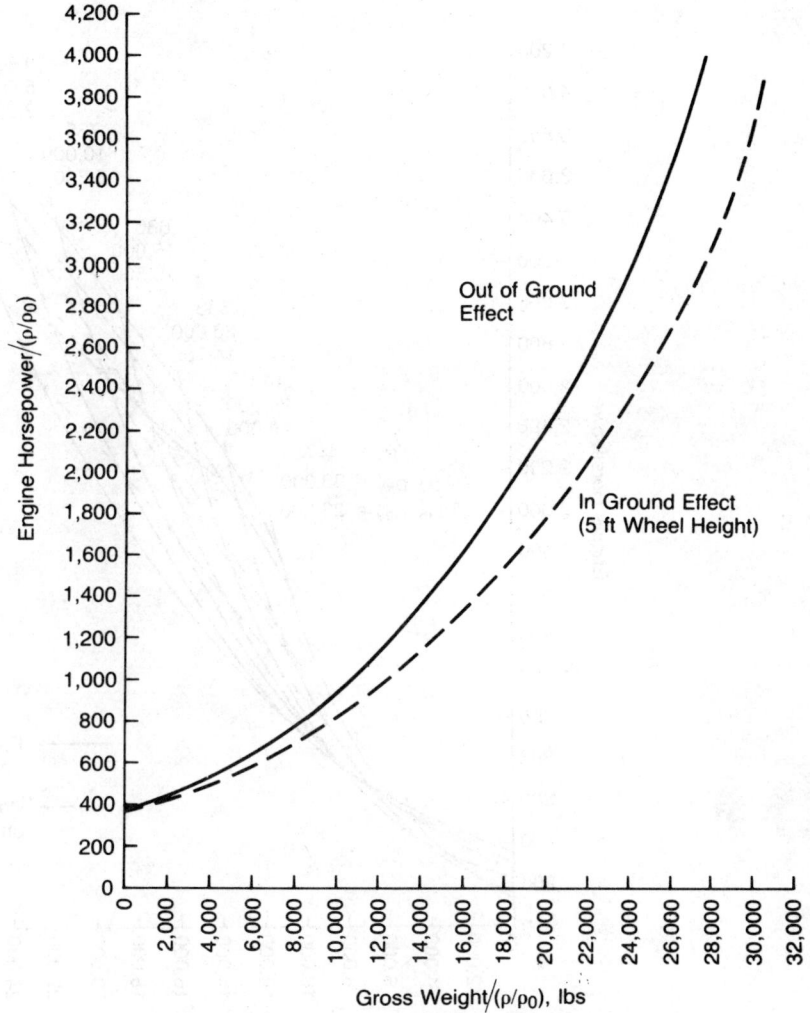

FIGURE 4.32   Normalized Engine Hover Power Required

VERTICAL CLIMB

A momentum method for computing the power required in a vertical climb in excess of that required to hover at the same conditions was derived in Chapter 2. The equation is:

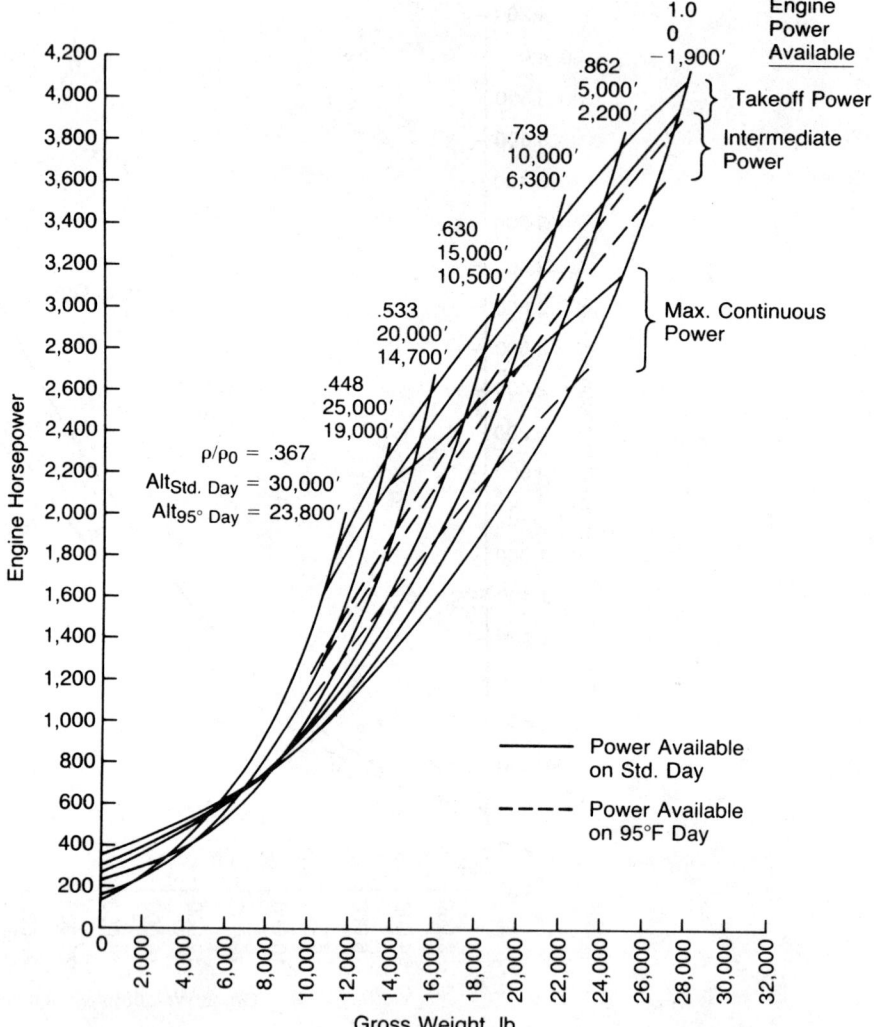

FIGURE 4.33 **Hover Performance out of Ground Effect**

$$\Delta\text{h.p.} = \frac{1}{550}\left\{\left[\text{G.W.}(v_{1_c} + V_c) + 4\left(\frac{D_v}{\text{G.W.}}\right)_{\text{hov.}}\frac{\rho}{2}(v_{1_c} + V_c)^3 A_M + (\Delta A_z C_D)\frac{\rho}{2}V_c^3\right]\right.$$

$$\left. - \left[\text{G.W.}(v_{1_{\text{hov.}}}) + 4\left(\frac{D_v}{\text{G.W.}}\right)_{\text{hov.}}\frac{\rho}{2}v_{1_{\text{hov.}}}^3 A_M\right]\right\}\left\{1 + \frac{v_{1_{\text{hov.}T}}}{(\Omega R)_M}\frac{R_M}{l_T}\right\}$$

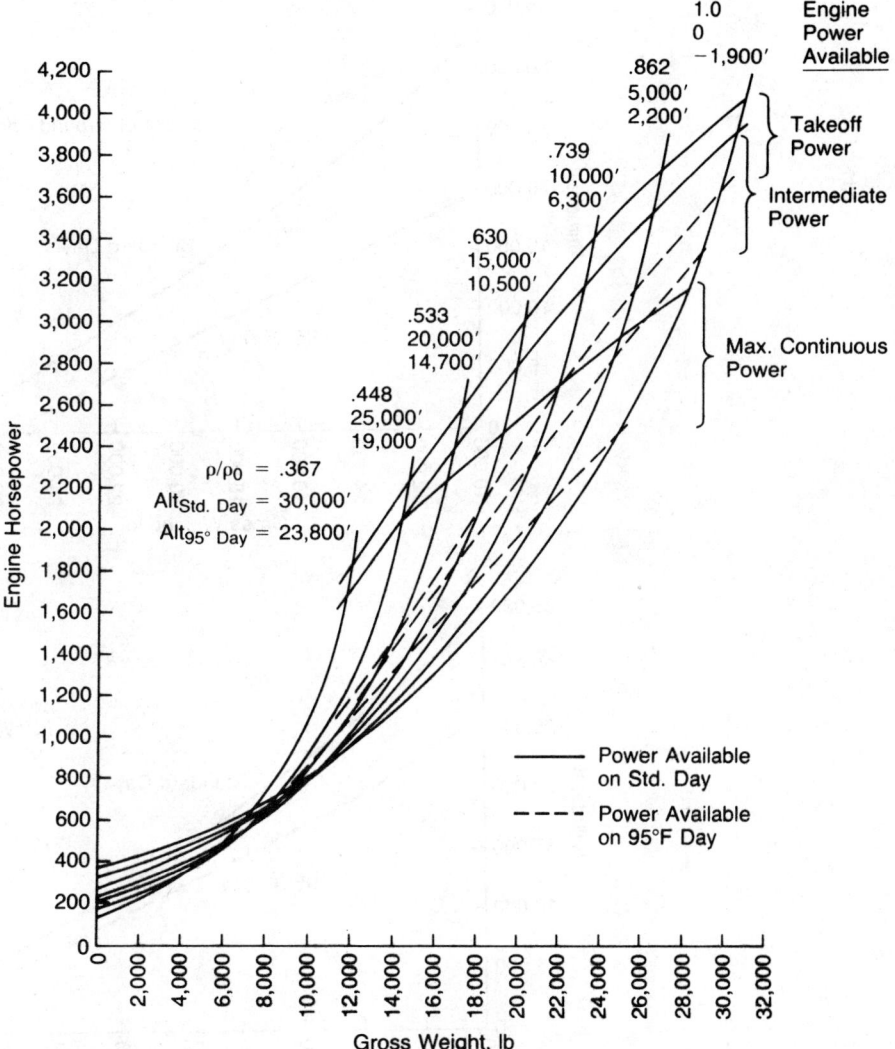

FIGURE 4.34   Hover Performance in Ground Effect at 5 Foot Wheel Height

where

$$v_{1_c} + V_c = \frac{V_c}{2} + \sqrt{\left(\frac{V_c}{2}\right)^2 + v_{1_{hov.}}^2}$$

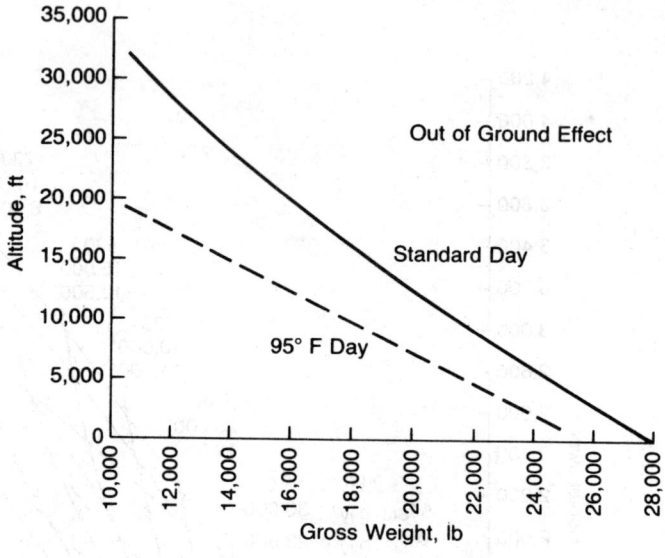

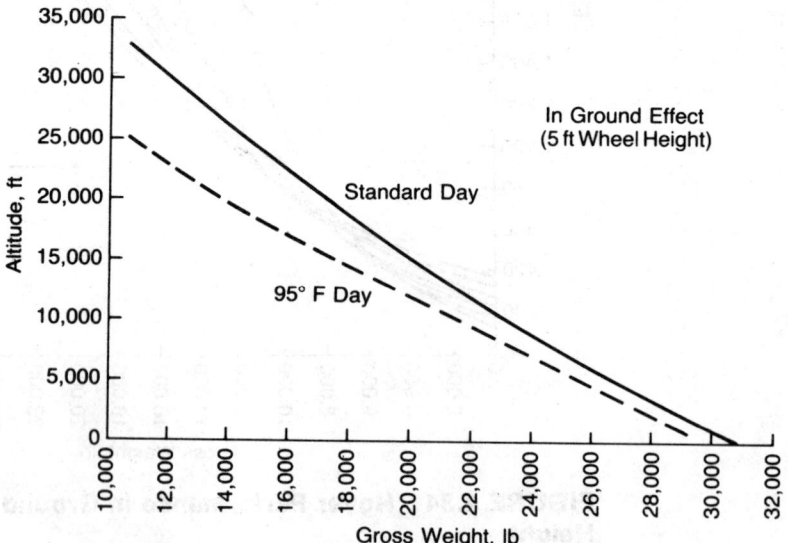

FIGURE 4.35   Hover Ceiling with Takeoff Power

This equation has been evaluated for the example helicopter by equating it to the excess power available and the resultant vertical rate of climb as a function of altitude is shown on Figure 4.36.

Figure 4.37 shows the same calculations as a function of gross weight at two conditions of usual interest: sea level, standard; and 4,000 ft, 95°F.

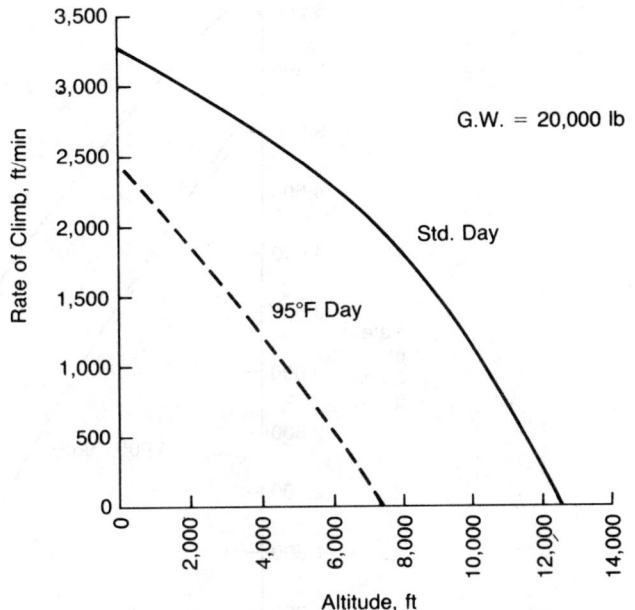

FIGURE 4.36   Vertical Rate of Climb with Takeoff Power

FORWARD FLIGHT PERFORMANCE

Power Required

Charts representing the nondimensional forward flight performance of isolated rotors and the iterative procedure for determining the trim conditions for the entire helicopter were developed in Chapter 3. These charts and procedures have been used for the example helicopter with appropriate accessory and transmission losses to calculate the engine power required in level forward flight for several gross weights. The results are presented in Figure 4.38. The basic curves show the power required assuming an advancing tip Mach number of 0.7 and the supplementary curves give the additional compressibility corrections as a function of temperature and gross weight based on actual advancing tip Mach number and the method of Figure 3.43, including the secondary effect of increased tail rotor power due to increased main rotor torque.

Also included in the figure are lines representing "upper and lower stall limits" where the increments of the profile torque coefficient, C_Q/σ_0, at the most

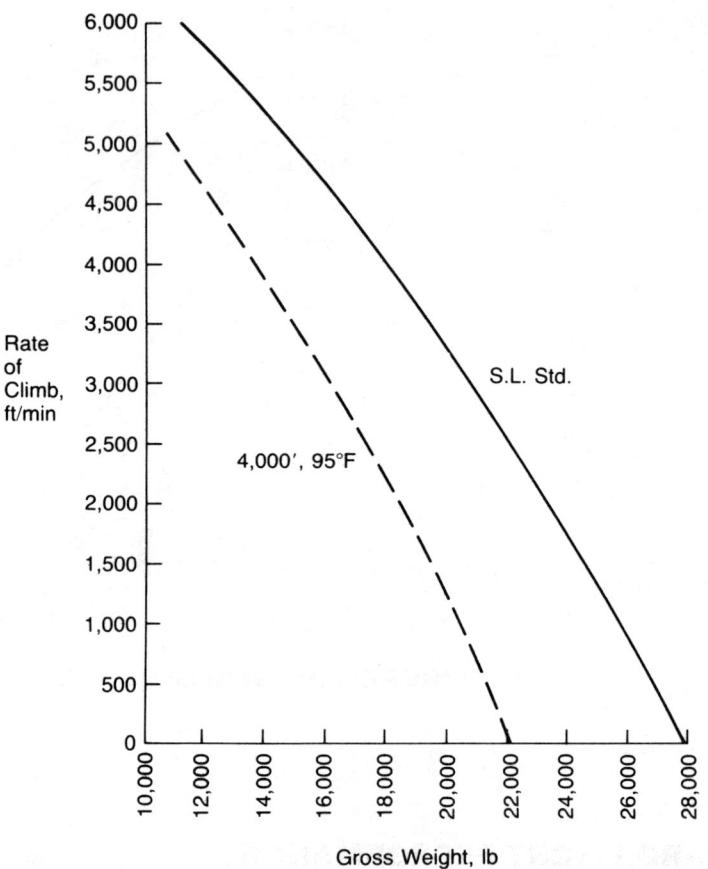

FIGURE 4.37 Vertical Climb Capability with Takeoff Power

critical azimuth position are 0.004 and 0.008, respectively. These limits are from the isolated rotor charts of Chapter 3, corrected for the twist effect as outlined in the discussion of those charts. The presence of these lines is not truly a limit, but simply a warning that many rotors suffer from high blade loads, high control loads, high vibration, and/or erratic flapping due to retreating blade stall in this regime. On the other hand, many rotors have proved relatively stall-tolerant to much higher levels. The aerodynamicist should assume that the rotor designers and dynamicists will provide such a rotor and that he can ignore the stall limits for purposes of estimating performance. An optimistic discussion of the aerodynamic capability of rotors will be found in reference 4.25.

The power required curves of Figure 4.38 as shown apply directly to sea level, standard, conditions but can also be used at other conditions, as follows:

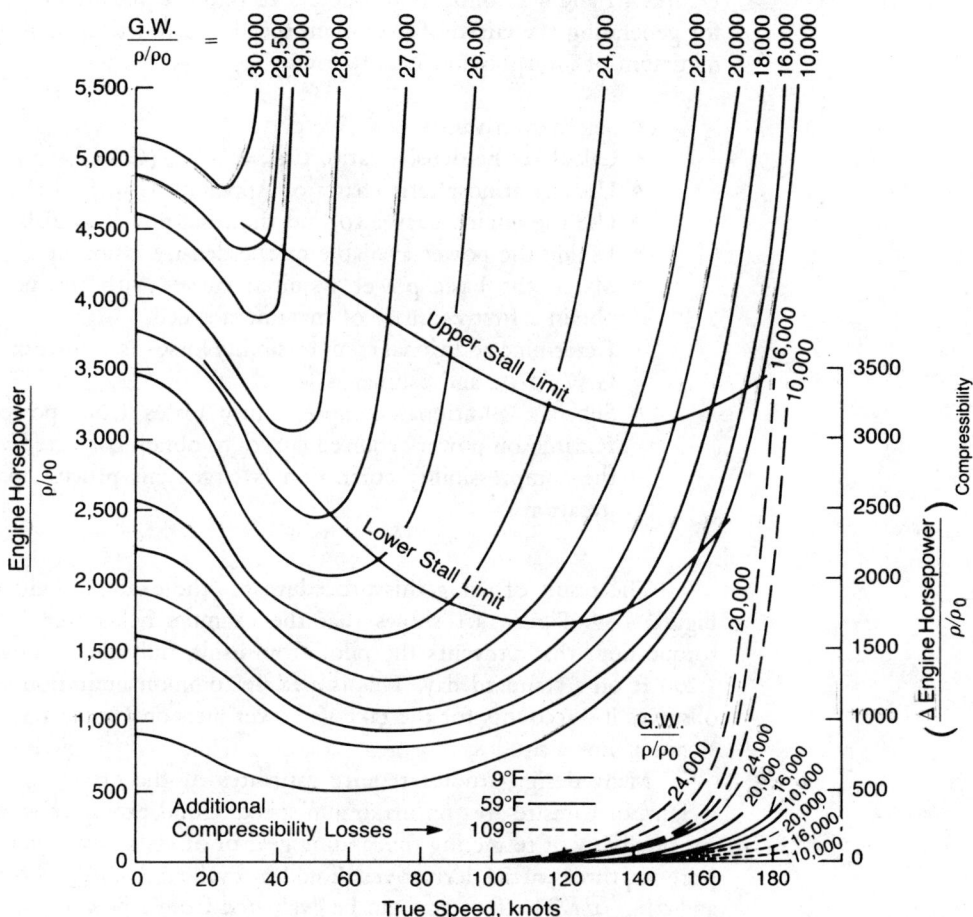

FIGURE 4.38 Normalized Engine Power Required in Forward Flight

			Example Helicopter, GW = 20,000 lb, V = 140 knots				
Altitude *(ft)*	*Temp.* *(deg F)*	*Density* *ratio* (ρ/ρ_0)	$\dfrac{G.W.}{(\rho/\rho_0)}$	$\dfrac{h.p._{eng}}{\rho/\rho_0}$ *(no comp.)*	$\dfrac{\Delta h.p._{eng}}{\rho/\rho_0}$ *(comp)*	$\dfrac{Total}{\dfrac{h.p._{eng.}}{\rho/\rho_0}}$	$h.p._{eng.}$
0	59	1.0	20,000	2,330	20	2,350	2,350
5,000	20	0.9	22,222	3,500	150	3,650	3,285

Maximum Speed

The power-required curves of Figure 4.38 and the power-available curves of Figures 4.1 and 4.2 can be used together to find the maximum speed. A procedure for generating the curve of maximum speed as a function of altitude requiring a minimum of interpolation is as follows:

- Select even values of $G.W./\rho/\rho_0$.
- Calculate the density ratio, $\rho/\rho_0 = G.W./(G.W./\rho/\rho_0)$.
- Use the atmospheric charts of Appendix C to find the altitude.
- Use the engine curves to find the total power available.
- Divide the power available by the density ratio, $h.p._{available}/\rho/\rho_0$.
- Match the basic power required curves with the power available to obtain a first estimate of maximum speed.
- Determine additional compressibility losses as a function of temperature, $G.W./\rho/\rho_0$, and estimated V_{max}.
- Subtract additional compressibility losses from power available and rematch on power required curves to obtain corrected value of V_{max}. (If the compressibility correction is large, this process may require some iteration.)

The result of using this procedure for the example helicopter is given in Figure 4.39. Figure 4.1 shows that the example helicopter has a transmission torque limit that prevents the pilot from using full takeoff power below about 1,200 ft on a standard day. This is a fairly common limitation and—if the pilot observes it—accounts for the takeoff power lines on Figure 4.39 having different slopes at low altitudes.

Many design studies require estimates of the effects of installed power, weight, or parasite area on maximum speed. Good examples would be studies of the feasibility of retracting the landing gear or of installing larger engines. For this purpose, three partial derivatives should be evaluated: $\partial V_{max}/\partial h.p.$, $\partial V_{max}/\partial G.W.$, and $\partial V_{max}/\partial f$. The first two can be evaluated from a power-required plot such as Figure 4.38. For the example helicopter at initial conditions of 20,000 lb, 160 knots, and 3,920 h.p. (intermediate installed power rating), the derivatives at sea level are:

$$\frac{\partial V_{max.}}{\partial h.p.} \doteq 0.009 \text{ K/h.p.}$$

$$\frac{\partial V_{max.}}{\partial G.W.} \doteq 0.005 \text{ K/lb}$$

A study of Figure 4.38 shows that these derivatives are strongly dependent on the initial trim point.

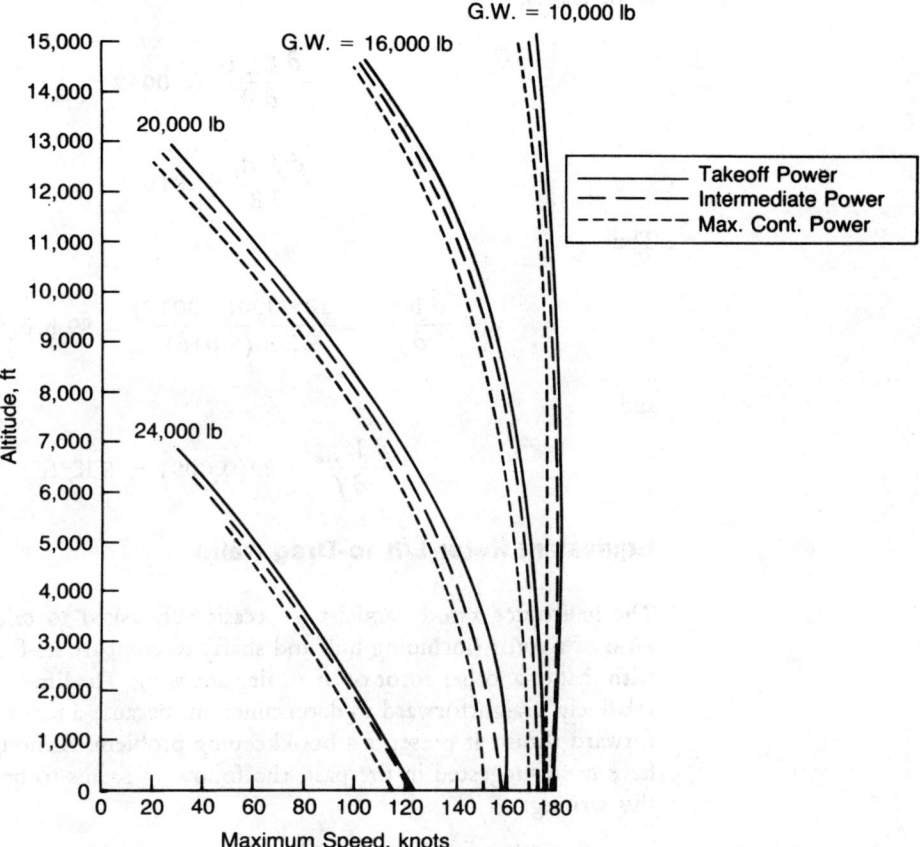

FIGURE 4.39   Maximum Speed for Example Helicopter, Standard Day

The parasite power area derivative can be found from the isolated rotor charts by noting that:

$$\frac{\partial V_{\text{max.}}}{\partial f} = \frac{\partial \text{h.p.}}{\partial f} \frac{\partial V_{\text{max.}}}{\partial \text{h.p.}}$$

where

$$\frac{\partial \text{h.p.}}{\partial f} = \frac{\dfrac{\rho A_b (\Omega R)^3}{550} \dfrac{\partial C_Q/\sigma}{\partial \theta}}{A_b \dfrac{\partial f/A_b}{\partial \theta}}$$

The trim condition is near $\mu = 0.4$ and $C_T/\sigma = 0.085$. From the isolated rotor charts:

$$\frac{\partial\, C_Q/\sigma}{\partial\,\theta} \doteq .0012$$

$$\frac{\partial\, f/A_b}{\partial\,\theta} \doteq .016$$

Thus

$$\frac{\partial\ \text{h.p.}}{\partial\, f} = \frac{285,000(0.0012)}{240(0.016)} \doteq 89\ \text{h.p./ft}^2$$

and

$$\frac{\partial\, V_{\text{max.}}}{\partial\, f} = 89(0.009) \doteq .8\ \text{K/ft}^2$$

Equivalent Rotor Lift-to-Drag Ratio

The helicopter aerodynamicist is occasionally asked to calculate the lift-to-drag ratio of a rotor (including hub and shaft) to compare its forward flight efficiency with that of another rotor or of an airplane wing. The lift-to-drag ratio of a wing is relatively straightforward to determine; but because a rotor provides both lift and forward thrust, it presents a bookkeeping problem. Although several procedures have been suggested in the past, the following seems to be generally accepted at this writing:

Equivalent lift: The vertical component of rotor thrust.

Equivalent drag: The difference between the main rotor power divided by forward speed and the parasite drag of the rest of the helicopter (not including main rotor hub and mast.)

Figure 4.40 shows the results of this procedure applied to the example helicopter.

Collective and Cyclic Pitch for Trim

A fallout of the calculations for power required is the collective and cyclic pitch associated with the trim points. Figure 4.41 shows these for level flight at sea level at the example helicopter's design gross weight.

Cruising Flight

There is no single accepted definition of cruise speed. Depending on the situation, it may mean the speed for maximum range, the speed for 99% of maximum range,

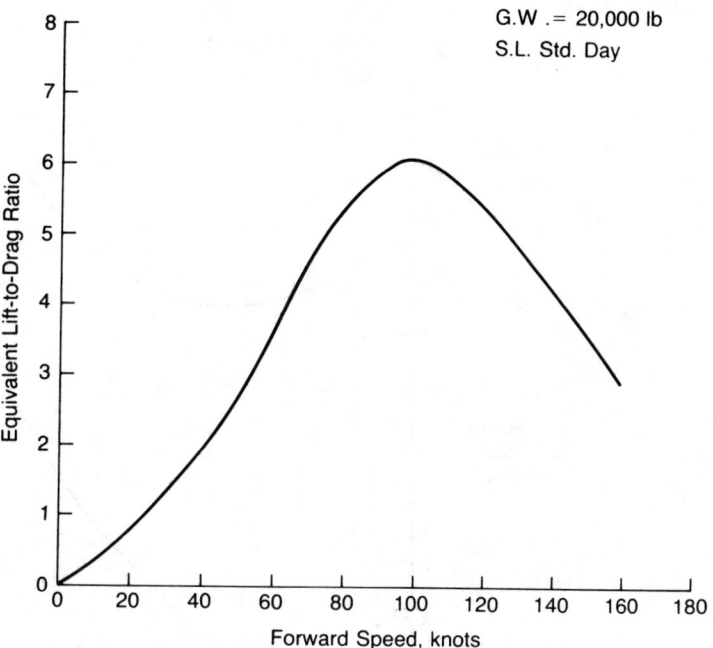

FIGURE 4.40   Equivalent Rotor Lift-to-Drag Ratio for Example Helicopter

the speed at maximum continuous power, or any other continuous speed required to do a specific mission.

The speed for maximum range may be determined using a plot of fuel flow vs. forward speed such as the one for the example helicopter on Figure 4.42. The speed for maximum range is the speed at which a ray through the origin is tangent to the fuel flow curve—for this case, 114 knots. Note that this is slightly faster than the speed for maximum equivalent rotor lift-to-drag ratio in Figure 4.40. The reason is that a turbine engine is more efficient at high power than at low power because of the fuel flow that must be used just to keep the gas generator spinning, regardless of the power output. The effects of head- and tailwinds are also shown. A headwind requires a higher cruise speed than no wind, and a tailwind a lower speed. The specific range, S.R., is the distance flown while burning one unit of fuel. It is generally expressed in nautical miles (N.M.) per pound of fuel and is determined as:

$$\text{S.R.} = \frac{\text{Ground speed}}{\text{Fuel flow}}, \frac{\text{N.M./hr}}{\text{lb/hr}}, \frac{\text{N.M.}}{\text{lb}}$$

If one knows the engine power required as in Figure 4.38 and the engine fuel consumption characteristics as in Figure 4.3, the specific range can be determined

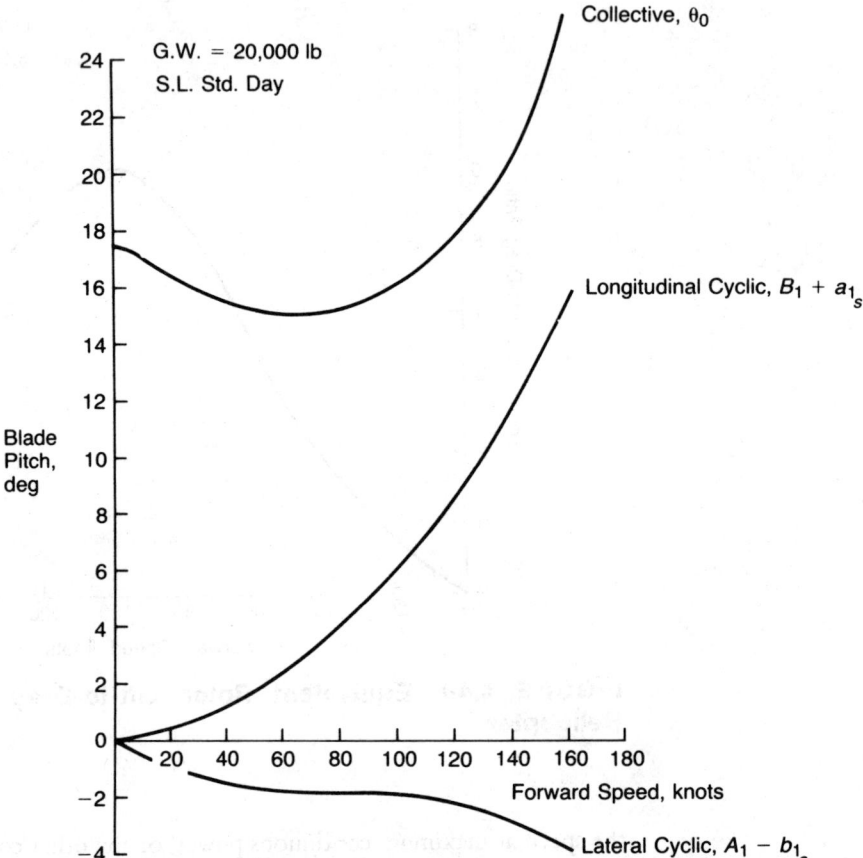

FIGURE 4.41   Collective and Cyclic Pitch for Level Flight

as a function of gross weight, altitude, and forward speed. (*Note*: For multiengine helicopters, divide the power required by the number of engines before going to the fuel flow curves.) The calculation has been done for the example helicopter, and the results are plotted on Figure 4.43. At low gross weight, maximum range is obtained at high altitude whereas at high gross weight it is obtained at sea level. At low gross weight and low altitude, the average blade element angle of attack is below the condition for the maximum lift-to-drag ratio. For this case, a higher value of C_T/σ obtained by flying at a higher altitude is beneficial. This is analogous to the dependency of hover Figure of Merit on C_T/σ, which was discussed in Chapter 1. The better specific range at low gross weight and high altitude can be important during a long flight in which the helicopter is allowed to drift up as fuel is used. One word of caution: if the cruise altitude is expected to be over 15,000 ft, oxygen should be provided for the crew.

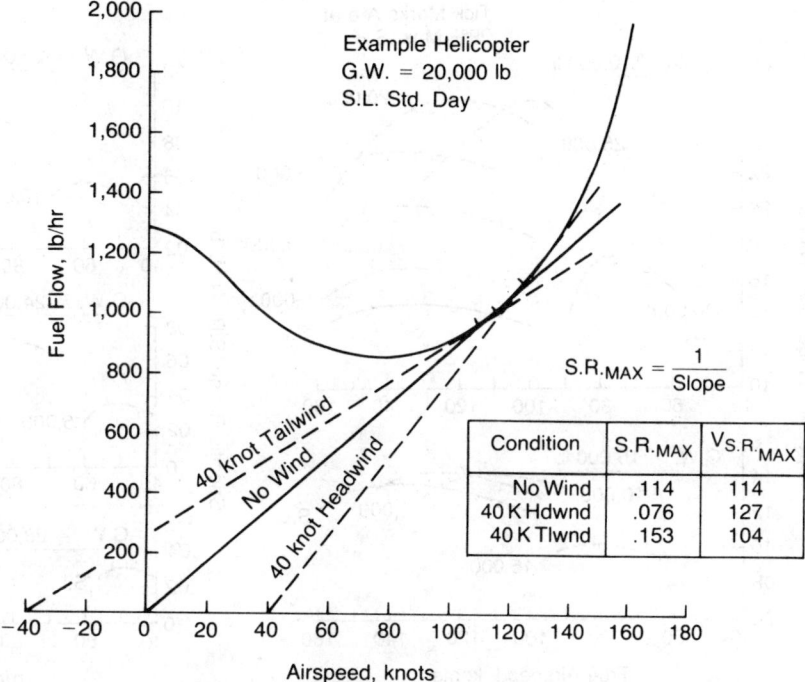

FIGURE 4.42   Graphic Determination of Maximum Specific Range

The speed for maximum range corresponds to the maximum specific range, but it is common to use the speed to the right of the peak where the specific range is 99% of maximum, the rationale being that the mission time can be shortened with little sacrifice of economy by flying at this speed.

Figure 4.44 is a cross-plot of Figure 4.43 showing the effects of gross weight and altitude on the 99% maximum specific range value. The cruise performance is better with one engine than with two. This is due to the characteristics of a turbine engine, which is more efficient near full power than at partial power since a large part of its energy is consumed internally in driving the compressor. (This effect is not so pronounced on reciprocating engines.) The single engine advantages are generally not used because of the risk of not being able to restart the dead engine in an emergency. It does, however, give the pilot an option—for example, in stretching the range at the end of a long over-water flight.

The distance traveled at the speed for 99% maximum specific range while burning a given amount of fuel can be found by integrating the area under the specific range curves of Figure 4.44, since:

$$\text{Dist.} = \int_{G.W._{end}}^{G.W._{start}} (S.R.)\, d G.W. \quad \text{N.M.}$$

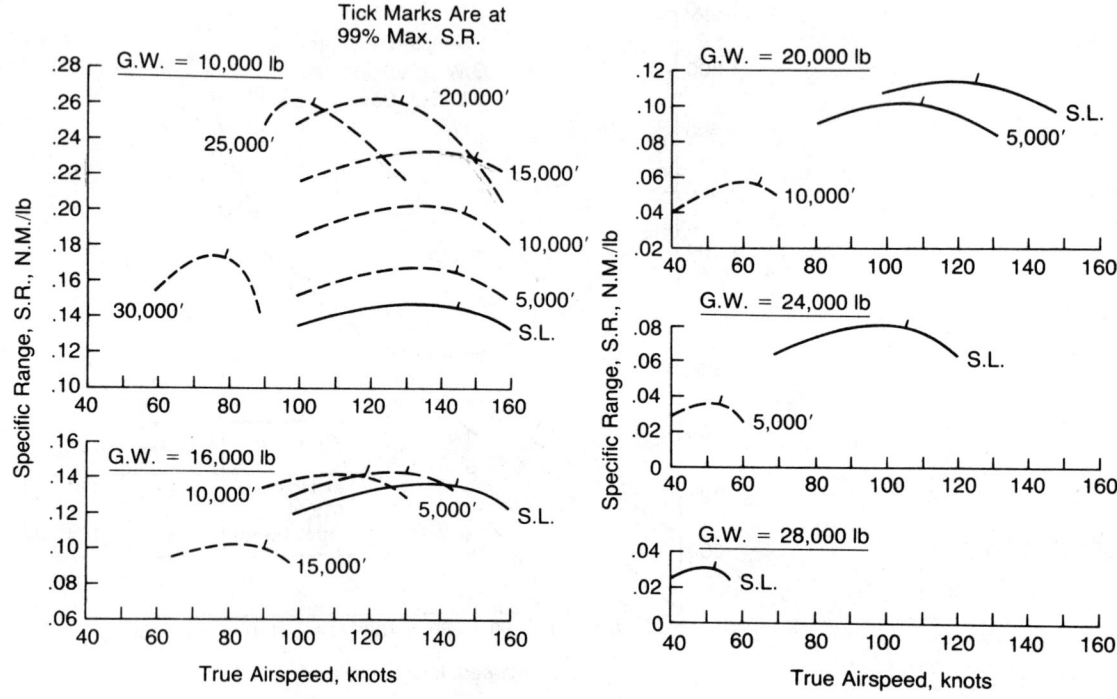

FIGURE 4.43   Specific Range Characteristics, Example Helicopter, Standard Day, No Wind

Payload-Range

A payload-range analysis is done to provide an indication of the usefulness of the helicopter as a load-carrying vehicle. Figure 4.45 shows the payload-range curve for the example helicopter at two takeoff gross weights. The conditions are listed on the figure and are typical for studies of this type. The curves are based on the evaluation of two fundamental equations:

$$\text{Range} = \int_{G.W._{\text{ldng.}}}^{G.W._{\text{T.O.}}} (S.R.) dG.W., \text{ N.M.}$$

where $G.W._{\text{lndg.}} = G.W._{\text{T.O.}} - (\text{Expended fuel} + \text{WUTO fuel}) \text{ lb}$

and $\text{Payload} = (G.W._{\text{T.O.}} - G.W._{\text{min.OP.}}) - (\text{Expended fuel} + \text{WUTO fuel} + \text{Reserves} + \text{Aux. fuel tank wt.}) \text{ lb}$

(WUTO stands for warmup and takeoff.)

The figure shows that as a transport aircraft the example helicopter, taking off at 20,000 pounds, can carry its design payload of 6,600 pounds—consisting of 30 passengers with baggage averaging 220 pounds apiece—for 330 nautical miles at

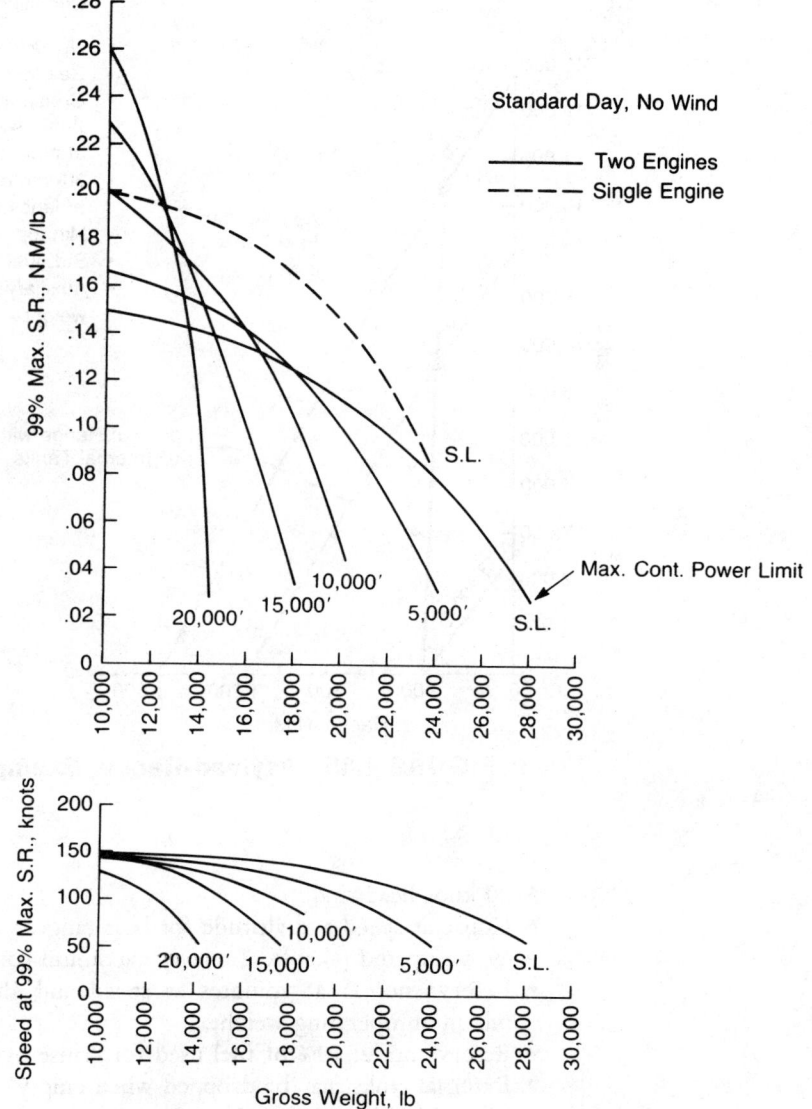

FIGURE 4.44   Cruise Conditions, Example Helicopter

sea level with adequate reserves. Offloading payload and replacing with fuel in auxiliary tanks can extend the range significantly.

Maximum Ferry Range

The calculation of the maximum ferry range uses the same general procedures as for the payload-range curves. Typical special considerations that might be used, however, are:

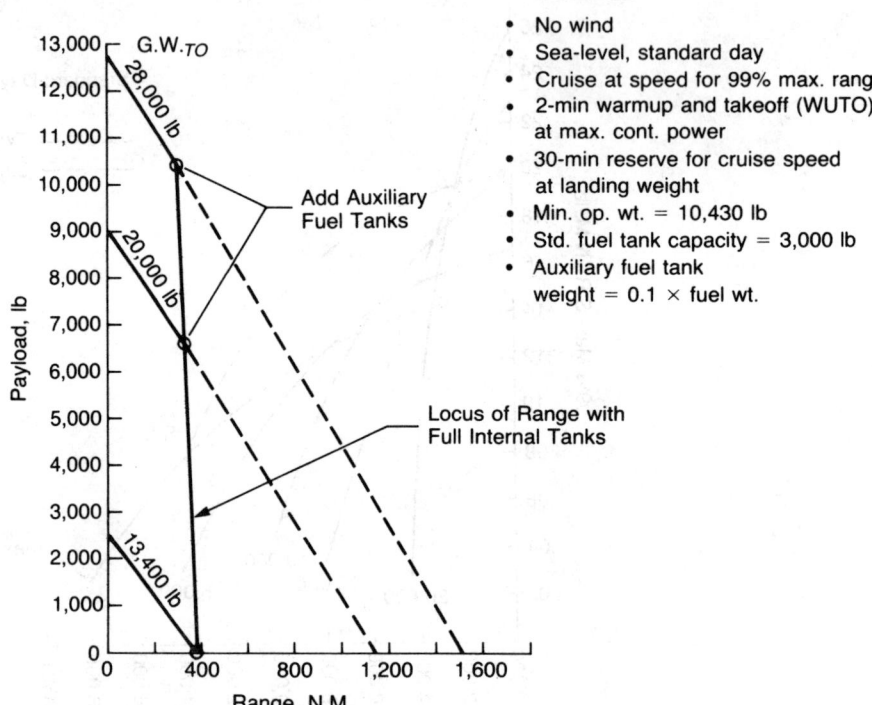

Conditions

- No wind
- Sea-level, standard day
- Cruise at speed for 99% max. range
- 2-min warmup and takeoff (WUTO) at max. cont. power
- 30-min reserve for cruise speed at landing weight
- Min. op. wt. = 10,430 lb
- Std. fuel tank capacity = 3,000 lb
- Auxiliary fuel tank weight = 0.1 × fuel wt.

FIGURE 4.45   Payload-Range, Example Helicopter

- 20-knot headwind.
- Cruise at speed and altitude for best range.
- Warm up and takeoff: 2 min at maximum continuous power.
- Reserve no. 1: 45 minutes at speed and altitude for best range at minimum operating weight.
- Reserve no. 2: 10% of fuel used for cruise over 3 hours.
- External tanks may be dropped when empty.
- Noncritical items of equipment may be left behind (weapons, passenger seats, etc.).
- Special items of equipment may be required (life rafts, oxygen, etc.).

For illustration, the maximum ferry range of the example helicopter will be determined using these conditions. The assumed takeoff gross weight is 28,000 lb, which is almost the maximum that can be flown in level flight on maximum continuous power.

The Group Weight Statement for the ferry mission in Appendix A can be used to find the minimum operating weight as:

Empty weight	10,000 lb
Crew	360
Unusable fuel	30
Engine oil	40
Auxiliary cabin tank	481
Survival kits	100
Lift rafts	50
Oxygen equipment	200
Minimum operating weight	11,261 lb

The total usuable fuel for the ferry mission is contained in five tanks:

Forward internal	1,485 lb
Aft internal	1,485
Cabin auxiliary	4,809
First external	4,091
Second external	4,091
Total	15,961 lb

Each external tank weighs 409 pounds and is assumed to have an equivalent flat plate area of 2 square feet.

Figure 4.46 shows the specific range plot based on these conditions. The maximum range is calculated by integrating under the curve from the mission start weight, which is the takeoff gross weight minus fuel for warmup and takeoff, and the landing weight, which is the minimum operating weight plus reserves. The reserves consist of two amounts:

Reserve no. 1 = Fuel for 45 minutes at minimum operating weight.

Reserve no. 2 = .1/1.1 [Total fuel − (Fuel for first 3 hr + WUTO + reserve no. 1)]

For this example, reserve no. 1 is 420 lb when calculated at 11,261 lb, 20,000 ft altitude, and 120 knots air speed. The fuel required for the first 3 hours was found by calculating the distance and time represented by burning 1,000-pound increments of fuel from takeoff. In the first 3 hours, 4,400 lb of fuel will be used. Using a warm-up and takeoff (WUTO) fuel allowance of 56 lb, reserve no. 2 becomes 1,008 lb. Thus the landing weight is 12,689 lb, and the takeoff weight is 27,944 lb. Integrating the envelope on Figure 4.46 between these two weights and accounting for dropping the external tanks when empty gives a maximum ferry range of 1,289 nautical miles.

In this example, the optimum altitude is sea level until more than half the fuel is burned off. At that time a slow climb is initiated, which reaches 15,000 feet at the end of the mission. The extra fuel required for this climb can be calculated and subtracted from the fuel available for cruise; but unless it is a significant quantity it can be ignored by assuming that the same amount of fuel is saved during the descent at the end of the flight.

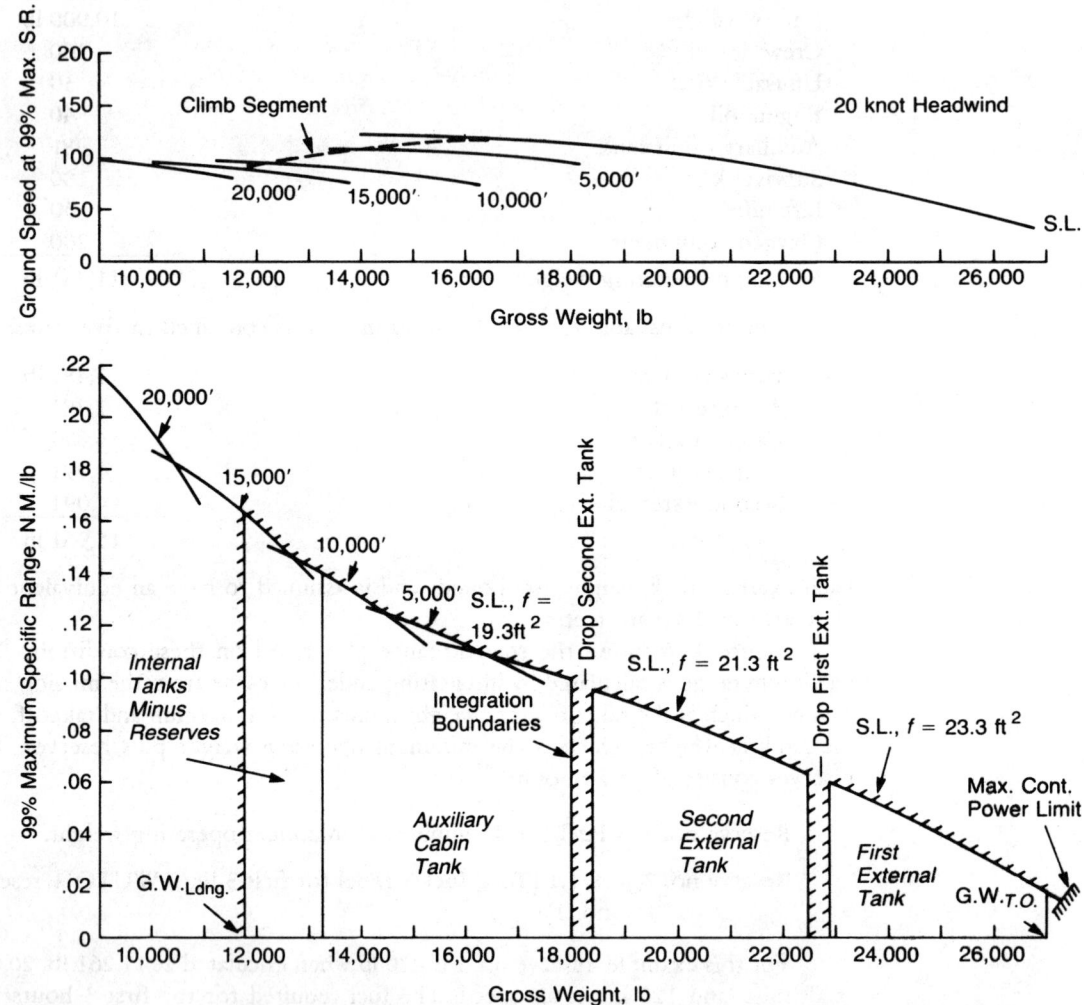

FIGURE 4.46   Ferry Mission Conditions, Example Helicopter

Loiter Flight

Many military missions require the helicopter to loiter while observing or waiting for some specific action. For these mission legs, the optimum speed is at the bottom of the fuel flow curve of Figure 4.42, and the endurance at any speed is a function of the specific endurance, S.E., whose units are hr/lb:

$$S.E. = \frac{1}{\text{Fuel flow}} , \frac{1}{\text{lb/hr}} , \frac{\text{hr}}{\text{lb}}$$

Figure 4.47 shows the calculated specific endurance curves for the example helicopter. Here again, the advantage of flying on one engine is evident. This might become important to a pilot who is forced to wait for a change in the weather or for a landing site to be prepared. The time spent in loiter flight while burning a given amount of fuel can be found by integrating under the specific endurance curves of Figure 4.47, since:

$$t = \int_{G.W._{end}}^{G.W._{start}} (S.E.)dG.W. \text{ hr}$$

This equation can be used to generate payload-endurance curves similar to payload-range curves.

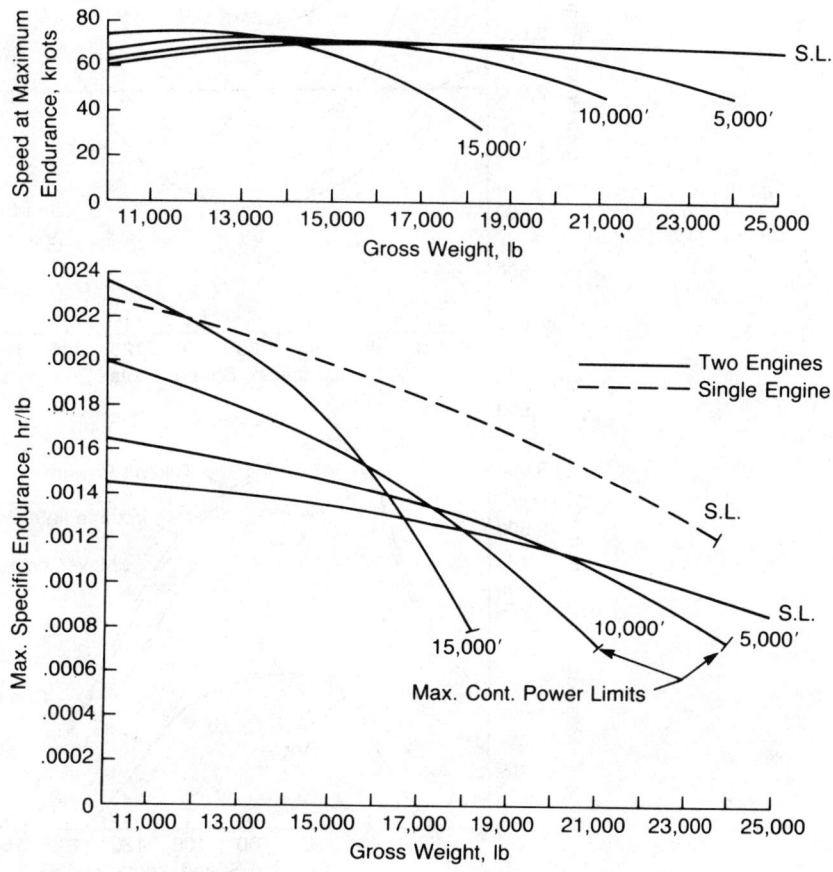

FIGURE 4.47   Maximum Endurance Conditions, Example Helicopter

CLIMB IN FORWARD FLIGHT

Rates of Climb

The power required to climb at a given flight path angle, γ, can be determined by using the charts of Chapter 3 just as for level flight but modifying the flat plate

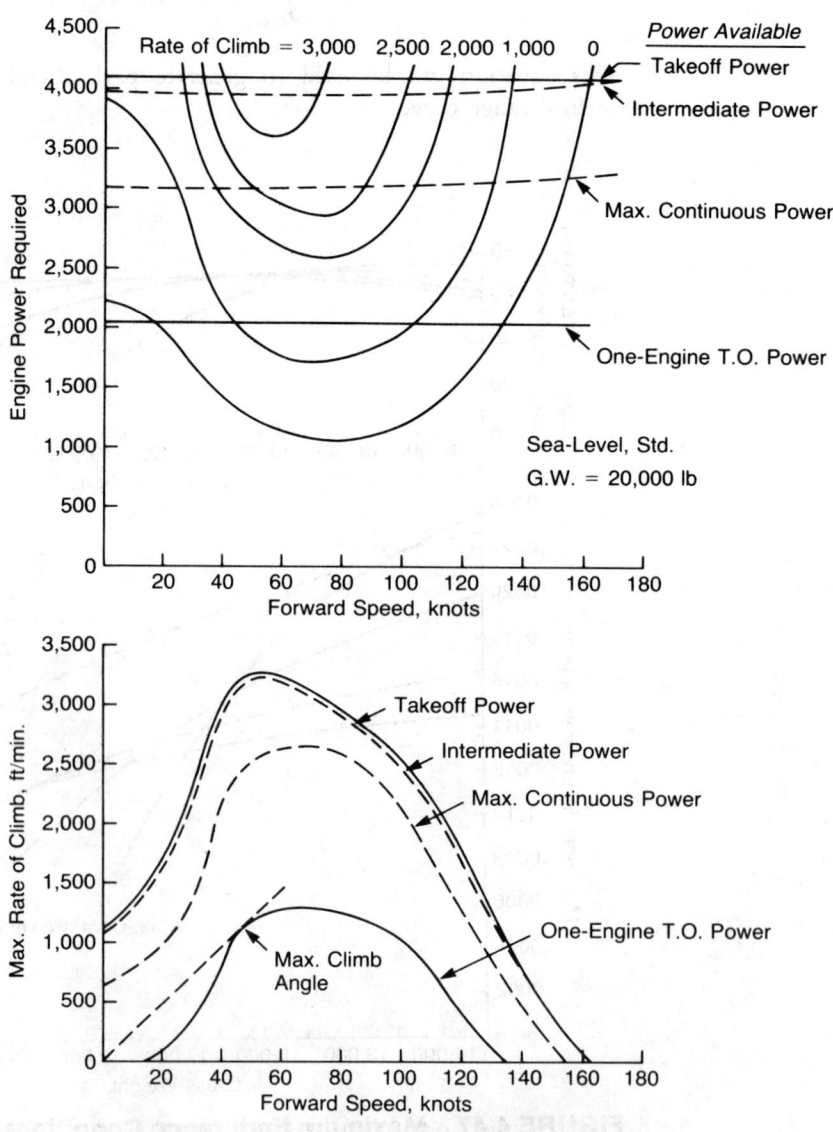

FIGURE 4.48   Climb in Forward Flight

area to account for the rearward component of gross weight along the flight path:

$$f_{climb} = f + \frac{G.W. \sin \gamma}{q}$$

Figure 4.48 shows the results of the calculations for the example helicopter at rates of climb from zero to 3,000 ft/min at sea level conditions. A cross-plot at the various ratings gives corresponding rates of climb as a function of forward speed in the lower part of the figure. Carrying out the same analysis at several altitudes allows us to find the maximum rates of climb as a function of altitude, as in Figure 4.49.

Angle of Climb

A figure like the bottom part of Figure 4.48 can also be used to determine the maximum angle of climb, which might be of interest in clearing an obstacle in the flight path. In this case the maximum angle is 90° when both engines are running, since the helicopter can climb vertically. With one engine, however, it cannot even hover, but it can climb at 1,100 ft/min at 48 knots, which gives a climb angle of 12.7 degrees. A similar analysis can be made when the gross weight or the altitude is too high for hover out of ground effect with two engines.

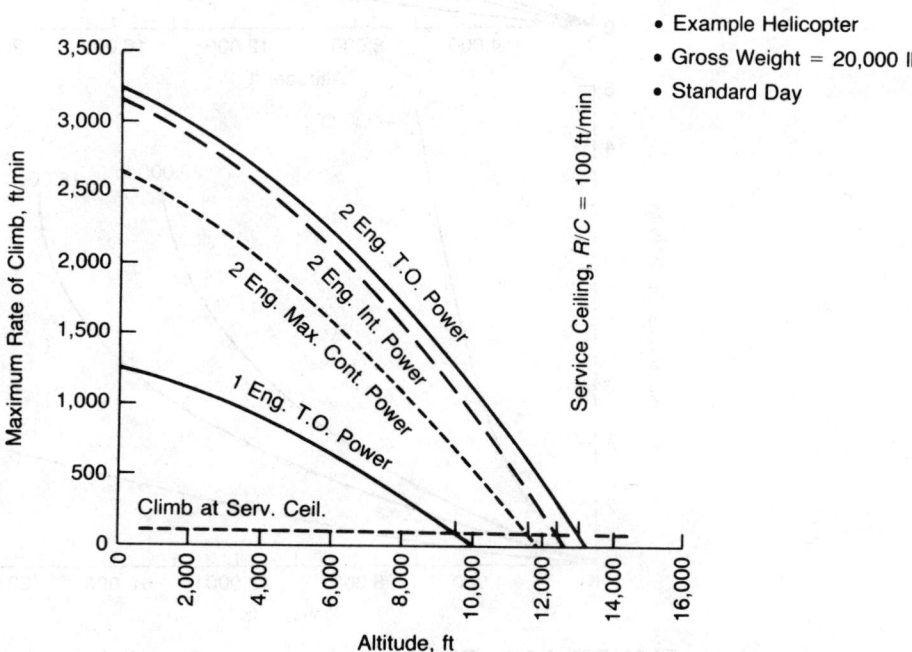

FIGURE 4.49   Maximum Rate of Climb in Forward Flight

Time and Distance to Climb

The minimum time to climb to a given altitude can be calculated using the maximum rates of climb on Figure 4.49 to determine the time required to climb in 1,000-ft increments. The summation of this process is shown on Figure 4.50. Knowing the speed corresponding to the climb rates makes it possible to find the distance required to climb. These values will later be useful in doing mission analyses.

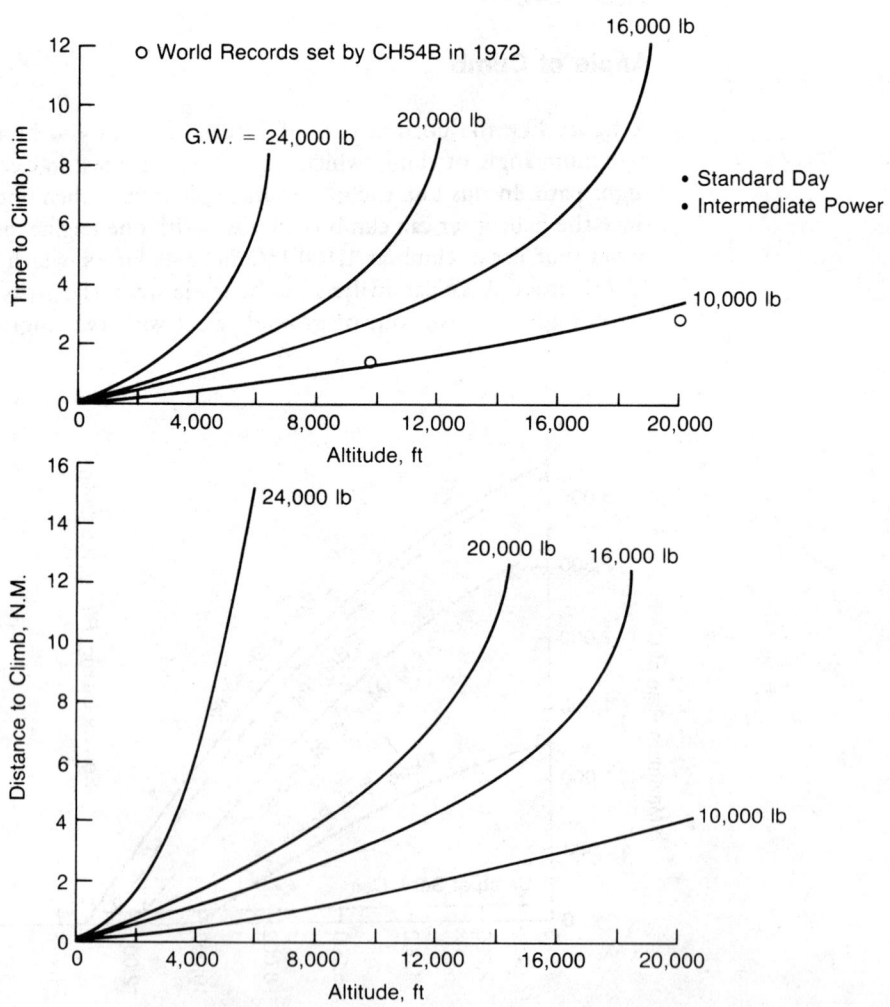

FIGURE 4.50   Time and Distance Required for Climb in Forward Flight

Two world records for time-to-climb are spotted in Figure 4.50. These were set in 1972 with a Sikorsky CH-54B and were still valid in 1985.

Forward Flight Ceilings

The altitude at which the rate of climb is zero is the *absolute ceiling*, and the altitude at which the rate of climb is 100 ft/min is the *service ceiling*. Both these ceilings can be found from Figure 4.49 for a gross weight of 20,000 lb. The same analysis at other gross weights can be used to obtain the plot of ceiling as a function of gross weight, as in Figure 4.51.

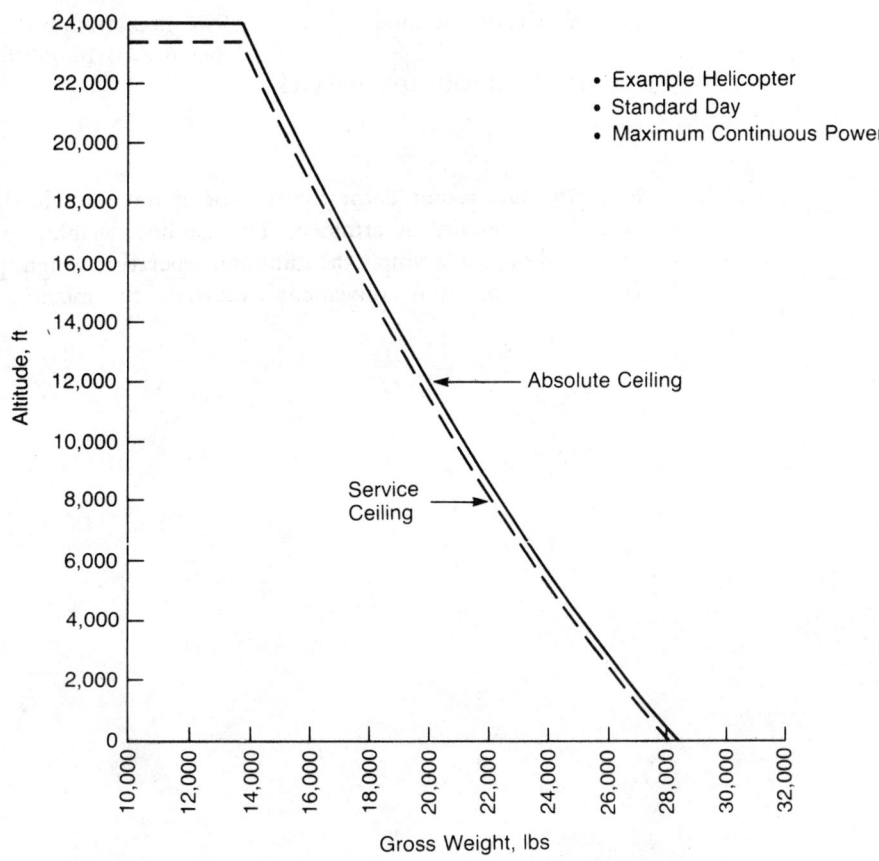

FIGURE 4.51   Absolute and Service Ceiling

ANALYSIS OF THE MILITARY-TYPE MISSION

A military mission may be considerably more complicated than simply flying from one place to another. A typical mission might include the following segments:

1. Warm-up and takeoff   Maximum continuous power, 2 min, S.L.
2. Climb   Intermediate power, S.L. to 5,000 ft
3. Cruise outbound   Maximum continuous power, 90 N.M. minus the distance required to climb 5,000 ft
4. Loiter   20 min S.L.
5. Dash   Intermediate power, 10 min, S.L.
6. Hover OGE   10 min S.L.
    Land and discharge 6,600 lb of payload
7. Warm-up and takeoff   Maximum continuous power, 2 min, S.L.
8. Climb   Intermediate power, S.L. to 10,000 ft
9. Cruise inbound   99% maximum S.R., 100 N.M. minus climb distance to 10,000 ft
10. Land with 10% reserves

Since the fuel required for the mission is not known, the takeoff gross weight cannot be initially determined. The landing weight, however, can be closely estimated since it is simply the minimum operating weight plus a small fuel reserve. For this reason, it is convenient to analyse the mission in reverse as on Table 4.4.

TABLE 4.4
Analysis of a Military Mission

Segment	Altitude	Power Rating	G.W. at End of Segment (Add Fuel from Last Column)	Average G.W. (Estimated)	Segment Distance, N.M.	Segment Time, Hr	Speed, Knots	Average Power per Eng.	Specific Range	Specific Endurance	Fuel Flow per Engine	Total Fuel Flow	Fuel for Segment
9. Cruise	10,000	—	10,730*	11000	96	—	—	—	.192[a]	—	—	—	510
8. Climb	0–10,000	Int.	11,240	11280	2[b]	.033[b]	—	1800[c]	—	—	900[d]	1800	59
7. WUTO	S.L.	Max. Cont.	11,299	—	—	.033	—	1600[c]	—	—	840[d]	1680	56
6. Hover OGE	S.L.	—	17,955**	18000	—	.167	0	940[e]	—	—	590[d]	1180	194
5. Dash	S.L.	Int.	18,149	—	10	.060	166[f]	2080[c,h]	—	—	1040[d]	2080	125
4. Loiter	S.L.	—	18,274	—	—	.333	—	—	—	.00121[g]	—	—	275
3. Cruise	5,000	Max. Cont.	18,549	—	88	.667	132[f]	1530[c,h]	—	—	785[d]	1570	1047
2. Climb	0–5,000	Int.	19,596	—	2[b]	.028[b]	—	1900[c]	—	—	950[d]	1900	53
1. WUTO	S.L.	Max. Cont.	19,649	—	—	.033	—	1600[c]	—	—	840[d]	1680	56

Est. T.O.G.W. = 19,705   (36 lb too high due to initial reserve estimate)

Mission Fuel = 2,375

Total Fuel = $\dfrac{2375}{0.9}$ = 2,639

Reserve = 264

Min. Op. Wt. = 10,430
Payload = 6,600
Total Fuel = 2,639
T.O.G.W. = 19,669

*Reserve estimated at 300 lb
**6,600 lb of payload accounted for

Source Figures

a. 4.44   e. 4.33
b. 4.50   f. 4.39
c. 4.1   g. 4.47
d. 4.3   h. 4.2

337

EXAMPLE HELICOPTER CALCULATIONS

HOW TO'S

The following items can be evaluated by the methods of this chapter.

CHAPTER 5

Special Performance Problems

INTRODUCTION

The previous chapters described methods for estimating the performance of the helicopter as an aircraft that hovers or simply flies from one place to another. Frequently, however, it is necessary to analyze the performance from other aspects. The aspects that will be discussed in this chapter are:

- Turns and pullups
- Autorotation
- Maximum accelerations
- Maximum decelerations
- Optimum takeoff procedure at high gross weights
- Return-to-target maneuver
- Towing
- Aerobatic maneuvers

TURNS AND PULLUPS

Load Factor Relationships

Conditions in a steady turn or in a symmetrical pullup are similar to conditions in level flight except that the rotor thrust is significantly higher than the gross weight and the rotor has a pitching velocity that relieves the retreating tip angle of attack. The rotor thrust must overcome the vector sum of weight and centrifugal force, as shown in Figure 5.1

$$T = \sqrt{(G.W.)^2 + (C.F.)^2}$$

The ratio of rotor thrust to gross weight is known as the load factor, n:

$$n_{turn} = \frac{T}{G.W.} = \sqrt{1 + \left(\frac{C.F.}{G.W.}\right)^2}$$

or, in terms of the bank angle:

$$n_{turn} = \sqrt{1 + \tan^2 \Phi} = \frac{1}{\cos \Phi}$$

The centrifugal force is:

$$C.F. = \frac{G.W.}{g} R_{turn} \omega^2$$

where ω is the rate of turn and R_{turn} is the radius of the turn. But

$$R_{turn} \omega = V$$

so that, based on the rate of turn and the speed, the load factor becomes

$$n_{turn} = \sqrt{1 + \left(\frac{V\omega}{g}\right)^2}$$

and

$$R_{turn} = \frac{V^2}{g \sqrt{n^2 - 1}}$$

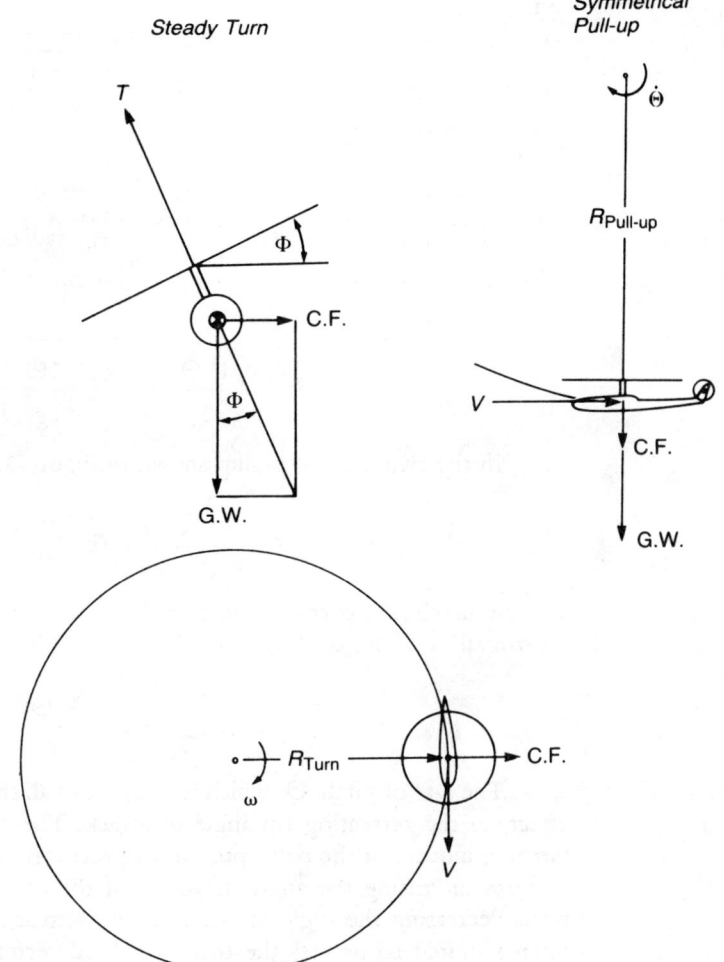

FIGURE 5.1   Conditions in a Turn and in a Symmetrical Pullup

Yet another form of the equation can be written in terms of the rate of pitch, $\dot{\Theta}$, by noting that

$$\dot{\Theta} = \omega \sin \Phi$$

(in a vertical bank the rate of pitch would be equal to the rate of turn), but

$$\omega = \frac{g}{V} \sqrt{n^2 - 1} \text{ radian/sec}$$

and

$$\sin \Phi = \frac{\sqrt{n^2 - 1}}{n}$$

Thus

$$\dot{\Theta} = \frac{g}{V}\left(\frac{n^2 - 1}{n}\right) \text{ radian/sec}$$

or

$$n_{\text{turn}} = \frac{V\dot{\Theta}}{2g} + \sqrt{\left(\frac{V\dot{\Theta}}{2g}\right)^2 + 1}$$

In the symmetrical pullup shown in Figure 5.1, the centrifugal force is

$$\text{C.F.} = \frac{\text{G.W.}}{g}\,\dot{\Theta}^2 R_{\text{pullup}}$$

and the maximum corresponding load factor, as the aircraft passes through the horizontal position, is:

$$n_{\text{pullup}} = 1 + \frac{V\dot{\Theta}}{g}$$

The rate of pitch, $\dot{\Theta}$, which is inherent in all these maneuvers, has a relieving effect on the retreating tip angle of attack. The rotor, which is producing the pitching motion of the helicopter, must precess itself nose up as a gyroscope; this requires increasing the angle of attack of the advancing blade with cyclic pitch while decreasing the angle of attack of the retreating blade. The change in cyclic pitch required to precess the rotor can be determined by setting the change in aerodynamic moment from trim equal to the gyroscopic moment. If the pitching velocity is constant, with no pitching acceleration, then no additional hub moment is required except for a small one due to damping moments from the fuselage and horizontal stabilizer. The analysis of the rotor with pitching velocity will be found in Chapter 7, "Rotor Flapping Characteristics." From that analysis:

$$\Delta B_1 = -\frac{16}{\gamma\Omega}\,\dot{\Theta} \text{ radians}$$

For the example helicopter in a turn at 115 knots with a load factor of 1.2, the pitch rate is 0.08 rad/sec and the decrease of cyclic pitch required due to this rate is 0.2 degrees, which decreases the retreating angle of attack by this amount and thus allows the rotor to develop somewhat more thrust before stalling than it could in static conditions such as in a wind tunnel test.

The power required during a steady turn may be estimated from the level flight characteristics by using an effective gross weight equal to the actual gross weight multiplied by the load factor. Using this method and the information in Figure 4.38 of Chapter 4, the example helicopter requires a total power of 3,170 h.p. in a 1.2-g turn at 115 knots, compared to 1,470 h.p. for level flight at the same speed. Sometimes for demonstration purposes, it is permissible to lose some speed and some altitude during a "steady" turn. In this way, a significant amount of power can be extracted from the changes in kinetic and potential energy in order to demonstrate higher load factors within the installed power limitations. Assuming that a speed, ΔV, is lost during a 180° turn, the change in kinetic energy is:

$$\Delta \text{K.E.} = \frac{\text{G.W.}}{g} V_{\text{avg.}} \Delta V$$

and, combining equations derived earlier, the time to make the 180° turn is:

$$t_{180°} = \frac{\frac{\pi}{2} V_{\text{avg.}}}{g \sqrt{n^2 - 1}}$$

Thus the gain in power is:

$$\Delta \text{h.p.}_{\Delta V} = 2 \left(\frac{\text{G.W.} \sqrt{n^2 - 1}}{550 \, \pi} \right) \Delta V$$

Similarly, the power associated with a loss of altitude during the same 180° turn is:

$$\Delta \text{h.p.}_{\Delta h} = 2 \left(\frac{\text{G.W.} \sqrt{n^2 - 1}}{550 \, \pi} \right) \frac{g \Delta h}{V_{\text{avg.}}}$$

For the example helicopter in a 1.5-g, 180° turn starting at 115 knots, ending at 100 knots, and losing 50 ft of altitude, the increment of available power due to the change of speed is 656 h.p. and due to the change in altitude, 230 h.p.

For analysis of nap-of-the-earth flight, it is necessary to consider load factors less than 1 developed during a pushover as would be used to fly over the top of a hill and into the valley beyond. For this case:

$$n = 1 - \frac{\text{C.F.}}{\text{G.W.}}$$

or

$$n = 1 - \frac{\dot{\Theta}^2 R_{\text{pushover}}}{g}$$

but

$$\dot{\Theta} = \frac{V}{R_{\text{pushover}}}$$

so that:

$$n = 1 - \frac{V^2}{g R_{\text{pushover}}}$$

This can be used to calculate the load factor associated with flying over a hill with a given radius of curvature.

Producing Maximum Load Factors

The pilot who has been asked to demonstrate the maximum load factor capability of a helicopter will build up the maneuver until he reaches one of the following limits:

- Maximum engine power
- Maximum stick displacement
- Unacceptable level of vibration
- High nose-up attitude or pitch rate from which recovery is uncertain
- Indication of abnormally high loads in the rotor or the control system
- Aircraft instability
- Ominous change in the rotor noise level
- Sudden rotor out-of-track condition

Some of these limits—such as reaching maximum engine power—are straightforward and can be predicted by methods already developed in previous chapters. Others, however, are a function of the structural and dynamic characteristics of the rotor, the control system, and the remainder of the helicopter and of the pilot's willingness to subject himself to uncomfortable or potentially dangerous flight conditions. There is as yet no analytical method for predicting the maximum attainable load factor when these latter considerations are involved, but there are enough experimental data to provide some insight into the problem. A convenient nondimensional representation of the maximum thrust capability is a plot of C_T/σ versus μ such as Figure 5.2. (*Note*: if the blade is not of constant chord, the definition of *thrust-weighted solidity* given in Chapter 1 should be used in this analysis.) The plot has three boundaries depending on the flight condition.

The *transient* boundary can be achieved momentarily in flight or continuously in a wind tunnel at high rotor angles of attack and corresponds to every blade element operating at its maximum lift coefficient. Test results indicate that this boundary is in the neighborhood of $C_T/\sigma = 0.17$, which is equivalent to an average lift coefficient of over 1. Values of this magnitude or higher have been

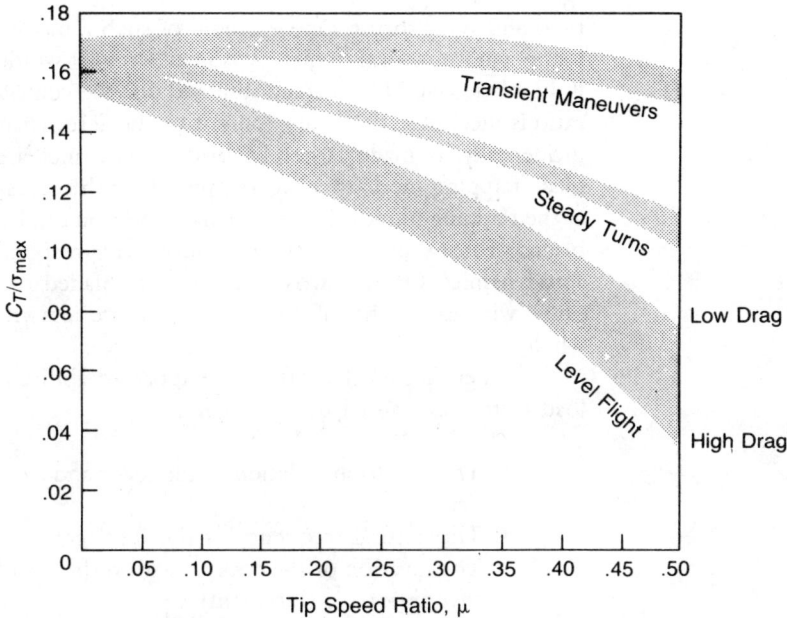

FIGURE 5.2   Rotor Thrust Capability

reported: in flight maneuvers by references 5.1 and 5.2; in autorotative flares by reference 5.3; and in wind tunnel tests by references 5.4 and 5.5. The theoretical limit of the transient capability is a value of C_T/σ of about 0.3, which could be achieved on a heavily stalled rotor at a very high angle of attack, where the contribution of the drag acting on each blade element parallel to the shaft is producing most of the rotor force. Other boundaries in Figure 5.2 represent level flight and steady turns and are primarily based on the flight experience reported in references 5.6 and 5.7. For this reason, these boundaries probably reflect some of the nonaerodynamic limits listed in the beginning of this section. These limits vary from helicopter to helicopter and determine how heavily loaded a rotor may be before it gets into stall-related troubles. For example, it is common to generate high oscillating loads in the control system when the blade goes into stall on the retreating side. The resultant change in aerodynamic pitching moment can twist the blade nose down to an unstalled condition, only to have it spring back into stall. Depending on the torsional natural frequency of the blade and control system, each blade may go through several cycles of this oscillation—sometimes called *stall flutter*—while passing through the retreating side. A theoretical and experimental investigation of this problem is reported in reference 5.8 in which it is shown that control loads are higher when the torsional natural frequency ratios are between 5 and 12 per revolution than for frequencies on either side of this range.

Another possible source of high vibration and high loads is the excitation of those blade-bending modes for which the forcing function increases with tip speed ratio and rotor thrust. One example of such a mode is the second blade flapping mode, whose natural frequency is near 3/rev (it would be exactly $\sqrt{2\pi}$ or 2.51/rev if the blade acted like a chain stiffened only be centrifugal forces). As the tip speed ratio is increased, the 3/rev content of the aerodynamic forcing function—which, incidentally, is made larger by blade twist—increases, thus exciting the second blade flapping mode to a high amplitude with correspondingly high vertical shear at the flapping hinges. This particular vibration can be especially severe for a three-bladed rotor since all three blades will respond in unison to the 3/rev aerodynamics. On the two-, four-, or five-bladed rotor, the blades will be out of phase with each other and thus will produce less vibration in the helicopter as a whole.

Other factors that affect the apparent ability of the rotor to develop high load factors as limited by vibration are:

- The vibration isolation of the rotor and transmission from the rest of the helicopter;
- The natural frequency of the fuselage;
- The location of the cockpit with respect to fuselage nodal points (points that appear to stand still)

Thus it may be seen that there can be no single simple method for predicting the maximum load factor that will subject the pilot to unacceptable levels of vibration.

An exploration of the aerodynamic limits with a wind tunnel model is reported in reference 5.9. Figure 5.3 shows the limit of test points achieved at a tip speed ratio of 0.6. These limits are well beyond those normally accepted and were not limited by aerodynamic or structural phenomena but by straightforward mechanical interferences on flapping and cyclic pitch control on that particular model. Reference 5.9 uses these test results to predict high rotor performance for future rotors. A continuation of this wind tunnel study is reported in reference 5.10, which concludes that a 225-knot helicopter is feasible. The bottom portion of Figure 5.3 summarizes the maximum rotor thrust capability as measured on the wind tunnel model during that project.

AUTOROTATION

The questions most asked of the helicopter aerodynamicist concerning auto-rotation are:

- How much does the rotor speed decay following a power failure before the pilot can react?
- What is the minimum steady rate of descent?
- How far can the helicopter fly following the power failure while the pilot looks for a suitable landing spot?

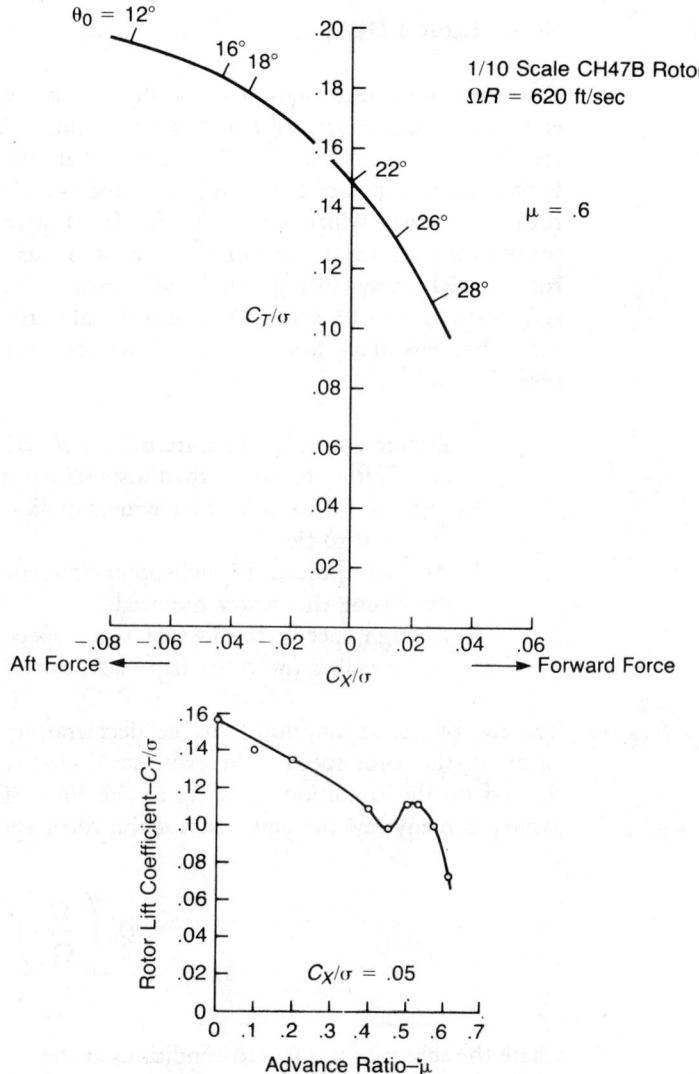

FIGURE 5.3 Envelope of Test Points for Model Rotor in Wind Tunnel

Source: McHugh & Harris, "Have We Overlooked the Full Potential for the Conventional Rotor?" *JAHS* 21-3, 1976; McHugh, "What are the Lift and Propulsive Force Limits at High Speed for the Conventional Rotor?" AHS 34th Forum, 1978

- What are the boundaries of the "Deadman's Curve"?
- What is the minimum touchdown speed?
- How does this helicopter compare with others?

These questions can be answered using the following analytical methods.

Rotor Speed Decay

The entry into autorotation is usually an emergency—or at least a simulated emergency—maneuver, so there is a finite time delay between the power loss and the pilot's corrective action. The decay of rotor speed during this period is a function of the power required and of the rotor's level of kinetic energy. If the rotor were on a whirl tower, the decelerating torque could be assumed to be proportional to the square of the instantaneous rotor speed. Measurements of rotor speed decay during simulated power failures in flight such as those in references 5.11, 5.12, and 5.13, however, indicate that the actual rate of decay is somewhat less than would be predicted by this assumption. The reasons are several:

1. During a simulated failure using a *throttle chop*, the engine torque decays at a finite rate rather than instantaneously.
2. In some helicopters, the engine supplies a small amount of torque even after a throttle chop.
3. At low speeds, the helicopter immediately begins to descend, thus decreasing the power required.
4. At high speeds, the loss of rotor speed increases the tip speed ratio, which makes the rotor flap back, thus decreasing the power required.

The use of the assumption that the decelerating torque is proportional to the square of the rotor speed is therefore somewhat conservative, but the degree will depend on the conditions existing at the time of the engine failure. Using the assumption anyway, the equation for the rotor speed decay is:

$$\dot{\Omega} = -\frac{Q_0 \left(\dfrac{\Omega}{\Omega_0}\right)^2}{J}$$

where the subscript 0 refers to conditions at the time of the power failure, and J is the total effective polar moment of inertia of the drive system, including the main rotor, the tail rotor, and the transmission referred to the main rotor speed; that is,

$$J = J_M + \left(\frac{\Omega_T}{\Omega_M}\right) J_T + J_{\text{trans, effective}}$$

For the example helicopter, this gives:

$$J = 11,600 + \left(\frac{100}{21.67}\right) 25 + 20 = 11,735 \text{ slug ft}^2$$

An equation for the rotor speed time history can be derived by integrating the equation for the rate of decay:

$$\int_0^t \frac{1}{\Omega^2}\, \dot{\Omega}\, dt = -\int_0^t \frac{Q_0}{J\Omega_0^2}\, dt$$

or

$$\int_{\Omega_0}^{\Omega} \frac{1}{\Omega^2}\, d\Omega = -\int_0^t \frac{Q_0}{J\Omega_0^2}\, dt$$

Carrying out the integration gives:

$$\frac{\Omega}{\Omega_0} = \frac{1}{1 + \dfrac{Q_0 t}{J\Omega_0}}$$

It is convenient to rewrite this as a function of the time during which all the rotor's kinetic energy would be dissipated at the initial horsepower and rotor speed:

$$t_{\text{K.E.}} = \frac{\frac{1}{2}J\Omega_0^2}{550\,\text{h.p.}_0}$$

or

$$t_{\text{K.E.}} = \frac{\frac{1}{2}J\Omega_0}{Q_0}$$

Using this definition, the rotor decay equation becomes:

$$\frac{\Omega}{\Omega_0} = \frac{1}{1 + \dfrac{t}{2t_{\text{K.E.}}}}$$

The equation has been evaluated for several values of $t_{\text{K.E.}}$ and the results are plotted in Figure 5.4. The figure applies to any flight condition from hover to maximum speed. The example helicopter has a value of $t_{\text{K.E.}}$ of 1.2 seconds with full engine power. Thus for flight in this condition, the rotor speed would decay 30%

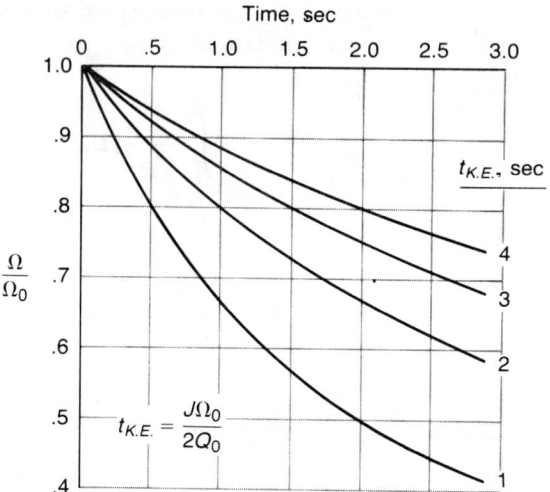

FIGURE 5.4   Rotor Speed Decay Following Power Failure

in the first second following a sudden failure of both engines. Failure of one engine from a full power condition would result in a decay of 17% in the first second if the other engine did not increase its power.

Steady Rate of Descent in Autorotation

The isolated rotor charts of Chapter 3 can be used with the method outlined there to calculate the steady rate of descent in autorotation. The results for the example helicopter are plotted in Figure 5.5 as a function of forward speed (the value at zero forward speed is from the example calculation in Chapter 2). The speed for minimum rate of descent is at the bottom of the curve, but the speed for minimum angle of descent—or maximum glide distance—is at a higher speed where a ray from the origin is tangent to the curve.

The figure shows that the example helicopter can make approaches down paths as steep as 9° at any speed; but if it is required to come down a steeper path, the allowable speed range is limited by the inability to dissipate fast enough the potential energy above that required just to maintain autorotation. The figure also shows that very steep approaches at intermediate rates of descent would be complicated by the possibility of power settling in the vortex/ring state, as discussed in Chapter 2.

As a practical matter, there are two requirements for safely doing steep approaches, especially with limited visibility: an airspeed (or possibly ground speed) indicator, which is accurate at low speeds, and an adequate field of view from the cockpit to the intended touchdown point.

In airplane studies, the ratio of forward speed to rate of descent in a glide is used to define the lift-to-drag ratio; since:

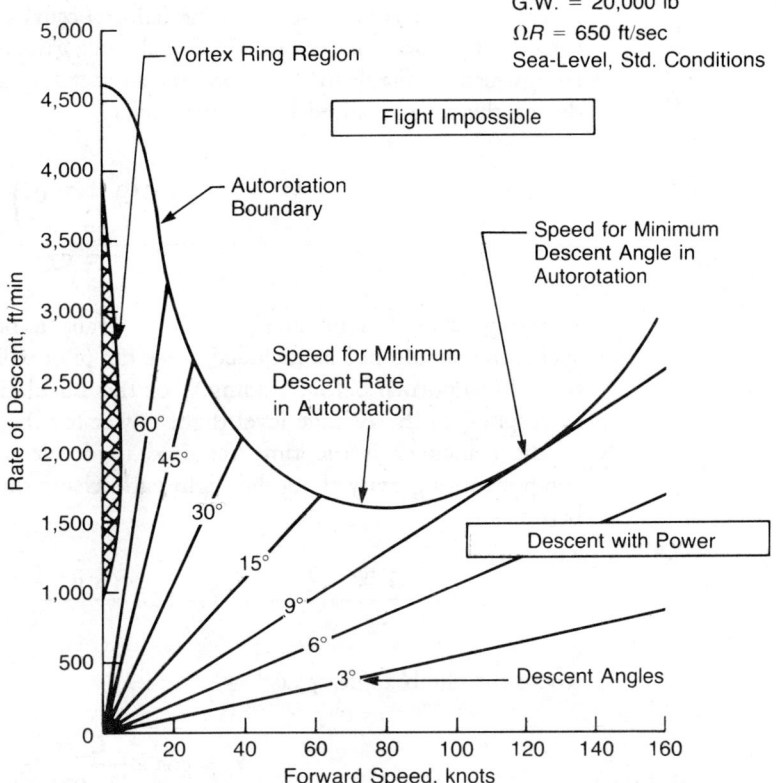

G.W. = 20,000 lb
ΩR = 650 ft/sec
Sea-Level, Std. Conditions

FIGURE 5.5   Rate of Descent in Autorotation, Example Helicopter

$$\frac{L}{D} = \frac{V_{\text{fwd.}}}{V_{\text{descent}}}$$

For the example helicopter the maximum lift-to-drag ratio based on this definition is 6.3. This is lower than would be expected for a comparable airplane. The discrepancy can be charged to the drag of the rotor hubs and to the fact that, over a large portion of the rotor disc, the blade elements are not operating at their angles of attack for maximum airfoil lift-to-drag ratio, whereas on an airplane each wing element can be operating near its optimum angle of attack.

Glide Distance

Following a power failure, the pilot has a limited zone in which to select a suitable landing spot. The radius of this zone is equal to the maximum glide distance from the altitude at which he enters autorotation. This altitude is either the altitude he had at the time of the power failure or the altitude to which he can zoom. The zoom maneuver is possible when the failure occurs at a forward speed, V_0, higher than the autorotational speed, V_1. The altitude gained can be related to the kinetic energy made available by decreasing the air speed modified by the power dissipated during the time required by the maneuver:

$$\Delta h = \frac{\dfrac{G.W.}{2g}(V_0^2 - V_1^2) - 550 \displaystyle\int_{t_0}^{t_1} \text{h.p.}\, dt, \text{ ft}}{G.W.}$$

(Although some kinetic energy might be available because of a decrease in rotor speed, this should not be included, since the pilot will want to get the rotor speed back up to normal before landing.) For this calculation, the power is assumed to correspond to the average level flight power for the two conditions and the time of the maneuver is the time required to decelerate from V_0 to V_1 using the component of gravity along the flight path. Using these assumptions, the equation becomes:

$$\Delta h = \frac{G.W.}{2g}(V_0^2 - V_1^2) - 550\,\frac{(\text{h.p.}_0 + \text{h.p.}_1)}{2}\,\frac{(V_0 - V_1)}{g \sin \gamma_c}$$

where the climb angle, γ_c, is:

$$\gamma_c = \cos^{-1}\frac{C_w/\sigma}{(C_T/\sigma)_{\max}}$$

A conservative value of $(C_T/\sigma)_{\max}$ for this calculation is 0.12.

The extra glide distance due to the zoom maneuver is:

$$\Delta d = 60\Delta h\,\frac{V_1}{R/D},\ \text{ft}$$

The analysis has been done for the example helicopter assuming a complete power failure at 160 knots, and the results are plotted in Figure 5.6. The power required was taken from Figure 4.24 and the rate of descent from Figure 5.5. It may be seen that the best autorotative speed is about 87 knots, which is closer to the speed for minimum rate of descent than to the speed for minimum angle of descent.

Generating the Deadman's Curve or Height-Velocity Diagram

At some combinations of altitude and forward speed, it is impossible to demonstrate safe autorotative landings at a vertical touchdown speed within the design limits of the landing gear. The boundaries of these combinations define the

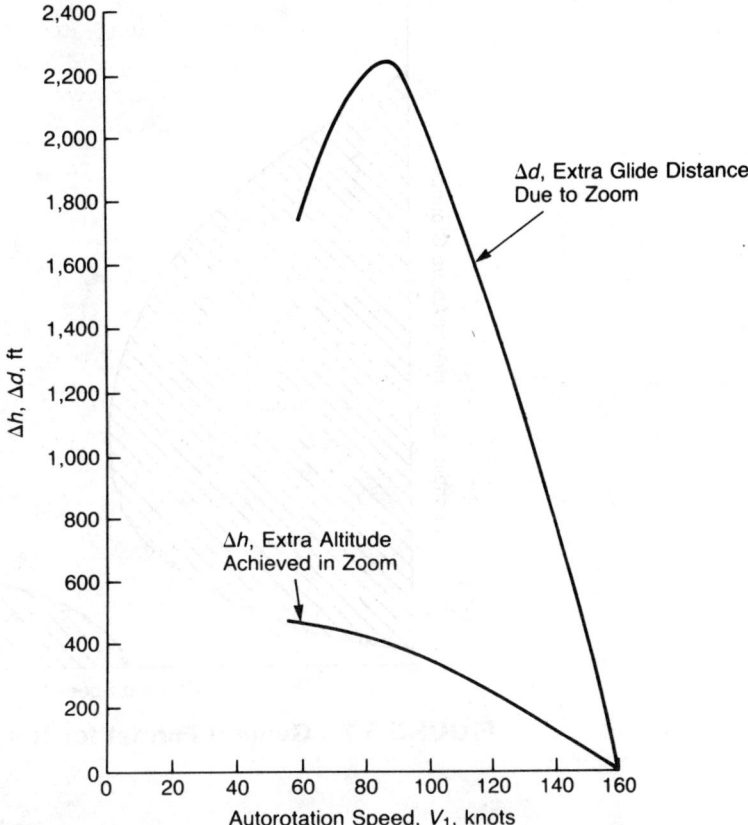

FIGURE 5.6   Benefits of Zoom Maneuver for Example Helicopter with Power Failure at 160 Knots

height-velocity diagram or Deadman's Curve. An unscaled height-velocity diagram is shown in Figure 5.7. Since the actual ability to make a safe landing depends on the interaction between the helicopter and the pilot, the height-velocity diagram can only be accurately determined in flight test. Prior to these tests, however, a first approximation can be obtained using a combination of empirical and analytical considerations.

The helicopter aerodynamicist should be aware that there are two different sets of ground rules used in determining the height-velocity diagram. When certificating a civil helicopter, the United States Federal Aviation Agency (FAA) specifies a pilot delay time following the power failure of 1 second along the upper boundary and no pilot time delay along the lower boundary. The United States military branches, on the other hand, specify a 2-second pilot delay at all points in order to define an *operational* height-velocity diagram. A method for calculating the diagram using FAA flight test data is given in reference 5.3. This method will be

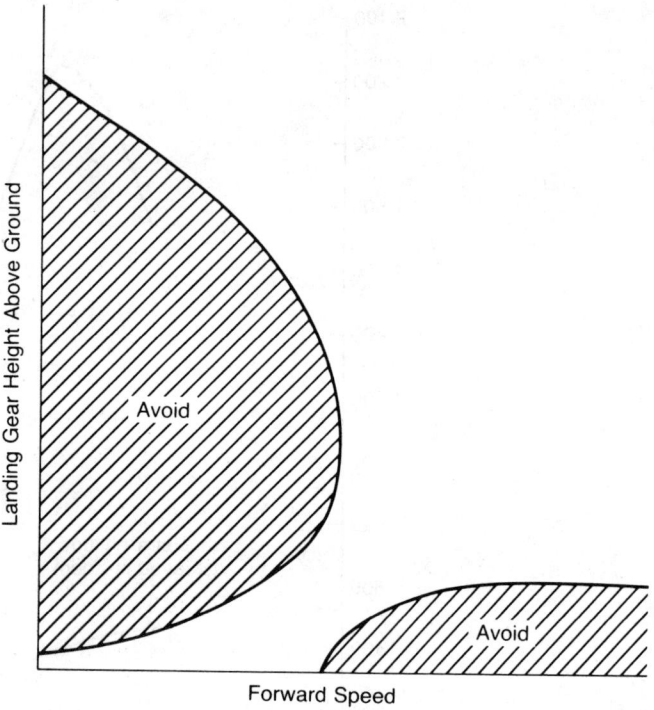

FIGURE 5.7   General Format for Height-Velocity Curve

outlined along with modifications to make it suitable for generating military operational height-velocity diagrams.

The method uses the generalized, nondimensional height-velocity diagram shown in Figure 5.8, which was generated from test data using three single-engine, single-rotor helicopters flown by skilled test pilots in a series of FAA flight test programs. To establish the diagram for a given helicopter, three unique heights and one velocity must be found: h_{lo}, h_{hi}, h_{cr}, and V_{cr}. The low hover height, h_{lo}, can be calculated by assuming that the entire maneuver is done at a vertical rate of descent equal to the landing gear sink speed, V_{LG}, and lasts as long as the kinetic energy associated with rotor speed can provide power equivalent to that required for hover in ground effect without exceeding a value of C_T/σ of 0.2. The resultant equation based on the analysis of reference 5.3 is:

$$h_{lo} = \frac{V_{L.G.} J \Omega_0^2 \left[1 - \dfrac{C_w/\sigma}{0.2} \right]}{1,100 \ \text{h.p.}_{IGE}}$$

where C_w/σ is the value of C_T/σ based on gross weight. Since the tail rotor is not required to balance the main rotor torque during this maneuver, the power required should not include tail-rotor-induced power. A study of height-velocity

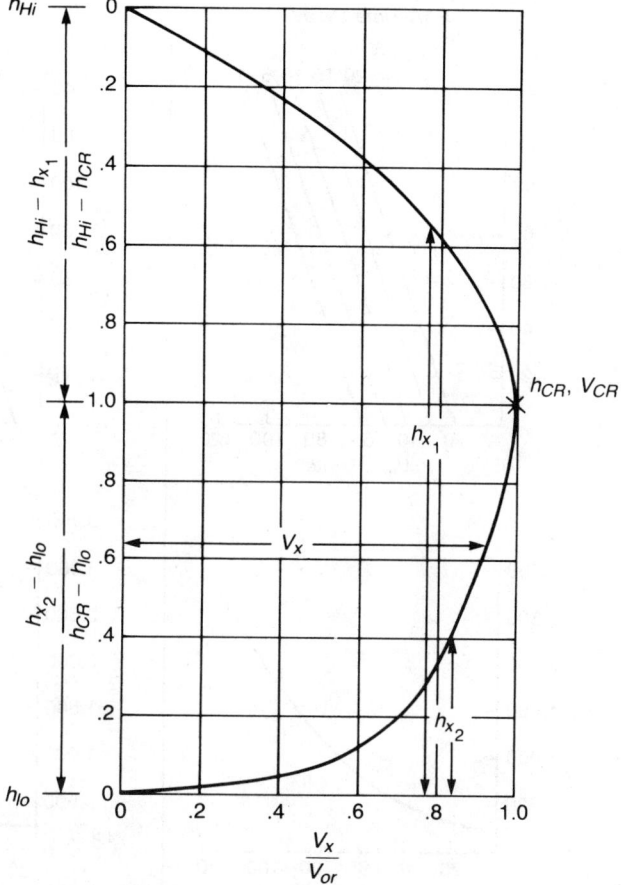

FIGURE 5.8   Generalized Nondimensional Height-Velocity Curve for Single-Engine Helicopters

Source: Pegg, "An Investigation of the Height-Velocity Diagram Showing Effects of Density, Altitude, and Gross Weight," NASA TND-4336, 1968.

curves generated by the military test agencies indicate that this equation can be used for military as well as FAA-type time delays.

The analysis of flight test data for the three helicopters at several gross weights each indicated that the critical velocity, V_{CR}, at the nose of the height-velocity curve is a function of the speed for minimum power, V_{min}, and the gross weight. The relationships deduced from this test data are plotted in the top half of Figure 5.9. The results, however, cannot be considered to be universally valid. The spread in gross weights for each of the test helicopters was only about 20%, and in this range the speed for minimum power did increase. At higher gross weights this trend would be expected to reverse as a result of blade stall effects, as indicated in Figure 4.24 of Chapter 4. A solution to this dilemma that preserves the method of

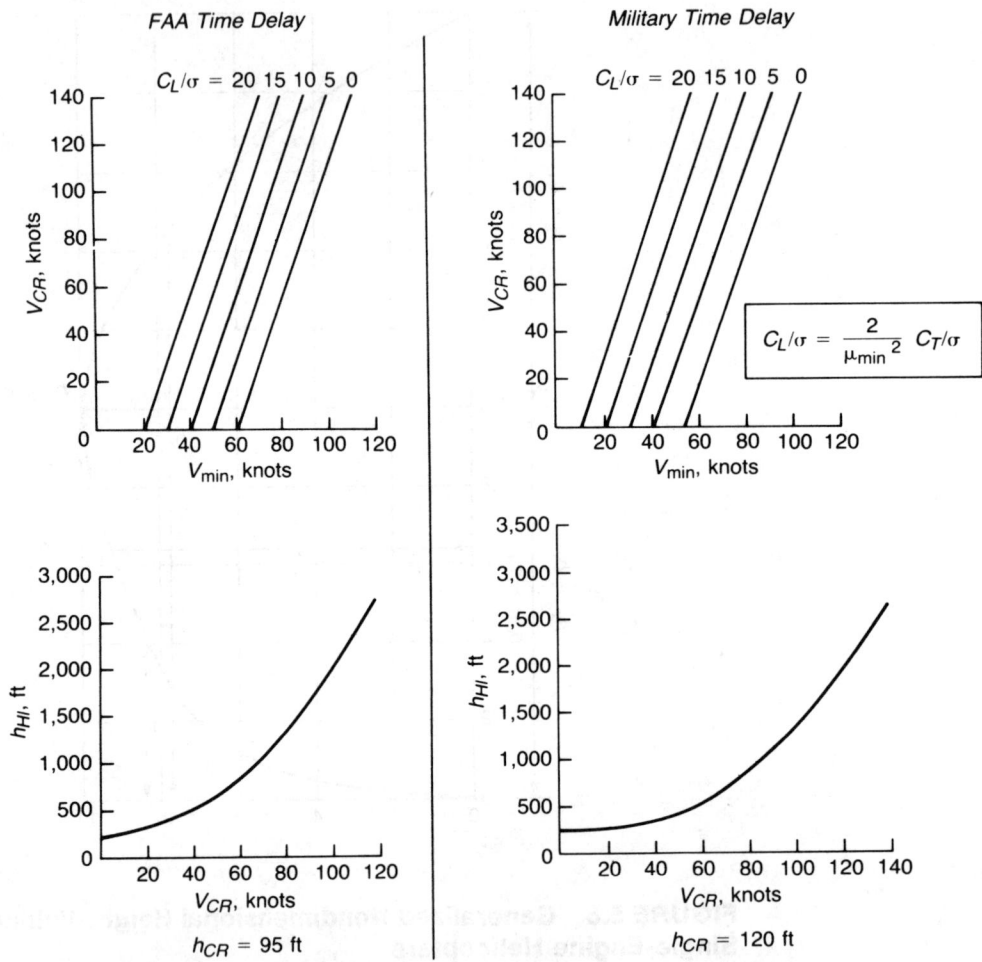

FIGURE 5.9   Parameters for Height-Velocity Diagram

reference 5.3 is to calculate V_{min} from energy considerations that do not include the effects of stall. In Chapter 3 an equation was derived for the main rotor power using energy methods:

$$\left(\frac{h.p.}{\rho/\rho_0}\right)_M = \frac{\left(\dfrac{G.W.}{\rho/\rho_0}\right)^2}{1{,}100\ e\ \rho_0 AV} + \frac{\rho_0 fV^3}{1{,}100} + \frac{\rho_0 A_b(\Omega R)^3 C_d/8}{550}\left[1 + 3\left(\frac{V}{\Omega R}\right)^2\right]$$

Differentiating with respect to V and equating to zero gives the conditions for minimum power:

$$\frac{d\left(\dfrac{\text{h.p.}}{\rho/\rho_0}\right)_{\text{M}}}{dV} = \frac{-\left(\dfrac{\text{G.W.}}{\rho/\rho_0}\right)^2}{1{,}100\ e\rho_0 A V_{\text{min}}^2} + \frac{3}{1{,}100}\ \rho_0 f V_{\text{min}}^2 + \frac{6}{550}\ \rho_0 A_b(\Omega R)C_d/8\ V_{\text{min}} = 0$$

or

$$V_{\text{min}}^4 + \frac{1}{2}\frac{A_b}{f}\ \Omega R\ C_d\ V_{\text{min}}^3 = \frac{\left(\dfrac{\text{G.W.}}{\rho_0}\right)^2}{3(\rho/\rho_0)^2 efA}\ ;\ (V_{\text{min}}\ \text{in ft/sec})$$

where the Induced Efficiency Factor, e, can be estimated from the low μ portion of Figure 3.7 of Chapter 3. The equation can be solved by trial and error procedures. Once V_{min} (in knots) has been determined, V_{CR} can be found from Figure 5.9, which is based on a similar figure in reference 5.3. The left-hand side of this figure is valid for the time delay used by the FAA—1 second along the top of the height-velocity curve and no delay along the bottom. The right-hand side applies to the military delay of 2 seconds at all points on the height-velocity curve and was guided by the published data on the Bell AH-1G as tested by the Army and reported in reference 5.11.

The high hover height, h_{hi}, as a function of V_{CR} is plotted at the bottom of Figure 5.9 again for both types of time delay based on the test data of references 5.3 and 5.11.

The critical height, h_{CR}, at the nose of the height-velocity diagram can be taken as 95 ft for all single-engine helicopters using the FAA time delay according to reference 5.3. This should be raised to 120 ft for the military time delay.

The failure of one engine on a multiengined helicopter is, of course, less of a problem than on a single-engined helicopter. For this case, the low hover height is:

$$h_{\text{lo}} = \frac{V_{LG} J\Omega_0^2\left[1 - \sqrt{\dfrac{C_w/\sigma}{.2}}\right]}{1{,}100(\text{h.p.}_{\text{IGE}} - \text{h.p.}_{\text{avail.}})}$$

The study of multiengined helicopters of reference 5.14 recommends that the critical speed, V_{CR}, be taken as half the speed at which the remaining power can maintain a rate of descent equal to the landing gear design sink speed—that is, half the speed at which:

$$\frac{550}{\text{G.W.}}\ (\text{h.p.}_{\text{req.}} - \text{h.p.}_{\text{avail.}}) = V_{\text{L.G.}}\ \text{ft/sec}$$

This method of determining V_{CR} should be used for the FAA time delay, but for the military 2-second time delay the critical speed is undoubtedly higher. For want of a more precise value, it is suggested that the critical speed for this case be taken as equal to the speed at which the remaining power can maintain a rate of descent equal to the landing gear design sink speed.

The critical height, h_{CR}, for multiengined helicopters can be assumed to be 50 ft or h_{lo}, whichever is higher. Until more data are available regarding single-engine height-velocity curves on multiengined helicopters, it may be assumed that there is no difference between the values of h_{CR} that would be obtained using the FAA and military-type time delays.

The methods described here have been used to construct height-velocity diagrams for the example helicopter at sea-level standard conditions and at 4,000 ft, 95° F, for both the FAA and the military time delays. The results are shown in Figure 5.10. At sea level the example helicopter can hover in ground effect with one engine out at its normal gross weight, so there is no single-engine failure envelope for this condition. At 4,000 ft, 95° F, however, the helicopter cannot hover on one engine; thus there is a single-engine envelope, as shown on the bottom set of curves of Figure 5.10.

The high-speed portion of the height-velocity diagram in Figure 5.7 is simply a warning that a power failure at high speed and close to the ground is a dangerous situation. No analytical method has been developed for predicting this portion of the diagram, and some presentations omit it entirely. Two considerations regarding the high-speed portion are worth noting. First, for this flight condition, the pilot can be assumed to be alert and able to react quickly to a power failure. Second, for most helicopters, when the rotor slows down at constant collective pitch, the increase in tip speed ratio causes the rotor to flap back so that a pitchup is started, which tends to keep the rotor speed from decaying further and, at high speeds, results in the automatic start of a climb. The height of this portion of the diagram depends only on what is considered prudent. The trend with time has been to decrease the height as shown from the values given in pilot handbooks for the following helicopters:

Helicopter	Height (ft)	Date Certificated
Hiller 12E	75	1959
Bell 47 J-2	50	1959
Bell OH-4A	15	1963
Hughes 500	5	1964

Minimum Touchdown Speed

The touchdown occurs at the end of the autorotative landing flare in which the helicopter has been brought from steady autorotation with moderate forward and vertical velocities to a condition with little or no velocity in either direction. An idealized flare maneuver is illustrated in Figure 5.11. It starts with a cyclic flare at

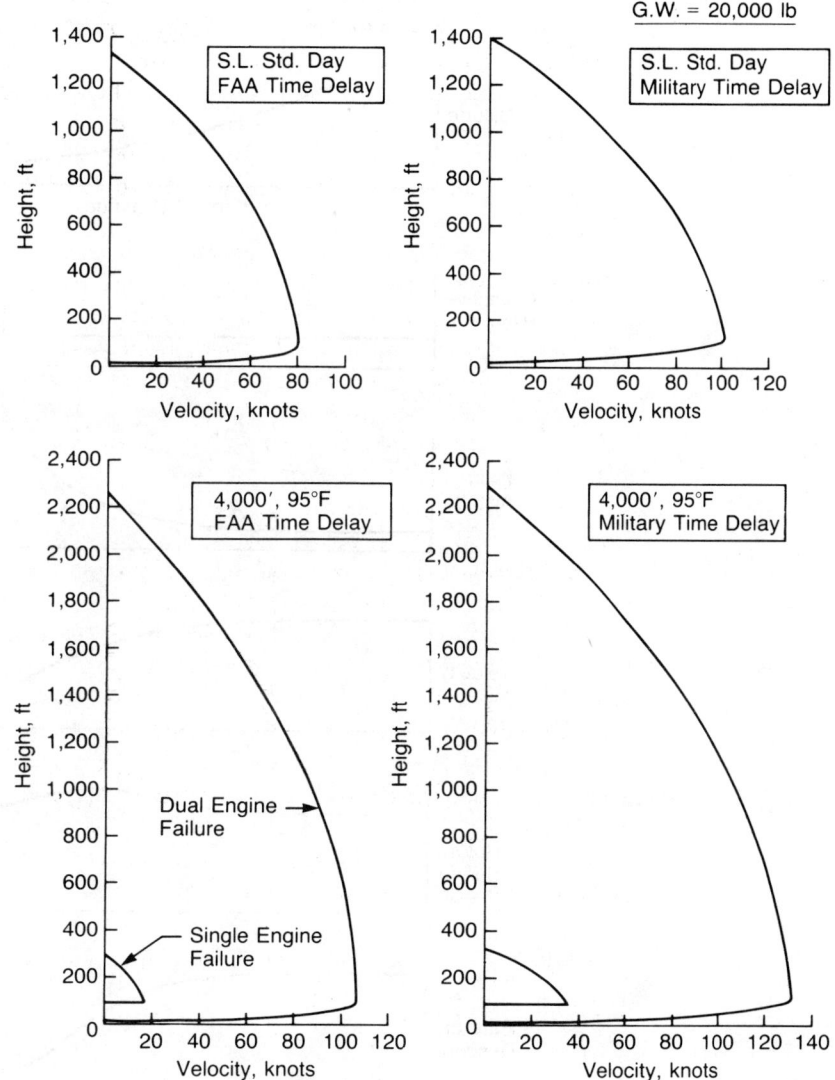

FIGURE 5.10   Height-Velocity Diagrams for Example Helicopter

constant collective pitch in which increased rotor thrust and its aft tilt are used to decrease both the vertical and the horizontal velocity components. At the end of this cyclic flare, the aircraft should be near the ground with its vertical component zero—or within the design sink speed of the landing gear—and with its horizontal velocity corresponding to autorotation at the angle of attack to which the rotor has been pitched. The maximum safe flare angle is the highest angle from which

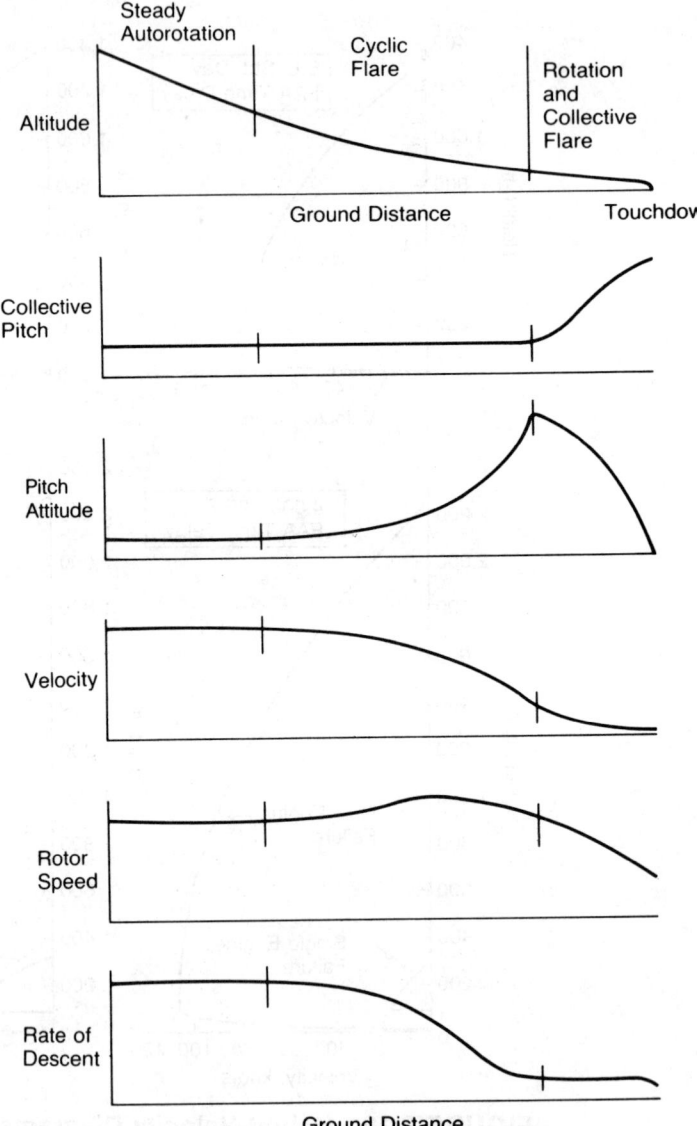

FIGURE 5.11   Idealized Flare Maneuver

the helicopter can be subsequently rotated nose down to a level attitude during the time that the rotor energy can be used to develop hovering thrust. (Two qualifications should be noted: The flare angle may be limited in an actual case if the pilot fears that he may lose sight of the ground in a machine without good downward visibility; and some helicopters do not have to flare to a level attitude if they have a tail wheel or skid structurally designed to take high loads.) The final nose-down rotation and collective flare are done during the time required to use up the rotor energy and should result in both the vertical and horizontal velocity components ending up as low as possible.

A method for estimating the final touchdown velocity is as follows:

• Calculate the maximum allowable tip path plane angle of attack at the end of the cyclic flare as a function of the maximum nose-down pitch rate, $\dot{\Theta}_{max}$, as limited by the longitudinal control power of the helicopter and the time available for the maneuver:

$$\alpha_{TPP_{max}} = \dot{\Theta}_{max}\, \Delta t, \text{ deg}$$

where the maximum pitch rate is given by the equation:

$$\dot{\Theta}_{max} = \frac{\gamma\Omega}{16}\, \Delta B_1, \text{ deg/sec}$$

where B_1 is the forward cyclic pitch available. See Chapter 7 for the derivation of this equation.

The time for the maneuver is:

$$\Delta t = \frac{J\Omega_0^2 \left[1 - \dfrac{C_W/\sigma}{C_T/\sigma_{max}} \right]}{1,100 \text{ h.p.}_{OGE}}, \text{ sec}$$

In this equation, the maximum value of C_T/σ at the end of the maneuver can be taken from hover charts such as those in Chapter 1. If the resultant flare angle is more than 45°, use 45° as the value.

• Use Figure 5.12 to find the tip speed ratio, μ_{auto}, at which autorotation can be sustained at the maximum flare angle while still developing a vertical component of rotor thrust equal to the gross weight. Figure 5.12 was produced from the isolated rotor charts of Chapter 3 by letting

$$C_T/\sigma_{auto} = \frac{C_W/\sigma}{\cos \alpha_{TPP}}$$

• Find the minimum touchdown velocity in knots as:

$$V_{min_{T.D.}} = \frac{1}{1.69} \left[\mu_{auto}(\Omega R)_{normal} - \frac{g}{2} \tan \alpha_{TPP}\, \Delta t \right], \text{ knots}$$

where the term $(g/2) \tan \alpha_{TPP}\Delta t$ is the decrease of forward speed during the nose-

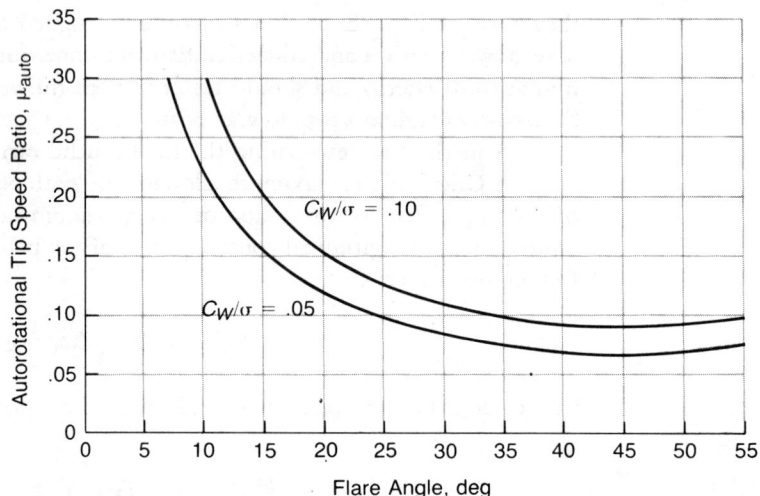

FIGURE 5.12   Conditions for Autorotation at End of Cyclic Flare

down rotation to the horizontal. Applying this process to the example helicopter gives:

$$\dot{\Theta}_{max} = 88 \text{ deg/sec (assuming } \Delta B_1 = 8 \text{ deg)}$$

$$\Delta t = 1.25 \text{ sec}$$

$$\alpha_{TPP_{max}} = 100 \text{ deg (use } 45° \text{ instead)}$$

$$\mu_{auto} = 0.085$$

$$V_{min_{T.P.}} = 21 \text{ knots}$$

This method assumes an ideal flare maneuver in which a skilled pilot does the right thing at the right time. It is actually a difficult maneuver in which the pilot must simultaneously satisfy the equations of motion for vertical forces, horizontal forces, and pitching moments with only his cyclic and collective controls in order to end the maneuver within narrow limits of height above the ground, rate of descent, and forward speed. Reference 5.13 states the problem thus: "Pilot apprehension is a factor because of ground proximity and rate of closure."

Although the initial rate of descent does not enter into the calculation directly, it does affect the pilot's chances of achieving the ideal flare. The higher the initial rate of descent, the less time he has to correct mistakes in control inputs. A study made with a B-25 airplane showed that satisfactory deadstick landings could be made up to rates of descent of 2,500 ft/min. Above that, the quality of the landings decreased. Presumably, helicopters have a similar limit.

Even though the military have had a requirement for many years that the minimum touchdown speed be 15 knots or less, there is a lack of actual test data on the maneuver, so the method has not yet been checked against an actual flare.

Autorotative Indices

The ability of the pilot to make a safe entry into autorotation and a safe flare from autorotation depends on both his skill and the physical characteristics of the helicopter. Some helicopters have been found to be reasonably forgiving of sloppy piloting; others are dangerous even for skilled test pilots. This situation leads to the desire to quantify the autorotative characteristics with some sort of simple index number that will indicate how a given helicopter compares to other helicopters in this regard. Since both the entry into autorotation and the flare from autorotation have to do with the kinetic energy stored in the rotor and the rate it is dissipated, a logical index is the equivalent hover time, or the time that the stored kinetic energy could supply the power required to hover before stalling. This and several other indices are discussed in reference 5.15 but the equivalent hover time appears to give the best correlation with the qualitative opinions of test pilots. In Reference 5.15 the equivalent hover time is called t/k. In this book it will be called t_{equiv} and defined as:

$$t/k = t_{equiv} = \frac{J\Omega^2 \left(1 - \frac{C_W/\sigma}{.8C_T/\sigma_{max}} \right)}{1,100 \ \text{h.p.}_{OGE}} \ \text{seconds}$$

Although reference 5.15 does not specify what value to use for C_T/σ_{max}, a reasonable approach would be to use the maximum calculated value from isolated rotor hover charts such as those in Chapter 1. Figure 5.13 is taken from reference 5.15 and represents the correlation of pilot opinion and equivalent hover time for several Bell helicopters. The figure indicates that the design goal for single-engine helicopters should be at least 1.5 seconds in order to be considered satisfactory. It is not yet clear what the corresponding goal should be for multiengined helicopters. For what it is worth, the twin-engine example helicopter has an equivalent hover time of 0.8 seconds. A later study reported in reference 5.16 results in a simple autorotative index for the landing flare in the form:

$$AI = \left[\frac{J\Omega^2}{\text{G.W.}} \right] \left[\frac{\rho}{\rho_0} \frac{1}{\text{D.L.}} \right], \ \text{ft}^3/\text{lb}$$

where the first term corresponds to the altitude to which all the rotor kinetic energy could lift the aircraft, and the second term is the penalty associated with altitude and disc loading. The study as applied to Sikorsky helicopters indicates that an index as low as 60 is satisfactory for single-engine helicopters and 25 for twin-engined. The calculated autorotative index for the example helicopter is 39.

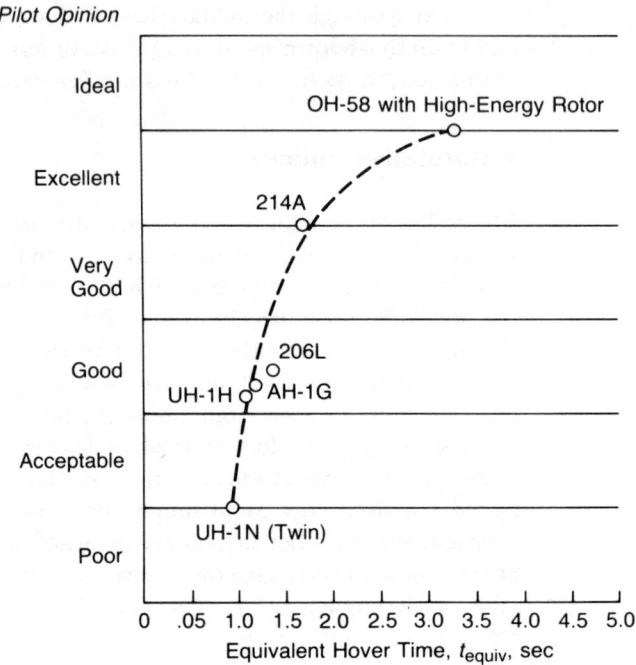

FIGURE 5.13   Autorotative Ratings of Several Bell Helicopters

Source: Wood, "High Energy Rotor System," 32nd AHS Forum, 1976.

MAXIMUM ACCELERATION

The ability to increase speed rapidly is important for many types of operations. The maximum level flight acceleration capability is primarily a function of the excess power available. At hover, the maximum acceleration is achieved when the maximum available rotor thrust is tilted until the vertical component is equal to the gross weight.

For this situation, the acceleration is:

$$\text{acc}_{\text{hover}} = 32.2 \sqrt{\left(\frac{T_{\text{max}}}{\text{G.W.}}\right)^2 - 1}, \text{ ft/sec}^2$$

where the maximum thrust is taken as equal to the maximum hover gross weight from a hover ceiling plot such as Figure 4.35. From hover, the equation can be used for accelerations rearward and sideward as well as forward. The acceleration capability in forward flight varies from the hover value to zero at V_{max}. For speeds between hover and V_{max}, the acceleration capability can be computed using the forward flight charts of Chapter 3 with the following procedure:

- Assume that $C_T/\sigma = C_W/\sigma$.
- For even values of tip speed ratio, find θ_0 at $C_Q/\sigma_{\text{max. avail. to main rotor}}$.
- Find α_{TPP} from charts.
- Calculate

$$C_T/\sigma = \frac{C_W/\sigma}{\cos \alpha_{\text{TPP}}} .$$

- Find new θ_0 at $C_Q/\sigma_{\text{max. avail.}}$.
- Find f/A_b corresponding to C_T/σ and θ_0.
- Convert tip speed ratio into forward speed and dynamic pressure, q.
- Calculate the acceleration capability from the equation:

$$\text{acc} = \frac{32.2\,A_b\,q}{\text{G.W.}} [f/A_{b_{\text{calc}}} - f/A_{b_{\text{actual}}}],\ \text{ft/sec}^2$$

Figure 5.14 shows the acceleration capability of the example helicopter at its design gross weight and at sea level standard conditions.

MAXIMUM DECELERATION

At speeds near hover, the deceleration capability is equal and opposite to the acceleration capability but in high speed flight, the capability may be limited by

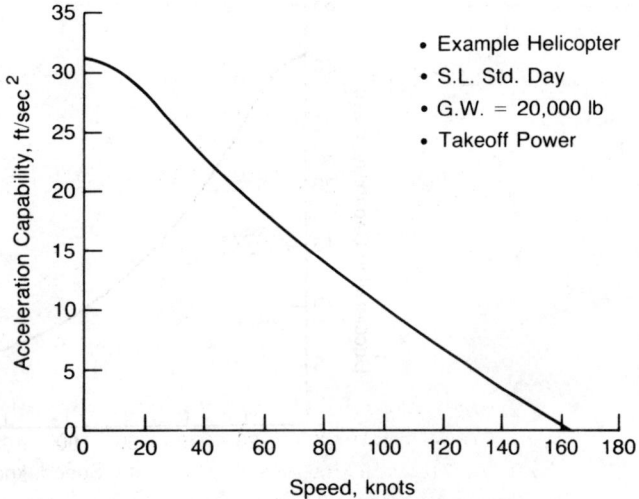

FIGURE 5.14   Maximum Longitudinal Acceleration Capability

rotor autorotation at some overspeed limit usually specified by structural design to be 10–20% over normal rotor speed. The procedure for calculating the deceleration capability under this limitation using the rotor charts of Chapter 3 are, as follow:

- Assume that $C_T/\sigma = C_W/\sigma$.
- For even values of tip speed ratio, find θ_0 at $C_Q/\sigma = 0$.
- Find α_{TPP} from charts.
  Calculate

$$C_T/\sigma = \frac{C_W/\sigma}{\cos \alpha_{TPP}}$$

- Find new θ_0 at $C_Q/\sigma = 0$.
- Convert C_T/σ to T at applicable tip speed.
- Convert C_H/σ to H at applicable tip speed.
- Convert tip speed ratio into speed and dynamic pressure, q.
- Calculate the deceleration capability from the equation:

$$\text{decel} = 32.2 \left(\frac{fq + H + T\alpha_{TPP}}{\text{G.W.}} \right) \text{ ft/sec}^2$$

Figure 5.15 shows the decelerative capability of the example helicopter at its design gross weight and at sea-level standard conditions.

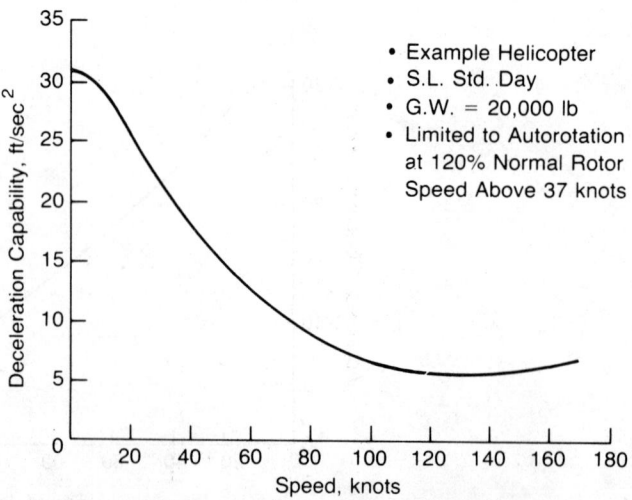

FIGURE 5.15 Maximum Longitudinal Deceleration Capability in Level Flight

OPTIMUM TAKEOFF PROCEDURE AT HIGH GROSS WEIGHTS

A study of the takeoff maneuver of helicopters too heavily loaded to hover out of ground effect is reported in reference 5.17. The conclusions of that study are that the shortest—and safest—takeoffs are achieved by accelerating into forward flight with as much ground effect as possible until reaching a *rotation speed* and then climbing out at that speed. An equation for the distance required to accelerate from hover to a given forward speed can be derived by considering that the acceleration capability is linear with speed. That this is a satisfactory assumption is shown by Figure 5.14. If x is the distance, then $\dot{x}$ is speed and $\ddot{x}$ is acceleration. Between hover and V_{max} (or $\dot{x}_{max}$), the equation for acceleration is:

$$\ddot{x} = \ddot{x}_{HIGE} + \frac{d\ddot{x}}{d\dot{x}} \dot{x}$$

where $d\ddot{x}/d\dot{x}$ is a negative number.

The solution to this differential equation is:

$$x_{acc} = \frac{\ddot{x}_{HIGE}}{-\dfrac{d\ddot{x}}{d\dot{x}}} \left[t - \frac{1}{\dfrac{d\ddot{x}}{d\dot{x}}} \left(e^{\frac{d\ddot{x}}{d\dot{x}} t} - 1 \right) \right] \quad \text{ft}$$

or

$$x_{acc} = \dot{x}_{max} \left[t - \frac{1}{\dfrac{d\ddot{x}}{d\dot{x}}} \left(e^{\frac{d\ddot{x}}{d\dot{x}} t} - 1 \right) \right] \quad \text{ft}$$

and

$$\dot{x} = \dot{x}_{max} \left[1 - e^{\frac{d\ddot{x}}{d\dot{x}} t} \right] \quad \text{ft/sec}$$

Combining these last two equations gives the equation for the distance, x, required to accelerate to the speed, $\dot{x}$:

$$x_{acc} = \dot{x}_{max} \left[\frac{\ln\left(1 - \dfrac{\dot{x}}{\dot{x}_{max}} \right) + \dfrac{\dot{x}}{\dot{x}_{max}}}{d\ddot{x}/d\dot{x}} \right] \quad \text{ft}$$

Once the helicopter is accelerated to the rotation speed, $\dot{x}_{rot}$, a climb is started. The additional distance to climb over an obstacle with height h is:

$$x_{CL} = \frac{60\, h\, \dot{x}_{rot}}{R/C}, \text{ ft}$$

where an approximation to the rate of climb from momentum considerations can be used:

$$R/C = \frac{33,000\, \Delta h.p.}{G.W.}, \text{ ft/min}$$

where

$$\Delta h.p. = h.p._{avail} - h.p._{level\, flt@\dot{x}_{rot}}$$

The total distance required for the maneuver is thus:

$$x_{tot} = x_{acc} + x_{CL} = \dot{x}_{max} \left[\frac{\ln\left(1 - \dfrac{\dot{x}_{rot}}{\dot{x}_{max}}\right) + \dfrac{\dot{x}_{rot}}{\dot{x}_{max}}}{d\ddot{x}/d\dot{x}} \right] + \frac{h\, \dot{x}_{rot}\, G.W.}{550\, \Delta h.p.}, \text{ ft}$$

The optimum rotation speed depends on the height of the obstacle to be cleared. For a low obstacle the rotation speed will be low, but for a high obstacle (such as a mountain) the speed will be that for maximum climb angle, which could be found from a plot of rate of climb versus forward speed as in Figure 4.48 as the speed at which a ray from the origin is tangent to the curve. Calculations have been done for the example helicopter at a gross weight of 28,000 lb—a weight just above that at which hover out of ground effect is possible on a sea-level standard day according to Figure 4.35. Figure 5.16 shows the results of the calculations: first the distance required to clear a 50-ft obstacle as a function of the rotation speed, and then the optimum rotation speed and minimum distance as a function of obstacle height.

RETURN-TO-TARGET MANEUVER

A rather complex indication of the maneuverability of a military helicopter is the return-to-target maneuver. In this maneuver, the helicopter passes over a spot (target) in forward flight and then tries to return to that same spot in the shortest possible time. The most critical ground rule is that the whole maneuver must be done at constant altitude. This simulates a nap-of-the-earth maneuver in which the "target" might be hostile and capable of destroying the helicopter if given the opportunity. Shorter times could be achieved if the maneuver were done with a zoom and a dive.

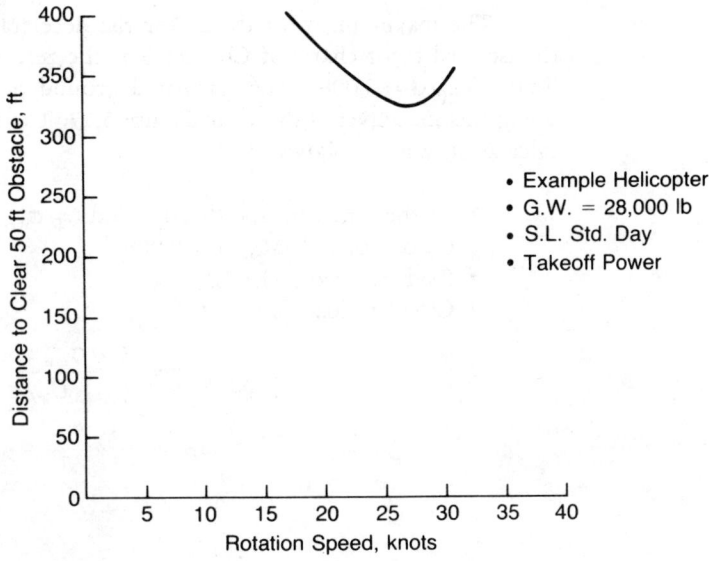

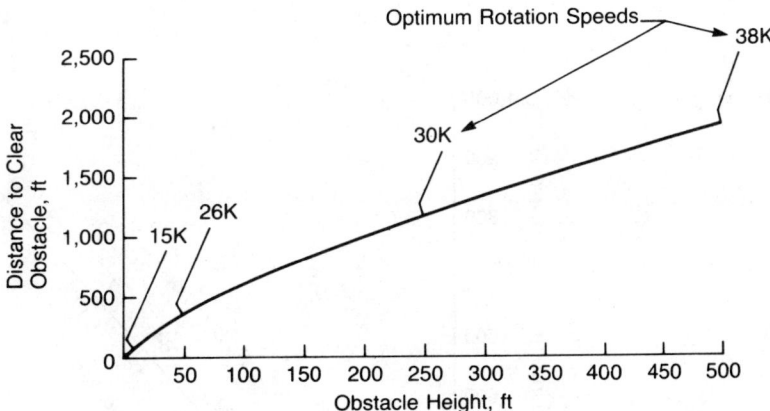

FIGURE 5.16   Conditions for Takeoff at High Gross Weight

The optimum return-to-target maneuver at constant altitude consists of a combined deceleration and banked turn until the flight path is pointed back at the target followed by a level flight acceleration. A discussion of the maneuver in reference 5.18 points out that the deceleration could be increased by sideslipping to increase the parasite drag, but that flight tests using this technique revealed the danger of the pilot becoming disoriented and losing the target. For this reason, calculations should be done assuming a zero sideslip turn during the banked deceleration phase.

The maximum rotor thrust for the deceleration phase may be taken from the isolated rotor charts of Chapter 3 as the zero torque value at the upper stall limit ($\Delta C_Q/\sigma = 0.008$). The calculated ground path for the example helicopter doing this maneuver is shown in Figure 5.17. The procedure for the step-by-step calculation was as follows:

- At the initial tip speed ratio, find C_T/σ_{max} from Chapter 3 rotor chart at $C_Q/\sigma = 0$, and $\Delta C_Q/\sigma = 0.008$.
- Find α_{TPP} from charts.
- Calculate load factor:

$$n = \frac{C_T/\sigma_{max} \cos \alpha_{TPP}}{C_W/\sigma}$$

- Example Helicopter
- Initial Speed = 115 k
- G.W. = 20,000 lb
- Sea Level, Std. Day
- Acceleration Done at Takeoff Power Rating
- Deceleration Done in Autorotation at 100% rpm

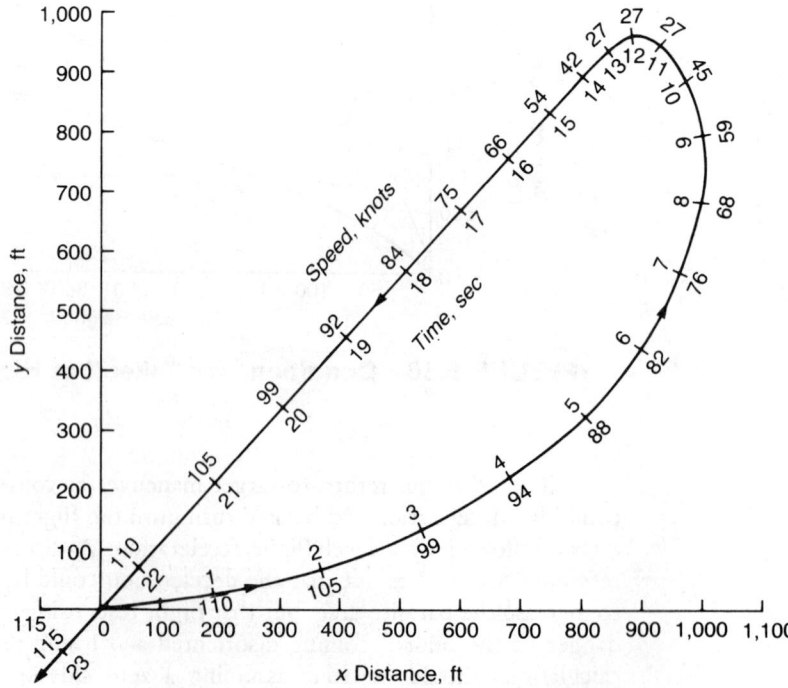

FIGURE 5.17   Return-to-Target Maneuver

- Calculate radius of turn:

$$R_T = \frac{V^2}{g \sqrt{n^2 - 1}} \text{, ft}$$

- Calculate deceleration:

$$\dot{V} = -32.2 \left(\frac{fq + H}{\text{G.W.}} + \frac{C_T/\sigma_{max} \sin \alpha_{TPP}}{C_W/\sigma} \right), \text{ ft/sec}^2$$

- At end of time increment, Δt, calculate:

$$\Delta \psi = \frac{57.3 \, V \Delta t}{R_T} \text{, deg}$$

$$\Delta x = V \Delta t \cos \psi \text{, ft}$$
$$\Delta y = V \Delta t \sin \psi \text{, ft}$$

- Repeat the procedure using the appropriate rotor chart as speed decreases until ψ is in the third quadrant and

$$\tan \psi = \frac{y}{x}$$

- If the speed drops below the autorotative limit as defined in Figure 5.12, a powered, steady turn at this speed should be used with the corresponding load factor taken as the ratio of maximum gross weight to actual gross weight from a power required curve such as Figure 4.24 at the appropriate engine power rating.
- Use an acceleration curve such as Figure 5.14 to continue the flight path until the origin is reached.

TOWING

Helicopters are occasionally used for towing in special situations such as minesweeping, rescue, or salvage operations. The maximum towline tension that can be maintained is a function of the maximum rotor thrust and the angle the towline makes with the horizon. For most towing operations, the speed will be slow enough that hover conditions can be assumed to apply. The equation for the towline tension can be derived from the balance of forces acting on the helicopter as shown in Figure 5.18. When the equations for the vertical and the horizontal components of the forces are solved simultaneously, the ratio of towline tension to gross weight is:

$$\frac{\text{Tension}}{\text{G.W.}} = -\sin \gamma + \sqrt{\sin^2 \gamma + \left[\left(\frac{T_{max}}{\text{G.W.}}\right)^2 - 1\right]}$$

Figure 5.18 shows this ratio as a function of γ for several ratios of maximum thrust to gross weight. The example helicopter at sea-level conditions can develop a maximum net rotor thrust of 27,800 lb out of ground effect, as shown in Figure 4.35. If it is flown at a gross weight of 17,000 lb and the towline is kept as flat as possible, it can maintain a towline tension of 22,000 lb.

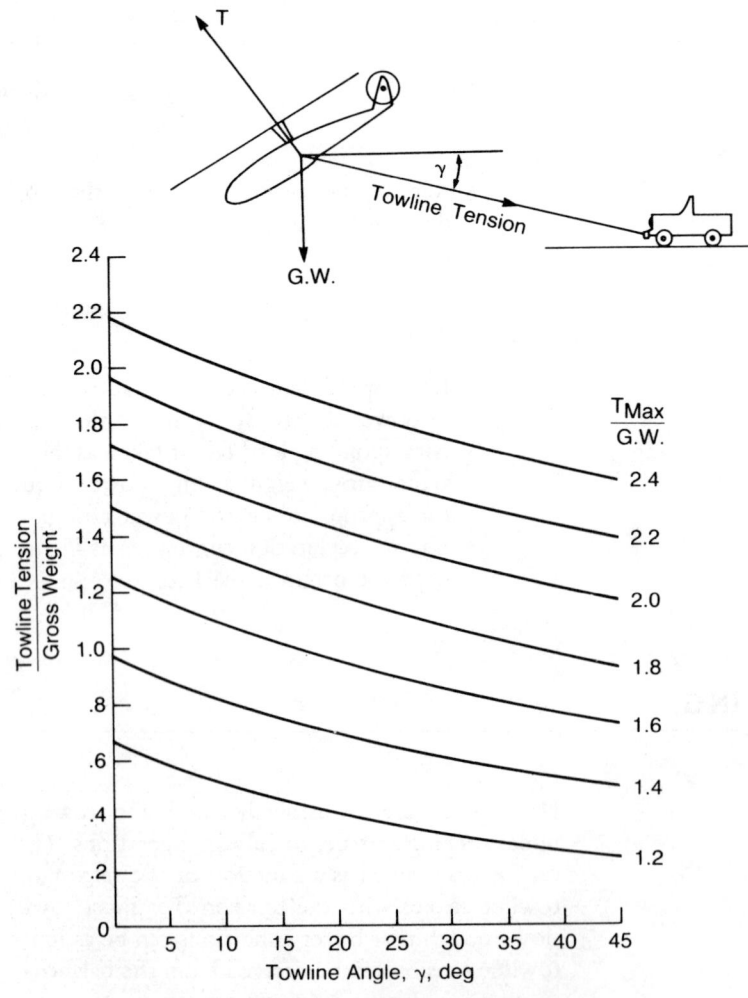

FIGURE 5.18   Towing Capability

AEROBATIC MANEUVERS

Although aerobatic maneuvers are not considered to be normal helicopter flight conditions, loops and rolls are quite possible and have been done by a variety of designs.

Loops

A well-done loop does not put excessive loads on any of the aircraft components. Figure 5.19 shows a loop being done by an airplane, which, having a separate

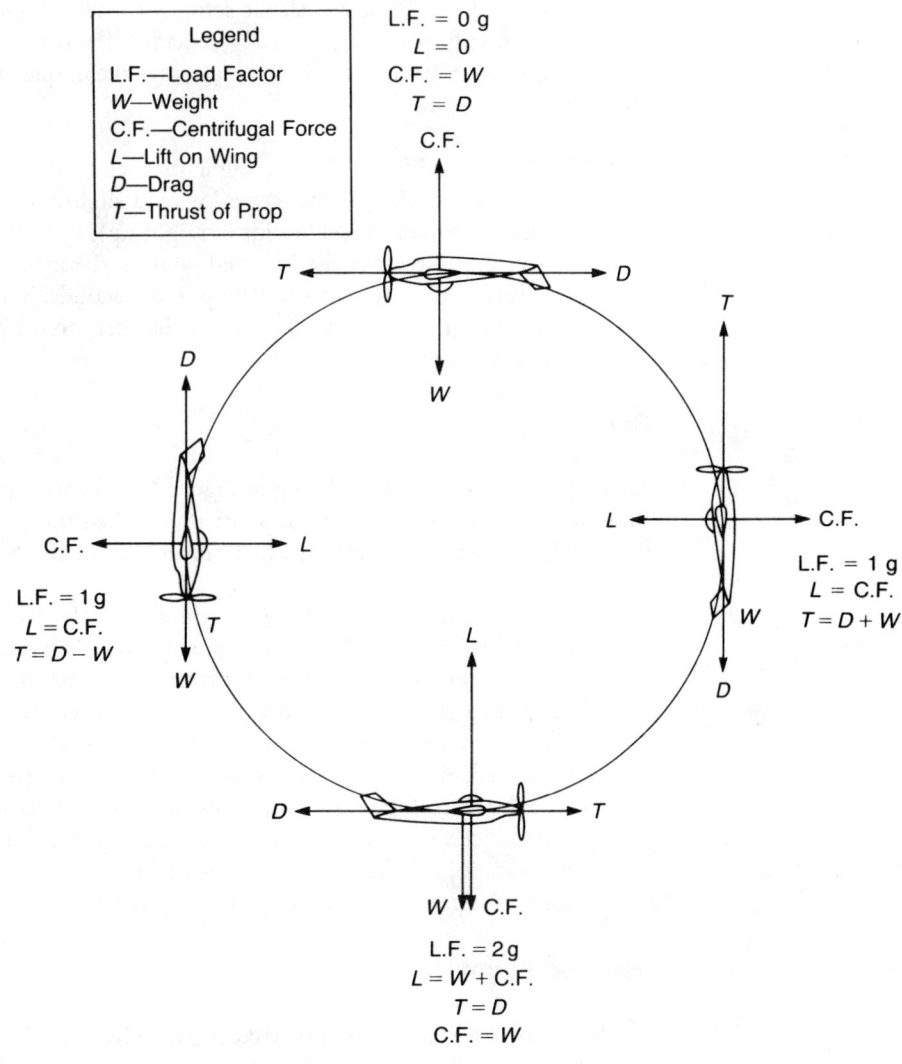

FIGURE 5.19   An Ideal Loop

propulsion device, is easier to visualize than a helicopter. The loop shown is idealized in that a constant speed is being maintained and the flight path is a perfect circle. The highest load factor is at the bottom of the loop, where it is only 2 g. Although no airplane or helicopter can do such a perfect loop, nevertheless any loop can be considered to be a relatively mild maneuver as far as loads go.

There is, however, a problem with respect to control. At the top of the loop, where the rotor thrust is zero or at least very low, all helicopters have reduced control power in pitch and roll; some—those with teetering rotors—may have none at all. If the pilot wants to make a cyclic correction in this situation, he might be surprised by how far he has to move the stick in order to get a response. The rotor will respond readily to the cyclic pitch and may tilt further than the designers made provision for. This is the classic setup for mast bumping on teetering rotors and for droop stop pounding on fully articulated rotors.

Reference 5.19 describes the piloting technique required to loop the Sikorsky S-67:

> The loop is initiated from a slight dive at approximately 175 knots. The cyclic is pulled aft and collective lowered slightly to limit control loads. As the aircraft passes the 90° point (straight up), collective is added to maintain positive g. Airspeed at the inverted point in the maneuver averages 50 knots. The average time to execute a loop is 21 seconds. The load factor range for the maneuver runs from 2.5 plus g at the entry to 0.7 g inverted to 2.5 plus g during the recovery.

Rolls

A roll is similar to a loop, as shown in Figure 5.20. Most aerobatic airplanes rely on substantial fuselage sideforce to support them when the wings are straight up and down. The Sikorsky description of the maneuver in the S-67 is:

> The roll maneuver is conducted to the right only to eliminate the problem of interference between the collective stick, pilot's left leg, and the cyclic stick. Generally the maneuver is started from 150 knots in level flight. The aircraft is pulled to 20° nose up and the pitch rate is reduced to a minimum. As the airspeed reaches 130 knots, full right and a slight amount of aft cyclic are introduced. As the aircraft reaches the 270° point (three-fourths of the way through the roll) lateral cyclic is returned to neutral and additional aft cyclic is introduced to counteract the nose tucking that initiates at approximately the 270° point. The roll takes an average of 6 seconds to complete and the load factor ranges from 0.8 g to 1.7 g for the maneuver.

Inverted Flight

Could a helicopter do steady inverted flight? Theoretically, yes; practically, no. A rotor could produce enough negative thrust to support the helicopter's weight if it

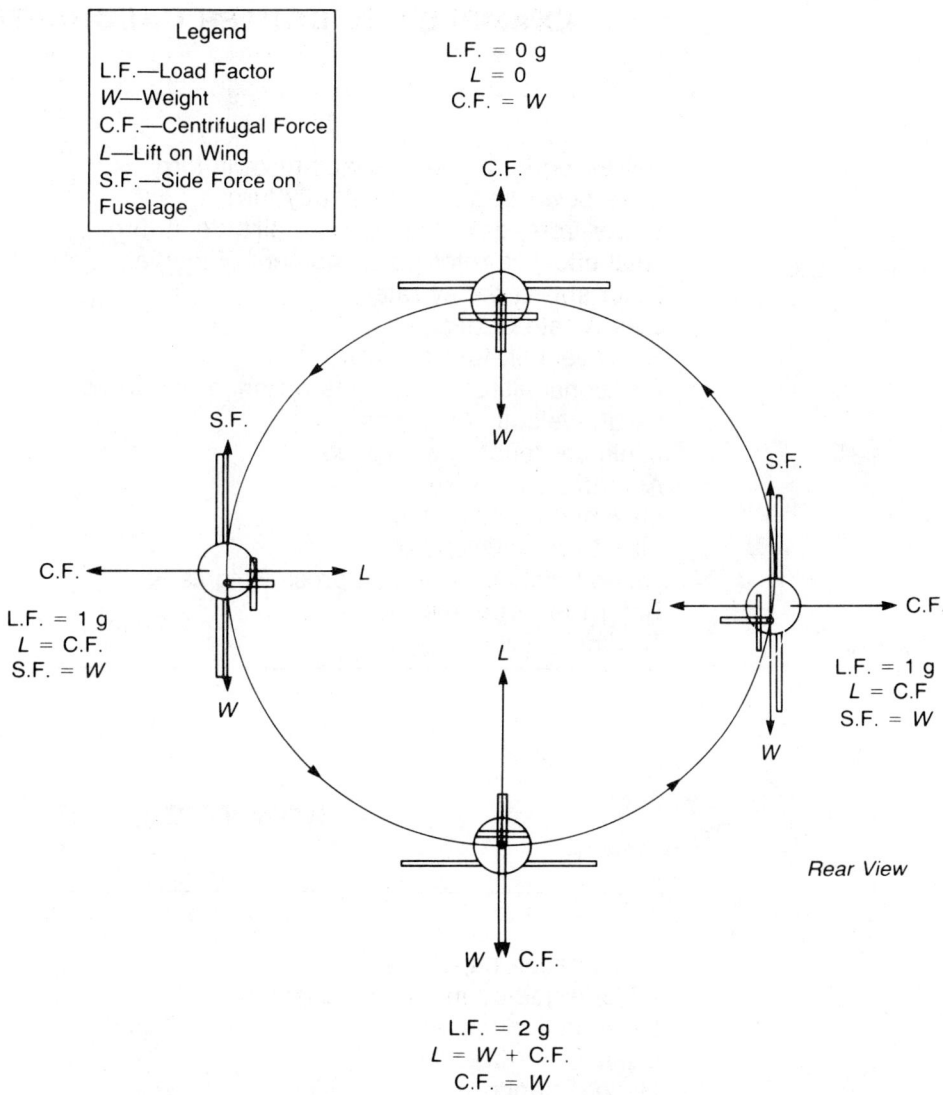

FIGURE 5.20   An Ideal Roll

were designed with enough negative collective pitch range. Some radio-controlled helicopter models have demonstrated this capability. No actual helicopters are rigged in this manner, for two reasons. First, it would require a collective control system with twice the normal travel and thus would require more space and weight than a normal system. Second, it would lose the important safety feature of having the down collective stop approximately corresponding to the right position for autorotation.

EXAMPLE HELICOPTER CALCULATIONS

HOW TO'S

CHAPTER 6

Airfoils for Rotor Blades

INTRODUCTION

The interest of the helicopter aerodynamicist in airfoils is either for the analysis of an existing rotor or for the design of a new one. In the first case, he may acquire the data he needs either directly, from two-dimensional wind tunnel tests or from whirl tower tests of a rotor with the specific airfoil, or indirectly, from test results of similar airfoils modified by empirical or theoretical means. For the design of a new helicopter, he may select one of the many airfoils already investigated or design an entirely new airfoil to incorporate characteristics he considers desirable. Since a blade with a good airfoil costs little or no more to build than one with a poor airfoil, there is strong motivation for improving airfoils even if the expected performance benefits are relatively small. A good airfoil for a rotor has:

- High maximum static and dynamic lift coefficients to allow flight at high tip speed ratios and/or at high maneuver load factors.
- A high drag divergence Mach number to allow flight at high advancing tip Mach numbers without prohibitive power losses or noise.
- Low drag at moderate lift coefficients and Mach numbers to minimize the power in normal flight conditions.

- A low pitching moment to minimize blade torsion moments and control loads.
- An aft aerodynamic center position to minimize the nose ballast required to balance the blade.
- Enough thickness for efficient structure.
- Easy-to-manufacture contours.

Unfortunately, because many of these requirements are conflicting, the choice of the best airfoil is not easy. (This situation is not unique to helicopters. For even such simple aircraft as sailplanes, there is no single airfoil that is considered optimum by a consensus of designers.) It is, however, possible to use wind tunnel results to show the effects of various physical parameters on the static aerodynamic characteristics, and to a lesser extent on the dynamic aerodynamic characteristics, for use either in the analysis or in the design process. For a discussion of airfoils in general, especially as viewed by the airplane aerodynamicist, the reader is referred to reference 6.1.

LIFT CHARACTERISTICS

Maximum Static Lift Coefficient at Low Mach Numbers

A large body of test data exists for airfoils at essentially zero Mach number. Although these data are thus not strictly applicable to rotor blades, the trends demonstrated can be shown to apply at the Mach numbers at which the retreating blade operates in forward flight and are, therefore, of interest. The maximum static lift that can be developed by an airfoil has been found to be related to the type of stall characteristic of the airfoil. Three types of low speed stall have been identified. They are discussed in references 6.2 and 6.3 and are shown pictorially in Figure 6.1. The three types of stall are:

- *Thin airfoil stall*: Caused by separation of the laminar boundary layer at the nose that produces a bubble whose outer surface is laminar. At moderate angles of attack, the flow reattaches and then may become turbulent before reaching the trailing edge. As the angle of attack is increased, the reattachment point moves aft producing the characteristic *long separation bubble* until at stall it covers the entire top surface of the airfoil. The characteristics of thin airfoil stall are that it occurs at low Reynolds numbers, has a gentle lift stall but a sharp moment stall, may show a slight jog in the lift coefficient at about 50% of maximum lift, has hysteresis in both lift and moment, and adding leading edge roughness produces either the same maximum lift or an increase.

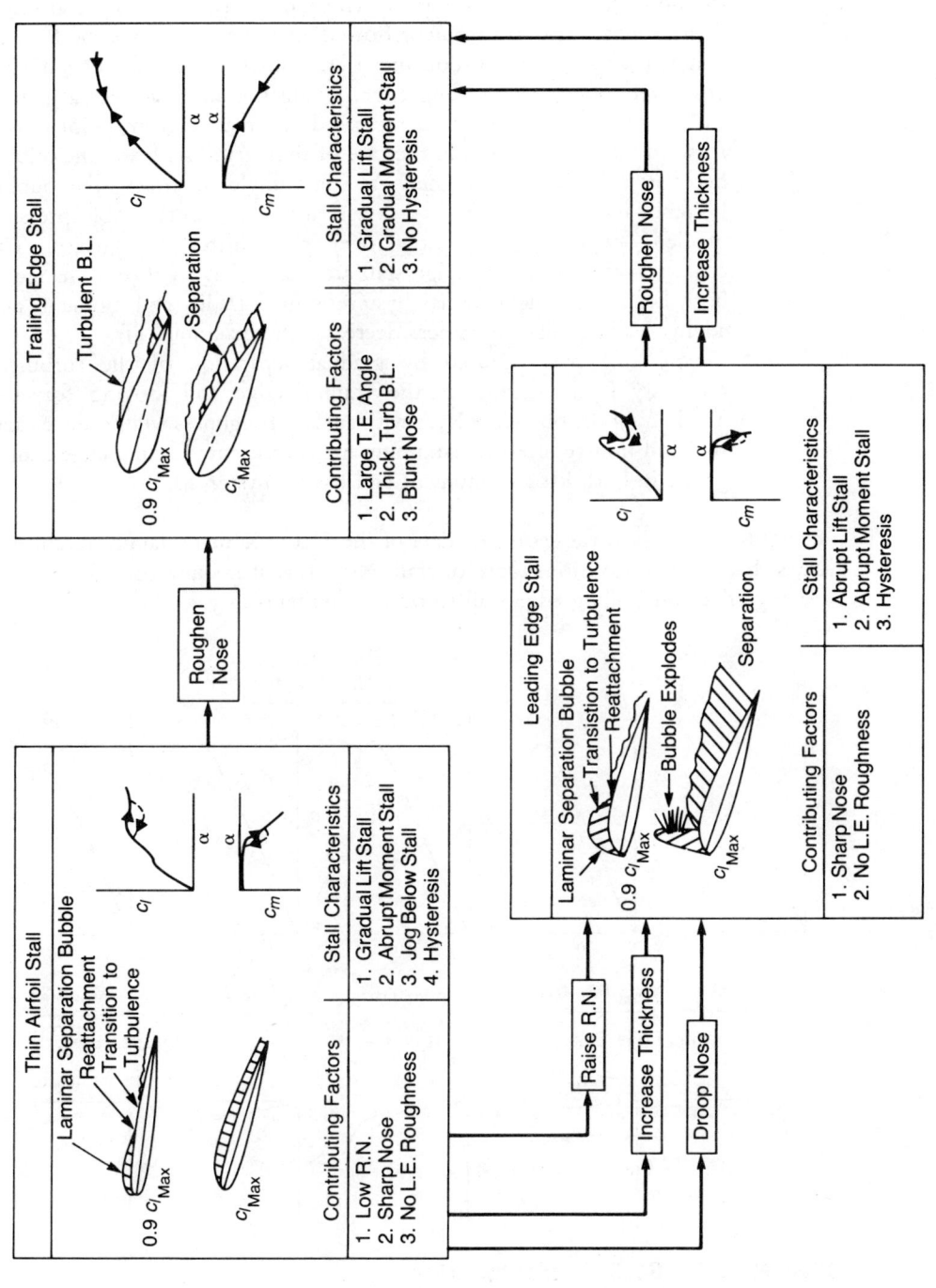

FIGURE 6.1 Types of Stall

- *Leading edge stall*: Caused by separation of the laminar boundary layer at the nose that produces a bubble whose outer surface has a transition from laminar flow to turbulent flow. This transition causes the flow to reattach quickly, thus producing a *short separation bubble*. The bubble effectively blunts the leading edge, giving the air molecules a gentler path. As the angle of attack is increased, the reattachment point moves forward rather than aft as in the case of thin airfoil stall, and the bubble becomes narrower but higher. At some angle of attack, the bubble becomes unstable and bursts—as a result of the unfavorable pressure gradient—separating the flow over the entire top surface. The characteristics of leading edge stall are that it has an abrupt change in both lift and moment, it has hysteresis in both lift and moment, and adding leading edge roughness decreases the maximum lift.
- *Trailing edge stall*: Caused by gradual separation of the turbulent boundary layer starting at the trailing edge and moving forward. Thickening the boundary layer with surface roughness will produce early stall and thus reduce the maximum lift coefficient. Trailing edge stall is gentle in both lift and moment and has no hysteresis.

Figure 6.2 shows how the measured data of the 63-OXX airfoil family exhibit the clues that typify the various types of stall. Note that it is quite common for the leading edge and trailing edge stall to occur simultaneously.

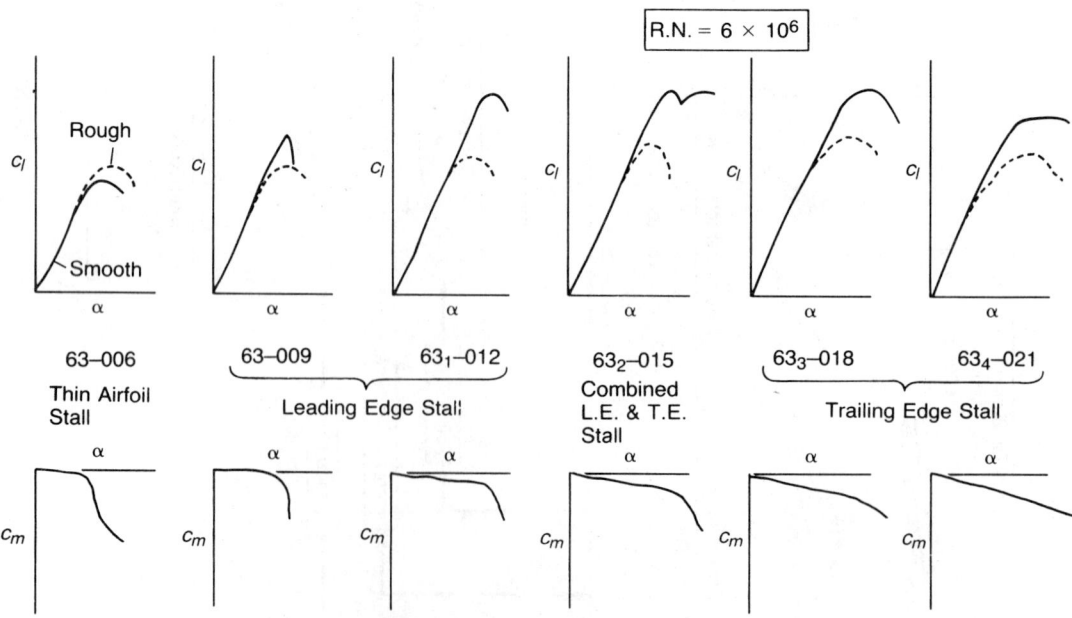

FIGURE 6.2 Stall Characteristics

Source: Abbott & Von Doenhoff, *Theory of Wing Sections* (New York: Dover, 1959).

The pressure distribution over an airfoil reflects the local velocity. As the air accelerates over the nose, the pressure decreases (almost always plotted upside down on a plot of pressure distribution). Behind the nose, the air must decelerate to reach the free-stream velocity at the trailing edge, thus causing the pressure to rise. This region is called the *pressure recovery region* and is characterized by an unfavorable pressure gradient. The air can accelerate at almost any rate, but it can decelerate at only a limited rate; that is, it can maintain only an unfavorable pressure gradient up to a certain value, which will depend on the type and thickness of the boundary layer. Attempts to make it decelerate too rapidly will lead to separation and to the establishment of a more comfortable path away from the airfoil surface. This is illustrated by the sequence of Figure 6.3, which shows an

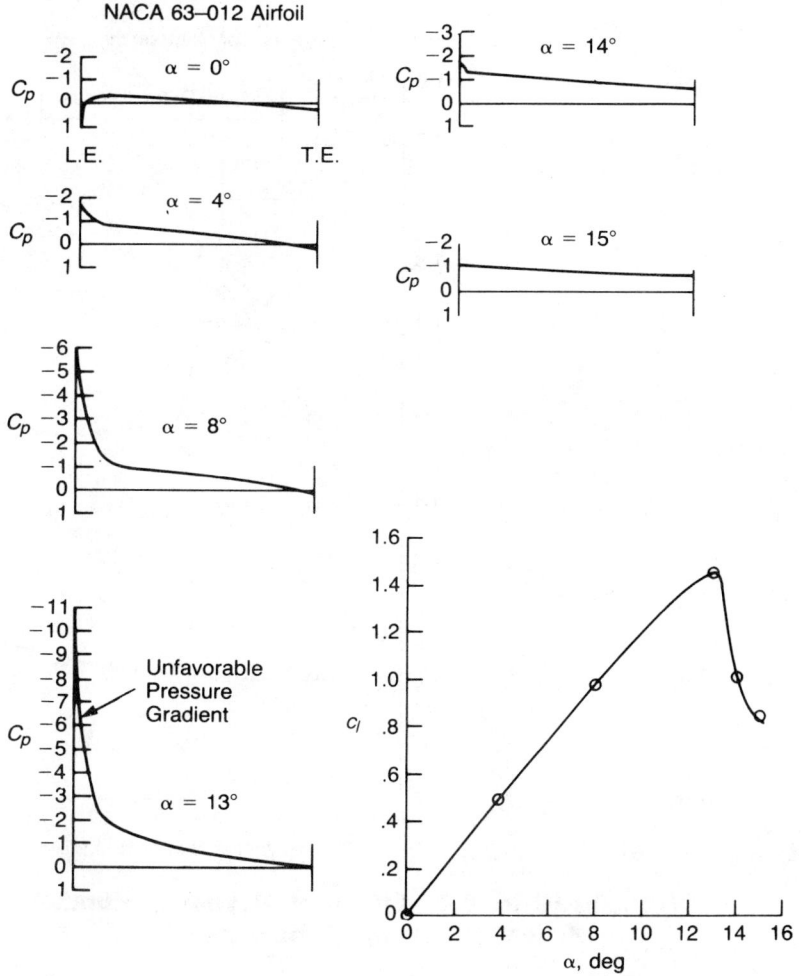

FIGURE 6.3   Upper Surface Pressure Distributions through Stall

Source: McCullough & Gault, "Examples of Three Types of Stall," NACA TN 2502, 1951.

airfoil undergoing leading edge stall during which the increasingly more unfavorable pressure gradient finally causes the leading edge bubble to burst, thus destroying the pressure peak and causing the pressure level over the aft portion of the airfoil to go to a nearly constant value.

Both the type of stall and the maximum lift coefficient are affected by the Reynolds number, as shown in Figure 6.4 from reference 6.3. The same airfoil may give different two-dimensional test results depending on the test Reynolds numbers. For this reason, the same airfoil may also give different results when installed on rotor blades with different chords and tip speeds. Figure 6.5 shows how Reynolds numbers vary. The characteristics of the three thinner airfoils of Figure 6.4 are the results of two trends: (1) at low Reynolds numbers a laminar

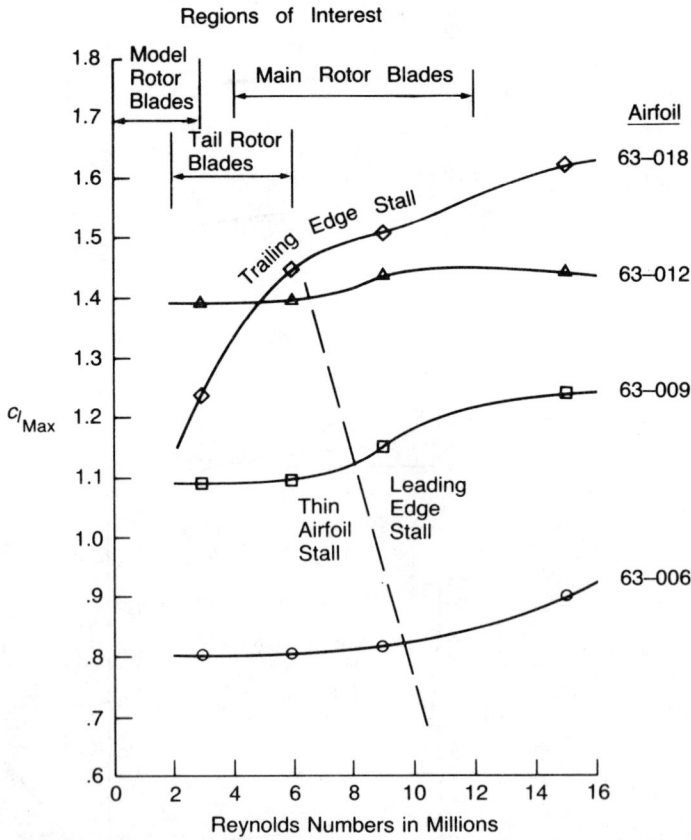

FIGURE 6.4 Effect of Reynolds Number on Maximum Lift Co-efficient at Low Mach Numbers

Source: Loftin & Bursnall, "The Effects of Variations in Reynolds Number between 3.0 × 10^6 and 25.0 × 10^6 upon the Aerodynamic Characteristics of a Number of NACA 6-Series Airfoil Sections," NACA TR 964, 1950.

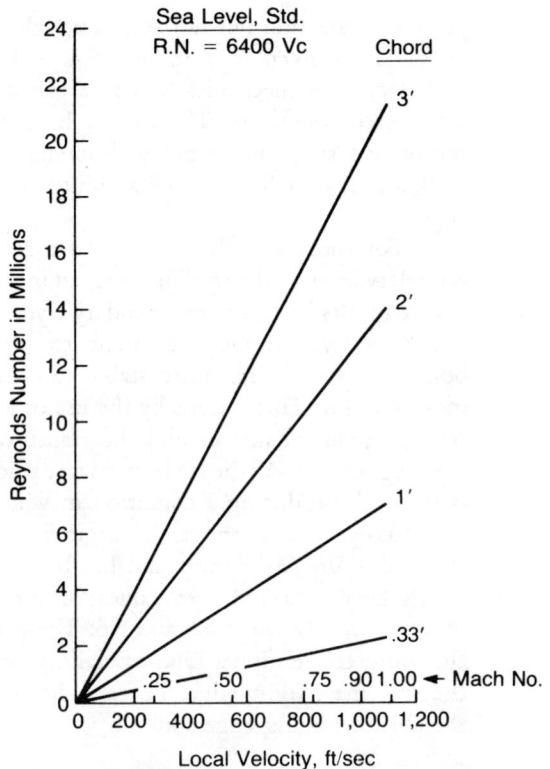

FIGURE 6.5   Reynolds Number Conditions at the Blade Element

boundary layer resists natural transition to a turbulent boundary layer; and (2) a laminar boundary layer will tend to separate more easily under the influence of an unfavorable pressure gradient than will a turbulent boundary layer. More specifically, the thinner airfoils exhibit thin airfoil stall at low Reynolds numbers. As long as the outer surface of the separation bubble remains laminar, increases in Reynolds numbers have little effect on the maximum lift. At some Reynolds numbers, however, the outer surface of the bubble undergoes transition from laminar to turbulent flow, which allows reattachment to occur closer to the nose and thus delays total separation. This is characteristic of leading edge stall. As the Reynolds number is increased further, the transition point and the reattachment point move further forward thus increasing the amount of chord which is influenced by the relatively stable turbulent boundary layer. When the transition point moves to the forward edge of the bubble, the beneficial effects have been exhausted and no further rise in the maximum lift takes place. Designers of low-speed airfoils have developed a technique for changing the type of stall from thin airfoil to leading edge or from leading edge to trailing edge when it is advantageous to do so. This technique involves careful use of a contour change that produces a slightly unfavorable pressure gradient that is just steep enough to promote transition but not steep enough to cause separation. A discussion of this technique is given by reference 6.4. A less sophisticated method is to trip the boundary layer mechanically with a transition strip consisting of a finite step or distributed roughness. The danger here is that if the tripping procedure is too severe, the resultant turbulent boundary layer will start out with a large initial thickness that will weaken its ability to withstand separation near the trailing edge.

For those airfoils that stall as a result of the separation of the turbulent boundary layer at the trailing edge, an increase in Reynolds number is beneficial in that it results in a thinner boundary layer with respect to the chord. This thinner boundary layer is more resistant to separation. The concept of the thinner boundary layer being more stable can also be used to design airfoils for high maximum lift. This is done by the use of a concave pressure distribution, with the steepest gradient just behind the transition point where the boundary layer can best negotiate it. As the boundary layer thickens, the unfavorable pressure gradient is reduced, producing a condition in which the boundary layer over the trailing edge has everywhere the same margin from separation. Such a pressure distribution is called a *Stratford recovery* and has been used for the two airfoils of Figure 6.6 which have achieved test values of maximum lift coefficient of over 2.2, as reported in references 6.5 and 6.6. These airfoils stall abruptly, unlike those with the more conventional type of trailing edge stall that progresses gradually from the extreme trailing edge. There is, however, little lift hysteresis as there is with the abrupt leading edge stall.

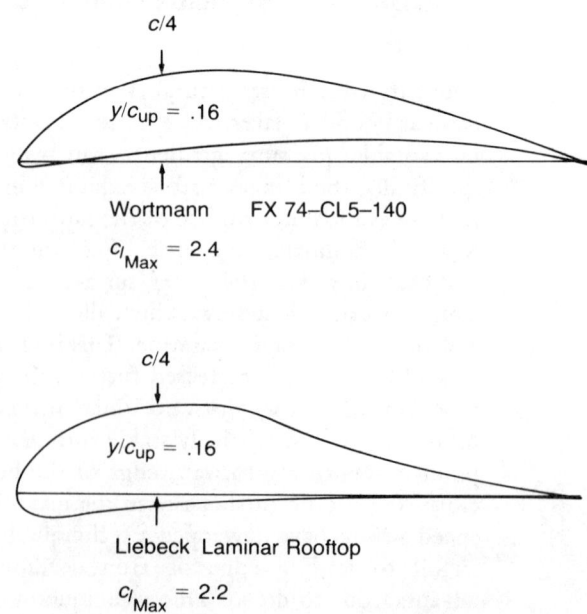

FIGURE 6.6   High Lift Airfoils

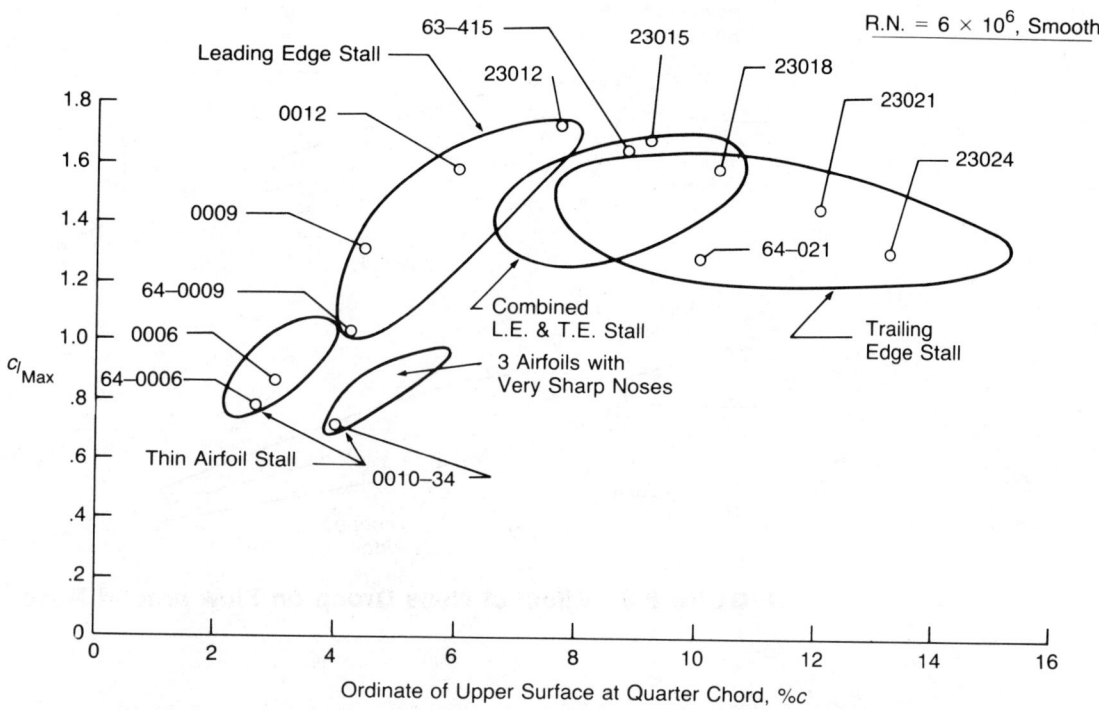

FIGURE 6.7   Maximum Lift Coefficient at Low Mach Numbers

Source: Abbott & von Doenhoff, *Theory of Wing Sections* (New York: Dover, 1959).

Reference 6.1 presents results of the testing of 118 airfoils in the NACA low-turbulence, two-dimensional wind tunnel at test Reynolds numbers of 3 to 9 million, corresponding to the Reynolds numbers existing at the retreating tips of rotor blades with about 1- to 3-foot chords. A convenient summary of the stall characteristics of these airfoils is obtained by plotting $c_{l_{max}}$ against the ordinate of the upper surface at the 25% chord station. The types of stall fall into separate envelopes, as shown in Figure 6.7. It may be seen that for those airfoils that have thin airfoil or leading edge stall, the maximum lift coefficient is almost directly proportional to the ordinate at the 25% chord station. Modifying one of these airfoils by extending the trailing edge without changing the leading edge will not significantly increase the maximum lift capability of the blade, since the maximum lift coefficient will decrease as the chord is increased. For those airfoils that stall at the trailing edge, the maximum lift coefficient is nearly constant and is relatively independent of the ordinate of the upper surface.

A study of Figure 6.7 indicates several potential methods for increasing the maximum lift coefficient. One obvious method is to increase the thickness ratio as in the 0006, 0009, 0012 series. Another method is to introduce forward camber, or "droop snoot," as in going from the 0012 to the 23012. This improvement comes from modifying the path from the stagnation point to the upper surface, as shown

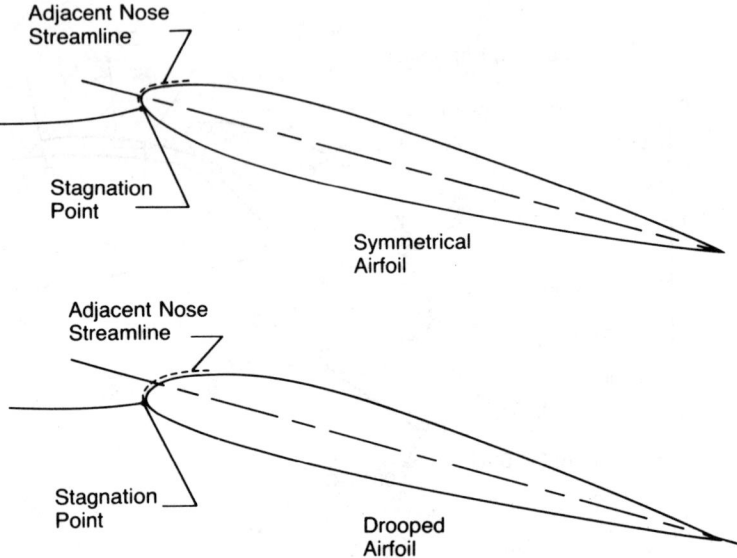

FIGURE 6.8   Effect of Nose Droop on Flow around Nose

in Figure 6.8. Because of the less violent changes in curvature and direction experienced as the molecules travel over the nose, the local velocities are reduced. This decreases the centrifugal force on the air, delaying the formation of the laminar separation bubble and also decreasing the magnitude of the deceleration required as the air goes toward the trailing edge (where it must slow to free stream velocity), thus decreasing the unfavorable pressure gradient. Figure 6.9 shows pressure distributions of a six-series airfoil and of its drooped nose modification from reference 6.4. The modification resulted in a 40% increase in maximum lift. Reference 6.7 reports on tests of families of airfoils produced by drooping the noses of NACA four-digit symmetrical airfoils. The results of these tests are shown in Figure 6.10, which indicates that drooping the nose is indeed an effective method of increasing the maximum lift coefficient. In the airplane industry, airfoils with drooped noses such as the NACA 23012 have bad reputations for abrupt stall. It appears that this is actually a characteristic of what would otherwise be considered a very good airfoil, which achieves its maximum lift coefficient by maintaining attached flow on both the leading and trailing edges longer than other airfoils do. When the flow does separate, the resultant stall is abrupt. This characteristic is not significant on rotors, however, because stall conditions are entered gradually starting with a small portion of the retreating side and also because the stall becomes less abrupt as a result of compressibility effects at Mach numbers about 0.3 or 0.4, as discussed in the next section.

A third potential for improvement is indicated in a negative way in Figure 6.7 by noting that three airfoils with very sharp noses lie below the envelope. This

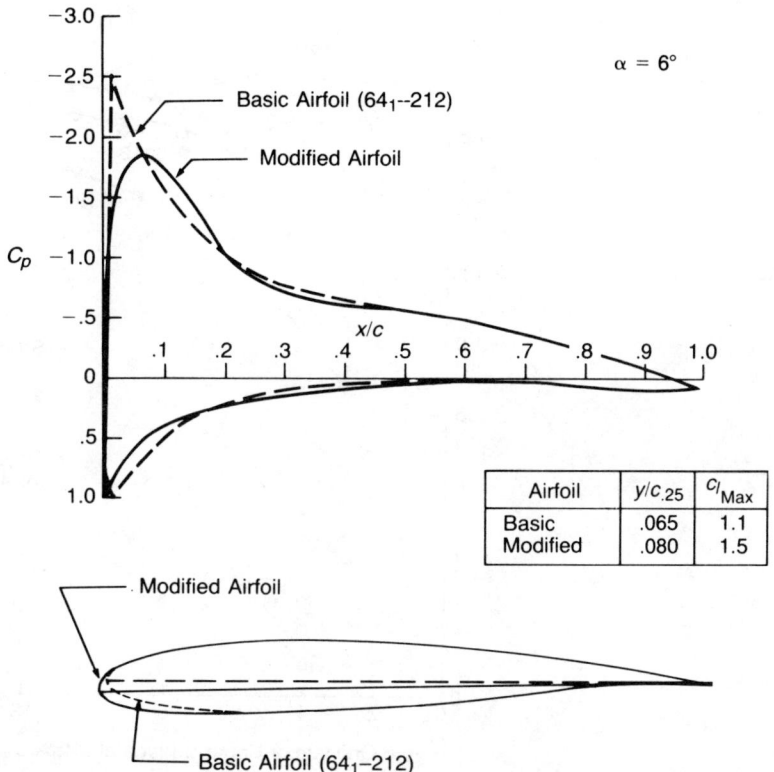

FIGURE 6.9   Effect of Droop Nose Modification on Pressure Distribution

Source: Hicks, Mendoza, & Bandettini, "Effects of Forward Contour Modification on the Aerodynamic Characteristics of the NACA 64,-212 Airfoil Section," NASA TM X-3293, 1975.

leads to the speculation that airfoils with very blunt noses might lie above. Tests reported in reference 6.8 have shown that blunting can produce a small but measurable improvement if done carefully. Figure 6.11 shows both good and disappointing results of increasing the nose radius. The disappointing result, from reference 6.9, is due to too sudden a change in curvature. In simple terms, the curvature of the surface governs the velocity of the air over the airfoil; thus the change in curvature governs the acceleration. If a sudden change in curvature from high to low demands a higher deceleration than the air can readily accomplish, it will separate. Most successful airfoils have gradual changes in curvature—at least in the first 10% of chord—and any modifications aimed at increasing the maximum lift coefficient should maintain this characteristic.

The final potential that can be inferred from the trends of Figure 6.7 is that the maximum lift coefficient could be increased if the trailing edge stall could be delayed. This path of development leads to the high lift airfoils of Figure 6.6.

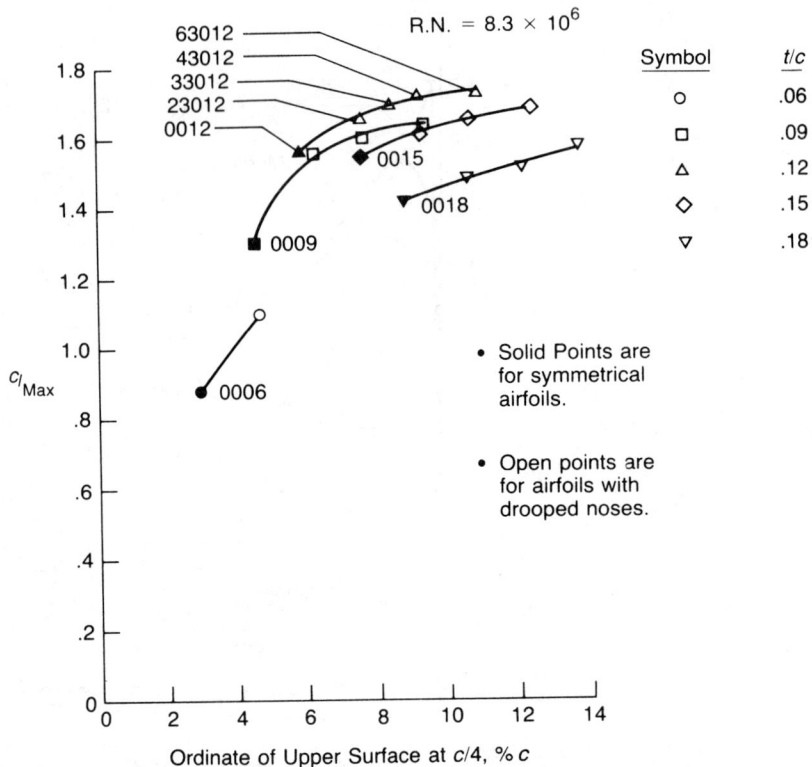

FIGURE 6.10   Effect of Nose Droop on Maximum Lift Coefficient

Source: Jacobs, Pinkerton, & Greenberg, "Tests of Related Forward-Camber Airfoils in the Variable-Density Wind Tunnel," NACA TR 610, 1937.

Maximum Static Lift Coefficient at High Mach Numbers

The test results discussed so far are in the Reynolds number range of interest to blade designers but were obtained at essentially zero Mach number, whereas the blade elements that are likely to stall on an actual rotor operate at Mach numbers from 0.25 to 0.5. The effects of compressibility on the lift characteristics of the NACA 0012 airfoil are shown in Figure 6.12 from reference 6.10. Even at relatively low Mach numbers, the local velocity over the surface can exceed the speed of sound, giving *supercritical* or *mixed* flow as indicated by the hatched lines. For this condition, the air returns to subsonic flow before reaching the trailing edge by passing through a shock wave. If the velocity is only slightly supercritical, the shock wave will be weak and its primary effect on maximum lift will be to thicken the boundary layer and thus increase the tendency toward early trailing

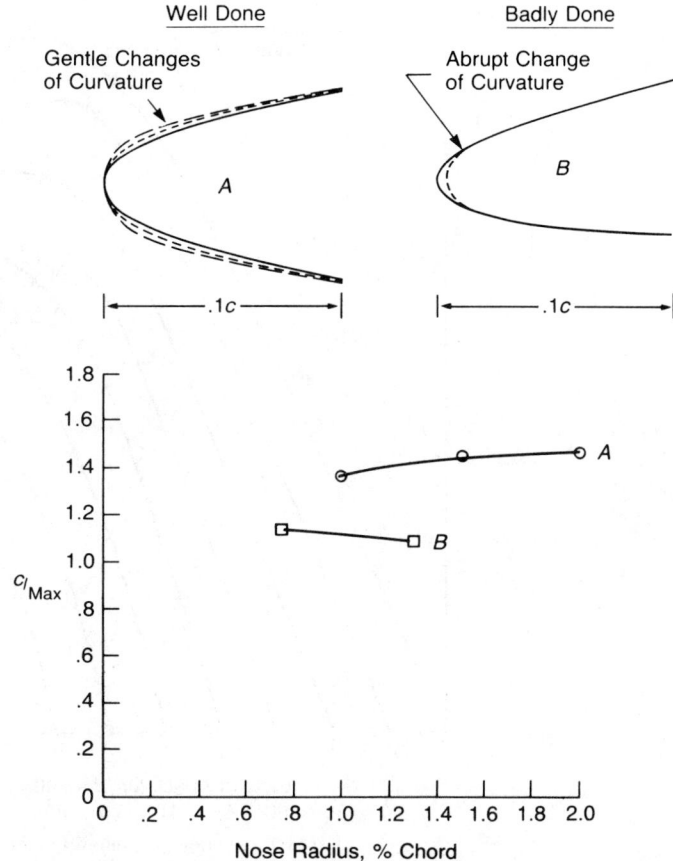

FIGURE 6.11   Effect of Nose Radius on Maximum Lift

edge stall. This is shown by the change in stall characteristic between 0.3 and 0.4 Mach number. For high supercritical velocities, the shock wave may be strong enough to cause the boundary layer to separate entirely, thus producing *shock stall*. Figure 6.13 shows measured pressure distributions for selected angles of attack at 0.3, 0.4, and 0.65 Mach numbers. It may be seen that at the lower Mach numbers the large pressure peak generated at maximum lift is rapidly destroyed as the airfoil stalls. At higher Mach numbers the height of the pressure peak is limited by the inability of the flow to sustain local Mach numbers above about 1.4. At an angle of attack of 18° and a Mach number of 0.65, the 0012 has extreme shock stall. Despite this, its lift coefficient is high and is still rising, leading to the observation that the term *maximum lift coefficient* loses its significance in these circumstances. A good survey of the influence of Mach number on maximum lift is presented in reference 6.11.

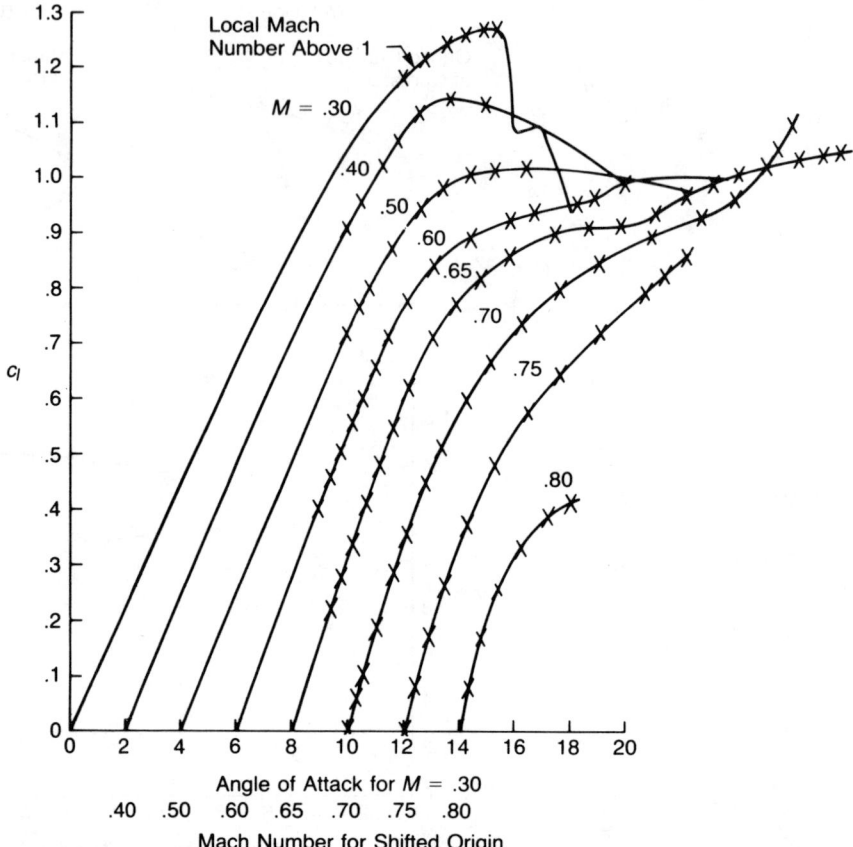

FIGURE 6.12   Effect of Mach Number on Lift Characteristics of 0012 Airfoil

Source: Lizak, "Two-Dimensional Wind Tunnel Tests of an H-34 Main Rotor Airfoil Section," USA TRECOM TR 60-53, 1960.

Most modern jet transports cruise at speeds such that a shock wave stands just ahead of the quarter-chord of the wing. The shadow of this shock wave can sometimes be seen when the wing is pointed toward or away from the sun.

Even at high Mach numbers, variations in Reynolds numbers are significant. Wind tunnel tests of several six-series airfoils at various Reynolds numbers and Mach numbers are reported in reference 6.12, and the results for one of these, the NACA 64-210, are shown in Figure 6.14. Since Mach number and Reynolds number are directly related for a given chord, lines of constant chord can be plotted across the family of curves. For sea-level, standard conditions:

$$M = \frac{.14 \text{ R.N.}/10^6}{c}$$

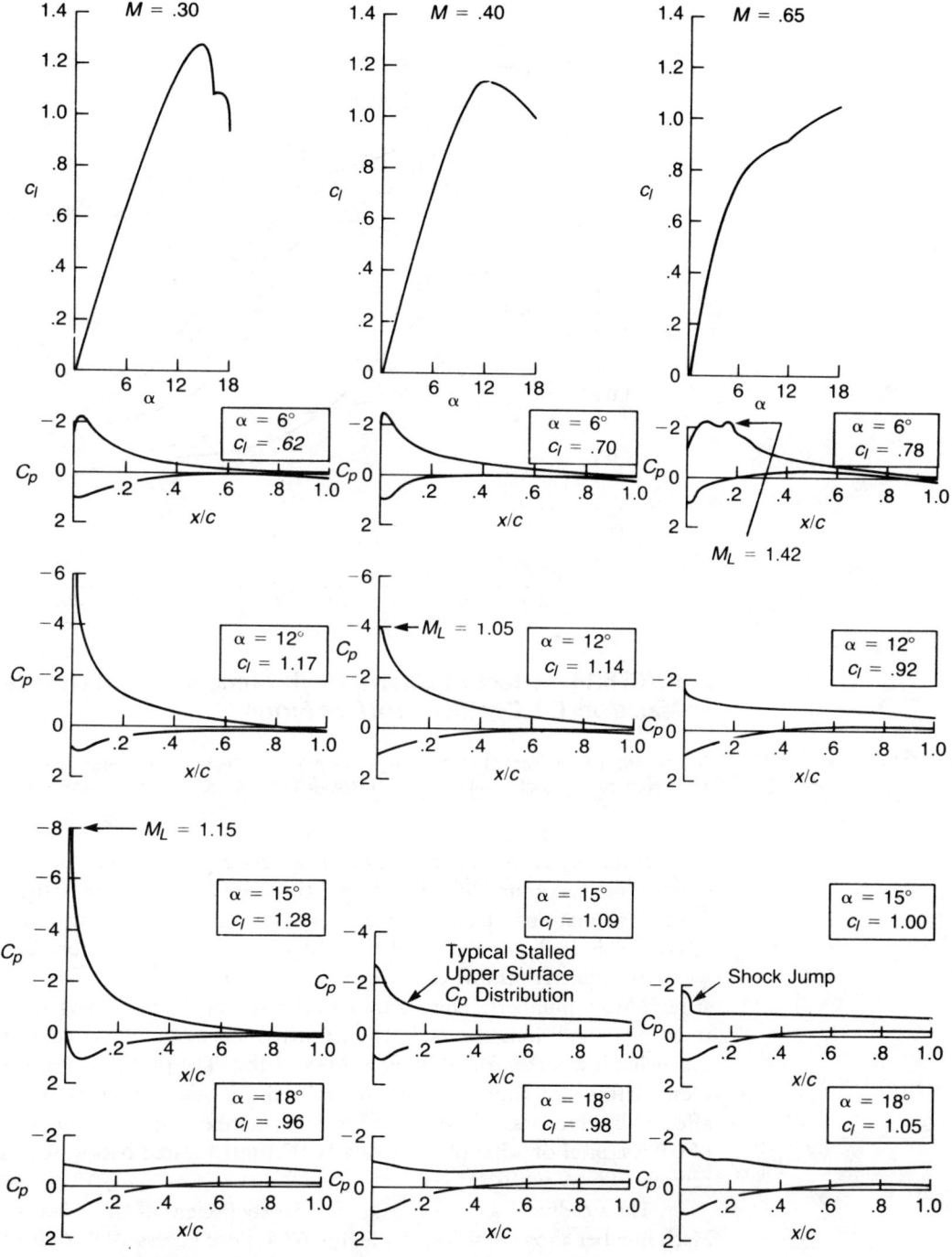

FIGURE 6.13 Effect of Mach Number on Lift Coefficient and Pressure Distribution of NACA 0012 Airfoil

Source: Lizak, "Two-Dimensional Wind Tunnel Tests of an H-34 Main Rotor Airfoil Section," USA TRECOM TR 60-53, 1960.

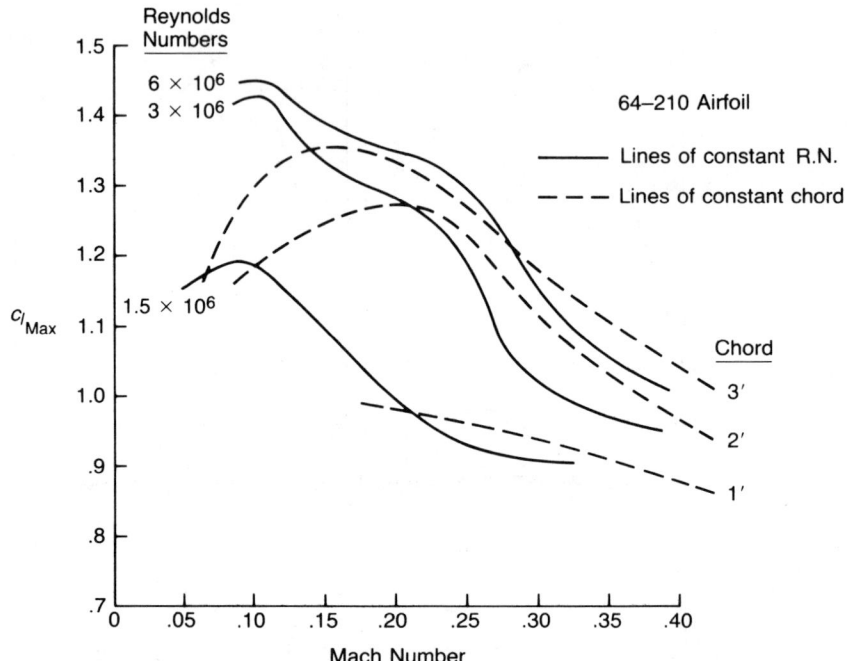

FIGURE 6.14   Effect of Chord, Mach Number, and Reynolds Number, on Maximum Lift Coefficient of One Airfoil

Source: Racisz, "Effects of Independent Variations of Mach Number and Reynolds Number on the Maximum Lift Coefficients of Four NACA 6-Series Airfoil Sections," NACA TN 2824, 1952.

It may be seen that the chord has a strong influence on the effect of Mach number on maximum lift. The larger the chord, the higher is the detrimental effect. As a matter of fact, tests with very small chord models may show beneficial effects. This is illustrated by the test data on Figure 6.15 from reference 6.13 where the effective chord varied from 5 inches at low Mach numbers to 3.8 inches at high Mach numbers. This figure also shows a rather surprising result—that for these test conditions the 64A010 airfoil has a maximum lift coefficient that is somewhat lower than the thinner 64A006 airfoil. The probable reason is that both these airfoils are experiencing pure thin airfoil stall, which is not significantly affected by thickness ratio at low Reynolds numbers. A maximum lift coefficient of 0.8 is typical of a flat plate or of a NACA 0012 tested backwards, according to references 6.14 and 6.15.

The steady reduction of maximum lift coefficient of large chord airfoils with Mach number above .3 shown on Figure 6.14 is not necessarily a trait of all airfoils. The type of airfoil known as *supercritical* or *peaky* can actually experience an increase in maximum lift coefficient with Mach number. A supercritical airfoil is one on which the nose is shaped such that the strength of the shock wave is reduced by slowing the air ahead of it through a favorable arrangement of expansion and

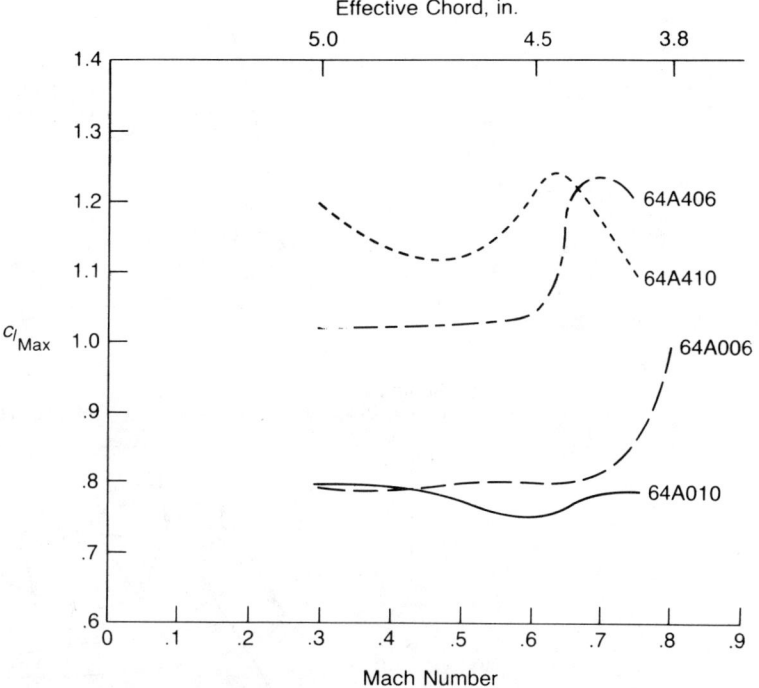

FIGURE 6.15   Maximum Lift Coefficients of Two-Dimensional Airfoil Models with Small Chords

Source: Stivers, "Effects of Subsonic Mach Number on the Forces and Pressure Distributions on Four NACA 64A-Series Airfoil Sections at Angles of Attack as High as 28°, NACA TN 3162, 1954.

compression waves, as shown in Figure 6.16. The initial expansion wave is generated by the high nose peak on the pressure distribution, which is reflected from the sonic line as a compression wave and then from the airfoil surface as another compression wave. If the shock wave is located in the region influenced by the second compression wave, its strength will be reduced as a result of the lower local velocity, and its ability to produce shock stall by separating the boundary layer will be correspondingly reduced. If the shock wave lies at some other location, the shock stall may be more severe. Thus the favorable conditions exist only for certain combinations of angle of attack and Mach number, and a penalty may apply to operating at other combinations. The first half of Figure 6.17, from reference 6.16, shows that a 12% thick airfoil with supercritical characteristics has a higher maximum lift coefficient than a 12% thick drooped-nose airfoil only for Mach numbers above about 0.43. At low Mach numbers, the peak on the supercritical pressure distribution—and the corresponding unfavorable pressure gradient behind it—encourages early bursting of the laminar separation bubble. Results for the NACA 0012 from the same series of tests are also shown for comparison. It

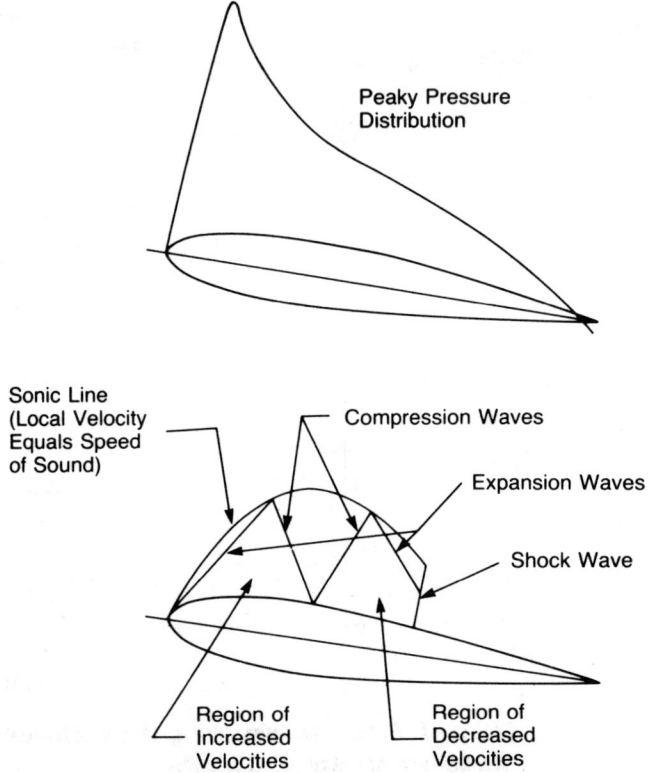

FIGURE 6.16   Features of a Peaky Supercritical Airfoil

may be seen that both of the cambered airfoils are better than the symmetrical airfoil, because the upper surface coordinate at the quarter chord is higher and the corresponding benefit shown in Figure 6.7 applies. Similar results for 15% thick airfoils are also shown in Figure 6.17 from reference 6.17. A complete discussion of supercritical airfoils is given in reference 6.18, and procedures for designing these airfoils for rotors are described in references 6.19 and 6.20.

Figure 6.18 shows the effect of Mach number on the measured maximum lift coefficients of several airfoils with nominal 2-foot chords as reported in references 6.10, 6.12, 6.21, and 6.22. Those airfoils that have good characteristics at low Mach numbers appear to retain their relative advantage in the Mach number range of the retreating blade: from 0.25 to 0.5.

Some caution should be exercised in drawing conclusions on the relative merits of the various airfoils shown in Figure 6.18, since the results appear to be affected by the wind tunnel that produced them. Figure 6.19 shows test results for the NACA 0012 airfoil obtained from several tunnels as reported in references 6.9, 6.10, 6.22, 6.23, 6.24, 6.25 and 6.26. It is possible that the difference between tunnels is associated with the thickness of the boundary layer on the side walls and

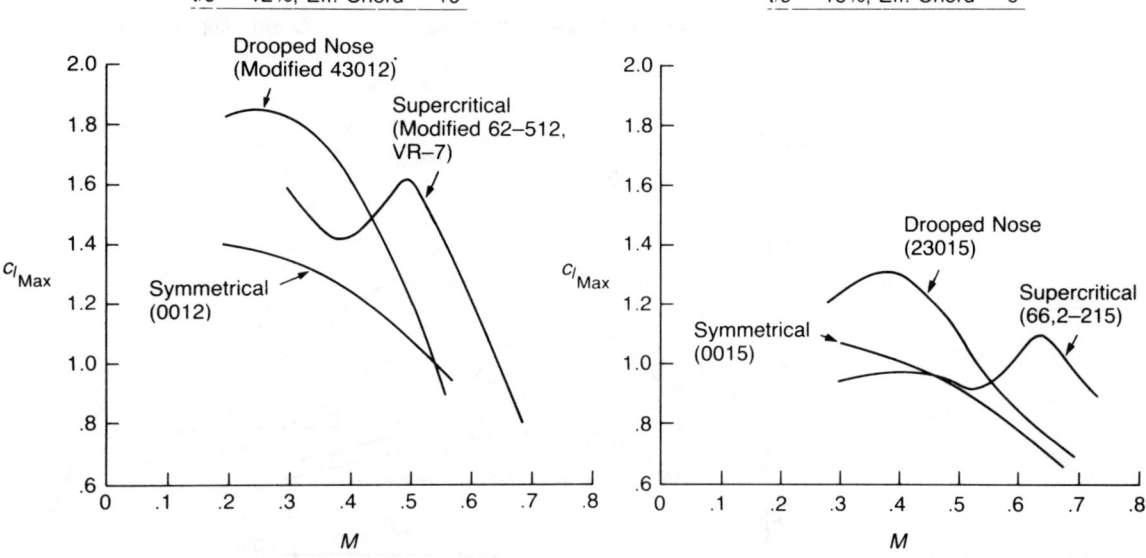

FIGURE 6.17   Maximum Lift of Symmetrical, Drooped Nose, and Supercritical Airfoils

Sources: Benson, Dadone, Gormont, & Kohler, "Influence of Airfoils on Stall Flutter Boundaries of Articulated Helicopter Rotors," *JAHS* 18-1, 1973; Graham, Nitzberg, & Olson, "A Systematic Investigation of Pressure Distributions at High Speeds over Five Representative NACA Low-Drag and Conventional Airfoil Sections," NACA TR 832, 1945.

the amount of the model affected. In any case, the differences should be resolved for the good of both the helicopter aerodynamicist and the wind tunnel engineer.

Maximum Dynamic Lift Coefficient

The airplane people discovered some time ago that when a wing's angle of attack is increased rapidly, it can momentarily generate a higher maximum lift coefficient than it could if the angle of attack were increased slowly. This also applies to helicopter blades. A good review of the phenomenon, which is referred to as *dynamic stall*, will be found in reference 6.27. Figure 6.20 shows this dynamic overshoot as measured during ramp-type angle-of-attack increases in the two-dimensional tests reported in reference 6.28. The overshoot can be related to the change in angle of attack during the time required for the airfoil to travel one chord length. For airfoils that stall first at the leading edge, the dynamic overshoot is attributed to two effects: the delay in the separation of the boundary layer, and the momentary existence of a vortex shed at the leading edge after the boundary layer does separate. These effects are discussed in reference 6.29. The delay in

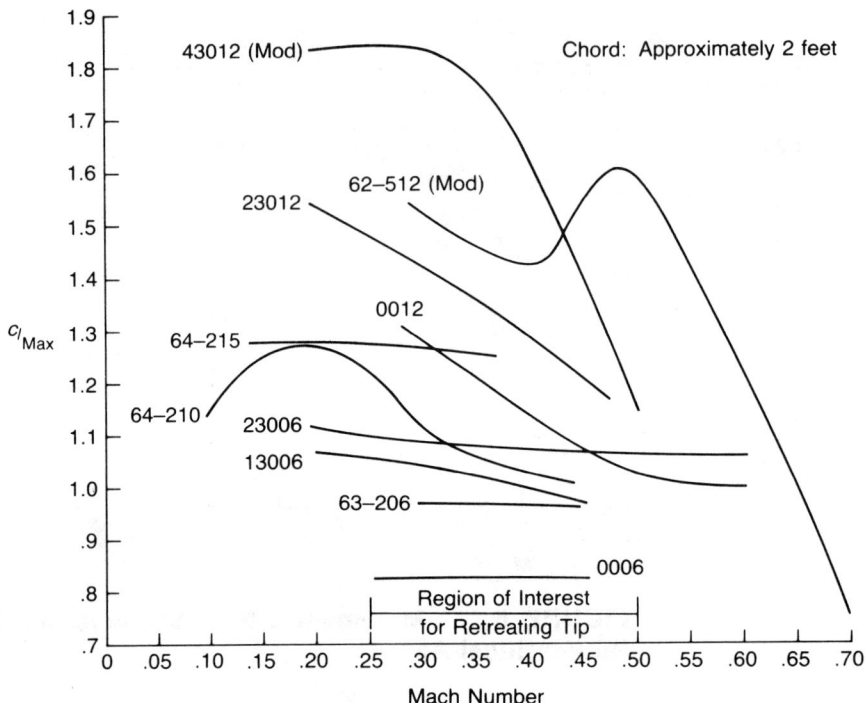

FIGURE 6.18   Effect of Mach Number on Maximum Lift

Source: Lizak, "Two-Dimensional Wind Tunnel Tests of an H-34 Main Rotor Airfoil Section," USA TRECOM TR 60-53, 1960; Davenport & Front, "Airfoil Sections for Helicopter Rotors—A Reconsideration," AHS 22nd Forum, 1966; Racisz, "Effects of Independent Variations of Mach Number and Reynolds Numbers on the Maximum Lift Coefficients of Four NACA Six-Series Airfoil Sections," NACA TN 2824, 1952.

separation corresponds to the finite time required for the aft edge of the separation bubble to move forward to its bursting position. This time delay is lengthened if the airfoil is pitching nose up, for two reasons. First, the motion raises the position of the stagnation point, as shown in Figure 6.21, and produces an effect similar to drooping the nose, which was shown to be beneficial for static stall. Second, the nose-up pitching motion causes the boundary layer to develop a fuller and more stable profile, which resists separation. A quantitative evaluation of the second effect is given in reference 6.30. If the airfoil has plunging motion instead of pitching motion, the maximum angle of attack occurs while the airfoil is descending, and thus the nose droop and pressure gradient effects are detrimental instead of beneficial. Nevertheless, an airfoil in plunge still exhibits a dynamic overshoot, though not as great as that of the airfoil in pitch.

Even after the leading edge separates, the airfoil can momentarily still generate high lift as a result of a vortex that is shed at the leading edge at the

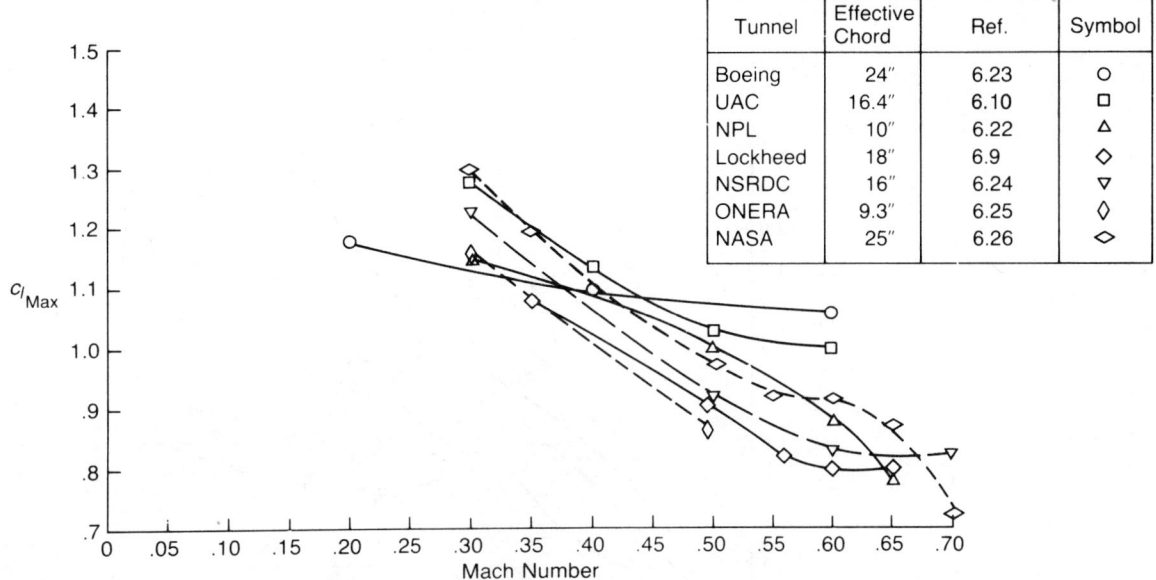

Tunnel	Effective Chord	Ref.	Symbol
Boeing	24″	6.23	○
UAC	16.4″	6.10	□
NPL	10″	6.22	△
Lockheed	18″	6.9	◇
NSRDC	16″	6.24	▽
ONERA	9.3″	6.25	◊
NASA	25″	6.26	◇

FIGURE 6.19   Comparison of NACA 0012 Wind Tunnel Data

instant of stall. The vortex travels back over the top of the airfoil at approximately half of the free stream velocity, according to reference 6.31, carrying with it a low-pressure wave that accounts for the very large lift coefficients shown in Figure 6.20.

Airfoils that stall first at the trailing edge also exhibit a dynamic overshoot, though considerably less than those airfoils that have leading edge stall. For example, reference 6.32 shows that a 16% thick airfoil, which would be expected to have trailing edge stall, has approximately half the dynamic overshoot of a 9% thick airfoil, which would be expected to have leading edge stall. The favorable effect of pitching motion for trailing edge stall is apparently the thinning of the boundary layer near the nose, which has a beneficial effect extending to the trailing edge.

Wind tunnel tests of oscillating, two-dimensional airfoils are reported in References 6.23, 6.33, 6.34, and 6.35. The first set of tests used modified NACA 0012 and 23010 airfoils and oscillated them in sinusoidal pitch and plunge motions at Mach numbers from 0.2 to 0.6; the second set used modified NACA 0006 and 13006 airfoils oscillating only in pitch through the same Mach number range; and the third set used a NACA 0012 in sinusoidal and sawtooth pitch oscillations at a Mach number of about 0.3. The tests of reference 6.35 used several modern airfoils, but at low Mach numbers. Some of the primary effects of varying the test parameters are shown in Figure 6.22. The first set of comparisons shows that oscillations entirely below stall or entirely above stall have only small dynamic effects, but that oscillation through stall produces a hysteresis loop in which stall is

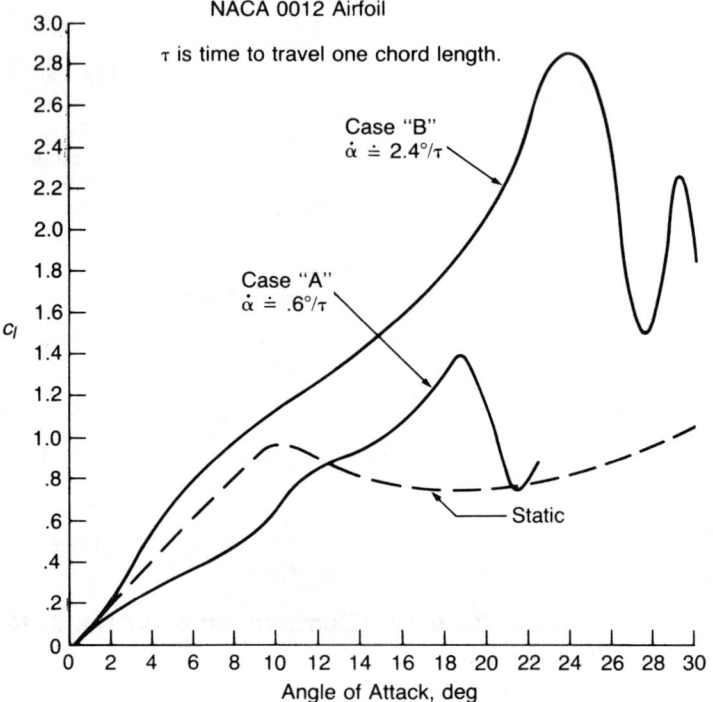

FIGURE 6.20   Lift Characteristics with Ramp Changes in Angle of Attack

Source: Ham & Garelick, "Dynamic Stall Considerations in Helicopter Rotors," *JAHS* 13-2, 1968.

reached late on the upstroke and is induced early on the downstroke. Reference 6.36 suggests that the lower limit of stalled lift on the downstroke is approximately the static maximum coefficient of a flat plate. Note that the lift coefficient above the stall angle of attack in Figure 6.22 does not go below 0.6. The second set of comparisons shows that the dynamic overshoot is a function of the frequency of oscillation. The frequency is expressed in terms of the reduced frequency, k:

$$k = \omega \, \frac{c/2}{V}$$

where ω is the frequency of oscillation in radians per second, $c/2$ is the semichord, and V is the local velocity. Physically, k is the portion of the oscillation cycle, in radians, which occurs during the time the air travels half of a chord length over the airfoil. For a blade element, the velocity is the tangential velocity, U_T. In order to

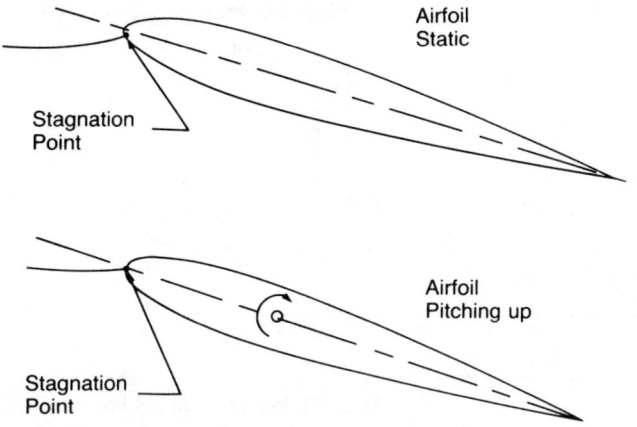

FIGURE 6.21 Effect of Pitch Motion on Location of Stagnation Point

obtain an understanding of the approximate magnitude of k, assume a rotor in hover is undergoing a once-per-rev pitch change. Then the reduced frequency at the tip is one-half the amount of azimuth subtended by the chord:

$$k_{\text{tip}} = \frac{1}{2(R/c)}$$

or, for the example helicopter with a 30-ft radius and a 2-foot chord:

$$k_{\text{tip}} = 0.033$$

The reduced frequency will be higher for inboard blade elements, for blade elements on the retreating blade in forward flight, or for blade elements that are being subjected to higher frequencies because they are coming close to a series of vortices shed by the tips of previous blades or being affected by blade torsional oscillations. The effect of increasing the Mach number is shown in the third set of comparisons in Figure 6.22. At low Mach numbers, the airfoil has leading edge stall; thus changes in conditions at the nose are significant in determining the amount of dynamic overshoot. At higher Mach numbers, however, where stall is caused by separation behind a shock wave, the stall occurs before the nose conditions become critical, and thus the pitching motion produces less overshoot.

A blade element in flight experiences plunge motion where the leading and trailing edges have vertical velocity in the same direction, as well as pitch motion where the leading and trailing edges have vertical velocities in opposite directions. The equation for the local angle of attack is:

Effect of Mean Angle of Attack, α_0

$M = .4, k = .12$

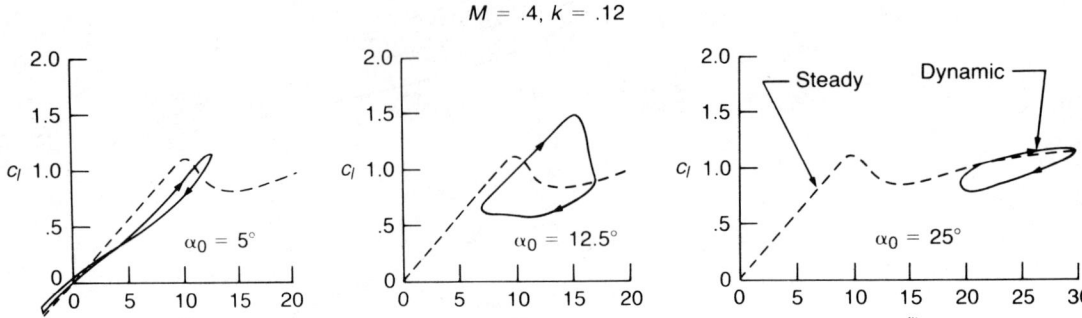

Effect of Reduced Frequency, k

$M = .4, \alpha_0 = 12.5°$

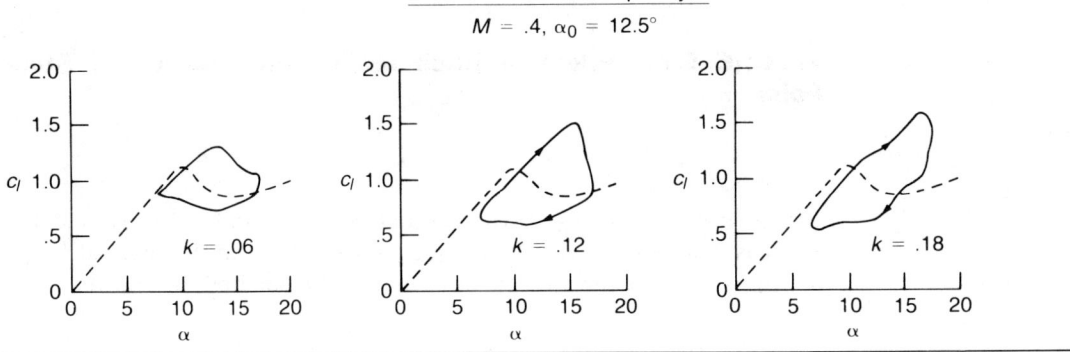

Effect of Mach Number, M

$k = .12$

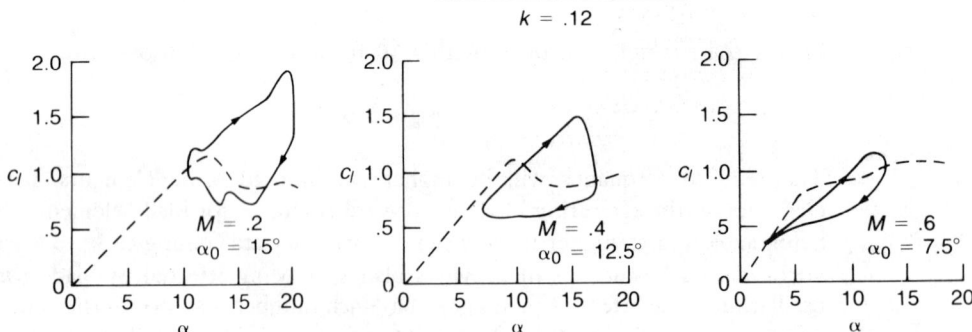

FIGURE 6.22   Effects of Test Parameters on Dynamic Overshoot of NACA 0012 Airfoil Oscillating in Pitch

Source: Liiva, Davenport, Gray, & Walton, "Two-Dimensional Tests of Airfoils Oscillating Near Stall," USAAVLABS TR 68-13, 1968.

$$\alpha = \theta + \tan^{-1} \frac{U_P}{U_T}$$

and thus the rate of change of angle of attack is:

$$\dot{\alpha} = \dot{\theta} + \overline{\tan^{-1} \frac{U_P}{U_T}}$$

where the first term is the rate of pitch (up) and the second is the rate of plunge (down). It is of interest to note that in the third quadrant, where the blade is pitching up, it is also plunging up as a result of the effects of coning and of the longitudinal gradient of induced velocity. Figure 6.23 shows that for the NACA 0012 airfoil, the dynamic overshoot corresponding to plunge is considerably less than that due to pitch.

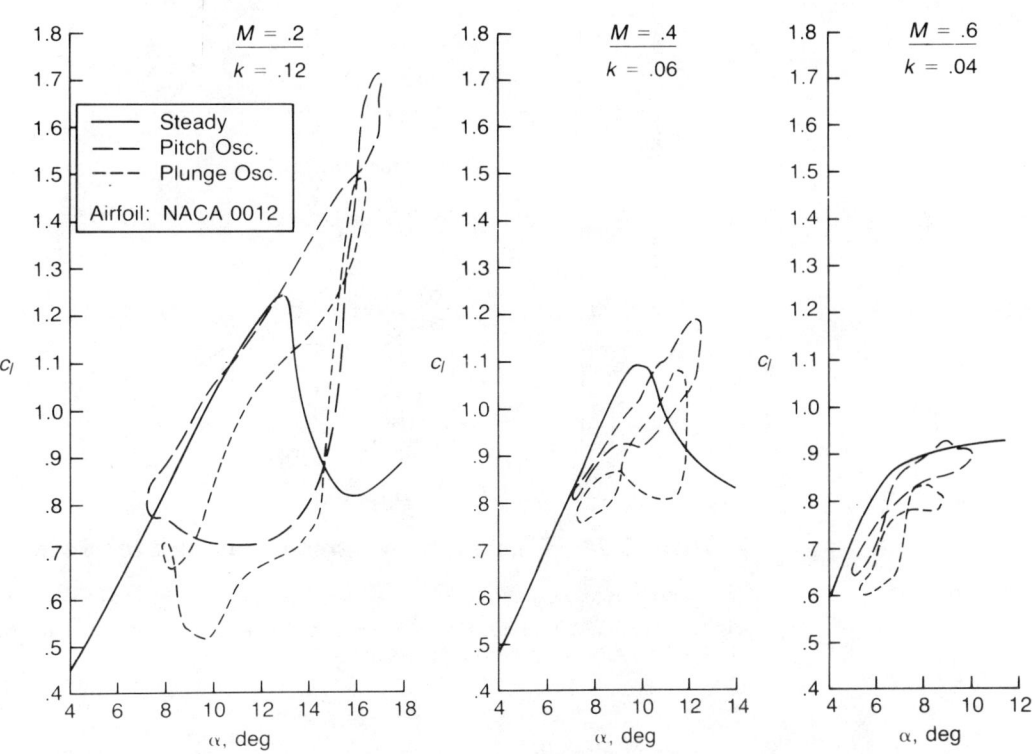

FIGURE 6.23   Effects of Pitch and Plunge Oscillations through Stall

Sources: Liiva et al., "Two-Dimensional Tests of Airfoils Oscillating Near Stall," USAAVLABS TR 68-13, 1968: Gray & Liiva, "Wind Tunnel Tests of Thin Airfoils Oscillating Near Stall," USAAVLABS TR 68-89, 1969.

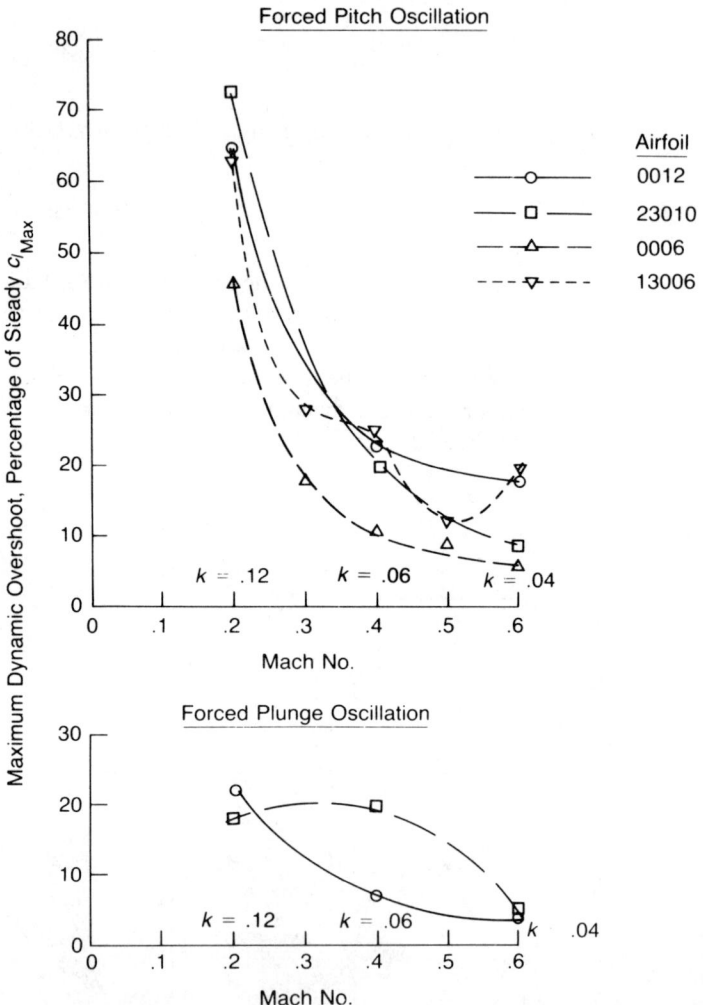

FIGURE 6.24   Maximum Dynamic Overshoot for Several Airfoils

Sources: Liiva, Davenport, Gray, & Walton, "Two-Dimensional Tests of Airfoils Oscillating Near Stall," USAAVLABS TR 68-13, 1968; Gray & Liiva, "Wind Tunnel Tests of Thin Airfoils Oscillating Near Stall," USAAVLABS TR 68-89, 1969.

A limited indication of the effect of the airfoil physical parameters on dynamic overshoot is given by Figure 6.24, which shows test values of dynamic overshoot in both pitch and plunge as a percentage of the static maximum lift coefficient for the four airfoils of references 6.23 and 6.33.

Another set of dynamic test results for a number of modern helicopter airfoil sections oscillating in pitch at a Mach number of 0.3 is reported in reference 6.35. Figure 6.25 summarizes the results. One conclusion that can be drawn is that

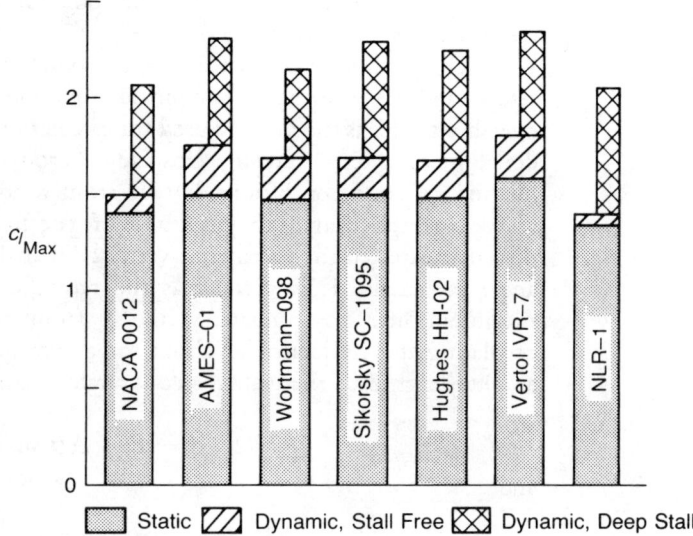

$c_{l_{Max}}$

Static   Dynamic, Stall Free   Dynamic, Deep Stall

FIGURE 6.25   Dynamic Stall Characteristics of Several Helicopter Airfoils

Source: McCrosky et al., "Dynamic Stall on Advanced Airfoil Sections," *JAHS* 26-3, 1981.

the dynamic overshoot in lift coefficient is nearly independent of the airfoil shape, varying from 0.7 to 0.8 for this group.

A number of ways have been suggested for representing the dynamic overshoot in actual practice. Several sophisticated analytical methods are given in references 6.37, 6.38, 6.27, and 6.39. These use potential flow and boundary-layer equations to predict the separation angle. Although they might be suitable for predicting the performance of an oscillating airfoil in a wind tunnel, they are much too complicated to use in a rotor analysis program. Three simpler methods based on empirical studies of oscillating airfoil wind tunnel data have been used in rotor programs and, at this writing, all must be considered valid, although none of them has undergone rigorous comparison against a wide range of rotor experimental data.

The method of reference 6.40 as expanded in references 6.31 and 6.41 calculates the overshoot of angle of attack as a function of the pitch rate, which is taken as the rate of change of the calculated angle of attack. The method is based on a series of analogies backed up by the selected test results. It does suffer, however, from a confusing writeup and a lack of explicitly stated correlation factors.

The method of references 6.34 and 6.42 is based on the observation that the overshoot is a function not only of the angle of attack and its velocity, but of the acceleration as well. Tabulated influence factors as a function of these three parameters are stored in the computer, or curve-fitting equations are used based on analysis of test data. Hints for extrapolation to test conditions or airfoils not tested are given in reference 6.42.

The method of reference 6.43 is also based on test data. It uses the angle of attack and its velocity in empirical equations to approximate the dynamic overshoot. This is the method used for calculations in this book and is explained in detail in Chapter 3 under "Unsteady Aerodynamics and Yawed Flow." This method relies on conclusions derived from wind tunnel results that the angle of attack corresponding to the maximum lift coefficient in a dynamic situation can be directly related to the parameter: $\sqrt{c\dot{\alpha}/2V}$. For this purpose, the rate of change of angle of attack, $\dot{\alpha}$, is defined as the rate the airfoil had when it reached its maximum lift. To evaluate $\sqrt{c\dot{\alpha}/2V}$ from wind tunnel data of an airfoil oscillating at a frequency, ω, about some average angle of attack, α_0, through an amplitude, $\pm \Delta\alpha$, the instantaneous angle of attack is:

$$\alpha = \alpha_0 + \Delta\alpha \sin \omega t$$

and

$$\dot{\alpha} = \Delta\alpha\omega \cos \omega t$$

At the instant of stall:

$$\omega t_{stall} = \sin^{-1}\left(\frac{\alpha_{stall} - \alpha_0}{\Delta\alpha}\right)$$

thus

$$\dot{\alpha} = \Delta\alpha\omega \cos \sin^{-1}\left(\frac{\alpha_{stall} - \alpha_0}{\Delta\alpha}\right)$$

From the definition of the reduced frequency, k, previously used:

$$\frac{c}{2V} = \frac{k}{\omega}$$

thus

$$\sqrt{\frac{c\dot{\alpha}}{2V}} = \sqrt{\left| k\Delta\alpha \cos \sin^{-1}\left(\frac{\alpha_{stall} - \alpha_0}{\Delta\alpha}\right) \right|}$$

The test data for four different oscillating airfoils tabulated in references 6.23 and 6.33 include all of the factors required to evaluate $\sqrt{c\dot{\alpha}/2V}$. (Note that the only usable test points are those for which the maximum lift was reached before the maximum angle of attack, since at that point the rate was zero.) Strictly speaking, the dynamic angle of overshoot is the difference between the measured angle of attack at maximum lift for the dynamic and static test conditions, but sometimes the lift curve slope is reduced at high angles of attack so that the magnitude of the angle of attack overshoot is much greater than that corresponding to the overshoot of maximum lift coefficient. For use in the analysis, which assumes that the increase in maximum lift coefficient is directly proportional to the overshoot of angle of attack, an effective stall angle has been defined based on the measured maximum lift coefficient; the slope of the lift curve; and, in the case of the cambered section, the angle of zero lift:

$$\alpha_{\text{stall}_{\text{eff}}} = \frac{c_{l_{\max}}}{a} + \alpha_{\text{LO}}, \text{deg}$$

In Figure 6.26, the data for the V0011 and the V23010–1.58 are plotted. The slope of the lines through the data points is the function, γ, where:

$$\alpha_{\text{stall}_{\text{eff}}} - \alpha_{\text{stall}_{\text{static}}} = \Delta\alpha = \gamma \sqrt{\frac{c\dot{\alpha}}{2V}} \text{ radians}$$

The values of γ for pitching oscillations from Figure 6.26 have been plotted on Figure 6.27 as a function of Mach number. Both the values and the trend with Mach number are different from those indicated by reference 6.43 which analysed the same test data. The differences arise primarily from the fact that reference 6.43 plots actual stall angles at which maximum lift is achieved, whereas in Figure 6.26 the effective stall angle is plotted. Another difference is apparently in the interpretation of the test data for Mach number of 0.6. This can be illustrated by examining the dynamic stall hysteresis loop at $M = 0.6$ in the lower right-hand corner of Figure 6.22. The angle of attack overshoot can be referenced either to the initial static stall at about 6° or to the final static stall at about 15°. In the case of the first, there is considerable angle of attack overshoot; in the case of the latter, none at all as was apparently assumed in the studies of references 6.43 and 6.44. The method used in this book was based on the 6° stall angle as being a more realistic value in light of how the results are used in the analysis.

Confession: The forward flight charts of Chapter 3 were computed before this study was made. The gamma function used was similar to the lines defined by the solid points (and dotted lines) in figure 6.27. Fortunately, the results are about the same for the blade tips at the tip speed ratio range of 0.25 to 0.35, which includes most conventional high-speed flight.

A comparison of the overshoot during pitch oscillations and plunge motions is given in reference 6.45 for $M = 0.4$. The general conclusion is that at this Mach number the gamma functions are not significantly different for the two types of motion, and thus they do not have to be treated differently in the analysis.

DRAG CHARACTERISTICS

Drag Divergence Mach Number

The advancing tip operates at nearly zero lift coefficient at high speed, so the drag characteristics of the airfoil in this condition are important. As the free stream

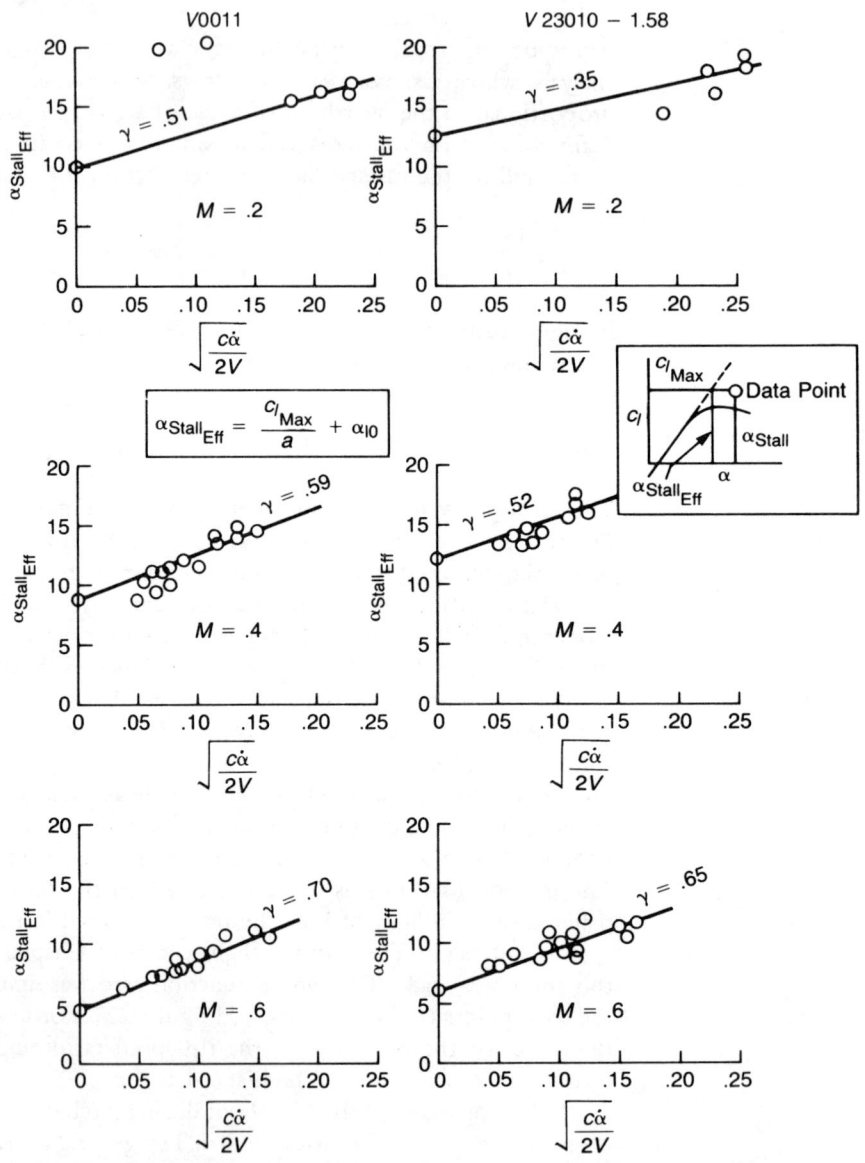

FIGURE 6.26 Effective Stall Angle for Oscillating Airfoils

Source: Liiva, Davenport, Gray, & Walton, "Two-Dimensional Tests of Airfoils Oscillating Near Stall," USAAVLABS TR 68-13, 1968.

Mach number is increased toward and beyond its critical Mach number, the local velocity on the surface first reaches the speed of sound and then exceeds it, until at some speed and at some position on the airfoil a shock wave is formed through which the velocity decreases to a subsonic value. Tests show that if the shock is weak and close to the nose, there is no significant drag penalty; but as the shock

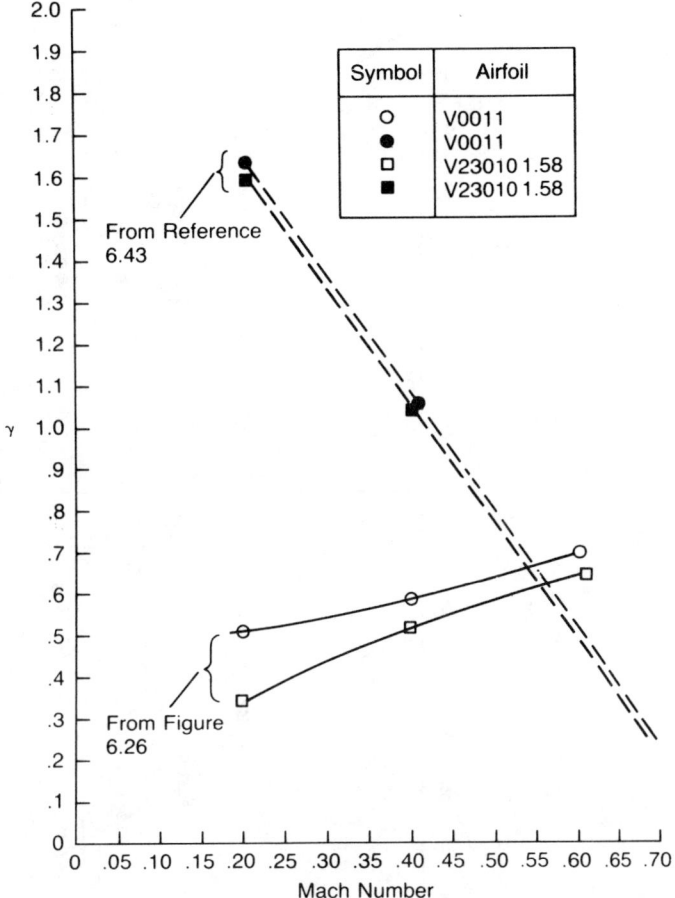

Symbol	Airfoil
○	V0011
●	V0011
□	V23010 1.58
■	V23010 1.58

FIGURE 6.27   Gamma Function as Affected by Mach Number

passes beyond the crest or, for this condition, the position of maximum thickness, the drag increases rapidly as a result of the momentum loss through the shock (wave drag) and in some cases as a result of separation of the boundary layer (shock stall). The free stream Mach number at which the drag coefficient increases significantly is known as the *drag divergence Mach number.* In many studies, it is defined as the Mach number at which the drag coefficient rises at the rate of 0.1 per unit Mach number. For helicopter applications, however, where the magnitude of the drag change may be more significant than the rate of increase, it is more appropriate to define it as the Mach number at which the drag coefficient is twice its incompressible value. Using this definition, the measured two-dimensional drag divergence Mach numbers of a number of airfoils at zero lift as reported in references 6.1, 6.13, 6.17, 6.21, 6.46, 6.47, 6.48, 6.49, and 6.50 have been compiled and are plotted in Figure 6.28a as a function of thickness ratio. It may be seen that the symmetrical airfoils of all families form a reasonably tight grouping, but that

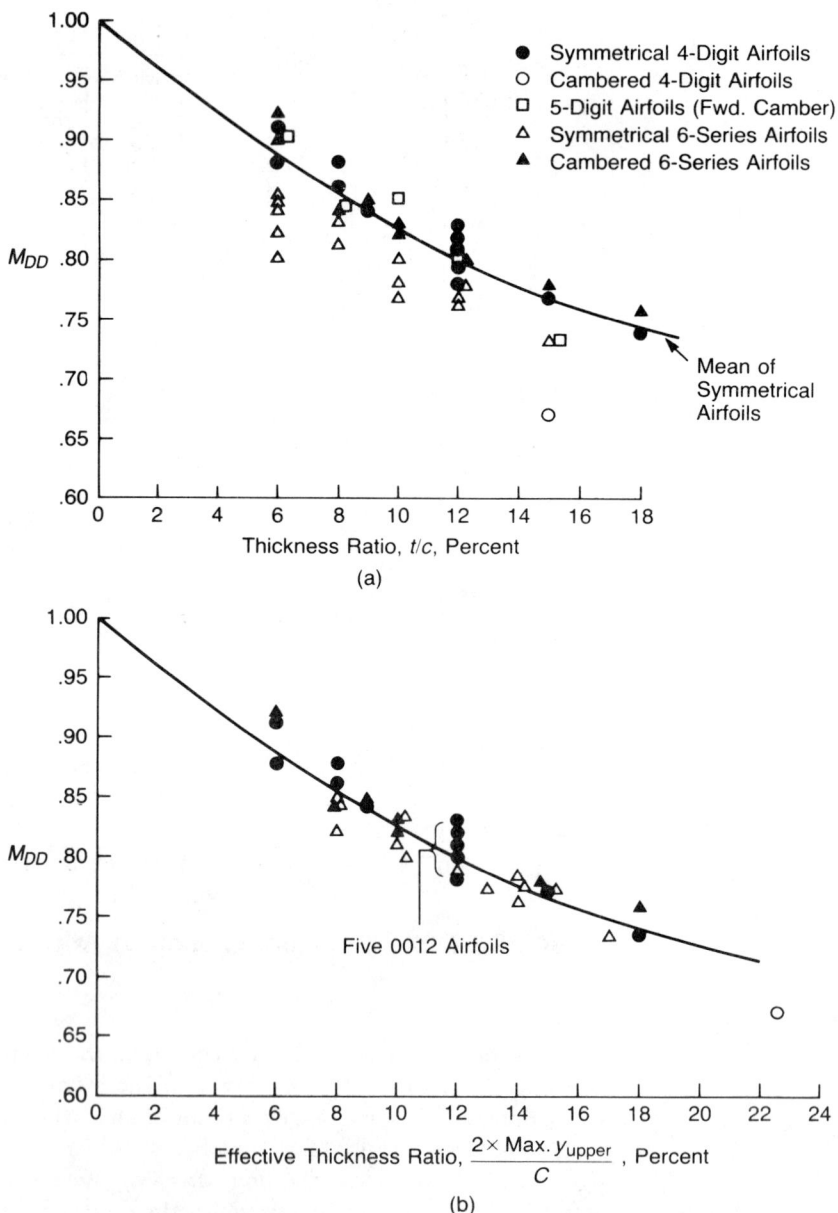

FIGURE 6.28   Drag Divergence Mach Numbers at Zero Lift

Sources: Sipe & Gorenberg, "Effect of Mach Number, Reynolds Number, and Thickness Ratio on the Aerodynamic Characteristics of NACA 63A-Series Airfoil Sections," USATRECOM TR 65-28, 1965; Van Dyke, "High-Speed Subsonic Characteristics of 16 NACA Six-Series Airfoil Sections," NACA TN 2670, 1952; Wilson & Horton, "Aerodynamic Characteristics at High and Low Subsonic Mach Numbers of Four NACA Six-Series Airfoil Sections at Angles of Attack from −2° to 31°," NACA RM 876 (L53620), 1953; Stivers, "Effects of Subsonic Mach Numbers on the Forces and

the six-series cambered airfoils generally have lower drag divergence Mach numbers. Even these airfoils can be brought into the grouping if the physical parameter is taken as twice the maximum upper ordinate instead of the thickness ratio, as shown in Figure 6.28b.

Just as with the maximum lift coefficient, the scatter of the points in Figure 6.28 represents differences in airfoils and differences in tunnels. Some airfoils have beneficial supercritical characteristics—either by design or by accident—and thus have measurably higher drag divergence Mach numbers than average. For other airfoils, the supercritical characteristics might be detrimental if an expansion wave is located ahead of the shock instead of a compression wave. The difference that a tunnel can make is shown by 5 points for the NACA 0012 airfoil.

A conclusion that can be drawn from Figure 6.28 is that for the same thickness ratio, aft camber as used in the six-series airfoils lowers the drag divergence Mach number, but that forward camber as used in the five-digit airfoils has little effect. This conclusion, however, must be qualified. If the nose is drooped so far that the lower surface becomes concave, the drag curve at zero lift can have the distinct characteristic shown for the 33008 airfoil in Figure 6.29 from reference 6.9. This is known as a "creepy" drag rise and is apparently due to a shock wave formed on the lower surface that locally separates the boundary layer in the concave region. Even airfoils with moderate forward camber may exhibit significant drag creep at small negative angles of attack such as might exist on the advancing tip at high speeds.

Drag at Moderate Angles of Attack and Mach Numbers

For conditions in which there is no separation and no significant compressibility effects, the drag of the airfoil is primarily due to skin friction, whose coefficient is a function of the local velocities on the surface and whether the boundary layer is laminar or turbulent. Figure 6.30, from reference 6.1, shows the effect of Reynolds number on the calculated minimum drag coefficient for fully laminar and for fully turbulent boundary layers. Also shown are wind tunnel results for several airfoils. Laminar flow can usually be maintained only on the nose of the airfoil where the flow is accelerating. Thus no airfoil can take full advantage of laminar flow; but some, such as the six-series airfoils, keep the flow accelerating further back than others and thus have a larger region of laminar boundary layer. As a matter of fact,

Pressure Distributions on Four NACA 64A-Series Airfoil Sections at Angles of Attack as High as 28°," NACA TN 3162, 1954; Abbott & Von Doenhoff, *Theory of Wing Sections* (New York: Dover, 1959); Wiesner & Kohler, "Tail Rotor Design Guide," USAAMRDL TR 73-99, 1973; Davenport & Front, "Airfoil Sections for Helicopter Rotors—A Reconsideration," AHS 22nd Forum, 1966; Graham, Nitzberg, & Olson, "A Systematic Investigation of Pressure Distributions at High Speeds over Five Representative NACA Low-Drag and Conventional Airfoil Sections," NACA TR 832, 1945; Gothert, "Airfoil Measurements in the DVL High-Speed Wind Tunnel," NACA TM 1240.

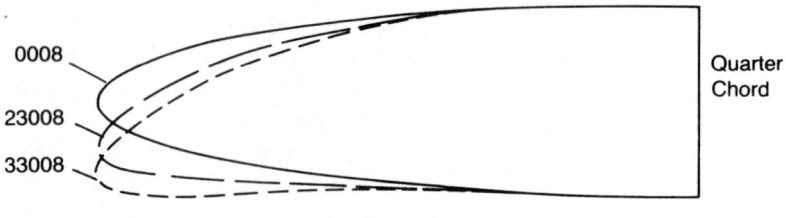

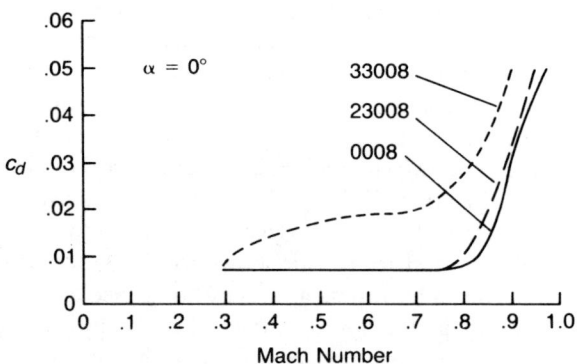

FIGURE 6.29   Effect of Excessive Nose Droop

Source: Prouty, "A State-of-the-Art Survey of Two-Dimensional Airfoil Data," *JAHS* 20-4, 1975.

the six-series airfoils were originally known as *laminar flow* airfoils and were especially designed to take advantage of this effect. Wind tunnel tests of these airfoils showed that, compared to other airfoils, the drag coefficient had only a small rise as the angle of attack was increased until at some angle of attack most of the boundary layer suddenly became turbulent. This resulted in the *drag bucket* illustrated in Figure 6.31. Subsequent use of these airfoils on airplanes in the 1940s and 1950s proved disappointing from a drag standpoint, since dirt, bugs, and surface imperfections triggered premature transition to turbulent flow and nullified the promising characteristics that had been measured on carefully shaped and polished wind tunnel models.

There is some indication—primarily based on the observations of reference 6.51—that a rotor blade, even one with leading edge erosion, can maintain laminar flow more easily than a wing, possibly because built-in surface imperfections are usually less and also because pitting is less detrimental than protrusions.

Even though the six-series airfoils were disappointing from a skin friction standpoint, it was found that some of them had lower drag characteristics for certain combinations of lift coefficients and Mach numbers than did the older airfoils. Figure 6.32 shows drag data for two 15% thick airfoils, a NACA 23015 and a 66.2-215 taken from reference 6.17. At low lift the drag characteristics of the two airfoils are similar, but at higher lift the six-series airfoil has higher drag at low Mach numbers but dips to lower drag at high Mach numbers. At the time

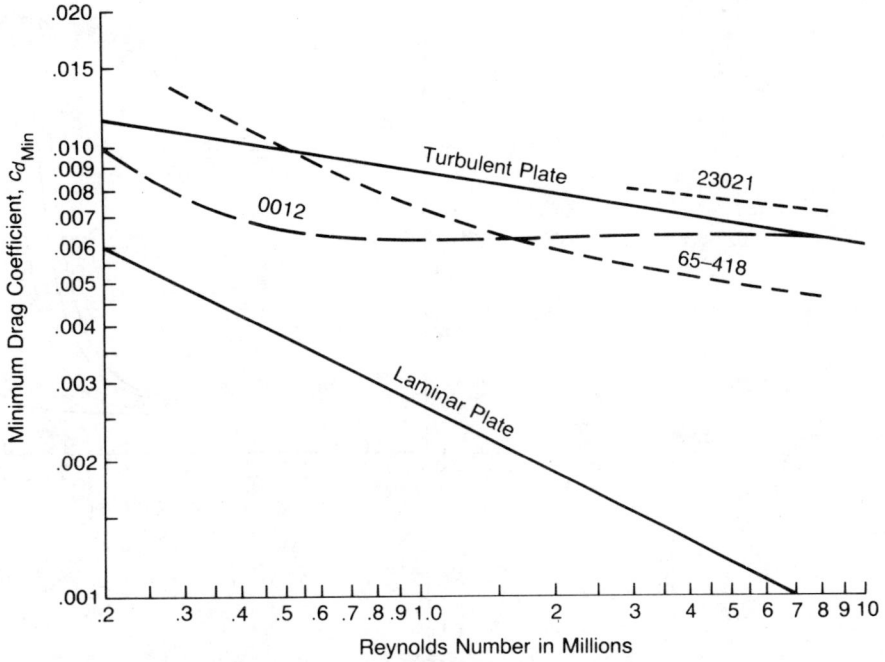

FIGURE 6.30   Minimum Drag Coefficients

Source: Abbott & Von Doenhoff, *Theory of Wing Sections* (New York: Dover, 1959).

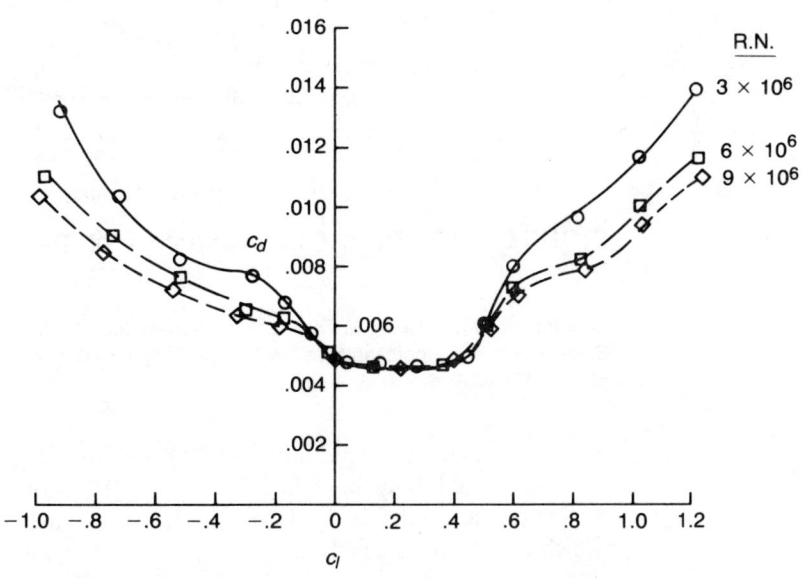

FIGURE 6.31   NACA 63-210 Drag Characteristics

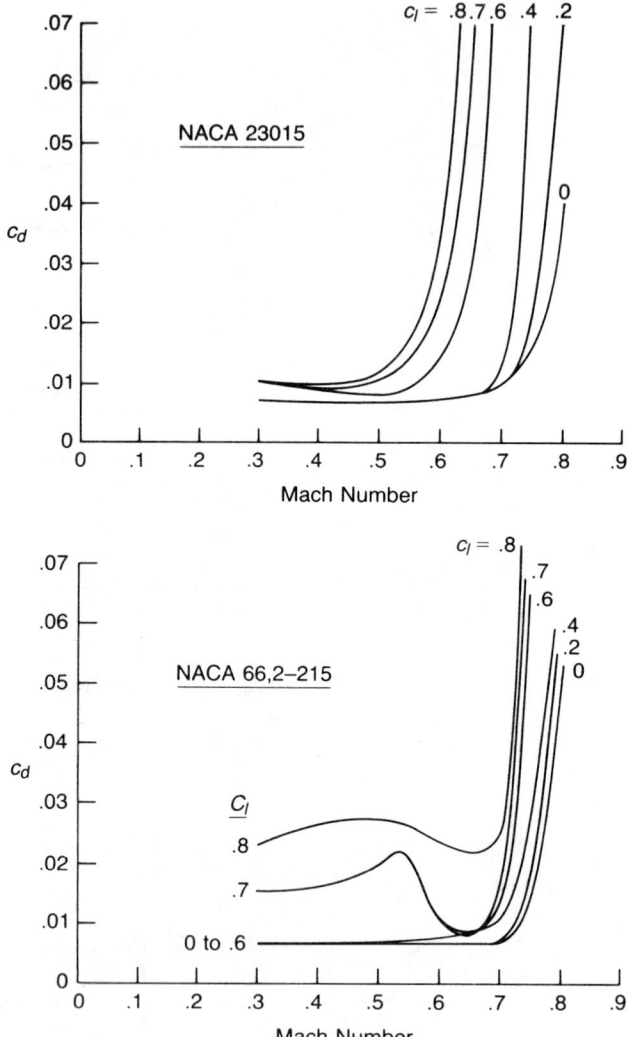

FIGURE 6.32   Drag Characteristics of Two 15 Percent Thick Airfoils

Source: Graham, Nitzberg, & Olson, "A Systematic Investigation of Pressure Distributions at High Speeds over Five Representative NACA Low-Drag and Conventional Airfoil Sections," NACA TR 832, 1945.

these test results were published (1945), the low drag could not be explained and the dips could only be called "peculiar." We now know that these dips are characteristic of the class of airfoils that has come to be known as *supercritical*, which benefit from an advantageous pattern of expansion and compression waves, as shown in Figure 6.16. It is typical of all these airfoils that the region of low drag is limited to a narrow range of operating conditions. The design task for the airplane aerodynamicist is to design airfoils in which this region coincides with the cruise Mach number and lift coefficient of the airplane. Since the helicopter rotor experiences a much wider range of operating conditions than a wing, it is more of a challenge to adapt this concept to a rotor blade; attempts to do this are outlined in references 6.52 and 6.53. Some whirl tower tests of rotors with six-series airfoils, or modifications of this family, indicate a small but measurable advantage in hover over the older airfoils. These tests are reported in references 6.16, 6.54, and 6.55. It is not known whether these benefits were due to increased laminar flow or to supercritical characteristics. Figure 6.33 shows a comparison of two-dimensional data backing up these whirl tower results.

Comparison of one airfoil with another with respect to drag characteristics should be done at the same Reynolds number if possible. Figure 6.34 shows the results of tests on a single airfoil at several Reynolds numbers. It may be seen that at the same Mach number and lift coefficient, the coefficient of drag decreases as the Reynolds number—or chord—is increased. The effect is present in other sets of wind tunnel data, though usually not as dramatically as in this case. It appears that from the standpoint of minimum profile power, a rotor with a few wide-chord blades is better than one with many narrow-chord blades. It is also evident that those computer programs that use tabulated values of aerodynamic coefficients may introduce sizable errors unless the data are based on the Reynolds number corresponding to the chord of the blade. When new tests in pressurized wind tunnels are being planned, consideration should be given to running at

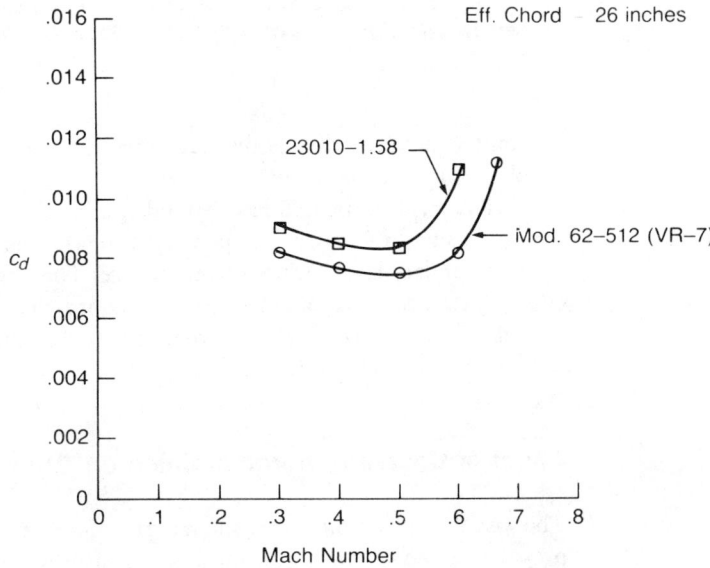

FIGURE 6.33   Drag of Two Arifoils at 0.6 Lift Coefficient

Source: Benson, Dadone, Gormont, & Kohler, "Influence of Airfoils on Stall Flutter Boundaries of Articulated Helicopter Rotors," *JAHS* 18-1, 1973.

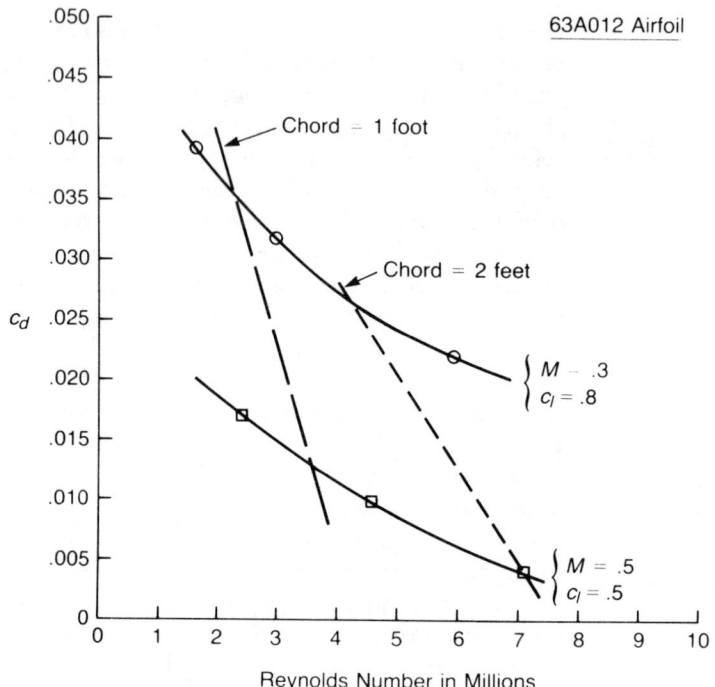

FIGURE 6.34   Effect of Reynolds Number on Drag Coefficient

Source: Sipe & Gorenberg, "Effect of Mach Number, Reynolds Number, and Thickness Ratio on the Aerodynamic Characteristics of NACA 63A-Series Airfoil Sections," USATRECOM TR 65-28, 1965.

constant equivalent chords by matching the test Reynolds number to the Mach number whenever possible.

The airplane people have found that a wing can have a blunt trailing edge without a significant base drag penalty if the trailing edge thickness is less than that of the boundary layer on the upper surface. This can result in an allowable trailing edge thickness of about 1% of the chord, as used on the Lockheed S3-A antisubmarine airplane. This feature might be considered for those blades that need high chordwise stiffness.

Effect of Unsteady Aerodynamics on Drag

The oscillating airfoil experiments of references 6.23 and 6.33 used surface-pressure surveys, which are good for measuring instantaneous lift and pitching moment but are not usable to establish the instantaneous drag. Fortunately, another set of tests reported in reference 6.56 used strain gauge balances instead of pressure measurements and thus measured the instantaneous drag. Figure 6.35 shows typical lift and drag loops. It may be seen that the delay in drag corresponds fairly well to the overshoot in lift and thus can be characterized by the same increase in the stall angle of attack.

PITCHING-MOMENT CHARACTERISTICS

Influence of Airfoil Type on Pitching Moments

During the autogiro era of rotary wing flight, some frightening moments resulted from extreme blade twisting and high control loads with cambered blade airfoil

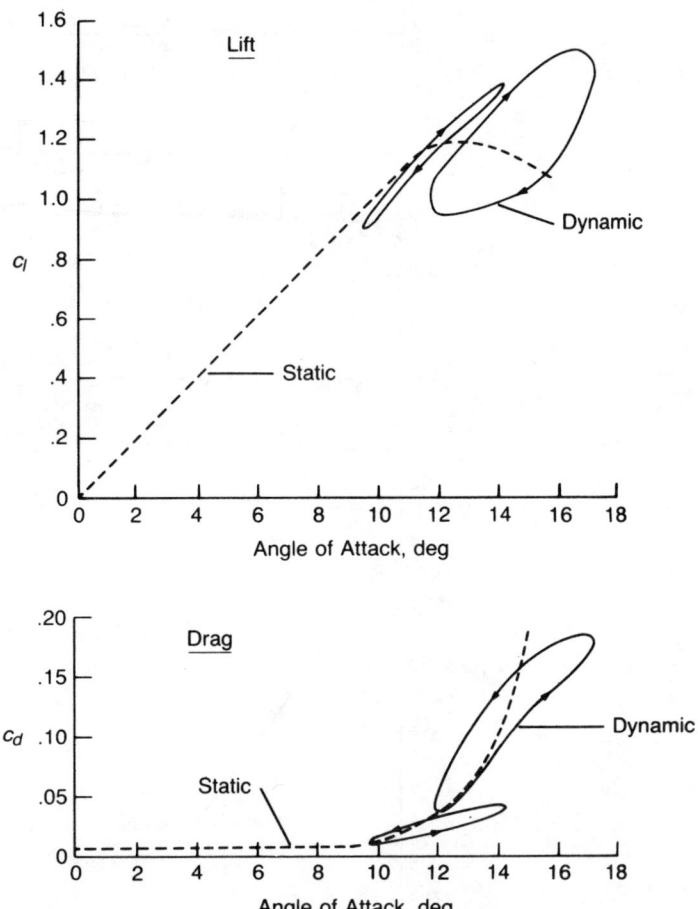

FIGURE 6.35   Measured Lift and Drag Dynamic Characteristics

Source: Philippe & Sagner, "Aerodynamic Forces Computation and Measurement on an Oscillating Aerofoil Profile with and without Stall," AGARD CP 111, 1972.

sections that had high aerodynamic pitching moments. In one case, an autogiro with airfoil sections that had nose-down pitching moments entered an unscheduled high-speed dive as the aerodynamic moments twisted the advancing blade nose down and produced effective forward cyclic pitch, even though the pilot had his stick all the way back! Experiences like this led to a period of almost exclusive use of low-moment, symmetrical airfoils. The development of stiffer blades and control systems has alleviated the problem somewhat, so that cambered airfoils with some inherent pitching moment can again be considered. It is noteworthy, however, that even symmetrical airfoils can produce substantial moments at high angles of attack and high Mach numbers. This is illustrated in Figure 6.36, from

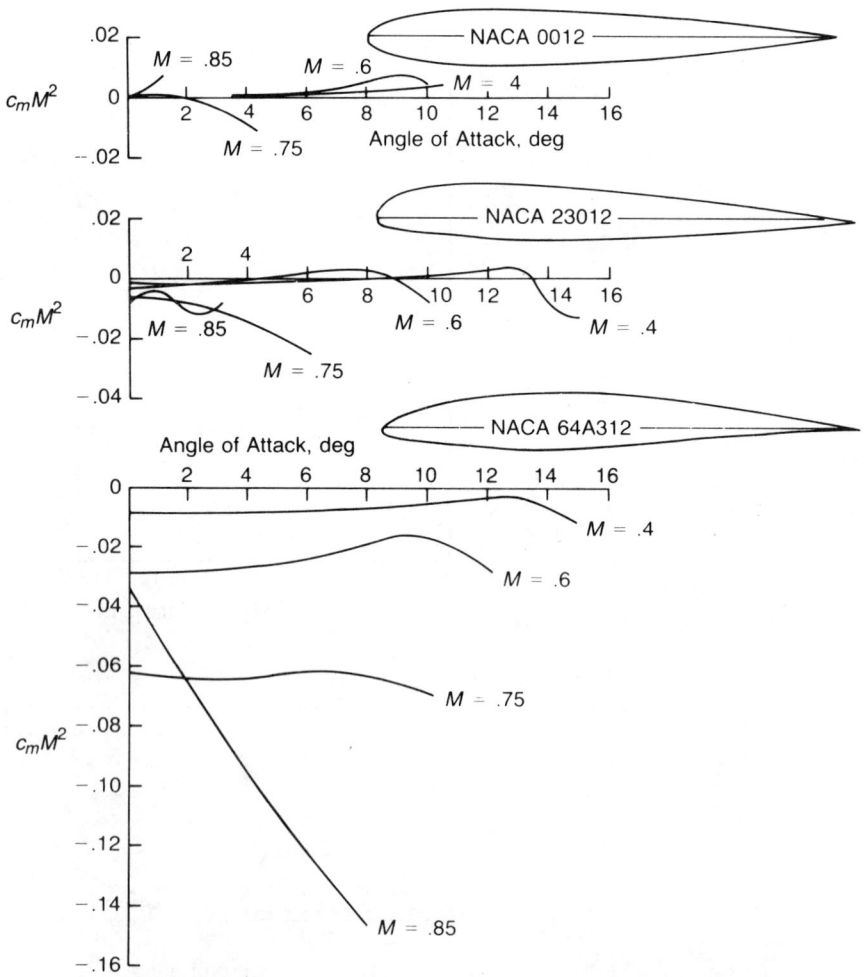

FIGURE 6.36   Comparison of Pitching Moment Functions

Source: Dadone, "Helicopter Design Datcom," Vol. I. "Airfoils," USAAMRDL CR 76-2, 1976.

reference 6.57, which shows the product of the pitching-moment coefficient and Mach number squared for the symmetrical NACA 0012 airfoil within the Mach number range encountered by rotors. (This presentation has been chosen since the product is directly proportional to the actual moment generated at the blade element.) Also shown are two cambered airfoils, one with forward camber and one with aft camber. It may be seen that forward camber has only a small effect on the pitching moment, whereas aft camber has a large effect.

The amount of aerodynamic pitching moment that can be tolerated in a given helicopter rotor depends on the structural and dynamic characteristics of the blades, hub, and control system. Thus the experience of various helicopter manufacturers has been different. Vertol has found that for their rotors, the value of c_{m_0}—the pitching moment coefficient at zero angle of attack and low Mach number—should be slightly positive (about 0.01) to maintain a satisfactory control system oscillatory load level. On the other hand, the use of a highly cambered airfoil with a c_{m_0} of about -0.07 on a Hughes tail rotor, reported in reference 6.58, produced only a small increase in oscillatory pitch link loads compared to the symmetrical airfoil it replaced, since the loads due to the aerodynamic pitching moments were out of phase with the existing pitch link loads. Tests of a model rotor using both a NACA 0012 airfoil and the same airfoil with the aft 20% deflected down 5° are reported in reference 6.59. It was found that for cases in which the retreating blade was stalled, the control loads were lower for the blade with the modified trailing edge than for the blade with the standard NACA 0012 airfoil, although below stall the opposite was true. It appears that the component of control and blade loads produced by the various modes of blade deflection can be as high as those due to aerodynamic pitching moments on the airfoil, and that the two can either add or subtract depending on the particular flight condition and the dynamic characteristics of the blade.

At the time of this writing, the accepted limit for c_{m_0} for main rotors is:

$$|c_{m_0}| < 0.02$$

It can be expected that more experience with cambered airfoils on a variety of rotors will lead to the establishment of more rational limits.

Trailing Edge Tabs

Reduction of pitching moments on a given airfoil can be achieved by using a reflex trailing edge either as an integral part of the airfoil design or as an add-on flat tab. Several wind tunnel tests of airfoils with tabs have been made. Figure 6.37 summarizes the results presented in references 6.9, 6.10, 6.21, and 6.49 in terms of the effect of tab deflection on both c_{m_0} and $c_{l_{max}}$ as a function of the ratio of tab to airfoil chord. The test results for pitching-moment sensitivity correlate well with those obtained from thin airfoil theory as given in reference 6.1. It may be seen that the larger tabs are desirable for their ability to produce a pitching moment

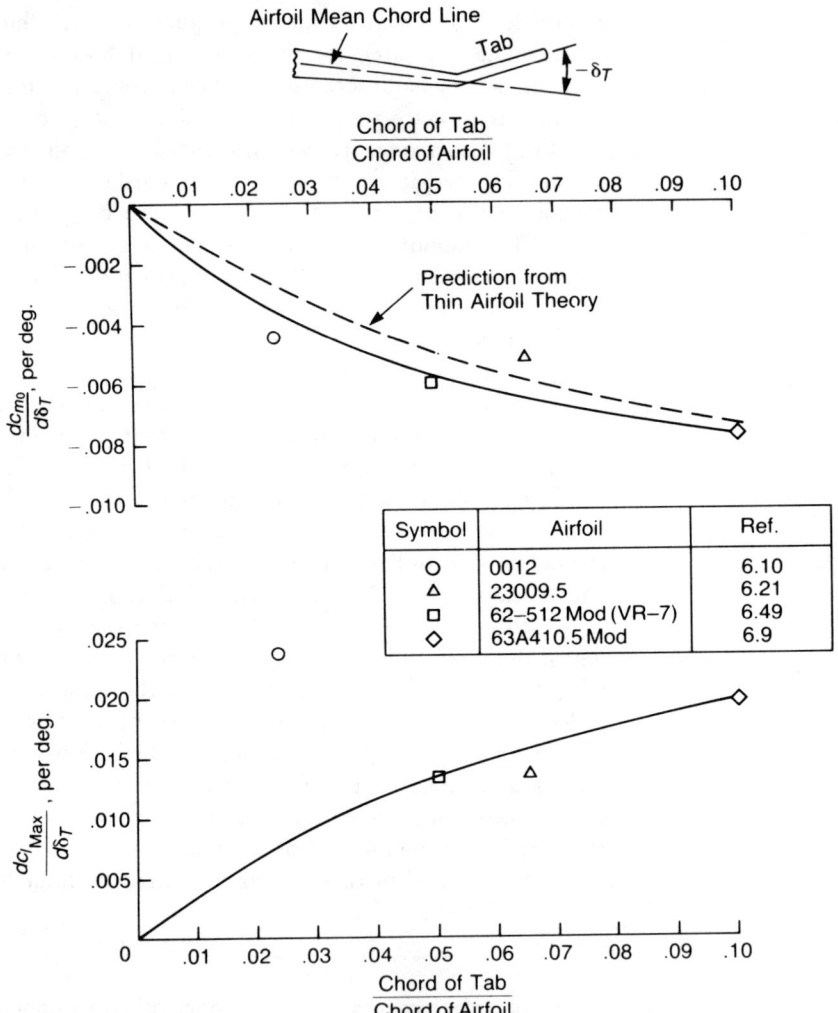

FIGURE 6.37   Effects of Trailing Edge Tabs

while minimizing the penalty to maximum lift coefficient. Drag penalties of tabs appear to be small—in most cases not much more than the accuracy of the drag measurement.

Position of the Aerodynamic Center

The position of the aerodynamic center is important in designing the blade to be free of flutter. The more aft the aerodynamic center is, the less weight will have to be installed in the nose to achieve the proper balance. It has been estimated that moving the aerodynamic center back 2% of the chord can result in a 10% saving in total blade weight. The positions of the aerodynamic center of NACA airfoils are listed in reference 6.1. For the four- or five-digit airfoils, the aerodynamic center is on or ahead of the quarter chord; for the six-series airfoils, the aerodynamic center is behind the quarter chord. For a given thickness ratio, the position of the aerodynamic center is roughly proportional to the enclosed trailing edge angle, as shown in Figure 6.38. The addition of a flat tab to the trailing edge is an effective way to decrease the trailing edge angle and to move the aerodynamic center aft. The measured effect of tabs on several airfoils is as follows:

Basic Airfoil	c_t/c	a.c. Position for Basic Airfoil, %C	Measured a.c. Position, %C	Reference
0012	.024	25	26	6.10
62-512 (VR-7) Mod	.050	27	29	6.49
23012	.050	24	25.8	6.60
23012	.100	24	26	6.60
63A410.5 Mod	.100	25	27	6.9

Pitching Moments at High Mach Numbers

When the speed at a blade element is high enough that shock waves are formed, the resulting effects can drastically change the pitching-moment characteristics.

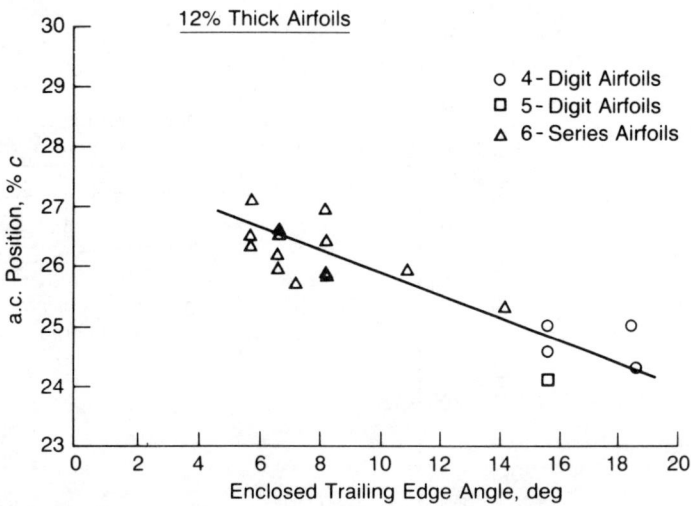

FIGURE 6.38   Effect of Trailing Edge Angle on Position of Aerodynamic Center

Source: Abbott & Von Doenhoff, *Theory of Wing Sections* (New York: Dover, 1959).

One of the most important changes is the so-called Mach tuck, a sudden nose-down pitching moment that was encountered on airplanes as their dive speeds first approached transonic conditions. When this occurs on the advancing tip of a helicopter rotor, it may produce high enough loads in the blades and control system to effectively limit the maximum allowable forward speed. For most airfoils that have been tested in two-dimensional wind tunnels, the Mach tuck characteristic comes slightly after drag divergence. This does not eliminate the Mach tuck problem, however, since drag divergence may be exceeded in many flight conditions.

Aft-cambered airfoils generally have a worse Mach tuck problem than do forward-cambered airfoils, as illustrated by the pitching-moment coefficient at zero lift shown in Figure 6.39, which is based on curves presented in reference 6.57. The basic source of the nose-down pitching moment is the change in the shape of the pressure distribution as the flow becomes supercritical and shock waves are established, first on the top surface and then on the bottom. This change is illustrated in Figure 6.40. Since the changes in pressure distribution patterns are caused by the shock waves, any reductions in shock wave strength, such as those obtained by supercritical airfoil design techniques, will be beneficial in decreasing the Mach tuck problem.

Trailing edge tabs that are used to adjust the pitching moment at low Mach numbers appear to retain most of their effectiveness at high Mach numbers, according to wind tunnel data presented in reference 6.57.

Symmetrical airfoils, of course, do not exhibit Mach tuck at zero lift, but they do produce compressibility-related pitching moments when developing some

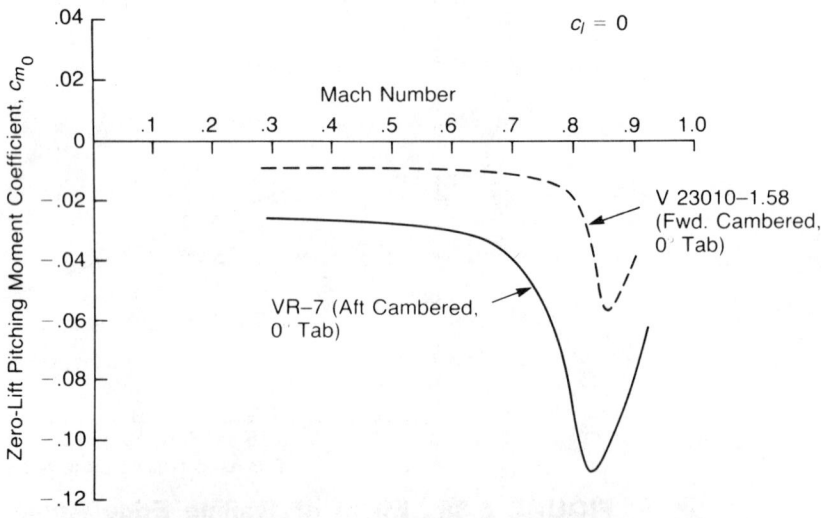

FIGURE 6.39   Mach Tuck Characteristics of Two Airfoils

Source: Dadone, "Helicopter Design Datcom," Vol. I. "Airfoils," USAAMRDL CR 76-2, 1976.

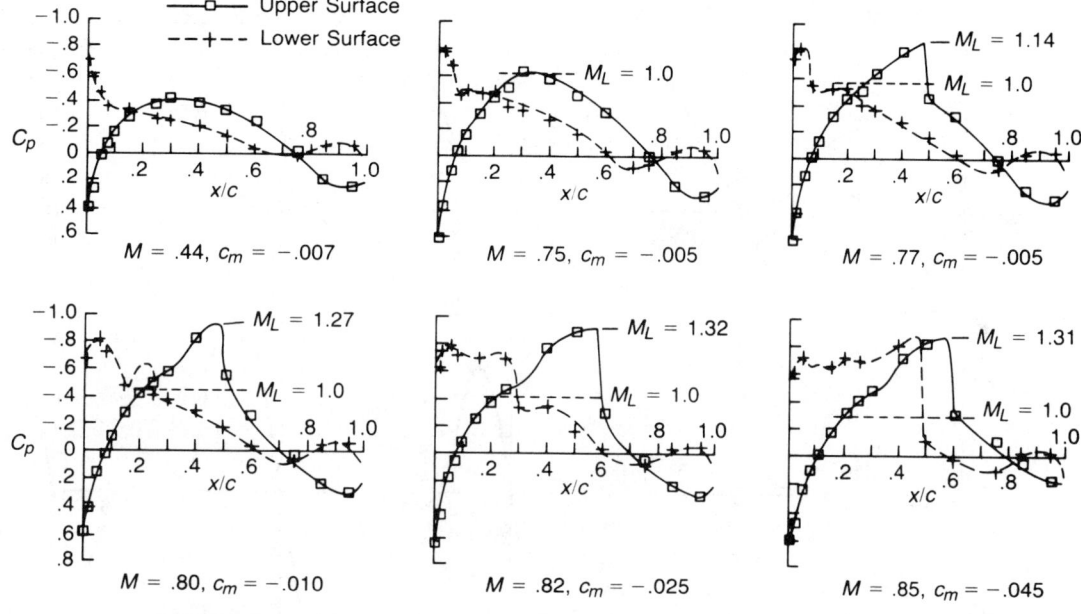

FIGURE 6.40 Pressure Distributions at Zero Lift—Modified 63A410.5 Airfoil

Source: Hughes, unpublished document.

lift. The presence and relative position of shock waves on both the upper and lower surfaces at about $M = 0.9$ causes the pitching moment to vary with angle of attack as the shock waves shift position. This produces a definite reversal in pitching-moment characteristics for small angles of attack such that the pitching moment generates an unstable blade twist—that is, an increased angle of attack twists the blade nose up and vice versa. This characteristic was identified in reference 6.61 as the cause of a significant dynamic problem involving an out-of-track phenomenon occurring every other rotor revolution on the Sikorsky NH-3A compound helicopter when the advancing tip Mach number exceeded 0.92. Figure 6.41 from reference 6.62 shows wind tunnel results for the NACA 0012 in the Mach number region from 0.80 to 0.96. It may be seen that at $M = 0.90$, even the lift curve slope exhibits a reversal. This odd behavior is not limited to symmetrical airfoils. Reference 6.57 shows that the VR-7 also has reversals in both pitching moment and lift at a Mach number of 0.82 and a c_l of about -0.3.

Dynamic Effects on Pitching Moment

As with lift and drag, the change in pitching moment due to stall is delayed if the angle of attack is rapidly increasing as the airfoil goes through its static stall angle

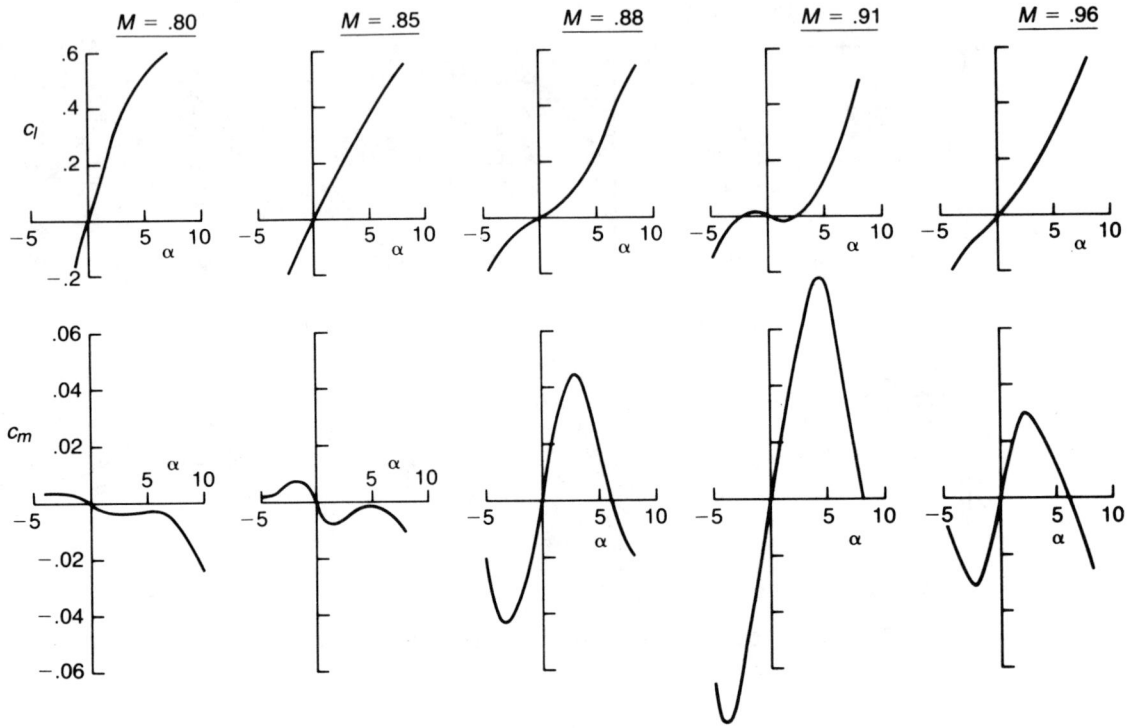

FIGURE 6.41   Transonic Characteristics of NACA 0012 Airfoil

Source: Prouty, "Aerodynamics" column, *Rotor & Wing International*, Vol. 18, no. 9 (August), 1984.

of attack. Since the dynamic pressure on the retreating blade is small, any pitching-moment characteristics associated with stall would be expected to be of little importance to the aerodynamicist in terms of their effects on performance or stability. The dynamicist, however, has recognized that the delay in the pitching-moment break while going up through stall, and the corresponding delay in returning to unstalled conditions on the downstroke, can lead to *negative damping*, which may excite the torsional vibration mode of the blade. If the direction of change in pitching moment is nose down while the angle of attack is changing nose up, or vice versa, the damping is positive. The top portion of Figure 6.42 illustrates this with a simple schematic representing an airfoil model mounted on a shaft through a moment balance and restrained by a damper. If the inertia of the model is neglected, the moment measured by the balance is just that caused by the damper, and a plot of moment versus angle of attack as the shaft is oscillated will trace out a counterclockwise hysteresis loop. Thus a plot of aerodynamic pitching moment from an oscillating airfoil test will show positive damping for regions enclosed by counterclockwise loops, but negative damping for regions enclosed by clockwise loops. The lower portion of Figure 6.42 from reference 6.63 shows the hysteresis curves for a NACA 0012 as it oscillates ±6° about mean angles of 0°,

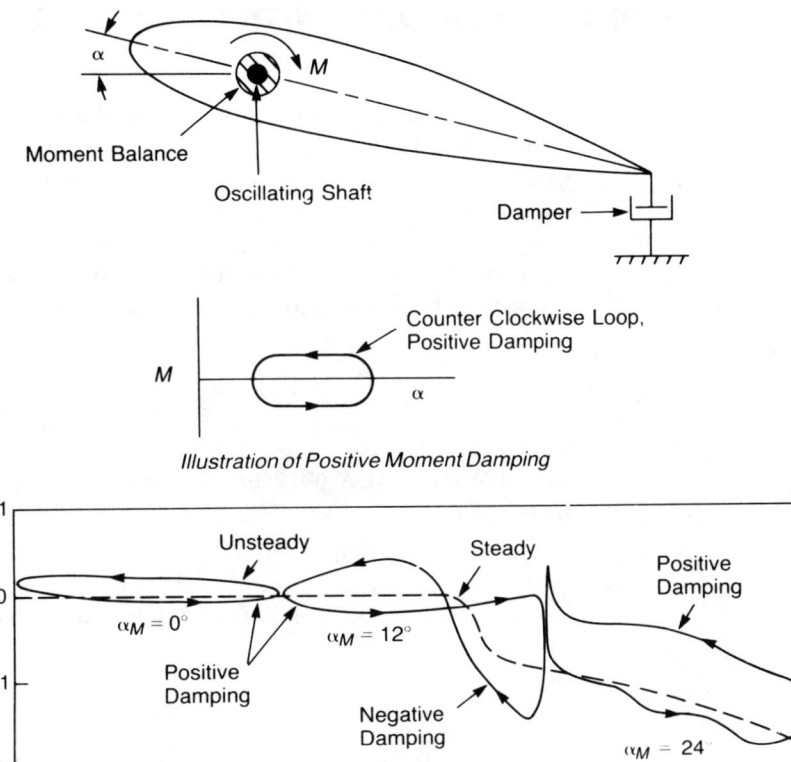

FIGURE 6.42   Pitching Moments of Oscillating Airfoil

12°, and 24°. The data show positive damping for the low and high mean angles representing nonstalled and fully stalled conditions, but show both positive and negative damping through the stall region. The negative damping region represents a condition where energy is being extracted from the airstream and put into the oscillating blade. This is the source of *stall flutter*, which, if allowed to persist, could result in serious structural problems. Fortunately, helicopters generally subject their blades to the stalling conditions only during a small portion of the retreating side. For this reason, only a few cycles of oscillation can occur before the angle of attack drops below stall and has positive damping. The phenomenon manifests itself on some helicopters in the form of high control system loads and vibration levels. Detailed discussions of these aspects of stall flutter may be found in references 6.64 and 6.65.

REPRESENTING AIRFOIL DATA WITH EQUATIONS

Airfoil lift and drag characteristics as a function of angle of attack and Mach number are often used in a computer as a bivalue table in conjunction with some table look-up scheme. There are occasions, however, when it is desirable to convert the airfoil data to equation form for use in simple hand analysis, in a small computer with limited storage capacity, or even in a large computer for quick comparisons of rotor performance using different airfoils. The simplest analytical expressions based on ignoring the effects of stall and compressibility are:

$$c_l = a\alpha$$

$$c_d = c_{d_0} + c_{d_1}\alpha + c_{d_2}\alpha^2$$

For the NACA 0012 airfoil data of reference 6.65, these equations can be written:

$$\left.\begin{array}{c} c_l = 0.1\alpha \\ c_d = 0.0081 + [-0.25\alpha + 0.12\alpha^2] \times 10^{-3} \end{array}\right\} \alpha \text{ in degrees}$$

or

$$\left.\begin{array}{c} c_l = 6\alpha \\ c_d = 0.0081 - 0.014\alpha + 0.4\alpha^2 \end{array}\right\} \alpha \text{ in radians}$$

(Note that the drag equations are unsymmetrical with respect to angle of attack and will give unrealistically low values of the drag coefficient for negative angles of attack. It is suggested that, if the negative angle of attack region is of importance in a specific analysis, the drag equations be rewritten in terms of only even powers of α. A cambered airfoil does not have zero lift at zero angle of attack. This can be accounted for by writing the equations in terms of $\alpha - \alpha_{LO}$ or by redefining the angle of attack such that it is zero at zero lift.)

A procedure for writing equations for the lift and drag coefficients for cases in which compressibility and stall cannot be ignored is given next.

For purposes of illustration, the lift and drag data synthesized from whirl tower tests of a rotor with the NACA 0012 airfoil as published in reference 6.66 will be used. These data are shown as solid lines in Figure 6.43. Both the lift and the drag coefficients of the NACA 0012 airfoil change characteristics at a Mach number of about 0.725, where compressibility effects are first evident. Because of this, two sets of equations must be written for the two separate Mach number regimes.

Lift Coefficient below M = .725
Theoretically, the slope of the lift curve should follow the Prandtl-Glauert relationship:

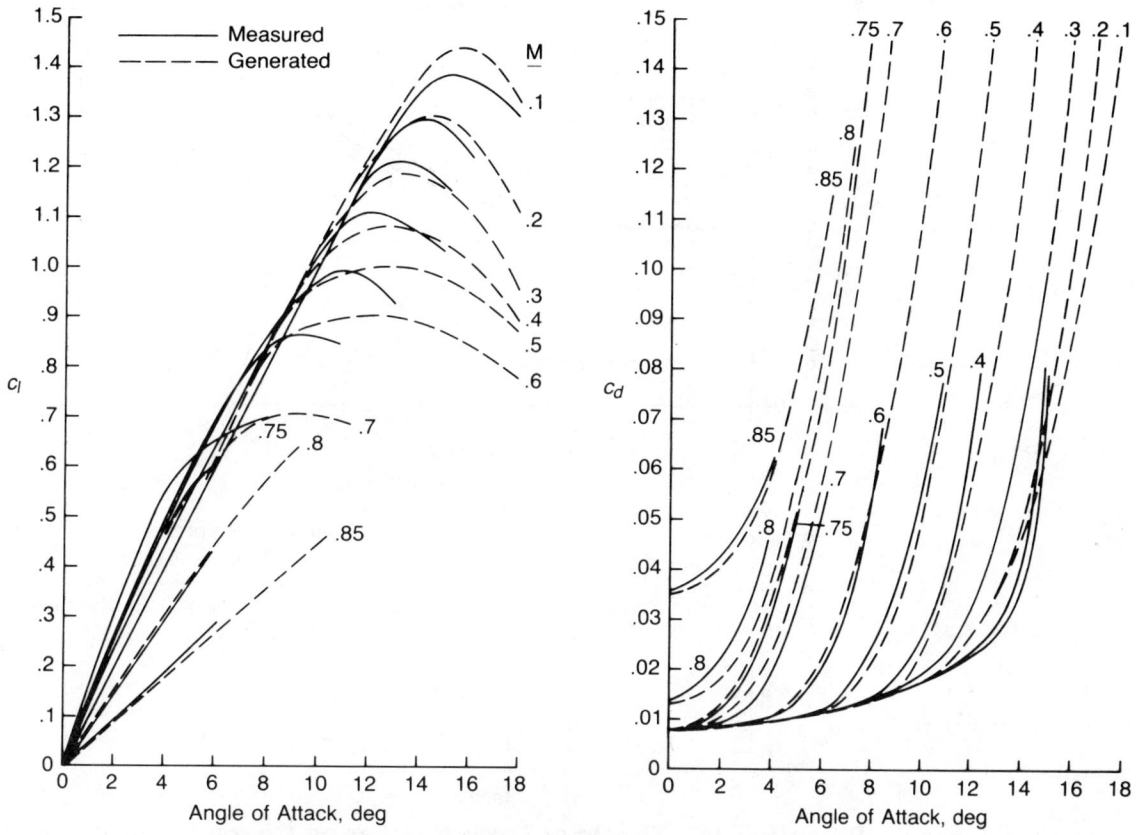

FIGURE 6.43   Comparison of Measured and Generated Characteristics of NACA 0012 Airfoil

Source: Carpenter, "Lift and Profile-Drag Characteristics of an NACA 0012 Airfoil Section as Derived from Measured Helicopter-Rotor Hovering Performance," NACA TN 4357, 1958.

$$a = \frac{a_0}{\sqrt{1 - M^2}}$$

Figure 6.44 shows, however, that the slope is somewhat lower, being better fit by the equation:

$$a = \frac{.1}{\sqrt{1 - M^2}} - .01M \text{ (per degree)}$$

In the linear region below stall, the lift coefficient is:

$$c_l = \left(\frac{0.1}{\sqrt{1 - M^2}} - 0.01M \right) \alpha$$

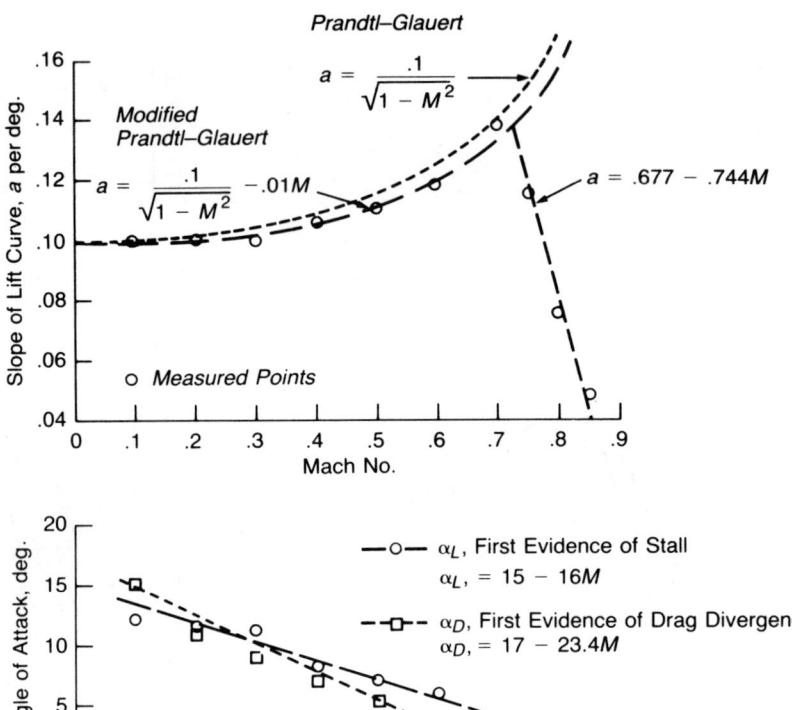

FIGURE 6.44   Effects of Mach Number on Characteristics of NACA 0012 Airfoil

The angle of attack at which the lift coefficient first shows the effects of stall will be defined as α_L and is a function of Mach number. For the NACA 0012, α_L is essentially linear with Mach number, as shown in Figure 6.44. An approximate equation is:

$$\alpha_L = 15 - 16M$$

Above α_L, the lift coefficient can be represented by:

$$c_{l_{\alpha>\alpha_L}} = a\alpha - K_1(\alpha - \alpha_L)^{K_2}$$

The exponent, K_2, at any Mach number is obtained by plotting the difference between $a\alpha$ and the measured lift coefficient versus $(\alpha - \alpha_L)$ on log-log paper. The slope of the straight line faired through the points is the value of the

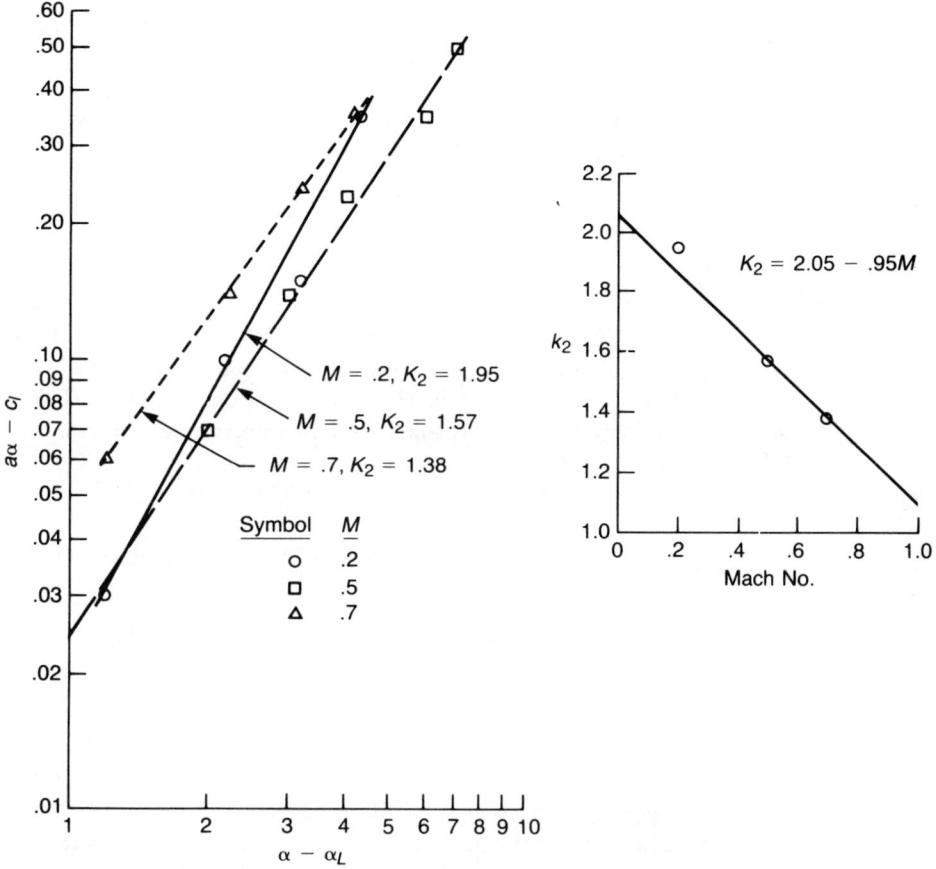

FIGURE 6.45   Evaluation of Equation for K_2

exponent to use. For the NACA 0012, this procedure, as illustrated in Figure 6.45, gives:

M	K_2
0.2	1.95
0.5	1.57
0.7	1.38

These points lie nearly on a straight line, whose equation—favoring the highest Mach number points—is:

$$K_2 = 2.05 - 0.95\,M$$

The coefficient, K_1, at any Mach number is found by evaluating the equation for c_l at the angle of attack corresponding to the maximum lift coefficient:

$$K_1 = \frac{a\alpha_{c_{l_{max}}} - c_{l_{max}}}{(\alpha c_{l_{max}} - \alpha_L)^{K_2}}$$

for the NACA 0012 airfoil:

M	K_1
0.2	.0233
0.5	.0257
0.7	.0497

When K_1 is plotted against Mach number, as in the top portion of Figure 6.46, it is seen to be a constant plus some power of M. Again using the log-log paper technique, as in the lower portion of Figure 6.46, the exponent can be evaluated and an approximate equation can be generated:

$$K_1 = 0.0233 + 0.342\, M^{7.15}$$

Thus the lift coefficient below 0.725 Mach number and above stall is:

$$c_{l_M} < 0.725,\ \alpha > \alpha_L = \left(\frac{0.1}{\sqrt{1 - M^2}} - 0.01M\right)\alpha$$

$$- (0.0233 + 0.342M^{7.15})(\alpha - 15 + 16M)^{(2.05 - 0.95M)}$$

Lift Coefficient above 0.725 Mach Number
The slope of the lift curve of the NACA 0012 breaks at 0.725 Mach number. The slope above this value is shown in Figure 6.44 to be nearly a straight line, represented by:

$$a = 0.677 - 0.744\, M$$

In order to satisfy the experimental lift characteristics by equations, the following constants were evaluated by the same methods used at the lower Mach numbers:

$$\alpha_L = 3.4$$
$$K_1 = 0.575 - 0.144(M - 0.725)^{0.44}$$
$$K_2 = 2.05 - 0.95\, M$$

Thus:

$$c_{l_{M>.725,\alpha>3.4}} = (0.677 - 0.744M)\alpha$$
$$- [0.0575 - 0.144(M - 0.725)^{0.44}]\,[\alpha - 3.4]^{(2.05 - 0.95M)}$$

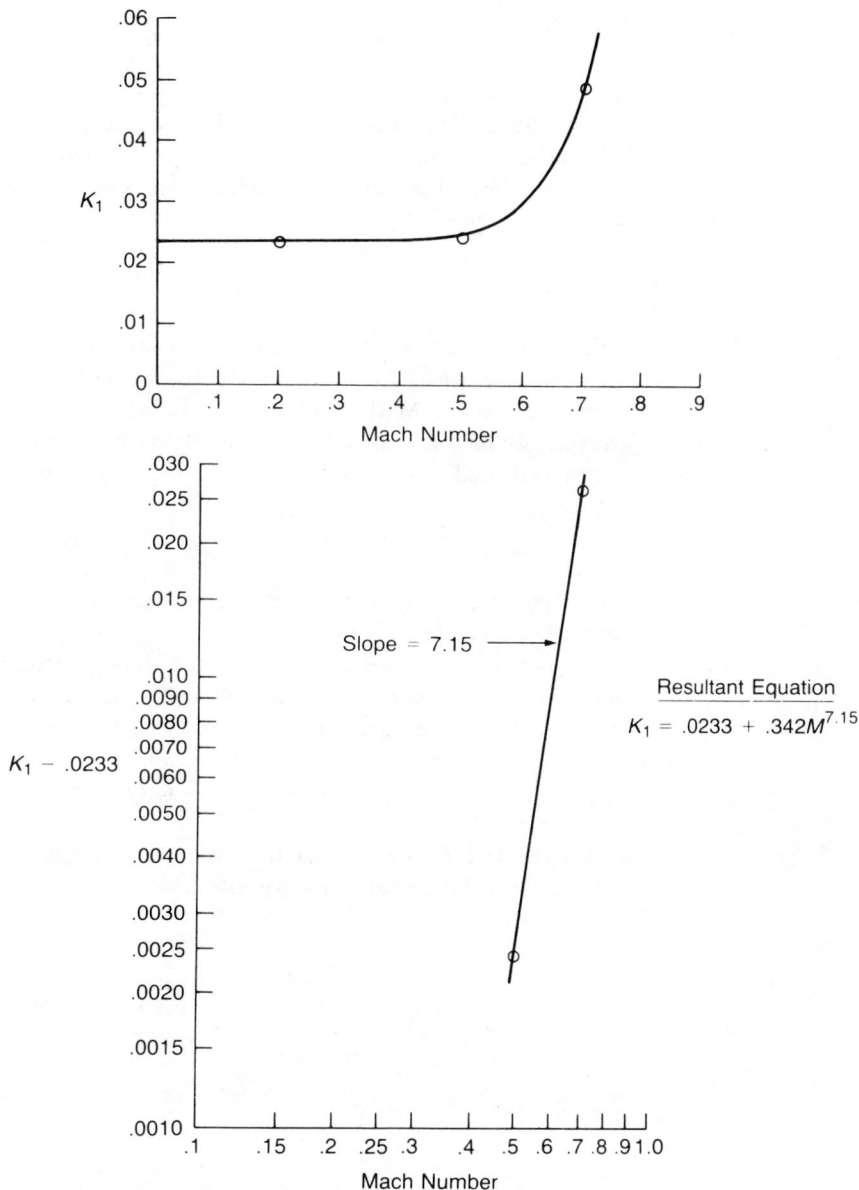

FIGURE 6.46   Evaluation of Equation for K_1

Figure 6.43 shows the correlation of measured lift coefficient and the generated lift coefficient.

Drag Coefficient below Drag Divergence
At the lowest test Mach number, the incompressible drag coefficient can be represented by a power series of the form:

$$c_{d_{\text{incomp}}} = c_{d_0} + c_{d_1} + c_{d_2}\alpha^2 + \ldots + c_{d_n}\alpha^n$$

The coefficients can be evaluated by selecting n control points and solving a set of n simultaneous equations. Before selecting the control points, the angle, α_D, at which the individual drag curves break away from the incompressible curve should be established. For the 0012 airfoil, this angle, as shown in Figure 6.44, is approximately:

$$\alpha_D = 17 - 23.4\, M$$

(Note that the line represented by this equation goes through the drag divergence Mach number of 0.725 at $\alpha = 0$.) The control points should include $\alpha = 0$ and α_D for the lowest test Mach number and $n - 2$ other points in between. For the 0012 a satisfactory fit was obtained with the 0.1 Mach number test data using a five-term series evaluated at 0, 2, 6, 10, and 14.7 degrees. This gives (with α in degrees):

$$c_{d_{\text{incomp}}} = 0.0081 + (-350\alpha + 396\alpha^2 - 63.3\alpha^3 + 3.66\alpha^4) \times 10^{-6}$$

(Warning: too many terms in the series may introduce large fluctuations of the curve between the control points).

For Mach numbers above 0.1, the drag coefficient breaks away from the incompressible value as α exceeds α_D. The curves have the characteristics represented by the equation:

$$c_d = c_{d_{\text{incomp}}} + K_3(\alpha - \alpha_D)^{K_4}$$

where K_3 and K_4 are evaluated in the same manner in which K_1 and K_2 were evaluated in the equation for the lift coefficient. For the NACA 0012 airfoil:

M	K_3	K_4
0.3	0.00071	2.60
0.5	0.00063	2.48
0.7	0.00064	2.57

Using average values gives:

$$K_3 = 0.00066$$
$$K_4 = 2.54$$

Thus:

$$c_{d_{M<.725,\alpha>\alpha_D}} = c_{d_{\text{incomp}}} + 0.00066\,(\alpha - 17 + 23.4M)^{2.54}$$

Drag Coefficient above Drag Divergence
For Mach numbers above drag divergence, another term must be added to account for the drag increment at zero angle of attack. The equation becomes:

$$c_d = c_{d_{\text{incomp}}} + K_3(\alpha - \alpha_D)^{K_4} + K_5(M - M_{DD})^{K_6}$$

For the NACA 0012, the coefficients and exponents that give the best fit are:

$$\alpha_D = 0$$
$$K_3 = 0.00035$$
$$K_4 = 2.54$$
$$K_5 = 21$$
$$K_6 = 3.2$$

Thus:

$$c_{d_{M>.725}} = c_{d_{\text{incomp}}} + 0.00035\alpha^{2.54} + 21(M - 0.725)^{3.2}$$

The correlation of the drag coefficient generated by this procedure with the test data is shown in Figure 6.43.

Equations Suited for Forward Flight Analysis

The foregoing procedure is adequate for hover performance methods and has been used to prepare the hover charts at the end of Chapter 1. For forward flight, however, the equations should be modified somewhat. Since negative angles of attack are possible on the advancing tip, the equation for the incompressible drag coefficient should be written in terms of even powers of α so that it is symmetrical about $\alpha = 0$. For the NACA 0012, the four-term series on α in degrees has been found to give satisfactory representation:

$$c_{d_{\text{incomp}}} = 0.0081 + (65.8\alpha^2 - 0.226\alpha^4 + 0.0046\alpha^6) \times 10^{-6}$$

A second modification accounts for the fact that some of the inboard elements on the retreating side will be subjected to angles of attack up to 360°, although at low Mach numbers. Figure 6.47 shows measured NACA 0012 lift and drag coefficients from 0° to 180° taken from reference 6.15. Equations for the lift coefficient can be written by dividing the angle of attack range up into segments:

Segment	Lift Coefficient
$\alpha < 20°$	c_l = generated coefficient
$20° < \alpha < 161°$	$c_l = 1.15 \sin 2\alpha$
$161° < \alpha < 173°$	$c_l = -0.7$
$173° < \alpha < 187°$	$c_l = 0.1(\alpha - 180°)$
$187° < \alpha < 201°$	$c_l = 0.7$
$201° < \alpha < 340°$	$c_l = 1.15 \sin 2\alpha$
$340° < \alpha < 360°$	c_l = generated coefficient

Similarly, the drag coefficient is represented by:

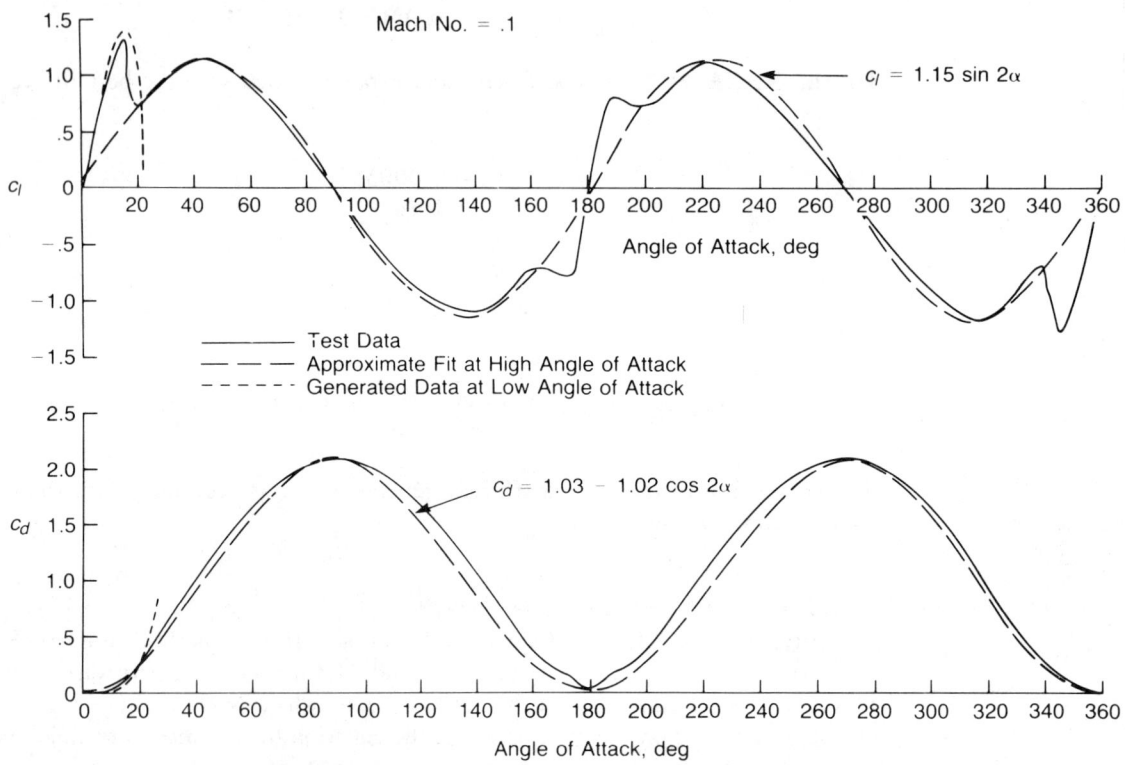

FIGURE 6.47   Lift and Drag Characteristics of NACA 0012 Airfoil through 360 Degrees

Segment	Drag Coefficient
$\alpha < 20°$	$c_d = $ generated coefficient
$20° < \alpha < 340°$	$c_d = 1.03 - 1.02 \cos 2\alpha$
$340° < \alpha < 360°$	$c_d = $ generated coefficient

As a further illustration, the results of applying the method to a cambered airfoil are presented in Figures 6.48, 6.49, and 6.50 for lift, drag, and pitching-moment coefficients. The airfoil was a modified 63A410.5 tested in the Lockheed transonic wind tunnel. (The aerodynamic center of this airfoil is approximately at 27% chord, so moments are referenced to this point.)

SOURCES OF TWO-DIMENSIONAL AIRFOIL DATA

Airfoil data in the region of angles of attack, Mach numbers, and Reynolds numbers of interest to the rotor aerodynamicist have been obtained from both

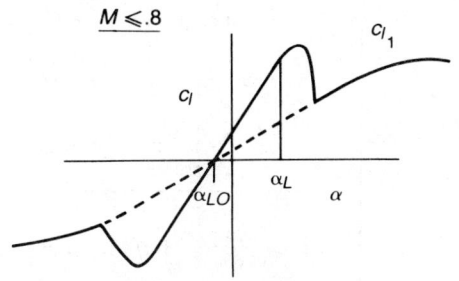

$\underline{M \leqslant .8}$

for $|\alpha| < \alpha_L$ (below Stall)

$c_l = a(\alpha - \alpha_{LO})$

for $|\alpha| > \alpha_2$ (above Stall)

$c_l = a(\alpha - \alpha_{LO}) - K_1 |\alpha - \alpha_L|^{k_2}$ (sign α)

$a = \dfrac{.1}{\sqrt{1 - M^2}} - .025M$

$\alpha_{LO} = -1.2 + 1.125M - \dfrac{.009}{.81 - M}$

$\alpha_L = 11(1 - 2M^4)$

$K_1 = .13(1 - M)$

$K_2 = 1.3 + .5M^4$

$c_{l_1} = 1.15(1 + M)^{.6} \sin 2(\alpha - \alpha_{LO})$

if $|c_l| < |c_{l_1}|$ then $c_l = c_{l_1}$

$\underline{M > .8}$

for $\alpha < 2°$

$c_l = a(\alpha - \alpha_{LO})$

for $\alpha > 2°$

$c_l = a(\alpha - \alpha_{LO} - K_1(|\alpha| - 2)^{K_2} \left[\text{sign}(\alpha - 2) \right]$

$a = .1466$

$\alpha_{LO} = -1.2 + 88.9(M - .8)^2$

$K_1 = .026 + .67(M - .8) - 4.93(M - .8)^2$

$K_2 = 1.505 - 6.18(M - .8) + 41.7(M - .8)^2$

if $K_1 < 0$ then $K_1 = 0$

if $|\alpha| > 20$ then $c_l = c_{l_1}$

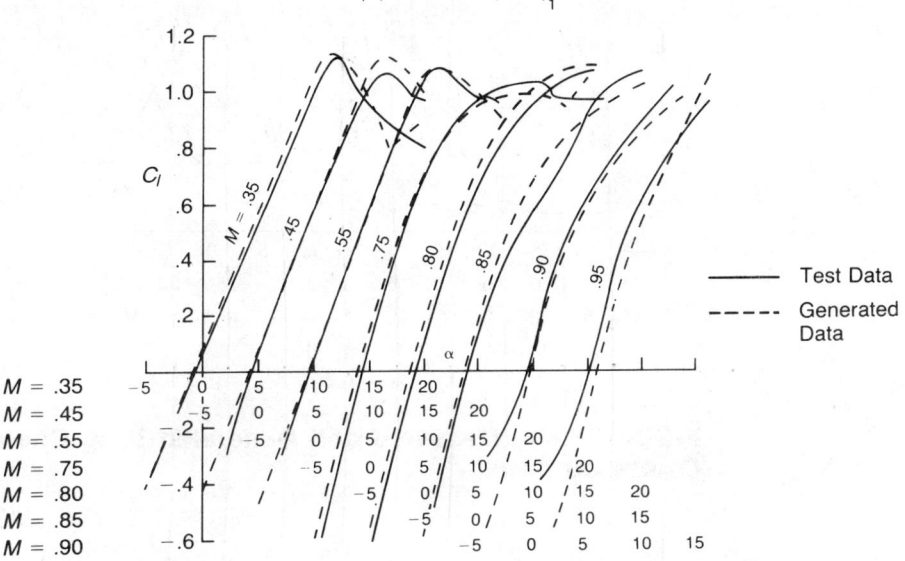

FIGURE 6.48 Measured and Generated Lift Characteristics of a Cambered Airfoil

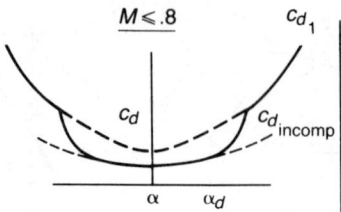

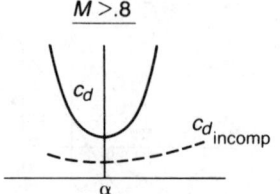

for $|\alpha| < |\alpha_d|$ (below comp. effects)

$$c_d = c_{d_{incomp}}$$

for $|\alpha| > |\alpha_d|$ (above comp. effects)

$$c_d = c_{d_{incomp}} + K_3 |\alpha - \alpha_d|^{2.58}$$

$$\alpha_d = 14 - 30\,M^3$$

if $\alpha_d < -1.2$ then $\alpha_d = -1.2$

$$K_3 = .01 - .021\,M^4$$

$$c_{d_{incomp}} = .0081 + (65.8\alpha^2 - .226\alpha^4 + .0046\alpha^6)/1{,}000{,}000$$

$$c_{d_1} = 1.05 + .05M - 1.04 \cos 2\alpha$$

if $c_d > c_{d_1}$ then $c_d = c_{d_1}$

$$c_d = c_{d_{incomp}} + .19(M - .8)^{.73} + K_3|\alpha - \alpha_{LO}|^{2.58}$$

$$\alpha_{LO} = -1.2 + 88.9(M - .8)^2$$

$$K_3 = .0004 + .09(M - .8)^3$$

if $|\alpha| > 20$ then $c_d = c_{d_1}$

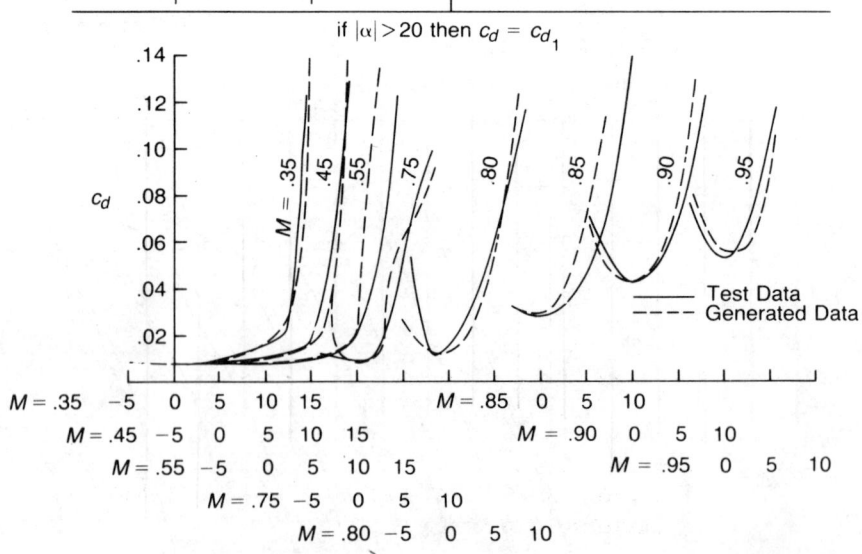

FIGURE 6.49   Measured and Generated Drag Characteristics of a Cambered Airfoil

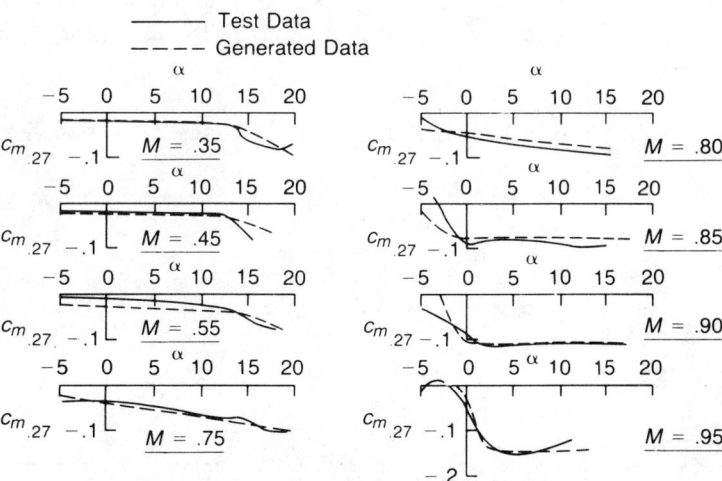

$M \leqslant .8$

$c_m = c_{m_0} + K_7 \alpha$

$c_{m_0} = -.02 - .04M^3$

$K_7 = -.0002 - .01M^4$

$c_{m_1} = -.55 \sin(\alpha - 10)$

if $c_m > c_{m_1}$ then $c_m = c_{m_1}$

$M > .8$

for $\alpha > \alpha_m$

$c_m = c_{m_c} + K_{13} \alpha$

for $\alpha < \alpha_m$

$c_m = c_{m_c} + K_{13}\alpha + K_{14}(\alpha_m - \alpha)^2$

if $c_m > 0$ then $c_{m_1} = 0$

$\alpha_m = -4 + 53.3(M - .8)$

$c_{m_c} = .49 - .663M$

$K_{13} = \dfrac{-.0043}{10^{100(M - .8)}}$

$K_{14} = .025(m - .67)$

if $|\alpha| > 20$ then $c_m = c_{m_1}$

——— Test Data
– – – Generated Data

FIGURE 6.50   Measured and Generated Moment Characteristics of a Cambered Airfoil

whirl tower and two-dimensional wind tunnel tests. The whirl tower method has the advantage that three-dimensional effects are accounted for, but these tests are relatively expensive. Two-dimensional wind tunnel tests are relatively inexpensive but do not include any three-dimensional effects. The airfoil data that are readily available at this time and are considered suitable for rotor analysis are summarized in Table 6.1. Much of the data is plotted in a consistent manner in reference 6.57.

TABLE 6.1
Sources of Two-Dimensional Airfoil Data

Airfoil	Angle of Attack Range	Mach No. Range	Effective Chord, In.	Test Apparatus	Date	Ref
0006	−10 to 19	.2 to .90	7	Boeing Tunnel	1965	6.57
0012	0 to 16	.1 to .85	16	Langley Whirl Tower	1958	6.66
0012	0 to 16	.3 to .80	16	UAC Wind Tunnel	1960	6.10
0012	0 to 15	.2 to .7	1.5	UARL 4' Rotor Rig	1972	6.67
0012	0 to 28	.1 to .95	2.7	UARL 9' Rotor Rig	1961	.6.68
0012	0 to 20	.2 to .85	21	Bell Rotor in 40 × 80 Tunnel	1965	6.69
0012	0 to 16	.3 to 1.08	16	NSRDC Tunnel	1977	6.24
0012	0 to 23	.2 to .6	24	Boeing Tunnel	1968	6.23
0012	−4 to 18	.35 to .9	3, 4, 14	Langley 6 × 28 Tunnel	1980	6.70
0012	0 to 12	.3 to .85	10	NPL Tunnel	1968	6.22,6.57
0012	−3 to 11	.35 to .89	20	Langley 6 × 28 Tunnel	1977	6.71
0015	0 to 17	.1 to .78	11	Langley Whirl Tower	1958	6.72
0015	0 to 16	.3 to .85	6	Ames Tunnel	1945	6.17,6.57
4415	0 to 15	.3 to .85	6	Ames Tunnel	1945	6.17,6.57
V13006-0.7	−10 to 20	.2 to .9	7	Boeing Tunnel	1965	6.57
V(1.9)3009-1.25	−2 to 16	.2 to .9	6	Boeing Tunnel	1965	6.57
SA 13109-1.58	0 to 12	.3 to .89	8	Onera Tunnel	1965	6.57
23010-1.58	0 to 24	.2 to .6	24	Boeing Tunnel	1968	6.23,6.57
23010-1.58, Refx	0 to 24	.4	24	Boeing Tunnel	1968	6.23
23012	−4 to 15	.4 to .85	5	ARA Tunnel	1971	6.57
23012	−2 to 11	.35 to .90	4	Langley 6 × 19 Tunnel	1971	6.71
23112	0 to 15	.2 to .7	1.5	UARL 4' Rotor Rig	1972	6.67
23015	−10 to 20	.3 to .85	6	Ames Tunnel	1945	6.17,6.57
V43012-1.58	−10 to 20	.2 to .7	7	Boeing Tunnel	1971	6.57
63A009, 12, 15, 18	−5 to 29	.26 to .94	8 to 24	Lockheed Tunnel	1965	6.46
63-015	0 to 14	.3 to .75	13	Langley Whirl Tower	1956	6.73
63-206, 8, 10, 12						
64-206, 8, 10, 12	−6 to 12	.3 to .9	6	Ames Tunnel	1952	6.47
65-206, 8, 10, 12						
66-206, 8, 10, 12						
64A(4.5)08	−4 to 16	.4 to .96	6	Boeing Tunnel	1969	6.57
64A608	−6 to 12	.4 to .96	6	Boeing Tunnel	1969	6.57
64A312	−2 to 15	.4 to .90	6	Boeing Tunnel	1969	6.57

438

Airfoil						
64A(4.5)12	−4 to 14	.4 to .96	6	Boeing Tunnel	1969	6.57
64A612	−6 to 13	.4 to .90	6	Boeing Tunnel	1969	6.57
64A516	−4 to 14	.4 to .80	6	Boeing Tunnel	1969	6.57
64-006, 8, 10, 12	−2 to 31	.3 to .9	6 and 11	Langley Tunnel	1953	6.48
64A006-406 }	−2 to 28	.3 to .92	6	Ames Tunnel	1954	6.13
64A010, 410 }	−4 to 16	.3 to .9	16	NSRDL Tunnel	1977	6.24
65-213						
65-006 }						
64-009 }						
64-210	−2 to 13	.1 to .47	12 to 36	Langley Tunnel (Lift Only)	1952	6.12
64-215 }						
64-A010 (Mod)	−2 to 14	.3 to .9	6	Ames Tunnel	1956	6.8
65-215 }	0 to 16	.3 to .85	6	Ames Tunnel	1945	6.17
66-215 }						
FX69-H-098	−4 to 16	.3 to .78	18	UAC Tunnel	1973	6.53, 6.57
FX69-H-098	−4 to 13	.35 to .90	20	Langley 6 × 28 Tunnel	1977	6.71
NPL 9615	−2 to 13	.3 to .85	10	NPL Tunnel	1968	6.22
NPL 9626 }	0 to 13	.3 to .75	10	NPL Tunnel	1969	6.74
NPL 9627 }						
NPL 9660	−2 to 13	.3 to .85	10	NPL Tunnel	1973	6.57
NACA-CAMBRE	−2 to 16	.3 to .9	8	Onera Tunnel	—	6.57
VR-7	−10 to 20	.3 to .92	29	Boeing Tunnel	1971	6.57
VR-7.1	−10 to 20	.2 to .71	29	Boeing Tunnel	1971	6.57
VR-8	−10 to 20	.2 to .95	29	Boeing Tunnel	1971	6.57
NLR-1(7223-62)	−2 to 11	.35 to .85	4	Langley 6 × 19 Tunnel	1977	6.71
NLR-1	−10 to 20	.2 to .9	25	Boeing Tunnel	1977	6.75
SC-1095	−4 to 8	.3 to 1.1	16	NSRDL Tunnel	1977	6.24
SC-1095-R8	−4 to 18	.35 to .9	3, 4, 9, 15	Langley 6 × 28 Tunnel	1980	6.70
DBLN-518	−5 to 17	.3 to .8	16	NSRDL Tunnel	1977	6.24
BHC-540	−3 to 12	.35 to .89	20	Langley 6 × 28 Tunnel	1977	6.71
SC-1095	−4 to 18	.35 to .88	3, 4, 14	Langley 6 × 28 Tunnel	1980	6.70
RC-10(N)-1	−4 to 13	.33 to .87	15, 25	Langley 6 × 28 Tunnel	1981	6.76
RL(1)-10	−4 to 14	.35 to .89	24	Langley 6 × 28 Tunnel	1981	6.77
RC(1)-10MOD1	"	"	"	"	"	6.77
RC(1)-10MOD2	"	"	"	"	"	6.77
RC(3)-08, 10, 12	"	"	"	"	1982	6.78
A-1	−2 to 14	.2 to .84		Ames 2 × 2 Tunnel	1980	6.79

The existence of several sets of published data for the NACA 0012 airfoil presents an opportunity for a comparison of test variables. Figure 6.51 shows the significant lift and drag characteristics of four sets of data as a function of angle of attack and Mach number. The comparison shows that the maximum lift coefficient and the drag coefficient are both strongly influenced by the chord of the test airfoil. This is a Reynolds number effect, which has been demonstrated in other tests.

Line	Source	Eff. Chord	RN at 600 ft/sec	Ref.
————	Whirl Tower	16″	5.05×10^6	6.66
—·—·—	Model Whirl Tower	1.5″	$.47 \times 10^6$	6.67
— — —	2-D Tunnel	9″	2.84×10^6	6.66
– – –	2-D Tunnel	20″	6.30×10^6	6.80

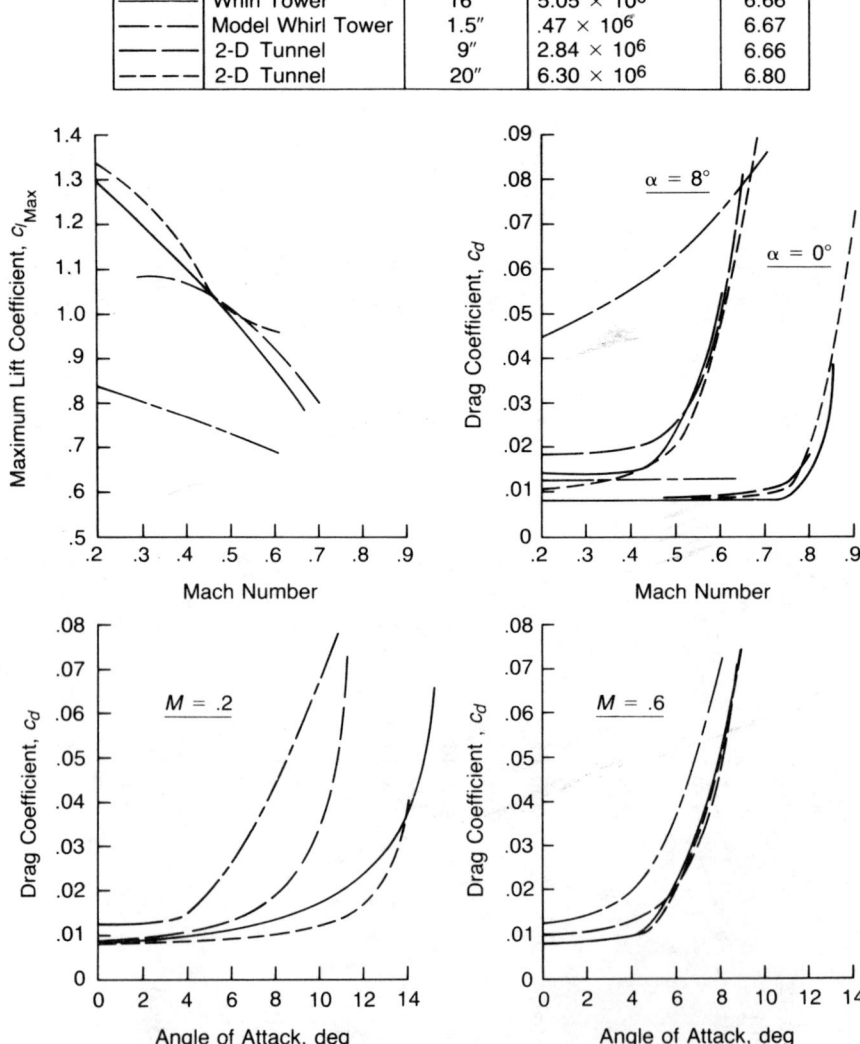

FIGURE 6.51   Comparison of NACA 0012 Airfoil Data

EXAMPLE HELICOPTER CALCULATIONS

HOW TO'S

CHAPTER 7

Rotor Flapping Characteristics

QUALITATIVE DISCUSSION OF FLAPPING

Before deriving the helicopter equations of equilibrium and motion, it seems well to develop a general understanding of rotor flapping, a primary factor in helicopter stability and control analysis.

As might be expected, the stability and control characteristics of a helicopter are different from those of an airplane primarily as a rotor is different from a wing. Whereas the wing remains more or less rigidly attached to the airplane airframe, the rotor tip path plane tilts easily with respect to the helicopter airframe in response to changing flight conditions and control inputs. This tilt produces changes in forces and moments at the top of the rotor shaft. A hingeless blade may have no single point at which flapping occurs, but an *effective hinge offset* can be determined that will give it the same stability and control characteristics as a blade with an actual mechanical hinge at that point. This concept will be used to eliminate any consideration of blade structural stiffness in the following analyses.

Several different flapping characteristics will be discussed in a cause-and-effect manner and then mathematically derived from the laws of physics. These characteristics are summarized in Table 7.1 for rotors with and without hinge

TABLE 7.1
Rotor Flapping Characteristics (for Counterclockwise Rotation)

Change in Condition	With No Hinge Offset	With Hinge Offset (Or Hingeless Rotor)
	Result	
Shaft tilted in vacuum.	Tip path plane remains in original position.	Tip path plane aligns itself perpendicular to shaft.
Shaft tilted in air in hover.	Tip path plane aligns itself perpendicular to shaft.	Tip path plane aligns itself perpendicular to shaft.
Longitudinal pitch (B_1) increased in hover.	Tip path plane tilts down in front by exactly B_1.	Tip path plane tilts down in front approximately by B_1 and tilts slightly down to right.
Lateral cyclic pitch (A_1) increased in hover.	Tip path plane tilts down to right by exactly A_1.	Tip path plane tilts down to right approximately by A_1 and tilts slightly up in front.
Forward speed increased.	Tip path plane tilts back.	Tip path plane tilts back and slightly down to left.
Shaft tilted back in forward flight.	Tip path plane tilts back further than change in shaft tilt.	Tip path plane tilts back further than change in shaft tilt and slightly down to left.
Collective pitch increased in forward flight.	Tip path plane tilts back.	Tip path plane tilts back and slightly down to left.
Coning increased in forward flight.	Tip path plane tilts down to right.	Tip path plane tilts down to right and slightly down in back.
Longitudinal pitch increased in forward flight.	Tip path plane tilts down in front slightly more than B_1.	Tip path plane tilts down in front approximately by B_1 and tilts slightly down to right.
Lateral pitch increased in forward flight.	Tip path plane tilts down to right by exactly A_1.	Tip path plane tilts down to right approximately by A_1 and tilts slightly up in front.
Steady nose-up pitch rate.	Longitudinal tilt of tip path plane lags shaft, tilts down to left.	Longitudinal tilt of tip path plane lags shaft, lateral tilt may be either right or left.
Steady right roll rate.	Lateral tilt of tip path plane lags shaft, tilts up in front.	Lateral tilt of tip path plane lags shaft, longitudinal tilt may be either up or down.
Sideslip to right.	Tip path plane tilts down to left and down in front.	Tip path plane tilts down to left and down in front.
Sideslip to left.	Tip path plane tilts down to right and down in front.	Tip path plane tilts down to right and down in front.

444

offset. (There are few modern rotors with individual blades hinged with no offset but two-bladed teetering rotors fall into this category for the purposes of this discussion.)

All these flapping characteristics can be explained on the basis that at the flapping hinge (or at an effective flapping hinge in the case of a hingeless rotor) the summation of moments produced by aerodynamic, centrifugal, weight, inertial, and gyroscopic forces must be zero.

Shaft Tilted in a Vacuum

If a rotor with no hinge offset is operating in a vacuum, there are no aerodynamic forces; only centrifugal forces acting in the plane of rotation. These can produce no moments about the flapping hinges. If the shaft is tilted, no changes in moments will be produced and the rotor disc will remain in its original position, as shown in Figure 7.1. If, on the other hand, the rotor has hinge offset, the centrifugal forces acting in the plane of rotation will produce moments about the hinges that will force the blades to align themselves perpendicular to the shaft.

Shaft Tilted in Air in Hover

If the shaft is tilted while the rotor is hovering in air, aerodynamic forces will be generated that will force the tip path plane to align itself perpendicular to the shaft whether the rotor has hinge offset or not. The sequence of steps leading to this is shown in Figure 7.2.

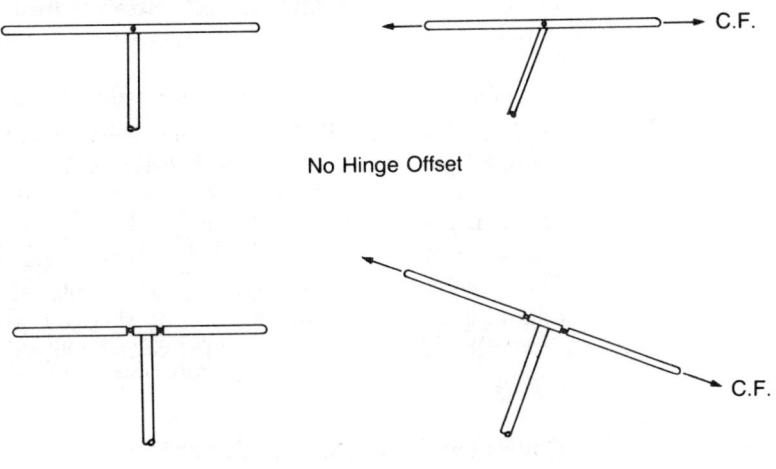

No Hinge Offset

With Hinge Offset

FIGURE 7.1   Effects of Shaft Tilt in Vacuum

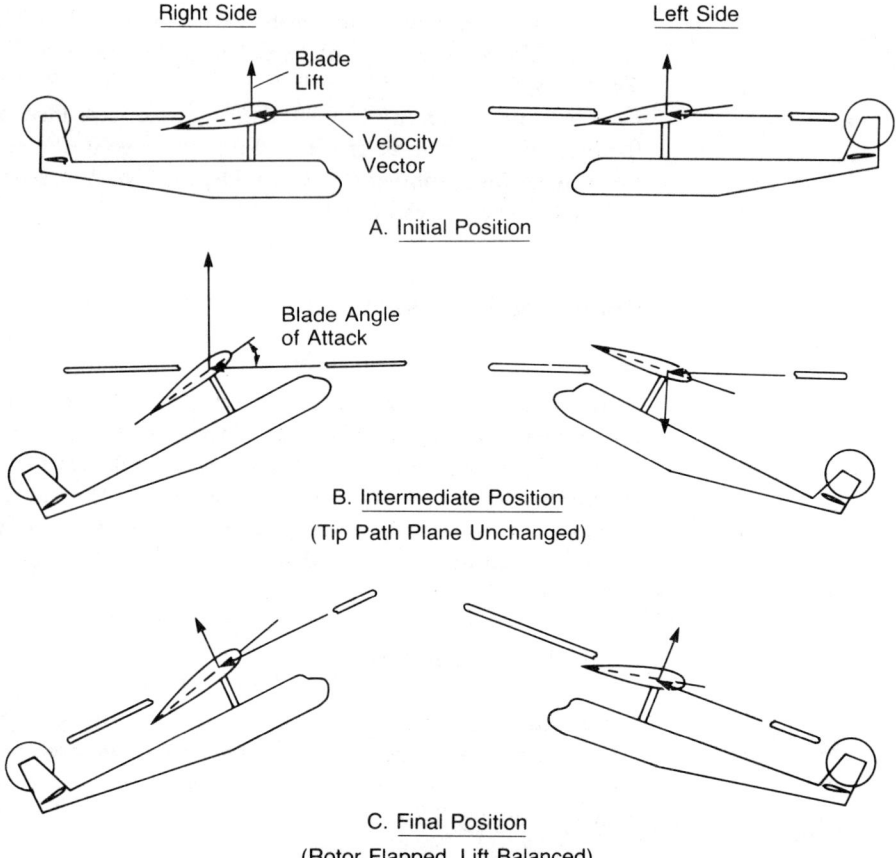

Right Side   Left Side

Blade Lift

Velocity Vector

A. Initial Position

Blade Angle of Attack

B. Intermediate Position
(Tip Path Plane Unchanged)

C. Final Position
(Rotor Flapped, Lift Balanced)

FIGURE 7.2 Shaft Tilted in Hover

First, there is the tilt of the shaft alone as the rotor disc acts as a gyroscope and remains in its original plane. Since the blade feathering is referenced to the shaft, however, the angle of attack of the right-hand blade is increased and that of the left-hand blade decreased by the same amount. This causes the rotor to flap until it is perpendicular to the shaft, where it will again be in equilibrium with a constant angle of attack around the azimuth and the moments will be balanced. This alignment is very rapid, usually taking less than one rotor revolution following a sudden tilt. Because of this, the flapping motion in hover has practically no effect on the stability of the helicopter in terms of holding a given attitude.

Cyclic Pitch Change in Hover

In Chapter 3 it was shown that a hinged rotor blade without offset is a system in resonance; that is, its natural frequency is identically equal to its rotational frequency. A characteristic of a system in resonance is that its maximum response follows by exactly a quarter of a cycle its maximum force input. Thus, if cyclic pitch is applied to a rotor of this type, it will have its maximum flapping amplitude 90° later. Although the derivation of Chapter 3 ignored the existence of aerodynamic forces, they do not change the phase lag from 90°. The aerodynamic forces only add damping to the flapping motion and, as shown in Figure 7.3, the phase angle is always 90° for any damping level. For a zero offset rotor in hover, a 1° change in cyclic pitch will result in a 1° change in flapping. The mathematical derivation of this will be given later; from an intuitive point of view, however, it is because the rotor's stable condition is with no cyclic angle of attack variations with respect to its tip path plane without regard to the relative position between the shaft and the tip path plane. Thus the rotor flaps just enough to cancel out the initial cyclic pitch input and to return it to its initial hover angle of attack configuration with respect to its tip path plane.

If the rotor has hinge offset, the phase angle is somewhat less than 90° and the flapping is not quite numerically equal to the cyclic pitch. This is because as offset is increased, the restoring moment due to centrifugal forces increases faster than the moment of inertia about the flapping hinge; as a result, the flapping natural frequency is higher than the rotational frequency. In short, the system is no longer in resonance. The frequency ratio will be less than unity and the phase angle will be less than 90°, as shown in Figure 7.3. The total magnitude of the flapping

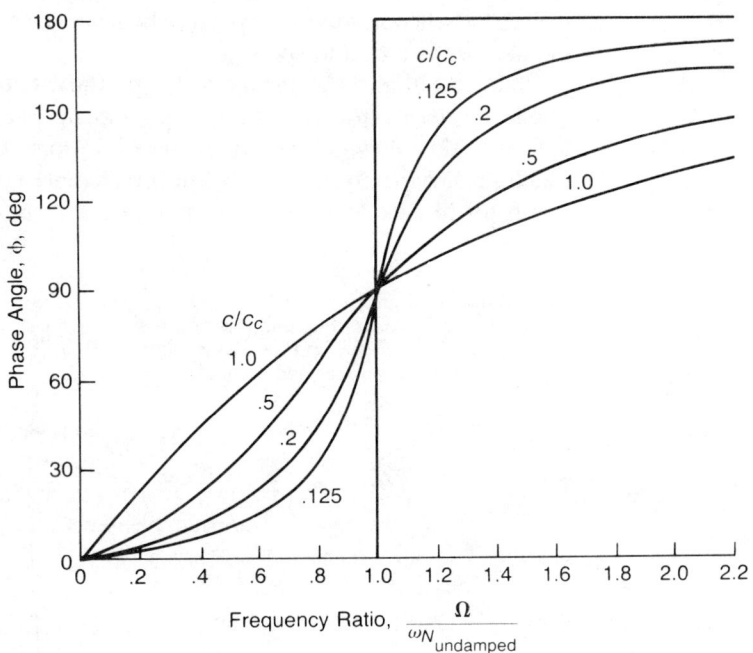

FIGURE 7.3 Phase Angle as Function of Frequency Ratio and Damping

will be slightly less than for a rotor with no hinge offset because of the restraint provided by the nonflapping inner portions of the blades.

Change of Forward Speed

One of the most important of the rotor's flapping characteristics is that caused by a change in speed. To illustrate, let us examine a lifting rotor in a wind tunnel as the tunnel is started. The tunnel speed adds to the velocity on the advancing blade, thus increasing its lift, but subtracts from the velocity on the retreating blade thus decreasing its lift. The advancing blade accelerates up about its flapping hinge, but at the same time the blade is being rotated toward the nose. As a result, the advancing blade reaches its maximum upward flapping angle over the nose and the retreating blade reaches its maximum downward flapping angle over the tail. The rotor trims itself to an equilibrium position when the flapping velocities on the advancing and retreating blades are just enough to reduce the angles of attack to compensate for the change in dynamic pressure. Thus the flapping will increase as the tunnel speed increases. The magnitude of the flapping is a function of the lift of the rotor. If the blades initially had no lift, the unbalanced dynamic pressure would cause no flapping. The rearward tilt of the rotor thrust vector produces a nose-up pitching moment with respect to the helicopter's center of gravity, as shown in Figure 7.4. The resultant change in pitching moment as a function of forward speed is known as *speed stability* and is one of the most important differences between a helicopter and an airplane, which has no corresponding change in pitching moment with respect to speed.

In free flight, the change in longitudinal flapping with increasing speed is stabilizing since it produces a nose-up moment that causes the helicopter to pitch up and to slow down to its original speed. In some cases the effect of a horizontal stabilizer carrying positive lift, or the interference effects of the front rotor on the rear rotor of a tandem rotor helicopter, can overpower the natural speed stability

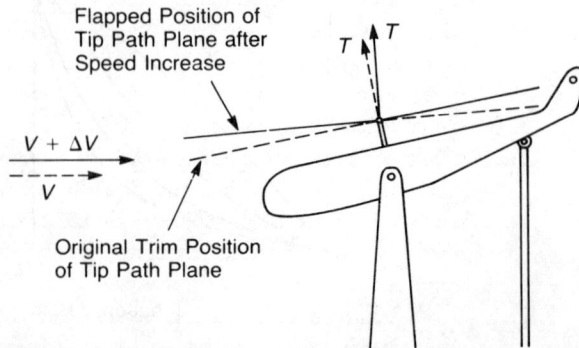

FIGURE 7.4 Change of Flapping Due to Change of Speed

of the rotors and produce negative speed stability. In this case, an increase in speed produces a nose-down pitching moment. This causes the helicopter to go into a dive in which the speed and the nose-down pitching moment increase as a pure divergence. This characteristic is, of course, undesirable from a flying-qualities standpoint; but pilots can learn to fly even such unstable aircraft, or the characteristic can be changed by methods which will be discussed. It will later be shown that in all flight conditions too much positive speed stability is as bad as too much negative speed stability and that in hover the optimum condition is one of neutral speed stability. A manifestation of positive speed stability is the requirement for the pilot to move the cylic stick forward as he increases speed to keep the helicopter trimmed. If the speed stability is negative, the pilot will push the stick forward to initiate an increase in speed; but when he finally trims at the new speed, the stick will be further aft than when he started.

Shaft Tilted in Forward Flight

If the shaft is tilted laterally in forward flight, the effect is the same as it is in hover—the tip path plane follows the shaft, and the flapping with respect to the shaft remains unchanged. If, however, the shaft is tilted longitudinally, the nonuniformity of the velocity distribution produces a different situation.

Figure 7.5 illustrates this with a rotor that, for simplicity's sake, starts from a condition of zero lift on the advancing and retreating blades. Following a sudden nose-up tilt, the immediate result is the same as in hovering: the advancing blade receives a increase in angle of attack and the retreating blade a decrease—producing an unbalanced lift that causes the tip path plane to flap nose up. In forward flight, however, when the blade flaps nose up until it is perpendicular to the shaft, the forces are not yet balanced. The unbalance is due to the forward flight velocity vector. The airflow coming at the rotor as a result of its forward speed modifies both the local angle of attack and the local velocity. Both blades have positive angles of attack, with the angle on the retreating blade actually being the greater, as shown in Figure 7.5c. The lift on the retreating blade, however, is less than on the advancing blade because the lift is proportional to the product of the angle of attack and the square of the local velocity. This causes the rotor to flap past the perpendicular to the shaft to a more nose-up position where the forces are in balance, as shown in Figure 7.5d. The magnitude of the excessive flapping is approximately proportional to the square of the forward speed. The result is negative angle of attack stability, since the aft flapping generates a nose-up pitching moment about the helicopter's center of gravity that tends to cause a further increase in the shaft angle of attack. It is an undesirable characteristic, but it can be compensated for rather easily with a horizontal stabilizer of reasonable size.

From this illustration it may be seen that increasing the angle of attack of the shaft also increases the rotor thrust, just as increasing the angle of attack of a wing increases its lift.

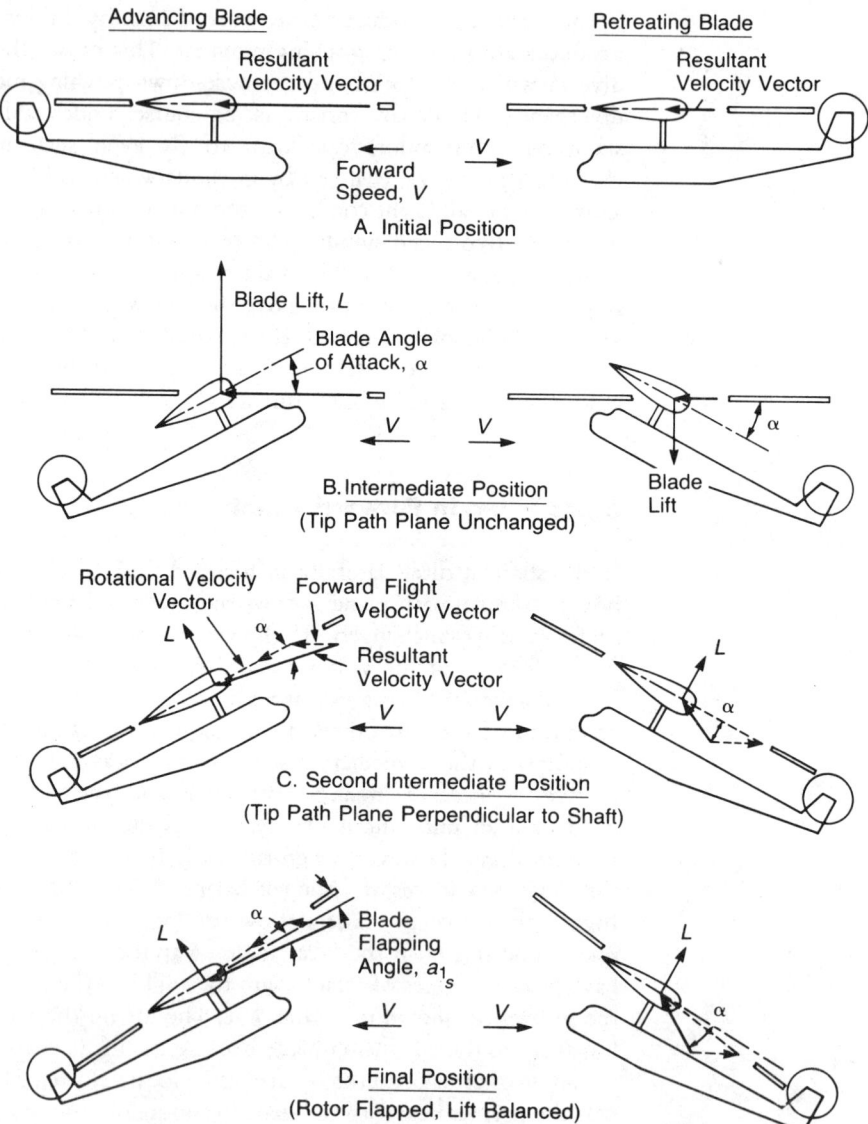

Advancing Blade

Resultant
Velocity Vector

Forward
Speed, V

Retreating Blade

Resultant
Velocity Vector

A. Initial Position

Blade Lift, L

Blade Angle
of Attack, α

B. Intermediate Position
(Tip Path Plane Unchanged)

Blade
Lift

Rotational Velocity
Vector

Forward Flight
Velocity Vector

Resultant
Velocity Vector

C. Second Intermediate Position
(Tip Path Plane Perpendicular to Shaft)

Blade
Flapping
Angle, a_{1_s}

D. Final Position
(Rotor Flapped, Lift Balanced)

FIGURE 7.5   Shaft Tilted in Forward Flight

Collective Pitch Change in Forward Flight

Another longitudinal flapping effect is that caused by an increase in collective pitch in forward flight. Both the advancing and retreating blades receive the same change in angle of attack; but the advancing blade, being at a higher dynamic pressure, develops more additional lift than the retreating blade. Both blades flap up a quarter of a revolution later, but the advancing blade flaps up more than the retreating blade, resulting in both an increase in coning and a net rearward tilt of the tip path plane. This type of flapping is sometimes noticed by the pilot when he decreases collective pitch when entering autorotation. In this case, of course, it is a nose-down tilt of the tip path plane that results. It is also partially responsible for the trend of cyclic stick position with speed. From hover to approximately the speed for minimum power, the collective pitch required for trim decreases, thus causing the tip path plane to want to tilt forward. An aft cyclic stick motion is required to keep the helicopter in trim. This effect may be larger that the aft flapping caused by the increase in forward speed; and, as a consequence, the trimmed stick position may initially move aft or have an unstable gradient. At high speeds, where collective pitch is increasing, the gradient will almost always be stable.

Lateral Flapping in Forward Flight

Steady lateral flapping, like steady longitudinal flapping, is caused by asymmetric airloads. In this case, the asymmetry of airloads is on the blades at $\psi = 0°$ and $\psi = 180°$ and is caused by coning. In forward flight, the coning causes the blade over the nose to be affected by a velocity component of forward speed that is more up with respect to the blade than for the blade over the tail. This asymmetry of vertical velocity at the blades at $\psi = 0°$ and $\psi = 180°$ produces a corresponding asymmetrical angle of attack and airload distribution that causes maximum response 90° later and thus lateral flapping that is upward on the retreating side. The magnitude of this flapping is a function of the coning and tip speed ratio. Because of it, the pilot must move his stick toward the advancing blade as well as forward in order to keep the helicopter in trim as he increases speed.

Cyclic Pitch Change in Forward Flight

Just as in hover, flapping is produced in forward flight by changes in cyclic pitch. for a rotor with no hinge offset, the flapping occurs 90° after the cyclic pitch input. Because of the distribution of the aerodynamics, a 1° change in lateral cyclic pitch, A_1, will cause the rotor to flap down to the right by exactly 1°, but a 1° change in the longitudinal pitch, B_1, will cause the rotor to flap down in front by slightly more than 1°. Thus the response to longitudinal control is somewhat more sensitive in forward flight than in hover, but the response to lateral control remains the same. If the rotor has hinge offset, the maximum response is less than 90° after the input and the magnitude of the response is a function of the offset, as will be shown by the flapping equations.

Flapping That Is Caused by Pitch and Roll Angular Velocities

A factor that is important to helicopter flying qualities is the damping moment the rotor produces when the helicopter is subjected to pitch or roll angular velocities by external means such as gusts. This damping is produced by the tilt of the tip path plane, which lags behind the motion of the shaft by an amount that is proportional to the rate of pitch or roll, as illustrated in Figure 7.6. It has already been shown how the aerodynamics on the blade causes the tip path plane to tend to stabilize itself in an equilibrium position with respect to the shaft. A rotor attached to a shaft that is continuously tilting, therefore, will follow the shaft. The rotor disc may be considered to be a gyro that must be precessed by a moment applied 90° before the direction of tilt. The moment can only be generated by aerodynamic means, which in turn can only be produced by asymmetric flapping velocities. To maintain a steady nose-up pitch rate, for example, the airload must be higher on the blade at $\psi = 90°$ than on the blade at $\psi = 270°$: this asymmetry is generated by a downward flapping velocity—with respect to the shaft—at $\psi = 90°$ and an upward flapping velocity at $\psi = 270°$. The maximum flapping amplitude with respect to the shaft, therefore, is down at $\psi = 180°$ and up at $\psi = 0°$, and the

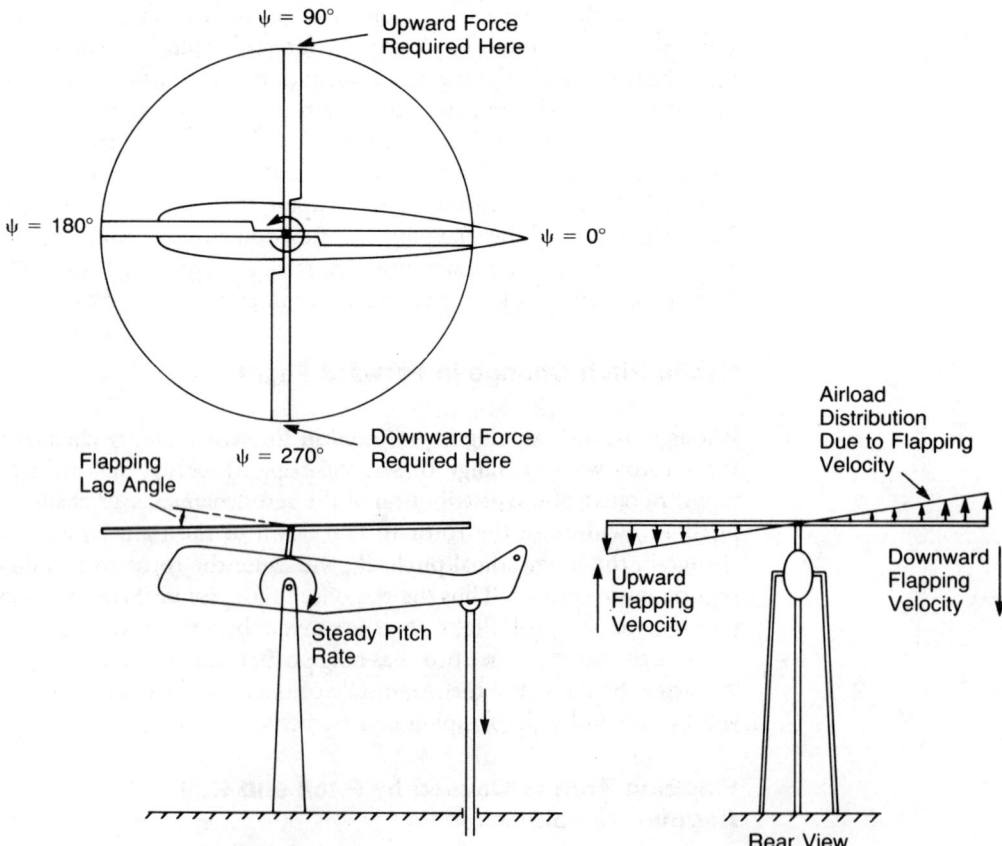

FIGURE 7.6   Flapping Due to a Steady Pitch Rate

tip path plane follows the motion of the shaft at a lag angle that is proportional to the rate of pitch and the rotor moment of inertia.

During a steady nose-up pitching maneuver, some lateral flapping also is generated as a result of the decreased angle of attack at $\psi = 180°$ and the increase at $\psi = 0°$ caused by the rate of pitch itself. This difference in angle of attack is compensated for by lateral flapping that produces enough flapping velocity at these two blade positions to cancel out the effect of the pitch rate. The production of both a lateral and a longitudinal flapping by a pure pitch rate is a source of cross-coupling, which applies equally to rotors with and without hinge offset. The cross-coupling ratio, b_1/a_1, is a function of blade inertia, being highest for light blades.

Consider the converse case in which the steady pitch rate is being produced by the pilot in a deliberate maneuver instead of by external means. Since the rate is steady (no acceleration), no hub moment is needed if we ignore the damping moments generated by the airframe. Under this condition, the longitudinal and lateral cyclic pitch required to maintain the maneuver will be exactly equal to the longitudinal and lateral flapping that would have been produced by a pitch rate caused by external means. This is a consequence of the equivalence of flapping and feathering explained in Chapter 3.

Dihedral Effect

In addition to all of the effects discussed above, the rotor will also respond to changes in sideslip. This is because blade flapping is produced by conditions referenced to the flight path rather than to whatever orientation the fuselage might have at the time.

Imagine the helicopter of Figure 7.7 in forward flight with no sideslip and with the rotor trimmed perpendicular to the shaft. If the flight direction is suddenly changed so that the helicopter is flying directly to the right without changing fuselage heading or control settings, the blade over the tail becomes the advancing blade and the one over the nose the retreating blade. Since the cyclic pitch no longer corresponds to trim conditions, the rotor will flap down on the left side because of the asymmetrical velocity distribution—thus producing a rolling moment to the left.

In practice, of course, sideslip angles are less than the 90° used for illustration, but the trend is the same—the helicopter tends to roll away from the approaching wind. This is the same characteristic found on airplanes with dihedral (both wings slanted up) and is known as the *positive dihedral effect* on rotors even though the source is different.

It is a desirable characteristic that helps the pilot. With negative dihedral, a sideslip would tend to roll the aircraft into an ever-tightening spiral dive. Too much positive dihedral, on the other hand, also can be undesirable, as will be later pointed out in the discussion of lateral-directional stability in Chapter 9. Positive

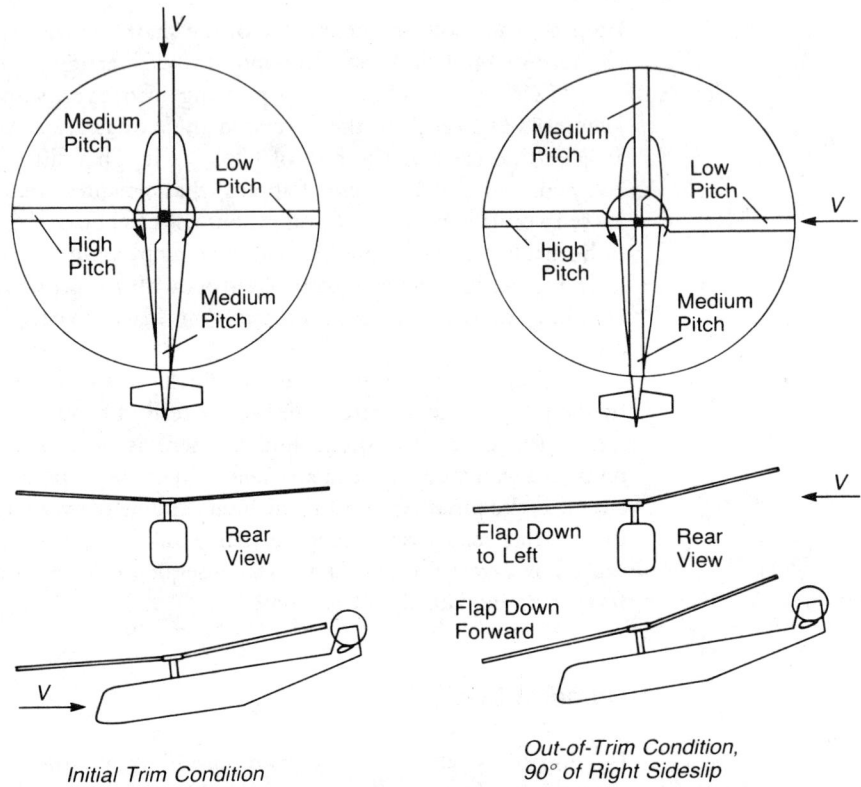

FIGURE 7.7   Effect of Sideslip on Rotor Flapping

dihedral manifests itself during flight as a lateral stick displacement required in the direction of sideslip to maintain equilibrium.

The rolling moment due to the dihedral effect is also accompanied by a pitching moment. Again going back to Figure 7.7 and the helicopter when it is flying directly to the right, it may be seen that the blade pointing in the direction of flight was originally the advancing blade and had a low pitch. It still has a low pitch that will cause the rotor to flap down over the nose, producing a nose-down pitching moment.

Similarly, during flight to the left, the blade pointing in the direction of flight has a high pitch and will cause the rotor to flap up over the tail—also producing a nose-down moment. Thus steady sideslip in either direction requires aft stick displacement, an effect that does not exist on an airplane.

Somewhat surprisingly, if the same analysis is made on a rotor turning clockwise when viewed from above, the pitching moment direction is unchanged—nose-down for sideslip in either direction. This pitching effect is not always observable in flight, since other pitching moments may be generated by changes in airflow conditions on the horizontal stabilizer and tailboom as they move out from behind the fuselage during sideslip.

Gust Response

The blade flapping is also responsible for a helicopter behaving better in gusty air than an airplane does. This is because the rotor blades flap individually in response to the gusts, allowing the rest of the helicopter to have a relatively smooth ride. The wing of an airplane, on the other hand, transmits its unsteady loading directly into the fuselage. This *gust alleviation* feature has been demonstrated by flying helicopters and airplanes of the same size in formation through gusty air, as reported in reference 7.1. The recording instrumentation showed that the helicopter had a smoother ride. The comparison is similar to that of an automobile with independent wheel suspension compared to one on which the wheels are rigidly mounted to the chassis.

FLAPPING EQUATIONS IN HOVER INCLUDING THE EFFECT OF HINGE OFFSET

Frequency Ratio

As already discussed, the addition of hinge offset changes the characteristics of the rotor from being a system in resonance to one whose natural frequency is higher than the rotational frequency. An analysis of this system in hover gives equations for the frequency ratio, damping, phase lag, and cross-coupling as a function of the hinge offset. From Figure 7.8, it may be seen that the increment of moment about the hinge due to centrifugal force perpendicular to the shaft is:

$$\Delta M_{\text{C.F.}} = m\,\Delta r\,\Omega^2 r(r - e)\beta$$

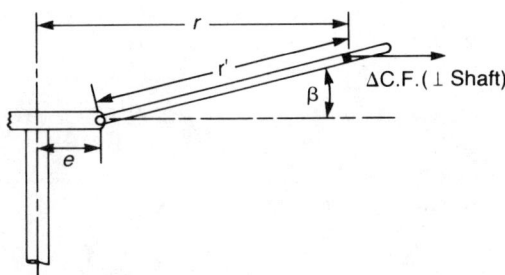

FIGURE 7.8   Geometry of a Flapping Blade

Since it is the portion of the blade outside the flapping hinge that is being analyzed, it will prove convenient to redefine the distance to the blade element by introducing a new parameter, r', such that:

$$r' = r - e$$

Then:

$$\Delta M_{\text{C.F.}} = m\Delta r\Omega^2(r' + e)r'\beta$$

and the total moment is:

$$M_{\text{C.F.}} = \Omega^2\beta \int_0^{R-e} m(r' + e)r'dr'$$

but

$$\int_0^{R-e} mr'^2dr' = I_b$$

and

$$\int_0^{R-e} mr'dr' = M_b/g$$

where I_b and M_b are the moment of inertia and the static moment of the blade about the flapping hinge. Thus:

$$M_{\text{C.F.}} = \Omega^2\beta(I_b + eM_b/g)$$

The natural frequency of the flapping motion is defined by the equation:

$$\omega_n = \sqrt{\frac{k}{I_b}}$$

where

$$k = \frac{M_{\text{C.F.}}}{\beta} = \Omega^2(I_b + eM_b/g)$$

so that the frequency ratio is:

$$\frac{\omega_n}{\Omega} = \sqrt{1 + \frac{eM_b}{gI_b}}$$

If the blade has a constant mass distribution, m, then:

$$I_b = m \int_0^{R-e} r'^2 dr = \frac{mR^3}{3}\left(1 - \frac{e}{R}\right)^3$$

and

$$\frac{M_b}{g} = m \int_0^{R-e} r' dr = \frac{mR^2}{2}\left(1 - \frac{e}{R}\right)^2$$

so that

$$\frac{\omega_n}{\Omega} = \sqrt{1 + \frac{\dfrac{3}{2}\dfrac{e}{R}}{1 - \dfrac{e}{R}}}$$

For the example helicopter, the hinge offset is 5% of the radius, and thus the corresponding first flapping frequency ratio is 1.04.

The equation can be rearranged for use in determining the effective hinge offset of a hingeless rotor. For these rotors, the value of ω_n/Ω is generally calculated separately for the purpose of performing rotor dynamic analyses. Once this parameter is known at the operational rotor speed, the effective hinge offset ratio can be found as:

$$\left(\frac{e}{R}\right)_{\text{eff}} = \frac{2\left[\left(\dfrac{\omega_n}{\Omega}\right)^2 - 1\right]}{1 + 2\left(\dfrac{\omega_n}{\Omega}\right)^2}$$

Damping Ratio

Since the ratio of rotational frequency to natural frequency for a rotor with hinge offset is lower than unity, the phase angle will be less than 90°, as can be seen from Figure 7.3. How much less depends on the damping ratio, c/c_{crit}, where c_{crit} is critical damping. The concept of critical damping comes from the study of a single pendulum-spring-damper system that has the differential equation:

$$I\ddot{\beta} + c\dot{\beta} + k\beta = 0$$

The general solution to this equation is:

$$\beta = \beta_0 e^{\left(-\frac{c/2}{I} + \sqrt{\left(\frac{c/2}{I}\right)^2 - \frac{k}{I}}\right)t}$$

If the term under the radical is negative, the motion will be oscillatory. If it is positive, the motion will have a pure convergence with no oscillations. Critical damping is the value of c that makes the radical term zero—that is, that lies on the boundary between oscillatory and nonoscillatory motion:

$$\left(\frac{c_{crit}/2}{I}\right)^2 = \frac{k}{I}$$

or

$$c_{crit} = 2I\,\omega_{n_{undamped}}$$

The damping of the blade, c, may be evaluated from the aerodynamic hinge moment due to flapping velocity:

$$M_A = \int_0^{R-e} r'\,\frac{\rho}{2}\,acr'\dot{\beta}\,(r' + e)\,\Omega dr$$

(Reader, please note that italic c stands for chord and roman c for damping; and that italic e stands for hinge offset and roman e for the base of natural logarithms.)

or
$$c = \frac{\partial M_A}{\partial \dot{\beta}} = \frac{c\rho a R^4}{8}\left(1 - \frac{e}{R}\right)^4 \Omega\left[\frac{1 + \frac{1}{3}\frac{e}{R}}{1 - \frac{e}{R}}\right]$$

A useful nondimensional parameter relating the inertia and aerodynamic characteristics of a blade is the Lock number, γ, which was already introduced in Chapter 1.

$$\gamma = \frac{c\rho a R^4}{I_\ell}$$

For most blades, γ is between 6 and 10. Adding a large tip weight such as a jet engine may lower it to as low as 2. The damping written in terms of the Lock number is:

$$c = \frac{I_b\gamma\Omega}{8}\left(1 - \frac{e}{R}\right)^4\left[\frac{1 + \frac{1}{3}\frac{e}{R}}{1 - \frac{e}{R}}\right]$$

The damping ratio is:

$$c/c_{crit} = \frac{\dfrac{I_b \gamma \Omega}{8} \left(1 - \dfrac{e}{R}\right)^4 \left[\dfrac{1 + \dfrac{1}{3}\dfrac{e}{R}}{1 - \dfrac{e}{R}}\right]}{2 I_b \omega_n}$$

or

$$c/c_{crit} = \frac{\gamma}{16} \left(1 - \frac{e}{R}\right)^4 \left[\frac{1 + \dfrac{1}{3}\dfrac{e}{R}}{1 - \dfrac{e}{R}}\right] \frac{1}{\omega_n/\Omega}$$

The factor, $\gamma/16$, is a universal parameter that will arise again when rotor damping as a whole is discussed.

Cross-Coupling

Knowing the ratio of rotational frequency to the undamped natural frequency and the damping ratio for the rotor with offset flapping hinges, the lag between the maximum aerodynamic input and the maximum flapping amplitude can be found from Figure 7.3. For the example helicopter, the Lock number, γ, is 8.1; the hinge offset is 0.05; and the frequency ratio is 1.04. Thus the damping ratio, c/c_{crit}, is 0.42. From Figure 7.3, the phase angle, ϕ, is 86°. The difference between this angle and 90° represents one source of flapping cross-coupling that exists on a rotor with hinge offset but not on a rotor without offset. The magnitude of the coupling is:

$$\frac{b_{1_s}}{a_{1_s}} = -\cot\phi$$

where a_{1_s} and b_{1_s} are shown in Figure 7.9.

The equation for the coupling can also be written directly. From the analysis of vibrating linear systems, the equation for the phase angle is:

$$\phi = \cos^{-1} \frac{\left[\left(\dfrac{\omega_n}{\Omega}\right)^2 - 1\right]}{\sqrt{\left[\left(\dfrac{\omega_n}{\Omega}\right)^2 - 1\right]^2 + 4\left(\dfrac{c}{c_{crit}}\right)^2 \left(\dfrac{\omega_n}{\Omega}\right)^2}}$$

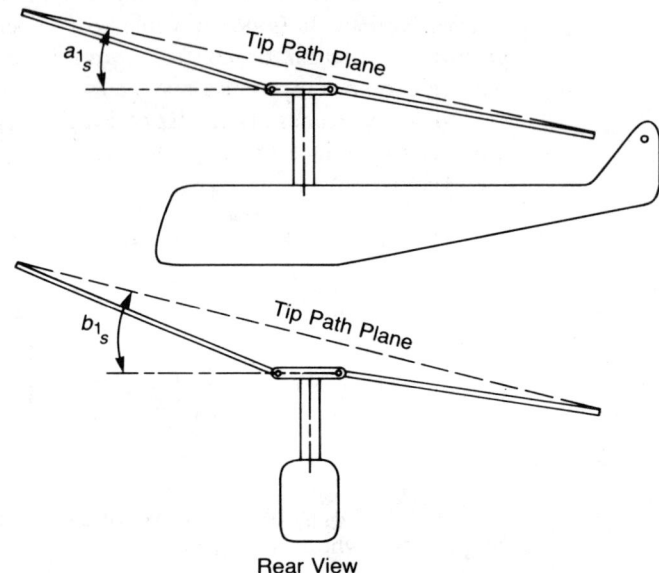

FIGURE 7.9   Longitudinal and Lateral Flapping

and thus:

$$\frac{b_{1_s}}{a_{1_s}} = -\cot\phi = \frac{-\left[\left(\dfrac{\omega_n}{\Omega}\right)^2 - 1\right]}{2\,\dfrac{c}{c_{crit}}\dfrac{\omega_n}{\Omega}}$$

Substituting in the expressions for the frequency and damping ratios gives the equation for coupling.

$$\frac{b_{1_s}}{a_{1_s}} = -\frac{\dfrac{12}{\gamma}\dfrac{e}{R}}{\left[1 + \dfrac{1}{3}\dfrac{e}{R}\right]}$$

For the example helicopter, this means that a 1° change in longitudinal flapping will be accompanied by a lateral flapping of −.07°. If the pilot wishes to use cyclic pitch to tilt the tip path plane nose up in a purely longitudinal direction, he will have to move the stick slightly to the right as he pulls it aft in order to cancel out the left roll that would otherwise be generated. This type of cross-coupling is sometimes called *acceleration cross-coupling* because it is associated with the rotor moments, which provide the initial acceleration during a maneuver. Once steady rates are established in the maneuver, the cross-coupling changes to *rate cross-*

coupling, as will be shown later. Because these two types of cross-coupling have different values, it is not possible to compensate exactly for both by a simple mechanical rotation of control inputs between the stick and the swashplate, although some compromise may be used in an attempt to minimize both.

It is sometimes of interest to calculate the increment in cyclic angle of attack required to produce 1° of flapping. The change in angle of attack experienced by the blade is:

$$\Delta\alpha = \frac{\dot{\beta}\,r'}{\Omega(r' + e)}$$

or at $\psi = 180°$

$$\Delta\alpha = \frac{-b_{1_s}r'}{r' + e}$$

but, as was just shown:

$$b_{1_s} = \frac{-\dfrac{12}{\gamma}\dfrac{e}{R}}{1 + \dfrac{1}{3}\dfrac{e}{R}}\,a_{1_s}$$

thus:

$$\Delta\alpha = \left(\frac{\dfrac{12}{\gamma}\dfrac{e}{R}}{1 + \dfrac{1}{3}\dfrac{e}{R}}\right)\left(\frac{r'}{r' + e}\right)a_{1_s}$$

For the blade of the example helicopter at the 75% radius station, this gives:

$$\frac{\Delta\alpha}{a_{1_s}} = 0.07$$

which says that 1° of cyclic flapping is maintained with only 0.07° change in angle of attack. If the hinge offset had been zero, the flapping could have been sustained with no cyclic change in angle of attack.

Blade Time Constant

The expression for blade damping derived several paragraphs ago allows the blade time constant to be determined. The time constant is the time required by a system to achieve 63% of its final amplitude following the application of a step forcing function. For the blade subjected to a step moment, M_{st}, the non-oscillatory part of the displacement, β_{n-0} is:

$$\beta_{n-0} = \frac{M_{st}}{k}\left[1 - e^{\frac{-c/2}{I_b}t}\right]$$

where M_{st}/k is the final value of β when time is very large. If this is evaluated at the time at which

$$\frac{c/2}{I_b} t = 1$$

then

$$\beta_{n-0} = \frac{M_{st}}{k}\left[1 - \frac{1}{e}\right] = .63\,\frac{M_{st}}{k}$$

The time that makes this true is called the time constant of the system:

$$t_{.63} = \frac{I_b}{c/2}\ \sec$$

But from one of the previous equations:

$$t_{.63} = \frac{I_b}{c/2} = \frac{\dfrac{16}{\gamma\Omega}}{\left(1 - \dfrac{e}{R}\right)^4\left(\dfrac{1 + \dfrac{1}{3}\dfrac{e}{R}}{1 - \dfrac{e}{R}}\right)}$$

The amount of azimuth required for the blade to achieve 63% of its response is thus:

$$\psi_{.63} = \Omega t_{.63} = \frac{917}{\gamma\left(1 - \dfrac{e}{R}\right)^3\left(1 + \dfrac{1}{3}\dfrac{e}{R}\right)},\ \deg$$

For the example helicopter, this equation gives an *azimuth constant* of 130°. This relatively fast response, compared to the response of the entire helicopter, will later be used to justify a simplifying assumption when writing the helicopter equations of motion.

FLAPPING EQUATIONS IN FORWARD FLIGHT

The complete equations for rotor flapping in forward flight can be derived by equating the increments of hinge moment due to aerodynamic, centrifugal, inertial, and weight forces to zero. The resultant equations will be similar to those derived for the trim values of cyclic pitch in Chapter 3 and are based on similar assumptions:

- Aerodynamic forces are considered to act from the hinge to the tip.
- The reverse flow region is ignored.
- The airfoil lift characteristics are linear and free of stall and compressibility effects.
- The blade motion consists of only coning and first harmonic flapping.
- Small-angle assumptions are valid.

The use of these assumptions has been found to be justified for conventional helicopters flying within conventional flight envelopes. For special applications, these assumptions can be relaxed at the expense of simplicity.

Since the blade motion consists only of coning and first harmonic flapping, the centrifugal forces may be assumed to lie in the plane of rotation of the blade element, as shown in Figure 7.10. This assumption also eliminates all inertial moments due to flapping from the analysis. The increment of moment about the flapping hinge due to centrifugal force is:

$$\Delta M_{\text{C.F.}} = -\Delta \text{C.F.} \, h$$

where

$$\Delta \text{C.F.} = m \Delta r (r' + e) \Omega^2$$

and

$$h = r' a_0 + \frac{e}{r' + e} (r' \beta - r' a_0)$$

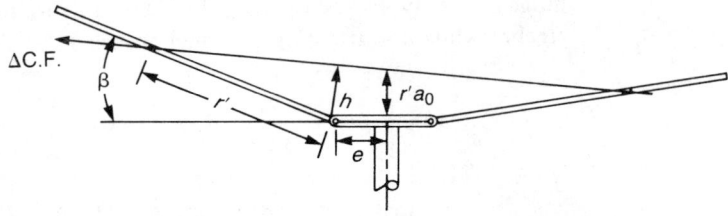

FIGURE 7.10   Geometry of Flapping

but

$$\beta = a_0 - a_{1_s} \cos \psi - b_{1_s} \sin \psi$$

where a_0, a_{1_s}, and b_{1_s} are the Fourier coefficients of the flapping angle about the offset flapping hinge. Thus:

$$\Delta M_{\text{C.F.}} = -mr'(r' + e)\Omega^2 \left[a_0 - \frac{e}{r' + e} (a_{1_s} \cos \psi - b_{1_s} \sin \psi) \right] \Delta r$$

Integrating out from the hinge gives:

$$M_{\text{C.F.}} = -\Omega^2 \left\{ a_0 \int_0^{R-e} mr'(r' + e)dr' - (a_{1_s} \cos \psi + b_{1_s} \sin \psi) e \int_0^{R-e} mr'dr \right\}$$

or

$$M_{\text{C.F.}} = -\Omega^2 \left\{ a_0 \left(I_b + e \frac{M_b}{g} \right) - (a_{1_s} \cos \psi + b_{1_s} \sin \psi) e \frac{M_b}{g} \right\}$$

The aerodynamic contribution to the hinge moment is:

$$\Delta M_A = r' \frac{\rho}{2} U_T^2 a\alpha c \Delta r'$$

where, from blade element theory:

$$U_T = \Omega R \left[\frac{r' + e}{R} + \mu \, sin \, \psi \right]$$

$$\alpha = \theta + \frac{U_P}{U_T}$$

$$\theta = \theta_0 + \frac{r'}{R} \theta_1 - A_1 \cos \psi - B_1 \sin \psi$$

(Note that for this derivation, the blade twist, θ_1, is measured from the flapping hinge to the tip instead of from the center of rotation. This helps cut down on the algebra while not affecting the final answer.)

$$U_P = \Omega R \left[\mu\alpha_s - \frac{v_1}{\Omega R} \left(1 + \frac{r'}{R} \cos \psi \right) - \frac{r'}{R} (a_{1_s} \sin \psi - b_{1_s} \cos \psi) \right.$$

$$\left. - \mu (a_0 - a_{1_s} \cos \psi - b_{1_s} \sin \psi) \cos \psi \right]$$

(In Chapter 3, where rotor performance was the subject, the inflow velocity, U_P, was written as a function of the angle of attack of the tip path plane. For the study of rotor flapping, however, it is better to write it as a function of the shaft angle of attack as here.)

The expression for M_A can be obtained by using these equations and integrating out the blade from the hinge to the tip. Since only constant and first harmonic flapping are to be studied, the moment will be written in terms of the constant, sine, and cosine terms by throwing out all higher harmonics such as $\sin 2\psi$ and the like.

$$M_A = M_{A_{\text{const}}} + M_{A_{\text{sine}}} \sin \psi + M_{A_{\text{cosine}}} \cos \psi$$

In order to do this, it is necessary to use the following trignometric identities:

$$\sin \psi \cos \psi = \frac{\sin 2\psi}{2}$$

$$\sin^2 \psi = \frac{1}{2} - \frac{\cos 2\psi}{2}$$

$$\cos^2 \psi = \frac{1}{2} + \frac{\cos 2\psi}{2}$$

$$\sin^3 \psi = \frac{3}{4} \sin \psi - \frac{1}{4} \sin 3\psi$$

$$\sin \psi \cos^2 \psi = \frac{1}{4} \sin \psi + \frac{1}{4} \sin 3\psi$$

$$\sin^2 \psi \cos \psi = \frac{1}{4} \cos \psi - \frac{1}{4} \cos 3\psi$$

After making these substitutions and performing the required algebra, the coefficients of the constant, sine, and cosine components of the aerodynamic hinge moment can be written. They are:

$$M_{A_{\text{const}}} = \frac{\rho}{2} ac(\Omega R)^2 \left(1 - \frac{e}{R}\right)^2 R^2 \left\{\frac{\theta_0}{4}\left[1 + \mu^2 + \frac{2}{3}\frac{e}{R} + \frac{1}{3}\left(\frac{e}{R}\right)^2\right]\right.$$

$$+ \theta_1\left[\frac{1}{5} + \frac{\mu^2}{6}\left(1 - \frac{e}{R}\right) - \frac{1}{10}\left(\frac{e}{R}\right) - \frac{1}{15}\left(\frac{e}{R}\right)^2 - \frac{1}{30}\left(\frac{e}{R}\right)^3\right]$$

$$\left. + \left(\mu\alpha_s - \frac{v_1}{\Omega R}\right)\left(\frac{1}{3} + \frac{1}{6}\frac{e}{R}\right) - B_1\mu\left(\frac{1}{3} + \frac{1}{6}\frac{e}{R}\right) + \mu\frac{e}{R}\frac{a_{1_s}}{4}\right\}$$

$$M_{A_{\text{sine}}} = \frac{\rho}{2} ac(\Omega R)^2 \left(1 - \frac{e}{R}\right)^2 R^2 \left\{ 2\theta_0\mu \left[\frac{1}{3} + \frac{1}{6}\frac{e}{R}\right] \right.$$

$$+ 2\theta_1\mu \left[\frac{1}{4} - \frac{1}{6}\frac{e}{R} - \frac{1}{12}\left(\frac{e}{R}\right)^2\right]$$

$$- B_1 \left[\frac{1}{4} + \frac{3}{8}\mu^2 + \frac{1}{6}\frac{e}{R} + \frac{1}{12}\left(\frac{e}{R}\right)^2\right]$$

$$+ \frac{\mu}{2}\left(\mu\alpha_s - \frac{v_1}{\Omega R}\right) - a_{1_s} \left[\frac{1}{4} - \frac{\mu^2}{8} - \frac{1}{6}\frac{e}{R} - \frac{1}{12}\left(\frac{e}{R}\right)^2\right] \right\}$$

$$M_{A_{\text{cosine}}} = \frac{\rho}{2} ac(\Omega R)^2 \left(1 - \frac{e}{R}\right)^2 R^2 \left\{ -A_1 \left[\frac{1}{4} + \frac{\mu^2}{8} + \frac{1}{6}\frac{e}{R} + \frac{1}{12}\left(\frac{e}{R}\right)^2\right] \right.$$

$$+ b_{1_s} \left[\frac{1}{4} + \frac{\mu^2}{8} - \frac{1}{6}\frac{e}{R} - \frac{1}{12}\left(\frac{e}{R}\right)^2\right]$$

$$- \frac{v_1}{\Omega R}\left[\frac{1}{4} - \frac{1}{6}\frac{e}{R} - \frac{1}{12}\left(\frac{e}{R}\right)^2\right] - \mu a_0 \left[\frac{1}{3} + \frac{1}{6}\frac{e}{R}\right] \right\}$$

For most rotors, the hinge offset ratio, e/R, will be small enough that it can reasonably be eliminated from those terms inside the square brackets.

The final contribution to the hinge moment that must be considered is that due to the blade weight, M_W. But:

$$M_W = -M_b = -\int_0^{R-e} mgr'dr'$$

The summation of the constant portions of the various contributions to the hinge moment can be used to give the equation for coning at the flapping hinge as it did in Chapter 3 for the rotor without hinge offset:

$$M_{\text{C.F.}_{\text{const}}} + M_{A_{\text{const}}} + M_{W_{\text{const}}} = 0$$

which gives:

$$a_0 = \frac{\dfrac{1}{6}\rho a c R^4 \left(1 - \dfrac{e}{R}\right)^2}{I_b + e\dfrac{M_b}{g}}\left[\theta_0\left(\frac{3}{4} + \frac{3}{4}\mu^2\right)\right.$$

$$\left. + \theta_1\left(\frac{3}{5} + \frac{\mu^2}{2}\right) + \mu(\alpha_s - B_1) - \frac{v_1}{\Omega R}\right] - \frac{M_b}{\Omega^2\left(I_b + e\dfrac{M_b}{g}\right)}$$

The equation for C_T/σ for a rotor with hinge offset may be derived as:

$$C_T/\sigma = \left(1 - \frac{e}{R}\right)\frac{a}{4}\left[\theta_0\left(\frac{2}{3} + \mu^2\right) + \theta_1\left(\frac{1}{2} + \frac{\mu^2}{2}\right) + \mu(\alpha_s - B_1) - \frac{v_1}{\Omega R}\right]$$

Thus, for all practical purposes:

$$a_0 = \frac{\dfrac{2}{3}\rho c R^4\left(1 - \dfrac{e}{R}\right)C_T/\sigma}{I_b + e\dfrac{M_b}{g}} - \frac{M_b}{\Omega^2\left(I_b + e\dfrac{M_b}{g}\right)}$$

For blades with uniform mass distribution, the coning becomes:

$$a_0 = \frac{\dfrac{2}{3}\gamma C_T/\sigma}{a}\left[\frac{\left(1 - \dfrac{e}{R}\right)^2}{1 + \dfrac{1}{2}\dfrac{e}{R}}\right] - \frac{3}{2}\frac{gR}{(\Omega R)^2}\left[\frac{1}{1 + \dfrac{1}{2}\dfrac{e}{R}}\right], \text{ radians}$$

(Compare this to the equation for coning on page 171.)
The sine components of the flapping moments must also be zero:

$$M_{\text{C.F.}_{\text{sine}}} + M_{A_{\text{sine}}} = 0$$

or

$$\Omega^2 b_{1_s} e\frac{M_b}{g} + \frac{\gamma I_b}{2}\Omega^2\left(1 - \frac{e}{R}\right)^2\left[\frac{2}{3}\theta_0\mu + \frac{1}{2}\theta_1\mu - B_1\left(\frac{1}{4} + \frac{3}{8}\mu^2\right)\right.$$

$$\left. + \frac{\mu}{2}\left(\mu\alpha_s - \frac{v_1}{\Omega R}\right) - a_{1_s}\left(\frac{1}{4} - \frac{\mu^2}{8}\right)\right] = 0$$

Similarly, the cosine equation becomes:

$$\Omega^2 a_{1_s} e \frac{M_b}{g} + \frac{\gamma I_b}{2} \Omega^2 \left(1 - \frac{e}{R}\right)^2 \left[-A_1 \left(\frac{1}{4} + \frac{\mu^2}{8}\right) - \frac{\mu a_0}{3} - \frac{1}{3} \frac{v_1}{\Omega R} \right.$$

$$\left. + b_{1_s} \left(\frac{1}{4} + \frac{\mu^2}{8}\right) \right] = 0$$

These two equations can be solved simultaneously to give two equations for a_{1_s} and b_{1_s}. In the process of performing the algebra, the following approximations have been used:

$$\frac{e M_b/g}{I_b} = \frac{\frac{3}{2} \frac{e}{R}}{1 - \frac{e}{R}}$$

$$\frac{v_1}{\Omega R} = C_T/\sigma \frac{\sigma}{2\mu}$$

$$a_0 = \frac{2}{3} \gamma \frac{C_T/\sigma}{a} \frac{\left(1 - \frac{e}{R}\right)^2}{1 + \frac{1}{2} \frac{e}{R}}$$

The equation for longitudinal flapping is:

$$a_{1_s} = \frac{\frac{8}{3} \theta_0 \mu + 2\theta_1 \mu - B_1 \left(1 + \frac{3}{2} \mu^2\right) + 2\mu \left(\mu \alpha_s - C_T/\sigma \frac{\sigma}{2\mu}\right)}{\left(1 - \frac{\mu^2}{2}\right) + \frac{144 \left(\frac{e}{R}\right)^2}{\gamma^2 \left(1 - \frac{e}{R}\right)^6 \left(1 + \frac{\mu^2}{2}\right)}}$$

$$+ \frac{12 \left(\frac{e}{R}\right)}{\gamma \left(1 - \frac{e}{R}\right)^3 \left(1 + \frac{\mu^2}{2}\right)} \left[\frac{A_1 \left(1 + \frac{\mu^2}{2}\right) + \frac{4}{3} C_T/\sigma \left(\frac{\frac{2}{3} \frac{\mu \gamma}{a}}{1 + \frac{3}{2} \frac{e}{R}} + \frac{6}{\mu}\right)}{\left(1 - \frac{\mu^2}{2}\right) + \frac{144 \left(\frac{e}{R}\right)^2}{\gamma^2 \left(1 - \frac{e}{R}\right)^6 \left(1 + \frac{\mu^2}{2}\right)}} \right]$$

For most applications, the second term in each denominator is small with respect to the first term. Taking it as negligible, the equation becomes:

$$a_{1_s} = \frac{\frac{8}{3}\theta_0\mu + 2\theta_1\mu - B_1\left(1 + \frac{3}{2}\mu^2\right) + 2\mu\left(\mu\alpha_s - C_T/\sigma\,\frac{\sigma}{2\mu}\right)}{1 - \frac{\mu^2}{2}} + \frac{12\left(\dfrac{e}{R}\right)}{\gamma\left(1 - \dfrac{e}{R}\right)^3\left(1 + \dfrac{\mu^4}{4}\right)}$$

$$\times\left[A_1\left(1 + \frac{\mu^2}{2}\right) + \frac{4}{3}C_T/\sigma\left(\frac{\frac{2}{3}\dfrac{\mu\gamma}{a}}{1 + \dfrac{3}{2}\dfrac{e}{R}} + \frac{\sigma}{2\mu}\right)\right]$$

The first term may be thought of as the basic flapping that is independent of hinge offset. The second term represents cross-coupling due to hinge offset and is significant when the determination of cross-coupling is a primary objective of the analysis. The similar equation for the lateral flapping, b_{1_s}, is:

$$b_{1_s} = A_1 + \frac{\frac{4}{3}C_T/\sigma\left(\dfrac{\frac{2}{3}\dfrac{\mu\gamma}{a}}{1 + \dfrac{3}{2}\dfrac{e}{R}} + \dfrac{\sigma}{2\mu}\right)}{1 + \dfrac{\mu^2}{2}}$$

$$+ \frac{12\left(\dfrac{e}{R}\right)}{\gamma\left(1 - \dfrac{e}{R}\right)^3\left(1 - \dfrac{\mu^4}{4}\right)} \times \left[\frac{8}{3}\theta_0\mu + 2\theta_1\mu - B_1\left(1 + \frac{3}{2}\mu^2\right) + 2\mu\left(\mu\alpha_s - C_T/\sigma\,\frac{\sigma}{2\mu}\right)\right]$$

The primary use of these equations in the analysis of stability and control is to yield flapping derivatives as a function of changes in flight conditions and control inputs. These flapping derivatives can then be converted into pitch and roll moment derivatives using the relationships between moments and flapping. The evaluation of the derivatives will be discussed in detail in Chapter 9.

FLAPPING DUE TO PITCH AND ROLL VELOCITIES

The flapping associated with pitch and roll velocities was discussed earlier without deriving the equations. The derivations can now be made by using the same techniques as were used to analyze flapping in steady flight. For this derivation, the moment at the hinge must still be zero, but an extra contribution, due to gyroscopic effects, must be considered:

$$M_{\text{C.F.}} + M_A + M_W + M_{\text{gyro}} = 0$$

The equation previously derived for the moment due to centrifugal forces is valid for this case. Since we are interested in only the sine and cosine components, it can be written:

$$M_{C.F.} = \Omega^2 (a_{1_s} \cos \psi + b_{1_s} \sin \psi) e \frac{M_b}{g}$$

The increment of hinge moment due to the aerodynamics is:

$$\Delta M_A = r' \frac{\rho}{2} U_T^2 a \alpha c \Delta r'$$

For this increment we need only consider the change in local angle of attack caused by the angular rates and the associated flapping. The result can then be superimposed on the flapping due to the other independent variables such as collective pitch and shaft angle of attack. The change in the local angle of attack is caused by the components of vertical velocity due to the pitch rate, q; the roll rate, p; and the flapping angles and flapping velocities associated with these rates, shown in Figure 7.11.

The local angle of attack to be used is:

$$\alpha = \frac{U_P}{U_T}$$

where:

$$U_P = (r' + e)(q \cos \psi + p \sin \psi) - r'\Omega(a_{1_s} \sin \psi - b_{1_s} \cos \psi)$$
$$+ V(a_{1_s} \cos \psi + b_{1_s} \sin \psi) \cos \psi$$

and:

$$U_T = \Omega(r' + e) + V \sin \psi$$

Substituting these equations into the moment equation, integrating out the blade, and discarding all but the first harmonic sine and cosine terms, as was done in the analysis of flapping in steady flight, gives:

$$M_A = \frac{\rho}{8} ac(\Omega R)^2 \left(1 - \frac{e}{R}\right)^2 R^2 \left\{ \left[\frac{p}{\Omega} - \left(1 - \frac{\mu^2}{2}\right) a_{1_s} \right] \sin \psi \right.$$
$$\left. + \left[\frac{q}{\Omega} + \left(1 + \frac{\mu^2}{2}\right) b_{1_s} \right] \cos \psi \right\}$$

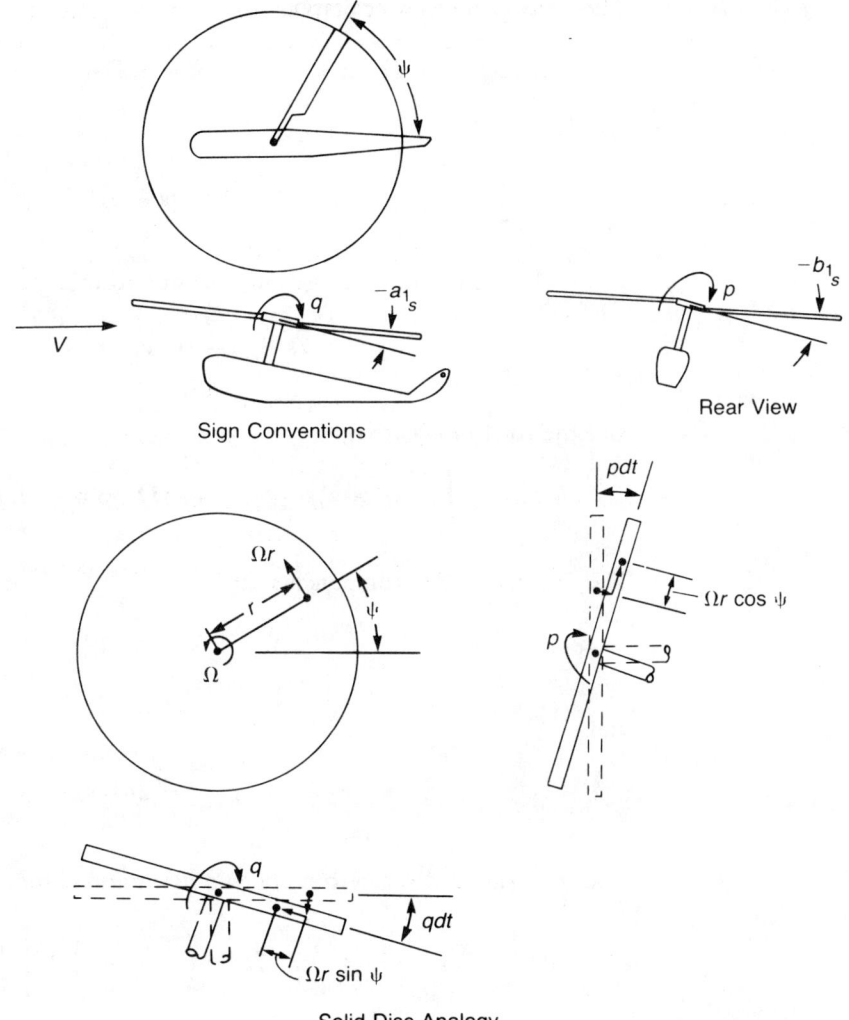

FIGURE 7.11   The Rotor with Pitch and Roll Velocities

The hinge moment due to gyroscopic forces can be found by examining the upward acceleration of a blade element perpendicular to the tip path plane, assuming that the tip path plane is being pitched and rolled as the solid disc of Figure 7.11.

The vertical velocity at the blade element is made up of four components: two due to the pitch and roll rates without rotation and two due to rotation on a constantly pitching and rolling disc:

$$V_{\text{gyro}} = -qr \cos \psi - pr \sin \psi + \Omega r \sin \psi \, qdt - \Omega r \cos \psi \, pdt$$

The corresponding acceleration is:

$$a_{\text{gyro}} = qr\Omega \sin \psi - pr\Omega \cos \psi + \Omega rq \sin \psi - p\Omega r \cos \psi$$

or:

$$a_{\text{gyro}} = 2qr\Omega \sin \psi - 2pr\Omega \cos \psi$$

The increment of hinge moment due to this acceleration is:

$$\Delta M_{\text{gyro}} = r(-a_{\text{gyro}}m)dr'$$

and the total moment is:

$$M_{\text{gyro}} = -\int_0^{R-e} (r' + e)m \left[2q(r' + e)\Omega \sin \psi - 2p(r' + e)\Omega \cos \psi\right]dr$$

The sine and cosine components are:

$$M_{\text{gyro}_{\text{sine}}} = -2q\Omega I_b$$

and

$$M_{\text{gyro}_{\text{cosine}}} = 2p\Omega I_b$$

The total sine and cosine components that must be zero are:

$$M_{\text{sine}} = \Omega^2 e \frac{M_b}{g} b_{1_s} + \frac{\rho}{8} ac(\Omega R)^2 \left(1 - \frac{e}{R}\right)^2 R^2 \left[\frac{p}{\Omega} - \left(1 - \frac{\mu^2}{2}\right)a_{1_s}\right]$$

$$-2q\Omega I_b = 0$$

$$M_{\text{cosine}} = \Omega^2 e \frac{M_b}{g} a_{1_s} + \frac{\rho}{8} ac(\Omega R)^2 \left(1 - \frac{e}{R}\right)^2 R^2 \left[\frac{q}{\Omega} + \left(1 + \frac{\mu^2}{2}\right)b_{1_s}\right]$$

$$+ 2p\Omega I_b = 0$$

These two equations can be solved simultaneously, using the same techniques as were used for steady flapping, to give equations for longitudinal and lateral flapping, a_{1_s} and b_{1_s}, as a function of the pitch and roll rates, q and p.

$$a_{1_s} = \frac{-\dfrac{16}{\gamma}\left(\dfrac{q}{\Omega}\right)}{\left(1-\dfrac{e}{R}\right)^2} + \left(\dfrac{p}{\Omega}\right)}{1-\dfrac{\mu^2}{2}} + \frac{\dfrac{12}{\gamma}\dfrac{e}{R}}{\left(1-\dfrac{e}{R}\right)^3}\left[\dfrac{-\dfrac{16}{\gamma}\left(\dfrac{p}{\Omega}\right)}{\left(1-\dfrac{e}{R}\right)^2} - \left(\dfrac{q}{\Omega}\right)\right]}{1-\dfrac{\mu^4}{4}}$$

$$b_{1_s} = \frac{\dfrac{-\dfrac{16}{\gamma}\left(\dfrac{p}{\Omega}\right)}{\left(1-\dfrac{e}{R}\right)^2} - \left(\dfrac{q}{\Omega}\right)}{1+\dfrac{\mu^2}{2}} - \frac{\dfrac{12}{\gamma}\dfrac{e}{R}}{\left(1-\dfrac{e}{R}\right)^3}\left[\dfrac{-\dfrac{16}{\gamma}\left(\dfrac{q}{\Omega}\right)}{\left(1-\dfrac{e}{R}\right)^2} + \left(\dfrac{p}{\Omega}\right)\right]}{1-\dfrac{\mu^4}{4}}$$

These terms can be added to the previously derived equations for flapping in steady flight. In each of these equations, the first term may be thought of as the basic flapping due to pitch and roll velocities and the second term as the cross-coupling due to hinge offset, which may safely be ignored unless cross-coupling is a primary objective of the analysis. It is of some interest to note that for a rotor in hover with zero hinge offset and only a pitch rate, the flapping is:

$$a_{1_s} = -\left(\frac{16}{\gamma}\right)\left(\frac{q}{\Omega}\right)$$

and

$$b_{1_s} = -\left(\frac{q}{\Omega}\right)$$

That is, for a nose-up pitch rate, the tip path plane will lag behind longitudinally and tilt down to the left laterally. For a more or less conventional Lock number of 8, the lateral tilt will be one-half the longitudinal lag angle. This is *rate cross-coupling* as contrasted to *acceleration cross-coupling*, discussed earlier.

If the pitch rate is being produced by the pilot in a deliberate maneuver using cyclic pitch, both the longitudinal and lateral flapping will be essentially zero and the trim value of lateral cyclic pitch will be approximately half of the longitudinal cyclic pitch. In this case, the longitudinal cyclic pitch, B_1, will be negative, and the lateral cyclic pitch, A_1, will be positive—as required to prevent a left roll.

Two other sources of cross-coupling may be present during pitching maneuvers in forward flight. The first is due to the change in coning as rotor thrust changes during the maneuver. For a nose-up pitching maneuver—or pull-up—the increase in coning causes the rotor to tilt down to the right, which is the opposite to the left tilt caused by nose-up pitch rate just discussed. A second possible source is associated with the sideslip that will occur if the pilot holds his pedals fixed during a pull-up. The sequence of events is as follows: The helicopter slows down; the tail rotor thrust decreases, allowing the helicopter to sideslip such that the flight path is to the left of the nose (for counterclockwise main rotor rotation); the rotor longitudinal flapping aligns itself with the new flight path and, in so doing, produces a tilt of the tip path plane down to the right. This is the source of the *rotor dihedral effect* since it generates a rolling moment in the same sense that wing dihedral generates a rolling moment on an airplane. Because the various coupling effects have different characteristics, the magnitude and even the direction of roll during pitching maneuvers—or the lateral cyclic pitch required to suppress roll—will depend on the physical parameters of the helicopter, on the flight conditions throughout the maneuver, and on the pilot's actions on the rudder pedals. To gain some insight into the coupling, assume that a helicopter is in a steady turn with no sideslip. If the lateral flapping is to be eliminated, the change in lateral cyclic pitch must be that due to both the pitch rate and that due to coning. The pitch rate effect is:

$$\Delta A_{1_{\text{pitch rate}}} = \frac{q}{\Omega}$$

From Chapter 5:

$$q = \frac{g}{V} \frac{(n^2 - 1)}{n}$$

The coning effect can be found from the analysis made in Chapter 3, where it was shown that:

$$A_1 - b_{1_s} = -\frac{\left(\frac{4}{3} \mu a_0 + \frac{v_1}{\Omega R} \right)}{1 + \frac{\mu^2}{2}}$$

Since both a_0 and v_1 are direct functions of rotor thrust:

$$\Delta A_{1_{\text{maneuver}}} = (n - 1) A_{1_{\text{level flight}}}$$

For illustration, assume the example helicopter in a steady zero sideslip turn at 115 knots with a load factor of 1.5.

For this maneuver:

$$q = 0.14 \text{ rad/sec}$$
$$A_{1_{\text{level flight}}} = -2.2°$$
$$\Omega = 21.67 \text{ rad/sec}$$

Thus

$$\Delta A_1 = 57.3 \left(\frac{0.14}{21.67} \right) + 0.5(-2.2) = 0.37 - 1.10 = -0.73°$$

In other words, the pilot will have to hold left stick to prevent roll to the right because the effect of coning is more than the effect of pitch rate.

The pitching rate in this maneuver would be expected to produce a change in longitudinal flapping, but this flapping must be suppressed with longitudinal cyclic pitch in order to maintain zero pitching moments about the center of gravity (ignoring any damping effects of the fuselage and horizontal stabilizer until Chapter 9). The cyclic pitch required to do this is:

$$\Delta B_1 = \frac{1}{\partial a_{1_r}/\partial B_1} \frac{16}{\gamma} \frac{q}{\Omega}$$

From the trim equation of longitudinal flapping:

$$\frac{\partial a_{1_r}}{\partial B_1} = - \frac{\left(1 + \frac{3}{2} \mu^2 \right)}{1 - \frac{\mu^2}{2}}$$

Thus

$$\Delta B_1 = - \frac{\left(1 - \frac{\mu^2}{2} \right)}{1 + \frac{3}{2} \mu^2} \frac{16}{\gamma} \frac{q}{\Omega}$$

For the example maneuver:

$$\Delta B_1 = -57.3 \left[\frac{1 - \frac{0.3^2}{2}}{1 + \frac{3}{2}(.3)^2} \right] \frac{16}{8.1} \frac{(0.14)}{21.67} = -0.62°$$

This means that the pilot will experience an almost one-to-one coupling when performing this maneuver.

MOMENTS PRODUCED BY FLAPPING

The moments about the aircraft's center of gravity due to rotor flapping are produced by the couple at the hub as a result of the *rotor stiffness*, the tilt of the thrust vector perpendicular to the tip path plane, and the rotor's inplane force. From Figure 7.12:

$$M_{C.G.} = \frac{dM_M}{da_{1_s}} a_{1_s} + (Ta_{1_s} + H_{a_{1_s}=0})h_m + Tl_m$$

The rotor stiffness is produced by the vertical component of blade centrifugal force acting at the hinge offset; from Figure 7.10:

$$d\text{C.F.}_{\text{vert}} = m\Delta r(r' + e)\Omega^2 \frac{(\beta - a_0)r'}{(r' + e)}$$

$$\therefore \text{C.F.}_{\text{vert}} = \Omega^2(\beta - a_0) M_b/g$$

For pure longitudinal flapping:

$$\beta - a_0 = -a_{1_s} \cos \psi$$

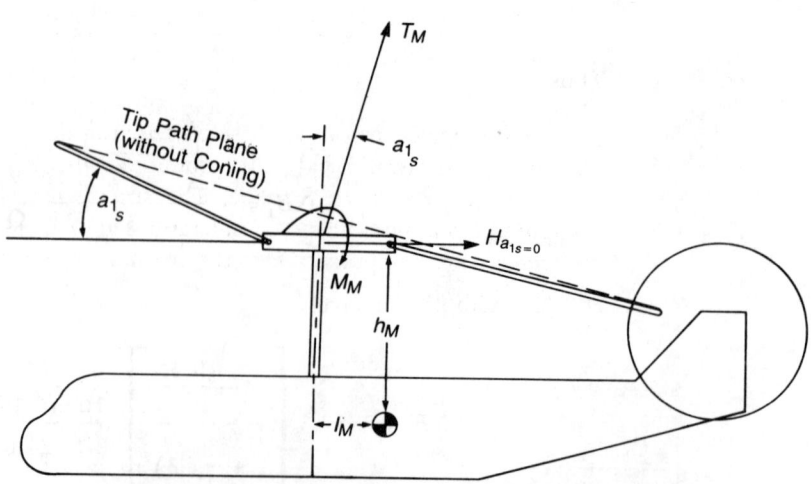

FIGURE 7.12   Moments Produced by Flapping

and the total rotor moment in the pitch direction for b blades is:

$$M_M = eb\Omega^2 a_{1_s} M_b/g \frac{1}{2\pi} \int_0^{2\pi} \cos^2 \psi \, d\psi$$

where M_b = First static moment

or

$$M_M = \frac{1}{2} eb\Omega^2 a_{1_s} M_b/g$$

For uniform blade mass distribution, m, in slugs per foot:

$$\frac{dM_M}{da_{1_s}} = \frac{1}{4} \left(\frac{e}{R} \right) bmR(\Omega R)^2$$

or

$$\frac{dM_M}{da_{1_s}} = \frac{3}{4} \left(\frac{e}{R} \right) \frac{A_b \rho R(\Omega R)^2 a}{\gamma}$$

(Note: This equation could have also been derived by using the vertical shear at the hinge location associated with the incremental inertia forces acting normal to the blade elements due to the flapping motion.) For the example helicopter at sea level:

$$\frac{dM_M}{da_{1_s}} = \frac{3}{4} \frac{(0.05)(240)(0.002378)(30)(650)^2(6)}{8.1} = 200,940 \text{ ft lb/rad}$$

For hingeless rotors whose blades are cantilevered from the shaft, the rotor stiffness can be visualized as being produced by blade bending moments as pairs of blades are bent into an S shape. The stiffness of a hingeless rotor is usually computed from mode shape considerations at the same time the flapping natural frequencies are computed. Once the stiffness is known, an effective hinge offset can be determined:

$$\left(\frac{e}{R} \right)_{\text{eff}} = \frac{1}{\dfrac{3}{4} \dfrac{b\Omega^2 I_b}{dM_M/da_{1_s}}}$$

For the hingeless rotor used on the Lockheed AH-56A, the effective hinge offset was approximately 13% of the radius.

H-FORCE DUE TO FLAPPING

For simple analyses, it is often satisfactory to assume that the total rotor vector is perpendicular to the tip path plane. It is more correct, however, to account for an effect of inflow that modifies this assumption. The geometry of this phenomenon is illustrated by a simple example in Figure 7.13. Here a helicopter is hovering with some nose-up rotor flapping, due in this case to the center of gravity being ahead of the shaft. The thrust vectors on two typical blade elements on the right and left blades are equal and tilted symmetrically with respect to the tip path plane but not to the shaft. Their contributions to the H-force, which is defined as being perpendicular to the shaft, are:

$$\Delta H_{\text{right}} = \Delta T_R \sin(a_{1_s} + \phi)$$

and

$$\Delta H_{\text{left}} = \Delta T_L \sin(a_{1_s} - \phi)$$

or

$$\Delta H = \Delta T_{\text{total}} \sin a_{1_s} \cos \phi$$

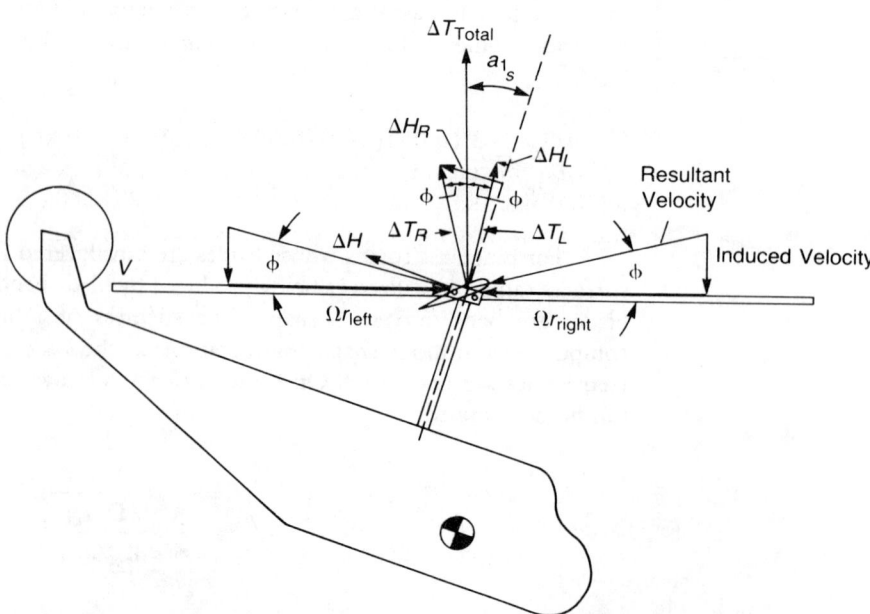

FIGURE 7.13 Illustration of Inflow Effect on H-Force

Thus, when the tip path plane is tilted with respect to the shaft, the H-force is decreased by the cosine of the inflow angle in this simple example.

The effect can be derived for the entire rotor from the first equation for C_H/σ derived in Chapter 3. Differentiating with respect to a_{1_s} gives:

$$\frac{\partial C_H/\sigma}{\partial a_{1_s}} = \frac{a}{8}\lambda' + \frac{a}{4}\left[\left(\frac{2}{3} + \mu^2\right)\theta_0 + \left(\frac{1}{2} + \frac{\mu^2}{2}\right)\theta_1 + \lambda' - \mu(B_1 + 2a_{1_s})\right]$$

but the second term, for all practical purposes, is the equation for C_T/σ, so that:

$$\frac{\partial C_H/\sigma}{\partial a_{1_s}} = C_T/\sigma + \frac{a}{8}\lambda'$$

Thus, as the helicopter goes faster and λ' becomes more and more negative, the effect of the tilt of the thrust vector is reduced and in some extreme conditions might even change sign. This has the effect of reducing the damping that one might expect from rotor flapping.

This effect was first pointed out in reference 7.2. The equation can be written in the format of that report by assuming that the thrust coefficient is a function of the pitch at the three-quarter radius position:

$$C_T/\sigma = \frac{a}{4}\left(\frac{2}{3}\theta_{.75} + \lambda'\right)$$

Solving for λ' and substituting in the equation for $(\partial C_H/\sigma)/\partial a_{1_s}$ gives:

$$\frac{\partial C_H/\sigma}{\partial a_{1_s}} = \frac{3}{2}C_T/\sigma\left(1 - \frac{a}{18}\frac{\theta_{.75}}{C_T/\sigma}\right)$$

This is a convenient form to use in hovering, but the one with λ' is recommended for forward flight. The reduction in the effect of the H-force due to flapping is proportional to the inflow which has to be compensated for by collective pitch.

For the example helicopter, the moment in foot-pounds produced about the center of gravity per radian of flapping is:

Flight Condition	Contribution of Hub Couple	Contribution of Rotor Force	Total
Hover	200,940	73,500	274,440
115 knots	200,940	123,000	323,940
160 knots	200,940	108,000	308,940

A derivation of the rotor Y-force equation shows that the effect is the same as with the H-force; that is:

$$\frac{\partial C_Y/\sigma}{\partial b_{1_s}} = C_T/\sigma + \frac{a}{8}\lambda'$$

EXAMPLE HELICOPTER CALCULATIONS

HOW TO'S

The following items can be evaluated by the methods in this chapter.

CHAPTER 8

The Helicopter in Trim

EQUATIONS OF EQUILIBRIUM

Like any aircraft in steady flight, the helicopter must be in equilibrium with respect to three forces and three moments acting along and around three orthogonal axes through its center of gravity. The analysis can be based on one of three possible systems of axes: wind axes, stability axes, or body axes. These systems are distinguished from each other by relating them to the three types of balance systems that are used in wind tunnels as shown in Figure 8.1. If the balance system is always aligned with the tunnel center line, the measurements will be in the wind axes system in which the X axis points along the line of flight in both the side and top views. If the balance system is mounted on a yaw table that rotates to produce sideslip conditions, the measurements will be in the stability axes system. For this case, the X axis will be aligned with the flight path in the side view but with the body in the top view. If the balance system is contained in the body of the model, the forces and moments will be measured in the body axes system and the X axis will line up with the body in both the side and top views. Although each system is valid, there are two reasons for using the body axes system in helicopter analysis. First, the other systems lose their significance in hover. Second, many helicopters are equipped with stability augmentation systems using gyros or bobweights whose displacements are measured with respect to the airframe, and the analysis of the effects of these devices is easiest in the body axes system.

(One of the minor annoyances that may have to be faced with the body axis system is the reluctance of the design department to define waterlines as being perpendicular to the rotor shaft. It is more likely that the designers will lay out the fuselage using some such arbitrary line as a cabin floor for a waterline and then tilt the rotor shaft forward to obtain a level floor and a streamline fuselage attitude at

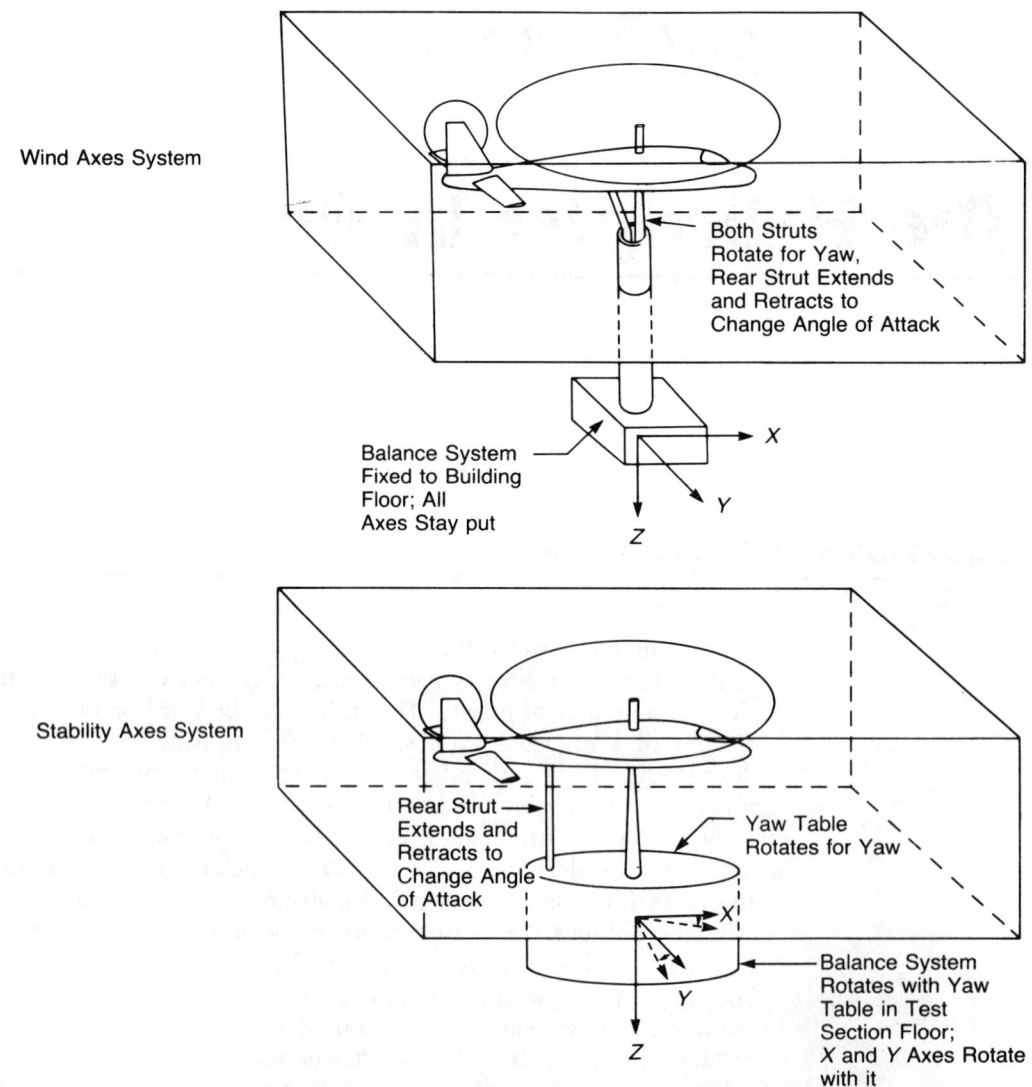

Wind Axes System

Both Struts
Rotate for Yaw,
Rear Strut Extends
and Retracts to
Change Angle of Attack

Balance System
Fixed to Building
Floor; All
Axes Stay put

Stability Axes System

Rear Strut
Extends and
Retracts to
Change Angle
of Attack

Yaw Table
Rotates for Yaw

Balance System
Rotates with Yaw
Table in Test
Section Floor;
X and *Y* Axes Rotate
with it

FIGURE 8.1 Wind Tunnel Balance Systems Using Various Axes Systems

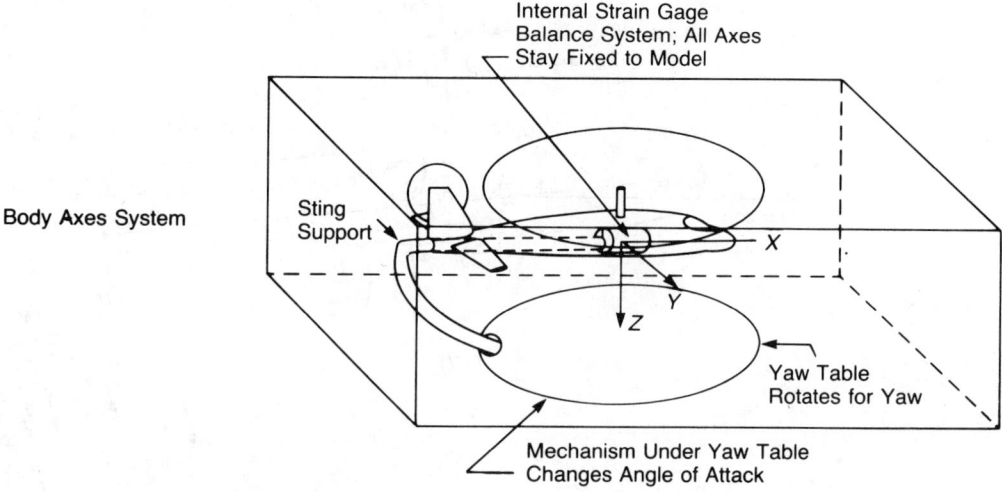

Internal Strain Gage
Balance System; All Axes
Stay Fixed to Model

Body Axes System

Sting
Support

X

Y

Z

Yaw Table
Rotates for Yaw

Mechanism Under Yaw Table
Changes Angle of Attack

FIGURE 8.1 (cont.)

the cruise condition. This produces unnecessary bookkeeping complications for the aerodynamicist in keeping track of center-of-gravity positions, moments of inertia, and horizontal stabilizer angle of attack. It would be helpful if the designers would use the shaft as their primary reference and then tilt the cabin floor and draw the fuselage contours for minimum drag. The finished helicopter would look the same but would be easier to analyze. This is the scheme that has been used for the example helicopter, but the analytical procedures that follow allow for those unfortunate cases where the shaft is tilted with respect to the body axis defined as a designer's waterline.)

The aerodynamic moments and forces in the body axis system that are acting on the helicopter are shown in Figure 8.2. They are due to the main rotor, the tail rotor, the horizontal stabilizer, the vertical stabilizer, and the fuselage. (For some helicopters, add a wing and/or a propeller.)

The six equations of equilibrium are:

Longitudinal
force (forward)
$$X_M + X_T + X_H + X_V + X_F = \text{G.W. } \sin \Theta$$

Lateral
force (right)
$$Y_M + Y_T \qquad + Y_V + Y_F = -\text{G.W. } \sin \Phi$$

Vertical
force (down)
$$Z_M + Z_T + Z_H + Z_V + Z_F = -\text{G.W. } \cos \Theta$$

Rolling moment
(down to right)
$$R_M + Y_M h_M + Z_M y_M + Y_T h_T + Y_V h_V + Y_F h_F + R_F = 0$$

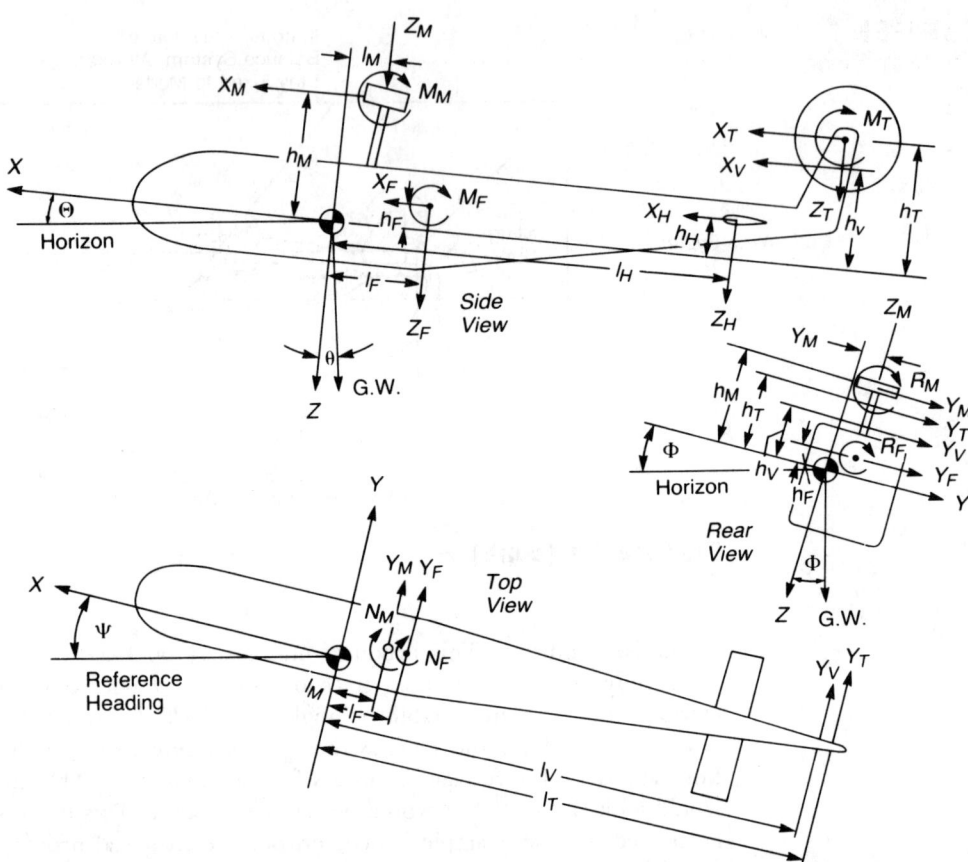

FIGURE 8.2   Forces and Moments Acting on Helicopter in Trim

Pitching moment
(nose-up)

$$M_M - X_M h_M + Z_M l_M + M_T - X_T h_T + Z_T l_T - X_H h_H + Z_H l_H$$
$$- X_V h_V + M_F + Z_F l_F - X_F h_F = 0$$

Yawing moment
(nose to right)

$$N_M - Y_M l_M - Y_T l_T - Y_V l_V + N_F - Y_F l_F = 0$$

When the equilibrium equations were developed for airplanes many years ago, all engineering analysis and reports were hand-lettered, and the three moments were written in script: $\mathscr{L}$, $\mathscr{M}$, and $\mathscr{N}$. Nowadays, typewriters and word processors don't make script characters. This makes a special problem with rolling moment, since L is used for total lift and, if also used for total rolling moment, causes confusion, especially in distinguishing C_L, meaning lift coefficient, from C_L meaning roll moment coefficient. For that reason, R is used in this book for rolling moment.

ELEMENTS OF THE TRIM EQUATIONS

Main Rotor

The main rotor as the primary source of forces and moments on the helicopter is shown in Figure 8.3 and is described by the following equations for a rotor whose advancing blade is on the right side:

$$X_M = -H_{a_{1_s}=0_M} \cos(a_{1_{s_M}} + i_M) + T_M \sin(a_{1_{s_M}} + i_M)$$

$$Y_M = T_M \cos b_{1_{s_M}} - H_{a_{1_s}=0_M} \sin \beta$$

$$Z_M = -T_M \cos(a_{1_{s_M}} + i_M)$$

$$R_M = \left(\frac{dR}{db_{1_{s_M}}}\right) b_{1_{s_M}} + Q_M \sin(i_M + a_{1_{s_M}})$$

$$M_M = \left(\frac{dM}{da_{1_{s_M}}}\right) a_{1_{s_M}} - Q_M \sin(b_{1_{s_M}})$$

$$N_M = Q_M \cos(i_M + a_{1_{s_M}}) \cos b_{1_{s_M}}$$

The equations contain several types of terms, as listed in Table 8.1.

TABLE 8.1
Main Rotor Parameters

Term	Symbol	Type
Thrust	T_M	unknown
Longitudinal flapping	$a_{1_{s_M}}$	unknown
Lateral flapping	$b_{1_{s_M}}$	unknown
Sideslip angle	β	unk. or known
Rotor H-Force	$H_{a_{1_s}=0_M}$	trim
Rotor torque	Q_M	trim
Hub rolling moment stiffness	$(dR/db_{1_s})_M$	phys. parameter
Hub pitching moment stiffness	$(dM/da_{1_s})_M$	phys. parameter

An example calculation will be done later to illustrate the use of these terms.

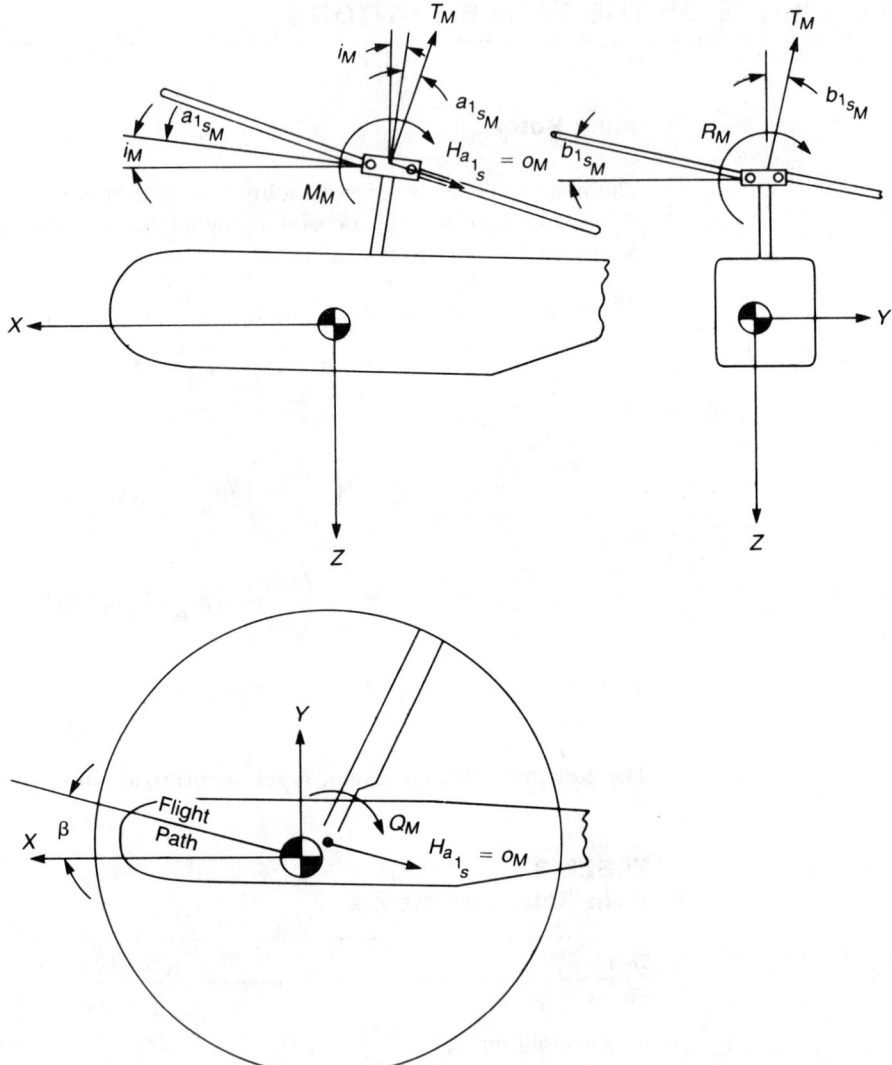

FIGURE 8.3   Main Rotor Forces and Moments

Tail Rotor

The tail rotor contributes primarily to the side force and yawing moment equations of equilibrium. Nevertheless, a complete analysis will also account for small contributions to the other four equations of equilibrium. Figure 8.4 shows the geometric relationships that govern the individual tail rotor forces and moments.

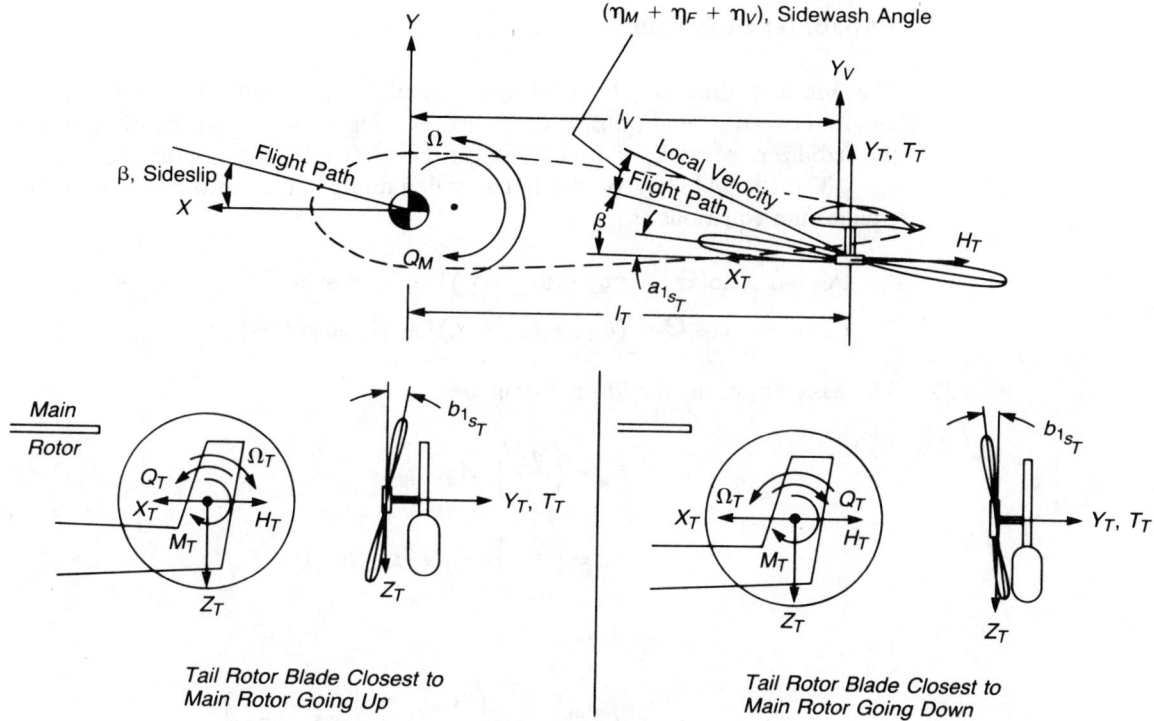

FIGURE 8.4   Aerodynamic Conditions at the Tail Rotor

$$X_T = -H_T$$

$$Y_T = T_T = \frac{Q_M}{l_T} - Y_V \frac{l_V}{l_T}$$

$$Z_T = b_{1_{S_T}} T_T \text{ (for tail rotor blade closest to main rotor going up)}$$

or

$$Z_T = -b_{1_{S_T}} T_T \text{ (for tail rotor blade closest to main rotor going down)}$$

$$M_T = -Q_T \text{ (for tail rotor blade closest to main rotor going up)}$$

or

$$M_T = Q_T \text{ (for tail rotor blade closest to main rotor going down)}$$

Once the tail rotor thrust to provide the antitorque forces that the vertical stabilizer does not provide is known, the other tail rotor parameters, $b_{1_{S_T}}$, H_T, and Q_D, can be calculated by the methods of Chapter 3.

Horizontal Stabilizer

The lift and drag of the horizontal stabilizer play important roles in the longitudinal trim conditions of the helicopter. Figure 8.5 shows the conditions at the stabilizer, especially the relationships that define its angle of attack.

The lift and drag of the horizontal stabilizer enter into the longitudinal equilibrium equations as:

$$X_H = L_H \sin[\Theta - (\varepsilon_{M_H} + \varepsilon_{F_H} + \gamma_c)] - D_H \cos[\Theta - (\varepsilon_{M_H} + \varepsilon_{F_H} + \gamma_c)]$$

$$Z_H = -L_H \cos[\Theta - (\varepsilon_{M_H} + \varepsilon_{F_H} + \gamma_c)] - D_H \sin[\Theta - (\varepsilon_{M_H} + \varepsilon_{F_H} + \gamma_c)]$$

The basic equations for lift and drag are:

$$L_H = \left(\frac{q_H}{q}\right) q A_H C_{L_H}$$

$$L_H = \left(\frac{q_H}{q}\right) q A_H [a(\alpha - \alpha_{LO})]_H$$

and

$$D_H = \left(\frac{q_H}{q}\right) q A_H \left\{ \frac{C_{L_H}^2}{\pi \text{A.R.}} (1 + \delta_i) + C_{D_0} \right\}_H$$

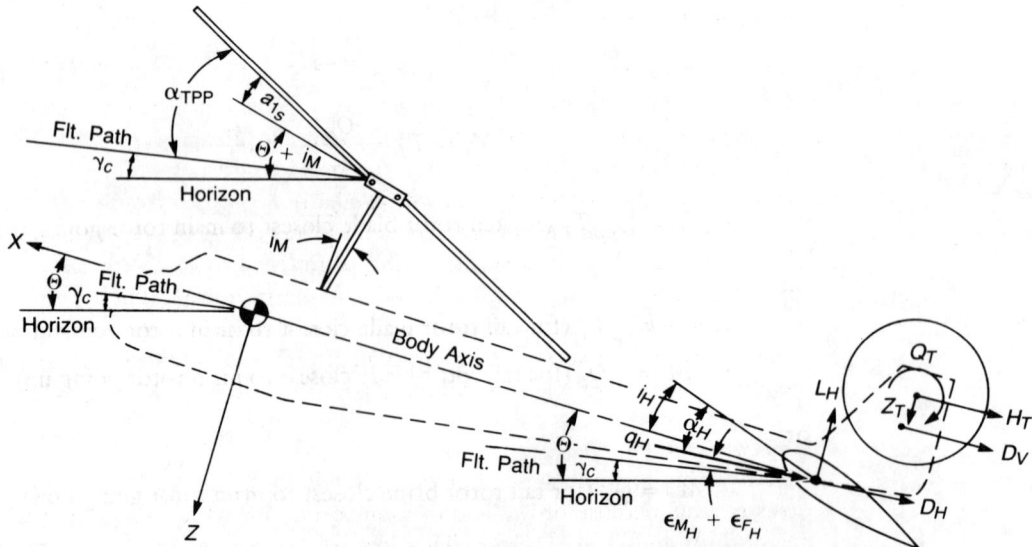

Note on subscripts: The first subscript refers to the element generating the effect. The second subscript refers to the element receiving the effect.

FIGURE 8.5 Longitudinal Component Forces and Moments

By examination of Figure 8.5, the angle of attack of the chord line of the stabilizer is:

$$\alpha_H = \Theta + i_H - (\varepsilon_{M_H} + \varepsilon_{F_H}) - \gamma_c, \text{ radians}$$

or

$$\alpha_H = \alpha_{TPP} - a_{1_s} - i_M + i_H - (\varepsilon_{M_H} + \varepsilon_{F_H}), \text{ radians}$$

These equations contain several types of parameters that must be evaluated. The first are the geometric terms, which are defined by the configuration description: stabilizer area, A_H; stabilizer aspect ratio, AR_H; incidence of the main rotor shaft with respect to the body axis, i_M; incidence of the chord line of the horizontal stabilizer, i_H; and angle of zero lift of the airfoil used on the stabilizer, α_{LO}. Three more parameters that may be found indirectly from the configuration description are the slope of the lift curve, a_H, which is a function of aspect ratio and sweep of the stabilizer; the span-efficiency factor, δ_{i_H}; and the profile drag coefficient, $C_{D_{0_H}}$. Some methods for evaluating these will be given. The second set of parameters consists of the flight conditions: dynamic pressure, q, and the angle of climb, γ_c, which for most analyses are known beforehand. The third type are the environmental conditions in which the stabilizer operates: the dynamic pressure ratio q_H/q, and the downwash angles induced by the main rotor and fuselage, ε_{M_H} and ε_{F_H}. These terms must be evaluated by some more-or-less empirical methods based either on studies of previous designs or on wind tunnel tests of the configuration analyzed. Both types of methods will be discussed. The last set of parameters are the trim conditions, Θ, α_{TPP}, and a_{1_s}, which are to be solved for in the procedure to be outlined later.

Slope of Lift Curve

Evaluation of the slope of the lift curve of the horizontal stabilizer, a_H, as a function of aspect ratio and sweep of the mid-chord line can be made from Figure 8.6, which has been adapted from reference 8.1. Test data for several stabilizers and wings of Bell helicopters are shown on Figure 8.7, taken from reference 8.2.

Many modern helicopters have end plates on their horizontal stabilizers in the form of small vertical surfaces. End plates tend to block the flow around the tip from bottom to top and thus reduce the tip vortex. The result is that the effective span—or the aspect ratio—is increased, with a corresponding increase in the slope of the lift curve. The increase in effective aspect ratio as a function of the height-to-span ratio is given in Figure 8.8, which was taken from reference 8.3.

Dynamic Pressure Ratio

The dynamic pressure at the horizontal stabilizer, q_H, is usually less than the free-stream value because of the loss of momentum due to the air passing around the main rotor hub and fuselage. This loss, of course, is greatest for the inboard regions of the stabilizer and is less outboard. For preliminary design purposes, Figure 8.9 can be used to obtain an estimate of what the average dynamic pressure

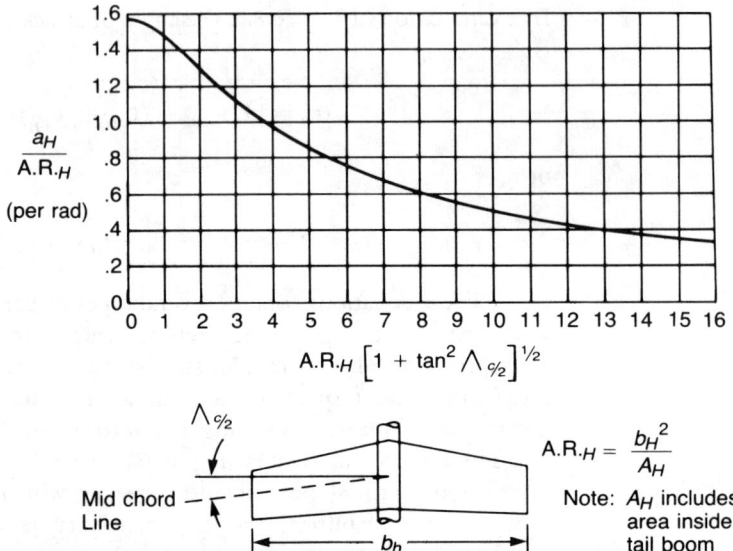

FIGURE 8.6 Slope of Lift Curve for Horizontal Stabilizer

Source: Hoak, "USAF Stability and Control Datcom," 1960.

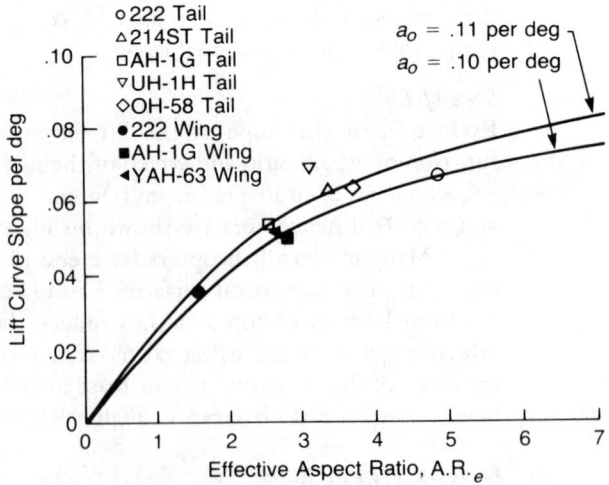

FIGURE 8.7 Lift Curve Slope for Helicopter Wings and Horizontal Tails

Source: Harris et al., "Helicopter Performance Methodology at Bell Helicopter Textron," AHS 35th Forum, 1979.

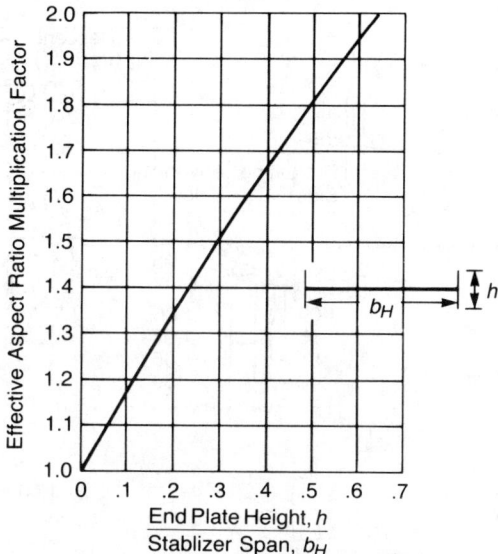

FIGURE 8.8   Effect of End Plates on Effective Aspect Ratio

Source: Hoerner & Borst, *Fluid-Dynamic Lift*, published by Mrs. Hoerner.

ratio might be for a given configuration. One set of flight test data from reference 8.4 and three sets obtained during the powered wind tunnel tests reported in references 8.5, 8.6, and 8.7 are presented. Note that the presence of the rotor wake has a significant effect on the dynamic pressure ratio, making the distribution unsymmetrical and producing some values above unity, especially behind the advancing side of the rotor disc. The effect of the rotor can be even more dramatic for those speed conditions where the horizontal stabilizer is immersed in the rotor wake. As an illustration of this effect, Figure 8.10 presents the results of wind tunnel and flight tests reported in reference 8.8. The fact that the increase in dynamic pressure can be higher than the disc loading is a result of the distortion of the wake in low-speed flight discussed in Chapter 3 under "Correlation of Flapping with Test Results."

If a wind tunnel model of a new design is being tested, the dynamic pressure ratio can be determined indirectly. This is done by holding the model at a constant angle of attack and varying the incidence of the horizontal stabilizer (or, as is more common, cross-plotting data from several pitch runs with different incidence settings). The change in measured pitching moment can then be used to evaluate the product of dynamic pressure ratio and the slope of the lift curve for the horizontal stabilizer:

$$a_H \left(\frac{q_H}{q} \right) = - \frac{\dfrac{\Delta M}{\Delta i_H}}{l_H q A_H}$$

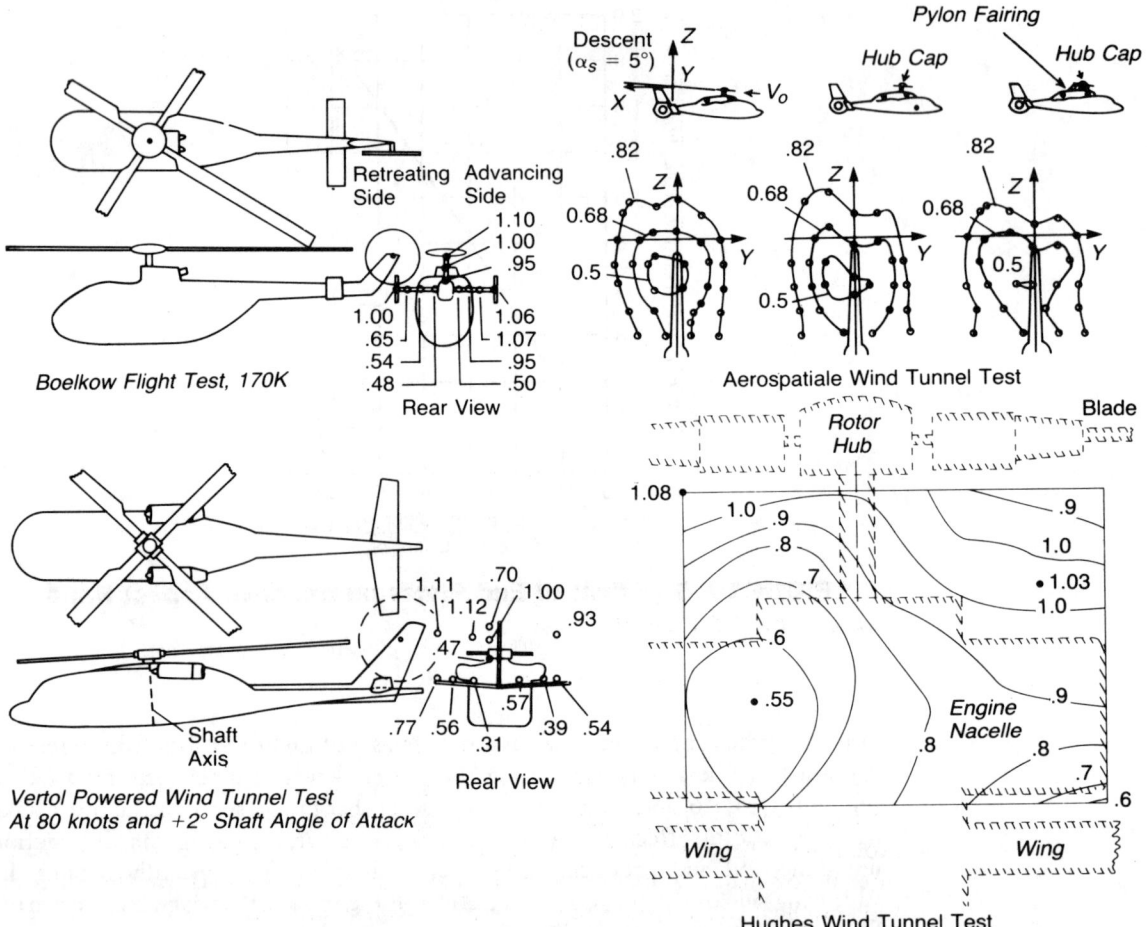

FIGURE 8.9   Measured Dynamic Pressure Ratios at Empennage

This product will be valid for the stabilizer geometry tested and can be used directly in this form in the analysis. For design studies in which alternate stabilizers are being investigated, the ratio, q_H/q can be found by assuming that the value of a_H found from Figure 8.6 is valid.

Downwash Angle
The downwash angle at the horizontal stabilizer is generated by both the main rotor and the fuselage—although the latter effect is usually small unless the helicopter has a wing or has widely spaced engine nacelles with a combined span greater than the span of the stabilizer. The downwash due to the rotor can be estimated from the results of downwash surveys made in a wind tunnel and reported in reference 8.9. Figure 8.11 presents the test measurements, at a tip speed ratio of 0.23, which may be taken as typical of forward flight, for several

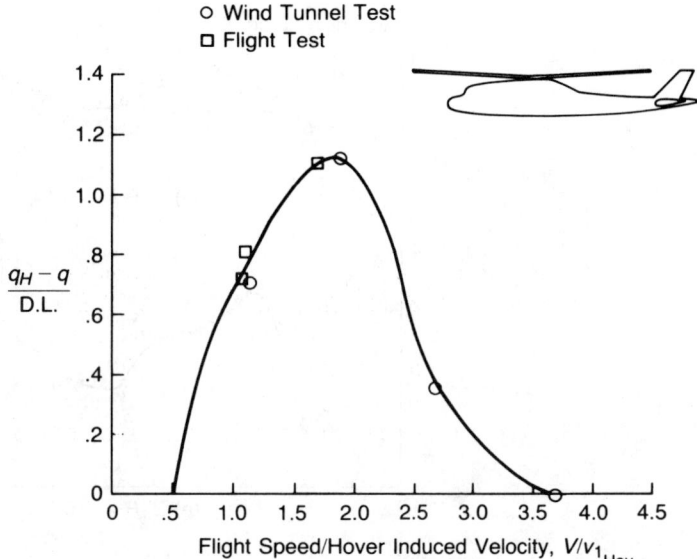

○ Wind Tunnel Test
□ Flight Test

FIGURE 8.10   Dynamic Pressure Increase at Horizontal Stabilizer at Low Forward Speed

Source: Blake & Alansky, "Stability and Control of the YUH-6IA," *JAHS* 22-1, 1977.

vertical locations and two longitudinal locations. It may be seen that the induced velocity ratio, v_H/v_1, exceeds 2—the maximum theoretical value—in some locations. The downwash angle due to the main rotor at the horizontal stabilizer, can be estimated from the curve of Figure 8.11 that most closely matches the longitudinal and vertical position of the stabilizer with respect to the main rotor. Using a mean value of v_H/v_1 that corresponds to the span of the stabilizer, the downwash angle is:

$$\varepsilon_{M_H} = \frac{v_H}{v_1} \frac{v_1}{V}$$

or

$$\varepsilon_{M_H} = \frac{v_H}{v_1} \frac{\text{D.L.}}{4q}$$

As an illustration, the horizontal stabilizer of the example helicopter is located at $X'/R = -1.08$ and $Z'/R = +0.3$, and the semispan/radius ratio is 0.2. From Figure 8.11, the effective value of v_H/v_1 is 1.5. At 115 knots and design gross weight, the corresponding downwash angle is 0.06 radians or 3.5°.

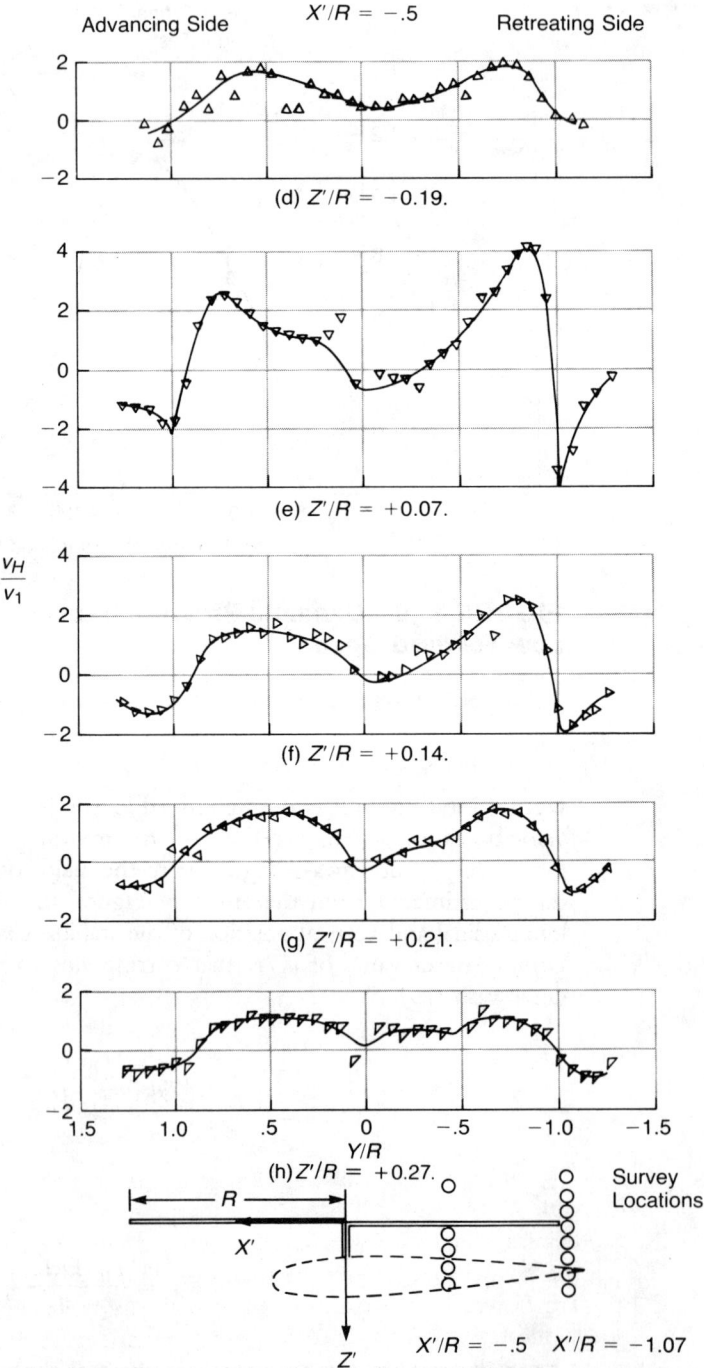

FIGURE 8.11   Downwash Behind a Rotor

Source: Herson & Katzoff, "Induced Velocities near a Lifting Rotor with Non-Uniform Disc Loading," NACA TR 1319, 1957.

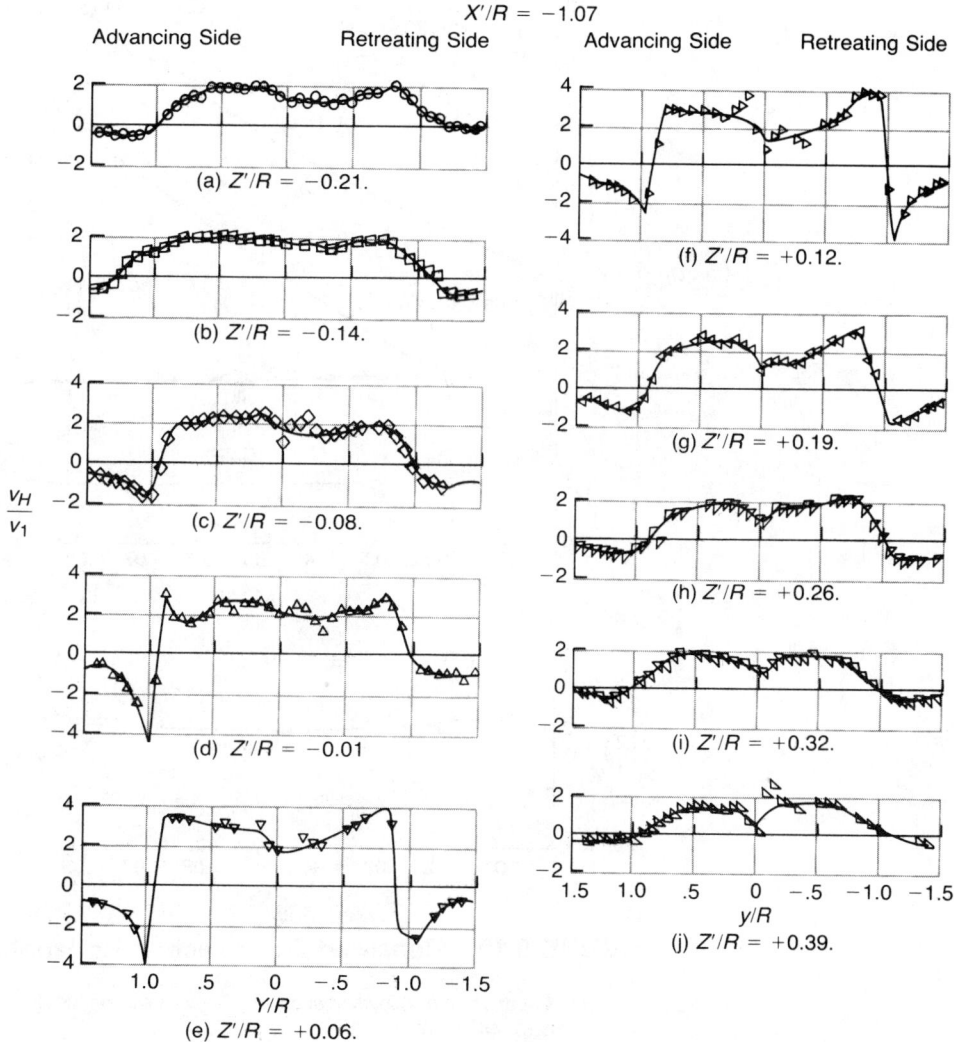

$$\frac{v_H}{v_1}$$

$X'/R = -1.07$

(a) $Z'/R = -0.21$.

(b) $Z'/R = -0.14$.

(c) $Z'/R = -0.08$.

(d) $Z'/R = -0.01$

(e) $Z'/R = +0.06$.

(f) $Z'/R = +0.12$.

(g) $Z'/R = +0.19$.

(h) $Z'/R = +0.26$.

(i) $Z'/R = +0.32$.

(j) $Z'/R = +0.39$.

FIGURE 8.11 (cont.)

Figure 8.11 also shows an effect that was not recognized as significant at the time of the test but has been recognized since then. It is the higher downwash behind the advancing side than behind the retreating side for locations just behind the hub. When helicopters with high disc loading and large horizontal stabilizers are flown, it is found that they have a coupling of pitch with sideslip as the stabilizer moves either to a high downwash region behind the advancing side or to a low downwash region behind the retreating side. Discussions of this effect are contained in references 8.10 and 8.11.

Another wind tunnel test in which the downwash angle at the horizontal stabilizer was measured directly is reported in reference 8.12. In these tests, a free-

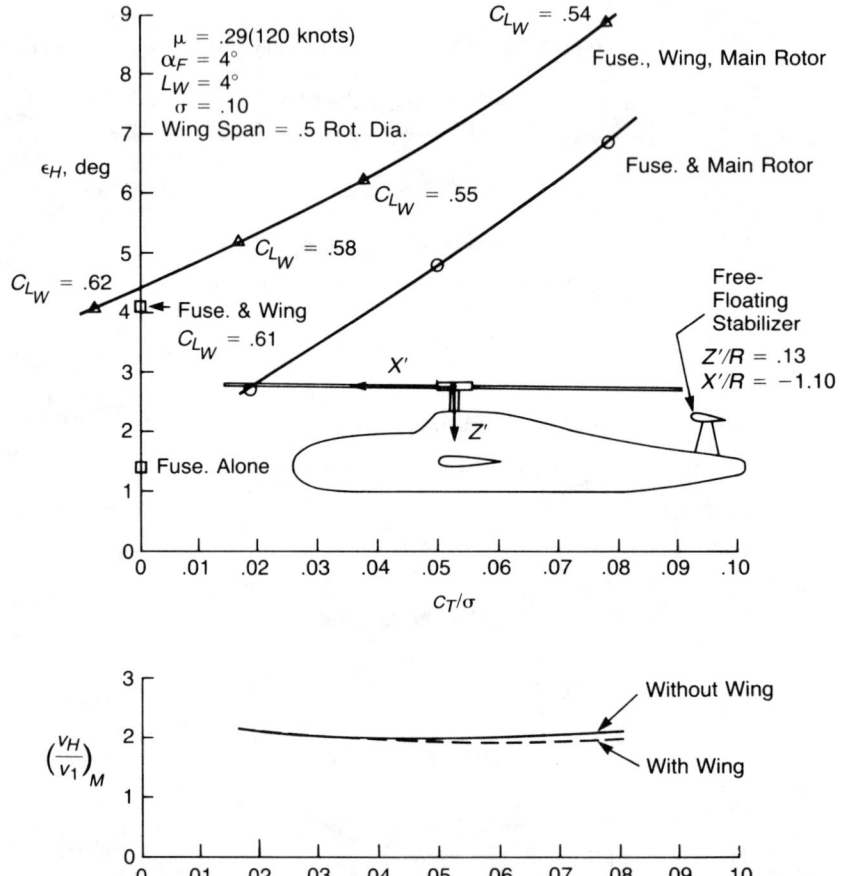

FIGURE 8.12   Measured Downwash at Horizontal Stabilizer

Source: Bain & Landgrebe, "Investigation of Compound Helicopter Interference Effects," USAAVLABS TR 67-44, 1967.

floating stabilizer was used as a flow vane. Figure 8.12 shows the measured downwash angle for various combinations of fuselage, wing, and rotor at constant fuselage angle of attack as the rotor thrust was varied with collective pitch. The downwash angles due to rotor with and without the wing have been converted into induced-velocity ratios on the bottom portion of the figure by using the equations:

$$\left(\frac{v_H}{v_1}\right)_{\text{without wing}} = \frac{1}{57.3}\left[\frac{\varepsilon_{M+F_H} - \varepsilon_{F_H}}{C_T/\sigma \, \frac{\sigma}{2\mu^2}}\right]$$

and

$$\left(\frac{v_H}{v_1}\right)_{\text{with wing}} = \frac{1}{57.3}\left\{\frac{\varepsilon_{M+F+W_H} - (\varepsilon_{F+W_H})\left(\dfrac{C_{L_W}}{C_{L_{W_{C_T/\sigma=0}}}}\right)}{C_T/\sigma\,\dfrac{\sigma}{2\mu^2}}\right\}$$

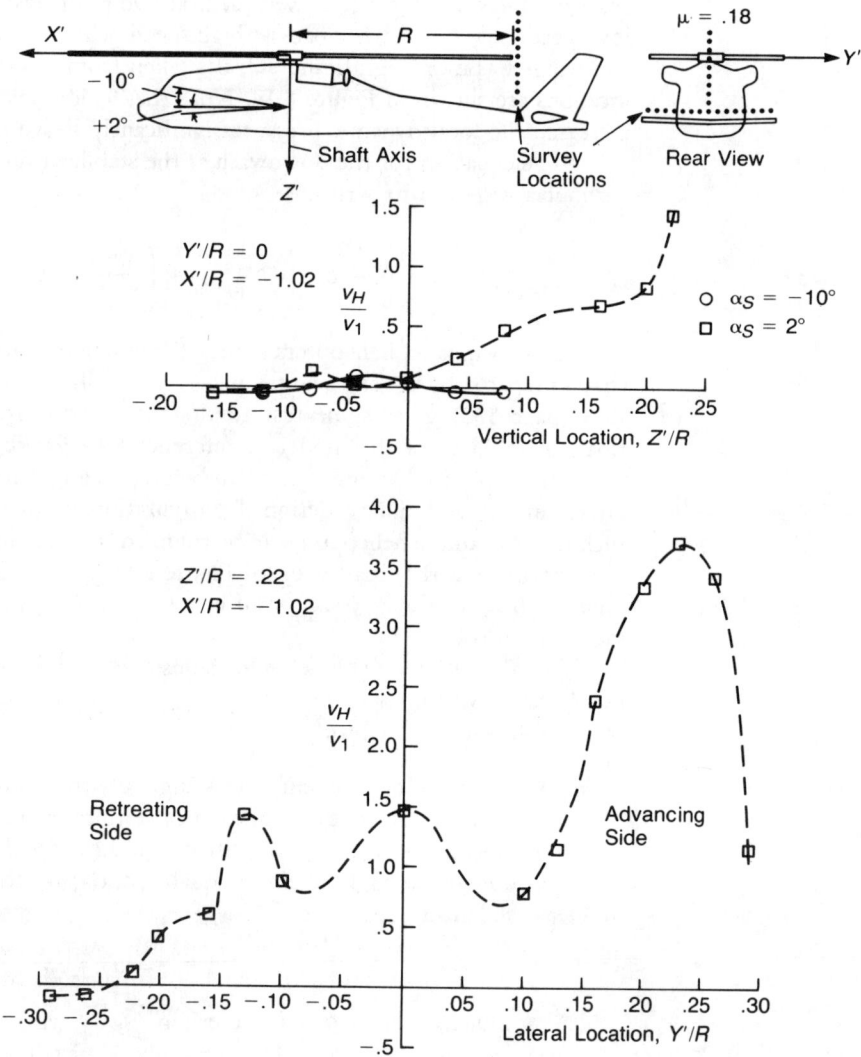

FIGURE 8.13   Rotor-Induced Downwash at Empennage

Source: Sheridan, "Interactional Aerodynamics of the Single Rotor Helicopter Configuration" USARTL TR 78-23A, 1978.

It may be seen that for this configuration, the rotor-induced velocity ratio is essentially the fully developed value of 2.0, which would also be read from Figure 8.11 for this stabilizer position, and that the presence of the wing has essentially no effect. Figure 8.13 shows yet another set of downwash measurements from the wind tunnel tests of the powered model of reference 8.5. Again the asymmetry of the lateral distribution is evident.

A more recent wind tunnel test produced the data on Figure 8.14 for the flow conditions at the stabilator position of the Hughes AH-64. The powered model used a rotor from a previous test that was somewhat small for the size of the fuselage. Thus the flow survey was made at two positions: one close to the rotor at low speeds, and one further back at high speed, where the influence of the fuselage and wings was more significant. Results taken from reference 8.7 for both survey locations are shown in Figure 8.14. Note that at low test speeds, the rotor wake increased the local dynamic pressure significantly above that of the wind tunnel.

The equation for the downwash at the stabilizer due to the fuselage with or without a wing can be written:

$$\varepsilon_{F_H} = \varepsilon_{F_{\alpha_F=0_H}} + \left(\frac{d\varepsilon_{F_H}}{d\alpha_F} \right) \alpha_F$$

Since wings on helicopters are relatively small compared to the fuselage, the charts prepared by the airplane people are generally not directly applicable. Figure 8.15 (page 500) gives some test results for the downwash as measured by the floating stabilizer of the model of reference 8.12 for the fuselage alone and for three different-sized wings. These can be used as a rough guide for estimating the effect during preliminary design. Configurations with external engine nacelles, such as the example helicopter, can be assumed to have almost the same downwash characteristics as the small wing of Figure 8.15. In another wind tunnel test, this time on the Sikorsky S-76, reported in reference 8.13, the value of $d\varepsilon_{F_H}/d\alpha_F$ was measured as 0.15.

If a wind tunnel model with adjustable horizontal stabilizer incidence is being tested without a rotor, the fuselage-induced downwash can be determined by the following procedure:

- Plot pitching moment versus angle of attack with stabilizer off and with stabilizer on at several incidence settings, as in Figure 8.16 (page 502).
- At each intersection the lift of the horizontal stabilizer is zero so the downwash angle must be equal to the geometric angle of attack of the stabilizer:

$$\varepsilon_{F_H} = \alpha_F + i_H$$

- Plot ε_{F_H} versus α_F and determine $\varepsilon_{F_{\alpha_F=0_H}}$ and $(d\varepsilon_F/d\alpha_F)_H$

Span Efficiency Factor
Since the stabilizer lift distribution is usually strongly affected by flow irregularities coming back from the fuselage and rotor, it is not certain that

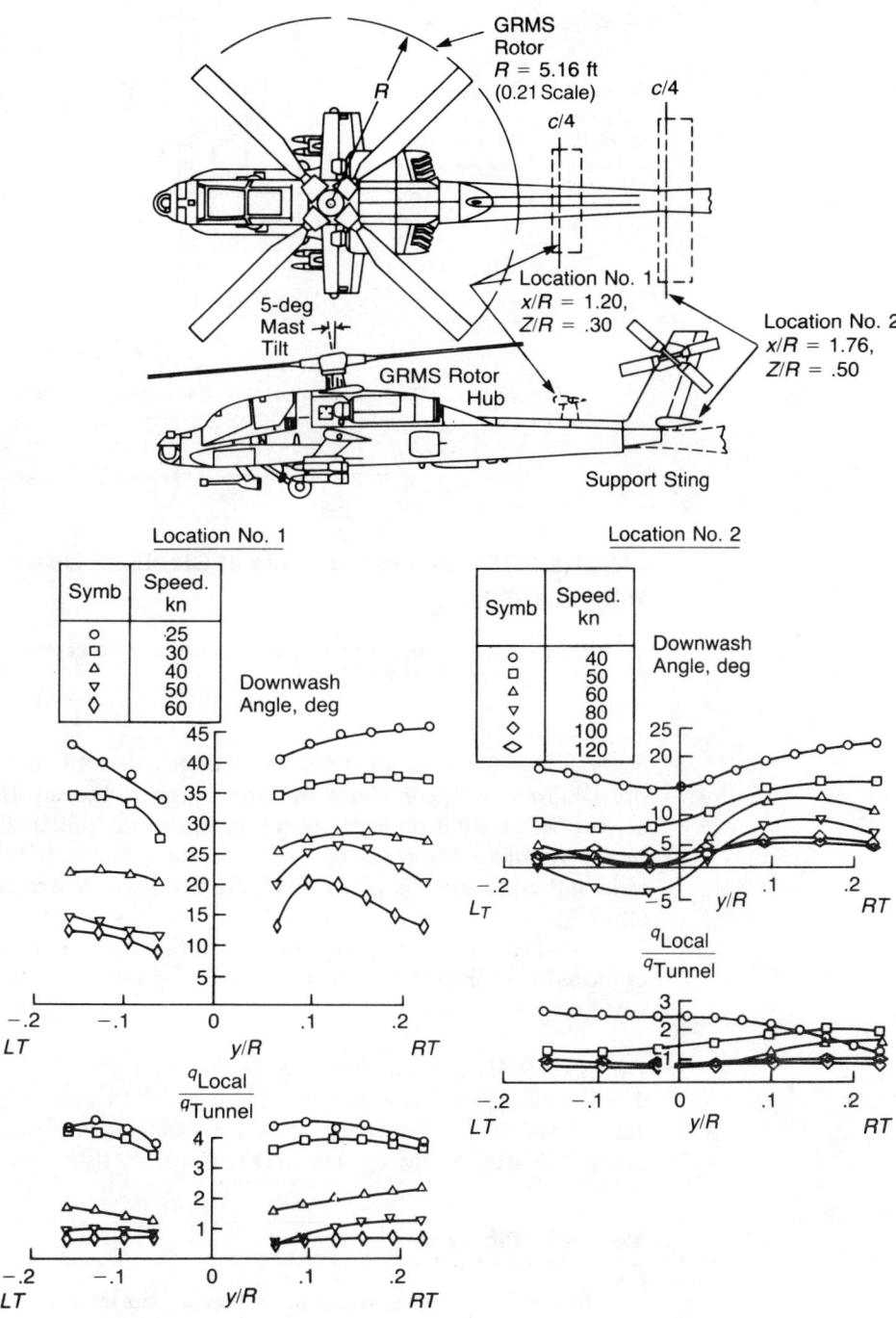

FIGURE 8.14   Measured Downwash Angles

Source: Logan, Prouty, & Clark, "Wind Tunnel Tests of Large and Small Scale Rotor Hubs and Pylons," USAVRADCOM TR-80-D-21, 1981.

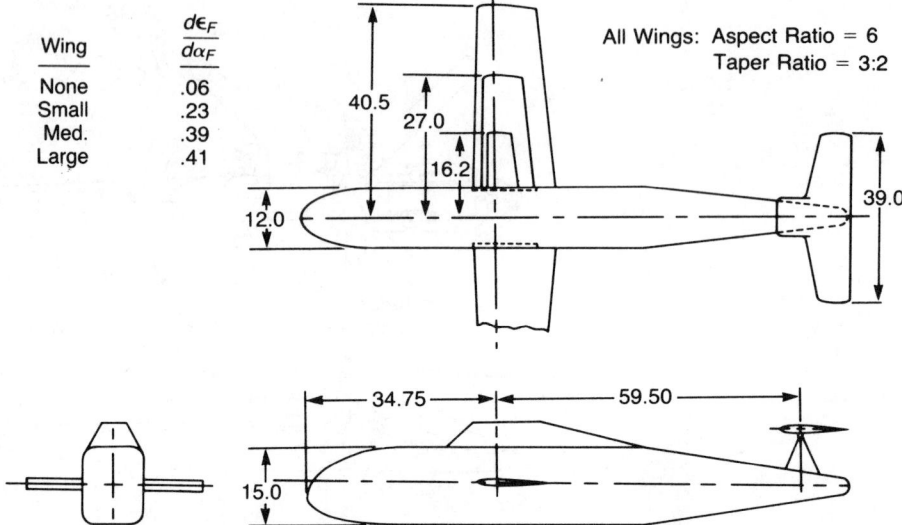

Wing	$\dfrac{d\epsilon_F}{d\alpha_F}$
None	.06
Small	.23
Med.	.39
Large	.41

All Wings: Aspect Ratio = 6
Taper Ratio = 3:2

FIGURE 8.15   Downwash Ratio at Stabilizer Due to Fuselage with or without Wing

Source: Bain & Landgrebe, "Investigation of Compound Helicopter Aerodynamic Interference Effects," USAAVLABS TR 67-44, 1967.

methods used for wings are directly applicable unless the configuration is such that the stabilizer can be considered to be operating in clean air. However, if values are needed before wind tunnel tests are done, the wing method is the only method readily available. The theoretical span-efficiency factor, δ_i as a function of aspect ratio and taper ratio is given in Figure 8.17, which was taken from reference 8.14.

The zero lift profile drag coefficient, C_{D_0}, of the horizontal stabilizer can be estimated from Figure 6.30 of Chapter 6. For this case, the Reynolds number is based on the chord.

Horizontal Stabilizer Characteristics of the Example Helicopter
The preceding methods can be used to make an estimate of the aerodynamic characteristics of a horizontal stabilizer from its physical parameters. For the example helicopter, these parameters and characteristics are shown in Table 8.2.

Vertical Stabilizer

The forces on the vertical stabilizer play a primary role in the yawing moment equilibrium equation but also appear as small participants in some of the others. Figure 8.18 shows the geometric relationships that are used in evaluating these forces.

TABLE 8.2
Horizontal Stabilizer—Example Helicopter

Physical Parameters			*Aerodynamic Characteristics*			
Parameter	*Symbol*	*Value*	*Characteristic*	*Symbol*	*Value*	*Source*
Span	b_H	9 ft	Slope of Lift Curve	a_H	4.0/rad	Figure 8.6
Root chord	c_{r_H}	2.34 ft	Dynamic Pressure Ratio	q_H/q_H	.6	Figure 8.9
Tip chord	c_{t_H}	1.66 ft	Rotor Downwash Ratio	v_H/v_{1_H}	1.5	Figure 8.11
Area	A_H	18 sq ft	Fuse. Downwash Ratio	$d\epsilon/d\alpha_F$	.23	Figure 8.15
Taper ratio	λ_H	.71	Span Efficiency Factor	δ_{i_H}	.02	Figure 8.17
Half chord sweep angle	$\Lambda c/2_H$	13°	Zero Lift Drag Coefficient	$C_{D_{0_H}}$	.0064	Figure 6.30
Aspect ratio	$A.R._H$	4.5				
Incidence	i_H	−3°				
Horizontal distance from main rotor	x/R_H	−1.08				
Vertical distance from main rotor	z'/R_H	.3				
Zero lift angle of attack	α_{LO_H}	0°				

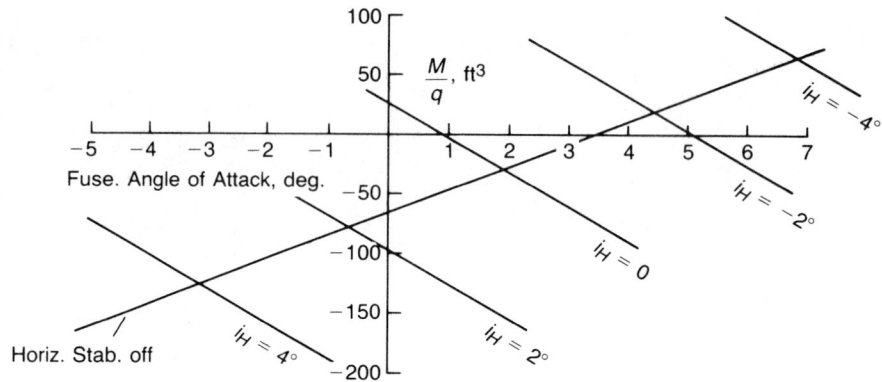

$$\epsilon_F = \alpha_F + i_H \text{ at intersections}$$

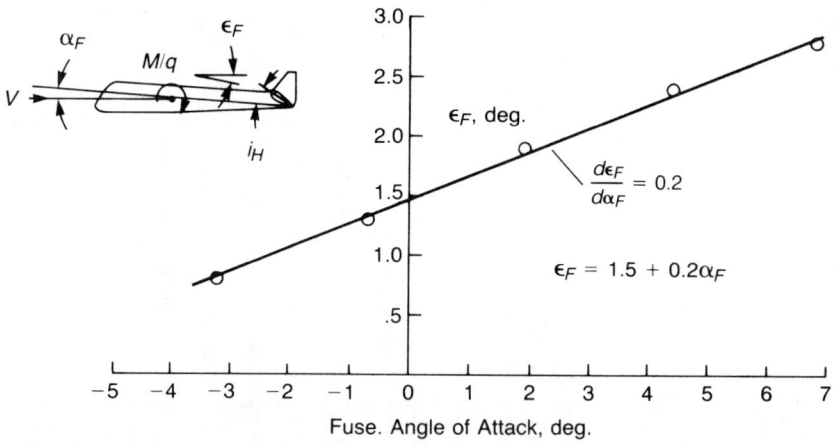

FIGURE 8.16   Wind Tunnel Results Used to Calculate Fuselage-Induced Downwash Ratio at Horizontal Stabilizer

The equations for the forces on the vertical stabilizer are:

$$X_V = -D_V \cos[\beta + \eta_{M_V} + \eta_{T_V} + \eta_{F_V}] - L_V \sin[\beta + \eta_{M_V} + \eta_{T_V} + \eta_{F_V}]$$

$$Y_V = L_V \cos[\beta + \eta_{M_V} + \eta_{T_V} + \eta_{F_V}] - D_V \sin[\beta + \eta_{M_V} + \eta_{T_V} + \eta_{F_V}]$$

$$Z_V = X_V \sin[\Theta - (\varepsilon_{M_V} + \varepsilon_{F_V} + \gamma_c)]$$

where β is the sideslip angle and η is the sidewash angle at the vertical stabilizer induced by the other components of the helicopter.

The basic equations for lift and drag of the vertical stabilizer are:

$$L_V = q\left(\frac{q_V}{q}\right)A_V C_{L_V}$$

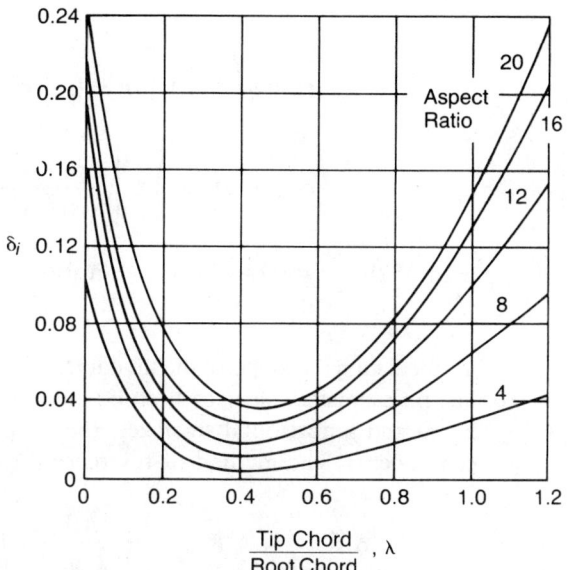

FIGURE 8.17   Chart for Determining Span Efficiency Factor

Source: Pope, *Basic Wing and Airfoil Theory*, McGraw-Hill, 1951.

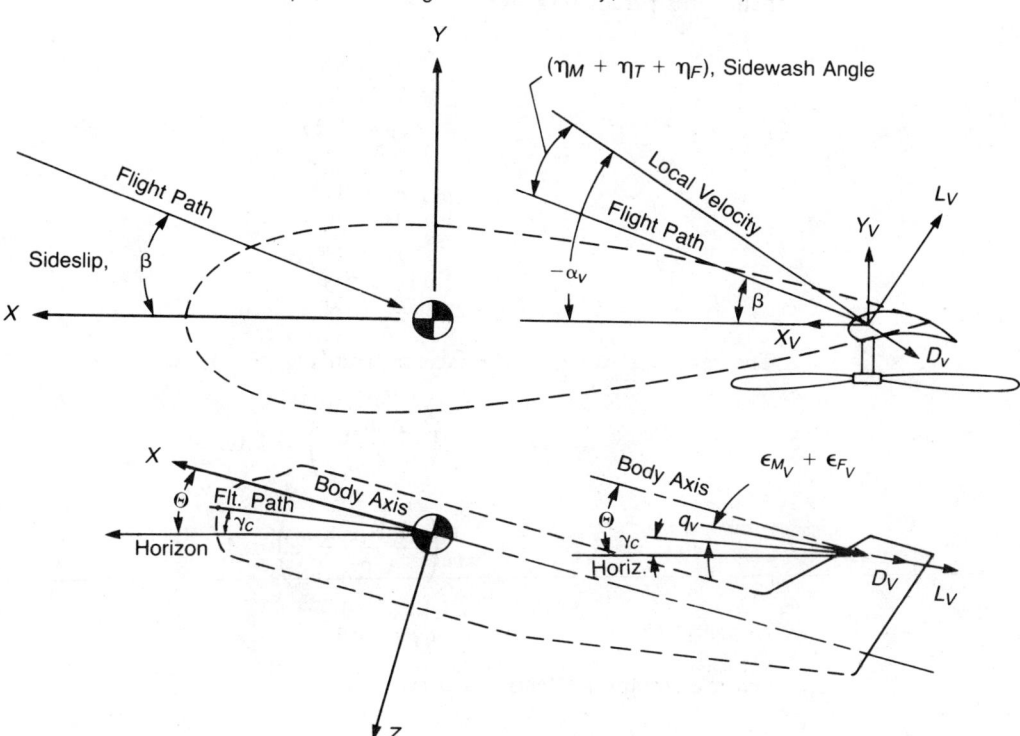

FIGURE 8.18   Aerodynamic Conditions at Vertical Stabilizer

or:

$$L_V = q\left(\frac{q_V}{q}\right) A_V \left[a(-\beta - \alpha_{LO} - \eta_M - \eta_T - \eta_F) \right]_V$$

$$D_V = q\left(\frac{q_V}{q}\right) A_V \left[\frac{C_L^2(1+\delta_i)}{\pi A.R.} + C_{D_0} \right]_v + \Delta D_{V_{int}}$$

Each of the new terms in these equations will be discussed separately.

Slope of Lift Curve

In calculating the slope of the lift curve, the effective aspect ratio of the surface must first be estimated. Since this can be influenced by the end-plating effects of the tail boom and a horizontal stabilizer, the estimating process must account for these components. The method of reference 8.1 makes use of the following equation:

$$A.R._{V_{eff}} = A.R._{V_{geo}} \left\{ \left(\frac{A.R._V}{A.R._{V+B}}\right) \left[1 + K_H \left(\frac{A.R._{V+B+H}}{A.R._{V+B}}\right) \right] \right\}$$

where the three factors; $(A.R._V/A.R._{V+B})$, $(A.R._{V+B+H}/A.R._{V+B})$, and K_H are given as functions of geometric parameters in Figure 8.19. As an illustration, the example helicopter parameters are:

$$S_V = 33$$
$$b_V = 7.7$$
$$c_V = 4.25$$
$$A.R._{V_{geo}} = 1.8$$
$$\lambda_V = .21$$
$$2r_1 = 1.5$$
$$S_H = 18$$
$$x = 2.5$$
$$Z_H = 0$$

The resultant values of the factors from Figure 8.19 are:

$$\left(\frac{A.R._V}{A.R._{V+B}}\right) = 1.05$$

$$\left(\frac{A.R._{V+B+H}}{A.R._{V+B}}\right) = 1.1$$

$$K_H = .64$$

and the resultant effective aspect ratio is:

$$A.R._{V_{eff}} = 3.2$$

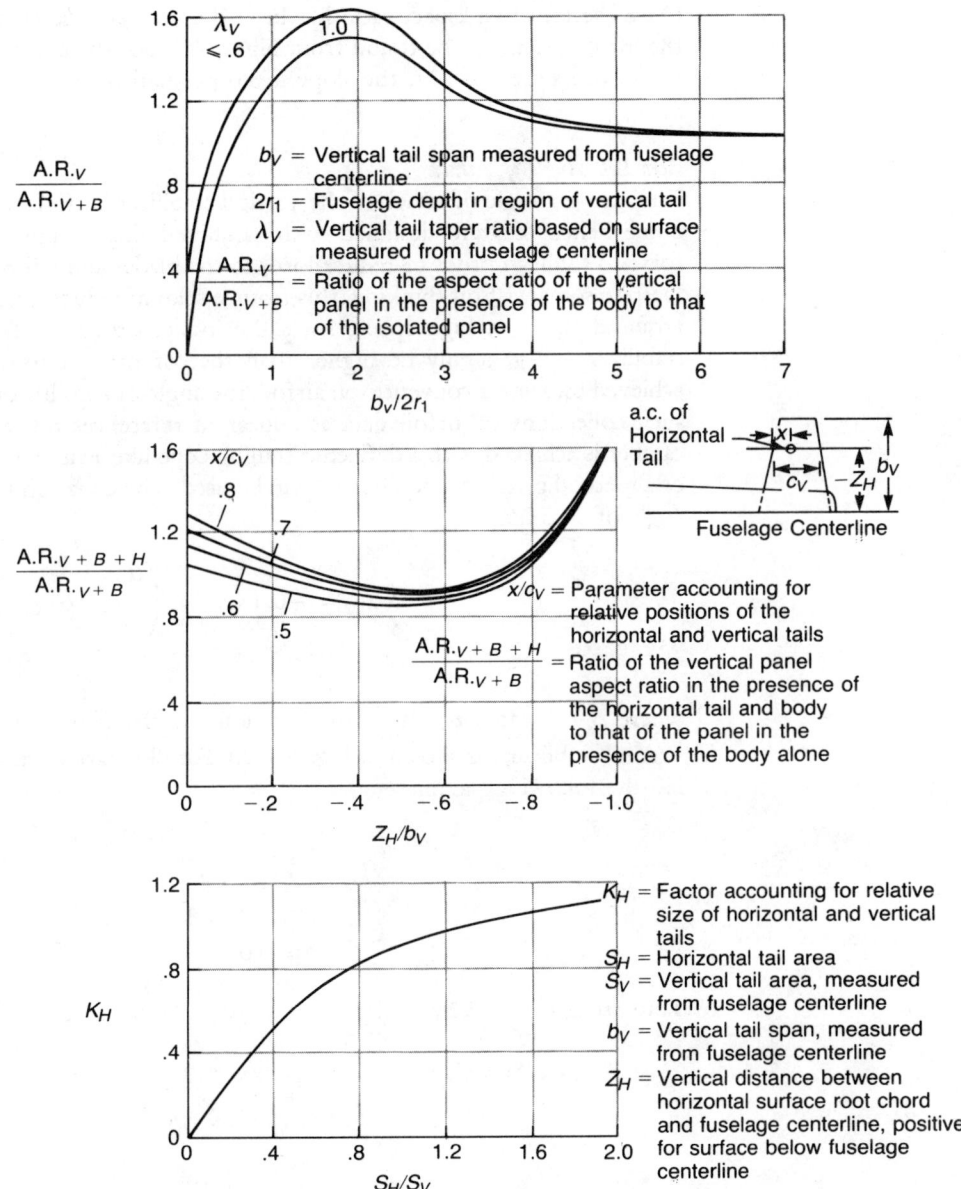

FIGURE 8.19 Charts for Determining Effective Aspect Ratio of Vertical Stabilizer

Source: Hoak, "USAF Stability and Control Datcom," 1960.

Once the effective aspect ratio has been determined, the corresponding slope of the lift curve, a_v, can be found from Figure 8.6. For the example helicopter with a half-chord sweep of 27°, the slope is 3.0 per radian.

Zero Lift Angle of Attack

The zero lift angle of attack of the vertical stabilizer is affected by built-in camber. Some helicopters are designed with cambered fins to unload the tail rotor in forward flight in order to reduce loads in the blades and tail rotor control system. Despite some wishful thinking, this feature seldom reduces the total engine power required to fly at high speed, since the induced drag of the cambered vertical stabilizer will generally be higher than that of the tail rotor. If the camber is achieved by using a conventional airfoil, the angle of zero lift can be obtained from such collections of airfoil data as appear in references 8.1, 8.15, or 8.16. If the camber is achieved with a deflected trailing edge like a rudder, a simplified version of a method given in reference 8.1 can be used. The equation for the change in the angle of zero lift is:

$$\Delta\alpha_{LO} = -0.14 c_{l_{\delta_{\text{theory}}}} \left(\frac{\alpha_{\delta_{C_L}}}{\alpha_{\delta_{c_l}}} \right) \delta$$

where $c_{l_{\delta_{\text{theory}}}}$ and $(\alpha_{\delta_{C_L}}/\alpha_{\delta_{c_l}})$ are functions of the geometric parameters of the vertical stabilizer, as shown in Figure 8.20. For illustration, the example helicopter has the following parameters:

$$t/c = .12$$
$$c_f/c = .2$$
$$\text{A.R.}_{V_{\text{eff}}} = 3.2$$
$$\delta = 10°$$

Thus from Figure 8.20,

$$c_{l_{\delta_{\text{theory}}}} = 3.7$$
$$(\alpha_\delta)_{c_l} = -.55$$
$$\frac{\alpha_{\delta_{C_L}}}{\alpha_{\delta_{c_l}}} = 1.12$$

Thus

$$\Delta\alpha_{LO} = -.14(3.7)(1.12)(10) = -5.8°$$

Sidewash Angle

The sidewash angle at the vertical stabilizer is produced by lateral velocities induced by the main rotor, by the tail rotor, and by the sidewash caused by the

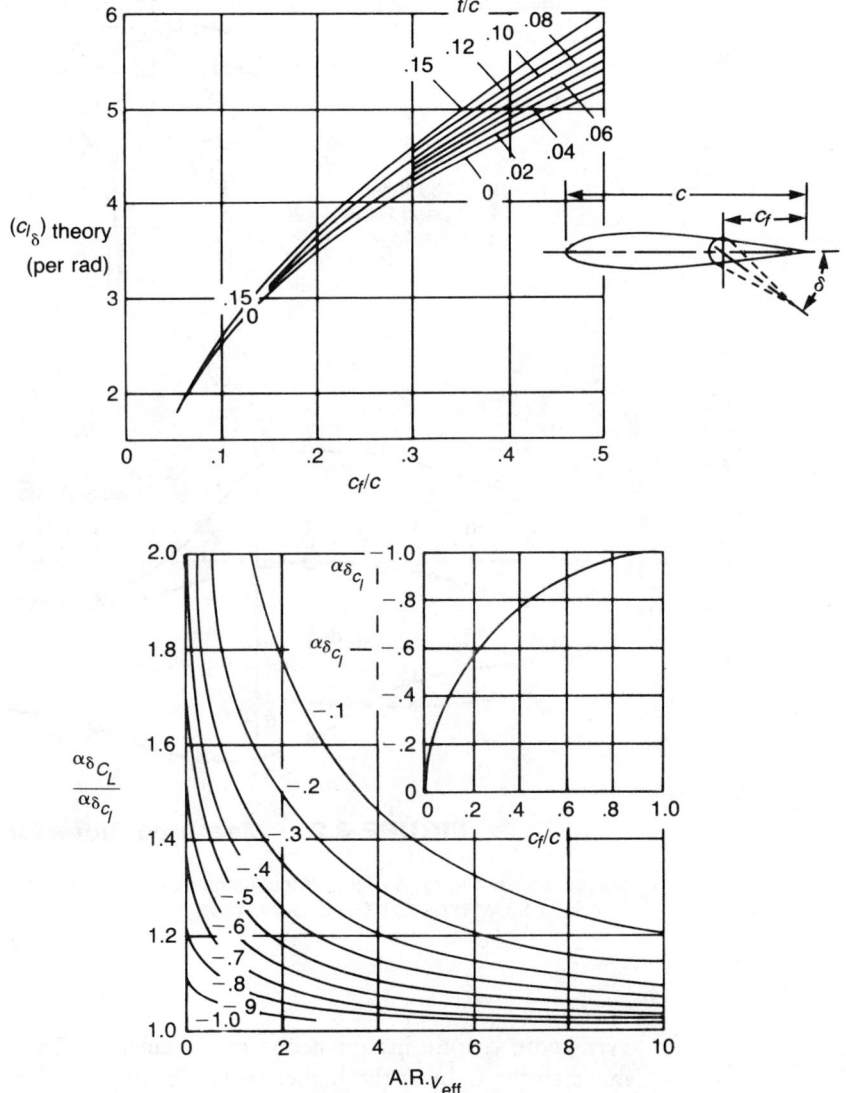

FIGURE 8.20   Charts for Determining Angle of Zero Lift for Vertical Stabilizer with Deflected Trailing Edge

Source: Hoak, "USAF Stability and Control Datcom," 1960.

fuselage in sideslip. There has been less attention paid to sidewash at the empennage than to downwash, so analysis will have to rely primarily on estimates. One set of available test data from reference 8.7 is given in Figure 8.21 for the Hughes AH-64 (without wings). The top figure shows the rather chaotic pattern existing at the empennage. (Note: These vectors are based on averages of some

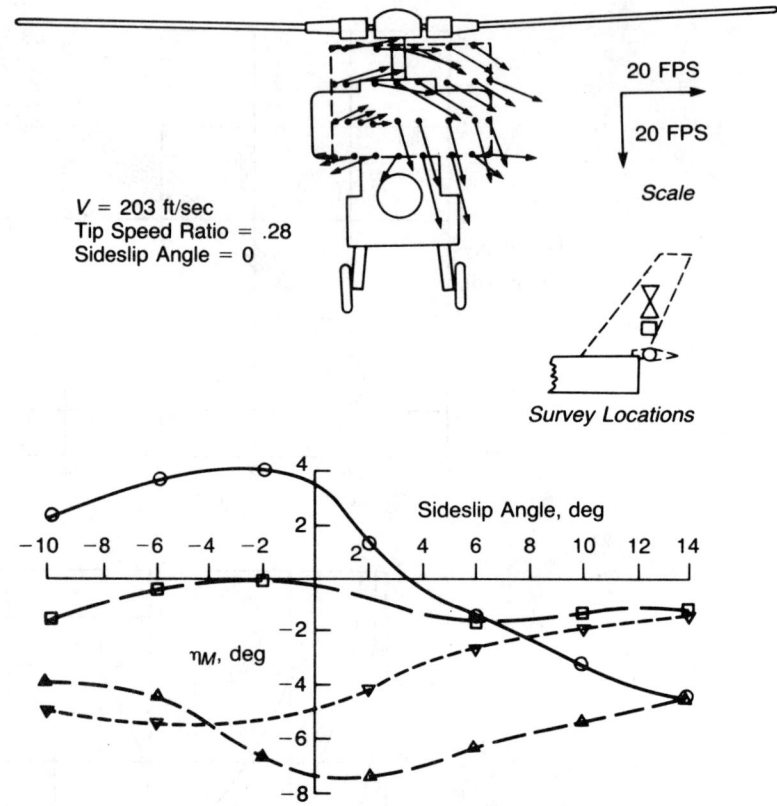

FIGURE 8.21   Measured Sidewash Angles

Source: Logan, Prouty, & Clark, "Wind Tunnel Tests of Large- and Small-Scale Rotor Hubs and Pylons," USAAVRADCOM TR-80-D-21, 1981.

even more chaotic instantaneous measurements.) The tendency of the flow to move to the right at the higher survey locations is due to the swirl in the main rotor wake. The lowest row of measurements is probably reflecting the effects of vortices generated by the nacelles.

The sidewash angle induced by the tail rotor may be assumed to be a function of the momentum value of induced velocity. Unless there is an appreciable separation distance, the effect may be assumed to be the same as that of the main rotor on the fuselage; that is:

$$\eta_T = \left[\frac{\text{D.L.}}{4\left(\frac{q_V}{q}\right)q} \right]_V \text{ radians}$$

where

$$\text{D.L.} = \frac{1}{A_T}\left[\frac{Q_M}{l_T} - L_V\frac{l_V}{l_T}\right]$$

In most cases, the sidewash induced by a sideslipping fuselage will be small enough to ignore. This is another way of saying that there is not yet much data on this effect. For those analyses that must at least give the appearance of completeness, it may be assumed that:

$$\frac{d\eta_f}{d\beta} = \frac{d\varepsilon_f}{d\alpha} = 0.06$$

where this value is taken from Figure 8.15 for the downwash effect of the body alone. The final approximation is thus:

$$\eta_f = 0.06\beta$$

Drag Parameters

In the equation for the drag of the vertical stabilizer, there are three new terms that must be evaluated; δ_i; C_{D_0}, and $\Delta D_{v_{\text{int}}}$. The first two also occurred in the equation for the drag of the horizontal stabilizer and can be found by the same methods using Figures 8.17 and Figure 8.19. For the vertical stabilizer of the example helicopter, the required parameters have already been calculated for use in determining the slope of the lift curve. They are:

$$\lambda_V = .21$$
$$\text{A.R.}_{V_{\text{eff}}} = 3.2$$

From Figure 8.17:

$$\delta_i \doteq .01$$

and from Figure 6.30 of Chapter 6, at a Reynolds number of 2×10^6 corresponding to a flight speed of 150 knots:

$$C_{D_0} = 0.0064$$

The vertical stabilizer and tail rotor have a mutual interference sometimes called *biplane effect*, which results in the combination having a slightly higher induced drag than if they were isolated components producing the same side force and thrust. Using the methods derived for biplanes in reference 8.17, an equation for the additional interference drag can be written in helicopter terms:

$$\Delta D_{V_{\text{int}}} = \frac{8}{\pi}\frac{\left|\dfrac{T_T}{2R_T}\ \dfrac{Y_V}{b_V}\right|}{q}K_{\text{int}}$$

where K_{int} is the interference factor, which is a function of the tail rotor radius, the span of the vertical stabilizer, and the separation distance between the two as shown in Figure 8.22. For the purposes of this estimate, the vertical stabilizer force, Y_V, can be assumed to be the value calculated assuming no tail rotor interference effects; thus:

$$T_T = \frac{Q_M}{l_T} - Y_V \frac{l_V}{l_T}$$

Vertical Stabilizer Characteristics of the Example Helicopter

Just as was done for the horizontal stabilizer, the aerodynamic characteristics of the vertical stabilizer of the example helicopter can be estimated from its physical parameters (Table 8.3).

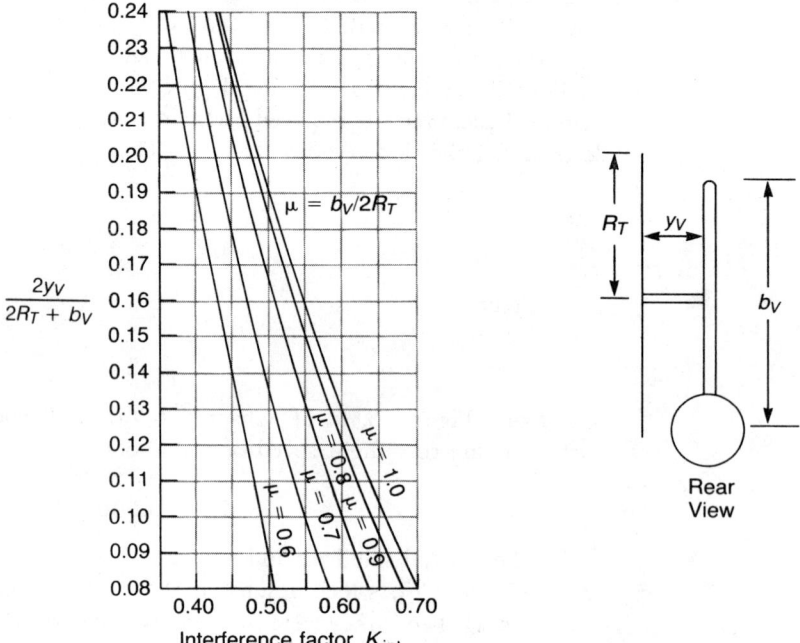

FIGURE 8.22   Interference Factor between Vertical Stabilizer and Tail Rotor

Source: Von Mises, *Theory of Flight*, McGraw-Hill, 1944.

TABLE 8.3
Vertical Stabilizer—Example Helicopter

Physical Parameters			Aerodynamic Characteristics			
Parameter	Symbol	Value	Characteristic	Symbol	Value	Source
Span	b_V	7.7 ft	Dynamic press. ratio	q_V/q	.6	Fig. 8.9
Root chord	c_{r_V}	7.08 ft	Eff. aspect ratio	$A.R._{eff_V}$	3.2	Fig. 8.19
Tip chord	c_{t_V}	1.49 ft	Slope of lift curve	a_V	3.0	Fig. 8.6
Average chord	c_{avg_V}	4.3 ft	Zero lift ang. of attack due to rudder	$\Delta\alpha_{LO_V}$	−5.8°	Fig. 8.20
Area	A_V	33 sq ft	Avg. rotor sidewash angle	η_{MV}	−3°	Fig. 8.21
Taper ratio	λ_V	.21	Fuse. sidewash derivative	$\dfrac{d\eta_F}{d\beta_V}$	.06	Fig. 8.15
Half chord sweep angle	$\Lambda_{c/2_V}$	27°				
Geometric aspect ratio	$A.R._V$	1.8	Span efficiency factor	δ_{iV}	.01	Fig. 8.17
Fuselage depth	r_1	.75 ft	Zero lift drag coefficient	$C_{D_{OV}}$	.0064	Fig. 6.30
Horizontal position of a.c. of horizontal stabilizer	x	2.5 ft	Interference drag factor	K_{int}	.4	Fig. 8.22
Vertical position of a.c. of horizontal stabilizer	z	0 ft				
Rudder angle	δ	10°				
Thickness ratio	t/c_V	.12				
Flap/chord ratio	c_f/c	.2				
Tail rotor radius	R_T	6.5 ft				
Dist. between T.R. and vert. stab.	y_V	2 ft				

Fuselage

Conditions at the fuselage are shown in Figure 8.23. The various fuselage effects enter into the equations of equilibrium as:

$$X_F = -D_F \cos[\Theta - \gamma_C - \varepsilon_{MF}] + L_F \sin[\Theta - \gamma_C - \varepsilon_{MF}]$$
$$Y_F = \text{S.F.}_F \cos \beta - D_F \sin \beta$$
$$Z_F = -L_F \cos[\Theta - \gamma_C - \varepsilon_{MF}] - D_F \sin[\Theta - \gamma_C - \varepsilon_{MF}]$$
$$M_F = q(M/q)_F, \ N_F = q(N/q)_F, \ R_F = q(R/q)_F$$

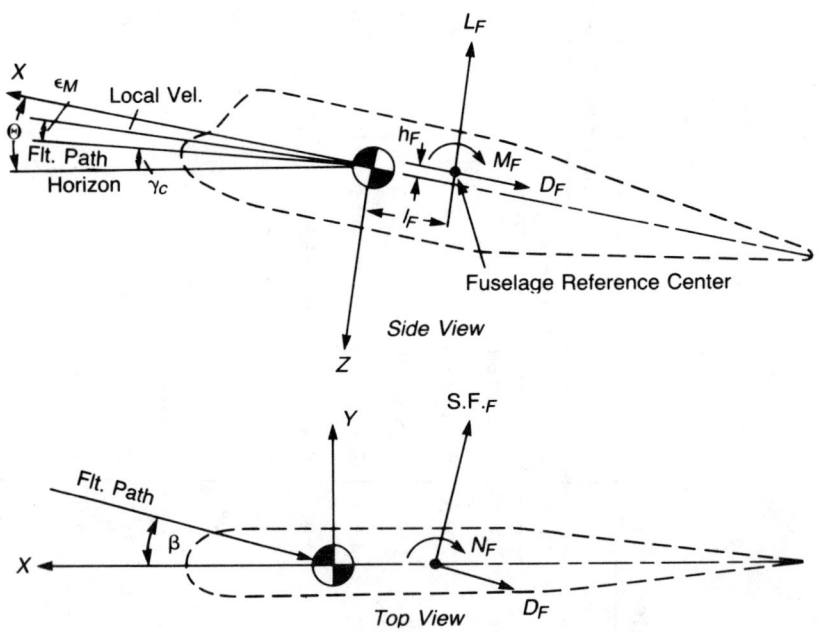

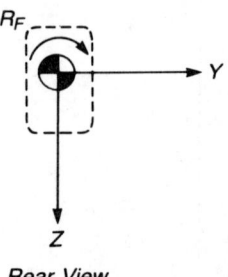

FIGURE 8.23 Aerodynamic Conditions at the Fuselage

Lift, Drag, and Pitching Moments

Methods for estimating fuselage drag were outlined in Chapter 4. Methods for estimating fuselage lift and pitching moment prior to wind tunnel tests can be done either with the airplane methods of reference 8.1 or by direct comparison with wind tunnel results of previous helicopter fuselages. Figure 8.24 shows a compilation of the lift and pitching moment data as a function of angle of attack for several rather typical single-rotor helicopter fuselages in full-scale wind tunnel tests reported in references 8.18 and 8.19. The data has been nondimensionalized by dividing by the product of maximum fuselage width and length for lift data and the product of maximum fuselage width and fuselage length squared for moment data.

Another set of data reduced to coefficient form is given on Figure 8.25 for several fuselages of Bell helicopters. This data was taken from reference 8.2 and uses a different nondimensionalizing scheme than Figure 8.24.

The angle of attack of the fuselage is

$$\alpha_F = \Theta - \gamma_c - \varepsilon_{M_F}$$

where

$$\varepsilon_{M_F} = \frac{v_F}{v_1} = \frac{T_M}{4qA_M}$$

The angle for zero lift can be assumed to be the same as for zero moment which in lieu of wind tunnel tests may be taken as the angle the aerodynamic *chordline* of the fuselage makes with the body axis.

Sideforce and Yawing and Rolling Moments

The fuselage will produce side forces, yawing moments, and rolling moments as a function of sideslip angle. If the calculations are being done without the benefit of wind tunnel tests, the fuselage sideforce (in wind axes) and the yawing moment can be estimated using the same procedures outlined for fuselage lift and pitching moment above. The rolling moment of the fuselage due to sideslip is caused by its *dihedral effect* and can be either slightly positive or slightly negative. A method for estimating this moment using the physical parameters of the fuselage and wing is given in reference 8.1, but since flight test experience has shown that the dihedral effect is strongly, but mysteriously, influenced by the interference of the main rotor wake on the tail rotor and the vertical stabilizer, the estimation of fuselage dihedral effect has a very low priority, so the method will not be repeated here.

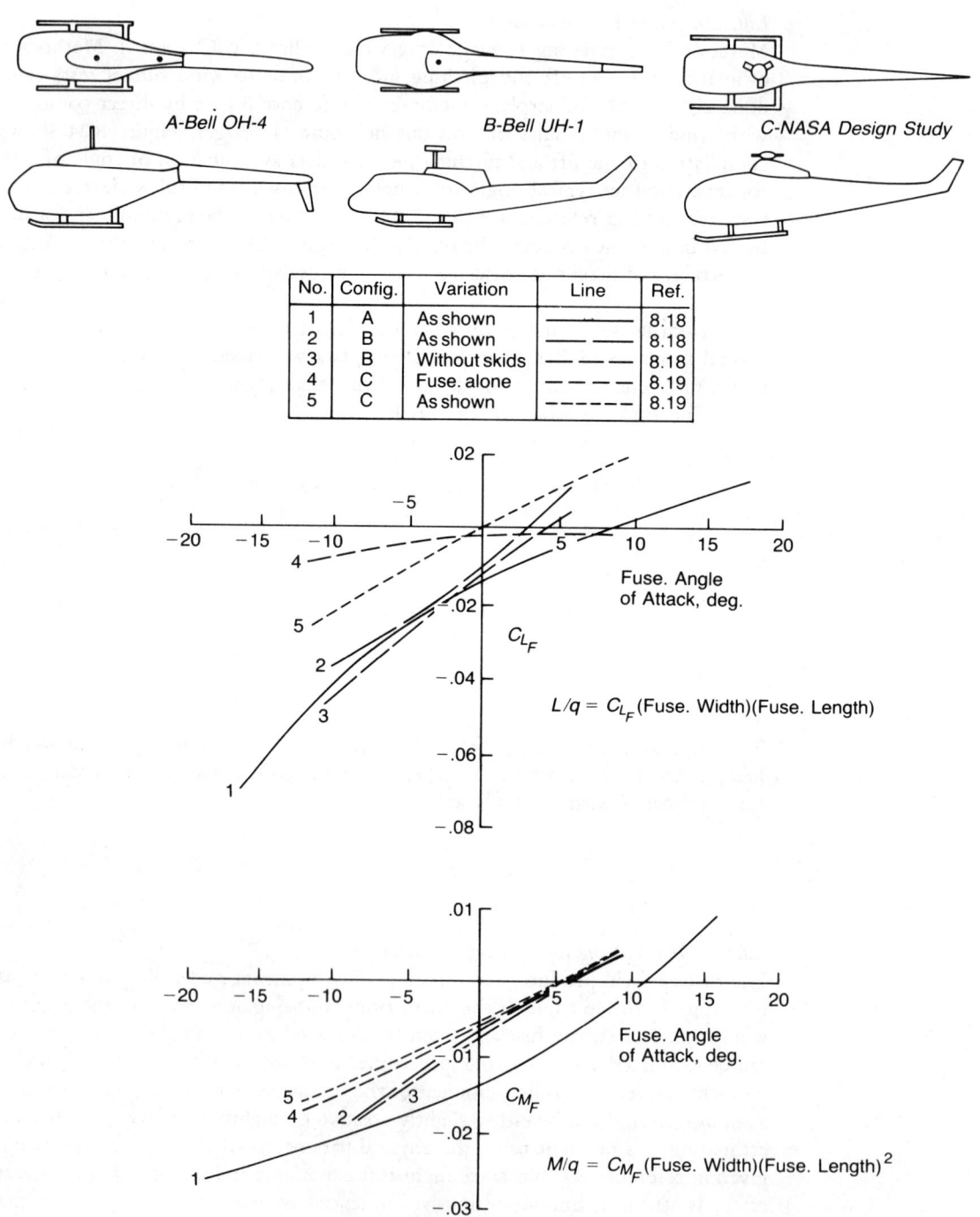

No.	Config.	Variation	Line	Ref.
1	A	As shown	———————	8.18
2	B	As shown	— — —	8.18
3	B	Without skids	— — — —	8.18
4	C	Fuse. alone	– – – –	8.19
5	C	As shown	- - - - -	8.19

$$L/q = C_{L_F}(\text{Fuse. Width})(\text{Fuse. Length})$$

$$M/q = C_{M_F}(\text{Fuse. Width})(\text{Fuse. Length})^2$$

FIGURE 8.24 Fuselage Aerodynamics

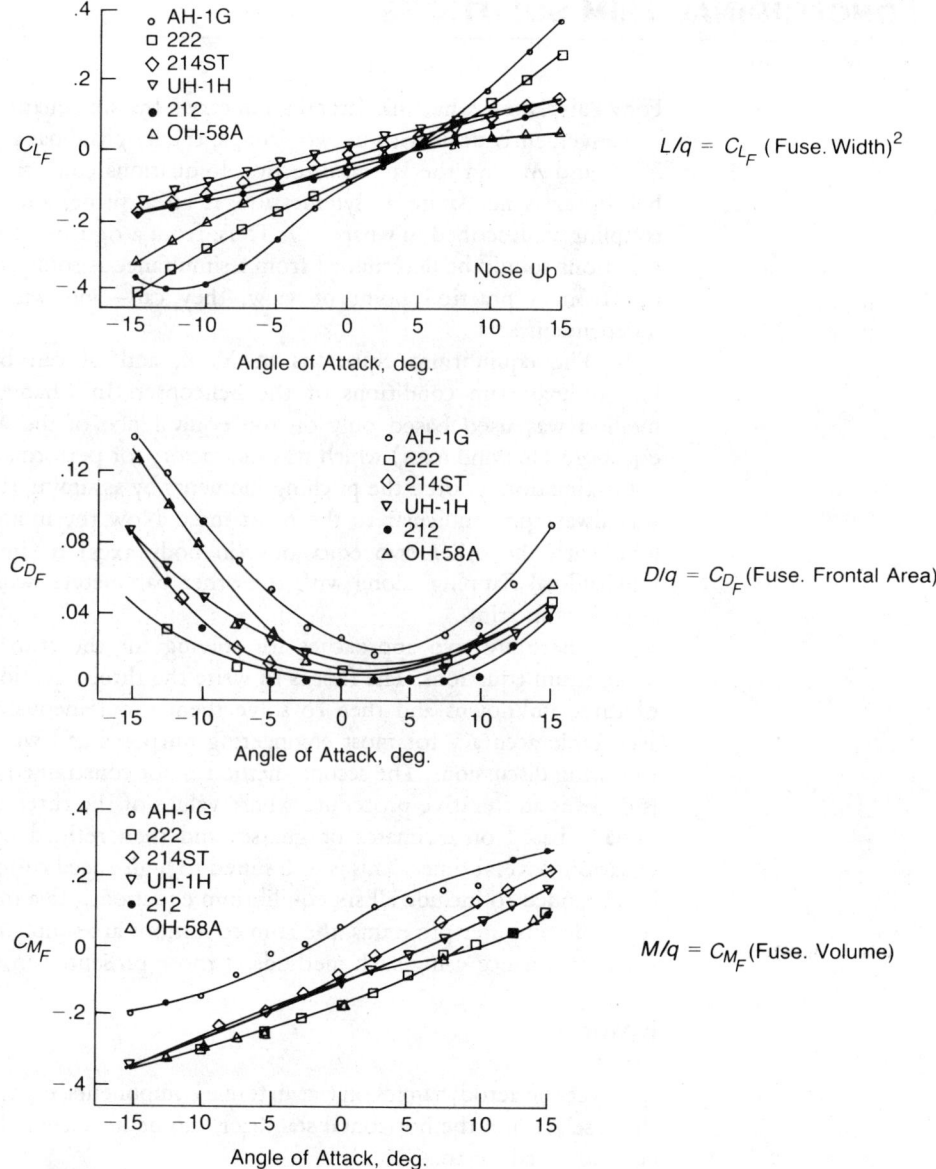

FIGURE 8.25   Aerodynamic Characteristics of Several Helicopter Fuselages

Source: Harris et al., "Helicopter Performance Methodology at Bell Helicopter Textron," AHS 35th Forum, 1979.

LONGITUDINAL TRIM SOLUTIONS

For an airplane—which has lateral symmetry—the six equations of equilibrium can be conveniently dealt with in two groups; the longitudinal equations consisting of X, Z, and M; and the lateral-directional equations consisting of Y, R, and N. A helicopter is not quite as symmetrical as an airplane, and there is some cross-coupling as described in Chapter 7. Thus, from a rigorous point of view, the trim equations should be determined from a simultaneous solution of all six equations; but from a practical point of view, they can—and will—be treated as two independent sets.

The equilibrium equations in X, Z, and M can be used to find the longitudinal trim conditions of the helicopter. In Chapter 3 an approximate method was used based only on the equivalence of the X and Z equilibrium equations (in wind axes) which was satisfactory for performance calculations. That approximation ignored the pitching moments by assuming that the tip path plane was always perpendicular to the rotor mast. Now the moment equation will be used with the other two equations (in body axes) to find the magnitude of longitudinal flapping, along with the other parameters, which must exist if the aircraft is in trim.

There are two approaches for solving for the trim conditions from the equilibrium equations. The first is to write the three equations as linear functions of three unknowns and then to solve them simultaneously. This method gives acceptable accuracy for most engineering purposes and will be illustrated in the following discussions. The second method is not constrained to linear functions. It is done as an iterative procedure where values of the three unknowns are chosen initially based on estimates or guesses and then refined by going through the equations several times. This is well suited to high-speed computers and can readily be expanded to include all six equilibrium equations. As a matter of fact, in these types of computer programs, the trim conditions are simply fallouts of computing the performance using such methods as those presented in Chapter 3.

Hover

In hover the aerodynamics on the airframe components, except for vertical drag on the fuselage and the horizontal stabilizer, can be neglected. The three equilibrium equations reduce to:

$$X_M = \text{G.W. } \sin \Theta$$

$$Z_M + Z_H + Z_F = -\text{G.W. } \cos \Theta$$

$$M_M - X_M h_M + Z_M l_M + M_T + Z_H l_H + Z_F l_F = 0$$

The three independent variables that are convenient for this analysis are Θ, T_M, and a_{1_s}. Rewriting each equation in terms of these variables and using small-angle assumptions gives:

$$-T_M(i_M + a_{1_s}) = \text{G.W.} \; \Theta$$

$$-T_M + T_M \left(\frac{D_V}{\text{G.W.}}\right)_H + T_M \left(\frac{D_V}{\text{G.W.}}\right)_F = -\text{G.W.}$$

$$\left(\frac{dM_M}{da_{1_s}}\right) a_{1_s} + T_M(i_M + a_{1_s})h_M - T_M l_M \pm Q_T + T_M \left(\frac{D_V}{\text{G.W.}}\right)_H l_H + T_M \left(\frac{D_V}{\text{G.W.}}\right)_F l_F = 0$$

(Note: the $\pm$ sign in front of Q_T is in recognition of the fact that tail rotors can rotate in either direction.)

All the constants in these equations are either physical parameters defined by the configuration description or can be calculated by methods already presented in previous chapters.

For the example helicopter:

$$i_M = 0, \; \text{G.W.} = 20{,}000 \text{ lb}$$

$$\left(\frac{D_V}{\text{G.W.}}\right)_H = 0, \; \left(\frac{D_V}{\text{G.W.}}\right)_F = 0.042$$

$$-Q_T = -\left(\frac{550 \; hp}{\Omega}\right)_T = -990 \text{ ft lb (Chapter 4, Figure 4.30, O.G.E.)}$$

$$h_M = 7.5 \text{ ft}, \; l_M = -0.5 \text{ ft}, \; l_F = -0.5 \text{ ft}$$

$$\frac{dM_M}{da_{1_s}} = 200{,}940 \text{ ft lb/rad (Chapter 7)}$$

Even though the first and third equations contain nonlinear terms as products of T_M and a_{1_s}, the fact that the second equation provides a unique value of T_M means that the set can be easily solved. The solution gives:

$$T_M = 20{,}877 \text{ lb}$$
$$\Theta = 0.026 \text{ rad}(1.4 \text{ deg})$$
$$a_{1_s} = -0.025 \text{ rad}(-1.5 \text{ deg})$$

Forward Flight

In forward flight, the aerodynamics of the airframe components cannot be neglected; but if they are considered to be either constants or linear functions of the independent variables, the procedure is essentially the same as for hover. The elements required for the analysis are listed in Table 8.4. The three longitudinal equilibrium equations in forward flight can be linearized to simplify their solution. This is done by using initial trim values of some of the parameters as constants to

TABLE 8.4
Longitudinal Equilibrium Equations in Forward Flight

Component	Source	Equation	Results for Example Helicopter at 115 Knots
X-Equilibrium Equation			
X_M	Main rotor H-force	$-\bar{H}_{a_{1_s}=0_M} - \bar{T}_M a_{1_{s_M}} - T_M i_M$	$145 - 20{,}606\,a_{1_{s_M}}$
X_T	Tail rotor H-force	$-\bar{H}_T$	-40
X_H	Tilt of horiz. stab. lift	$+\bar{L}_H\left[\left(\Theta - \gamma_c - \left(\dfrac{v_H}{v_1}\right)\dfrac{T_M}{4qA_M} - \varepsilon_{\alpha_F=0_H}\right) - \left(\dfrac{d\varepsilon_F}{d\alpha_F}\right)_H\left(\Theta - \gamma_c - \dfrac{T_M}{4qA_M}\right)\right]$	7 $-235\Theta + .0008\,T_M$
	Horiz. stab drag	$-\bar{L}_H a_H\left[\Theta - \gamma_c + i_H - \alpha_{L=0_H} - \left(\dfrac{v_H}{v_1}\right)\dfrac{T_M}{4qA_M} - \varepsilon_{F\alpha_F=0_H} - \left(\dfrac{d\varepsilon_F}{d\alpha_F}\right)_H\left(\Theta - \gamma_c - \dfrac{T_M}{4qA_M}\right)\right]\left(\dfrac{1+\delta_{i_H}}{\pi A.R._H}\right) - \dfrac{q_H}{q}\,qA_H C_{D_{0_H}}$	-10 $+68\Theta - .0002\,T_M$
X_V	Vert. stab. drag	$-\bar{D}_V$	-58
	Horiz. tilt of vert. stab. lift	$-\bar{L}_V(\bar{\eta}_{M_V} + \bar{\eta}_{T_V})$	2
X_F	Fuselage drag	$-\bar{D}_F$	-811
	Tilt of fuse. lift	$+\bar{L}_F\left(\Theta - \gamma_c - \dfrac{T_M}{4qA_M}\right)$	$-283\Theta + .0006\,T_M$
	Tilt of gross weight	$-G.W.\cdot\Theta$	$-20{,}000\Theta$

Equilibrium $= 0$

$$-765 - 20{,}606\,a_{1s_M} - 20{,}450\Theta + .0012\,T_M = 0$$

Z-Equilibrium Equation

Z_M — Main rotor thrust

$$-T_M \qquad\qquad -T_M$$

Z_T — Tilt of tail rotor thrust

$$+b_{1s_T}\bar{T}_T \qquad\qquad -4$$

Z_H — Horiz. stab. lift

$$-\left(\frac{q_H}{q}\right) q A_H a_H \left[\Theta - \gamma_c + i_H - \alpha_{LO_H} - \left(\frac{v_H}{v_1}\right)\frac{T_M}{4qA_M} - \varepsilon_{F\alpha_{F=0_H}} - \left(\frac{d\varepsilon_F}{d\alpha_F}\right)_H \left(\Theta - \gamma_c - \frac{T_M}{4qA_M}\right)\right]$$

$$149 \qquad\qquad -1497\Theta + .0048\,T_M$$

Z_H — Tilt of horiz. stab. drag

$$-\bar{D}_H \left[\Theta - \gamma_c - \left(\frac{v_H}{v_1}\right)\frac{T_M}{4qA_M} - \varepsilon_{F\alpha_{F=0_H}} - \left(\frac{d\varepsilon_F}{d\alpha_F}\right)_H \left(\Theta - \gamma_c - \frac{T_M}{4qA_M}\right)\right]$$

$$-13\Theta + .0000\,T_M$$

Z_V — Tilt of vert. stab. X-force

$$-[\bar{D}_V + \bar{L}_V(\bar{n}_{M_V} + \bar{n}_{T_V})]\left[\Theta - \gamma_c - \left(\frac{v_V}{v_1}\right)\frac{T_M}{4qA_M} - \varepsilon_{F\alpha_{F=0_V}} - \left(\frac{d\varepsilon_F}{d\alpha_F}\right)_V \left(\Theta - \gamma_c - \frac{T_M}{4qA_M}\right)\right]$$

$$1 \qquad\qquad -43\Theta + .0001\,T_M$$

Z_F — Fuselage lift

$$-q\left[\left(\frac{L}{q}\right)_{\alpha_{F=0_F}} + \left(d\frac{\frac{L}{q}}{d\alpha_F}\right)\left(\Theta - \gamma_c - \frac{T_M}{4qA_M}\right)_F\right]$$

$$68 \qquad\qquad -3375\Theta + .0066\,T_M$$

Z_F — Tilt of fuselage drag

$$-\bar{D}_F\left[\Theta - \gamma_c - \frac{T_M}{4qA_M}\right]$$

$$811\Theta + .0016\,T_M$$

TABLE 8.4 (continued)

Component	Source	Equation	Results for Example Helicopter at 115 Knots
	Gross weight	+ G.W.	20,000
	Equilibrium	= 0	$20214 - 5739\Theta - .9869T_M$
M-Equilibrium Equation			
M_M	Main rotor hub moment	$\left(\dfrac{d_M}{da_{1_s}}\right)_M a_{1_{s_M}}$	$200,940a_{1_{s_M}}$
$-X_M b_M$	Main rotor H-force	$-X_M b_M$	$-1,087 + 154,545a_{1_{s_M}}$
$Z_M l_M$	Main rotor thrust	$Z_M l_M$	$.5T_M$
M_T	Tail rotor torque	$-\bar{Q}_T$	-127
$-X_T b_T$	Tail rotor H-force	$+\bar{H}_T b_T$	240
$Z_T l_T$	Tilt of tail rotor thrust	$Z_T l_T$	-148

520

Symbol	Description	Value	Expression
$-X_H b_H$	Tilt of horiz. stab. lift and horiz. stab. drag	-5	$-250\Theta + .0008T_M$
$Z_H l_H$	Horiz. stab. lift and tilt of horiz. stab. drag	4917	$-49,830\Theta + .1614T_M$
$-X_V b_V$	Vert. stab. drag and horiz. tilt of vert. stab. lift	168	
M_F	Fuselage pitching moment	-7200	$80,100\Theta - .1574T_M$
$Z_F l_F$	Fuselage lift and tilt of fuselage drag	-34	$2092\Theta - .0041T_M$
$-X_F b_F$	Fuselage drag and tilt of fuselage lift	405	$141\Theta - .0003T_M$
Equilibrium	$= 0$		$-2,871 + 355,485a_{i_M} + 32,253\Theta + .5004T_M = 0$

$$M_F \quad q\left[\left(\frac{M_F}{q}\right)_{\alpha_F=0} + \frac{d\left(\dfrac{M_F}{q}\right)}{d\alpha_F}\left(\Theta - \gamma_c - \frac{T_M}{4qA_M}\right) \right]$$

reduce the system to three linear equations in the three unknowns: Θ, T_M, and $a_{1_{s_M}}$. In the equations of Table 8.4, the physical dimensions are, of course, known and the trim—or barred—terms for the aircraft components will already be known from previous performance calculations such as those done in Chapter 3. Table 8.4 also presents the numerical results for the example helicopter at 115 knots ($\mu = .3$). The elements that go into the equations for this example are tabulated in Table 8.5 which can serve as a check list for any helicopter.

Solving the three equations simultaneously yields the following solutions for the example helicopter:

$$T_M = 20,586 \text{ lb}$$
$$\Theta = -.016 \text{ rad} = -.9°$$
$$a_{1_{s_M}} = -.019 \text{ rad} = -1.1°$$

Besides the thrust, pitch attitude, and longitudinal flapping, the other trim value of interest is the longitudinal cyclic pitch. This can be determined from:

$$B_1 = (B_1 + a_{1_{s_M}}) - a_{1_{s_M}}$$

where the value of $(B_1 + a_{1_{s_M}})$ is found from the performance charts of Chapter 3. For the example helicopter at 115 knots, this was determined during the illustration of the method for the "Entire Helicopter in Level Flight" to be:

$$(B_1 + a_{1_{s_M}}) = 7.8°$$

and now, knowing the trim value of longitudinal flapping calculated above:

$$B_1 = 7.8 - (-1.1) = 8.9°$$

Approximate Method

The foregoing method is rigorous in accounting for the contributions of every aircraft component to the equations of equilibrium. A study of the numerical values in the example shows, however, that only a few terms dominate; the others have little effect on the final solutions. This observation leads to an approximate method involving solving the moment equation using only initial trim values to give the approximate value of the longitudinal flapping:

$$a_{1_{s_M}} = \frac{\bar{T}_M l_M - \bar{H}_{a_{1_{s_M}=0}} h_M + \bar{Q}_T - \bar{H}_T h_T + \bar{L}_H l_H - \bar{D}_H h_H - \bar{D}_V h_V + \bar{L}_F l_F - \bar{D}_F h_F - \bar{M}_F - \bar{T}_T b_{1_{s_T}} l_T}{\left(\dfrac{dM}{da_{1_s}}\right)_M + \bar{T}_M h_M}$$

TABLE 8.5
Elements of the Longitudinal Equilibrium Equations for the Example Helicopter at 115 Knots

Unknowns				Flight Conditions			
Unknown	Symbol	Units		Condition	Symbol	Units	Value
Rotor thrust	T_M	lb		Gross weight	G.W.	lb	20,000
Longitudinal flapping	$a_{1_{1_M}}$	rad		Climb angle	γ_c	rad	0
Fuselage attitude	Θ	rad		Dynamic pressure	q	lb/ft^2	45

Physical Dimensions (See Appendix A)				Initial Trim Forces (From Chap 3)			
Dimension	Symbol	Units	Value	Force	Symbol	Units	Value
Main rotor disc area	A_M	sq ft	2827	Main rotor thrust	$\bar{T}_M$	lb	20606
Main rotor shaft incidence	i_M	rad	0	Main rotor H-force	$\bar{H}_{a_{1_{=0_M}}}$	lb	−145
Main rotor long. offset	l_M	ft	−.5	Main rotor torque	$\bar{Q}_M$	ft lb	34726
Main rotor vert. offset	b_M	ft	7.5	Antitorque force	$\bar{Q}_M/l_T$	lb	939
Tail rotor long. offset	l_T	ft	37	Drag of H + V + F	$\bar{D}_H + \bar{D}_V + \bar{D}_F$	lb	886
Tail rotor vert. offset	b_T	ft	6	Lift of H + F	$\bar{L}_H + \bar{L}_F$	lb	−588
Horiz. stab. area	A_H	sq ft	18				
Horiz. stab. aspect ratio	A.R.$_H$	—	4.5	Resultant Component Trim Forces			
Horiz. stab. incidence	i_H	rad	−.052	Tail rotor thrust	T_T	lb	666
Horiz. stab. angle of zero lift	α_{LO_H}	rad	0	Tail rotor total H-force	H_T	lb	40
Horiz. stab. long. offset	l_H	ft	33	Tail rotor torque	Q_T	ft lb	127
Horiz. stab. vert. offset	b_H	ft	−1.5	Horiz. stab. lift	L_H	lb	−305
Vert. stab. long. offset	l_V	ft	35	Horiz. stab. drag	D_H	lb	17
Vert. stab. vert. offset	b_V	ft	3	Vert. stab. lift	L_V	lb	287
Fuselage long. offset	l_F	ft	−.5	Vert. stab. drag	D_V	lb	58
Fuselage vert. offset	b_F	ft	.5	Fuselage lift	L_F	lb	−283
				Fuselage drag	D_F	lb	811
				Fuselage moment	M_F	ft lb	−13140

TABLE 8.5 (continued)

Derived Parameters

Parameter	Symbol	Units	Source	Value
Main rotor stiffness	$(dM/da_{1_s})_M$	ft lb/rad	Calc, Chap 7	200940
Tail rotor lat. flapping	$b_{1_{y_T}}$	rad	Calc, based on chap 3	−.0054
Horiz. stab. dynamic pressure ratio	q_H/q	—	Table 8.2	0.6
Horiz. stab. slope of lift curve	a_H	C_L/rad	Table 8.2	4.0
Horiz. stab. span efficiency factor	δi_H	—	Table 8.2	.02
Horiz. stab. zero lift drag coefficient	$C_{D_{0_H}}$	—	Table 8.2	.0064
Horiz. stab. rotor induced velocity ratio	v_H/v_1	—	Table 8.2	1.5
Horiz. stab. fuselage induced velocity constant	$\varepsilon_{F\alpha_{F=0_H}}$	rad	Figure 8.12	.024
Horiz. stab. fuselage induced velocity slope	$(d\varepsilon_F/d\alpha_F)_H$	—	Figure 8.15	.23
Vert. stab. rotor induced velocity ratio	v_V/v_1	—	Figure 8.11	1.5
Vert. stab. fuselage induced velocity constant	$\varepsilon_{F\alpha_{F=0_V}}$	rad	Figure 8.12	.024
Vert. stab. fuselage induced velocity slope	$(d\varepsilon_F/d\alpha_F)_V$	—	Figure 8.15	.23
Vert. stab. sidewash angle from main rotor	η_{MV}	rad	Table 8.3	−.052
Vert. stab. sidewash angle from tail rotor	η_{TV}	rad	Calc.	.045
Fuselage lift constant	$(L/q_{\alpha_{F=0}})_F$	ft^2	Appendix A	−1.5
Fuselage lift slope	$(dL/q/d\alpha_F)_F$	ft^2/rad	Appendix A	75
Fuselage pitching moment constant	$(M/q_{\alpha_{F=0}})_F$	ft^3	Appendix A	−160
Fuselage pitching moment slope	$(dM/q/d\alpha_F)_F$	ft^3/rad	Appendix A	1780
Fuselage side force slope	$(dS.F./q/d\beta)_F$	ft^2/rad	Appendix A	−220
Fuselage rolling moment slope	$(dR/q/d\beta)_F$	ft^3/rad	Appendix A	230
Fuselage yawing moment slope	$(dNq/d\beta)_F$	ft^3/rad	Appendix A	−820

The numerical values called for in this equation were all given in Table 8.5. The result is:

$$a_{1_{s_M}} = -.019 \text{ rad} = -1.1°$$

This is exactly the same as the $-1.1°$ calculated by the more rigorous method. The pitch attitude, θ, can also be approximated since:

$$\alpha_{TPP} = \Theta + a_{1_s} + i_M \doteq -\left(\frac{\bar{H}_{a_{1_{s_M}}=0} + \bar{H}_T + \bar{D}_H + \bar{D}_V + \bar{D}_F}{\text{G.W.} - \bar{L}_H - \bar{L}_F}\right)$$

For the example calculation:

$$\Theta - a_{1_{s_M}} \doteq -.038 \text{ rad} = -2.2°$$

and thus

$$\Theta = -2.2 - 1.1 = -1.1°$$

which compares to the more exact value of $-0.9°$.

Speed Stability

The longitudinal trim equations can be used to evaluate speed stability, or, as it is sometimes confusedly called, "longitudinal static stability." The question here is whether with an inadvertent increase in speed with controls held fixed, the helicopter will pitch up and slow down, exhibiting speed stability, or pitch down and speed up in an unstable manner. The rotor flapping is always stabilizing, producing increased longitudinal flapping with increasing speed and a resulting nose-up moment, but the effect of the other components of the aircraft may be either stabilizing or destabilizing. The most important component in this regard is the horizontal stabilizer. If it is carrying a download at the initial trim condition, it should develop more of a download as speed is increased, thus producing a stabilizing nose-up pitching moment. Figure 8.26 summarizes the possibilities for several initial stabilizer loadings. In reality, the contribution of the horizontal stabilizer is modified by changes in angle of attack which accompany the changes in speed. For example, the helicopter's rate of descent that will exist at the higher-than-trim speed with the collective fixed at its trim value will increase the stabilizer's angle of attack. This is destabilizing. The draggier the helicopter, the more destabilizing is this effect, since a higher speed will require a higher rate of descent to achieve it. Another destabilizing change in angle of attack can be traced to the decrease in main rotor downwash at the horizontal stabilizer at the higher speed. This effect is especially significant on tandem rotor helicopters where the aft rotor is directly affected by what the front rotor is doing.

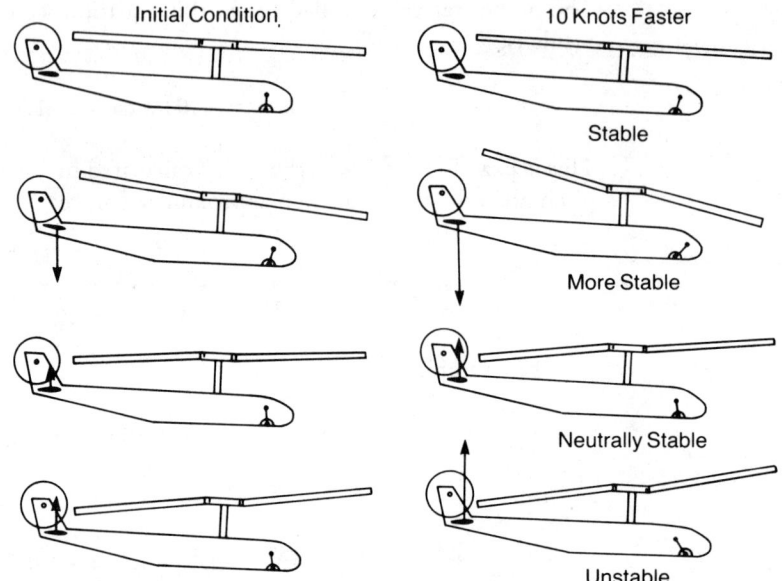

Initial Condition   10 Knots Faster

Stable

More Stable

Neutrally Stable

Unstable

FIGURE 8.26   Illustration of Possible Speed Stability Results

The magnitude of the stability—or the instability—in terms of the change in moment per unit change of speed can be found using the charts of Chapter 3 along with the longitudinal equilibrium equations. For this calculation, the equilibrium conditions are determined in level flight as has just been done and then recalculated at a slightly higher (or lower) speed assuming no change in collective pitch to represent an inadvertant speed change. A change, however, in tail rotor pitch is allowed to maintain zero sideslip as a pilot would do by instinct. For this recalculation, the seventh illustrative example, "Helicopter in Dive at Constant Collective Pitch" of Chapter 3 can be used to establish the new conditions including the angle of climb, γ_C (or dive, γ_D). The new trim condition for the example helicopter at 135 knots ($\mu = 0.35$) with the same collective pitch as at 115 knots ($\mu = 0.30$) are listed in Table 8.6.

Solving for the trim solutions as was done at 115 knots gives:

$$a_{1_{s_M}} = -.7°$$
$$\Theta = -2.3°$$
$$T_M = 20,496 \text{ lb}$$

From the chart of Chapter 3 for a tip speed ratio of 0.35 and the trim collective pitch and thrust coefficient:

$$a_{1_{s_M}} + B_1 = 8.6°$$

TABLE 8.6
Elements of the Longitudinal Equilibrium Equations for the Example Helicopter at 135 Knots

Flight Conditions				Initial Trim Forces			
Condition	Symbol	Units	Value	Force	Symbol	Units	Value
Gross weight	G.W.	lb	20,000	Main rotor thrust	$\bar{T}_M$	lb	20,618
Climb angle	γ_c	rad	−.049[a]	Main rotor H-Force	$\bar{H}_{a_{1_s}=0_M}$	lb	−72
Dynamic pressure	q	lb/ft^2	61.5	Antitorque force	$\bar{Q}_M/l_T$	lb	882
				Drag of $H + v + F$	$\bar{D}_H + \bar{D}_V + \bar{D}_F$	lb	1,193
				Lift of $H + F$	$L_H + L_F$	lb	−584
				Resultant Component Trim Forces			
				Tail rotor thrust	$\bar{T}_T$	lb	428
				Tail rotor total H-Force	$\bar{H}_T$	lb	34
				Tail rotor torque	$\bar{Q}_T$	ft-lb	164
	$\bar{n}_{M_V}$		−.052	Horiz. stab lift	L_H	lb	−334
	$\bar{n}_{T_V}$		.023	Horiz. stab. drag	$\bar{D}_H$	lb	16
	$b_{1_{s_T}}$		−.012	Vert. stab. lift	$\bar{L}_V$	lb	480
				Vert. stab. drag	$\bar{D}_V$	lb	128
				Fuselage lift	$\bar{L}_F$	lb	−250
				Fuselage drag	$\bar{D}_F$	lb	1,049
				Fuselage moment	$\bar{M}_F$	ft lb	−14,268

[a]Using method of Chapter 3.

Thus the longitudinal cyclic pitch is

$$B_1 = 8.6 + .7 = 9.3°$$

The comparable value at 115 knots was 8.9°. Thus the calculations have demonstrated that in this speed regime, the example helicopter has speed stability since the pilot must hold more forward cyclic stick to dive at 135 knots than to fly level at 115 knots. Had he not done so, the increase in speed would have resulted in a nose-up maneuver with a subsequent decrease in speed. Figure 8.27 shows the results of this analysis in terms of stick position. The rigging curves of Appendix A have been used to convert cyclic pitch to stick position.

The curvature of the collective-fixed speed sweep line on Figure 8.27 can be traced to the two destabilizing effects of the change in angle of attack discussed earlier. For this example, the dynamic pressure increased by 37% in going from 115 to 135 knots, but the download on the horizontal stabilizer increased only by 10%. This type of curvature is often seen in flight test results.

To a pilot, speed stability is seen as the change in stick position required to maintain a new speed. For example, on an unstable helicopter, an increase in speed

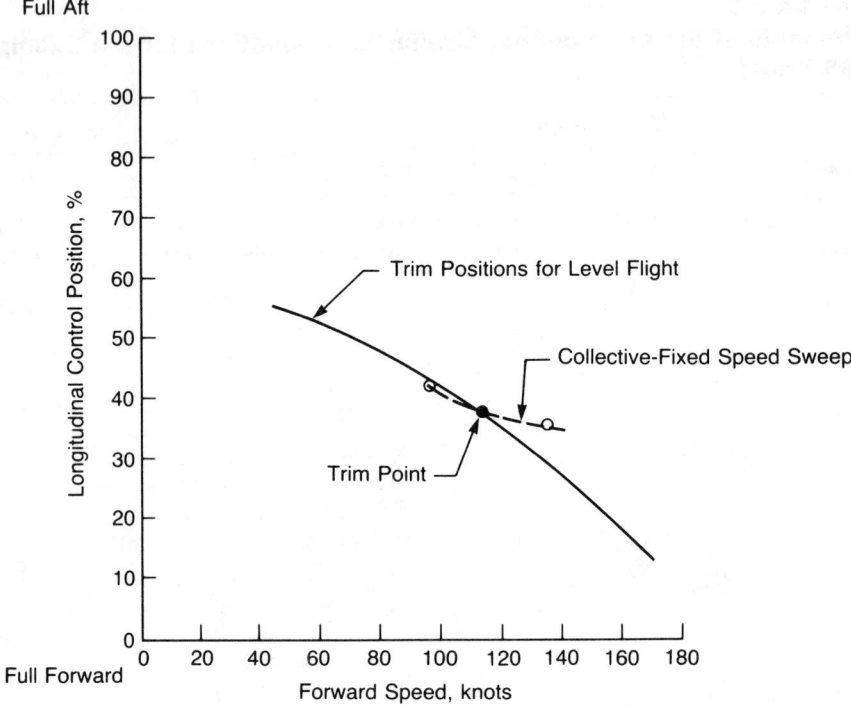

FIGURE 8.27   Speed Stability Calculation for Example Helicopter

will initially require a forward stick motion to accelerate; but when finally trimmed out at this new speed, the stick will be aft of its initial trim point. Such a helicopter is possible to fly, but will tend to wander off its trim point unless constantly corrected. This may not be much of a problem in normal flight, but it is generally considered unacceptable for instrument flight, where the cues are poor and the pilot has other things to worry about. As a general rule, pilots prefer a level of speed stability which is just slightly positive. Too much speed stability could result in running out of forward stick travel—or at least putting the stick into an uncomfortable far-forward position at high speeds.

It should be pointed out that the change in pitching moment with speed is unique to rotary-wing aicraft and does not normally exist for fixed-wings. For airplanes, any change in control position while changing speed is due only to the resulting change in angle of attack and not to the change in speed itself (except for high-speed airplanes, for which compressibility effects might be important). Airplanes also have inherent speed stability through another mechanism, however. For a given propeller pitch or jet power setting, the change in forward propulsive force is stabilizing; that is, if the airplane slows down, the propulsive force increases, thus accelerating the aircraft to its original trim speed—an automatic "cruise control." A rotary wing aircraft that has auxiliary propulsion, such as an autogiro or a compound helicopter, will also benefit from this inherent speed stability and need not be subjected to the analysis and tests that are necessary on a pure helicopter, which gets its propulsive force from tilt of the rotor.

Angle of Attack Stability

The stability with respect to angle of attack is evaluated in flight by trimming for level flight and then entering a steady, descending turn at the same speed without changing collective pitch. The extra rotor thrust required to support the weight and centrifugal force will come from the increased angle of attack produced by the rate of descent. If during the turn, the longitudinal control is aft of the level flight position, the helicopter is said to have positive *maneuver stability*, which is another way of saying that it has positive angle of attack stability. The procedure for calculating trim conditions can be used to investigate the sign and magnitude of this stability. The first step is to calculate the trim conditions in a dive at constant collective pitch (using the method outlined in Chapter 3) of a helicopter whose effective gross weight is equal to its actual gross weight multiplied by the load factor corresponding to the turn. The second step is to correct the calculated longitudinal cyclic pitch to account for the effect of pitch rate (using the procedure in Chapter 7). This process has been done for the example helicopter in a 1.3-g descending turn at 115 knots with collective held at the level flight value. The results are presented in Table 8.7.

The more forward position of the stick in the turn than in level flight is an indication that this particular helicopter does not have a large enough horizontal stabilizer to give it positive angle-of-attack stability. This might be considered to be a design deficiency that would be corrected in an actual program by increasing the stabilizer area or by installing a "black box" stability augmentation system that could use a signal from an angle of attack sensor to move the swashplate nose down or to change the stabilizer incidence nose up so that the pilot would have to use aft stick to compensate. For a helicopter with either inherent or artificially achieved

TABLE 8.7
Summary of Conditions for Example Helicopter at 115 Knots

Condition	Level Flight	Descending Turn
Speed, V, knots	115	115
Collective pitch, θ_0, deg	17.25	17.25
Load factor, n	1.0	1.3
Longitudinal flapping, a_1, deg	−1.1	−.5
Longitudinal cyclic pitch (uncorr.) B_1, deg	8.9	11.8
Longitudinal cyclic pitch (corr.) B_1, deg	8.9	11.4
Stick position from full foward, %	34	29

TABLE 8.8
Effect of Power Changes on Trim of Example Helicopter

Trim Conditions	Autorotation	Level	Climb
Climb angle, γ_c, deg	−9.2	0	9.7
Angle of attack of fuselage, α_F, deg	7.4	−3.3	−17.5
Angle of attack of horizontal stab., α_H, deg	.9	−7.9	−19.6
Pitching moment due to fuselage, M_F, ft-lb	2,925	−11,250	−30,600
Pitching moment due to horiz. stab., $−L_H l_H$, ft-lb	−1,008	8,845	21,944
Longitudinal flapping, a_{1_s}, deg	−3.3	−1.1	2.1
Total cyclic pitch and flapping, $(B_1 + a_{1_s})$, deg	5.2	7.7	11.5
Cyclic pitch, B_1, deg	8.5	8.8	9.4
Stick position, % from full forward	38	37	35

positive angle of attack stability, an inadvertant pitching motion which produced a nose-up angle of attack would result in a stabilizing nose-down control response without pilot action. Design decisions in this matter involve tradeoffs between the weight of an adequately sized stabilizer and the weight of an auxiliary stabilizing system. Also entering into the study are considerations of safety involving possible shutdowns or hardovers of the black box system.

Power Effects on Trim

If the shift in longitudinal control position while going from a full-power climb to steady autorotation at the same forward speed is excessive, pilots will complain. A typical minimum value written into the flying qualities specifications is 3 inches. The shift is primarily due to the change in aerodynamic pitching moments on the airframe—especially those contributed by the fuselage and the horizontal stabilizer as the angle of attack changes from one flight condition to the other. The calculations for the example helicopter in autorotation, level flight, and climb at 2,000 feet per minute—all at 115 knots—are given in Table 8.8.

For this helicopter, the change in $B_1 + a_{1_s}$ required to trim the rotor aerodynamically is almost equal to the change in a_{1_s} required to trim the airframe pitching moments so that practically no stick motion is needed when going from one trim condition to another. This is one benefit of having a small horizontal stabilizer. If it were large enough to produce positive angle of attack stability, the shift in control position would have been much greater.

LATERAL-DIRECTIONAL TRIM SOLUTIONS

The lateral-directional equilibrium equations are those involving Y-force, rolling moment, and yawing moment—the Y, R, and N equations. The aircraft components that are significant in these equations are the main rotor, tail rotor, vertical stabilizer, and the fuselage.

Hover

In hover, the airframe aerodynamic effects may be neglected so that only the two rotors enter into the equations. For example, Figure 8.28 shows the relationships that result in the equilibrium equations:

$$Y \qquad T_M b_{1_s} + T_T = -G.W. \, \Phi$$

$$R \qquad \left(\frac{dR_M}{db_{1_s}} + T_M h_M \right) b_{1_s} - T_M y_M + T_T h_T = 0$$

$$N \qquad Q_M - l_T T_T = 0$$

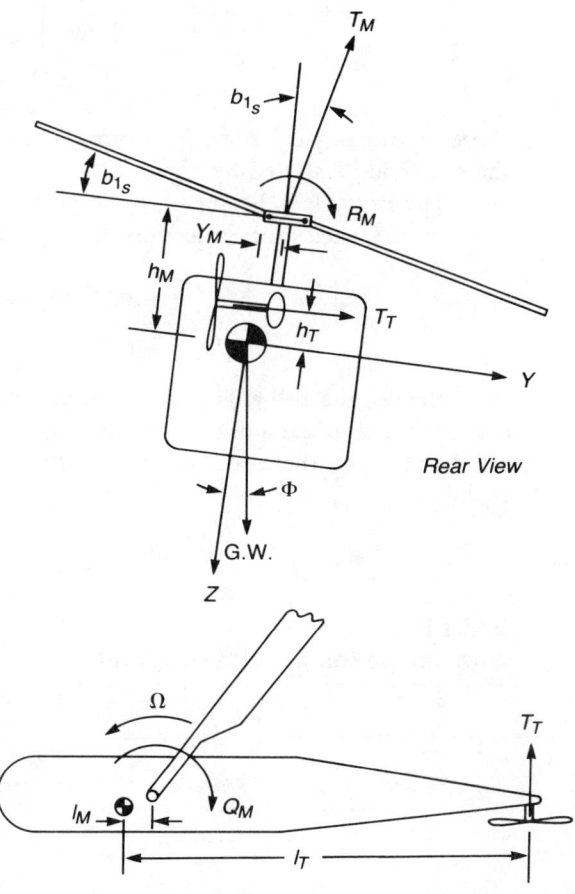

FIGURE 8.28   Lateral-Directional Components in Hover

where

$$T_M = \frac{\text{G.W.}}{1 - \left[\left(\dfrac{D_V}{\text{G.W.}} \right)_H + \left(\dfrac{D_V}{\text{G.W.}} \right)_F \right]}$$

Simultaneous solution of the Y and R equations gives:

$$b_{1_{s_M}} = \frac{T_M y_M - T_T b_T}{\dfrac{dR_M}{db_{1_s}} + T_M b_M} \text{ , rad}$$

and

$$\Phi = -\frac{T_T}{\text{G.W.}} - \frac{T_M}{\text{G.W.}} \left(\frac{T_M y_M - T_T b_T}{\dfrac{dR_M}{db_{1_s}} + T_M b_M} \right) \text{ , rad}$$

These equations yield some interesting results for special cases as presented in Table 8.9 and illustrated by Figure 8.29.

The example helicopter has the pertinent parameters given in Table 8.10 These values, used in the equations for b_{1_s} and Φ, give:

$$b_{1_s} = -.027 \text{ rad} = -1.5 \text{ deg}$$

$$\Phi = -0.050 \text{ rad} = -2.9 \text{ deg}$$

Besides the roll angle and the lateral flapping, the tail rotor pitch and its corresponding pedal position required to maintain heading in hover can be calculated from the conditions that satisfy the yawing moment equilibrium equation:

$$Q_M - l_T T_T = 0$$

TABLE 8.9
Special Cases of Trim in Hover

	Helicopter Parameters		Results	
Lat. C.G. offset, y_M	Rotor Type	Height of Tail Rotor, b_T	Φ	b_{1_s}
0	Teetering	b_M	0	$-T_T/\text{G.W.}$
0	Any	0	$-T_T/\text{G.W.}$	0
Any	Very rigid	Any	$-T_T/\text{G.W.}$	0

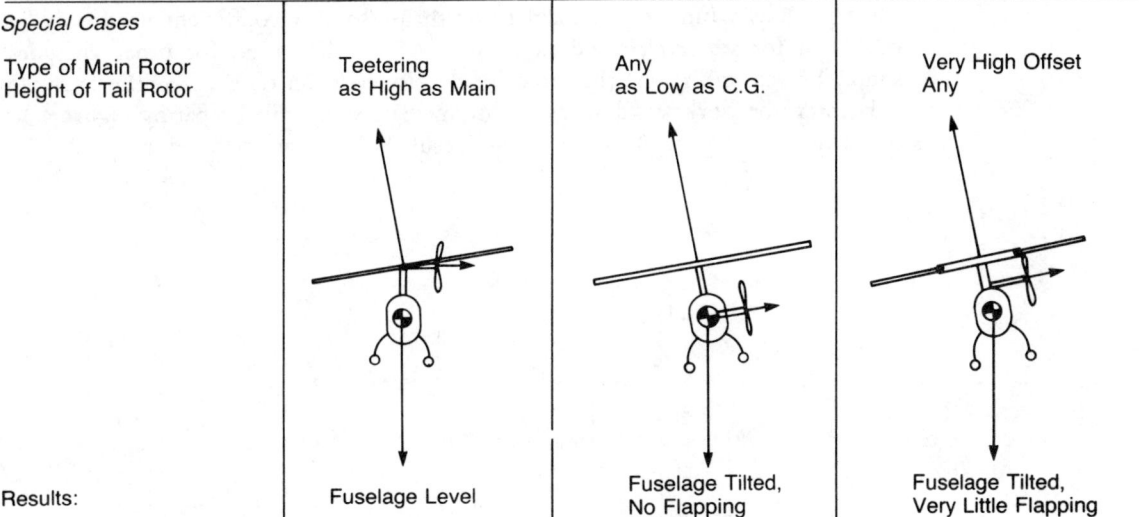

General Case

Type of Main Rotor—with Some Hinge Offset
Height of Tail Rotor—Between C.G. & Main Rotor

Results: Some flapping down to left,
some fuselage left tilt.

Special Cases

Type of Main Rotor
Height of Tail Rotor

| Teetering as High as Main | Any as Low as C.G. | Very High Offset Any |

Results:

| Fuselage Level | Fuselage Tilted, No Flapping | Fuselage Tilted, Very Little Flapping |

FIGURE 8.29   Lateral Tilt in Hover

TABLE 8.10
Example Helicopter Parameters

Physical Dimensions				Trim Forces in Hover (OGE)			
Dimension	Symbol	Units	Value	Force	Symbol	Units	Value
Main rotor lat. offset	y_M	ft	0	Gross weight	G.W.	lb	20,000
Main rotor vert. offset	b_M	ft	7.5	Main rotor thrust	T_M	lb	20,840
Main rotor stiffness	dR_M/db_{1_s}	ft lb / rad	200,940	Tail rotor thrust	T_T	lb	1,540
Tail rotor vert. offset	b_T	ft	6				

Before one does a numerical calculation, an interesting relationship can be obtained by rewriting the equation in terms of nondimensionalized coefficients, which gives:

$$\frac{C_T/\sigma_T}{C_Q/\sigma_M} = \frac{A_{b_M}}{A_{b_T}} \frac{(\Omega R)^2_M}{(\Omega R)^2_T} \frac{R_M}{(l_T - l_M)}$$

Thus for a given helicopter, this ratio is constant and independent of altitude. If the small effects of compressibility are ignored, the constancy of this ratio also defines another ratio that is independent of altitude: $\theta_{0_T}/C_T/\sigma_M$, since with this assumption and for given rotor geometries, C_T/σ_T is only a function of θ_{0_T} and C_Q/σ_M is only a function of C_T/σ_M. The practical application of this is that the tail rotor pitch as a function of main rotor thrust/solidity coefficient need only be calculated for sea-level, standard conditions and then used for other altitudes simply by accounting for the effect of density ratio on C_T/σ_M. For the example helicopter, the work was done as an intermediate step while preparing Figure 4.30 of Chapter 4. Figure 8.30 shows the result, which can be used to guide the

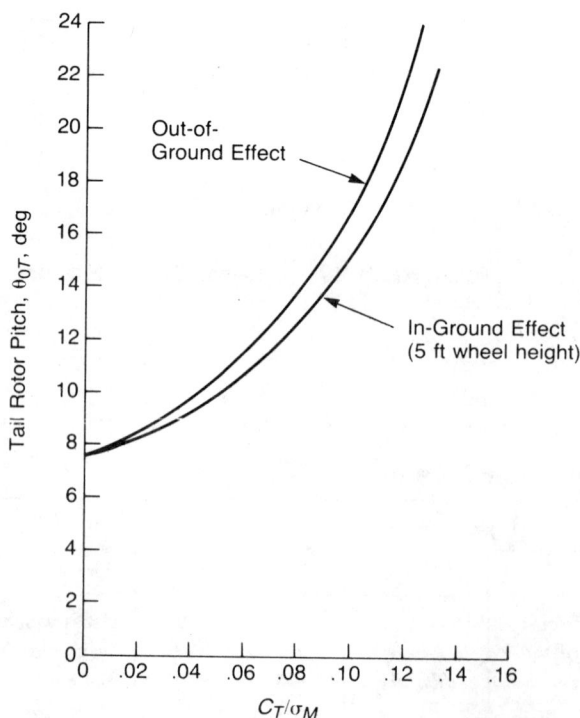

FIGURE 8.30   Tail Rotor Pitch Required in Hover

designers in choosing the maximum tail rotor pitch travel. For the example helicopter at its design gross weight of 20,000 lb, Figure 4.35 gives standard day hover ceilings of 12,400 ft out of ground effect and 15,000 ft in ground effect. Using Figure 8.30, it may be determined that, in each case, a tail rotor pitch of 22.2 degrees is required to hover. A prudent designer would, of course, allow some margin for maneuvering.

Forward Flight

In forward flight there are two natural trim conditions: zero bank angle and zero sideslip. Since very few helicopters have sideslip indicators, pilots tend to fly at zero bank angle, where they are most comfortable. In these cases, the pilots find the sideslip angle that makes the sideforces on the fuselage equal and opposite to the tail rotor thrust.

For test purposes, instrumented helicopters generally have sideslip indicators, and the pilot may be asked to fly at zero sideslip to minimize drag and maximize performance. In actual flight, of course, the helicopter is usually in a condition of some bank angle and some sideslip.

The relationships of bank angle versus sideslip angle are imbedded in the three equations of lateral-directional equilibrium of Table 8.11.

The solution of these equations results in relationships of b_{1_s}, Φ, and T_T as a function of the sideslip angle, β. The pilot can sense the roll angle, Φ, but not the other two variables. These show up indirectly as lateral and directional control positions. To calculate these positions, the sideslip is accounted for in the following manner:

Lateral:

$$A_1 = (A_1 + b_{1_s})_{\beta=0} - b_{1_s} + B_1 \sin \beta$$

Directional:

$$\alpha_{TPP_T} = \tan^{-1} \left(\frac{\lambda'}{\mu} + \frac{\sigma C_T/\sigma}{2\mu^2} \right)$$

$$\alpha_{TPP_T} = a_{1_{s_T}} - \beta$$

(Use the method of Chapter 3—"Tail Rotor"—to find θ_{0_T}.)

Using the rigging plots of Appendix A, the results for the example helicopter at 115 knots have been obtained and are plotted on Figure 8.31. These indicate that this helicopter has three types of positive stability: positive directional stability, positive dihedral effect, and a positive sideforce characteristic.

The amount of pedal position required to hold a sideslip angle indicates the magnitude of the directional stability and tail rotor control power. Although a

TABLE 8.11
Lateral-Directional Equilibrium Equations in Forward Flight

Y-Equilibrium Equation				
Component	Source		Equation	Results for Example Helicopter at 115 Knots
Y_M	Main Rotor Y-Force		$\bar{T}_M b_{1_{j_M}}$	$20606\, b_{1_{j_M}}$
	Main Rotor H-Force		$-(\bar{H}_{a_{1_s}=0_M} + \bar{T}_M \bar{a}_{1_{s_M}} + \bar{T}_M \dot{i}_M)\beta$	$+537\beta$
	Tail Rotor Thrust		T_T	T_T
Y_v	Vertical Stabilizer Lift		$q\left(\dfrac{q_V}{q}\right) A_V a_V \left\{ -\beta - \alpha_{LO_V} - M_{M_V} - \dfrac{T_T}{4q\left(\dfrac{q_V}{q}\right)A_T} - \left(\dfrac{dM_F}{d\beta}\right)\beta \right\}_v$	412 — $-2883\beta - .187T_T$
	Tilt of Vert. Stab. Drag		$-\bar{D}_V \left\{ \beta + \eta_{M_V} + \dfrac{T_T}{4q\left(\dfrac{q_V}{q}\right)A_T} + \left(\dfrac{d\eta_F}{d\beta}\right)\beta \right\}_v$	3 — $-61\beta - .004T_T$
Y_F	Fuselage Side Force		$q\left(\dfrac{dS.F./q}{d\beta}\right)\beta$	-9900β
	Tilt of Fuse. Drag		$-\bar{D}_F\beta$	
	Tilt of Gross Weight		$-G.W.\,\Phi$	$-20{,}000\Phi$
	Equilibrium		$= 0$	$415 + 20606b_{1_{j_M}} - 13068\beta + .809T_T - 20{,}000\Phi = 0$

R-Equilibrium Equation

Component	Source	Equation	Results for Example Helicopter at 115 Knots
R_M	Main Rotor Hub Moment	$\left(\dfrac{dR}{db_{1_{s_M}}}\right)b_{1_{s_M}}$	$+200940 b_{1_{s_M}}$
$Y_M b_M$	Main Rotor Y&H Forces		$154545 b_{1_{s_M}} + 4028\beta$
$Z_M J_M$	Main Rotor Thrust		0
$Y_T b_T$	Tail Rotor Thrust		$+6T_T$
$Y_V b_V$	Vertical Stab. Lift & Drag Tilt		$1245 \qquad -8682\beta \qquad -.57T_T$
$Y_F b_F$	Fuselage Side Force & Drag Tilt		-5355β
R_F	Fuselage Rolling Moment	$q\left(\dfrac{dR/q}{d\beta}\right)_F \beta$	10350β
Equilibrium		$= 0$	$1245 + 355485 b_{1_{s_M}} + 341\beta + 5.43T_T = 0$

N-Equilibrium Equation

Component	Source	Equation	Results for Example Helicopter at 115 Knots
N_M	Main Rotor Torque	Q_M	34726
$-Y_M l_M$	Main Rotor Y&H Forces		$12363 b_{1_{s_M}} + 228\beta$
$-Y_T l_T$	Tail Rotor Thrust		$-37T_T$
$-Y_V l_V$	Vertical Stab. Lift & Drag Tilt		$-14525 \qquad +101290\beta \qquad +6.68T_T$
$-Y_F l_F$	Fuselage Side Force & Drag Tilt		5355β
N_F	Fuselage Yawing Moment	$q\left(\dfrac{dN/q}{d\beta}\right)_F \beta$	-36900β
Equilibrium		$= 0$	$20201 + 12363 b_{1_{s_M}} + 69973\beta - 30.32T_T = 0$

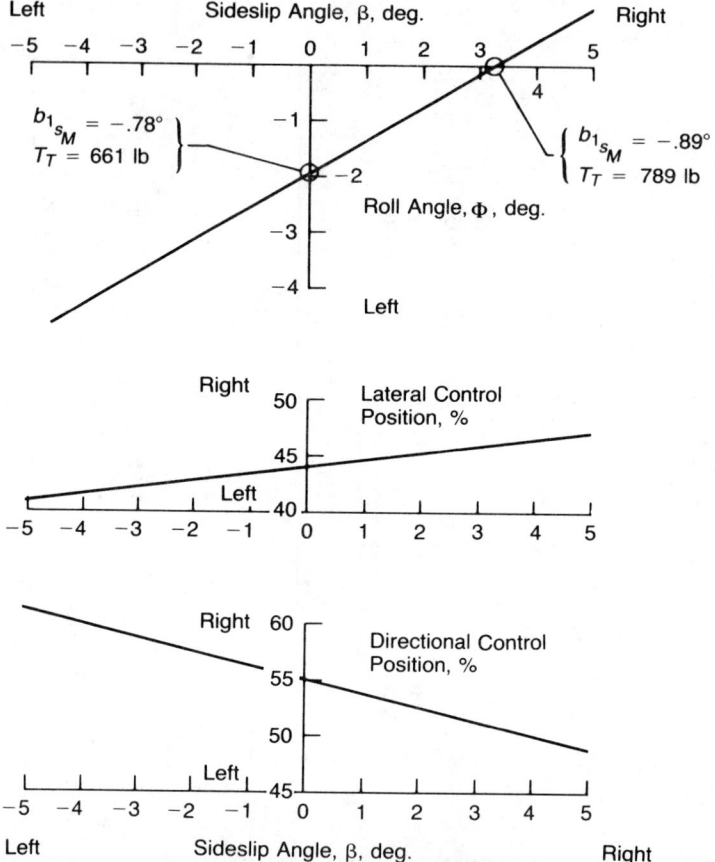

FIGURE 8.31   Trim Conditions in Sideslip; Example Helicopter at 115 Knots

helicopter might resist flying with sideslip because of high directional stability, only a little pedal displacement may be needed to hold sideslip if large changes in tail rotor thrust can be obtained with small pedal movements.

The change in lateral control position to hold a sideslip angle is an indication of the dihedral effect. Positive dihedral results in a roll to the left when the helicopter inadvertently slips to the right—just as on an airplane with both wings tilted up. Positive dihedral is desirable to insure dynamic lateral-directional stability.

The change in roll attitude to hold a given sideslip is an indication of the sideforce characteristics of the helicopter. Strong sideforce characteristics help the pilot make coordinated turns by giving him a "seat of the pants" feel whether he is skidding to the outside of the turn or sliding to the inside.

EXAMPLE HELICOPTER CALCULATIONS

HOW TO'S

CHAPTER 9

Stability and Control Analysis

INTRODUCTION

When designing a helicopter for good performance, the engineer deals almost entirely with laws of physics that are reasonably constant and more or less well understood. When designing for good stability and control, on the other hand, the engineer must also deal with the capabilities of pilots, which are variable and are only partially understood. The problem becomes one of machine-man matching. If the characteristics of the machine are not matched to the natural capabilities of the man, either the machine cannot be operated or the capabilities of the man must be upgraded by training. One example of the problem is the difference between the training and talent required to successfully ride a tricycle, a bicycle, and a unicycle. In this case, the increasing ability required is due primarily to the progressive deterioration in stability. Another example is the difference between attempting a docking maneuver with rowboat, a cabin cruiser, and an ocean liner. In this example, instability is no problem, but the ability to generate accelerations in the required directions becomes more and more degraded until, in the case of the ocean liner, the ability to dock without doing damage is completely inadequate and the job must be relinquished to tugboats.

Thus there are two important elements in machine-man matching: stability and response to control inputs. The best machine-man matching in the aircraft field involves an aircraft that simultaneously has high stability and a rapid and positive response to the pilot's control inputs. Many early helicopters were not good examples of machine-man matching. In general, a helicopter without special

541

stability augmentation provisions not only is unstable, but its response to control inputs is slow, with maximum results appearing some time after the pilot starts the control input. These characteristics give the pilot a combination of fear of the instability and impatience with the slow response. In many cases the student pilot overcontrols and actually finds himself contributing to the instability rather than damping it. A classic remark made by a student following his first attempt to hover was, "It's like riding a pogo stick over a floor covered with greasy ball bearings." Fortunately, pilots can be trained to fly even unstable helicopters and at the same time, the helicopter itself can be tamed by various means.

DEFINITIONS

Stability is the tendency of an object to return to its original conditions following a disturbance. A marble in a bowl is stable, but a marble balanced on top of an inverted bowl is unstable since once disturbed it will go away from its original position with ever increasing speed. The in-between case is a marble on a flat plate, which has no tendency either to return or to leave: it is thus neutrally stable. *Static stability* is measured by the force or moment per unit of displacement that acts to restore the object to its original position. *Dynamic stability* is measured by the time required to return to its original position following a unit displacement. The stability characteristics of a system can be categorized by the type of time history it has following a displacement. Figure 9.1 shows six types of time histories characteristic of aircraft.

Control is the ability to apply forces and moments to the aircraft to maintain it in a steady flight condition in gusty air or to perform a desired maneuver. Two terms are used to define control further: *control power* is the measure of the total moment or force available to the pilot for maneuvering from a steady trimmed flight condition or for compensating for large gust disturbances; *control sensitivity* is the measure of either aircraft acceleration or steady velocity produced by a unit of control motion. It is of importance in defining the precision of control.

A situation that exists in almost all types of aircraft is that increased stability of the basic airframe results in decreased controllability. As a general rule, pilots prefer controllability over stability, since it permits them to get out of tight spots that even very stable aircraft might get into. In this, they are following the lead of the Wright Brothers, whose "Flyer" was very unstable but also very controllable.

SPRING-WEIGHT-DAMPER SYSTEM

Almost all dynamic systems can be represented by combinations of springs, weights, and dampers. It is natural, therefore, to use this system to illustrate basic dynamic principles, since it can easily be visualized and even set up experimentally.

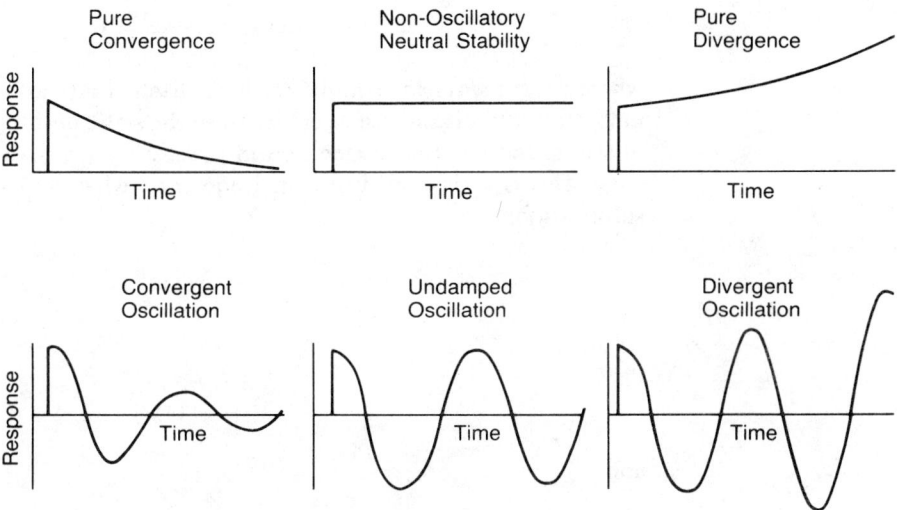

FIGURE 9.1   Possible Types of Time Histories

Figure 9.2 shows a simple spring-weight-damper system consisting of one of each of the elements. The variable of interest, x, is the distance between the instantaneous position of the center of the weight, x_i, and its static position, x_s, when there is no load in the spring. At any one time, the forces acting on the weight in the x direction are of three types: the spring force, F_S, the damper force, F_D, and the inertia force, F_I. In the absence of any external forces, the summation of these three forces must be zero:

$$F_S + F_D + F_I = 0$$

Referring to Figure 9.2, this equation can be written:

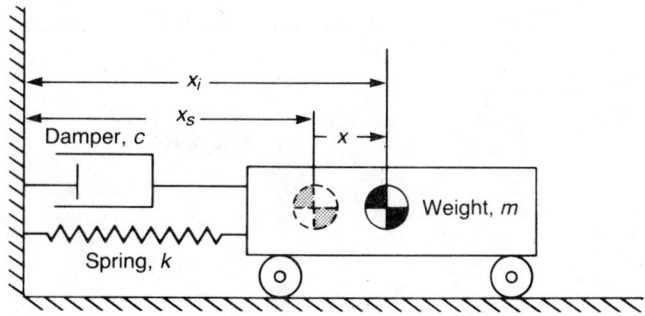

FIGURE 9.2   Spring-Weight-Damper System

$$kx + c\dot{x} + m\ddot{x} = 0$$

where k is the spring rate in lb/ft; c is the damper rate in lb/(ft/sec); m is the mass in lb/(ft/sec²); x is the displacement from the static position in ft; $\dot{x}$ is the velocity in ft/sec; and $\ddot{x}$ is the acceleration in ft/sec².

This is a classical differential equation, which can be solved by using the substitution:

$$x = x(s)e^{st}$$

so that

$$\dot{x} = sx(s)e^{st}$$

and

$$\ddot{x} = s^2x(s)e^{st}$$

When these are substituted into the equilibrium equation, the common factor $x(s)e^{st}$ can be eliminated from each term, leaving:

$$s^2 + \frac{c}{m}s + \frac{k}{m} = 0$$

which is a quadratic equation with two roots:

$$s_1 = -\left(\frac{c/2}{m}\right) + \sqrt{\left(\frac{c/2}{m}\right)^2 - \frac{k}{m}}$$

$$s_2 = -\left(\frac{c/2}{m}\right) - \sqrt{\left(\frac{c/2}{m}\right)^2 - \frac{k}{m}}$$

The general solution for x has both of these roots:

$$x = x_1(s)e^{\left[-\left(\frac{c/2}{m}\right) + \sqrt{\left(\frac{c/2}{m}\right)^2 - \frac{k}{m}}\right]t}$$

$$+ x_2(s)e^{\left[-\left(\frac{c/2}{m}\right) - \sqrt{\left(\frac{c/2}{m}\right)^2 - \frac{k}{m}}\right]t}$$

where $x_1(s)$ and $x_2(s)$ are constants, which in this case can be evaluated by two initial conditions at zero time:

at $t = 0$

$$x = x_0$$
$$\dot{x} = 0$$

thus

$$x_1(s) + x_2(s) = x_0$$

and

$$\left[-\frac{c/2}{m} + \sqrt{\left(\frac{c/2}{m}\right)^2 - \frac{k}{m}} \right] x_1(s) + \left[-\frac{c/2}{m} - \sqrt{\left(\frac{c/2}{m}\right)^2 - \frac{k}{m}} \right] x_2(s) = 0$$

Solving these two equations simultaneously gives:

$$x_1(s) = \frac{x_0}{2} \left[1 + \frac{\dfrac{c/2}{m}}{\sqrt{\left(\dfrac{c/2}{m}\right)^2 - \dfrac{k}{m}}} \right]$$

and

$$x_2(s) = \frac{x_0}{2} \left[1 - \frac{\dfrac{c/2}{m}}{\sqrt{\left(\dfrac{c/2}{m}\right)^2 - \dfrac{k}{m}}} \right]$$

The final solution is thus:

$$x = \frac{x_0}{2} e^{-\frac{c/2}{m}t} \left\{ \left[1 + \frac{\dfrac{c/2}{m}}{\sqrt{\left(\dfrac{c/2}{m}\right)^2 - \dfrac{k}{m}}} \right] e^{\sqrt{\left(\frac{c/2}{m}\right)^2 - \frac{k}{m}}\,t} \right.$$

$$\left. + \left[1 - \frac{\dfrac{c/2}{m}}{\sqrt{\left(\dfrac{c/2}{m}\right)^2 - \dfrac{k}{m}}} \right] e^{-\sqrt{\left(\frac{c/2}{m}\right)^2 - \frac{k}{m}}\,t} \right\}$$

This complicated-looking solution to an apparently simple system represents the six types of time histories shown in Figure 9.1 through various combinations of relative magnitudes of the constants, k, c, and m.

Case 1

$$c = 0, \frac{k}{m} > 0$$

$$x = \frac{x_0}{2}\left[e^{i\sqrt{\frac{k}{m}}t} + e^{-i\sqrt{\frac{k}{m}}t} \right]$$

The exponential terms can be written in terms of trignometric terms using Euler's equations:

$$e^{iz} = \cos z + i \sin z$$

and

$$e^{-iz} = \cos z - i \sin z$$

Thus

$$x = x_0 \cos \sqrt{\frac{k}{m}}\,t$$

This is a steady, or neutrally stable, oscillation with a half amplitude of x_0 and a frequency of $\sqrt{k/m}$ radians per second, or a period of $2\pi/\sqrt{(k/m)}$ seconds as shown in the sketch.

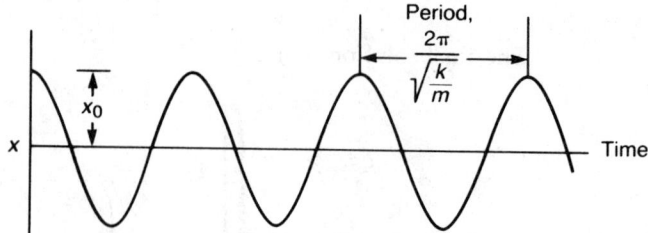

Case 2

$$c > 0, \frac{k}{m} > \left(\frac{c/2}{m}\right)^2$$

$$x = x_0 e^{-\frac{c/2}{m}t}\left[\cos\sqrt{\frac{k}{m} - \left(\frac{c/2}{m}\right)^2}\,t + \frac{\frac{c/2}{m}}{\sqrt{\frac{k}{m} - \left(\frac{c/2}{m}\right)^2}}\sin\sqrt{\frac{k}{m} - \left(\frac{c/2}{m}\right)^2}\,t \right]$$

or

$$x = x_0 e^{-\frac{c/2}{m}t} \cos\left(\sqrt{\frac{k}{m} - \left(\frac{c/2}{m}\right)^2}\, t - \frac{\dfrac{c/2}{m}}{\sqrt{\dfrac{k}{m} - \left(\dfrac{c/2}{m}\right)^2}}\right)$$

This is a damped, or stable, oscillation with a frequency of $\sqrt{(k/m) - [(c/2)/m]^2}$ radians per second whose envelope decays asymtotically. The rate of decay to half amplitude is given by the equation:

$$e^{-\frac{c/2}{m}t} = .5$$

or the time to half amplitude is:

$$t_{1/2} = -\frac{\ln .5}{\dfrac{c/2}{m}} = \frac{.693}{\dfrac{c/2}{m}}$$

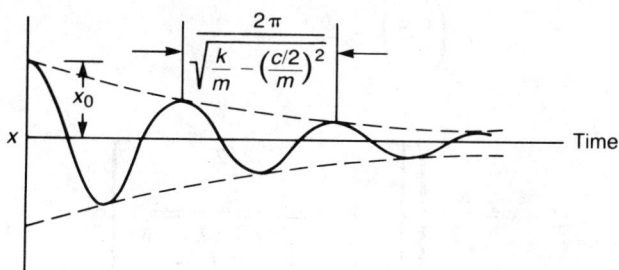

Case 3

$$c < 0, \quad \frac{k}{m} > \left(\frac{c/2}{m}\right)^2$$

(Note: Physically, there are no simple negative dampers, but in some systems a source of external energy gives the same effect.)

$$x = x_0 \, e^{\left|\frac{c/2}{m}\right| t} \cos \sqrt{\frac{k}{m} - \left(\frac{c/2}{m}\right)^2} \, t$$

This is a negatively damped, or unstable, oscillation, with a frequency of $\sqrt{(k/m) - [(c/2)/m]^2}$ radians per second, whose envelope expands to double its amplitude every $1.386 \, m/c$ seconds.

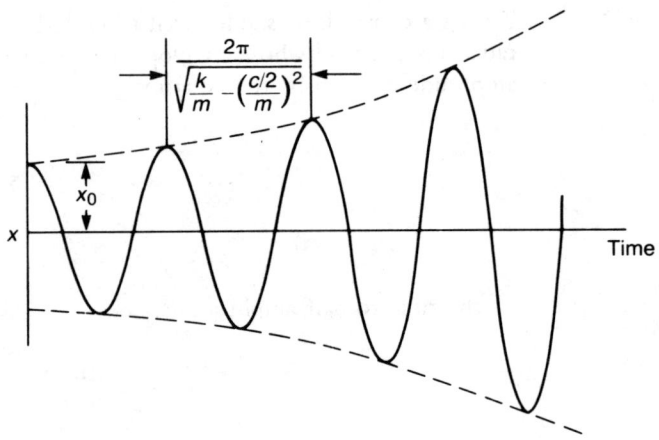

Case 4

$$c > 0, \quad \left(\frac{c/2}{m}\right)^2 > \frac{k}{m}$$

$$x = \frac{x_0}{2} \left\{ \left[1 + \frac{\frac{c/2}{m}}{\sqrt{\left(\frac{c/2}{m}\right)^2 - \frac{k}{m}}} \right] e^{\left[-\frac{c/2}{m} + \sqrt{\left(\frac{c/2}{m}\right)^2 - \frac{k}{m}} \right] t} \right.$$

$$\left. + \left[1 - \frac{\frac{c/2}{m}}{\sqrt{\left(\frac{c/2}{m}\right)^2 - \frac{k}{m}}} \right] e^{\left[-\frac{c/2}{m} - \sqrt{\left(\frac{c/2}{m}\right)^2 - \frac{k}{m}} \right] t} \right\}$$

In this case, the exponents of both e's are real and negative, so that the motion is a pure convergence with time as in the sketch.

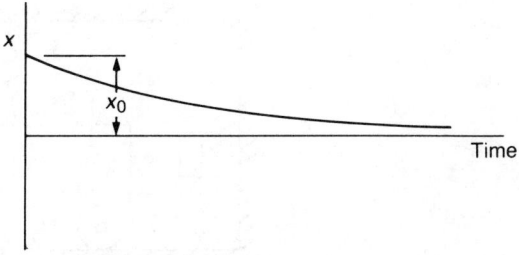

Note that the special subcase of $[(c/2)/m]^2 = k/m$ represents the dividing line between convergence with oscillation and convergence without oscillation. The value of damping that satisfies this condition is called *critical damping*. This concept allows a useful alternative form of the equilibrium equation to be written by defining a *damping ratio*, ζ, where

$$\zeta = c/c_{\text{crit}} = \frac{c}{2m\sqrt{\dfrac{k}{m}}}$$

But the *undamped natural frequency*, ω_0, is:

$$\omega_0 = \sqrt{\frac{k}{m}}$$

Thus

$$\frac{c}{m} = 2\zeta\omega_0$$

and the equilibrium equation becomes:

$$s^2 + 2\zeta\omega_0 s + \omega_0 = 0$$

The concept of critical damping is sometimes used as a criterion for dynamic systems that might go unstable under certain circumstances by specifying a damping in terms of the critical damping. For example, a value of ζ of 0.06, or 6% of critical damping, is often used as a goal.

Case 5a

$k < 0$

(Note: A simple system with a negative spring is a toggle switch poised at dead center.) Schematically, this system looks like this:

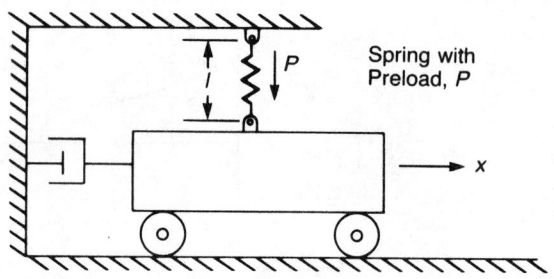

The effective spring constant, k, is:

$$k = -\frac{P}{l}$$

$$x = \frac{x_0}{2} \left\{ \left[1 + \frac{\frac{c/2}{m}}{\sqrt{\left(\frac{c/2}{m}\right)^2 + \left|\frac{k}{m}\right|}} \right] e^{\sqrt{1 + \frac{\left|\frac{k}{m}\right|}{\frac{c/2}{m}}}\,t} \right.$$

$$\left. + \left[1 - \frac{\frac{c/2}{m}}{\sqrt{\left(\frac{c/2}{m}\right)^2 + \left|\frac{k}{m}\right|}} \right] e^{-\frac{c/2}{m}\sqrt{\left(\frac{c/2}{m}\right)^2 + \left|\frac{k}{m}\right|}\,t} \right\}$$

Since at least one e has a positive real exponent, the resultant time history is a pure divergence:

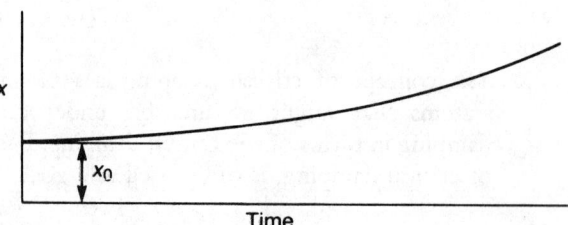

Case 5b

$$c < 0, \quad \left(\frac{c/2}{m}\right)^2 > \frac{k}{m}$$

$$x = \frac{x_0}{2}\left\{\left[1 - \frac{\left|\dfrac{c/2}{m}\right|}{\sqrt{\left(\dfrac{c/2}{m}\right)^2 - \dfrac{k}{m}}}\right]e^{\left|\frac{c/2}{m}\right|\sqrt{\left(\frac{c/2}{m}\right)^2 - \frac{k}{m}}\,t}\right.$$

$$\left. + \left[1 + \frac{\left|\dfrac{c/2}{m}\right|}{\sqrt{\left(\dfrac{c/2}{m}\right)^2 - \dfrac{k}{m}}}\right]e^{\frac{\left|\frac{c/2}{m}t\right|}{\sqrt{\left(\frac{c/2}{m}\right)^2 - \frac{k}{m}}}}\right\}$$

This solution has two *e*s with positive exponents, so it is also a pure divergence:

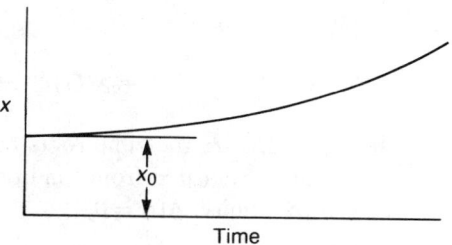

Case 6

$k = 0$

$$x = \frac{x_0}{2}\,e^{-\frac{c/2}{m}t}\left\{2e^{\frac{c/2}{m}t} + 0e^{-\frac{c/2}{m}t}\right\}$$

or

$$x = x_0$$

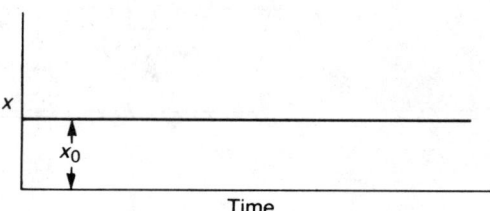

Note that this solution, based on zero initial velocity, is good whether the damping is positive or negative. If damping were negative, however, any initial velocity would result in a pure divergence.

Response to an External Force

Much of the analysis of helicopter flying qualities is involved with determining the response to a gust or to a control input. To illustrate the basic technique used for this type of analysis, let us suddenly impose a constant force on the system of Figure 9.3. Now the equation of equilibrium is:

$$kx + c\dot{x} + m\ddot{x} = F$$

After using the substitution for x, this equation becomes

$$k + cs + ms^2 = F(s)$$

The solution to this equation contains the roots of the homogeneous equation—that is, the equation with $F(s) = 0$, plus two new terms involving constants of integration:

$$x = x_1(s)e^{s_1 t} + x_2(s)e^{s_2 t} + C_1 + C_2 t$$

where s_1 and s_2 are the same roots obtained earlier. The constants $x_1(s)$, $x_2(s)$, C_1, and C_2 can be evaluated from conditions when time is equal to zero and when time is equal to infinity. At $t = 0$,

$$x = 0$$
$$\dot{x} = 0$$

for a system with positive damping. At $t = \infty$,

$$x = F/k$$

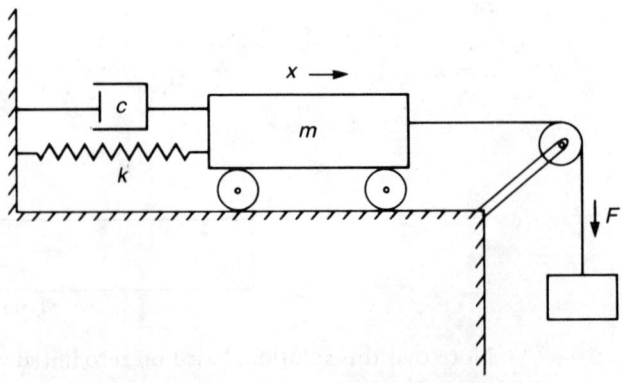

FIGURE 9.3   Forced System

Using the last condition first, it is obvious that C_2 must be zero and that since s_1 and s_2 have negative real parts for a system with positive damping:

$$x_{t=\infty} = C_1 = \frac{F}{k}$$

at zero time:

$$x_{t=0} = x_1(s) + x_2(s) = -\frac{F}{k}$$

$$\dot{x}_{t=0} = s_1 x_1(s) + s_2 x_2(s) = 0$$

Solving these two equations simultaneously and substituting for s_1 and s_2 gives:

$$x_1(s) = -\frac{F}{2k}\left[1 + \frac{\frac{c/2}{m}}{\sqrt{\left(\frac{c/2}{m}\right)^2 - \frac{k}{m}}} \right]$$

$$x_2(s) = -\frac{F}{2k}\left[1 - \frac{\frac{c/2}{m}}{\sqrt{\left(\frac{c/2}{m}\right)^2 - \frac{k}{m}}} \right]$$

Note that these equations are identical to those obtained for the free system except that $x_0/2$ has been replaced by $-F/2k$. The final solution is:

$$x = \frac{F}{2k}\left[2 - e^{-\frac{c/2}{m}t}\left\{ \left[1 + \frac{\frac{c/2}{m}}{\sqrt{\left(\frac{c/2}{m}\right)^2 - \frac{k}{m}}} \right] e^{\sqrt{\left(\frac{c/2}{m}\right)^2 - \frac{k}{m}}\,t} \right. \right.$$

$$\left. \left. + \left[1 - \frac{\frac{c/2}{m}}{\sqrt{\left(\frac{c/2}{m}\right)^2 - \frac{k}{m}}} \right] e^{-\sqrt{\left(\frac{c/2}{m}\right)^2 - \frac{k}{m}}\,t} \right\} \right]$$

A typical time history based on this equation for a system with a damped oscillation is as follows:

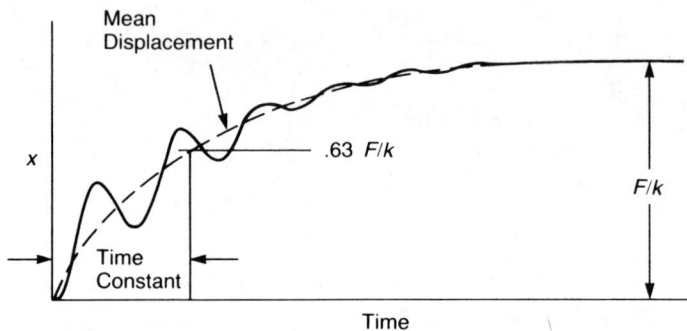

It is often desirable to describe the response in terms of the time required to achieve 63% of its final value. For this purpose, it is only necessary to work with real portions of the solution:

$$x = \frac{F}{k}\left[1 - e^{-\frac{c/2}{m}t}\right]$$

for

$$x = .63 \, F/k$$

$$e^{-\frac{c/2}{m}t_{.63}} = 1 - .63$$

or

$$\frac{c/2}{m}t_{.63} = 1.0$$

Thus

$$t_{.63} = \frac{m}{c/2} \sec$$

This is called the *time constant* of the system.

Multi-Degree-of-Freedom Systems

The system we have been using for illustration is known as a single-degree-of-freedom system since only the value of a single variable, x, is needed to describe its instantaneous position. Aircraft—and, indeed, most other dynamic systems—have more than a single degree of freedom. For example, an airplane has six degrees of freedom, representing three linear displacements—forward, upward, and sideward—and three angular displacements in pitch, roll, and yaw. A helicopter has at least three more, representing rotor coning, longitudinal flapping, and lateral flapping. For illustration of the principles, we will deal with a two-degree-of-freedom system obtained by adding one more spring, weight, and damper to our original system. Figure 9.4 shows the new system.

To define the instantaneous position of this system, we must know both x and y, so this is a two-degree-of-freedom system, and there are two simultaneous equations of equilibrium representing all the forces acting on both weights.

$$k_x x - k_y y + c_x \dot{x} - c_y \dot{y} + m_x \ddot{x} = 0$$

$$k_y y + c_y \dot{y} + m_y (\ddot{x} + \ddot{y}) = 0$$

Using the substitution,

$$x = x(s) e^{st}$$

$$y = y(s) e^{st}$$

the two equations of equilibrium become:

$$(k_x + sc_x + s^2 m_x) x(s) + (-k_y - sc_y) y(s) = 0$$

$$s^2 m_y x(s) + (k_y + sc_y + s^2 m_y) y(s) = 0$$

or in matrix form:

$$\begin{vmatrix} (k_x + sc_x + s^2 mx) & (-k_y - sc_y) \\ (s^2 m_y) & (k_y + sc_y + s^2 m_y) \end{vmatrix} \begin{vmatrix} x(s) \\ y(s) \end{vmatrix} = 0$$

The roots can be found by expanding the determinant and setting it to zero:

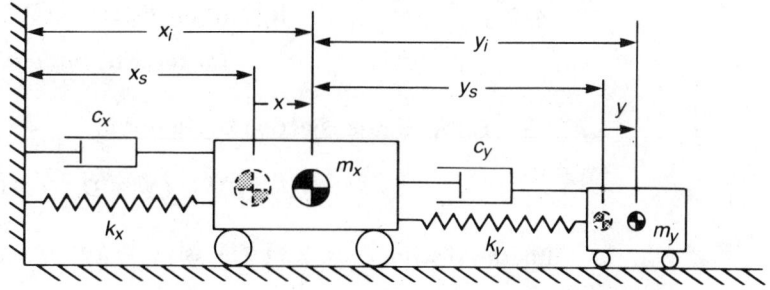

FIGURE 9.4   Two-Degree-of-Freedom System

$$s^4 m_x m_y + s^3[c_y m_x + m_y(c_x + c_y)] + s^2[k_y m_x + c_y c_x + m_y(k_x + k_y)]$$
$$+ s[k_y c_x + k_x c_y] + k_x k_y = 0$$

This is known as the *characteristic equation* of the system, and its roots are sometimes called *eigen values*. There is no simple analytical method for solving the quartic equation as there is for the quadratic that represents the single-degree-of-freedom system; but after the values of the spring, damper, and mass constants are inserted, it can be solved by numerical methods. Depending on the values of the constants, the four roots can be either four real numbers, two real numbers and one pair of complex numbers, or two pairs of complex numbers. The existence of two pairs of complex roots indicates that the system has two separate natural frequencies. The time histories for the displacement of either of the masses can independently have any of the characteristics of the six time histories shown in Figure 9.1.

Routh's Discriminant

It is not necessary to solve for the roots of the characteristic equation to determine whether the system is stable or unstable. Instead, we have only to examine the combination of coefficients known as *Routh's discriminant* (R.D.). For a cubic equation of the form:

$$As^3 + Bs^2 + Cs + D = 0$$

Routh's discriminant, R.D. (3), is:

$$\text{R.D.}(3) = BC - AD$$

For a quartic equation of the form:

$$As^4 + Bs^3 + Cs^2 + Ds + E = 0$$

Routh's discriminant, R.D. (4), is:

$$\text{R.D.}(4) = BCD - AD^2 - B^2E$$

(D must be positive)

For a quintic equation of the form:

$$As^5 + Bs^4 + Cs^3 + Ds^2 + Es + F = 0$$

Routh's discriminant, R.D. (5), is:

$$\text{R.D. }(5) = D(BC - AD)(BE - AF) - B(BF - AF)^2 - F(BC - AD)^2$$

Several tests of the discriminant are used to produce clues about the types of time history that are characteristic of the system:

If	*Then*
1. All coefficients are positive.	There can be no positive real root and thus no pure divergence.
2. R.D. is positive.	There can be no real part of complex roots and thus no unstable oscillation.
3. R.D. = 0	Neutrally stable.
4. R.D. is negative.	Unstable.
5. $D = 0$ (for cubic) $E = 0$ (for quartic) $F = 0$ (for quintic)	There will be one zero root, and one degree of freedom will have non-oscillatory neutral stability.
6. One of the coefficients is negative.	There will be either a pure divergence or an unstable oscillation.

EQUATIONS OF MOTION

The equilibrium equations of Chapter 8 can be converted into equations of motion by accounting for forces and moments corresponding to inertia effects associated with accelerations, either linear or angular, and combinations of velocities.

$$X_M + X_T + X_H + X_V + X_F = \text{G.W.} \sin \Theta + \frac{\text{G.W.}}{g}(\ddot{x} - \dot{y} r + \dot{z} q)$$

$$Y_M + Y_T + Y_V + Y_F = -\text{G.W.} \sin \Phi + \frac{\text{G.W.}}{g}(\ddot{y} + \dot{x} r - \dot{z} p)$$

$$Z_M + Z_T + Z_H + Z_V + Z_F = -\text{G.W.} \cos \Theta + \frac{\text{G.W.}}{g}(\ddot{z} - \dot{x} q + \dot{y} p)$$

$$R_M + Y_M b_M + Z_M y_M + Y_T b_T + Y_V b_V + Y_F b_F + R_F = I_{xx}\dot{p} - qr(I_{yy} - I_{zz})$$

$$M_M - X_M b_M + Z_M l_M + M_T - X_T b_T + Z_T l_T - X_H b_H + Z_H l_H - X_V b_V + M_F$$
$$+ Z_F l_F - X_F b_F = I_{yy}\dot{q} - pr(I_{zz} - I_{xx})$$

$$N_M - Y_M l_M - Y_T l_T - Y_V l_V + N_F - Y_F l_F = I_{zz}\dot{r} - pq(I_{xx} - I_{yy})$$

Figure 9.5 shows the sign convention used.

The centrifugal forces and the gyroscopic moments in the foregoing equations fall directly out of the analysis of the dynamics of a rigid body that can

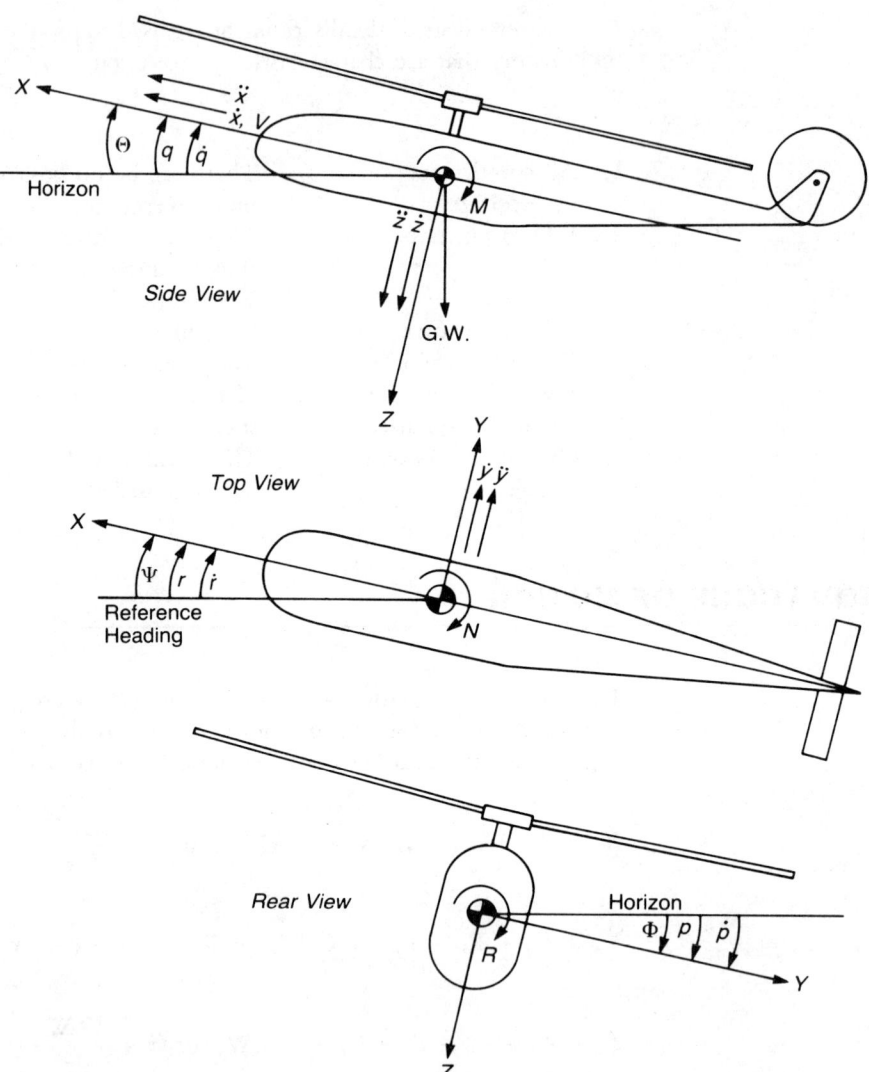

FIGURE 9.5   Parameters in Equations of Motion

simultaneously translate and rotate along and about all three axes. Many of us, however, need some graphical help to accept these terms. Figure 9.6 is an attempt to provide this help.

From a rigorous standpoint, the set of six equations of motion should be augmented with three more equations representing the coning, longitudinal flapping, and lateral flapping of the rotor, which is not attached very rigidly to the airframe. As was shown in Chapter 7, however, the time constant for the flapping of conventional rotor blades corresponds to one-quarter to one-half of a rotor

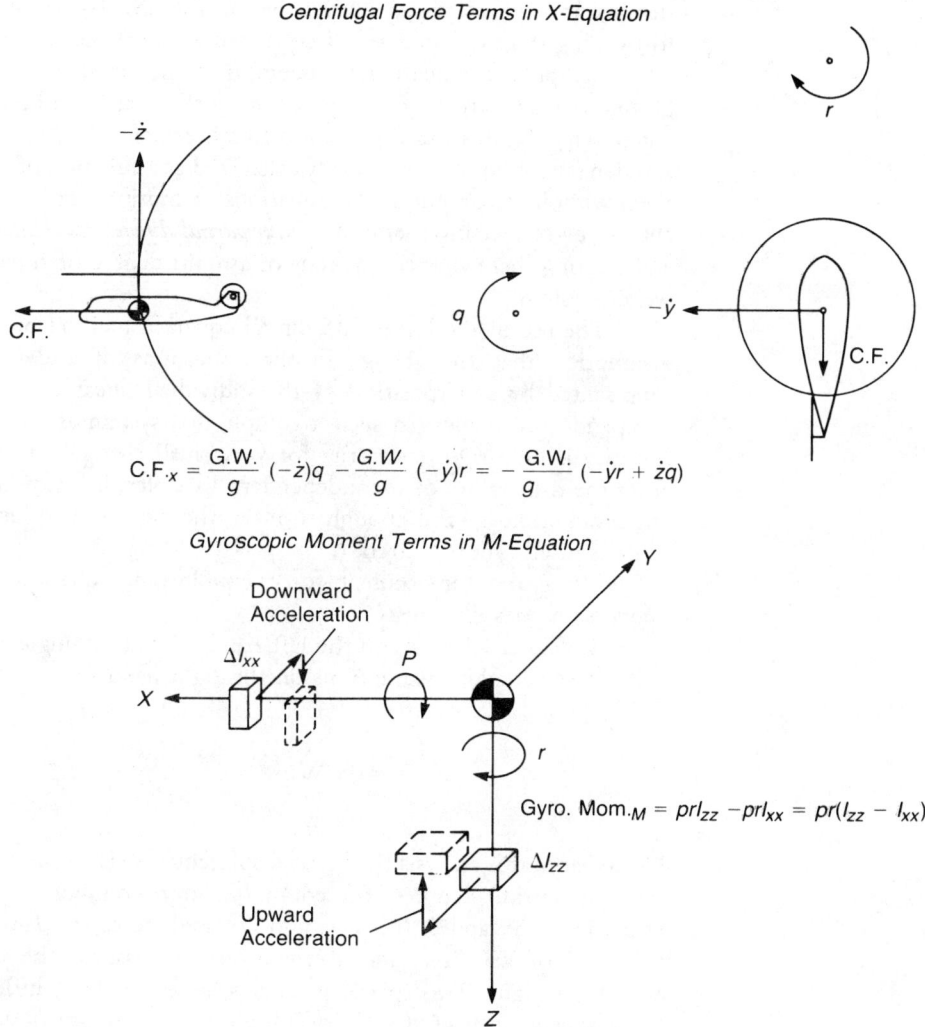

Centrifugal Force Terms in X-Equation

$$C.F._x = \frac{G.W.}{g}(-\dot{z})q - \frac{G.W.}{g}(-\dot{y})r = -\frac{G.W.}{g}(-\dot{y}r + \dot{z}q)$$

Gyroscopic Moment Terms in M-Equation

$$Gyro. \ Mom._M = prI_{zz} - prI_{xx} = pr(I_{zz} - I_{xx})$$

FIGURE 9.6   Illustration of Sources of Centrifugal and Gyroscopic Terms

revolution. This rapid response justifies the use of the *quasi-static* assumption, which eliminates blade motions as separate degrees of freedom and simulates replacing the rotor with a black box at the top of the mast, which essentially produces forces and moments instantaneously in response to changes in flight condition or control inputs.

There are some studies, however, in which the coning and cyclic flapping of the rotor on a short time basis cannot be ignored. These include the prediction of the immediate response to a gust or to a control input or the design of a high-gain stability and control augmentation system (SCAS). For these situations, methods for writing the expanded set of equations are outlined in reference 9.1.

Accepting the quasi-static assumption, the set of six equations is a perfectly general representation of the helicopter in flight and can be used in sophisticated computer programs and flight simulators in which the instantaneous values of the aerodynamic terms are continually calculated as a function of the flight conditions. Even without a computer, the equations of motion can be used in a meaningful analysis by converting them into *linear partial-differential small-perturbation equations* and treating them as the equations of a multi-degree-of-freedom spring-damper-weight system.

The use of the linear differential equation method of analysis relies on the assumption that the change in the conditions of a dynamic system can be represented by superposition of the individual linear effects of changes in the independent variables. In such a complicated system as an aircraft in flight, the linearity is only rigorously true for very small changes or "small perturbations" from the trim values of the independent variables; but experience has shown that the assumption is valid enough to make the method very useful for the stability analysis of all types of aircraft.

Using the longitudinal force equilibrium equation as an example, the conversion goes like this:

1. Gather all terms on the left-hand side into a single aerodynamic term, X, and use small-angle assumptions on the right-hand side.

$$X = \text{G.W.} \, \Theta + \frac{\text{G.W.}}{g} \, (\ddot{x} - \dot{y} \, r + \dot{z} \, q)$$

Note that in this analysis the linear displacements along the body axes, X, Y, and Z are used as prime degrees of freedom. It is more common to see the corresponding velocities $\dot{x}$, $\dot{y}$, and $\dot{z}$ (or u, v, and w) used, since the forces and moments are functions of velocities and accelerations but not of the displacements of the aircraft in flight. Two special problems, however, have influenced the choice of displacements instead of velocities: flight simulation of manual or automatic hover over a spot, and the dynamic analysis of a helicopter mounted in a wind tunnel on moderately flexible supports. Although neither of these problems will be addressed here, the present choice of variables allows them to be studied when they arise while putting practically no unnecessary complication into more routine studies.

2. Rewrite the left-hand side in terms of *stability derivatives* multiplied by pertinent changes in the six degrees of freedom and the four control inputs. Since the aerodynamics are affected only by velocities (with an exception to be covered later), only the first derivatives with respect to time appear in the equations, and all these variables as well as control motions are understood to now be changes from the initial trim conditions.

$$X = \frac{\partial X}{\partial \dot{x}} \dot{x} + \frac{\partial X}{\partial \dot{y}} \dot{y} + \frac{\partial X}{\partial \dot{z}} \dot{z} + \frac{\partial X}{\partial q} q + \frac{\partial X}{\partial p} p + \frac{\partial X}{\partial r} r + \frac{\partial X}{\partial \theta_{0_M}} \theta_{0_M}$$

$$+ \frac{\partial X}{\partial \theta_{0_T}} \theta_{0_T} + \frac{\partial X}{\partial A_1} A_1 + \frac{\partial X}{\partial B_1} B_1$$

3. Rewrite the right-hand side to linearize the products of variables:

$$\text{G.W.} \, \Theta + \frac{\text{G.W.}}{g} (\ddot{x} - \dot{y} \, r + \dot{z} \, q) =$$

$$\text{G.W.} \, \Theta + \frac{\text{G.W.}}{g} (\ddot{x} - \overline{\dot{y}} \, r - \dot{y} \, \overline{r} + \overline{\dot{z}} \, q + \dot{z} \, \overline{q})$$

where the bar refers to values existing in the initial trim condition. For most types of analysis that start with the helicopter in straight and level flight, the following initial velocities are zero:

$$\overline{q} = \overline{p} = \overline{r} = \overline{\dot{y}} = 0$$

In addition:

$$\overline{\dot{x}} = \overline{V}$$

and

$$\overline{\dot{z}} = \overline{V} \alpha_{\text{body}}$$

or, in level flight,

$$\overline{\dot{z}} = \overline{V} \Theta$$

4. In the equation for vertical forces, drop out the gross weight, G.W., since it is not a function of one of the degrees of freedom.

The six body equations of equilibrium converted to linear partial differential equations of motion for small perturbations from steady, level flight are:

$$X, \quad -\frac{\text{G.W.}}{g} \ddot{x} + \frac{\partial X}{\partial \dot{x}} \dot{x} + \frac{\partial X}{\partial \dot{y}} \dot{y} + \frac{\partial X}{\partial \dot{z}} \dot{z} + \left(\frac{\partial X}{\partial q} - \frac{\text{G.W.}}{g} \overline{V} \Theta \right) q - \text{G.W.} \Theta$$

$$+ \frac{\partial X}{\partial p} p + \frac{\partial X}{\partial r} r = -\frac{\partial X}{\partial \theta_{0_M}} \theta_{0_M} - \frac{\partial X}{\partial \theta_{0_T}} \theta_{0_T} - \frac{\partial X}{\partial A_1} A_1 - \frac{\partial X}{\partial B_1} B_1$$

$$Y, \quad \frac{\partial Y}{\partial \dot{x}} \dot{x} - \frac{\text{G.W.}}{g} \ddot{y} + \frac{\partial Y}{\partial \dot{y}} \dot{y} + \frac{\partial Y}{\partial \dot{z}} \dot{z} + \frac{\partial Y}{\partial q} q + \left(\frac{\partial Y}{\partial p} + \frac{\text{G.W.}}{g} \overline{V} \overline{\Theta} \right) p + \text{G.W.} \Phi$$

$$+ \left(\frac{\partial Y}{\partial r} - \frac{\text{G.W.}}{g} \overline{V} \right) r = -\frac{\partial Y}{\partial \theta_{0_M}} \theta_{0_M} - \frac{\partial Y}{\partial \theta_{0_T}} \theta_{0_T} - \frac{\partial Y}{\partial A_1} A_1 - \frac{\partial Y}{\partial B_1} B_1$$

$$Z, \quad \frac{\partial Z}{\partial \dot{x}}\dot{x} + \frac{\partial Z}{\partial \dot{y}}\dot{y} + \left(\frac{\partial Z}{\partial \ddot{z}} - \frac{G.W.}{g}\right)\ddot{z} + \frac{\partial Z}{\partial \dot{z}}\dot{z} + \left(\frac{\partial Z}{\partial q} + \frac{G.W.}{g}\bar{V}\right)q$$

$$+ \frac{\partial Z}{\partial p}p + \frac{\partial Z}{\partial r}r = -\frac{\partial Z}{\partial \theta_{0_M}}\theta_{0_M} - \frac{\partial Z}{\partial \theta_{0_T}}\theta_{0_T} - \frac{\partial Z}{\partial A_1}A_1 - \frac{\partial Z}{\partial B_1}B_1$$

$$R, \quad \frac{\partial R}{\partial \dot{x}}\dot{x} + \frac{\partial R}{\partial \dot{y}}\dot{y} + \frac{\partial R}{\partial \dot{z}}\dot{z} + \frac{\partial R}{\partial q}q - I_{xx}\dot{p} + \frac{\partial R}{\partial p}p + \frac{\partial R}{\partial r}r =$$

$$-\frac{\partial R}{\partial \theta_{0_M}}\theta_{0_M} - \frac{\partial R}{\partial \theta_{0_T}}\theta_{0_T} - \frac{\partial R}{\partial A_1}A_1 - \frac{\partial R}{\partial B_1}B_1$$

$$M, \quad \frac{\partial M}{\partial \dot{x}}\dot{x} + \frac{\partial M}{\partial \dot{y}}\dot{y} + \frac{\partial M}{\partial \ddot{z}}\ddot{z} + \frac{\partial M}{\partial \dot{z}}\dot{z} - I_{yy}\dot{q} + \frac{\partial M}{\partial q}q + \frac{\partial M}{\partial p}p + \frac{\partial M}{\partial r}r =$$

$$-\frac{\partial M}{\partial \theta_{0_M}}\theta_{0_M} - \frac{\partial M}{\partial \theta_{0_T}}\theta_{0_T} - \frac{\partial M}{\partial A_1}A_1 - \frac{\partial M}{\partial B_1}B_1$$

$$N, \quad \frac{\partial N}{\partial \dot{x}}\dot{x} + \frac{\partial N}{\partial \dot{y}}\dot{y} + \frac{\partial N}{\partial \dot{z}}\dot{z} + \frac{\partial N}{\partial q}q + \frac{\partial N}{\partial p}p - I_{zz}\dot{r} + \frac{\partial N}{\partial r}r =$$

$$-\frac{\partial N}{\partial \theta_{0_M}}\theta_{0_M} - \frac{\partial N}{\partial \theta_{0_T}}\theta_{0_T} - \frac{\partial N}{\partial A_1}A_1 - \frac{\partial N}{\partial B_1}B_1$$

Note that the terms associated with the control inputs—main rotor collective pitch, tail rotor collective pitch, and rotor cyclic pitch—have all been gathered on the right-hand side in order to separate free from forced motion. The aerodynamic terms may be seen to be functions only of velocities with the exceptions of $\partial Z/\partial \ddot{z}$ and $\partial M/\partial \ddot{z}$, which will later be shown to be associated with the time required for a change in rotor downwash to reach the horizontal stabilizer.

There are three methods for studying the flying qualities of the helicopter using these equations. The first method, discussed in the previous section, involves substituting the general solution in the form:

$$x = x(s)e^{st}$$

From this, the characteristic equation can be formed; evaluated by Routh's discriminant; and/or solved for its roots, which can be related to periods and damping. This classic method will be illustrated in the remainder of this chapter. A second method involves programming the equations on a so-called analog computer. This uses electronic components consisting of capacitors, resistors, and inductances to represent spring, damper, and mass terms, respectively, in the equations as if they were written for mechanical systems. Voltages within the circuits represent various degrees of freedom and can be used to produce time histories in response to simulated gusts or control motions. This method was in wide use from about 1955 to 1975, but has now largely been supplanted by the third method, using digital computers, which can both do the classical analysis and produce time histories more accurately than can the analog computers.

In any case, it is necessary first to write the equations of motion in numerical form by evaluating all the derivatives.

EVALUATION OF STABILITY DERIVATIVES

The stability derivatives can be evaluated by several methods. The most accurate is, naturally, also the costliest unless it is to be used over and over. That method consists of programming a computer to do the type of numerical calculations illustrated in Chapter 8 to solve first for the trim conditions and then for the changes in aerodynamic forces and moments due to sequential unit variations in each degree of freedom and control input. The heart of this type of program is the subroutine that analyzes the main and tail rotors. Programs with various levels of sophistication are operational in all the major helicopter companies and research organizations. A description of one will be found in reference 9.2.

A noncomputer means of obtaining values for the stability derivatives of the main and tail rotors consists of differentiating equations for the aerodynamic coefficients and flapping angles such as those found in Chapter 3. This procedure, widely used before computer technology matured, still has value for quick analysis and for verifying the more complex computer programs as they are being developed. A description of this method, along with results in the form of charts, is given in reference 9.3.

In the following pages, a third method will be used. It makes use of the rotor performance charts in Chapters 1 and 3 to determine changes to the non-dimensional rotor coefficients as basic parameters are varied.

In many analyses, the derivatives are presented as the ratio of the change in forces divided by the mass of the aircraft and the change in moments about each axis divided by the appropriate moment of inertia. That approach has two advantages: it makes the equations of motion look more compact, and it provides derivatives that can be compared with respect to magnitude from one aircraft to another even if they vary widely in size. This latter reason seems to be of more significance for fixed-wing aircraft than for helicopters, for which many different fuselages can be supported by rotors that are essentially alike. One disadvantage with this system is that these derivatives change magnitude with gross weight and with loadings that change the moments of inertia. In this book, we will leave the derivatives in their dimensional form in order to avoid making them any more abstract than they already are.

Rotor Derivatives near Hover

In the case of hover, only the aerodynamics of the two rotors are important, and the required derivatives can be simply evaluated using the charts and equations already presented in Chapters 1, 2, 3, and 7. The significant rotor derivatives are listed in Tables 9.1 through 9.4 with numerical values corresponding to the example helicopter. Table 9.1 gives the basic rotor derivatives. Tables 9.2 and 9.3 are separate tables for the dimensional derivatives of the main and tail rotors. Finally Table 9.4 is a summary table for the entire helicopter. These tables can serve as checklists for the analysis of any other helicopter.

TABLE 9.1
Basic Rotor Derivatives near Hover

Derivative	Equation	Values for Example Helicopter	
		Main Rotor	Tail Rotor
$\left(\dfrac{\partial \mu}{\partial \dot{x}}\right)_M, \left(\dfrac{\partial \lambda'}{\partial \dot{z}}\right)_M$	$\left(\dfrac{1}{\Omega R}\right)_M$	.00154	—
$\left(\dfrac{\partial \mu}{\partial \dot{x}}\right)_T, \left(\dfrac{-\partial \lambda'}{\partial \dot{y}}\right)_T$	$\left(\dfrac{1}{\Omega R}\right)_T$	—	.00154
$\dfrac{\partial a_{1_s}}{\partial \mu}$	$\dfrac{8}{3}\theta_0 + 2\theta_1 - 2\dfrac{v_1}{\Omega R}$	.34	.34
$\dfrac{\partial b_{1_s}}{\partial \mu}$	$\dfrac{4}{3}a_0$	.10	.05
$\dfrac{\partial C_T/\sigma}{\partial \theta_0}$	Take from hover charts of Chapter 1	.61	.50
$\dfrac{\partial C_Q/\sigma}{\partial \theta_0}$	Take from hover charts of Chapter 1	.078	.089
$\dfrac{\partial C_T/\sigma}{\partial \lambda'}$	$\dfrac{1}{\dfrac{8}{a} + \dfrac{\sqrt{\dfrac{\sigma}{2}}}{\sqrt{C_T/\sigma}}}$	.49	.44

TABLE 9.1 (continued)

Derivative	Equation	Values for Example Helicopter	
		Main Rotor	Tail Rotor
$\dfrac{\partial C_Q/\sigma}{\partial \lambda'}$	$-\dfrac{a}{4}\left(\theta_{.75} - 2\,\dfrac{v_1}{\Omega R}\right)$	$-.076$	$-.038$
$\dfrac{\partial a_{1_s}}{\partial q}, \dfrac{\partial b_{1_s}}{\partial p}$	$-\dfrac{16}{\gamma\Omega\left(1 - \dfrac{e}{R}\right)^2}$	$-.105$	0
$\dfrac{\partial a_{1_s}}{\partial p}, \dfrac{-\partial b_{1_s}}{\partial q}$	$-\dfrac{12\,\dfrac{e}{R}}{\gamma\Omega\left(1 - \dfrac{e}{R}\right)^3}$ $\dfrac{1}{\Omega}\left[1 - \dfrac{192\,\dfrac{e}{R}}{\gamma^2\left(1 - \dfrac{e}{R}\right)^5}\right]$	$.037$	0
$\dfrac{\partial a_{1_s}}{\partial A_1}, \dfrac{\partial b_{1_s}}{\partial B_1}$	$\dfrac{12\,\dfrac{e}{R}}{\gamma\left(1 - \dfrac{e}{R}\right)^3}$	$.086$	0
$\dfrac{\partial a_{1_s}}{\partial B_1}, \dfrac{-\partial b_{1_s}}{\partial A_1}$	$\dfrac{-1}{1 + \dfrac{144\left(\dfrac{e}{R}\right)^2}{\gamma^2\left(1 - \dfrac{e}{R}\right)^6}}$	$-.993$	0
$\dfrac{\partial C_H/\sigma}{\partial a_{1_s}}, \dfrac{\partial C_y/\sigma}{\partial b_{1_s}}$	$\dfrac{3}{2}\,C_T/\sigma\left(1 - \dfrac{a}{18}\,\dfrac{\theta_{.75}}{C_T/\sigma}\right)$	$.040$	$-.016$
$\dfrac{\partial M}{\partial a_{1_s}}, \dfrac{\partial R}{\partial b_{1_s}}$	$\dfrac{\dfrac{3}{4}\,\dfrac{e}{R}\,A_b\rho R(\Omega R)^2 a}{\gamma}$	$200{,}940$	$-$

TABLE 9.2
Main Rotor Derivatives near Hover

Derivative	Equation	Values for Example Helicopter
$\left(\dfrac{\partial X}{\partial \dot{x}}\right)_M$	$-\rho A_b (\Omega R)^2 \dfrac{\partial C_H/\sigma}{\partial a_{1_s}} \dfrac{\partial a_{1_s}}{\partial \mu} \dfrac{\partial \mu}{\partial \dot{x}}$	-5 lb/ft/sec
$\left(\dfrac{\partial X}{\partial \dot{y}}\right)_M$	$-\left(\dfrac{\partial Y}{\partial \dot{x}}\right)_M$	-1 lb/ft/sec
$\left(\dfrac{\partial X}{\partial q}\right)_M$	$-\rho A_b (\Omega R)^2 \dfrac{\partial C_H/\sigma}{\partial a_{1_s}} \dfrac{\partial a_{1_s}}{\partial q}$	$1{,}008$ lb/rad/sec
$\left(\dfrac{\partial X}{\partial p}\right)_M$	$-\rho A_b (\Omega R)^2 \dfrac{\partial C_H/\sigma}{\partial a_{1_s}} \dfrac{\partial a_{1_s}}{\partial p}$	-355 lb/rad/sec
$\left(\dfrac{\partial X}{\partial \theta_0}\right)_M$	$-\rho A_b (\Omega R)^2 (\bar{a}_{1_s} + i_M) \dfrac{\partial C_T/\sigma}{\partial \theta_0}$	$3{,}677$ lb/rad
$\left(\dfrac{\partial X}{\partial A_1}\right)_M$	$-\rho A_b (\Omega R)^2 \dfrac{\partial C_H/\sigma}{\partial a_{1_s}} \dfrac{\partial a_{1_s}}{\partial A_1}$	-825 lb/rad
$\left(\dfrac{\partial X}{\partial B_1}\right)_M$	$-\rho A_b (\Omega R)^2 \dfrac{\partial C_H/\sigma}{\partial a_{1_s}} \dfrac{\partial a_{1_s}}{\partial B_1}$	$9{,}531$ lb/rad
$\left(\dfrac{\partial Y}{\partial \dot{x}}\right)_M$	$\rho A_b (\Omega R)^2 \dfrac{\partial C_y/\sigma}{\partial b_{1_s}} \dfrac{\partial b_{1_s}}{\partial \mu} \dfrac{\partial \mu}{\partial \dot{x}}$	1 lb/ft/sec
$\left(\dfrac{\partial Y}{\partial \dot{y}}\right)_M$	$\left(\dfrac{\partial X}{\partial \dot{x}}\right)_M$	-5 lb/ft/sec
$\left(\dfrac{\partial Y}{\partial q}\right)_M$	$\left(\dfrac{\partial X}{\partial p}\right)_M$	-355 lb/rad/sec
$\left(\dfrac{\partial Y}{\partial p}\right)_M$	$-\left(\dfrac{\partial X}{\partial q}\right)_M$	$-1{,}008$ lb/rad/sec

TABLE 9.2 (continued)

Derivative	Equation	Values for Example Helicopter
$\left(\dfrac{\partial Y}{\partial \theta_0}\right)_M$	$\rho A_b (\Omega R)^2 \bar{b}_{1_s} \dfrac{\partial C_T/\sigma}{\partial \theta_0}$	$-3{,}971$ lb/rad
$\left(\dfrac{\partial Y}{\partial A_1}\right)_M$	$\left(\dfrac{\partial X}{\partial B_1}\right)_M$	$9{,}531$ lb/rad
$\left(\dfrac{\partial Y}{\partial B_1}\right)_M$	$-\left(\dfrac{\partial X}{\partial A_1}\right)_M$	825 lb/rad
$\left(\dfrac{\partial Z}{\partial \dot{z}}\right)_M$	$-\rho A_b (\Omega R)^2 \dfrac{\partial C_T/\sigma}{\partial \lambda'} \dfrac{\partial \lambda'}{\partial \dot{z}}$	-182 lb/ft/sec
$\left(\dfrac{\partial Z}{\partial \theta_0}\right)_M$	$-\rho A_b (\Omega R)^2 \dfrac{\partial C_T/\sigma}{\partial \theta_0}$	$-147{,}088$ lb/rad
$\left(\dfrac{\partial R}{\partial \dot{x}}\right)_M$	$\left(\dfrac{dR}{db_{1_s}}\right)_M \dfrac{\partial b_{1_s}}{\partial \mu} \dfrac{\partial \mu}{\partial \dot{x}} + \left(\dfrac{\partial Y}{\partial \dot{x}}\right)_M h_M$	39 ft-lb/ft/sec
$\left(\dfrac{\partial R}{\partial \dot{y}}\right)_M$	$-\left(\dfrac{\partial M}{\partial \dot{x}}\right)_M$	-143 ft-lb/ft/sec
$\left(\dfrac{\partial R}{\partial \dot{z}}\right)_M$	$\left(\dfrac{\partial Z}{\partial \dot{z}}\right)_M y_M$	0 ft-lb/ft/sec
$\left(\dfrac{\partial R}{\partial q}\right)_M$	$\left(\dfrac{dR}{db_{1_s}}\right)_M \dfrac{\partial b_{1_s}}{\partial q} + \left(\dfrac{\partial Y}{\partial q}\right)_M h_M$	$-10{,}097$ ft-lb/rad/sec
$\left(\dfrac{\partial R}{\partial p}\right)_M$	$\left(\dfrac{dR}{db_{1_s}}\right)_M \dfrac{\partial b_{1_s}}{\partial p} + \left(\dfrac{\partial Y}{\partial p}\right)_M h_M$	$-28{,}659$ ft-lb/rad/sec
$\left(\dfrac{\partial R}{\partial \theta_0}\right)_M$	$\left(\dfrac{\partial Y}{\partial \theta_0}\right) h_M + \left(\dfrac{\partial Z}{\partial \theta_0}\right) y_M$	$-29{,}783$ ft-lb/rad

continued

TABLE 9.2 (continued)

Derivative	Equation	Values for Example Helicopter
$\left(\dfrac{\partial R}{\partial A_1}\right)_M$	$\left(\dfrac{dR}{db_{1_s}}\right)_M \dfrac{\partial b_{1_s}}{\partial A_1} + \left(\dfrac{\partial Y}{\partial A_1}\right)_M h_M$	271,016 ft-lb/rad
$\left(\dfrac{\partial R}{\partial B_1}\right)_M$	$\left(\dfrac{dR}{db_{1_s}}\right)_M \dfrac{\partial b_{1_s}}{\partial B_1} + \left(\dfrac{\partial Y}{\partial B_1}\right)_M h_M$	23,468 ft-lb/rad
$\left(\dfrac{\partial M}{\partial \dot{x}}\right)_M$	$\left(\dfrac{dM}{da_{1_s}}\right)_M \dfrac{\partial a_{1_s}}{\partial \mu} \dfrac{\partial \mu}{\partial \dot{x}} - \left(\dfrac{\partial X}{\partial \dot{x}}\right)_M h_M$	143 ft-lb/ft/sec
$\left(\dfrac{\partial M}{\partial \dot{y}}\right)_M$	$\left(\dfrac{\partial R}{\partial \dot{x}}\right)_M$	39 ft-lb/ft/sec
$\left(\dfrac{\partial M}{\partial \dot{z}}\right)_M$	$\left(\dfrac{\partial Z}{\partial \dot{z}}\right)_M l_M$	91 ft-lb/ft/sec
$\left(\dfrac{\partial M}{\partial q}\right)_M$	$\left(\dfrac{dM}{da_{1_s}}\right)_M \dfrac{\partial a_{1_s}}{\partial q} - \left(\dfrac{\partial X}{\partial q}\right)_M h_M$	−28,659 ft-lb/rad/sec
$\left(\dfrac{\partial M}{\partial p}\right)_M$	$\left(\dfrac{dM}{da_{1_s}}\right)_M \dfrac{\partial a_{1_s}}{\partial p} - \left(\dfrac{\partial X}{\partial p}\right)_M h_M$	10,097 ft-lb/rad/sec
$\left(\dfrac{\partial M}{\partial \theta_0}\right)_M$	$-\left(\dfrac{\partial X_A}{\partial \theta_0}\right)_M h_M + \left(\dfrac{\partial Z}{\partial \theta_0}\right)_M l_M$	45,967 ft-lb/rad
$\left(\dfrac{\partial M}{\partial A_1}\right)_M$	$\left(\dfrac{dM}{da_{1_s}}\right)_M \dfrac{\partial a_{1_s}}{\partial A_1} - \left(\dfrac{\partial X}{\partial A_1}\right)_M h_M$	23,468 ft-lb/rad
$\left(\dfrac{\partial M}{\partial B_1}\right)_M$	$\left(\dfrac{dM}{da_{1_s}}\right)_M \dfrac{\partial a_{1_s}}{\partial B_1} - \left(\dfrac{\partial X}{\partial B_1}\right)_M h_M$	−271,016 ft-lb/rad
$\left(\dfrac{\partial N}{\partial \dot{z}}\right)_M$	$\rho A_b (\Omega R)^2 R \dfrac{\partial C_Q/\sigma}{\partial \lambda'} \dfrac{\partial \lambda'}{\partial \dot{z}}$	−847 ft-lb/ft/sec

TABLE 9.2 (continued)

Derivative	Equation	Values for Example Helicopter
$\left(\dfrac{\partial N}{\partial r}\right)_M$	$2\rho A_b(\Omega R)R^2 C_Q/\sigma$ (governed engine, counterclockwise rotation)	4,471 ft-lb/rad/sec
$\left(\dfrac{\partial N}{\partial \theta_0}\right)_M$	$\rho A_b(\Omega R)^2 R\,\dfrac{\partial C_Q/\sigma}{\partial \theta_0}$	564,242 ft-lb/rad

TABLE 9.3
Tail Rotor Derivatives near Hover

Derivative	Equation	Values for Example Helicopter
$\left(\dfrac{\partial Y}{\partial \dot{y}}\right)_T$	$\rho A_b(\Omega R)^2\,\dfrac{\partial C_T/\sigma}{\partial \lambda'}\,\dfrac{\partial \lambda'}{\partial \dot{y}}$	-13 lb/ft/sec
$\left(\dfrac{\partial Y}{\partial p}\right)_T$	$\left(\dfrac{\partial Y}{\partial \dot{y}}\right)_T h_T$	-78 lb/rad/sec
$\left(\dfrac{\partial Y}{\partial r}\right)_T$	$-\left(\dfrac{\partial Y}{\partial \dot{y}}\right)_T l_T$	481 lb/rad/sec
$\left(\dfrac{\partial Y}{\partial \theta_0}\right)_T$	$\rho A_b(\Omega R)^2\,\dfrac{\partial C_T/\sigma}{\partial \theta_0}$	9,746 lb/rad
$\left(\dfrac{\partial R}{\partial \dot{y}}\right)_T$	$\left(\dfrac{\partial Y}{\partial \dot{y}}\right)_T h_T$	78 ft-lb/ft/sec
$\left(\dfrac{\partial R}{\partial p}\right)_T$	$\left(\dfrac{\partial Y}{\partial p}\right)_T h_T$	-468 ft-lb/rad/sec
$\left(\dfrac{\partial R}{\partial r}\right)_T$	$\left(\dfrac{\partial Y}{\partial r}\right)_T h_T$	2,886 ft-lb/rad/sec

continued

TABLE 9.3 (continued)

Derivative	Equation	Values for Example Helic.
$\left(\dfrac{\partial R}{\partial \theta_0}\right)_T$	$\left(\dfrac{\partial Y}{\partial \theta_0}\right)_T b_T$	58,476 ft-lb/rad/sec
$\left(\dfrac{\partial M}{\partial \dot{y}}\right)_T$	$\rho A_b (\Omega R)^2 R \dfrac{\partial C_Q/\sigma}{\partial \lambda'} \dfrac{\partial \lambda'}{\partial \dot{y}}$	-7 ft-lb/ft/sec
$\left(\dfrac{\partial M}{\partial r}\right)_T$	$-\left(\dfrac{\partial M}{\partial \dot{y}}\right)_T l_T$	274 ft-lb/rad/sec
$\left(\dfrac{\partial M}{\partial \theta_0}\right)_T$	$\rho A_b (\Omega R)^2 R \dfrac{\partial C_Q/\sigma}{\partial \theta_0}$	11,276 ft-lb/rad
$\left(\dfrac{\partial N}{\partial \dot{y}}\right)_T$	$-\left(\dfrac{\partial Y}{\partial \dot{y}}\right)_T l_T$	481 ft-lb/ft/sec
$\left(\dfrac{\partial N}{\partial p}\right)_T$	$-\left(\dfrac{\partial Y}{\partial p}\right)_T l_T$	2,886 ft-lb/rad/sec
$\left(\dfrac{\partial N}{\partial r}\right)_T$	$-\left(\dfrac{\partial Y}{\partial r}\right)_T l_T$	$-17,797$ ft-lb/rad/sec
$\left(\dfrac{\partial N}{\partial \theta_0}\right)_T$	$-\left(\dfrac{\partial Y}{\partial \theta_0}\right)_T l_T$	$-360,602$ ft-lb/rad

Rotor Derivatives in Forward Flight

The procedure for evaluating the rotor derivatives in forward flight is similar to that used for hover except that more use is made of charts—in this case, those in Chapter 3. The method is based on the fact that for a given rotor, the forces and flapping conditions are uniquely determined by three independent variables, μ, θ, and λ'. Using this concept, the main and tail rotor derivatives may be written in terms of nondimensional rotor derivatives, physical parameters, and trim conditions.

TABLE 9.4
Total Derivatives in Hover

Component / Derivative	Main Rotor	Tail Rotor	Total
$\dfrac{\partial X}{\partial \dot{x}}$	−5		−5
$\dfrac{\partial X}{\partial \dot{j}}$	−1		−1
$\dfrac{\partial X}{\partial q}$	1,008		1,008
$\dfrac{\partial X}{\partial p}$	−355		−355
$\dfrac{\partial X}{\partial \theta_{o_M}}$	3,677		3,677
$\dfrac{\partial X}{\partial A_1}$	−825		−825
$\dfrac{\partial X}{\partial B_1}$	9,531		9,531

Component / Derivative	Main Rotor	Tail Rotor	Total
$\dfrac{\partial Y}{\partial \dot{x}}$	1		1
$\dfrac{\partial Y}{\partial \dot{j}}$	−5	−13	−18
$\dfrac{\partial Y}{\partial q}$	−355		−355
$\dfrac{\partial Y}{\partial p}$	−1,008	−78	−1,086
$\dfrac{\partial Y}{\partial r}$		481	481
$\dfrac{\partial Y}{\partial \theta_{o_M}}$	−3,971		−3,971
$\dfrac{\partial Y}{\partial A_1}$	9,531		9,531

TABLE 9.4 (continued)

Component / Derivative	Main Rotor	Tail Rotor	Total
$\dfrac{\partial Y}{\partial B_1}$	825		825
$\dfrac{\partial Y}{\partial \theta_{0_T}}$		9746	9,746
$\dfrac{\partial Z}{\partial \dot{z}}$	−182		−182
$\dfrac{\partial Z}{\partial \theta_{0_M}}$	−147,088		−147,088
$\dfrac{\partial R}{\partial \dot{x}}$	39		39
$\dfrac{\partial R}{\partial \dot{j}}$	−143	78	−65
$\dfrac{\partial R}{\partial q}$	−10,097		−10,097

Component / Derivative	Main Rotor	Tail Rotor	Total
$\dfrac{\partial R}{\partial p}$	−28,659	−468	−29,127
$\dfrac{\partial R}{\partial r}$		2,886	2,886
$\dfrac{\partial R}{\partial \theta_{0_M}}$	−29,783		−29,783
$\dfrac{\partial R}{\partial A_1}$	271,016		271,016
$\dfrac{\partial R}{\partial B_1}$	23,468		23,468
$\dfrac{\partial R}{\partial \theta_{0_T}}$		58,476	58,476

$\dfrac{\partial M}{\partial \dot{x}}$	143 143 −271,016
$\dfrac{\partial M}{\partial \dot{y}}$	39 −7 32 11,276 11,276
$\dfrac{\partial M}{\partial \dot{z}}$	91 91 481 481
$\dfrac{\partial M}{\partial q}$	−28,659 −28,659 −847 −847
$\dfrac{\partial M}{\partial p}$	10,097 10,097 2,886 2,886
$\dfrac{\partial M}{\partial r}$	274 274 4,471 −17,797 −13,326
$\dfrac{\partial M}{\partial \theta_{0_M}}$	45,967 45,967 564,242 564,242
$\dfrac{\partial M}{\partial A_1}$	23,468 23,468 −360,602 −360,602

$\dfrac{\partial M}{\partial B_1}$	−271,016
$\dfrac{\partial M}{\partial \theta_{0_T}}$	11,276
$\dfrac{\partial N}{\partial \dot{y}}$	481
$\dfrac{\partial N}{\partial \dot{z}}$	−847
$\dfrac{\partial N}{\partial p}$	2,886
$\dfrac{\partial N}{\partial r}$	−13,326
$\dfrac{\partial N}{\partial \theta_{0_M}}$	564,242
$\dfrac{\partial N}{\partial \theta_{0_T}}$	−360,602

TABLE 9.5
Rotor Derivatives in Forward Flight from Charts

Independent Variable	Trim Condition		Main Rotor					Tail Rotor				
	Dependent Variable											
	Main Rot.	Tail Rot.	C_T/σ	C_H/σ $(a_{1_s}=0)$	C_Q/σ	a_{1_s} (rad)	b_{1_s} (rad)	C_T/σ	C_H/σ $(a_{1_s}=0)$	C_Q/σ	a_{1_s} (rad)	b_{1_s} (rad)
μ	.30	.30	−.140	.008	−.005	.33	.05	−.070	.004	−.001	.12	.05
$\theta_{0_{\theta_t=-5°}}$	13.5°	7.1°	.46	−.04	.052	1.1	−.25	.659	.001	.005	.60	−.25
λ	−.023	.0051	.79	−.07	.010	1.2	−.59	1.04	−.002	−.026	.70	−.35

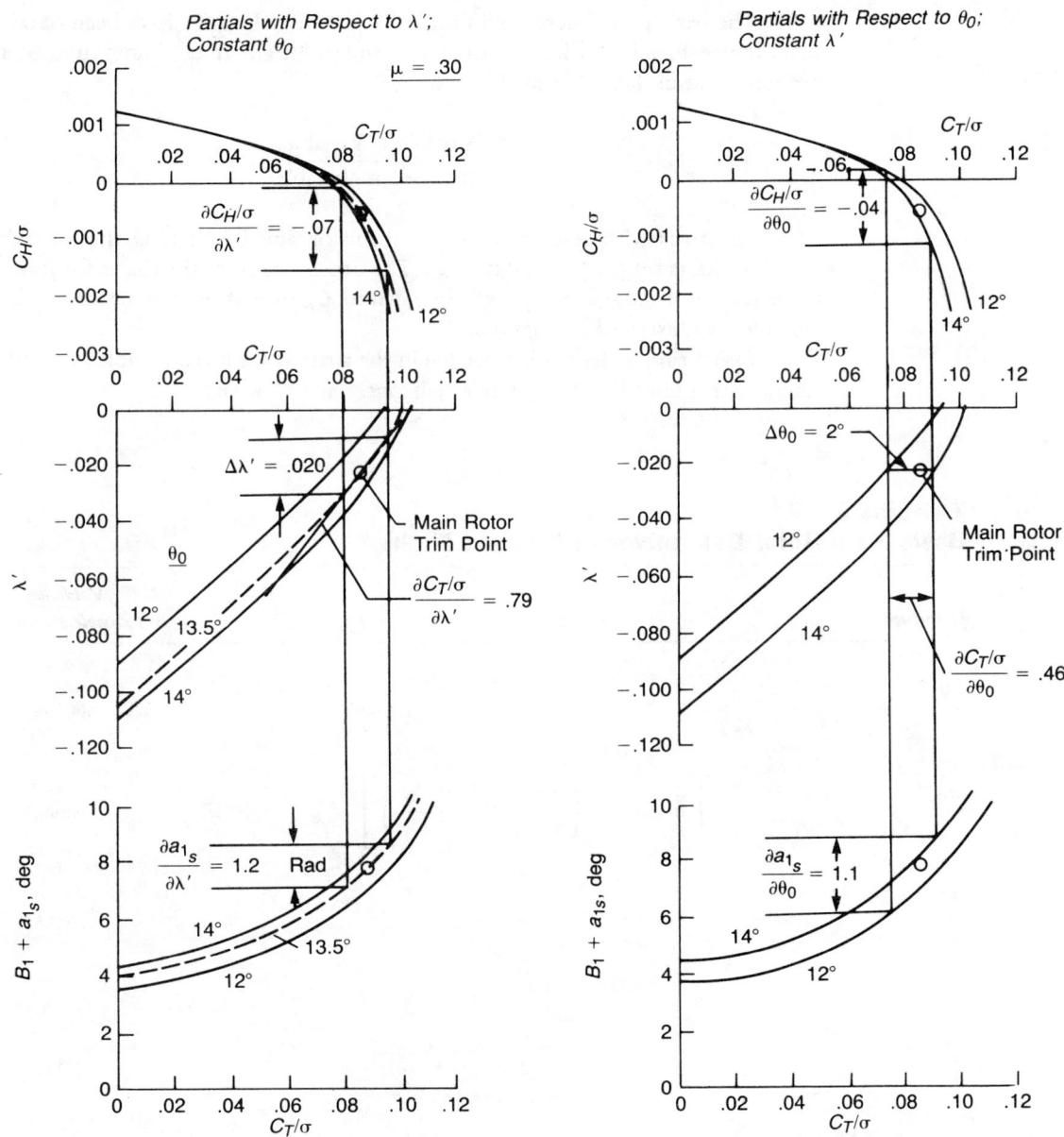

FIGURE 9.7 Illustration of Rotor Derivative Extraction from Performance Charts of Chapter 3

The basic partial derivatives for the example helicopter have been obtained from the performance charts of Chapter 3 in level flight at 115 knots ($\mu = .3$) and are tabulated in Table 9.5 in the form:

$$\frac{\partial \text{ Dependent Variable}}{\partial \text{ Independent Variable}}$$

The derivatives with respect to μ have been obtained by taking the difference between charts for $\mu = .25$ and $\mu = .35$. The others are from the charts for $\mu = .30$. Since the rotor parameters are plotted against C_T/σ, it is used as an intermediate variable, as illustrated by Figure 9.7.

Other simple derivatives needed in the analysis are listed in tables 9.6 and 9.7 along with values for the example helicopter at 115 knots.

TABLE 9.6
Basic Main Rotor Derivatives in Forward Flight

Derivative	Equation	Value for Example Helicopter
$\dfrac{\partial \mu}{\partial \dot{x}}$	$\dfrac{1}{\Omega R}$	.00154
$\dfrac{\partial \lambda'}{\partial \dot{x}}$	$\dfrac{1}{\Omega R}\left[\overline{\alpha_{TPP}} - \dfrac{\sigma}{2\mu}\left(\dfrac{\partial C_T/\sigma}{\partial \mu} - \dfrac{C_T/\sigma}{\mu}\right)\right]$	.000037
$\dfrac{\partial \lambda'}{\partial \dot{z}}$	$\dfrac{1}{\Omega R\left(1 + \dfrac{\partial C_T/\sigma}{\partial \lambda'}\dfrac{\sigma}{2\mu}\right)}$	.00138
$\dfrac{\partial \beta}{\partial \dot{y}}$	$\dfrac{1}{V}$	.00515
$\dfrac{\partial C_H/\sigma}{\partial a_{1_s}}, \dfrac{\partial C_Y/\sigma}{\partial b_{1_s}}$	$\overline{C_T}/\sigma + \dfrac{a}{8}\lambda'$	.069
$\dfrac{\partial a_{1_s}}{\partial q}$	$-\dfrac{16}{\gamma\Omega\left(1 - \dfrac{e}{R}\right)^2\left(1 - \dfrac{\mu^2}{2}\right)} - \dfrac{12\dfrac{e}{R}}{\gamma\Omega\left(1 - \dfrac{e}{R}\right)^3\left(1 - \dfrac{\mu^4}{4}\right)}$	$-.1098$

TABLE 9.6 (continued)

Derivative	Equation	Value for Example Helicopter
$\dfrac{\partial a_{1_s}}{\partial p}$	$\dfrac{1}{\Omega\left(1 - \dfrac{\mu^2}{2}\right)} - \dfrac{192\dfrac{e}{R}}{\gamma^2\Omega\left(1 - \dfrac{e}{R}\right)^5\left(1 - \dfrac{\mu^4}{4}\right)}$	.0396
$\dfrac{\partial a_{1_s}}{\partial A_1}$	$\dfrac{12\dfrac{e}{R}\left(1 + \dfrac{\mu^2}{2}\right)}{\gamma\left(1 - \dfrac{e}{R}\right)^3\left(1 - \dfrac{\mu^4}{4}\right)}$	.0905
$\dfrac{\partial a_{1_s}}{\partial B_1}$	$-\dfrac{\left(1 + \dfrac{3}{2}\mu^2\right)}{1 - \dfrac{\mu^2}{2}}$	−1.188
$\dfrac{\partial b_{1_s}}{\partial q}$	$-\dfrac{1}{\Omega\left(1 + \dfrac{\mu^2}{2}\right)} + \dfrac{192\dfrac{e}{R}}{\gamma^2\Omega\left(1 - \dfrac{e}{R}\right)^5\left(1 - \dfrac{\mu^4}{4}\right)}$	−.0354
$\dfrac{\partial b_{1_s}}{\partial p}$	$-\dfrac{16}{\gamma\Omega\left(1 - \dfrac{e}{R}\right)^2\left(1 + \dfrac{\mu^2}{2}\right)} - \dfrac{12\dfrac{e}{R}}{\gamma\Omega\left(1 - \dfrac{e}{R}\right)^3\left(1 - \dfrac{\mu^4}{4}\right)}$	−.101
$\dfrac{\partial b_{1_s}}{\partial A_1}$	1	1
$\dfrac{\partial b_{1_s}}{\partial B_1}$	$\dfrac{12\dfrac{e}{R}\left(1 + \dfrac{3}{2}\mu^2\right)}{\gamma\left(1 - \dfrac{e}{R}\right)^3\left(1 - \dfrac{\mu^4}{4}\right)}$	.0983
$\dfrac{dM}{da_{1_s}},\ \dfrac{dR}{db_{1_s}}$	$\dfrac{\dfrac{3}{4}\dfrac{e}{R}A_b\rho R(\Omega R)^2 a}{\gamma}$	200940

TABLE 9.7
Basic Tail Rotor Derivatives in Forward Flight

Derivative	Equation	Value for Example Helicopter
$\dfrac{\partial \mu}{\partial \dot{x}}$	$\dfrac{1}{\Omega R}$	.00154
$\dfrac{\partial \lambda'}{\partial \dot{x}}$	$\dfrac{1}{\Omega R}\left[\overline{a}_{1_s} - \dfrac{\sigma}{2\mu}\left(\dfrac{\partial C_T/\sigma}{\partial \mu} - \dfrac{C_T/\sigma}{\mu} \right) \right]$	.000169
$\dfrac{\partial \lambda'}{\partial \dot{y}}$	$-\dfrac{1}{\Omega R\left(1 + \dfrac{\partial C_T/\sigma}{\partial \lambda'} \dfrac{\sigma}{2\mu} \right)}$	$-.00121$
$\dfrac{\partial \beta}{\partial \dot{y}}$	$\dfrac{1}{V}$	.00515

By combining the preceding simple derivatives with rotor physical parameters and trim values, the total main rotor and tail rotor derivatives can be evaluated, as shown in tables 9.8 and 9.9.

TABLE 9.8
Main Rotor Derivatives in Forward Flight

Derivative	Equation	Value for Example Helicopter
$\left(\dfrac{\partial X}{\partial \dot{x}} \right)_M$	$-\rho A_b (\Omega R)^2 \left\{ \left[\dfrac{\partial C_H/\sigma}{\partial \mu} + \dfrac{\partial C_H/\sigma}{\partial a_{1_s}} \dfrac{\partial a_{1_s}}{\partial \mu} + (\overline{a}_{1_s} + i_M) \dfrac{\partial C_T/\sigma}{\partial \mu} \right] \dfrac{\partial \mu}{\partial \dot{x}} \right.$ $\left. + \left[\dfrac{\partial C_H/\sigma}{\partial \lambda'} + \dfrac{\partial C_H/\sigma}{\partial a_{1_s}} \dfrac{\partial a_{1_s}}{\partial \lambda'} + (\overline{a}_{1_s} + i_M) \right] \dfrac{\partial \lambda'}{\partial \dot{x}} \right\}$	-12 lb/ft/sec
$\left(\dfrac{\partial X}{\partial \dot{y}} \right)_M$	$+\rho A_b (\Omega R)^2 \overline{C}_T/\sigma (A_1 - \overline{b}_{1_s}) \dfrac{\partial \beta}{\partial \dot{y}}$	-6 lb/ft/sec
$\left(\dfrac{\partial X}{\partial \dot{z}} \right)_M$	$-\rho A_b (\Omega R)^2 \left[\dfrac{\partial C_H/\sigma}{\partial \lambda'} + \overline{C}_T/\sigma \dfrac{\partial a_{1_s}}{\partial \lambda'} + (\overline{a}_{1_s} + i_M) \right] \dfrac{\partial \lambda'}{\partial \dot{z}}$	-5 lb/ft/sec

TABLE 9.8 (continued)

Derivative	Equation	Value for Example Helicopter
$\left(\dfrac{\partial X}{\partial q}\right)_M$	$-\rho A_b(\Omega R)^2 \dfrac{\partial C_H/\sigma}{\partial a_{1_s}} \dfrac{\partial a_{1_s}}{\partial q} - \left(\dfrac{\partial X}{\partial \dot{x}}\right)_M h_M$	1937 lb/rad/sec
$\left(\dfrac{\partial X}{\partial p}\right)_M$	$-\rho A_b(\Omega R^2) \dfrac{\partial C_H/\sigma}{\partial a_{1_s}} \dfrac{\partial a_{1_s}}{\partial p} - \left(\dfrac{\partial X}{\partial \dot{y}}\right)_M h_M$	-659 lb/rad/sec
$\left(\dfrac{\partial X}{\partial \theta_0}\right)_M$	$-\rho A_b(\Omega R)^2 \left[\dfrac{\partial C_H/\sigma}{\partial \theta_0} + \dfrac{\partial C_H/\sigma}{\partial a_{1_s}} \dfrac{\partial a_{1_s}}{\partial \theta_0} + (\overline{a_{1_s}} + i_M) \dfrac{\partial C_T/\sigma}{\partial \theta_0} \right]$	-6727 lb/rad
$\left(\dfrac{\partial X}{\partial A_1}\right)_M$	$-\rho A_b(\Omega R)^2 \dfrac{\partial C_H/\sigma}{\partial a_{1_s}} \dfrac{\partial a_{1_s}}{\partial A_1}$	-1506 lb/rad
$\left(\dfrac{\partial X}{\partial B_1}\right)_M$	$-\rho A_b(\Omega R)^2 \dfrac{\partial C_H/\sigma}{\partial a_{1_s}} \dfrac{\partial a_{1_s}}{\partial B_1}$	18601 lb/rad
$\left(\dfrac{\partial Y}{\partial \dot{x}}\right)_M$	$\rho A_b(\Omega R)^2 \left[\dfrac{\partial C_Y/\sigma}{\partial b_{1_s}} \dfrac{\partial b_{1_s}}{\partial \mu} + \overline{b_{1_s}} \dfrac{\partial C_T/\sigma}{\partial \mu} \right] \dfrac{\partial \mu}{\partial \dot{x}}$	2 lb/ft/sec
$\left(\dfrac{\partial Y}{\partial \dot{y}}\right)_M$	$-\rho A_b(\Omega R)^2 \left[\overline{C_H}/\sigma + \overline{C_T}/\sigma(B_1 + \overline{a_{1_s}}) \right] \dfrac{\partial \beta}{\partial \dot{y}}$	-14 lb/ft/sec
$\left(\dfrac{\partial Y}{\partial \dot{z}}\right)_M$	$\rho A_b(\Omega R)^2 \dfrac{\partial b_{1_s}}{\partial \lambda'} \dfrac{\partial C_Y/\sigma}{\partial b_{1_s}} \dfrac{\partial \lambda'}{\partial \dot{z}}$	-13 lb/ft/sec
$\left(\dfrac{\partial Y}{\partial q}\right)_M$	$\rho A_b(\Omega R)^2 \dfrac{\partial C_Y/\sigma}{\partial b_{1_s}} \dfrac{\partial b_{1_s}}{\partial q} + \left(\dfrac{\partial Y}{\partial \dot{x}}\right)_M h_M$	-574 lb/rad/sec
$\left(\dfrac{\partial Y}{\partial p}\right)_M$	$\rho A_b(\Omega R)^2 \dfrac{\partial C_Y/\sigma}{\partial b_{1_s}} \dfrac{\partial b_{1_s}}{\partial p} + \left(\dfrac{\partial Y}{\partial \dot{y}}\right)_M h_M$	-1658 lb/rad/sec
$\left(\dfrac{\partial Y}{\partial \theta_0}\right)_M$	$\rho A_b(\Omega R)^2 \left[\dfrac{\partial C_Y/\sigma}{\partial b_{1_s}} \dfrac{\partial b_{1_s}}{\partial \theta_0} + \overline{b_{1_s}} \dfrac{\partial C_T/\sigma}{\partial \theta_0} \right]$	-5668 lb/rad

continued

TABLE 9.8 (continued)

Derivative	Equation	Value for Example Helicopter
$\left(\dfrac{\partial Y}{\partial A_1}\right)_M$	$\rho A_b (\Omega R)^2 \dfrac{\partial C_Y/\sigma}{\partial b_{1_s}} \dfrac{\partial b_{1_s}}{\partial A_1}$	16638 lb/rad
$\left(\dfrac{\partial Y}{\partial B_1}\right)_M$	$\rho A_b (\Omega R)^2 \dfrac{\partial C_Y/\sigma}{\partial b_{1_s}} \dfrac{\partial b_{1_s}}{\partial B_1}$	1635 lb/rad
$\left(\dfrac{\partial Z}{\partial \dot{x}}\right)_M$	$-\rho A_b (\Omega R)^2 \left[\dfrac{\partial C_T/\sigma}{\partial \mu} \dfrac{\partial \mu}{\partial \dot{x}} + \dfrac{\partial C_T/\sigma}{\partial \lambda'} \dfrac{\partial \lambda'}{\partial \dot{x}} \right]$	45 lb/ft/sec
$\left(\dfrac{\partial Z}{\partial \dot{z}}\right)_M$	$-\rho A_b (\Omega R)^2 \dfrac{\partial C_T/\sigma}{\partial \lambda'} \dfrac{\partial \lambda'}{\partial \dot{z}}$	−261 lb/ft/sec
$\left(\dfrac{\partial Z}{\partial r}\right)_M$	$-\dfrac{2}{\Omega} \rho A_b (\Omega R)^2 \overline{C_T/\sigma}$	−1846 lb/rad/sec
$\left(\dfrac{\partial Z}{\partial \theta_0}\right)_M$	$-\rho A_b (\Omega R)^2 \dfrac{\partial C_T/\sigma}{\partial \theta_0}$	−110919 lb/rad
$\left(\dfrac{\partial z}{\partial B_1}\right)_M$	$-\rho A_b (\Omega R)^2 \dfrac{\partial C_T/\sigma}{\partial \lambda'} \dfrac{\partial \lambda'}{\partial a_{1_s}} \dfrac{\partial a_{1_s}}{\partial B_1}$	67855 lb/rad
$\left(\dfrac{\partial R}{\partial \dot{x}}\right)_M$	$\left(\dfrac{dR}{a'b_{1_s}}\right)_M \left(\dfrac{\partial b_{1_s}}{\partial \mu}\right) \left(\dfrac{\partial \mu}{\partial \dot{x}}\right) + \left(\dfrac{\partial Y}{\partial \dot{x}}\right)_M h_M + \left(\dfrac{\partial Z}{\partial \dot{x}}\right)_M y_M$	31 ft lb/ft/sec
$\left(\dfrac{\partial R}{\partial \dot{y}}\right)_M$	$-\left(\dfrac{dR}{db_{1_s}}\right)_M \left(\overline{B_1 + a_{1_s}}\right) \dfrac{\partial \beta}{\partial \dot{y}} + \left(\dfrac{\partial Y}{\partial \dot{y}}\right)_M h_M$	− 246 ft lb/ft/sec
$\left(\dfrac{\partial R}{\partial \dot{z}}\right)_M$	$\left(\dfrac{dR}{db_{1_s}}\right)_M \left(\dfrac{\partial b_{1_s}}{\partial \lambda'}\right) \left(\dfrac{\partial \lambda'}{\partial \dot{z}}\right) + \left(\dfrac{\partial Y}{\partial \dot{z}}\right)_M h_M + \left(\dfrac{\partial Z}{\partial \dot{z}}\right)_A y_M$	−263 ft lb/ft/sec
$\left(\dfrac{\partial R}{\partial q}\right)_M$	$\left(\dfrac{dR}{db_{1_s}}\right)_M \left(\dfrac{\partial b_{1_s}}{\partial q}\right) + \left(\dfrac{\partial Y}{\partial q}\right)_M h_M$	−11418 ft lb/rad/sec
$\left(\dfrac{\partial R}{\partial p}\right)_M$	$\left(\dfrac{dR}{db_{1_s}}\right)_M \left(\dfrac{\partial b_{1_s}}{\partial p}\right) + \left(\dfrac{\partial Y}{\partial p}\right)_M h_M$	−32730 ft lb/rad/sec

TABLE 9.8 (continued)

Derivative	Equation	Value for Example Helicopter
$\left(\dfrac{\partial R}{\partial \theta_0}\right)_M$	$\left(\dfrac{dR}{db_{1_s}}\right)_M\left(\dfrac{\partial b_{1_s}}{\partial \theta_0}\right) + \left(\dfrac{\partial Y}{\partial \theta_0}\right)_M h_M + \left(\dfrac{\partial Z}{\partial \theta_0}\right) y_M$	-92745 ft lb/rad
$\left(\dfrac{\partial R}{\partial A_1}\right)_M$	$\left(\dfrac{dR}{db_{1_s}}\right)_M\left(\dfrac{\partial b_{1_s}}{\partial A_1}\right) + \left(\dfrac{\partial Y}{\partial A_1}\right)_M h_M$	325703 ft lb/rad
$\left(\dfrac{\partial R}{\partial B_1}\right)_M$	$\left(\dfrac{dR}{db_{1_s}}\right)_M\left(\dfrac{\partial b_{1_s}}{\partial B_1}\right) + \left(\dfrac{\partial Y}{\partial B_1}\right)_M h_M$	32014 ft lb/rad
$\left(\dfrac{\partial M}{\partial \dot{x}}\right)_M$	$\left(\dfrac{dM}{da_{1_s}}\right)_M\left[\dfrac{\partial a_{1_s}}{\partial \mu}\dfrac{\partial \mu}{\partial \dot{x}} + \dfrac{\partial a_{1_s}}{\partial \lambda'}\dfrac{\partial \lambda'}{\partial \dot{x}}\right] - \left(\dfrac{\partial X}{\partial \dot{x}}\right)_M h_M + \left(\dfrac{\partial Z}{\partial \dot{x}}\right)_M l_M$	182 ft lb/ft/sec
$\left(\dfrac{\partial M}{\partial \dot{y}}\right)_M$	$-\left(\dfrac{\partial X}{\partial \dot{y}}\right)_M h_M + \left(\dfrac{dm}{da_{1_s}}\right)\left(A_1 - \overline{b_{1_s}}\right)\dfrac{\partial \beta}{\partial \dot{y}}$	-15 ft lb/ft/sec
$\left(\dfrac{\partial M}{\partial \dot{z}}\right)_M$	$\left(\dfrac{dM}{da_{1_s}}\right)_M\dfrac{\partial a_{1_s}}{\partial \lambda'}\dfrac{\partial \lambda'}{\partial \dot{z}} - \left(\dfrac{\partial X}{\partial \dot{z}}\right)_M h_M + \left(\dfrac{\partial Z}{\partial \dot{z}}\right)_M l_M$	495 ft lb/ft/sec
$\left(\dfrac{\partial M}{\partial q}\right)_M$	$\left(\dfrac{dM}{da_{1_s}}\right)_M\left(\dfrac{\partial a_{1_s}}{\partial q}\right) - \left(\dfrac{\partial X}{\partial q}\right)_M h_M$	-36591 ft lb/rad/sec
$\left(\dfrac{\partial M}{\partial p}\right)_M$	$\left(\dfrac{dM}{da_{1_s}}\right)_M\left(\dfrac{\partial a_{1_s}}{\partial p}\right) - \left(\dfrac{\partial X}{\partial p}\right)_M h_M$	12900 ft lb/rad/sec
$\left(\dfrac{\partial M}{\partial \theta_0}\right)_M$	$\left(\dfrac{dM}{da_{1_s}}\right)_M\left(\dfrac{\partial a_{1_s}}{\partial \theta_0}\right) - \left(\dfrac{\partial X}{\partial \theta_0}\right)_M h_M + \left(\dfrac{\partial Z}{\partial \theta_0}\right)_M l_M$	326946 ft lb/rad
$\left(\dfrac{\partial M}{\partial A_1}\right)_M$	$\left(\dfrac{dM}{da_{1_s}}\right)_M\left(\dfrac{\partial a_{1_s}}{\partial A_1}\right) - \left(\dfrac{\partial X}{\partial A_1}\right)_M h_M$	29480 ft lb/rad
$\left(\dfrac{\partial M}{\partial B_1}\right)_M$	$\left(\dfrac{dM}{da_{1_s}}\right)_M\left(\dfrac{\partial a_{1_s}}{\partial B_1}\right) - \left(\dfrac{\partial X}{\partial B_1}\right)_M h_M$	-364158 ft lb/rad

continued

TABLE 9.8 (continued)

Derivative	Equation	Value for Example Helicopter
$\left(\dfrac{\partial N}{\partial \dot{x}}\right)_M$	$\rho A_b(\Omega R)^2 R\left[\dfrac{\partial C_Q/\sigma}{\partial \mu}\dfrac{\partial \mu}{\partial \dot{x}} + \dfrac{\partial C_Q/\sigma}{\partial \lambda'}\dfrac{\partial \lambda'}{\partial \dot{x}}\right]$	-53 ft lb/ft/sec
$\left(\dfrac{\partial N}{\partial \dot{z}}\right)_M$	$\rho A_b(\Omega R)^2 R\left(\dfrac{\partial C_Q/\sigma}{\partial \lambda'}\right)\dfrac{\partial \lambda'}{\partial \dot{z}}$	99 ft lb/ft/sec
$\left(\dfrac{\partial N}{\partial r}\right)_M$	$-\dfrac{2}{\Omega}\rho A_b(\Omega R)^2 R\overline{C_Q/\sigma}$	-39710 ft lb/rad/sec
$\left(\dfrac{\partial N}{\partial \theta_0}\right)_M$	$\rho A_b(\Omega R)^2 R\left(\dfrac{\partial C_Q/\sigma}{\partial \theta_0}\right)$	376161 ft lb/rad

TABLE 9.9
Tail Rotor Derivatives in Forward Flight

Derivative	Equation	Value for Example Helicopter
$\left(\dfrac{\partial Y}{\partial \dot{x}}\right)_T$	$\rho A_b(\Omega R)^2\left[\dfrac{\partial C_T/\sigma}{\partial \mu}\right]\dfrac{\partial \mu}{\partial \dot{x}}$	-2 lb/ft/sec
$\left(\dfrac{\partial Y}{\partial \dot{y}}\right)_T$	$\rho A_b(\Omega R)^2\left[\dfrac{\partial C_T/\sigma}{\partial \lambda'}\right]\dfrac{\partial \lambda'}{\partial \dot{y}}$	-24 lb/ft/sec
$\left(\dfrac{\partial Y}{\partial p}\right)_T$	$\left(\dfrac{\partial Y}{\partial \dot{y}}\right)_T h_T$	-147 lb/rad/sec
$\left(\dfrac{\partial Y}{\partial r}\right)_T$	$-\left(\dfrac{\partial Y}{\partial \dot{y}}\right)_T l_T$	907 lb/rad/sec
$\left(\dfrac{\partial Y}{\partial \theta_0}\right)_T$	$\rho A_b(\Omega R)^2\dfrac{\partial C_T/\sigma}{\partial \theta_0}$	12845 lb/rad

TABLE 9.9 (continued)

Derivative	Equation	Value for Example Helicopter
$\left(\dfrac{\partial R}{\partial \dot{x}}\right)_T$	$\left(\dfrac{\partial Y}{\partial \dot{x}}\right)_T b_T$	-13 ft lb/ft/sec
$\left(\dfrac{\partial R}{\partial \dot{y}}\right)_T$	$\left(\dfrac{\partial Y}{\partial \dot{y}}\right)_T b_T$	-147 ft lb/ft/sec
$\left(\dfrac{\partial R}{\partial p}\right)_T$	$\left(\dfrac{\partial Y}{\partial p}\right)_T b_T$	-882 ft lb/rad/sec
$\left(\dfrac{\partial R}{\partial r}\right)_T$	$\left(\dfrac{\partial Y}{\partial r}\right)_T b_T$	5442 ft lb/rad/sec
$\left(\dfrac{\partial R}{\partial \theta_0}\right)_T$	$\left(\dfrac{\partial Y}{\partial \theta_0}\right)_T b_T$	77070 ft lb/rad
$\left(\dfrac{\partial N}{\partial \dot{x}}\right)_T$	$-\left(\dfrac{\partial Y}{\partial \dot{x}}\right)_T l_T$	78 ft lb/ft/sec
$\left(\dfrac{\partial N}{\partial \dot{y}}\right)_T$	$-\left(\dfrac{\partial Y}{\partial \dot{y}}\right)_T l_T$	907 ft lb/ft/sec
$\left(\dfrac{\partial N}{\partial p}\right)_T$	$-\left(\dfrac{\partial Y}{\partial p}\right)_T l_T$	5439 ft lb/rad/sec
$\left(\dfrac{\partial N}{\partial r}\right)_T$	$-\left(\dfrac{\partial Y}{\partial r}\right)_T l_T$	-33559 ft lb/rad/sec
$\left(\dfrac{\partial N}{\partial \theta_0}\right)_T$	$-\left(\dfrac{\partial Y}{\partial \theta_0}\right)_T l_T$	-475265 ft lb/rad

Airframe Derivatives

Fortunately, the airframe derivatives are somewhat more straightforward to evaluate than the rotor derivatives. Sometime in the latter stages of the design process, wind tunnel results should be available for determining static derivatives directly. Up to that time, the equations of Chapter 8 can be used with the following methods.

Horizontal Stabilizer

Equations for the steady X and Z forces of the horizontal stabilizer were presented in Chapter 8. They can be used almost as is for obtaining derivatives (see tables 9.10 and 9.11). The exception is the addition of a term to the equation for stabilizer angle of attack to account for the time required for the main rotor downwash to reach the stabilizer during nonsteady conditions. The equation now becomes:

TABLE 9.10
Nondimensional Horizontal Stabilizer Derivatives

Derivative	Equation	Value for Example Helicopter
$\dfrac{\partial \gamma}{\partial \dot{z}}$	$-\dfrac{1}{V}$	$-.00515$
$\dfrac{\partial \varepsilon_{M_H}}{\partial \dot{x}}$	$\dfrac{v_H}{v_1} \dfrac{1}{4qA} \left[-\left(\dfrac{\partial Z}{\partial \dot{x}} \right)_M - \dfrac{2\bar{Z}_M}{V} \right]$	$-.00076$
$\dfrac{\partial \varepsilon_{M_H}}{\partial \dot{z}}$	$-\dfrac{v_H}{v_1} \dfrac{1}{4qA} \left(\dfrac{\partial Z}{\partial \dot{z}} \right)_M$	$.00072$
$\dfrac{\partial \varepsilon_{F_H}}{\partial \dot{z}}$	$\dfrac{d\varepsilon_F}{d\alpha_F} \left[\dfrac{1}{4qA} \left(\dfrac{\partial Z}{\partial \dot{z}} \right)_M - \dfrac{\partial \gamma}{\partial \dot{z}} \right]$	$.00108$
$\dfrac{\partial \alpha_H}{\partial \dot{x}}$	$-\dfrac{\partial \varepsilon_{M_H}}{\partial \dot{x}}$	$.00076$
$\dfrac{\partial \alpha_H}{\partial \dot{z}}$	$-\left[\dfrac{\partial \varepsilon_{M_H}}{\partial \dot{z}} + \dfrac{\partial \varepsilon_{F_H}}{\partial \dot{z}} + \dfrac{\partial \gamma}{\partial \dot{z}} \right]$	$.00335$
$\dfrac{\partial \alpha_H}{\partial \ddot{z}}$	$-\left(\dfrac{\alpha \varepsilon_{M_H}}{\partial \dot{z}} \right) \dfrac{l_H}{V}$	$-.00014$

TABLE 9.11
Horizontal Stabilizer Derivatives in Forward Flight

Derivative	Equation	Value for Example Helicopter
$\left(\dfrac{\partial X}{\partial \dot{x}}\right)_H$	$\dfrac{2}{V}X_H + \left(\dfrac{q_H}{q}\right)qA_H a_H \left\{ (\bar{\alpha}_H - \alpha_{LO})\left[1 - \dfrac{2a_H(1+\delta_i)}{\pi A.R.} \right] \right.$ $\left. + (\bar{\alpha}_H - i_H) \right\} \dfrac{\partial \alpha_H}{\partial \dot{x}}$	<1 lb/ft/sec
$\left(\dfrac{\partial X}{\partial \dot{z}}\right)_H$	$\left(\dfrac{q_H}{q}\right)qA_H a_H \left\{ (\bar{\alpha}_H - \alpha_{LO})\left[1 - \dfrac{2a_H(1+\delta_i)}{\pi A.R.} \right] + (\alpha_H - i_H) \right\} \dfrac{\partial \alpha_H}{\partial \dot{z}}$	-1 lb/ft/sec
$\left(\dfrac{\partial X}{\partial \ddot{z}}\right)_H$	$\left(\dfrac{\partial X}{\partial \dot{z}}\right)_H \dfrac{\dfrac{\partial \alpha_H}{\partial \ddot{z}}}{\dfrac{\partial \alpha_H}{\partial \dot{z}}}$	<1 lb/ft/sec^2
$\left(\dfrac{\partial Z}{\partial \dot{x}}\right)_H$	$\dfrac{2}{V}\bar{Z}_H - \left(\dfrac{q_H}{q}\right)qA_H a_H \left\{ 1 + \dfrac{a_H(1+\delta_i)}{\pi A.R.}\left[2(\bar{\alpha}_H - \alpha_{LO})(\bar{\alpha}_H - i_H) \right.\right.$ $\left.\left. + (\bar{\alpha}_H - \alpha_{LO})^2 \right] + C_{D_0} \right\} \dfrac{\partial \alpha_H}{\partial \dot{x}}$	1 lb/ft/sec
$\left(\dfrac{\partial Z}{\partial \dot{z}}\right)_H$	$-\left(\dfrac{q_H}{q}\right)qA_H a_H \left\{ 1 + \dfrac{a_H(1+\delta_i)}{\pi A.R.}\left[2(\bar{\alpha}_H - \alpha_{LO})(\bar{\alpha}_H - i_H) \right.\right.$ $\left.\left. + (\bar{\alpha}_H \alpha_{LO})^2 \right] + C_{D_0} \right\} \dfrac{\partial \alpha_H}{\partial \dot{z}}$	-10 lb/ft/sec
$\left(\dfrac{\partial Z}{\partial \ddot{z}}\right)_H$	$\left(\dfrac{\partial Z}{\partial \dot{z}}\right)_H \dfrac{\dfrac{\partial \alpha_H}{\partial \ddot{z}}}{\dfrac{\partial \alpha_H}{\partial \dot{z}}}$	<1 lb/ft/sec^2
$\left(\dfrac{\partial Z}{\partial q}\right)_H$	$\left(\dfrac{\partial Z}{\partial \dot{z}}\right)_H l_H$	-370 lb/rad/sec

continued

TABLE 9.11 (continued)

Derivative	Equation	Value for Example Helicopter
$\left(\dfrac{\partial M}{\partial \dot{x}}\right)_H$	$-\left(\dfrac{\partial X}{\partial \dot{x}}\right)_H b_H + \left(\dfrac{\partial Z}{\partial \dot{x}}\right)_H l_H$	27 ft-lb/ft/sec
$\left(\dfrac{\partial M}{\partial \dot{z}}\right)_H$	$-\left(\dfrac{\partial X}{\partial \dot{z}}\right)_H b_H + \left(\dfrac{\partial Z}{\partial \dot{z}}\right)_H l_H$	-327 ft-lb/ft/sec
$\left(\dfrac{\partial M}{\partial \ddot{z}}\right)_H$	$-\left(\dfrac{\partial X}{\partial \ddot{z}}\right)_H b_H + \left(\dfrac{\partial Z}{\partial \ddot{z}}\right)_H l_H$	14 ft-lb/ft/sec²
$\left(\dfrac{\partial M}{\partial q}\right)_H$	$\left(\dfrac{\partial Z}{\partial q}\right)_H l_H$	$-13,690$ ft-lb/rad/sec

$$\alpha_H = \Theta + i_H - (\varepsilon_{M_H} + \varepsilon_{F_H}) - \gamma_C - \left(\frac{\partial \varepsilon_{M_H}}{\partial \dot{z}}\right)\left(\frac{d\dot{z}}{dt}\right)\Delta t$$

where:

$$\frac{d\dot{z}}{dt} = \ddot{z}$$

and Δt is the time required for the air to go from the main rotor to the stabilizer, or:

$$\Delta t = \frac{l_H}{V}$$

Thus

$$\alpha_H = \Theta + i_H - (\varepsilon_{M_H} + \varepsilon_{F_H}) - \gamma_C - \left(\frac{\partial \varepsilon_{M_H}}{\partial \dot{z}}\right)\ddot{z}\frac{l_H}{V}$$

Vertical Stabilizer

The vertical stabilizer derivatives can be obtained using the equations from Chapter 8 and the same methods used for the horizontal stabilizer. The intermediate derivatives are shown in Table 9.12 and the dimensional derivatives in Table 9.13.

TABLE 9.12
Nondimensional Vertical Stabilizer Derivatives

Derivative	Equation	Value for Example Helicopter
$\dfrac{\partial \beta}{\partial \dot{y}}$	$\dfrac{1}{V}$	.00515
$\dfrac{\partial M_T}{\partial \dot{x}}$	$\dfrac{-1}{4\left(\dfrac{q_V}{q}\right)qA_V}\left[\left(\dfrac{\partial Y}{\partial \dot{x}}\right)_T - \dfrac{2T_T}{V}\right]$	.00034
$\dfrac{\partial \alpha_V}{\partial \dot{x}}$	$\dfrac{\partial \eta_T}{\partial \dot{x}}$	.00034
$\dfrac{\partial M_T}{\partial \dot{y}}$	$\dfrac{\left(\dfrac{\partial Y}{\partial \dot{y}}\right)_T}{4\left(\dfrac{q_V}{q}\right)qA_V}$	−.00167
$\dfrac{\partial M_F}{\partial \dot{y}}$	$\dfrac{\partial \eta_F}{\partial \beta}\dfrac{\partial \beta}{\partial \dot{y}}$	.00031
$\dfrac{\partial \alpha_V}{\partial \dot{y}}$	$-\left(\dfrac{\partial \beta}{\partial \dot{y}} + \dfrac{\partial \eta_T}{\partial \dot{y}} + \dfrac{\partial \eta_F}{\partial \dot{y}}\right)$	−.00379

TABLE 9.13
Vertical Stabilizer Derivatives in Forward Flight

Derivative	Equation	Value for Example Helicopter
$\left(\dfrac{\partial X}{\partial \dot{x}}\right)_V$	$\dfrac{2}{V}\left[\overline{X}_V + 2\Delta\overline{D_{V_{\text{int}}}}\right]$	<1 lb/ft/sec

continued

TABLE 9.13 (continued)

Derivative	Equation	Value for Example Helicopter
$\left(\dfrac{\partial X}{\partial \dot{y}}\right)_V$	$\dfrac{q_V}{q} q A_V a_V \left\{ (\bar{\alpha}_V - \alpha_{LO}) \left[1 - \dfrac{2a_V(1+\delta_i)}{\pi AR} \right] + \alpha_V \right\} \dfrac{\partial \alpha_V}{\partial \dot{y}}$ $+ \Delta \bar{D}_{V_{int}} \left[\dfrac{1}{\bar{Y}_V}\left(\dfrac{\partial Y}{\partial \dot{y}}\right)_V + \dfrac{1}{\bar{T}_T}\left(\dfrac{\partial Y}{\partial \dot{y}}\right)_T \right]$	-3 lb/ft/sec
$\left(\dfrac{\partial Y}{\partial \dot{x}}\right)_V$	$\dfrac{2}{V}\bar{Y}_V + \left(\dfrac{q_V}{q}\right) q A_V a_V \dfrac{\partial \alpha_V}{\partial \dot{x}}$	4 lb/ft/sec
$\left(\dfrac{\partial Y}{\partial \dot{y}}\right)_V$	$\dfrac{1}{1 - \left(\dfrac{\Delta \bar{D}_{V_{int}}}{\bar{Y}_V}\right)} \left\{ \left(\dfrac{q_V}{q}\right) q A_V a_V \dfrac{\partial \alpha_V}{\partial \dot{y}} + \Delta \bar{D}_{V_{int}} \left[-\dfrac{2}{V} + \dfrac{1}{\bar{T}_T}\left(\dfrac{\partial Y}{\partial \dot{y}}\right)_T \right] \right\}$	-14 lb/ft/sec
$\left(\dfrac{\partial Y}{\partial p}\right)_V$	$\left(\dfrac{\partial Y}{\partial \dot{y}}\right)_V h_V$	-42 lb/rad/sec
$\left(\dfrac{\partial Y}{\partial r}\right)_V$	$-\left(\dfrac{\partial Y}{\partial \dot{y}}\right)_V l_V$	490 lb/rad/sec
$\left(\dfrac{\partial R}{\partial \dot{x}}\right)_V$	$\left(\dfrac{\partial Y}{\partial \dot{x}}\right)_V h_V$	12 ft-lb/ft/sec
$\left(\dfrac{\partial R}{\partial \dot{y}}\right)_V$	$\left(\dfrac{\partial Y}{\partial \dot{y}}\right)_V h_V$	-42 ft-lb/ft/sec
$\left(\dfrac{\partial R}{\partial p}\right)_V$	$\left(\dfrac{\partial Y}{\partial p}\right)_V h_V$	-126 ft-lb/rad/sec
$\left(\dfrac{\partial R}{\partial r}\right)_V$	$\left(\dfrac{\partial Y}{\partial r}\right)_V h_V$	$1{,}470$ ft-lb/rad/sec
$\left(\dfrac{\partial N}{\partial \dot{x}}\right)_V$	$-\left(\dfrac{\partial Y}{\partial \dot{x}}\right)_V l_V$	105 ft-lb/ft/sec

TABLE 9.13 (continued)

Derivative	Equation	Value for Example Helicopter
$\left(\dfrac{\partial N}{\partial \dot{y}}\right)_V$	$-\left(\dfrac{\partial Y}{\partial \dot{y}}\right)_V l_V$	490 ft-lb/ft/sec
$\left(\dfrac{\partial N}{\partial p}\right)_V$	$-\left(\dfrac{\partial Y}{\partial p}\right)_V l_V$	1,470 ft-lb/rad/sec
$\left(\dfrac{\partial N}{\partial r}\right)_V$	$-\left(\dfrac{\partial Y}{\partial r}\right)_V l_V$	$-17{,}150$ ft-lb/rad/sec

Fuselage

The equations for fuselage derivatives can be derived from the Chapter 8 information just as the empennage derivatives were. The necessary intermediate derivatives are shown in Table 9.14, and the dimensional derivatives in Table 9.15.

TABLE 9.14
Nondimensional Fuselage Derivatives

Derivative	Equation	Value for Example Helicopter
$\dfrac{\partial \gamma}{\partial \dot{z}}$	$-\dfrac{1}{V}$	$-.00515$
$\dfrac{\partial \beta}{\partial \dot{y}}$	$\dfrac{1}{V}$	$.00515$
$\dfrac{\partial \varepsilon_{M_F}}{\partial \dot{x}}$	$\dfrac{v_F}{v_1}\dfrac{1}{4qA}\left[-\left(\dfrac{\partial Z}{\partial \dot{x}}\right)_M - \dfrac{2\bar{Z}_M}{V}\right]$	$-.00051$
$\dfrac{\partial \varepsilon_{M_F}}{\partial \dot{z}}$	$-\dfrac{v_F}{v_1}\dfrac{1}{4qA}\left(\dfrac{\partial Z}{\partial \dot{z}}\right)_M$	$.00048$
$\dfrac{\partial \alpha_F}{\partial \dot{x}}$	$-\dfrac{\partial \varepsilon_{M_F}}{\partial \dot{x}}$	$.00051$

continued

TABLE 9.14 (continued)

Derivative	Equation	Value for Example Helicopter
$\dfrac{\partial \alpha_F}{\partial \dot{z}}$	$-\left(\dfrac{\partial \varepsilon_{M_F}}{\partial \dot{z}} + \dfrac{\partial \gamma}{\partial \dot{z}} \right)$	.00467
$\dfrac{\partial f}{\partial \alpha_F}$		-2 ft²/rad
$\dfrac{\partial L/q}{\partial \alpha_F}$		74.5 ft²/rad
$\dfrac{\partial \text{S.F.}/q}{\partial \beta}$	From curves in Appendix A at trim conditions	-220 ft²/rad
$\dfrac{\partial M/q}{\partial \alpha_F}$		1780 ft³/rad
$\dfrac{\partial N/q}{\partial \beta}$		-820 ft³/rad
$\dfrac{\partial R/q}{\partial \beta}$		230 ft³/rad

TABLE 9.15
Fuselage Derivatives in Forward Flight

Derivative	Equation	Value for Example Helicopter
$\left(\dfrac{\partial X}{\partial \dot{x}} \right)_F$	$\dfrac{2}{V} \overline{X_F}$	-8 lb/ft/sec
$\left(\dfrac{\partial X}{\partial \dot{z}} \right)_F$	$\left(\overline{L_F} - q\,\dfrac{\partial f}{\partial \alpha_F} \right) \dfrac{\partial \alpha_F}{\partial \dot{z}}$	-2 lb/ft/sec
$\left(\dfrac{\partial Y}{\partial \dot{y}} \right)_F$	$\dfrac{1}{V} \left(q\,\dfrac{\partial \text{S.F.}/q}{\partial \beta} - \overline{D_F} \right)$	-55 lb/ft/sec

TABLE 9.15 (continued)

Derivative	Equation	Value for Example Helicopter
$\left(\dfrac{\partial Z}{\partial \dot{x}}\right)_F$	$\dfrac{2}{V}\overline{Z_F}$	-3 lb/ft/sec
$\left(\dfrac{\partial Z}{\partial \dot{z}}\right)_F$	$\left(-\overline{D_F} - q\,\dfrac{\partial L/q}{\partial \alpha_F}\right)\dfrac{\partial \alpha_F}{\partial \dot{z}}$	-19 lb/ft/sec
$\left(\dfrac{\partial R}{\partial \dot{y}}\right)_F$	$q\,\dfrac{\partial R/q}{\partial \beta}\dfrac{\partial \beta}{\partial \dot{y}}$	53 ft-lb/ft/sec
$\left(\dfrac{\partial M}{\partial \dot{x}}\right)_F$	$\dfrac{2}{V}\overline{M_F} + q\,\dfrac{\partial M/q}{\partial \alpha_F}\dfrac{\partial \alpha_F}{\partial \dot{x}}$	-94 ft-lb/ft/sec
$\left(\dfrac{\partial M}{\partial \dot{z}}\right)_F$	$q\,\dfrac{\partial M/q}{\partial \alpha_F}\dfrac{\partial \alpha_F}{\partial \dot{z}}$	374 ft-lb/ft/sec
$\left(\dfrac{\partial N}{\partial \dot{y}}\right)_F$	$q\,\dfrac{\partial N/q}{\partial \beta}\dfrac{\partial \beta}{\partial \dot{y}}$	-190 ft-lb/ft/sec

Total Derivatives

Table 9.16 summarizes the derivative contributions for all of the components for the example helicopter at 115 knots.

TABLE 9.16
Total Forward Flight Derivatives

Derivative \ Component	Main Rotor	Tail Rotor	Horizontal Stab.	Vert. Stab.	Fuselage	Total
$\dfrac{\partial X}{\partial \dot{x}}$	-12				-8	-20 lb/ft/sec
$\dfrac{\partial X}{\partial \dot{y}}$	-6			-3		-9 lb/ft/sec
$\dfrac{\partial X}{\partial \dot{z}}$	-5		-1		-2	-8 lb/ft/sec

continued

TABLE 9.16 (continued)

Derivative	Main Rotor	Tail Rotor	Horizontal Stab.	Vert. Stab.	Fuselage	Total
$\dfrac{\partial X}{\partial q}$	1,937					1,937 lb/rad/sec
$\dfrac{\partial X}{\partial p}$	−659					−659 lb/rad/sec
$\dfrac{\partial X}{\partial \theta_{0_M}}$	−6,727					−6,727 lb/rad
$\dfrac{\partial X}{\partial A_1}$	−1,506					−1,506 lb/rad
$\dfrac{\partial X}{\partial B_1}$	18,601					18,601 lb/rad
$\dfrac{\partial X}{\partial \theta_{0_T}}$		12,845				12,845 lb/rad
$\dfrac{\partial Y}{\partial \dot{x}}$	2	−2		4		4 lb/rad/sec
$\dfrac{\partial Y}{\partial \dot{y}}$	−14	−24		−14	−55	−107 lb/rad/sec
$\dfrac{\partial Y}{\partial \dot{z}}$	−13					−13 lb/rad/sec
$\dfrac{\partial Y}{\partial q}$	−574					−574 lb/rad/sec
$\dfrac{\partial Y}{\partial p}$	−1,658	−147		−42		−1,847 lb/rad/sec

TABLE 9.16 (continued)

Derivative \ Component	Main Rotor	Tail Rotor	Horizontal Stab.	Vert. Stab.	Fuselage	Total
$\dfrac{\partial Y}{\partial r}$		907		490		1,397 lb/rad/sec
$\dfrac{\partial Y}{\partial \theta_{0_M}}$	−5,668					−5,668 lb/rad
$\dfrac{\partial Y}{\partial A_1}$	16,638					16,638 lb/rad
$\dfrac{\partial Y}{\partial B_1}$	1,635					1,635 lb/rad
$\dfrac{\partial Y}{\partial \theta_{0_T}}$		12,845				12,845 lb/rad
$\dfrac{\partial Z}{\partial \dot{x}}$	45		1		−3	43 lb/ft/sec
$\dfrac{\partial Z}{\partial \dot{z}}$	−261		−10		−19	−290 lb/ft/sec
$\dfrac{\partial Z}{\partial q}$			−370			−370 lb/rad/sec
$\dfrac{\partial Z}{\partial r}$	−1,846					−1,846 lb/rad/sec
$\dfrac{\partial Z}{\partial \theta_{0_M}}$	−110,919					−110,919 lb/rad
$\dfrac{\partial Z}{\partial B_1}$	67,855					67,855
$\dfrac{\partial R}{\partial \dot{x}}$	31	−13		12		30 ft-lb/ft/sec
$\dfrac{\partial R}{\partial \dot{y}}$	−246	−147		−42	53	−382 ft-lb/ft/sec

continued

TABLE 9.16 (continued)

Derivative \ Component	Main Rotor	Tail Rotor	Horizontal Stab.	Vert. Stab.	Fuselage	Total
$\dfrac{\partial R}{\partial \dot{z}}$	−263					−263 ft-lb/ft/sec
$\dfrac{\partial R}{\partial q}$	−11,418					−11,418 ft-lb/rad/sec
$\dfrac{\partial R}{\partial p}$	−32,730	−882		−126		−33,738 ft-lb/rad/sec
$\dfrac{\partial R}{\partial r}$		5,442		1,470		6,912 ft-lb/rad/sec
$\dfrac{\partial R}{\partial \theta_{0_M}}$	−92,745					−92,745 ft-lb/rad
$\dfrac{\partial R}{\partial A_1}$	325,703					325,703 ft-lb/rad
$\dfrac{\partial R}{\partial B_1}$	32,014					32,014 ft-lb/rad
$\dfrac{\partial R}{\partial \theta_{0_T}}$		77,070				77,070 ft-lb/rad
$\dfrac{\partial M}{\partial \dot{x}}$	182		27		−94	115 ft-lb/ft/sec
$\dfrac{\partial M}{\partial \dot{y}}$	−15					−15 ft-lb/ft/sec
$\dfrac{\partial M}{\partial \dot{z}}$	495		−327		374	542 ft-lb/ft/sec
$\dfrac{\partial M}{\partial \ddot{z}}$			14			14 ft-lb/ft/sec²

TABLE 9.16 (continued)

Component Derivative	Main Rotor	Tail Rotor	Horizontal Stab.	Vert. Stab.	Fuselage	Total
$\dfrac{\partial M}{\partial q}$	−36,591		−13,690			−50,281 ft-lb/rad/sec
$\dfrac{\partial M}{\partial p}$	12,900					12,900 ft-lb/rad/sec
$\dfrac{\partial M}{\partial \theta_{0_M}}$	326,946					326,946 ft-lb/rad/sec
$\dfrac{\partial M}{\partial A_1}$	29,480					29,480 ft-lb/rad
$\dfrac{\partial M}{\partial B_1}$	−364,158					−364,158 ft-lb/rad
$\dfrac{\partial N}{\partial \dot{x}}$	−53	78		105		130 ft-lb/ft/sec
$\dfrac{\partial N}{\partial \dot{y}}$		907		490	−190	1,017 ft-lb/ft/sec
$\dfrac{\partial N}{\partial \dot{z}}$	99					99 ft-lb/ft/sec
$\dfrac{\partial N}{\partial p}$		5439		1470		6,909 ft-lb/rad/sec
$\dfrac{\partial N}{\partial r}$	−39,710	−33,559		−17,150		−90,419 ft-lb/rad/sec
$\dfrac{\partial N}{\partial \theta_{0_M}}$	376,161					376,161 ft-lb/rad
$\dfrac{\partial N}{\partial \theta_{0_T}}$		−475,265				−475,265 ft-lb/rad

EQUATIONS OF MOTION IN HOVER

Whether the helicopter is in hover or in forward flight, it is governed by the same basic equations of motion, but the fact that some of the derivatives are quite different strongly affects the behavior of the aircraft in the two flight regimes. It is therefore useful to study them separately. One of the advantages of starting with the hover regime is that the coupling between the longitudinal and the lateral-directional equations is weak enough to be ignored, at least for the purpose of instruction. It seems intuitively valid to consider a helicopter that is constrained by the pilot (or by an instructor pilot or by an autopilot) to allow motion in only selected degrees of freedom. For example, the pilot could use his pedals to hold heading and his lateral control to hold "wings level" so that only pitching, fore-and-aft, and up-and-down motions were permitted. This reduces the required equations of motion from six to three without doing too much violence to the concept of the hovering helicopter. (A further simplification to only two equations will later be shown to also produce useful results.)

For analysis of the longitudinal motion, the three required equations are:

$$-\frac{G.W.}{g}\ddot{x} + \frac{\partial X}{\partial \dot{x}}\dot{x} + \frac{\partial X}{\partial \dot{z}}\dot{z} + \frac{\partial X}{\partial q}q - G.W.\Theta = -\frac{\partial X}{\partial \theta_{0_M}}\theta_{0_M} - \frac{\partial X}{\partial B_1}B_1$$

$$\frac{\partial Z}{\partial \dot{x}}\dot{x} + \frac{\partial Z}{\partial \ddot{z}}\ddot{z} + \frac{\partial Z}{\partial \dot{z}}\dot{z} + \frac{\partial Z}{\partial q}q \qquad = -\frac{\partial Z}{\partial \theta_{0_M}}\theta_{0_M} - \frac{\partial Z}{\partial B_1}B_1$$

$$\frac{\partial M}{\partial \dot{x}}\dot{x} + \frac{\partial M}{\partial \ddot{z}}\ddot{z} + \frac{\partial M}{\partial \dot{z}}\dot{z} - I_{yy}\dot{q} + \frac{\partial M}{\partial q}q \qquad = -\frac{\partial M}{\partial \theta_{0_M}}\theta_{0_M} - \frac{\partial M}{\partial B_1}B_1$$

The numerical values for the stability derivatives of the example helicopter in hover have already been evaluated and could be put into the equations to generate the characteristic equation and to solve for the roots as a calculator exercise. A more rewarding learning process, however, is to operate on the equations in symbol form first, before resorting to numbers. Using this approach and substituting:

$$x = X(S)e^{st}$$
$$z = Z(S)e^{st}$$
$$\Theta = \Theta(S)e^{st}$$

the unforced equations in matrix form may be written:

$$\begin{vmatrix} \left(-\dfrac{\text{G.W.}}{g}s + \dfrac{\partial X}{\partial \dot{x}}\right) & \dfrac{\partial X}{\partial \dot{z}} & \dfrac{\partial X}{\partial q}s - \text{G.W.} \\[2em] \dfrac{\partial Z}{\partial \dot{x}} & \left(\left[\dfrac{\partial Z}{\partial \ddot{z}} - \dfrac{\text{G.W.}}{g}\right]s + \dfrac{\partial Z}{\partial \dot{z}}\right) & \dfrac{\partial Z}{\partial q}s \\[2em] \dfrac{\partial M}{\partial \dot{x}} & \left(\dfrac{\partial M}{\partial \ddot{z}}s + \dfrac{\partial M}{\partial \dot{z}}\right) & \left(-I_{yy}s^2 + \dfrac{\partial M}{\partial q}s\right) \end{vmatrix} \begin{vmatrix} \dot{x}\,(S) \\[2em] \dot{z}\,(S) \\[2em] \Theta(S) \end{vmatrix} = 0$$

Note that in hover, the following derivatives are zero:

$$\frac{\partial X}{\partial \dot{z}}, \ \frac{\partial Z}{\partial \dot{x}}, \ \frac{\partial Z}{\partial \ddot{z}}, \ \frac{\partial Z}{\partial q}, \ \frac{\partial M}{\partial \ddot{z}}$$

Also note that the expansion will yield terms with components consisting of:

$$\left(\frac{\partial M}{\partial q}\frac{\partial X}{\partial \dot{x}} - \frac{\partial M}{\partial \dot{x}}\frac{\partial X}{\partial q}\right)$$

For a single rotor helicopter, these terms cancel themselves out because:

$$\frac{\partial M}{\partial q} = \frac{dM}{da_{1_s}}\frac{\partial a_{1_s}}{\partial q}, \ \frac{\partial X}{\partial \dot{x}} = -\rho A_b(\Omega R)^2 \overline{C_T}/\sigma \, \frac{\partial a_{1_s}}{\partial \mu}\frac{\partial \mu}{\partial \dot{x}}$$

$$\frac{\partial M}{\partial \dot{x}} = \frac{dM}{da_{1_s}}\frac{\partial a_{1_s}}{\partial \mu}\frac{\partial \mu}{\partial \dot{x}}, \ \frac{\partial X}{\partial q} = -\rho A_b(\Omega R)^2 \overline{C_T}/\sigma \, \frac{\partial a_{1_s}}{\partial q}$$

The characteristic equation obtained by expanding the determinant is:

$$s^4 - \left[\frac{1}{\text{G.W.}/g}\left(\frac{\partial X}{\partial \dot{x}} + \frac{\partial Z}{\partial \dot{z}}\right) + \frac{1}{I_{yy}}\frac{\partial M}{\partial q}\right]s^3 + \frac{\partial Z}{\partial \dot{z}}\left[\frac{1}{(\text{G.W.}/g)^2}\frac{\partial X}{\partial \dot{x}}\right.$$

$$\left. + \frac{1}{(\text{G.W.}/g)I_{yy}}\frac{\partial M}{\partial q}\right]s^2 + \frac{g}{I_{yy}}\frac{\partial M}{\partial \dot{x}}s - \frac{g}{(\text{G.W.}/g)I_{yy}}\frac{\partial M}{\partial \dot{x}}\frac{\partial Z}{\partial \dot{z}} = 0$$

which for the example helicopter becomes:

$$s^4 + 1.02s^3 + .21s^2 + .12s + .034 = 0$$

The four roots are:

$$s_1 = -.89, \ s_2 = -.28, \ s_{3,4} = .076 \pm .360i$$

These represent two heavily damped pure convergences and an unstable oscillation whose period is:

$$P = \frac{2\pi}{.360} = 17.5 \text{ seconds}$$

and the time to double amplitude is:

$$t_{\text{double}} = \frac{.693}{.076} = 9.1 \text{ seconds}$$

Add Simplification

Nearly the same results are obtained if the helicopter is assumed to be constrained vertically so that the Z-Force equation can be eliminated. Now the equations are:

$$-\frac{\text{G.W.}}{g}\ddot{x} + \frac{\partial X}{\partial \dot{x}}\dot{x} + \frac{\partial X}{\partial q}q - \text{G.W.}\Theta = -\frac{\partial X}{\partial B_1}B_1$$

$$\frac{\partial M}{\partial \dot{x}}\dot{x} - I_{yy}\dot{q} + \frac{\partial M}{\partial q}q = -\frac{\partial M}{\partial B_1}B_1$$

It is often useful to see a more graphical representation of the system. One such is the block diagram of Figure 9.8. An even more graphic illustration is the mechanical analog of Figure 9.9, in which viscous dampers and screw jack actuated beam-rider weights are used to generate appropriate forces and moments.

The characteristic equation is:

$$s^3 - \left(\frac{1}{\text{G.W.}/g}\frac{\partial X}{\partial \dot{x}} + \frac{1}{I_{yy}}\frac{\partial M}{\partial q}\right)s^2 + \frac{g}{I_{yy}}\frac{\partial M}{\partial \dot{x}} = 0$$

or, for the example helicopter:

$$s^3 + .724s^2 + .115 = 0$$

whose roots are:

$$s_1 = -.87, \ s_{2,3} = .075 \pm .355i$$

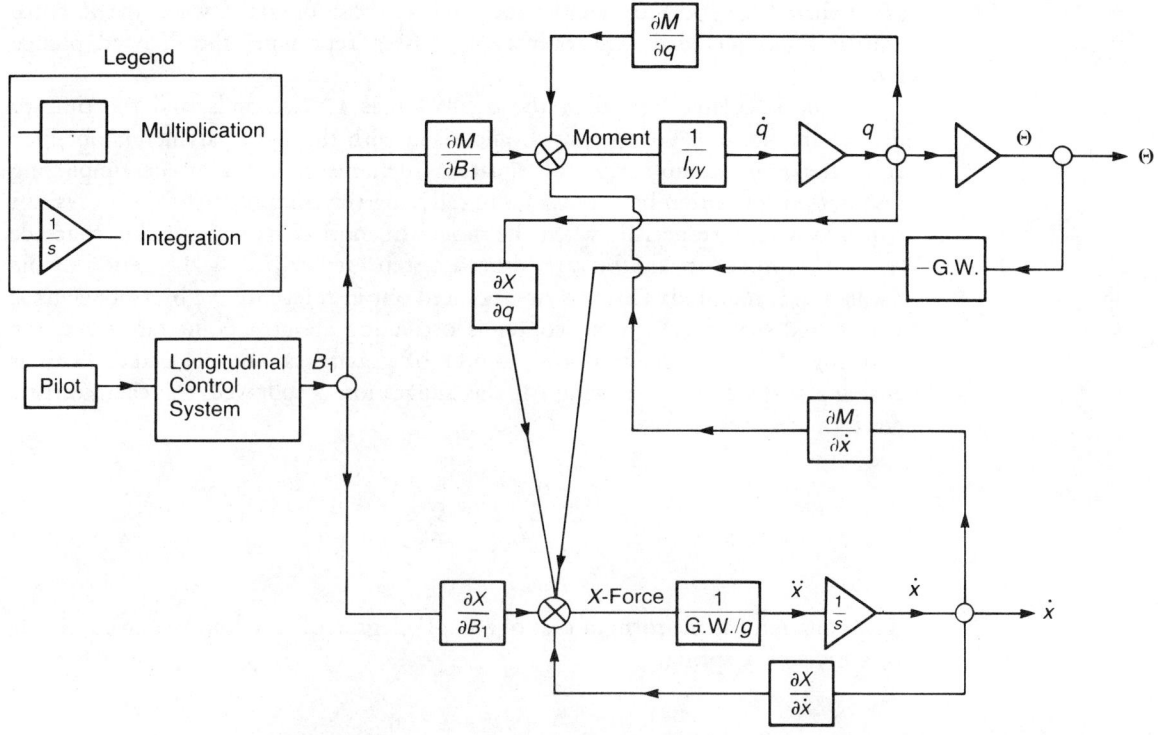

FIGURE 9.8   Block Diagram for Two-Degree-of-Freedom System Representing a Hovering Helicopter

$$\frac{\partial X}{\partial \dot{x}}\,\dot{x} + \frac{\partial X}{\partial q}\,q - \text{G.W.}\Theta = \frac{\text{G.W.}}{g}\,\ddot{x}$$

$$\frac{\partial M}{\partial \dot{x}}\,\dot{x} + \frac{\partial M}{\partial q}\,q = I_{yy}\dot{q}$$

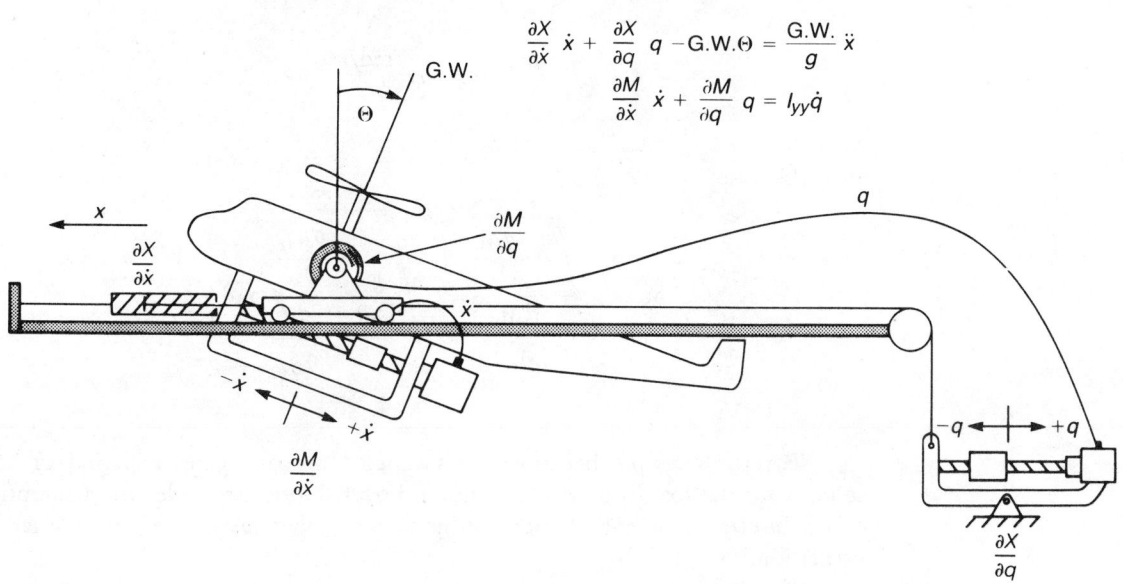

FIGURE 9.9   Mechanical Analog of Hovering Helicopter

Note that these roots are almost the same as those obtained when the Z-Force equation was included. The only root missing represents the damped plunge mode.

The calculated period of the oscillation is 17.7 seconds, and the time to double amplitude is 9.2 seconds. Comparison with the same parameters obtained from the more complete set of equations demonstrates that some simplifying assumptions can often be very useful in reducing the complexity of the analysis of dynamic systems, especially when the modes of motion are only weakly coupled.

The procedure can be carried even a step further if only the period of the oscillation is required. This was first pointed out in reference 9.4 by Hohenemser, who noted that since the helicopter is oscillating about a point far above, the moment of inertia about its own center of gravity can be neglected. (This is analogous to a child on a swing.) If this suggestion is followed, the characteristic equation reduces to:

$$-\frac{\partial M}{\partial q}s^2 + g\frac{\partial M}{\partial \dot{x}} = 0$$

This equation has the form of that of a single-degree-of-freedom system consisting of a mass on a spring:

$$ms^2 + k = 0$$

for which the natural frequency is:

$$\omega_N = \sqrt{\frac{k}{m}} \text{ rad/sec}$$

or in this case:

$$\omega_N = \sqrt{\frac{-g\dfrac{\partial M}{\partial \dot{x}}}{\dfrac{\partial M}{\partial q}}} = \sqrt{\frac{g}{\Omega R}\dfrac{\dfrac{\partial a_{1_s}}{\partial \mu}}{-\dfrac{\partial a_{1_s}}{\partial q}}} \text{ rad/sec}$$

For the example helicopter this simple approach gives a period of 15.7 seconds for the longitudinal oscillation in hover. Note that under the assumption of no fuselage moment of inertia, this value applies just as well to the lateral oscillation.

A generalization can be stated about the period of oscillation—it is essentially proportional to the square root of the rotor radius. The demonstration makes use of the following line of reasoning:

In hover

$$\frac{\partial a_{1_s}}{\partial \mu} = \frac{8}{3}\theta_0 + 2\theta_1 - 2\frac{v_1}{\Omega R}$$

and from Chapter 3, with the tip speed ratio, μ, set to zero:

$$C_T/\sigma = \frac{a}{16}\left[\frac{8}{3}\theta_0 + 2\theta_1 - 4\frac{v_1}{\Omega R}\right]$$

Thus with only a small white lie concerning the coefficient of the induced velocity ratio:

$$\frac{\partial a_{1_s}}{\partial \mu} \doteq \frac{16 C_T/\sigma}{a}$$

The damping derivative with zero hinge offset is:

$$\frac{\partial a_{1_s}}{\partial q} = \frac{16}{\gamma R}$$

When these are substituted into the equation for the period, we have:

$$P = \frac{2\pi\sqrt{R}}{\sqrt{g\,\dfrac{C_T/\sigma}{a}\gamma}}\ \sec$$

If it is assumed that most modern helicopters have nearly the same values of C_T/σ and Lock number, γ, as the example helicopter, the period can be roughly approximated by:

$$P = 3.2\sqrt{R}\ \sec$$

(It is interesting to note that the period of a pendulum is approximately $\sqrt{l/3}$ seconds and so, by analogy, the helicopter is swinging from a support 33 radii above it.)

A human can learn to control an unstable vehicle like a hovering helicopter as long as the period is significantly longer than the total time delay in perceiving an error, processing the information in the brain, and moving the appropriate control to correct the error. The period of very small one-man helicopters approaches the lower limit and of radio-controlled models generally goes below unless some means is used to change their characteristics. For this reason, most models use a version of the Hiller servorotor system that increases the damping by an order of magnitude and hence increases the period enough to make it compatible with the capabilities of the ground-based pilot.

What Routh's Discriminant Tells Us

Even without solving for the roots, the characteristic equation can be made to yield useful information by using Routh's discriminant (R.D.). For a cubic equation:

$$\text{R.D.}(3) = BC - AD$$

or in this case:

$$\text{R.D.}(3) = -\left(\frac{g}{I_{yy}} \frac{\partial M}{\partial \dot{x}} \right)$$

Table 9.17 lists the tests of Routh's discriminant.

TABLE 9.17
Tests of Routh's Discriminant

Test	Consequence	Application to Hovering Helicopter
1. All coefficients are positive.	No pure divergence	For most helicopters, both $\partial X/\partial \dot{x}$ and $\partial M/\partial q$ are negative, thus B is positive.
		$\dfrac{\partial M}{\partial \dot{x}}$ is positive, thus D is positive
2. R.D. is positive.	No unstable oscillation	Since $\partial M/\partial \dot{x}$ is generally positive, this test is failed.
3. R.D. = 0	Neutrally stable	Requires that $\partial M/\partial \dot{x}$ be zero
4. R.D. is negative.	Unstable	Generally true due to sign of $\partial M/\partial \dot{x}$
5. $D = 0$	Non-oscillatory neutral stability	Only true if $\partial M/\partial \dot{x} = 0$
6. One coefficient is negative.	Pure divergence or unstable oscillation	No negative coefficients for conventional helicopters

Note the strong role that the sign of the speed stability term, $\partial M/\partial \dot{x}$, plays in these tests, with a positive value leading to instability. (This is in contrast to the forward flight situation, where a positive sign on the speed derivative is usually necessary for stability, as will be discussed in the next section.) The equation for the derivative shows that it is made up of both hub stiffness (through offset flapping hinges, for instance) and the tilt of the thrust vector with a moment arm proportional to mast height. For a conventional helicopter, both effects have the same sign thus leading to Test 4 predicting an unstable oscillation. If the rotor were mounted under the helicopter, the contribution of the tilt of the thrust vector would change sign. For a teetering rotor with no hub stiffness, the sign of the entire derivative would be reversed as shown in Figure 9.10. Routh's discriminant would then satisfy Test 2, and no unstable oscillation could occur. Test 6, however, would indicate a pure divergence, represented by the tendency of this helicopter to drift off into translational flight.

Several helicopters have actually been built with the low rotor position. One was the De Lackner "stand on" configuration shown in Figure 9.11. This

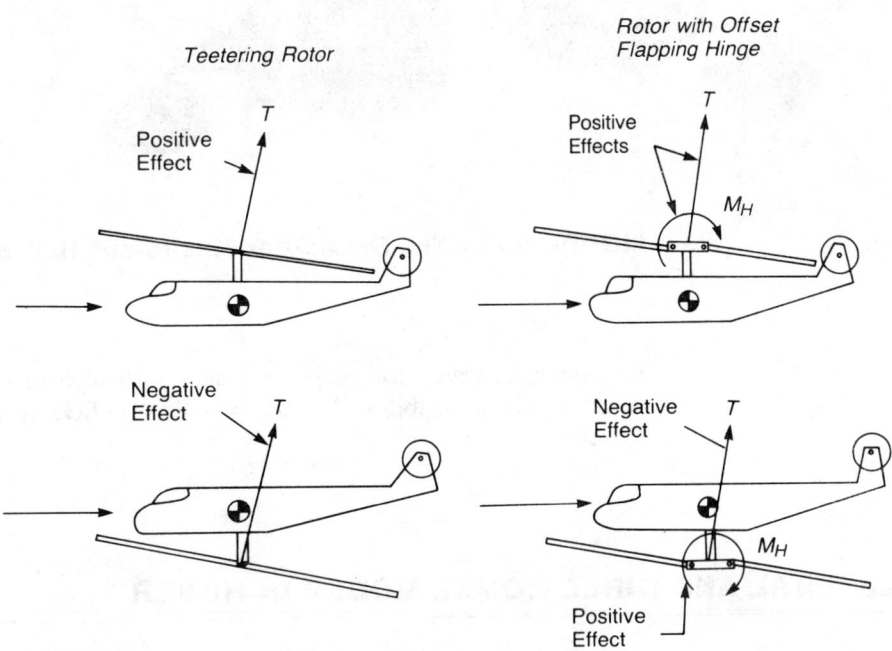

FIGURE 9.10   Effect of Rotor Location on Sign of Speed Stability Derivative

FIGURE 9.11   The DeLackner "Stand-on" Helicopter

helicopter, however, had hingeless blades with high inherent hub stiffness, and consequently its stability did not benefit significantly from its unique rotor location.

LATERAL AND DIRECTIONAL MODES IN HOVER

So far, this study has been limited to the longitudinal mode, but it is evident that the lateral mode in hover could have been treated in the same manner by using the moment of inertia in roll instead of pitch in the equations. For a simple analysis, it is again necessary to assume that the pilot holds heading by adjusting tail rotor thrust as required and that in this case longitudinal control is used to prevent any pitching motion. If the purpose of the analysis is to study cross-coupling problems, these assumptions will have to be dropped and, instead of simply using two or three equations, all six equations should be treated simultaneously. The method of solution is the same as for the more restricted situation. The mathematical manipulation to solve for the roots becomes very tedious if done by hand but very easy if turned over to a modern computer.

The Yaw Mode in Hover

There is another simple case, however, for which the motion can be assumed to be essentially decoupled, and that is hover motion about the yaw axis. In this case the helicopter can be considered to be a single-degree-of-freedom system representing a mass and damper combination. The yawing moment equation without control input is:

$$-I_{zz}\dot{r} + \frac{\partial N}{\partial r}r = 0$$

This has only one root representing a heavy damping:

$$s = \frac{\partial N}{\partial r}\frac{1}{I_{zz}}$$

or for the example helicopter:

$$s = -.38$$

which indicates a pure convergence that damps to half amplitude in 1.82 seconds.

CONTROL RESPONSE IN HOVER

The previous section developed the basis for the study of the stability characteristics of the hovering helicopter. We will now address its control characteristics. One useful piece of information in this regard is the *transfer function*, which relates the response of the helicopter to an individual control input.

The transfer function is obtained as the ratio between two determinants. The denominator is the determinant already used to generate the characteristic equation, and the numerator is identical except that the control column on the right-hand side of the equations of motion is substituted for the column representing the degree of freedom of interest. As an example, let us return to the analysis of the longitudinal degrees of freedom without plunge motion in hover and obtain the transfer function for pitch attitude, Θ, due to longitudinal cyclic pitch, B_1. The two pertinent equations of motion with only the nonzero derivatives retained are:

$$X \qquad -\frac{\text{G.W.}}{g}\ddot{x} + \frac{\partial X}{\partial \dot{x}}\dot{x} + \frac{\partial X}{\partial q}q - \text{G.W.}\Theta = -\frac{\partial X}{\partial B_1}B_1$$

$$M \qquad \frac{\partial M}{\partial \dot{x}}\dot{x} \qquad\qquad -I_{yy}\dot{q} + \frac{\partial M}{\partial q}q = -\frac{\partial M}{\partial B_1}B_1$$

The transfer function of pitch attitude due to longitudinal cyclic pitch in determinant form is:

$$\frac{\Theta(s)}{B_1(s)} = \frac{\begin{vmatrix} \left(-\dfrac{\text{G.W.}}{g}s + \dfrac{\partial X}{\partial \dot{x}}\right) & -\dfrac{\partial X}{\partial B_1} \\[2em] \dfrac{\partial M}{\partial \dot{x}} & -\dfrac{\partial M}{\partial B_1} \end{vmatrix}}{\begin{vmatrix} \left(-\dfrac{\text{G.W.}}{g}s + \dfrac{\partial X}{\partial \dot{x}}\right) & \left(\dfrac{\partial X}{\partial q}s - \text{G.W.}\right) \\[2em] \dfrac{\partial M}{\partial \dot{x}} & \left(-I_{yy}s^2 + \dfrac{\partial M}{\partial q}s\right) \end{vmatrix}}$$

or

$$\frac{\Theta(s)}{B_1(s)} = \frac{\dfrac{1}{I_{yy}}\dfrac{\partial M}{\partial B_1}s}{s^3 - \left(\dfrac{1}{\text{G.W.}/g}\dfrac{\partial X}{\partial \dot{x}} + \dfrac{1}{I_{yy}}\dfrac{\partial M}{\partial q}\right)s^2 + \dfrac{g}{I_{yy}}\dfrac{\partial M}{\partial \dot{x}}}$$

(Note that in the derivation for a single-rotor helicopter, some terms in the numerator cancel themselves out just as they did in the derivation of the characteristic equation, which becomes the denominator.)

The operation represented by s is differentiation with respect to time, so the transfer function of the pitch rate, $q(s)$ [or $\dot{\Theta}(s)$], can be written by multiplying the transfer function of pitch attitude by s:

$$\frac{q(s)}{B_1(s)} = \frac{\dfrac{1}{I_{yy}}\dfrac{\partial M}{\partial B_1}s^2}{s^3 - \left(\dfrac{1}{G.W./g}\dfrac{\partial X}{\partial \dot{x}} + \dfrac{1}{I_{yy}}\dfrac{\partial M}{\partial q}\right)s^2 + \dfrac{g}{I_{yy}}\dfrac{\partial M}{\partial \dot{x}}}$$

This equation can be made to produce a time history of pitch rate as a response to cyclic pitch. Most modern engineering computers now have canned programs for doing this, but it is of some interest to know that noncomputer methods exist, both for the historical perspective on how it was done in the "old days," and to make simple checks of computer results. The method that will be illustrated is the Heaviside Expansion, which for this application is:

$$\frac{q}{B_1} - \frac{N(0)}{D(0)} + \sum_{r=1}^{n} \frac{N(s)e^{st}}{\left[\dfrac{dD(s)}{ds}\right]s}$$

where $N(s)$ and $D(s)$ are, respectively, the numerator and the denominator of the transfer function, and s_r is a root of the characteristic equation. In this case for the example helicopter:

$$\left[\frac{dD(s)}{ds}\right]s = 3s^3 + 1.448s^2$$

and

$$\frac{q}{B_1} = -6.78\left[\frac{0}{.115} + \sum_{r=1}^{3}\frac{s^2 e^{st}}{3s^3 + 1.448s^2}\right]$$

or

$$\frac{q}{B_1} = -6.78\sum_{r=1}^{3}\frac{e^{st}}{3s + 1.448}$$

Using the three roots:

$$\frac{q}{B_1} = -6.78\left[\frac{e^{-.874t}}{3(-.874) + 1.448} + \frac{e^{(.075+.355i)t}}{3(.075 + .355i) + 1.448}\right.$$

$$\left. + \frac{e^{(.075-.355i)t}}{3(.075 - .355i) + 1.448}\right]$$

When the algebra is done and trignometric terms substituted for the complex variables, the result is:

$$\frac{q}{B_1} = 5.78e^{-.874t} - 6.85e^{.075t}\sin(20.34t + 57.54)$$

where the angles are in degrees. Figure 9.12 shows the time history obtained from this equation for the helicopter free both to pitch and to have horizontal translation.

If the helicopter had been mounted on trunions so that only pitching motion were permitted, the transfer function would reduce to:

$$\frac{q(s)}{B_1(s)} = \frac{\dfrac{\partial M}{\partial B_1}}{I_{yy}s - \dfrac{\partial M}{\partial q}}$$

The corresponding equation in terms of time is:

$$\frac{q}{B_1} = -\frac{\dfrac{\partial M}{\partial B_1}}{\dfrac{\partial M}{\partial q}}\left(1 - e^{\frac{\frac{\partial M}{\partial q}}{I_{yy}}t}\right)$$

This is a response that asymptotically approaches a steady value:

$$\left(\frac{q}{B_1}\right)_{t\to\infty} = -\frac{\dfrac{\partial M}{\partial B_1}}{\dfrac{\partial M}{\partial q}}\frac{\text{deg/sec}}{\text{deg of cyclic pitch}}$$

and has a time constant of:

$$t_{.63} = \frac{I_{yy}}{-\dfrac{\partial M}{\partial q}}\sec$$

For the example helicopter, this time history is plotted on Figure 9.12 along with the more unconstrained system. It may be seen that the two time histories are essentially identical during the first quarter cycle of the oscillation. After that point, the effects of horizontal translation become dominant.

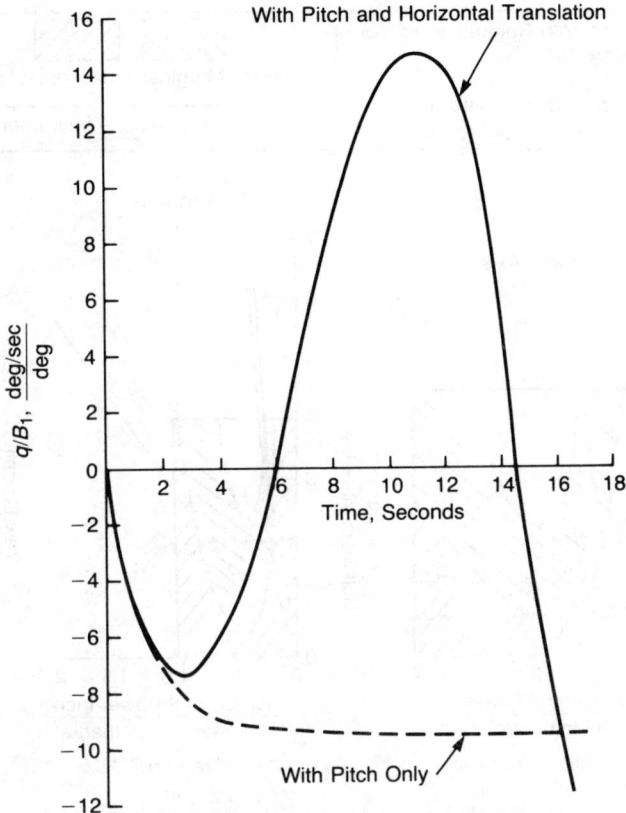

FIGURE 9.12 Response of Example Helicopter to Longitudinal Control Step in Hover

Guidelines for Response

Pilots have found that there are both maximum and minimum limits on the response to control motion for desirable flying qualities. If the response per inch of control motion is too small, the pilot will find the helicopter too sluggish; if the response is too large, he will complain of oversensitivity because even very small inadvertent control motions will produce large responses. (It is a well-documented observation, however, that pilot opinion changes with experience in a given helicopter design. What might be judged to be oversensitivity initially often later becomes sluggishness as the pilot becomes more experienced in the machine.) Many flight and simulator studies have been made to determine the limits. One of the first was done in the late 1950s with a small, variable stability helicopter making instrument landing system (ILS) approaches. This program is reported in reference 9.5 and the results are summarized in Figure 9.13 as regions on the plot

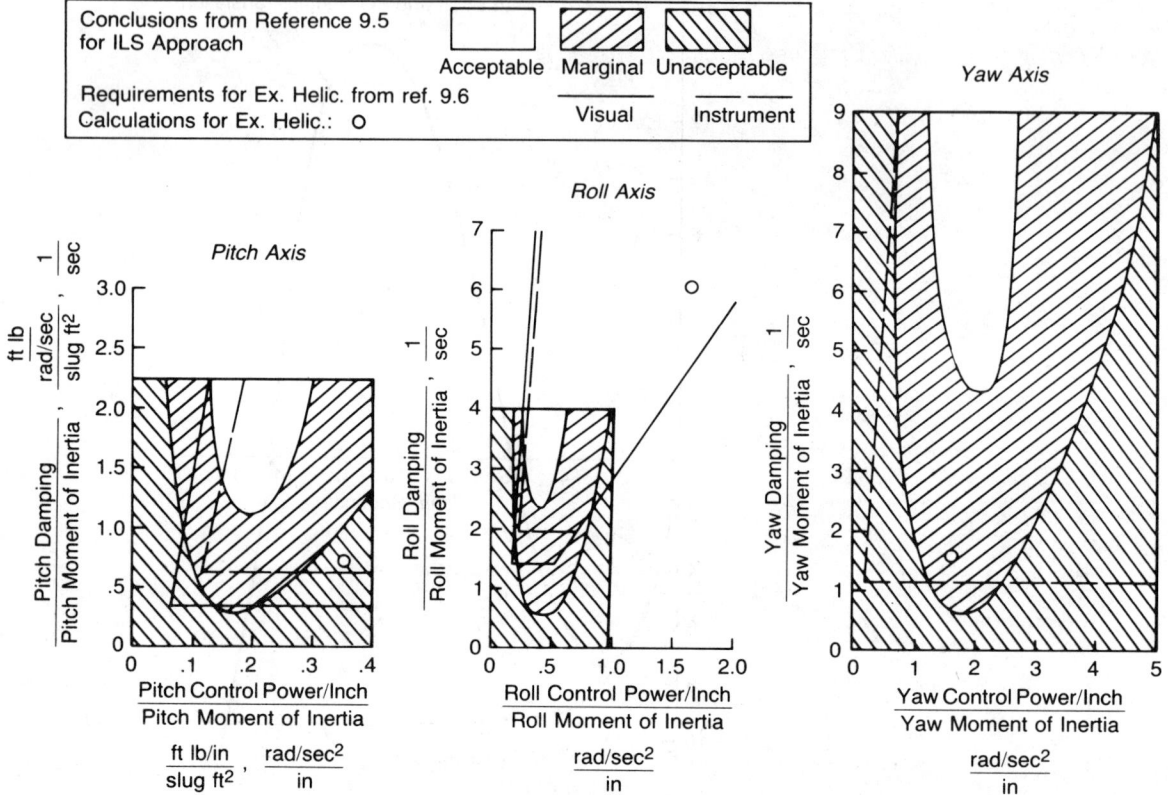

FIGURE 9.13   Control Power and Damping for Acceptable Handling Qualities

of damping versus control power, which produced varying degrees of satisfactory flying qualities.

At the time of these tests, it was felt that small helicopters should be more responsive than large helicopters, and this reasoning was used when the Military Specification for helicopter flying qualities, MIL-H-8501A—reference 9.6—was being written. These requirements specified the damping and the response to one-inch control steps for both visual and instrument flight conditions in all three axes. Paragraphs 3.2.13, 3.2.14, 3.3.5, 3.3.15, 3.3.18, 3.3.19, and 3.6.1.1 of reference 9.6 can be summarized as in Table 9.18.

A study of the results of later flight test programs such as those reported in references 9.7, 9.8, and 9.9 indicate that size is not really a factor and that all helicopters should have about the same control characteristics. (At the time of this writing, MIL-H-8501A is in the process of revision and will probably lose this size distinction.)

TABLE 9.18
Summary of MIL-H-8501A Response Requirements

Axis	Time, sec	Minimum Response (deg.)		Maximum Response (deg./sec)	Damping (ft-lb/rad/sec)	
		Visual	Instrument		Visual	Instrument
Longitudinal	1	$\dfrac{45}{\sqrt[3]{\text{G.W.} + 1{,}000}}$	$\dfrac{73}{\sqrt[3]{\text{G.W.} + 1{,}000}}$	—	$8I_{yy}^{.7}$	$15I_{yy}^{.7}$
Lateral	.5	$\dfrac{27}{\sqrt[3]{\text{G.W.} + 1{,}000}}$	$\dfrac{32}{\sqrt[3]{\text{G.W.} + 1{,}000}}$	20	$18I_{xx}^{.7}$	$25I_{xx}^{.7}$
Directional	1	$\dfrac{110}{\sqrt[3]{\text{G.W.} + 1{,}000}}$	$\dfrac{110}{\sqrt[3]{\text{G.W.} + 1{,}000}}$	—	$27I_{zz}^{.7}$ [a]	$27I_{zz}^{.7}$

[a] Not a requirement, only a preference.

The displacement requirements can be converted into combinations of the two parameters of Figure 9.13 by treating each of the moment equations of motion as single degrees of freedom. For example, the longitudinal equation reduces to:

$$\frac{\partial M}{\partial B_1} B_1 = I_{yy} \ddot{\Theta} - \frac{\partial M}{\partial q} \dot{\Theta}$$

The solution of this equation is:

$$\Theta = \frac{\dfrac{\partial M}{\partial B_1} B_1}{-\dfrac{\partial M}{\partial q}} \left[t + \frac{I_{yy}}{-\dfrac{\partial M}{\partial q}} \left(e^{\frac{\frac{\partial M}{\partial q}}{I_{yy}} t} - 1 \right) \right]$$

and the displacement at the end of one second is:

$$\Theta/\text{inch}_{1\,\text{sec}} = 57.3 \left\{ \frac{\dfrac{\text{Cont. Pow./Inch}}{\text{Inertia}}}{\dfrac{\text{Damping}}{\text{Inertia}}} \left[1 + \frac{1}{\dfrac{\text{Damping}}{\text{Inertia}}} \left(e^{-\frac{\text{Damping}}{\text{Inertia}}} - 1 \right) \right] \right\} \text{deg/inch}$$

For illustration, the response and damping requirements of reference 9.6 for the example helicopter are superimposed on the envelopes of Figure 9.13.

Calculated points for the example helicopter are also plotted. These indicate that this aircraft would satisfy the instrument flight requirements, while perhaps not being optimum for the ILS approach task on which Figure 9.13 was based unless equipped with some stability augmentation equipment.

Note that the response curves at zero damping go through a point obtained from the simple equation from high school physics:

$$s = \tfrac{1}{2}at^2$$

or in this case:

$$\left(\frac{\text{Cont. Pow./Inch}}{\text{Inertia}} \right)_{(D/I=0)} = 2\,\frac{\Theta/\text{Inch}}{t^2}$$

For very high damping, the terms inside the bracket approach unity, and the line becomes asymptotic to the ray defined by:

$$\left(\frac{\text{Cont. Pow./Inch}}{\text{Inertia}} \right)_{(D/I \to \infty)} = \Theta/\text{Inch}\,\frac{\text{Damping}}{\text{Inertia}}$$

Rays also take on another meaning related to steady velocity, since the equation for angular rate is:

$$q = \frac{\dfrac{\partial M}{\partial B_1}B_1}{-\dfrac{\partial M}{\partial q}}\left[1 - e^{\left(\frac{\partial M}{\partial q}/I_n\right)t}\right] \text{ radians/sec}$$

As time increases, the rate takes on its steady, constant value, which plots as a ray from the origin.

$$q/\text{inch} = 57.3\,\frac{\dfrac{\text{Cont. Pow./Inch}}{\text{Inertia}}}{\dfrac{\text{Damping}}{\text{Inertia}}}\ \text{degrees/sec/inch}$$

The fact that the displacement requirement is similar to a final rate line and that the damping-to-inertia ratio is the inverse of the time constant allows the maps of Figure 9.13 to be approximated in another format as combinations of the final rate and the time constant. This alternative format is shown in Figure 9.14 as simplified approximations of the boundaries that are generally accepted today as the result of several flight test and simulator studies such as those reported in references 9.10 and 9.11. This format is useful in that flight test data in the form of time histories following step control inputs can yield the information required to judge the flying qualities directly.

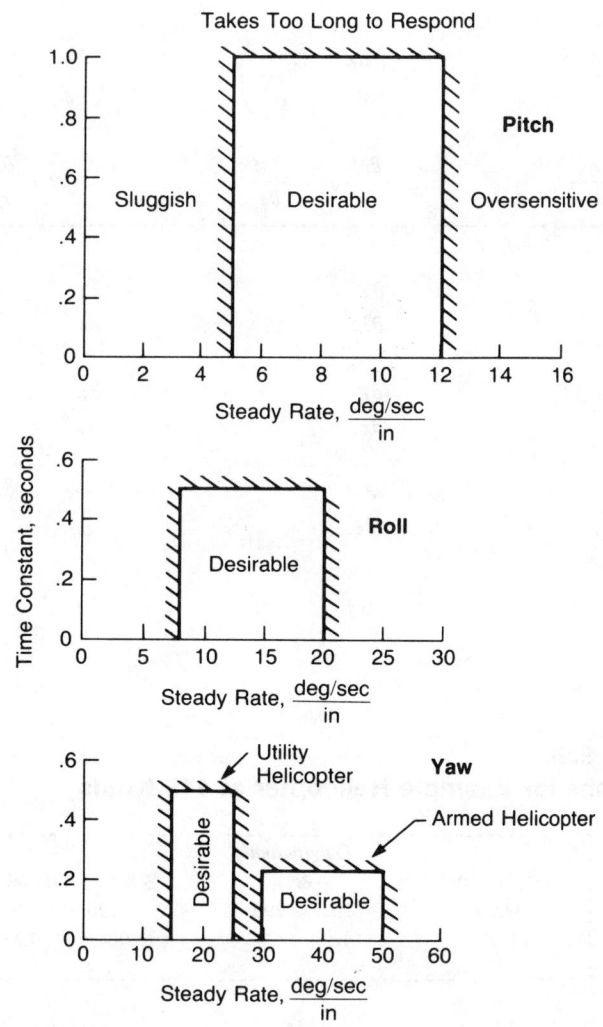

FIGURE 9.14 Alternate Presentation of Hover Maneuverability Requirements

TABLE 9.19
Equations of Motion

	X	Z	Θ	Y
	LONGITUDINAL			
X	$-\dfrac{\text{G.W.}}{g}s^2 + \dfrac{\partial X}{\partial \dot{x}}s$	$\dfrac{\partial X}{\partial \dot{z}}s$	$\left(\dfrac{\partial X}{\partial q} - \dfrac{\text{G.W.}}{g}\bar{V}\bar{\Theta}\right)s - \text{G.W.}$	$\dfrac{\partial X}{\partial \dot{y}}s$
Z	$\dfrac{\partial Z}{\partial \dot{x}}s$	$\left(\dfrac{\partial Z}{\partial \ddot{z}} - \dfrac{\text{G.W.}}{g}\right)s^2 + \dfrac{\partial Z}{\partial \dot{z}}s$	$\left(\dfrac{\partial Z}{\partial q} + \dfrac{\text{G.W.}}{g}\bar{V}\right)s$	$\dfrac{\partial Z}{\partial \dot{y}}s$
M	$\dfrac{\partial M}{\partial \dot{x}}s$	$\dfrac{\partial M}{\partial \ddot{z}}s^2 + \dfrac{\partial M}{\partial \dot{z}}s$	$-I_{yy}s^2 + \dfrac{\partial M}{\partial q}s$	$\dfrac{\partial M}{\partial \dot{y}}s$
Y	$\dfrac{\partial Y}{\partial \dot{x}}s$	$\dfrac{\partial Y}{\partial \dot{z}}s$	$\dfrac{\partial Y}{\partial q}s$	$-\dfrac{\text{G.W.}}{g}s^2 + \dfrac{\partial Y}{\partial \dot{y}}s$
R	$\dfrac{\partial R}{\partial \dot{x}}s$	$\dfrac{\partial R}{\partial \dot{z}}s$	$\dfrac{\partial R}{\partial q}s$	$\dfrac{\partial R}{\partial \dot{y}}s$
N	$\dfrac{\partial N}{\partial \dot{x}}s$	$\dfrac{\partial N}{\partial \dot{z}}s$	$\dfrac{\partial N}{\partial q}s$	$\dfrac{\partial N}{\partial \dot{y}}s$

TABLE 9.20
Equations for Example Helicopter at 115 Knots

	Longitudinal			Lateral-directional	
X	$-621s^2 - 20s$	$-8s$	$3,865s - 20,000$	$-9s$	$-659s$
Z	$43s$	$-621s^2 - 290s$	$120,273s$	0	0
M	$115s$	$14s^2 + 542s$	$-40,000s^2 - 50,281s$	$-15s$	$12,900s$
Y	$4s$	$-13s$	$-574s$	$-621s^2 - 107s$	$-3,775s + 20,000$
R	$30s$	$-263s$	$-11,418s$	$-382s$	$-5,000s^2 - 33,738s$
N	$130s$	$99s$	0	$1,017s$	$6,909s$

$$
\begin{array}{cc}
\Phi & \Psi
\end{array}
$$

$$
\begin{bmatrix}
\dfrac{\partial X}{\partial p}s & \dfrac{\partial X}{\partial r}s \\[2ex]
\dfrac{\partial Z}{\partial p}s & \dfrac{\partial Z}{\partial r}s \\[2ex]
\dfrac{\partial M}{\partial p}s & \dfrac{\partial M}{\partial r}s \\[2ex]
\left(\dfrac{\partial Y}{\partial p}+\dfrac{G.W.}{g}\bar{V}\bar{\Theta}\right)s + G.W.\left(\dfrac{\partial Y}{\partial r}-\dfrac{G.W.}{g}\bar{V}\right)s & \dfrac{\partial Y}{\partial r} \\[2ex]
-I_{xx}s^2+\dfrac{\partial R}{\partial p}s & \dfrac{\partial R}{\partial r}s \\[2ex]
\dfrac{\partial N}{\partial p}s & -I_{zz}s^2+\dfrac{\partial N}{\partial r}s
\end{bmatrix}
\begin{bmatrix}
x(s) \\[2ex] z(s) \\[2ex] \Theta(s) \\[2ex] y(s) \\[2ex] \Phi(s) \\[2ex] \Psi(s)
\end{bmatrix}
=
$$

LATERAL-DIRECTIONAL (enclosing the lower-left block of the left matrix)

Right-hand side:

$$
\begin{array}{cccc}
\theta_{0_M} & \theta_{0_T} & A_1 & B_1
\end{array}
$$

$$
\begin{bmatrix}
-\dfrac{\partial X}{\partial \theta_{0_M}} & -\dfrac{\partial X}{\partial \theta_{0_T}} & -\dfrac{\partial X}{\partial A_1} & -\dfrac{\partial X}{\partial B_1} \\[2ex]
-\dfrac{\partial Z}{\partial \theta_{0_M}} & -\dfrac{\partial Z}{\partial \theta_{0_T}} & -\dfrac{\partial Z}{\partial A_1} & -\dfrac{\partial Z}{\partial B_1} \\[2ex]
-\dfrac{\partial M}{\partial \theta_{0_M}} & -\dfrac{\partial M}{\partial \theta_{0_T}} & -\dfrac{\partial M}{\partial A_1} & -\dfrac{\partial M}{\partial B_1} \\[2ex]
-\dfrac{\partial Y}{\partial \theta_{0_M}} & -\dfrac{\partial Y}{\partial \theta_{0_T}} & -\dfrac{\partial Y}{\partial A_1} & -\dfrac{\partial Y}{\partial B_1} \\[2ex]
-\dfrac{\partial R}{\partial \theta_{0_M}} & -\dfrac{\partial R}{\partial \theta_{0_T}} & -\dfrac{\partial R}{\partial A_1} & -\dfrac{\partial R}{\partial B_1} \\[2ex]
-\dfrac{\partial N}{\partial \theta_{0_M}} & -\dfrac{\partial N}{\partial \theta_{0_T}} & -\dfrac{\partial N}{\partial A_1} & -\dfrac{\partial N}{\partial B_1}
\end{bmatrix}
\begin{bmatrix}
\theta_{0_M} \\[2ex] \theta_{0_T} \\[2ex] A_1 \\[2ex] B_1
\end{bmatrix}
$$

Numerical form:

Left column vector:

$$
\begin{bmatrix}
0 \\ -1{,}846s \\ 0 \\ -119{,}698s \\ 6{,}912s \\ -35{,}000s^2-90{,}419s
\end{bmatrix}
$$

$$
\begin{bmatrix}
x(s) \\ z(s) \\ \Theta(s) \\ y(s) \\ \Phi(s) \\ \Psi(s)
\end{bmatrix}
=
$$

	θ_{0_M}	θ_{0_T}	A_1	B_1	
$x(s)$	6,727	−12,845	1,506	−18,601	θ_{0_M}
$z(s)$	110,919	0	0	67,855	
$\Theta(s)$	−326,946	0	−29,480	364,158	θ_{0_T}
$y(s)$	5,668	−12,845	−16,638	−1,635	A_1
$\Phi(s)$	92,745	−77,070	−325,703	−32,014	
$\Psi(s)$	−376,161	475,265	0	0	B_1

FORWARD FLIGHT: THE COMBINED EQUATIONS OF MOTION

A fixed-wing aircraft is symmetrical, and in most flight conditions there is little coupling between its longitudinal degrees of freedom and its lateral-directional degrees of freedom. In the preceding discussion of a helicopter in hover, the same concept was used where it was assumed that there was no significant coupling. In forward flight, on the other hand, there are several obvious sources of cross-coupling, and it is not clear that they can be ignored. For a single-rotor shaft-driven helicopter, they include the yawing moment produced by main rotor torque as a function of both forward speed and rotor angle of attack; the pitching moment due to blade flapping during roll maneuvers and the rolling moment during pitch maneuvers; and the yawing moment caused by changes in tail rotor thrust during changes in forward speed. No similar sources of cross-coupling would be found on a fixed-wing aircraft. In the interest of rigorousness—if not of simplicity—the analysis will first be done on the combined equations of motion and then on the two uncoupled subsets.

The six equations of motion can be written in matrix form, as in Tables 9.19 and 9.20. The matrix has been so arranged that the longitudinal equations form a submatrix in the upper-left-hand corner while the lateral-directional equations are in the lower right. The other two corners represent the coupling between the primary submatrices. Table 9.20 gives the numerical matrices representing the example helicopter at 115 knots.

Expanding the left-hand determinant produces the coupled system's characteristic equation:

$$s^8 + 11.19s^7 + 38.70s^6 + 66.41s^5 + 38.18s^4 - 49.95s^3 - 6.90s^2 + 3.18s^2 + .341 = 0$$

In order of decreasing damping, the roots are:

$$-6.78, \ -2.39, \ -1.251, \ \pm 1.932i, \ -.253, \ -.096, \ .265, \ .564$$

The positive roots, of course, denote that the example helicopter is quite unstable—a discovery that should come as no surprise after the discussion in Chapter 8 of the inadequacy of its horizontal stabilizer area to give positive angle-of-attack stability. The roots in this form give no clue to which types of motion are stable and which are unstable. That information could be obtained from the equations with some available mathematical techniques, but for our purposes the same thing can be done by separately studying the longitudinal and the lateral-directional submatrices.

LONGITUDINAL EQUATIONS OF MOTION IN FORWARD FLIGHT

If only the longitudinal subset determinant is expanded for the example helicopter at 115 knots, the resultant characteristic equation is:

$$s^4 + 1.69s^3 - 2.00s^2 + .022s + .074 = 0$$

The Routh's discriminant is:

$$R.D. = -.29$$

According to the discriminant tests, the example helicopter is longitudinally unstable in this flight condition. The characteristic equation has four real roots, three of which match up well with three from the fully coupled solution:

Longitudinal subset (uncoupled):	$-2.49, -.180, .231, .752$
Full system (coupled):	$-2.39, -.253, .265, .564$

The positive roots produce a pure divergence, with the largest one governing and making the amplitude double in less than one second. As discussed earlier, the horizontal stabilizer of only 18 square feet on the example helicopter is not large enough. Many early helicopters had no stabilizers at all; and, although they could be flown by alert pilots in conditions giving them good cues, they were difficult to fly when the pilots were distracted or did not have a good view of the horizon. A flight test program using a variable-stability helicopter reported in reference 9.11 indicates that for flight on instruments, a time to double amplitude of less than about 8 seconds is unacceptable.

If this were an actual design program, the example helicopter would undoubtedly be given a bigger tail to put it on a more competitive footing with other modern designs. An alternative approach would involve an electronic auxiliary control system with various degrees of complexity to make up for the lack of inherent stability.

A guide to the resizing of the horizontal stabilizer can be generated as a *stability map* using two of the most important derivatives: one defining angle of attack stability, $\partial M / \partial \dot{z}$, and one defining speed stability, $\partial M / \partial \dot{x}$, as variables. The effect of combinations of these two derivatives on Routh's discriminant will define stable and unstable regions. The first step in preparing the stability map is to express the characteristic determinant as before, but leaving the two derivatives as variables:

$$D = \begin{vmatrix} -621s^2 - 20s & -8s & 3{,}865s - 20{,}000 \\ 43s & -621s^2 - 290s & 120{,}273s \\ \frac{\partial M}{\partial \dot{x}} s & 14s^2 + \frac{\partial M}{\partial \dot{z}} s & -40{,}000s^2 - 50{,}281s \end{vmatrix}$$

The resulting characteristic equation is:

$$s^4 + 1.69s^3 + \left(.65 - .0048\frac{\partial M}{\partial \dot{z}} - .00016\frac{\partial M}{\partial \dot{x}} \right)s^2$$

$$+ \left(.0227 - .00018\frac{\partial M}{\partial \dot{z}} + .00079\frac{\partial M}{\partial \dot{x}} \right)s$$

$$+ \left(.000056\frac{\partial M}{\partial \dot{z}} + .000372\frac{\partial M}{\partial \dot{x}} \right) = 0$$

and the corresponding equation for Routh's discriminant is:

$$\text{R.D.} = 10^{-6}\left[24{,}421 - 534\frac{\partial M}{\partial \dot{z}} - 220\frac{\partial M}{\partial \dot{x}} + 1.43\left(\frac{\partial M}{\partial \dot{z}}\right)^2 \right.$$

$$\left. - 6.08\left(\frac{\partial M}{\partial \dot{z}}\right)\left(\frac{\partial M}{\partial \dot{x}}\right) - .837\left(\frac{\partial M}{\partial \dot{x}}\right)^2 \right]$$

The loci of the two derivatives that make the discriminant vanish is the boundary between positive and negative stability. This is shown in Figure 9.15 along with the combinations that make the coefficient of the constant term, E, equal to zero. This defines the boundary between stable oscillations and unstable divergences. It is where:

$$\frac{\partial M}{\partial \dot{z}} = \frac{\partial M}{\partial \dot{x}}\frac{\frac{\partial Z}{\partial \dot{z}}}{\frac{\partial Z}{\partial \dot{x}}}$$

Also shown in Figure 9.15 is a boundary in the right unstable region between oscillations and divergences. This was determined by finding combinations of the two derivatives that made the roots of the characteristic equation switch from complex to real.

Note that the type of stability map of Figure 9.15 is unique to helicopters because for airplanes in trimmed level flight, the speed stability derivative is essentially zero. (Envision an airplane model in a wind tunnel with the elevator angle adjusted to make the model have no pitching moment. Then, unless compressibility is a factor, the moment will remain zero as the tunnel speed is changed. This is not true for a helicopter model whose rotor tip path plane will tilt as the tunnel speed is changed.)

One of the uses of the stability map is to predict the effect of increasing the area of the horizontal stabilizer. In this case, it may be seen that doubling the area would improve the longitudinal flying qualities by moving the example helicopter from a region of pure divergences to one of unstable oscillations, and that tripling the area would stabilize the aircraft. Stabilizer incidence can also be used to move the point on the map since it changes the speed stability parameter, $\partial M/\partial \dot{x}$, increasing it as incidence is decreased. It may be seen that the minimum increase in stabilizer area to achieve stability would involve increasing the incidence to take advantage of the corner of stability near the origin. In practice, of course, approaching the lower divergence boundary would introduce a risk of going unstable. Another consideration would be the possible problems of high oscillatory

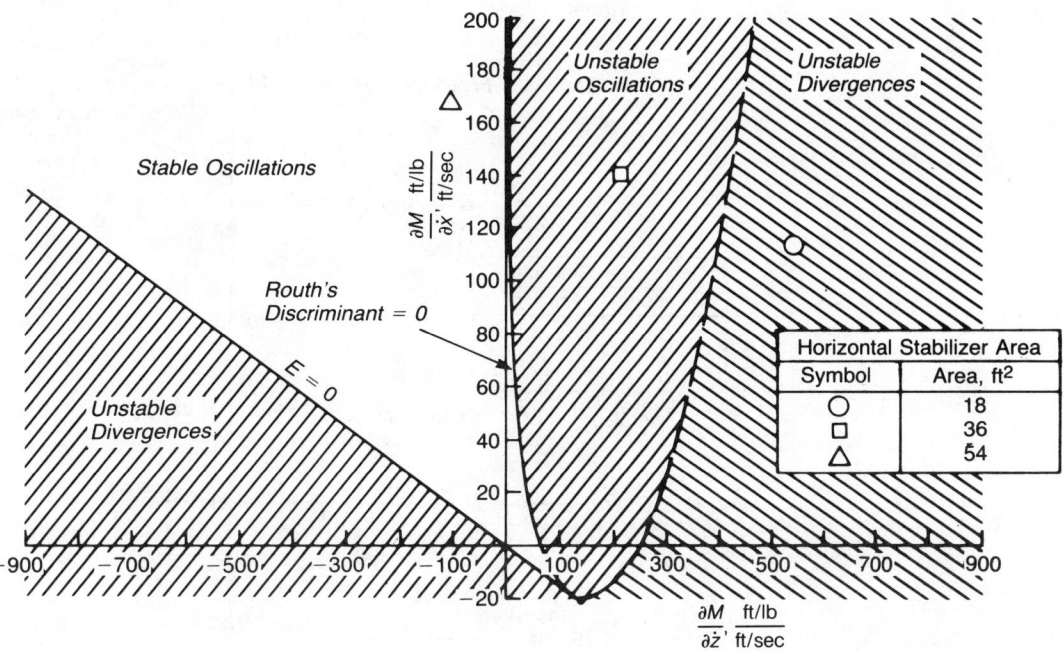

FIGURE 9.15   Longitudinal Stability Map for Example Helicopter at 115 Knots

blade loads if the big download on the stabilizer required excessive nose-down flapping to balance the helicopter.

Changing the size of the horizontal stabilizer will change many other derivatives in addition to the two on the stability map. The full effect is shown in a different format in the top portion of Figure 9.16 as the locus of roots of the characteristic equation as the stabilizer area is increased. Roots that have no imaginary components represent either pure divergences or convergences, and roots with imaginary components represent oscillations—unstable if they are in the right-hand plane.

Both helicopters and airplanes with enough stabilizer area to give positive angle-of-attack stability will exhibit oscillations in forward flight. The oscillation

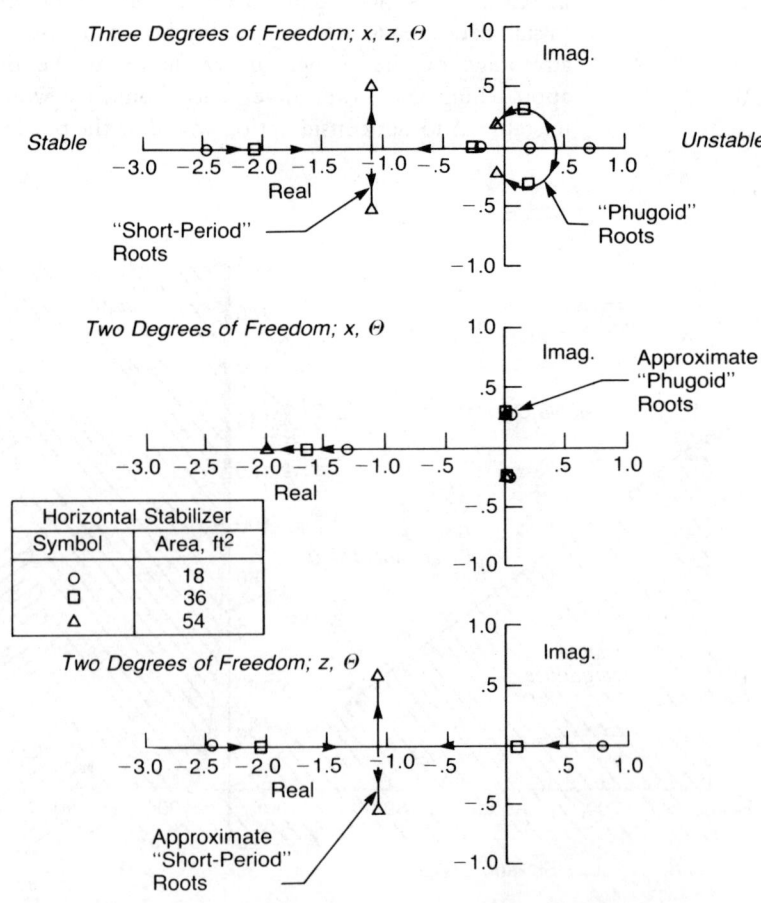

FIGURE 9.16 Root Locus Plots as Horizontal Stabilizer Area Is Increased

typically has a period of 10 to 30 seconds and primarily involves an interchange between forward speed and altitude—that is, between kinetic and potential energy at a nearly constant angle of attack.

This oscillation was first observed by W. F. Lanchester, a pioneer British aerodynamicist working with model gliders at about the same time that the Wrights were doing their first testing. Lanchester, in naming the motion, chose *phugoid*, based on a Greek verb that he thought meant "to fly." Actually the verb means "to flee," but we have happily used the word ever since.

Figure 9.17 shows the flight path of a helicopter following a brief encounter with a sharp-edged gust. The controls are held fixed so that the aircraft can demonstrate its inherent characteristics. It first shows its short-period response, which disappears rapidly because it is well damped in this example. The helicopter then goes into its phugoid motion, shown as slightly unstable in this illustration.

The assumption that the analysis can be based on the uncoupled equations has been shown in reference 9.12 to overestimate slightly the damping of the phugoid mode. This can be traced primarily to the omission of the coupling that gives pitching moments as a function of roll rate represented by the derivative, $\partial M/\partial p$. During an uncontrolled phugoid in flight, the helicopter will have a rolling motion phased with the pitching motion in such a way that the damping of the system is slightly reduced.

The imaginary component of the root is the frequency of the oscillation in radians per second. With a stabilizer area of 36 square feet, the example helicopter has a frequency of 0.31 radians per second or a period of just over 20 seconds. Even though this point is unstable, doubling in amplitude in about 4 seconds, it would still be considered satisfactory for visual flight but not for instrument flight. The acceptability of oscillations in the two flight regimes is indicated by the specifications found in paragraphs 3.2.11 and 3.6.1.2 of reference 9.6. In brief, they are as given in Table 9.21.

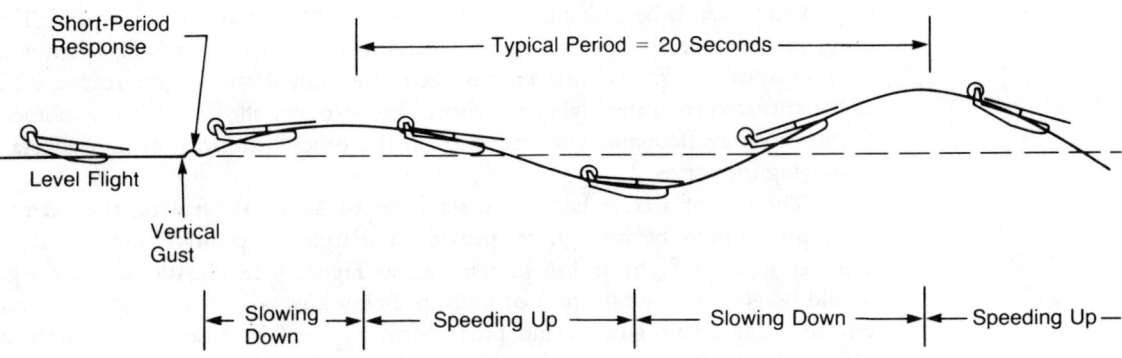

FIGURE 9.17 Longitudinal Motions

TABLE 9.21
Summary of MIL-H-8501A Stability Requirements

Period	Damping Requirement	
	Visual Flight	*Instrument Flight*
<5 sec	$\frac{1}{2}$ amplitude in 2 cycles	$\frac{1}{2}$ amplitude in 1 cycle
5–10 sec	At least lightly damped	$\frac{1}{2}$ amplitude in 2 cycles
10–20 sec	Not double in 10 sec	At least lightly damped
>20 sec	No requirement	Not double in 20 sec

The justification for not imposing a requirement in visual flight for periods above 20 seconds is that the time is so long that the pilot instinctively corrects for any instability with his normal control motions.

The 36-square-foot stabilizer on the example helicopter would allow the aircraft to satisfy the visual flight requirement, but a somewhat larger tail (or an auxiliary stability system) would be needed for instrument flight where the cues are not as good and the pilot has other duties that require his attention.

In an actual design project, of course, the empennage parameters should be selected to give good flying qualities not only at one flight condition but throughout the entire flight envelope as well. A full analysis of the example helicopter then would involve more than just the 115 knots at sea level that has been chosen for illustration. It is true, however, that a stabilizer area chosen to satisfy the requirements at one forward speed would not be dramatically different from the one that satisfies them at any other speed. This is because even though the upsetting effect on angle-of-attack stability of the rotor flapping is proportional to tip speed ratio squared—as can be seen from the equations for longitudinal flapping in Chapter 3—the correcting effect of the horizontal stabilizer is proportional to velocity squared, thus maintaining an overall balance for the entire helicopter.

It should be recognized, however, that a helicopter that is stable in level flight will probably be unstable at some higher load factor at the same speed. This is in contrast to an airplane, whose angle-of-attack stability is nearly invarient with wing angle of attack. The difference is due to the contribution of the rearward tilt of the thrust vector; the higher the thrust, the stronger the destabilizing moment due to nose-up flapping. The airframe, on the other hand, maintains a constant stabilizing influence.

The use of a large horizontal stabilizer of 54 square feet on the example helicopter would be enough to provide a margin of positive angle-of-attack stability in level flight at 115 knots; but, as Figure 9.18 illustrates, the margin would be gone in a 1.44-g turn or pull-up. Below this point, the helicopter, upon encountering an up-gust, would pitch down by itself because of the stabilizing effect of the large horizontal stabilizer. Above this point, however, the rotor would overpower the stabilizer and would pitch the helicopter nose-up unless prevented by an alert pilot.

Approximate Solutions for the Phugoid Mode

The observation that the phugoid involves little change in angle of attack allows its analysis to be made by eliminating the Z equation out of the set of the three longitudinal equations of motion. Retaining only the speed and pitch equations gives:

$$
\begin{vmatrix}
-\dfrac{\text{G.W.}}{g}\,s^2 + \dfrac{\partial X}{\partial \dot{x}}\,s & \left(\dfrac{\partial X}{\partial q} - \dfrac{\text{G.W.}}{g}\,\bar{V}\bar{\Theta}\right)s - \text{G.W.} \\[4ex]
\dfrac{\partial M}{\partial \dot{x}}\,s & -I_{yy}s^2 + \dfrac{\partial M}{\partial q}\,s
\end{vmatrix}
=
\begin{vmatrix}
x(s) \\[4ex]
\Theta(s)
\end{vmatrix}
$$

and the phugoid's characteristic equation may be written as:

$$
I_{yy}s^3 - \left(\frac{I_{yy}}{\text{G.W.}/g}\,\frac{\partial X}{\partial \dot{x}} + \frac{\partial M}{\partial q}\right)s^2 + \frac{1}{\text{G.W.}/g}\left[\frac{\partial X}{\partial \dot{x}}\,\frac{\partial M}{\partial q}\right.
$$

$$
\left. + \left(\frac{\partial X}{\partial q} - \frac{\text{G.W.}}{g}\,\bar{V}\bar{\Theta}\right)\frac{\partial M}{\partial \dot{x}}\right]s + g\,\frac{\partial M}{\partial \dot{x}} = 0
$$

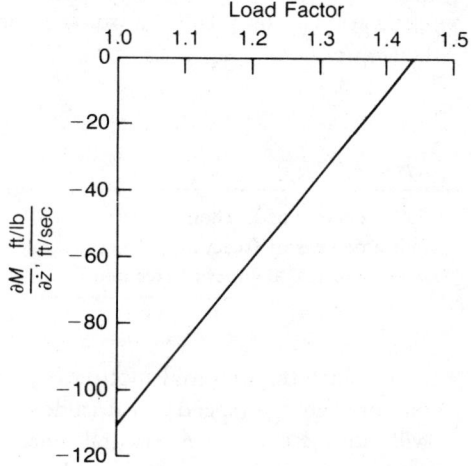

FIGURE 9.18 Angle of Attack Stability Derivative as Affected by Load Factor

For the example helicopter with the three different-sized horizontal stabilizers, the root loci of this mode are given in the second portion of Figure 9.16. It may be seen that although the phugoid period is approximately correct, the simplification has sacrificed reasonableness for the phugoid damping.

An even simpler way of calculating the phugoid frequency can be obtained by applying the same assumption used in hover—that since the pitching motion is occurring about a virtual center far away from the flight path, the aircraft's moment of inertia about its center of gravity can be ignored. This reduces the characteristic equation to:

$$s^2 - \frac{1}{\text{G.W.}/g}\left[\frac{\partial X}{\partial \dot{x}} + \left(\frac{\partial X}{\partial q} - \frac{\text{G.W.}}{g}\bar{V}\bar{\Theta}\right)\frac{\dfrac{\partial M}{\partial \dot{x}}}{\dfrac{\partial M}{\partial q}}\right]s - g\frac{\dfrac{\partial M}{\partial \dot{x}}}{\dfrac{\partial M}{\partial q}} = 0$$

where the natural frequency is approximately:

$$\omega_{\text{nat}} = \sqrt{-g\frac{\dfrac{\partial M}{\partial \dot{x}}}{\dfrac{\partial M}{\partial q}}} \quad \text{rad/sec}$$

which is the same equation as that derived for hover and gives results almost identical to those produced either by the full three-degree-of-freedom system or the two-degree-of-freedom system in which the moment of inertia is retained. For illustration, using the 36 square foot horizontal stabilizer on the example helicopter:

Equation	Calculated Natural Frequency, rad/sec	Period, sec
Full 3 degrees of freedom	0.31	20
Full 2 degrees of freedom	0.26	24
Approximate 2 degrees of freedom	0.27	23

Since the phugoid motion is primarily an interchange between kinetic and potential energy (speed and altitude), anything that dissipates energy in the process will add damping. A natural energy dissipator is parasite drag. Thus the aerodynamicist is faced with a dilemma because anything he does to clean up the aircraft will result in the phugoid having less damping.

The Short-Period Motion

Some of the roots in the root locus plot of the top portion of Figure 9.16 represent the *short-period mode*. As its name implies, the time associated with this mode is so short that it can be assumed that no speed change occurs while it is being excited. This allows us to reduce the three-degree-of-freedom analysis to two degrees of freedom for investigation of this mode. The equations in matrix form are:

$$\begin{vmatrix} \left[\left(\dfrac{\partial Z}{\partial \ddot{z}} - \dfrac{\text{G.W.}}{g}\right)s^2 + \dfrac{\partial Z}{\partial \dot{z}}\,s\right] & \left[\dfrac{\partial Z}{\partial q} + \dfrac{\text{G.W.}}{g}\,\bar{V}\right]s \\[4mm] \left[\dfrac{\partial M}{\partial \ddot{z}}\,s^2 + \dfrac{\partial M}{\partial \dot{z}}\,s\right] & \left[-I_{yy}s^2 + \dfrac{\partial M}{\partial q}\,s\right] \end{vmatrix} = \begin{vmatrix} z(s) \\[4mm] \Theta(s) \end{vmatrix}$$

And the characteristic equation of the short period mode is:

$$s^2 + \left[\frac{-I_{yy}\dfrac{\partial Z}{\partial \dot{z}} + \dfrac{\partial M}{\partial q}\left(\dfrac{\partial Z}{\partial \dot{z}} - \dfrac{\text{G.W.}}{g}\right) - \left(\dfrac{\partial Z}{\partial q} + \dfrac{\text{G.W.}}{g}\,\bar{V}\right)\dfrac{\partial M}{\partial \ddot{z}}}{I_{yy}\left(\dfrac{\text{G.W.}}{g} - \dfrac{\partial Z}{\partial \ddot{z}}\right)}\right]s$$

$$+ \left[\frac{\dfrac{\partial Z}{\partial \dot{z}}\dfrac{\partial M}{\partial q} - \left(\dfrac{\partial Z}{\partial q} + \dfrac{\text{G.W.}}{g}\,\bar{V}\right)\dfrac{\partial M}{\partial \dot{z}}}{I_{yy}\left(\dfrac{\text{G.W.}}{g} - \dfrac{\partial Z}{\partial \ddot{z}}\right)}\right] = 0$$

The bottom portion of Figure 9.16 shows the root locus of the short period as the stabilizer area is increased. It will be seen that these roots are very similar to the corresponding roots of the three-degree-of-freedom system plotted above them. A stability map for the short-period mode based on the two derivatives representing angle-of-attack stability, $\partial M/\partial \dot{z}$, and damping in pitch, $\partial M/\partial q$, is given in Figure 9.19. Damping in pitch is not dramatically increased (negative sign on $\partial M/\partial q$) by increasing stabilizer area, but it can be increased significantly by using a rate gyro that commands changes in main rotor cyclic pitch or in horizontal stabilizer incidence. Both these methods are used on modern military helicopters such as the Sikorsky UH-60 and the Hughes AH-64.

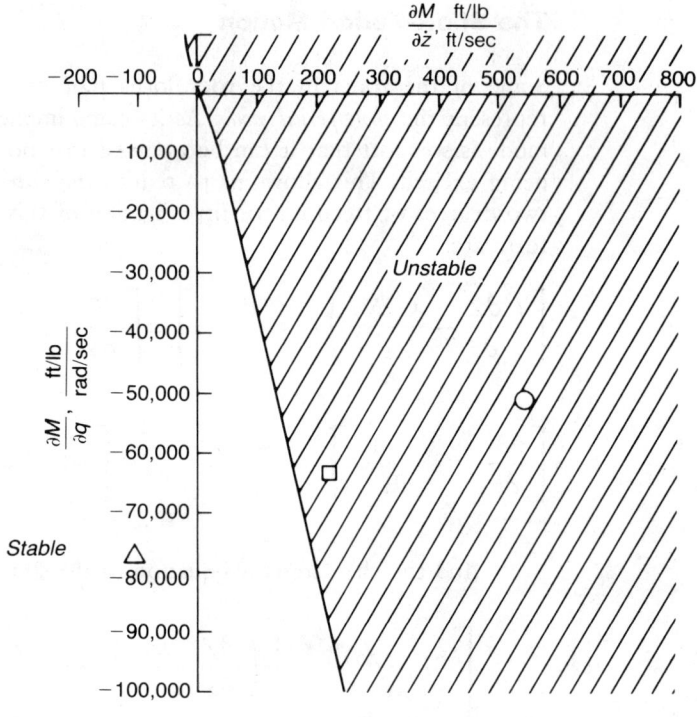

FIGURE 9.19   Short-Period Stability Map for Example Helicopter at 115 Knots

The characteristics of the short-period motion are addressed in paragraph 3.3.11.1 of reference 9.6. This requires that following an aft longitudinal control step, the time histories of normal acceleration and of pitch rate shall become concave downward within 2 seconds. A method for studying whether a given helicopter does or does not satisfy this requirement was first presented in reference 9.13. That analysis resulted in the criterion shown on Figure 9.20, which is similar to the stability map of Figure 9.19, and for the example helicopter at least would result in choosing about the same size stabilizer.

Yet another stabilizer-sizing study was reported in reference 9.14 as part of the development of the Boeing Vertol YUH-61. Here again the criterion is based on the short-period mode. The study made use of a ground-based simulator in which several pilots evaluated the acceptability of the longitudinal handling qualities as the last, or *spring*, term in the short-period characteristic equation was varied. This term is approximately:

$$\text{Short-period stability parameter} = \left(\frac{\frac{\partial Z}{\partial \dot{z}}}{\text{G.W.}/g} \right) \left(\frac{\frac{\partial M}{\partial q}}{I_{yy}} \right) - \bar{V} \frac{\frac{\partial M}{\partial \dot{z}}}{I_{yy}}$$

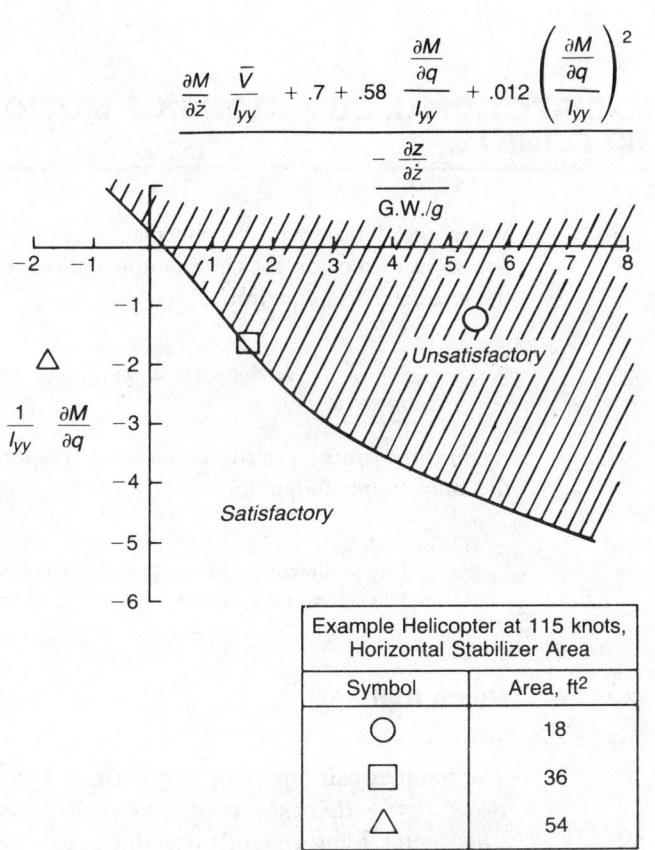

FIGURE 9.20   Parameters Determining Compliance with Two-Second Concave Downward Requirement of MIL-H-8501A

The airplane aerodynamicist would call this parameter the *maneuvering margin*. When it is positive, the aircraft will respond to a control input by going to a new steady-state flight condition; but when it is negative, there is no equilibrium condition that will satisfy it.

The simulator pilots were asked to assign a *Handling Qualities Rating* (sometimes referred to as a Cooper-Harper pilot rating, for its originators) for various values of the parameter. This rating system goes from 1, or "perfect," to 10, "completely unacceptable." A rating of 3.5 is considered to be the boundary between "acceptable as is" and "should be fixed." Figure 9.21 shows the results of the simulation study. Also shown are the three points representing the three horizontal stabilizer areas on the example helicopter. It may be seen that this criterion is more demanding than the others considered up to now, since in this case even the largest area is not quite acceptable.

LATERAL-DIRECTIONAL EQUATIONS OF MOTION IN FORWARD FLIGHT

When the determinant of the matrix subset representing the lateral-directional equations of motion for the example helicopter at 115 knots is expanded, it yields the characteristic equation:

$$s^4 + 9.50s^3 + 23.90s^2 + 39.35s + 5.06 = 0$$

Again three roots from this equation correspond closely to three of the roots from the fully coupled equation:

Lateral-directional subset (uncoupled): $-6.83, -1.266 \pm 1.924i, -.138$
Full system (coupled): $-6.78, -1.251 \pm 1.932i, -.096$

Dutch Roll

The complex pair represent an oscillation known as *Dutch roll* after the motion that two skaters with locked arms make as they travel down the canal. The rear view of a helicopter doing a slightly unstable Dutch roll is given in Figure 9.22 (page 630). In the case of the example helicopter, the calculated Dutch roll motion has a period of 3.3 seconds and is well damped. Many fixed-wing aircraft have unstable Dutch roll characteristics—especially at high altitude. You can observe how the autopilot is controlling this mode on a jet transport by watching the motion of the inboard ailerons, which will be going up and down in a more or less regular manner every 2 to 4 seconds.

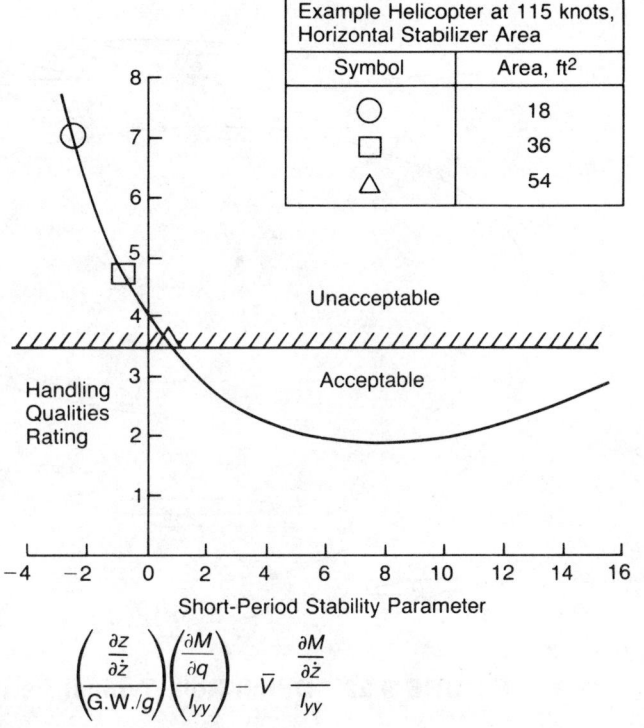

Example Helicopter at 115 knots, Horizontal Stabilizer Area	
Symbol	Area, ft^2
◯	18
▢	36
△	54

$$\left(\dfrac{\frac{\partial z}{\partial \dot z}}{\text{G.W.}/g}\right)\left(\dfrac{\frac{\partial M}{\partial q}}{I_{yy}}\right) - \bar V \; \dfrac{\frac{\partial M}{\partial \dot z}}{I_{yy}}$$

FIGURE 9.21   Boeing-Vertol Criteria for Stabilizer Sizing

Source: Blake & Alansky, "Stability and Control of the YUH-61A," *JAHS* 22-1, 1977.

Comparison of the roots from the uncoupled and the coupled systems for the example helicopter at 115 knots show little effect of the coupling on the Dutch roll mode. This is not to be taken as a general rule, however. In many cases, the solution of the fully coupled equations will show significantly different damping than the uncoupled subset. This can be traced primarily to the effect of angle of attack on rotor torque, as represented by the derivative, $\partial N/\partial \dot z$. The typical phasing of the relative motions is shown in Figure 9.22 with the helicopter pitching up as it yaws to the right. At low forward speed, rotor torque decreases with angle of attack, producing a positive damping effect; but at high speed, torque increases, giving negative yaw damping. The sign and magnitude of $\partial N/\partial \dot z$ can be obtained from the rotor performance charts in Chapter 3, specifically the charts of C_Q/σ versus C_T/σ for different values of collective pitch. At the trim values of μ, C_T/σ, and θ_0, an increase in C_T/σ caused by an increase in angle of attack (related to λ' in the next chart of the pair) will either have a negative slope, a positive slope, or be almost flat as it is for the example helicopter for $\mu = .3$, $C_T/\sigma = .085$, and $\theta_{0_{.75}} = 13.5°$. At higher speeds the collective pitch would be higher and the coupling would be more powerful, leading to a reduction in Dutch roll damping.

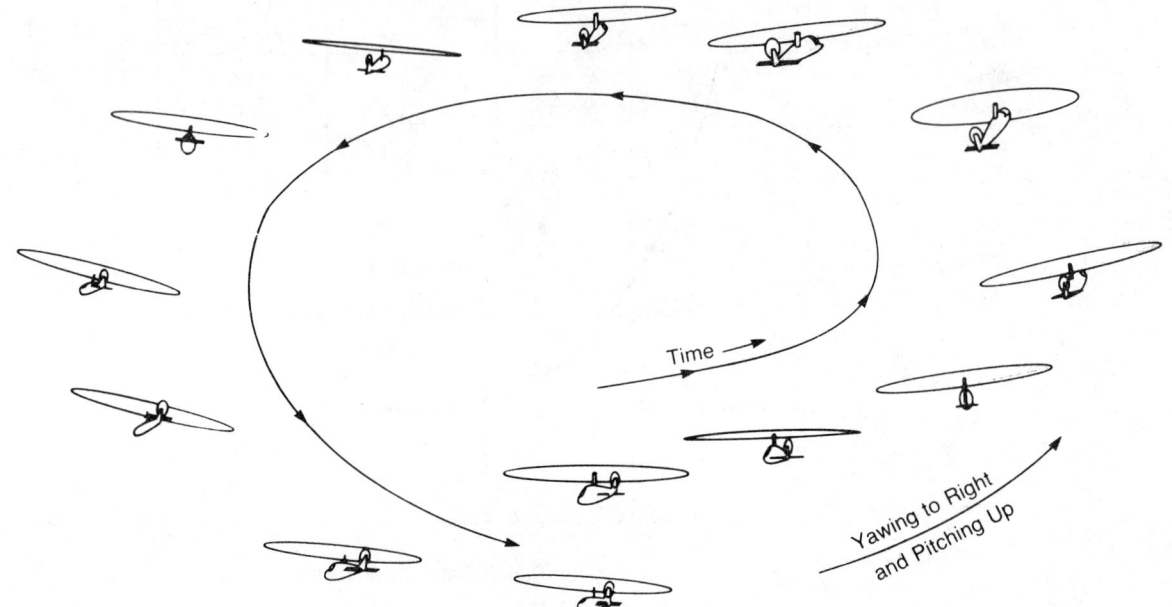

FIGURE 9.22   Dutch Roll, Typical Period, 3 seconds

Approximate equations for the Dutch roll roots can be derived by making some simplifying assumptions as outlined in reference 9.15 when discussing the Dutch roll characteristics of airplanes. The first assumption is that the aircraft is allowed to roll and yaw, but its center of gravity is constrained to follow a straight flight path. This is the same as eliminating all the side force contributions from aerodynamics and roll angle while retaining only the inertial terms in the Y equation.

$$Y, -\frac{\text{G.W.}}{g}\ddot{y} - \frac{\text{G.W.}}{g}\bar{V}r = 0$$

The lateral-directional determinant can thus be written in a simplified form as:

$$D = \begin{vmatrix} s & 0 & \bar{V} \\[2ex] \dfrac{\partial R}{\partial \dot{y}} & -I_{xx}s + \dfrac{\partial R}{\partial p} & \dfrac{\partial R}{\partial r} \\[2ex] \dfrac{\partial N}{\partial \dot{y}} & \dfrac{\partial N}{\partial p} & -I_{zz}s + \dfrac{\partial N}{\partial r} \end{vmatrix}$$

The characteristic equation is:

$$s^3 - \left(\frac{\dfrac{\partial N}{\partial r}}{I_{zz}} + \frac{\dfrac{\partial R}{\partial p}}{I_{xx}} \right) s^2 + \left(\frac{\dfrac{\partial R}{\partial p} \dfrac{\partial N}{\partial r} - \dfrac{\partial R}{\partial r} \dfrac{\partial N}{\partial p}}{I_{xx} I_{zz}} + \frac{\bar{V}}{I_{zz}} \frac{\partial N}{\partial \dot{y}} \right) s$$

$$+ \frac{\bar{V}}{I_{xx} I_{zz}} \left(\frac{\partial R}{\partial \dot{y}} \frac{\partial N}{\partial p} - \frac{\partial R}{\partial p} \frac{\partial N}{\partial \dot{y}} \right) = 0$$

The next assumption is that in the s^2 coefficient:

$$\frac{\dfrac{\partial N}{\partial r}}{I_{zz}} \ll \frac{\dfrac{\partial R}{\partial p}}{I_{xx}}$$

The final assumption is known as the Bairstow approximation, which says that for lightly damped systems represented by the cubic:

$$c_3 s^3 + c_2 s^2 + c_1 s + c_0 = 0$$

that

$$c_2 s^2 + c_0 \doteq 0, \quad \therefore s^2 = -\frac{c_0}{c_2}$$

Using this, the cubic becomes a quadratic:

$$c_2 s^2 + \left(c_1 - c_3 \frac{c_0}{c_2} \right) s + c_0 = 0$$

Incorporating all of these assumptions, the characteristic equation becomes:

$$s^2 - \frac{1}{I_{zz}} \left[\frac{\partial N}{\partial r} - \frac{\dfrac{\partial R}{\partial r} \dfrac{\partial N}{\partial p}}{\dfrac{\partial R}{\partial p}} + \bar{V} I_{xx} \frac{\dfrac{\partial R}{\partial \dot{y}} \dfrac{\partial N}{\partial p}}{\left(\dfrac{\partial R}{\partial p} \right)^2} \right] s$$

$$+ \frac{\bar{V}}{I_{zz}} \left[\frac{\partial N}{\partial \dot{y}} - \frac{\dfrac{\partial R}{\partial \dot{y}} \dfrac{\partial N}{\partial p}}{\dfrac{\partial R}{\partial p}} \right] = 0$$

For the example helicopter, this is:

$$s^2 + 2.61s + 5.23 = 0$$

and the two Dutch roll roots are:

$$s = -1.304 \pm 1.879i$$

which are essentially the same as obtained from the more complete equations. As a matter of fact, the damping and spring terms in this example are primarily due to the damping in yaw derivative, $\partial N/\partial r$, and the directional stability derivative, $\partial N/\partial \dot{y}$. Since this is probably true for most single-rotor helicopters, an approximation to the characteristic equation of the Dutch roll mode can be written:

$$s^2 - \frac{1}{I_{zz}} \frac{\partial N}{\partial r} s + \frac{\bar{V}}{I_{zz}} \frac{\partial N}{\partial \dot{y}} = 0$$

The roots for the example helicopter are:

$$s = -1.292 \pm 2.00i$$

Thus again little accuracy has been lost in the simplification. The period of the Dutch roll is approximately 3 seconds, which is fairly typical of both helicopters and airplanes of all sizes.

The success of the analysis of the Dutch Roll characteristics using either approximate or more "exact" methods is somewhat compromised by our poor understanding of the flow conditions in which the empennage and tail rotor actually operate. When evaluating the stability derivatives affected by these components, by neccessity we must assume a flow pattern that does not change drastically with small changes in flight conditions. A clue that this is not a good conclusion was illustrated in Figure 8.21. The measured flow distortion behind the rotor and engine installation of the Hughes AH-64 (and presumably, most other helicopters) can only be described as chaotic.

On the AH-64, small changes to the empennage had unexpectedly large effects on the damping of the Dutch roll mode with the stability augmentation system (SAS) turned off. (Turning the SAS on increased the damping in yaw by a factor of 3 and produced good damping even with the worst configuration.) A similar result is reported in reference 9.16, in which the Dutch roll was found to have much more positive damping in a right sideslip than in a left.

Spiral Stability

The roots of the lateral-directional equations for the example helicopter included a small negative real root that represented a time to half in amplitude of about 7 seconds. Reference 9.11 reports that if the time to double amplitude is less than 8 seconds, the helicopter is not satisfactory for flight on instruments.

The time history of the spiral mode is either a non-oscillatory convergence or a divergence. The sign of the constant term in the characteristic equation of the lateral-directional determinant determines which. The equation for E is:

$$E = \frac{g}{I_{xx}I_{zz}} \left[\frac{\partial R}{\partial \dot{y}} \frac{\partial N}{\partial r} - \frac{\partial N}{\partial \dot{y}} \frac{\partial R}{\partial r} \right]$$

The spiral mode will be unstable if E is negative. A variety of conditions could cause this to happen. The signs of the various derivatives should be noted:

Derivative	Name	Normal Sign	Value for Example Helicopter	Products
$\dfrac{\partial R}{\partial \dot{y}}$	Dihedral effect	−	−382 ft-lb/ft/sec	
$\dfrac{\partial N}{\partial r}$	Yaw damping	−	−90,419 ft-lb/rad/sec	34.54×10^6
$\dfrac{\partial N}{\partial \dot{y}}$	Directional stability	+	1,017 ft-lb/ft/sec	
$\dfrac{\partial R}{\partial r}$	Roll due to yaw rate	+	6,912 ft-lb/rad/sec	7.03×10^6
			$\dfrac{\partial R}{\partial \dot{y}} \dfrac{\partial N}{\partial r} - \dfrac{\partial N}{\partial \dot{y}} \dfrac{\partial R}{\partial r} = 27.51 \times 10^6$	

Stability Map

The airplane aerodynamicist illustrates the way important derivatives affect the spiral stability and the Dutch roll modes by plotting lines representing zero values for Routh's discriminant and the constant term, E, in the characteristic equation on a stability map on which the two axes represent the directional stability derivative, $\partial N/\partial \dot{y}$, and the dihedral derivative, $\partial R/\partial \dot{y}$. For airplanes, these two derivatives are significant since they can easily be modified during the design by changing the area of the vertical stabilizer and the wing dihedral. For helicopters, the direct relationships between these derivatives and easily changed geometric parameters are not so straightforward, especially in the case of the dihedral effect. As can be seen from the tabulation of the stability derivatives for the example helicopter,

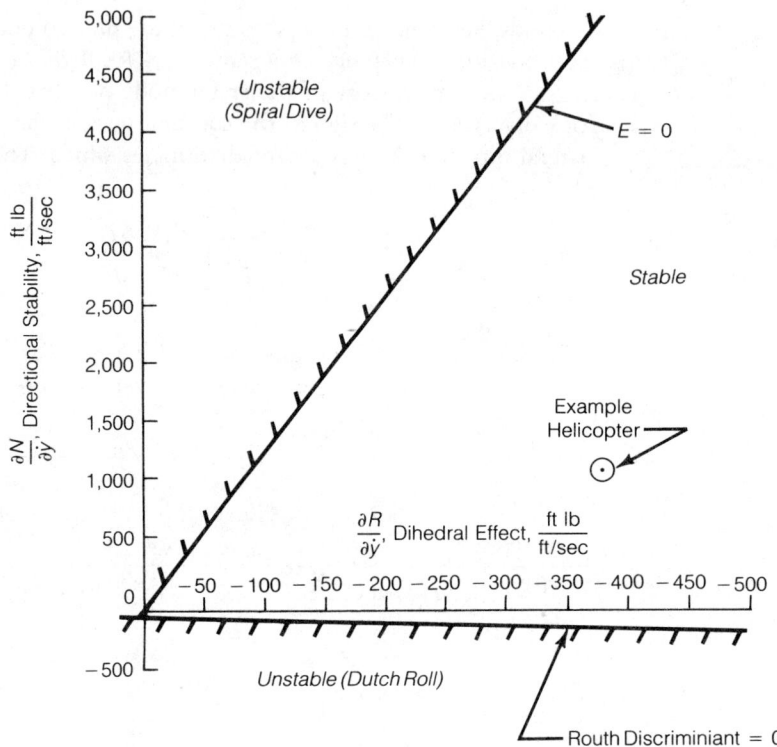

FIGURE 9.23   Lateral-Directional Stability Map for Example Helicopter at 115 Knots

contributions to $\partial R/\partial \dot{y}$ come from the main rotor, the tail rotor, the vertical stabilizer, and the fuselage (and from the horizontal stabilizer if it has dihedral.) The contributions are of nearly the same magnitude but of different signs, and none of them approach the power of the dihedral effect that can be obtained relatively easily with a wing. For this reason, the helicopter aerodynamicist is much more likely to use changes in the damping derivatives obtained by stability augmentation systems to improve the lateral-directional flying qualities.

Despite this qualification, the stability map gives a graphic illustration of the effects of the two dominant airplane-type derivatives as shown in Figure 9.23 for the example helicopter at 115 knots. In this case, the critical boundary is that associated with the constant term of the characteristic equation, E, being zero.

As one who flew model airplanes in his youth, I can attest to the fact that an effective way to cure spiral dives on hand-launched gliders is to reduce the directional stability by whittling away some of the area of the vertical stabilizer. A fix to do the same thing can be seen on the Bell 212 of Figure 9.24, which was

FIGURE 9.24   A Bell-212 Equipped with a Vertical Destabilizer

Source: Couresy *Rotor & Wing International*.

certificated for instrument flight by the FAA. On this aircraft, a vertical destabilizer is installed ahead of the center of gravity to reduce the directional stability. A more common fix is to install a auxiliary yaw damper either as an independent unit or as part of the stability augmentation system (SAS). Increased yaw damping can be used to cure either unstable Dutch roll or spiral dive.

EXAMPLE HELICOPTER CALCULATIONS

HOW TO'S

The following items can be evaluated by the methods of this chapter.

CHAPTER 10

Preliminary Design

INTRODUCTION

The preliminary design of a new helicopter is a team effort between the designer, the aerodynamicist, and the weight engineer, with help from other specialists. The effort progresses in cycles of iteration, at the end of which the design converges into its final form, leaving each member of the design team more or less satisfied.

The effort starts with a set of requirements established by the potential customer, by a marketing survey, or by some other means. The requirements that usually have the most influence on the design are:

1. Payload.
2. Range or endurance.
3. Critical hover or vertical climb condition.
4. Maximum speed.
5. Maximum maneuver load factor.

There are always design constraints, either formally stated or understood, that limit the design alternatives in some manner. Some of the most common involve:

1. Compliance with applicable safety standards.
2. Maximum disc loading.

3. Choice of engine from a list of approved engines.
4. Maximum physical size.
5. Maximum noise level.
6. Minimum one-engine-out performance.
7. Minimum autorotative landing capability.

The primary objective of the preliminary design team is to design the smallest, lightest, and least expensive helicopter that simultaneously satisfies all of the requirements and all of the constraints.

OUTLINE OF STEPS IN THE PRELIMINARY DESIGN PROCESS

Although every preliminary design team will have a different process for achieving its goals, the following steps are typical and can be used as a guide.

1. Guess at the gross weight and installed power on the basis of existing helicopters with similar performance.
2. Estimate the fuel required using a specific fuel consumption of 0.5 lb/h.p. hr for piston engines or 0.4 lb/h.p. hr for turbines applied to the installed power.

$$\text{Fuel} = sfc \times \text{h.p.}_{\text{installed}} \times \text{Mission time}$$

3. Calculate the useful load:

$$\text{U.L.} = \text{crew} + \text{payload} + \text{fuel}$$

4. Assume a value of the ratio U.L./G.W. based on existing helicopters and trends. Use Figure 10.1 for guidance.
5. Estimate gross weight as:

$$\text{G.W.} = \frac{\text{U.L.}}{\text{U.L./G.W.}}$$

and compare this value with the original estimate. Modify the estimate of installed power and fuel if the two gross weights are significantly different.
6. Assume a disc loading at the maximum allowable value or at the highest deemed practical, and lay out the configuration based on the rotor radius corresponding to this disc loading and to the estimated gross weight.
7. Make first design decisions for main rotor tip speed, solidity, and twist based on maximum speed or maneuverability requirements.

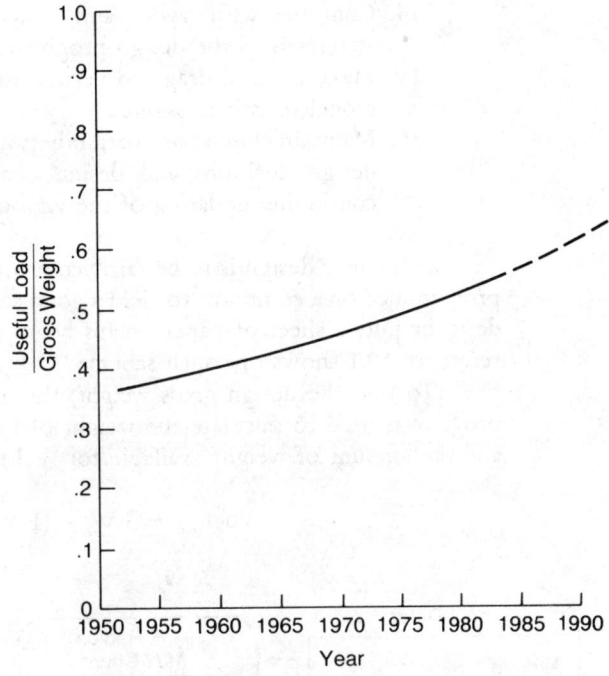

FIGURE 10.1   Historic Trend of Ratio of Useful Load to Gross Weight

8. Make first estimates of drag in forward flight and of vertical drag in hover.
9. Calculate installed power required to satisfy vertical rate of climb and maximum speed requirements at specified altitude and temperature.
10. Select engine, or engines, that satisfy both requirements and constraints. Decrease disc loading if necessary to use approved engine if the vertical rate of climb is the critical flight condition. Use retracting landing gear and other drag reduction schemes in order to use approved engine if high speed is the critical flight condition.
11. Recalculate the fuel required based on the design mission and on known engine characteristics.
12. Calculate group weights based on statistical methods modified by suitable state-of-the-art assumptions. If the resulting gross weight is different from the gross weight currently being used, return to the appropriate previous step.
13. Perform trade-off studies with respect to disc loading, tip speed, solidity, twist, taper, type and number of engines, and so on, to establish smallest allowable gross weight.

14. Continue with layout and structural design. Modify group weight statement as the design progresses.
15. Make detailed drag and vertical drag estimates based on drawings and model tests if possible.
16. Maintain close coordination between the team members to ensure that design decisions and design compromises are incorporated in the continuing updating of the various related tasks.

If the new design is to be fairly conventional, the first thirteen steps can be programmed on a computer to yield a good starting configuration even before the designer puts a sheet of paper on his board. Figure 10.2 slightly modified from reference 10.1 shows one such scheme.

To find the design gross weight, the input gross weight is varied and the program is used to calculate the weight of fuel required to perform the mission and the amount of weight available for fuel from the equation:

$$\text{Fuel}_{avail} = \text{G.W.} - (\text{E.W.} + \text{Payload})$$

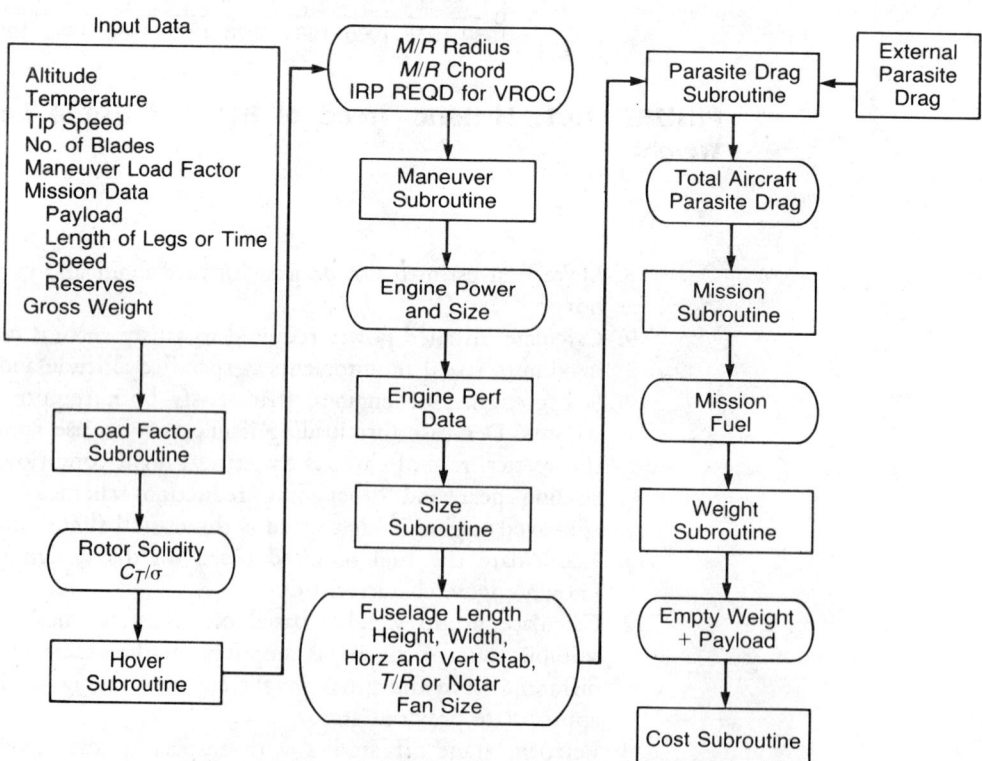

FIGURE 10.2 Block Diagram of Typical Computer Program for Initial Steps of Helicopter Preliminary Design

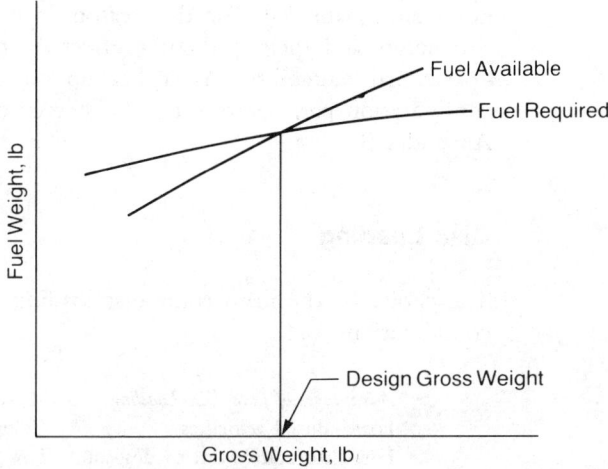

FIGURE 10.3   Result of Preliminary Design Study

The gross weight that makes the fuel available equal to the fuel required is the design gross weight as shown in Figure 10.3.

As a fallout of this process, the difference in the slopes of the two lines of fuel weight versus gross weight yields the *growth factor*—the change in gross weight that is forced by a 1-pound increase in the payload or the structural weight. The growth factor, G.F., is:

$$\text{G.F.} = \frac{1}{\dfrac{dF_{\text{avail.}}}{d\text{G.W.}} - \dfrac{dF_{\text{req.}}}{d\text{G.W.}}} \text{ , lb/lb}$$

where the two slopes are taken at the design gross weight. The denominator is always less than unity, so the growth factor is always greater than unity. Determining the magnitude of the growth factor in preliminary design gives the engineers an indication of the feasibility of their design. A growth factor of over 2 is an indication of serious trouble at this stage. After the helicopter is built, of course, a 1-pound increase in payload or structural weight is accepted as a 1-pound increase in gross weight, with a corresponding decrease in performance.

THE MAIN ROTOR

Although a program like the one just outlined is very useful, with or without a computer, it cannot be used to make all the necessary engineering judgments. The selection of some of the configuration parameters is influenced by considerations not easily quantified. For this reason, it is necessary to discuss the individual parameters and their qualitative effect on performance, size, weight, cost, and operational suitability. As a backup to these discussions, a tabulation of configuration parameters for a number of current helicopters will be found in Appendix B.

Disc Loading

The choice of the main rotor disc loading will be influenced by the following considerations:

Advantages of Low Disc Loading	*Advantages of High Disc Loadings*
Low induced velocities	Compact size
Low autorotative rate of descent	Low empty weight
Low power required in hover	Low hub drag in forward flight

In the early days of helicopter development, designers used low disc loadings because engines were heavy. With the development of the turbine engine, the consideration of engine weight became less important and designers chose higher disc loadings to take advantage of the potentials of smaller overall aircraft size and low empty weight. A possible limit to this trend, however, may be a specific customer requirement not to exceed a given disc loading. The total amount of air displaced by a hovering aircraft is a function of its gross weight, but the ability to "placer mine" the landing surface is a function of its disc loading. Thus interference with pilot visibility or the difficulty of concealment in snow or dust, or the ability to tumble equipment at some distance from the hover spot are all related to gross weight, but the ability to entrain gravel, dirt clods, or bushes into the recirculation pattern is related to the disc loading. The limits for operation are not yet well defined. Hovering aircraft with very high disc loadings (20 to 50 lb/ft^2) such as the Vertol Model 76 and the Ling-Temco-Vought XC-142, were proved to be quite satisfactory for operation from grass lawns or asphalt but completely unsuitable for operation above sand or gravel surfaces. It is evident that the intended use of the aircraft will have a bearing on the maximum "environmental" disc loading.

The autorotative rate of descent is a function of the disc loading, as shown in Chapters 2 and 3. The ability to autorotate is recognized as one of the inherent and desirable features of helicopters. Good autorotative capability is extremely important for single-engine helicopters since it is practiced extensively during pilot training, but even multiengine helicopters are required to demonstrate full power-off autorotations and landings by both the military and the FAA. Any rotor will autorotate if the rate of descent is high enough and, in theory at least, a successful landing can be made from any rate of descent if the stored energy in the rotor is sufficient. In practice, however, the pilot's chances of making a successful landing at high rates of descent are limited by his reaction time and his ability to judge the precise altitude at which to initiate the landing flare. The same problem applies to airplanes in deadstick landings. Tests made on a B-25 bomber showed that good landings could consistently be made by average pilots if the steady rate of descent in the glide was less than 2,500 ft/min, but that at higher rates of descent, increased pilot skill was required. Although these tests have not been repeated on helicopters, it is felt that the rate of descent of 2,500 ft/min is a valid division line between satisfactory and unsatisfactory for single-engine helicopters in which students will practice autorotations. For multiengined helicopters in which the autorotation capability will be demonstrated by skilled pilots, it is suggested that the boundary be raised to 3,500 ft/min. The energy methods of Chapter 3 can be used to establish the maximum disc loadings that correspond to these limits.

Tip Speed

Low tip speeds have the advantage of low noise and good hovering performance. High tip speeds have the advantage of low rotor and drive system weights and high stored energy for autorotative entries and flares.

One of the primary considerations that limits tip speeds on the high side is rotor noise. Tip speeds of more than about 750 ft/sec are considered to be excessively noisy. A lower limit is set by the requirement to store kinetic energy in the rotor in case of a power failure.

Avoiding advancing-tip compressibility and retreating-tip stall also limits the choices of rotor speed. It is generally accepted that advancing tip Mach numbers of more than about 0.92 will produce high blade loads as a result of the Mach tuck

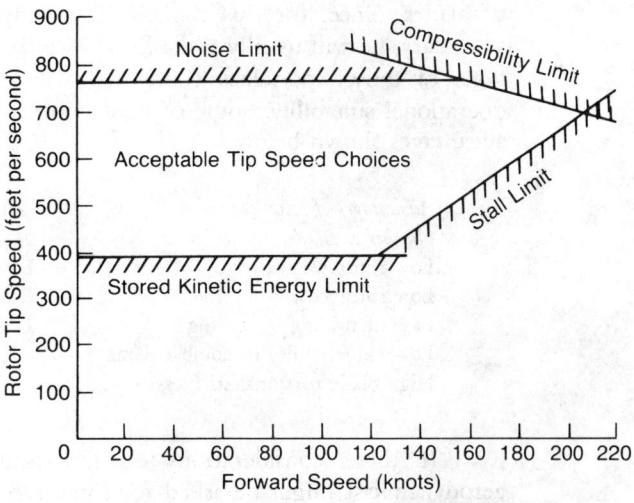

FIGURE 10.4 Constraints on Choice of Tip Speeds

phenomenon described in Chapter 6. It is also generally accepted that for conventional helicopters at maximum speed, the tip speed ratio limit should not exceed 0.5 to avoid retreating blade stall. Figure 10.4 shows how all these constraints limit the tip speed options available and why the maximum speed of "pure" helicopters is about 200 knots (as of this writing).

Solidity

Once values of disc loading and tip speed are selected, the solidity is the primary main rotor physical parameter to be chosen. There are three possible flight conditions that might establish solidity:

1. *Hover at high altitude and temperature*: The solidity is selected to achieve the maximum Figure of Merit. This criterion applies primarily to compound helicopters and flying cranes.
2. *Maximum speed*: The solidity is selected to prevent retreating blade stall at the design maximum speed.
3. *High load factors*: The solidity is selected to prevent retreating blade stall at the design maximum maneuverability requirement.

The basis for all these considerations will be found in Chapters 1, 3, and 5.

Number of Blades

Once the blade area has been selected, the rotor designer must chose the number of blades. Since, for this choice, the aerodynamic considerations are relatively secondary, he will usually make his choice based on the conflicting recommendations of those specialists concerned with vibration, noise, weight, cost, and operational suitability. Some of their concerns can be organized in terms of the advantages shown below.

Advantages of Low Number of Blades	*Advantages of High Number of Blades*
Low rotor weight	Low rotor-induced vibration
Low rotor cost	Ease of handling blades in the field
Ease of folding or storing	Less distinctive noise signature
Low vulnerability to combat damage	
High blade torsional stiffness	

If these considerations do not lead to an immediate decision, the aerodynamicist might be asked for some recommendations. Unfortunately, these will also be conflicting; both as applied to hover and to forward flight.

As discussed in Chapter 1, a performance penalty in hover is caused by blade-vortex interaction. For a small number of blades, this should be less because the tip vortex has a chance to get out of the way before the next blade passes by. This should provide a slight power advantage in hover. This advantage, however, might be lost if the small number of blades results in such a stubby shape that the tip loss region is a significant portion of the blade area. On the other hand, splitting the required blade area into many small chord blades might introduce Reynolds number penalties on maximum lift coefficient and skin friction if the chords are less than about 5 inches.

Some design teams limit the minimum aspect ratio of the blade outboard of the flapping hinge to about 12. This is done to ensure that the natural frequency of the second flapping mode will be below 3 cycles/revolution.

In forward flight, a potential advantage of a low number of blades—such as two—is in lower hub drag, as indicated by the hub drag survey presented in Chapter 4. Against this advantage is the disadvantage that the wake left by a rotor with a small number of blades is pulsating, which generates a higher induced power than the smoother wake left by a multibladed rotor. Some basis for this observation will be found in reference 10.2.

Twist

High blade twist produces good hovering performance and delays retreating blade stall at high forward speeds. but it also produces high vibration in forward flight as it causes the blades to bend as they go from the advancing to the retreating side. The effects on performance in hover and forward flight can be deduced from the discussions in Chapters 1 and 3, but the question of vibration is not easily quantified. At this writing, designers are generally selecting values of main rotor blade twist in the $-8°$ to $-14°$ range. If the dynamicists can find ways of reducing the vibration felt in the aircraft, the aerodynamicists will undoubtedly recommend higher values of twist.

Variations on the conventional twist distribution should be considered for special reasons. High twist that is good for hovering out of ground effect was shown to be too high for efficient hover in ground effect in reference 10.3. In addition, twist that is beneficial in powered flight is detrimental in autorotation. Thus the decision about what twist to use on a new design may depend on its projected use. In Chapter 1 there is a discussion of the tip vortex interference problem in hover. At least one helicopter design has attempted to deal with this by using a special nonlinear twist at the tip, as reported in reference 10.4. This twist distribution is shown in Figure 10.5. Yet another variation has been used to reduce the negative angle of attack in the reverse-flow region at high speed. This is shown in Figure 10.6 and is from reference 10.5.

Taper

From the earliest days of helicopter development, it was known that blade taper improved hovering performance by unloading the tips to achieve a more uniform

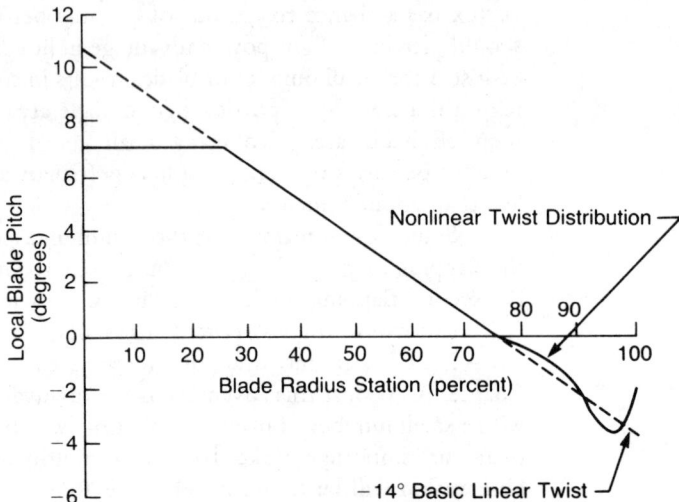

FIGURE 10.5   Twist Modification at Tip

Source: Arcidiacono & Zincone, "Titanium UTTAS Main Rotor Blade," *JAHS* 21-2, 1976.

induced velocity distribution across the disc. Because of this, many early helicopters had tapered blades. Generally, these blades were built like airplane wings with a spar, many ribs, and a covering of plywood and fabric. When blades began to be made of sheet metal, it was easier to use a constant chord and to rely on twist to make the induced velocity distribution uniform. The next innovation in blade construction was the use of fiberglass or another composite material. For these blades the fabrication engineers had no difficulty with taper, so it once again became feasible and has been used on several programs.

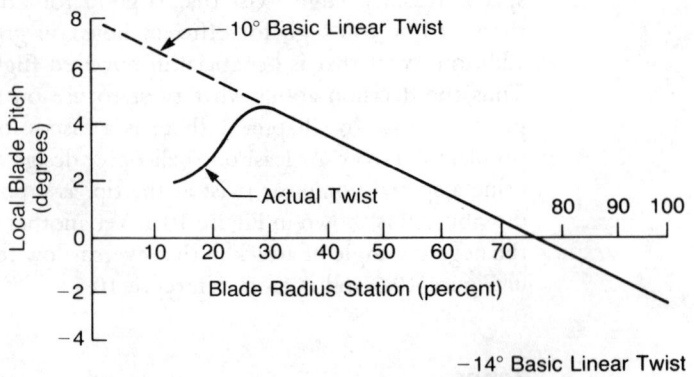

FIGURE 10.6   Twist Modification at Root

Source: Fradenburgh, "Aerodynamic Design of the Sikorsky S-76 Helicopter," *JAHS* 24-4, 1979.

Several factors, however, should be considered before deciding whether to use taper. For small helicopters, taper may drive the tip chord to such a small value that the tip airfoil will suffer penalties in drag and in maximum lift coefficient because of low Reynolds numbers. A second consideration has to do with tip weights. Most rotor blades have tip weights either to improve the autorotational capability or to control dynamic characteristics by placing the frequency of the second blade bending mode below three times per revolution (3/rev). Planform taper, especially when combined with thin tip airfoil sections, may not provide enough physical volume for the required weights. Finally, a tapered blade needs more area than a straight blade to produce the thrust required by a high load factor maneuver. Thus it may weigh more, which will subtract from whatever payload advantage resulted from its increased aerodynamic performance in hover and vertical climb. (Some preliminary studies indicate that perhaps inverse taper holds some promise in this regard.)

Airfoil Sections

The choice of an airfoil—or a series of airfoils—for the main rotor is another exercise in compromise. The ideal airfoil should simultaneously have a high maximum lift coefficient and a high drag divergence Mach number. A study of airfoil characteristics show that these two characteristics do not go together in any one airfoil. If the information in Chapter 6 does not provide enough guidance for this difficult decision, the blade designer will have to rely on the results of later analytical and experimental studies.

Tip Shape

Tip shapes other than square may be selected for a variety of reasons. The most common is the reduction of compressibility effects on the advancing tip at high forward speeds through the use of leading edge sweep. This, of course, is the same reason that jet transports use swept-back wings. There are two compressibility effects that have proved to be significant: the generation of high blade torsion and control loads through Mach tuck (discussed in Chapter 6) and the generation of noise by propagated shock waves. By sweeping the leading edge of the tip, both of these disturbing phenomena can be delayed to forward speeds above the helicopter's normal speed range.

Another motivation for non-square blade tips is to reduce the blade loads and noise generated when a blade passes through, or close to, the concentrated tip vortex left by a preceding blade. If the tip vortex could be spread out by using a special tip shape, the argument goes, the subsequent interaction should be less violent. At this writing, this remains a reasonable but as yet unverified hypothesis.

Swept-back tips have yet another potential advantage: *dynamic twist*. At high forward speeds, most blades with twist carry a nonproductive downward load on the advancing tip. If the tip is swept back, this download acts behind the structural axis and tends to twist the blade nose up, thus reducing the download and its aerodynamic penalty. On the retreating side, the upload on the swept tip twists the blade nose down and alleviates retreating blade stall. The effect also works in hover, where the upload increases the twist, which is beneficial for hover performance.

As with most concepts in helicopter aerodynamics, the use of a swept blade tip is not as straightforward as it might seem. Tests reported in reference 10.6 show a lag in the compressibility effects such that they peak after the blade enters the front half of the disc. This is where a straight blade naturally takes on a swept characteristic from the combination of rotational and forward speed—whereas the swept blade is being aerodynamically unswept by the same effects. Thus it is possible that the swept tip could suffer more from compressibility than the straight one in this region. This possibility has not prevented designers from using swept tips. Figure 10.6 shows several used on contemporary helicopters.

Unfortunately, nothing comes free. Swept tips complicate the structural design of the blade, doubly so if they must be replaceable in the field when damaged. The cost of designing, testing, and building these blades is significantly higher than for straight blades.

Collective and Cyclic Pitch Ranges

The main rotor pitch travels must be adequate to trim the helicopter throughout its flight envelope while leaving suitable margins for maneuvering. The methods of Chapters 3 and 8 can be used to calculate the trim values at the extreme flight conditions. Then these should be increased by 10 to 20 percent. The configuration lists of Appendix B show what designers of some existing helicopters have chosen.

The minimum collective setting must be that which will correspond to 100 percent rotor speed in autorotation at the most extreme combination of gross weight and altitude in the expected flight envelope. This is usually at the lowest gross weight and at sea level, since this is where the C_T/σ is the lowest.

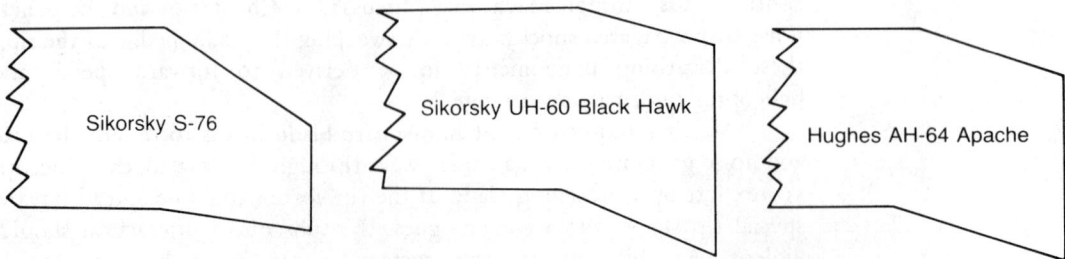

FIGURE 10.7   Production Main-Rotor Blade Tips

As a general rule, the maximum collective pitch is required when making a flare from autorotation to allow all the available rotor energy to be used as the rotor slows down.

The cyclic pitch requirement for trim is unsymmetrical. For example, the forward longitudinal cyclic pitch required at the maximum forward speed is much higher than the aft cyclic pitch required at the maximum allowable rearward speed. By recognizing this fact, the designer can save control system weight and space compared to designing for as much aft travel as forward. Similarly, the total lateral cyclic travel can be minimized by biasing toward the high-speed trim value.

Inertia

Inertia in the main rotor is valuable for two purposes: to prevent the rotor from decelerating too quickly following an unexpected loss of power, and to provide a source of energy for making a landing flare at the end of an autorotational descent. In either case, the level of rotor inertia is more important to a single-engine helicopter than to a multiengine machine, since a sudden power failure of two or more engines is unlikely and there is not the requirement to practice auto-rotational landings continuously as there is with single-engine helicopters.

Flight test experience with a number of single-engine helicopters has led to the conclusion that the rotor kinetic energy stored in the rotor at normal rotor speed should be sufficient to provide the equivalent of at least one and a half seconds of hover time to insure a satisfactory flare capability. It is believed that an equivalent hover time of 0.75 seconds is sufficient for a twin-engine helicopter. The methods for calculating the equivalent hover time are developed in Chapter 5.

Direction of Rotation

Most modern American helicopters have rotors that turn so that the advancing blade is on the right, but many European helicopters use the other rotation. At one time, it was thought that a pilot trained in one type would have trouble making the transition to the other because the pedal motion required to compensate for changes in rotor torque would be different. Today, enough pilots can fly both types on the same day that the argument seems no longer valid. The explanation is that the pilot instinctively does the right thing with his feet to hold heading. This is certainly true near the ground with good visibility, where the directional cues are very strong. Only at high altitude or in degraded visibility conditions might he revert, when making power changes, to pedal actions more appropriate to the helicopter he learned to fly than to the one he is now flying. It is thought that this should not be a strong factor and that the designer should feel free to choose the direction of rotation that will result in the lightest transmission, especially if the design also includes mechanical coupling between main and tail rotor pitch to minimize the pilot workload, as is true of many of the recent Sikorsky designs.

Shaft Tilt

For best performance in forward flight, the fuselage angle of attack should correspond to its minimum drag condition. Since the tip path plane angle of attack must be nose-down to produce the proper balance of forces, a fuselage whose longitudinal axis is parallel to the rotor may be generating more drag than it would if designed to fly more level. To minimize fuselage drag, many helicopters are built with nose-down shaft tilt. A typical value is $-5°$. Higher values might be desirable for very high speed helicopters, but the designer must consider the effects on hover attitude where the rotor must be horizontal. A helicopter with high shaft tilt would hover with the fuselage tilted far nose up, with possible effects on field of view and seat comfort.

The aerodynamicist and the weights engineer could save themselves considerable later work if they could convince the designer to make his layout drawings with the waterlines perpendicular to the rotor shaft. Drag optimization would then be done with *fuselage tilt* with respect to the waterlines instead of shaft tilt. This would simplify analyses in which both vertical and longitudinal positions of the center of gravity are used.

Flare Angle

A parameter that is affected by the nose-down shaft (or nose-up fuselage) tilt is the maximum fuselage angle that can be achieved during a landing flare without doing structural damage by striking the ground with the fuselage aft end or the tail rotor. Unless special provisions are made, such as a sturdy tail boom and a shock-absorbing tail landing gear, the maximum shaft flare angle is that reached as the main landing gear and the aft end of the tailboom or tail wheel/skid touch simultaneously. The higher this angle, the better will be the final deceleration capability. A survey of many helicopter side views such as those in Appendix B indicates that designers believe that a flare angle of at least 8° is desirable. Note that some designers—such as those of the Hughes AH-64—have preferred simply to make their tail booms sturdy enough to withstand a significant landing impact.

Hinge Offset

Once a hub design with hinge offset has been selected, the question becomes, "How much?" In most cases, the decision is based on mechanical or structural considerations rather than on performance or handling qualities. There is, however, some advice that can be given to the designer in this matter. Large hinge offsets produce penalties in the form of large, draggy hubs and the increased ability to accidentally roll the helicopter over on the ground. Small offsets, on the other hand, may produce marginal control power especially when the rotor is unloaded such as in pushovers to low load factors. A minimum offset can be defined to satisfy a flying quality requirement in the form of: "The control power during flight at zero load factor shall be no less than one half (or some other fraction) of the control power in level flight." A requirement such as this will help avoid "mast bumping" or "droop stop pounding" when the pilot tries to control the helicopter under conditions of low"g's".

In Chapter 7, it was shown that the control power is:

$$\frac{dM_{C.G.}}{da_{1_s}} = \frac{dM_M}{da_{1_s}} + hT$$

where the first term is the hub moment due to hinge offset:

$$\frac{dM_M}{da_{1_s}} = \frac{3}{4}\frac{e}{R}\frac{A_b\rho R(\Omega R)^2 a}{\gamma}$$

If the control power at zero load factor is to be a fraction, $1/1 + K$, of what it is in level flight, then:

$$\frac{\left(\dfrac{dM_{C.G.}}{da_{1_s}}\right)_0}{\left(\dfrac{dM_{C.G.}}{da_{1_s}}\right)_{level}} = \frac{1}{1 + K} = \frac{1}{1 + \dfrac{hG.W.}{\dfrac{dM_M}{da_{1_s}}}}$$

With a little algebraic manipulation, the required hinge offset ratio is:

$$\left(\frac{e}{R}\right)_{req} = \frac{\dfrac{4}{3}\dfrac{h}{R}\dfrac{\gamma}{a}C_T/\sigma_{level}}{K}$$

But both in hover and in forward flight, the coning is approximately:

$$a_0 = \frac{2}{3}\frac{\gamma}{a}C_T/\sigma_{level}, \text{ radians}$$

Thus the required hinge offset ratio can be related to the coning in level flight and to the rotor height above the center of gravity:

$$\left(\frac{e}{R}\right)_{req} = \frac{2}{K}\left(\frac{h}{R}\right)a_0$$

THE TAIL ROTOR

The main rotor, of course, is the dominant component of the helicopter and therefore receives most of the designer's attention. As several projects have shown, however, unfortunate decisions about tail rotors can jeopardize the operational success of the aircraft. Tail rotor parameters of some current helicopters are listed in Appendix B.

Location

The tail rotor can be installed either as a pusher or as a tractor. Test results given in Chapter 4 show that the pusher is more effective since it is less interfered with by the vertical fin. Nevertheless, there may be reasons for using the tractor installation. For example, on the Sikorsky UH-60, the designers decided to tilt the tail rotor shaft to take advantage of a vertical component of tail rotor thrust. To obtain adequate clearances as a pusher, the tail rotor shaft would have had to be very long. As a tractor, the installation was much lighter and more compact.

The longitudinal and vertical location of the tail rotor with respect to the main rotor affects the mutual interference in hover, sideward flight, and forward flight. Reference 10.7 discusses test results concerning these interferences in hover. As might be expected, these tests show minimum interferences when the gap between the main rotor and tail rotor discs is large.

Reference 10.8 cites experience indicating that left sideward flight was smoother and the Dutch roll made more stable when the tail rotor was raised on the Hughes AH-64 during its flight test development phase.

Direction of Rotation

Many designers in the past compromised the main rotor transmissions to obtain a "traditional" direction of rotation, although that mattered very little, while letting tail rotors turn in either direction even though that was an important decision. Experience gained during several recent development programs has proven that tail rotors should rotate with the blade closest to the main rotor going up to alleviate the instability and unsteadiness associated with the tail rotor vortex ring state in sideward flight (left sideward flight for helicopters with the "American" direction of main rotor rotation). A discussion of this subject will be found in Chapter 2. Figure 2.10 showed the dramatic improvement resulting from reversing the direction on the Lockheed AH-56. With the original—and wrong—direction, going from hover into left sideward flight not only required an unstable pedal travel but actually used up all the available control before reaching 15 knots. Not shown was the great unsteadiness that existed from 10 to 30 knots. When the rotation was reversed, the pedal displacement became stable, a good control margin remained, and flight was much steadier. Similar experiences can be cited by engineers at Bell, Westland, Aerospatiale, and Mil. This main rotor/tail rotor interference phenomenon is not well understood, and as yet there is no analysis that will provide quantitative results.

If steadiness in left sideward flight is not enough reason for selecting a preferred direction, perhaps noise reduction is. Tests performed at Westland and reported in reference 10.9 have shown that noise in forward flight is lower if the tail rotor blade on top is going aft since it is not slicing through the main rotor wake quite so violently.

Diameter

The choice of tail rotor diameter will be influenced by the following considerations:

Advantages of a Large Diameter	*Advantages of a Small Diameter*
Low power required in hover	Low tail rotor and drive system weight
High directional control power	Helps solve perpetual tail heavy problem
High stability in forward flight	Low hub drag

Although these conflicting considerations would seem to leave every designer on his own, a study of existing helicopters reported in reference 10.10 indicates a remarkably good relationship for the ratio of tail rotor diameter to main rotor diameter as a function of the main rotor disc loading. Figure 10.8 shows that small tail rotors are used with low main rotor disc loadings and big tail rotors are used with high disc loadings. There are probably three reasons for this:

1. With a high disc loading, the main rotor is requiring so much power that saving power with a big tail rotor is attractive.
2. With a low main rotor disc loading, the tail boom is very long and the difficulty of balancing the helicopter with the center of gravity near the main rotor is made easier by using a small tail rotor.
3. When looking at other helicopters, the observed trend becomes aesthetically pleasing.

The trend in Figure 10.8 is expressed approximately by the equation:

$$\frac{D_T}{D_M} = \frac{1}{7.15 - .27 \text{ D.L.}_M}$$

It may be significant that both the Hughes AH-64 and the Bell OH-58 as originally designed had directional control problems that were made less severe by increasing the diameter of the tail rotor to make it closer to the trend line.

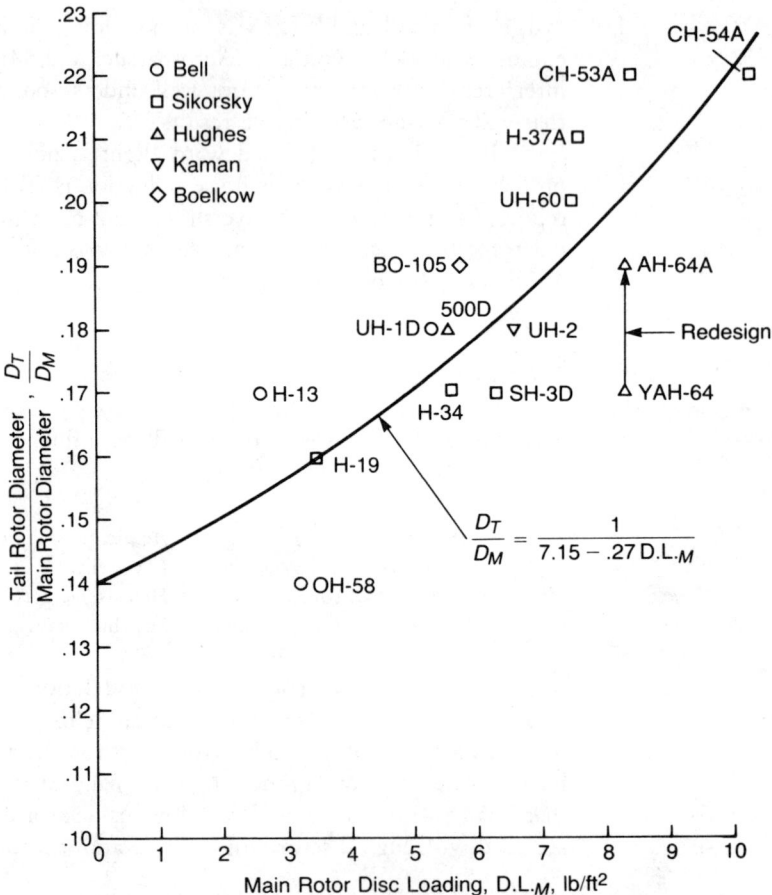

FIGURE 10.8   Tail Rotor Diameter Sizing Trend

Source: Wiesner & Kohler, "Tail Rotor Design Guide," USAAMRDL TR 73-99, 1973.

Tip Speed

The choice of tip speed for the tail rotor is usually based on just two conflicting considerations: a low tip speed minimizes noise, and a high tip speed minimizes weight. Aerodynamic factors are less important than they are on the main rotor. High tip speeds are not so likely to generate serious Mach tuck problems because the blades are stubbier and thus torsionally stiffer; and low tip speeds are not likely to produce retreating tip stall problems because the tail rotor operates in a nonpropulsive mode with a more benign angle of attack distribution. How designers have decided in the past is indicated by the tail rotor tip speeds used on contemporary helicopters as listed in Appendix B.

Solidity and Airfoil Section

With the diameter and tip speed chosen, the next task is to choose the tail rotor solidity. The criterion that most often dictates this is the ability of the tail rotor to balance the torque of the main rotor in a full-power vertical climb at a specified density altitude with at least a 10% thrust margin remaining for maneuvering before stalling. During the analysis of lateral-directional trim in hover in Chapter 8, it was shown that the required value of tail rotor thrust just to balance main rotor torque was:

$$T_T = \frac{Q_M}{l_T - l_M}$$

From this, the minimum tail rotor solidity can be obtained as:

$$\sigma_{T_{min}} = \frac{\left(\dfrac{Q_M}{\rho}\right)_{max}}{C_T/\sigma_{T_{max}} \pi R_T^2 (\Omega R)_T^2 (l_T - l_M)}$$

The critical condition may be either at low altitudes, where the power available is a maximum, or at high altitudes, where the density is low. The minimum solidity, of course, should be increased for design purposes to account for the 10% maneuvering margin and for any blockage effects of the fin, as discussed in Chapter 4. A further check should be made for right sideward flight (assuming "American" rotation of the main rotor) since the tail boom drag may be high as a result of main rotor wake impingement, as discussed in reference 10.11. The required blade area depends not only on the maximum thrust but also on the maximum usuable value of C_T/σ_T, which in turn depends on the airfoil section. It is generally true that compressibility effects on tail rotors are limited to relatively small power penalties, and so it is permissible to choose a thicker airfoil for the tail rotor than would be desirable for the main rotor, in order to take advantage of the higher maximum lift coefficient.

The charts of Chapter 1 give the maximum predicted values of C_T/σ for rotors with the NACA 0012 airfoil. The charts can be used directly if the actual airfoil has similar maximum lift characteristics, or they can be modified up or down to account for the difference between airfoils as measured in two-dimensional wind tunnel tests.

Number of Blades

Dividing the total blade area up into a finite number of blades is the next decision. The fewer the blades, the cheaper the tail rotor is to build and maintain; but if the result is a very stubby blade, high tip losses may penalize performance. Most designers will select the number of blades to satisfy an aspect ratio (radius/chord) criteria of 5 to 9.

Twist

High twist is beneficial on the main rotor for improving hover performance and for delaying retreating tip stall in forward flight. High twist also helps tail rotor hover performance; but, as mentioned in the discussion of tip speed, even untwisted tail rotor blades will not have high retreating tip angles of attack. Thus part of the usual advantage of high twist does not apply. A further consideration, reported in wind tunnel tests in reference 10.12 and in flight test observations in reference 10.8, indicate that untwisted rotor blades go through the vortex ring state more gracefully than twisted blades do. This then is a consideration regarding the ability of the new design to fly steadily in sideward flight.

Pitch Range

The pitch range designed into the tail rotor must provide for both maximum antitorque capability and adequate control in all flight regimes. On the negative side, the critical condition is directional control in autorotation. Experience indicates that a value of $-10°$ to $-15°$ at the 75% radius station is adequate for this. The tabulation of Appendix B lists values used on contemporary helicopters.

On the positive side, the collective pitch must be at least high enough to develop the maximum value of C_T/σ_T used earlier when calculating the required tail rotor solidity. A check should then be made for right sideward flight conditions to determine if this pitch is high enough to account for the increased inflow while still providing a thrust adequate for both antitorque and maneuvering. Choosing a higher value than necessary will cause a problem for the designer of the tail rotor drive system. During a fast hover turn with torque, the tail rotor pitch is very low. If the pilot quickly stops the turn by using maximum pitch, the angles of attack on the tail rotor blades will initially increase by that change before the final inflow pattern can establish itself. This angle may be enough to stall the tail rotor transiently, with a resulting high torque spike in its drive system. Designing for a torque that is caused by a maximum pitch higher than absolutely required for normal operation will increase the helicopter's weight and cost. A discussion of this problem faced during the development of the Hughes AH-64 will be found in reference 10.8.

Cant

Sikorsky has used tail rotors canted 20° down to the left on both the UH-60A and the CH-53E, as may be seen in Appendix B. Reference 10.13 cites the advantages of this concept as the more efficient use of hover power since the tail rotor thrust vector has an upward component, the ability to trim with a center-of-gravity position behind the main rotor, and the alleviation of unsteadiness in the vortex ring state in sideward flight. The primary disadvantage is the development of a longitudinal response with directional control inputs. This can be mechanically compensated for in one flight condition, but not in all. In weighing the advantages against the disadvantages, reference 10.13 concludes: "Although the UH-60A control and Automatic Flight Control System (AFCS) design solved the coupling problems associated with the canted tail rotor, the feeling is that unless forced into an aft c.g. problem by other constraints, the canted tail rotor should not be considered."

THE HORIZONTAL STABILIZER

A horizontal stabilizer is not absolutely required on a helicopter. Helicopters designed before 1960 were unlikely to have them but were nevertheless considered successful. A stabilizer does, however, make an order-of-magnitude improvement in the flying qualities in forward flight and is used routinely in modern designs.

The analytical methods of Chapters 8 and 9 can be used for guidance in choosing the area and fixed incidence that will provide acceptable static and dynamic longitudinal stability in high-speed forward flight at the most aft center-of-gravity position, but the stabilizer's effect on hover and low-speed flight should also be considered. The primary effect is the possibility of erratic longitudinal trim shifts when going from hover to forward flight if the main rotor wake impinges on the surface and produces a high download. There are three possible factors that might contribute to this problem: a large fixed incidence stabilizer located behind the main rotor wake in hover, a high rotor disc loading, and low control power. When one or more of these factors is significant enough to create a problem, designers have adopted three different approaches. One, favored by Bell, places the surface forward on the tail boom so that it is in the rotor wake in hover and thus does not experience a sudden change in download during the transition. The second method uses a *T-tail* with the horizontal stabilizer mounted at about the same height as the main rotor so that it does not feel the wake until high speeds are reached where the induced velocities are small. The third approach uses a movable incidence *stabilator* that can be aligned with the local flow during transition either by being permitted to float free or by being programmed as a function of flight parameters. A discussion of the development of the Hughes AH-64, which had both a T-tail and a stabilator during its development, can be found in reference 10.14. In this case, the incidence of the stabilator was programmed to be a function of air speed, collective pitch, and aircraft pitch rate, as shown in Figure 10.9. The air speed input is the primary means of changing incidence during the transition. The collective pitch input is used to reduce the upload in autorotation, and pitch

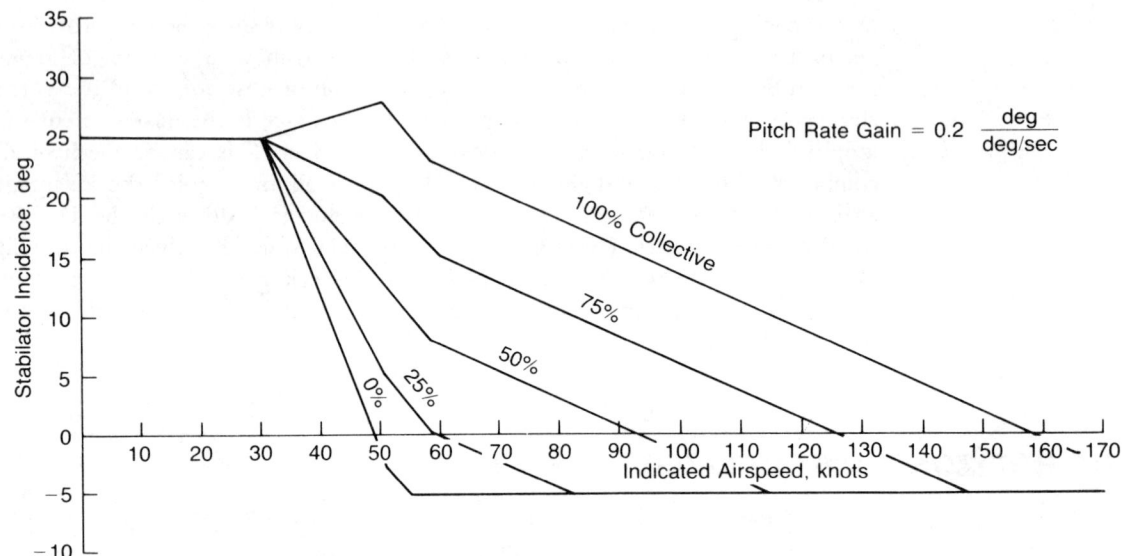

FIGURE 10.9   Stabilator Incidence Schedule for the Hughes AH-64

Source: Prouty & Amer, "The YAH-64 Empennage and Tail Rotor—A Technical History," AHS 38th Forum, 1982.

rate is used to make the surface into an "active control," which allows its area to be significantly less for the same level of dynamic stability than if the incidence were fixed.

A similar stabilator schedule for the Sikorsky UH-60 is found in reference 10.15.

THE VERTICAL STABILIZER

Just as a horizontal stabilizer is not absolutely necessary on a helicopter, neither is a vertical stabilizer, since the tail rotor alone can produce adequate directional stability. Nevertheless, most modern helicopters do have a vertical stabilizer. Depending on the helicopter, the vertical surface may be used to: streamline the tail rotor support, supplement the directional stability produced by the tail rotor, unload the tail rotor in forward flight, support a T-tail, or stabilize the fuselage in case the tail rotor drops off completely.

Because of the low dynamic pressure behind the fuselage and hub, as illustrated by the measured data of Figure 8.9, many designers use vertical surfaces on the ends of the horizontal stabilizer to put them into relatively clean air. In this configuration, they are also out of the way of the tail rotor's induced flow, and they increase the effectiveness of the horizontal stabilizer by acting as end plates. This configuration, however, is probably heavier than a centrally located fin.

As long as the vertical surface is there, it can be used to unload the tail rotor in forward flight by including camber or an offset incidence angle. The primary purpose is to increase the fatigue life of the tail rotor by minimizing the oscillatory flapping loads. Unloading the tail rotor in this manner may not save on total power, since now the induced drag of the vertical surface must be overcome by the main rotor. If the span of the vertical surface is much less than that of the tail rotor, the power tradeoff will probably be unfavorable.

Some recent specifications for combat helicopters have asked the designers to configure the aircraft so they could be flown home in case the tail rotor were completely shot off. In a sideslip, a big enough vertical surface could produce enough antitorque force to do this. Unfortunately, several development programs, such as those reported in references 10.8 and 10.13, have demonstrated that this much area can cause large blockage problems, especially in sideward flight. As a result no helicopter has this desirable characteristic at this writing.

THE WING

A wing is a convenient place to hang external stores or to carry fuel. From an aerodynamic standpoint, however, it is usually detrimental unless used in conjunction with a forward propulsion system in a compound helicopter configuration. The penalty in hover is due not only to the structural weight of the wing but also to its aerodynamic download. In high-speed forward flight, a lifting wing that is used to unload the rotor may actually increase the retreating blade angle of attack, leading to premature stall. This happens because the partially unloaded rotor must be tilted further forward to produce the required propulsive force that now must compensate for the drag of the wing in addition to the drag of the basic helicopter. This requires increased collective pitch to overcome the inflow and increased cyclic pitch to keep the rotor in trim. In many applications, these two effects will combine to increase the retreating tip angle of attack more than the decrease made by unloading the rotor. In some cases, if the requirement is for a high transient load factor during a pullup, then the addition of a wing may be justified.

If a wing is to be used, its incidence should be chosen so that at high speed it is operating at its best angle of attack. By analogy with biplanes, for minimum induced drag, the wing and the rotor should be sharing the lift such that their ratio is:

$$\frac{L_W}{T_M} = \left(\frac{b_W}{D_M}\right)^2$$

The wing should be located with its aerodynamic center behind the most aft center of gravity of the helicopter so that it acts as a stabilizer rather than a destabilizer.

A large wing may cause a problem in autorotation. If the wing supports a large portion of the gross weight, the rotor will be starved for thrust and will not be able to maintain autorotation. If this situation is possible, some means of reducing wing lift will have to be used, such as reverse flaps, spoilers, or incidence changes.

FORWARD PROPULSION SYSTEMS

Auxiliary propulsion in the form of propellers, ducted fans, or jet engines can be used to relieve the rotor of part or all of its propulsive requirement. The limit to this is the wingless autogiro, where no power at all is required by the rotor. The use of auxiliary propulsion to overcome drag reduces the needed forward tilt of the rotor, thus decreasing its collective and cyclic pitch requirements and relieving the high angle-of-attack pattern on the retreating side. This permits this type of aircraft to operate at very high tip speed ratios. Autogiros built in the 1930s routinely operated at tip speed ratios above 0.5, which is considered today to be the upper limit for "pure" helicopters.

An auxiliary propulsion system acting as a separate, controllable longitudinal force system can be used to change the aircraft pitch attitude at any speed. This is especially attractive for combat helicopters. In a hover, for example, this type of helicopter can be held either nose up or nose down to increase the usable field-of-fire. Another use is as a speed brake in forward flight. This is a valuable attribute where rapid and precise decelerations and descents are required to accomplish a mission.

The power to operate the propulsive devices depends on their thrust-to-power ratio. For propellers and ducted fans, this ratio is primarily a function of their diameter—the bigger, the better. Such a system does have its disadvantages, of course. Its weight subtracts from the payload that can be hovered; unless it is declutched in hover, it will reduce the power available to the main rotor, thus creating an even higher payload penalty.

COMPOUND HELICOPTERS

The compound helicopter is even more vulnerable than the pure helicopter to the generality that whatever helps the high speed capability hurts the hovering capability, and vice versa.

The complexity and weight penalties of compounding by using both a wing and an auxiliary propulsion system are justified only when the aircraft requirements combine higher speeds than can be achieved by a pure helicopter with helicopter-type performance in the lower speed regimes. If the rotor is completely unloaded, all the limitations based on blade element angles of attack are eliminated. At some forward speed, the advancing tip Mach number will approach drag divergence; at that point, however, the rotor can be slowed down to alleviate even this problem.

Despite the possibility of unloading the rotor completely, analysis shows that a lifting rotor acts as an efficient large-span wing compared to the typical wing that might be used on this type of configuration. For most effective use of the power, the main rotor should be loaded as highly as possible. This is done by adjusting the collective pitch to some value above flat pitch. In this sense, the collective control is something like the flap setting on an airplane.

WEIGHT ESTIMATES

A key actor in the preliminary design process is the weight engineer. He bases his estimates both on extensive previous experience and on good judgment about existing and future engineering trends. His primary tools are equations for each aircraft component that have been derived from weight data on previous aircraft subjected to a mathematical process known as *multiple linear regression*. This determines sensitivity with respect to every parameter that logically affects the weight of the component. The resulting equations, of course, are continually modified as more modern helicopters are added to the data base and as detail design of specific components is accomplished.

The following equations suitable for preliminary design weight estimates are based on those presented in references 10.16 and 10.17.

Main rotor blades	$W_{b_M} = 0.026 b^{.66} c R^{1.3} (\Omega R)^{.67}$
Main rotor hub and hinge	$W_{b_M} = 0.0037 b^{.28} R^{1.5} (\Omega R)^{.43} (.67 W_{b_M} + \dfrac{gJ}{R^2})^{.55}$
Stabilizer (horizontal)	$W_H = .72 A_H^{1.2} \text{A.R.}_H^{.32}$
Fin (Vertical stabilizer)	$W_V = 1.05 A_V^{.94} \text{A.R.}_V^{.53} (\text{no. of tail rotor gearboxes})^{.71}$
Tail rotor	$W_T = 1.4 R_T^{.090} \left(\dfrac{\text{Transmission h.p. rating}}{\Omega_M} \right)^{.90}$

Body (fuselage)

$$W_F = 6.9 \left(\frac{\text{G.W.}}{1,000} \right)^{.49} L_F^{.61} (Swet_F)^{.25}$$

Landing gear

$$W_{\text{L.G.}} = 40 \left(\frac{\text{G.W.}}{1,000} \right)^{.67} (\text{No. of wheel legs})^{.54}$$

(add 10% for retraction)

Nacelles

$$W_N = 0.041 (\text{Total engine wt})^{1.1} (\text{No. of engines})^{.24} + 0.33 (Swet_N)^{1.3}$$

Engine installation

$$W_{\text{eng}} = (\text{Installed wt. per eng.})(\text{No. of eng.})$$

Propulsion subsystems

$$W_{\text{P.SS}} = 2(W_{\text{eng}})^{.59} (\text{No. of eng.})^{.20}$$

Fuel system

$$W_{\text{F.S.}} = .43 (\text{cap. in gal.})^{.77} (\text{No. of tanks})^{.59}$$

Drive system

$$W_{\text{D.S.}} = 13.6 (\text{Transmission h.p. rating})^{.82} \left(\frac{\text{rpm}_{\text{eng}}}{1,000} \right)^{.037}$$

$$\times \left[\left(\frac{\text{Tail rotor h.p. rating}}{\text{Transmission h.p. rating}} \right) \left(\frac{\Omega_M}{\Omega_T} \right) \right]^{.068} \frac{(\text{No. of gearboxes})^{.066}}{\Omega_M^{.64}}$$

Cockpit controls

$$W_{\text{C.C.}} = 11.5 \left(\frac{\text{G.W.}}{1,000} \right)^{.40}$$

(Triple if no boost is used)

Systems controls (boosted)

$$W_{\text{S.C.}} = 36 b_M c_M^{2.2} \left(\frac{\Omega R_M}{1000} \right)^{3.2}$$

Auxiliary power plant

$$W_{\text{A.P.P.}} = 150$$
(1980 state-of-the-art)

Instruments

$$W_T = 3.5 \left(\frac{\text{G.W.}}{1,000} \right)^{1.3}$$

Hydraulics

$$W_{\text{hyd.}} = 37 b_M^{.63} c^{1.3} \left(\frac{\Omega R_M}{1,000} \right)^{2.1}$$

Electrical

$$W_{\text{EL.}} = \frac{9.6 (\text{Transmission h.p. rating})^{.65}}{\left(\dfrac{\text{G.W.}}{1,000} \right)^{.40}} - W_{\text{hyd}}$$

Avionics

$$W_{\text{av.}} = \begin{cases} 50 \ (\text{low}) \\ 150 \ (\text{avg}) \\ 400 \ (\text{high}) \end{cases}$$

Furnishings and equipment
$$W_{FE} = \begin{cases} 6\left(\dfrac{G.W.}{1,000}\right)^{1.3} \text{(low)} \\[2em] 13\left(\dfrac{G.W.}{1,000}\right)^{1.3} \text{(avg)} \\[2em] 23\left(\dfrac{G.W.}{1,000}\right)^{1.3} \text{(high)} \end{cases}$$

Air cond. & anti-ice
$$W_{AC\&AI} = 8\left(\dfrac{G.W.}{1,000}\right)$$

Manufacturing variation
$$W_{M.V.} = 4\left(\dfrac{G.W.}{1,000}\right)$$

The use of composite materials and advanced technology should result in a weight reduction for some of these components from trends based on the historical data base. A discussion of appropriate advanced technology factors can be found in references 10.1, 10.17, and 10.18.

The U.S. military has developed a standard weight reporting format known as a *Group Weight Statement*. The weight equations given earlier have been applied to the example helicopter and its Group Weight Statement is included in Appendix A.

BALANCING

The best position for the center of gravity on a single-rotor helicopter is slightly ahead of the main rotor shaft. Even when the designer has good intentions and achieves this goal during preliminary design, experience indicates that before the helicopter is ready for production, its average center of gravity (C.G.) will have drifted aft and settled down somewhere behind the shaft. In some cases, this aft C.G. position is forced by design considerations even during preliminary design. A classic example of this is the Sikorsky UH-60A. As explained in reference 10.13, this helicopter was limited in overall length by the requirement to load it into a C-130 without major disassembly. With the rotor sized by the vertical climb requirement, the air transportability requirement dictated a short nose. This and the desire to carry all the fuel behind the main cabin, led to a center of gravity range of more than 15 inches; all located aft of the main rotor. Sikorsky more or less satisfactorily solved the resultant trim problem by canting the tail rotor, as explained during the discussion of tail rotor design.

The longitudinal position of the center of gravity of the empty helicopter is calculated from the sum of the static moments about some arbitrary point contributed by each group that makes up the empty weight divided by that weight.

The arbitrary point should be selected ahead of and below the nose so that all parts of the aircraft will have positive locations. Figure 10.10 shows the scheme as it applies to the example helicopter. Since the center-of-gravity position with respect to the main rotor is of prime importance, it is convenient to chose the origin so that the hub falls on an easily remembered point. For the example helicopter, this has been chosen as Fuselage Station 300 and Waterline 200. The calculation that yields the center-of-gravity position for the empty helicopter is presented next.

Calculation for Center-of-Gravity Position of Empty Example Helicopter

Group	Group Weight (lb)	Fuselage Station (in.)	Moment (in.-lbs)
Rotor	1,466	300	439,800
Tail	281	725	203,725
Body	1,801	330	594,330
Alighting gear	539	220	118,580
Nacelle	207	310	64,170
Propulsion	2,773	305	845,765
Flight controls	205	240	49,200
Auxiliary power plant	150	250	37,500
Instruments	172	75	12,900
Hydraulic	88	280	24,640
Electrical	548	375	205,500
Avionics	400	150	60,000
Furnishings and equipment	1,130	290	327,700
Air cond & anti-ice	160	304	48,640
Manufacturing variation	80	300	24,000
Total Empty Weight	10,000	Total moment: 3,056,450	

$$\text{C.G. pos.}_{empty} = \frac{3,056,450}{10,000} = 305.6 \text{ fuselage station}$$

Note that as the design progresses, this calculation can become more and more precise by expanding it to account for the weight and location of each component of each group.

A plot of the longitudinal center of gravity position as a function of gross weight is known colloquially as the "C.G. potato." It is generated by loading items of the useful load into the empty helicopter in the most forward manner and then in the most aft manner. Figure 10.10 shows the location and weights of the useful load items of the example helicopter. The fuselage station for the center of gravity as the loading proceeds is obtained from the equation:

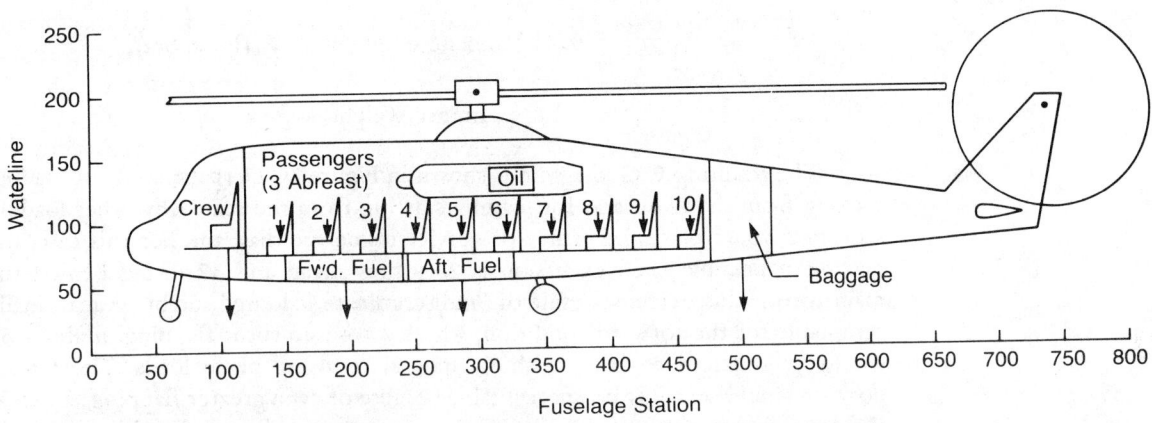

FIGURE 10.10   Location of Useful Load Items

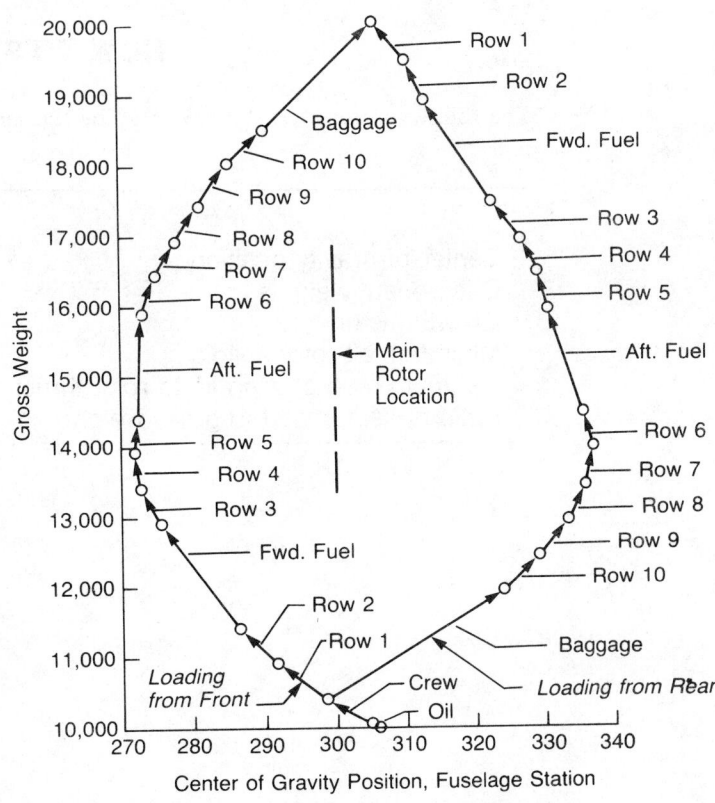

FIGURE 10.11   Center of Gravity Diagram for Example Helicopter

$$\text{Fuse. Sta. } n = \frac{\text{Total Moment, empty} + \sum_{n=1}^{N} W_n (\text{Fuse. Sta})_n}{\text{Empty Weight} + \sum_{n=1}^{N} W_n}$$

The resulting C.G. diagram is shown in Figure 10.11 (page 667) for loading starting from the front and for loading starting from the rear. Any other loading sequence would place the center of gravity inside the diagram. For this case, the maximum possible C.G. excursion is 28 inches ahead and 37 inches behind the main rotor. This extreme result of "indiscriminate loading" might be marginally acceptable for the forward condition, where a nose-up rotor flapping of about 6° would be required to trim, with substantial resulting blade loads. The far aft position would certainly be unacceptable because of even greater flapping as well as the destabilizing effect of the aft C.G. In operation, a helicopter of this size would probably be limited to not more than ±10 inches from the shaft, which means that the loading would have to be monitored and controlled by the crew.

HOW TO'S

The following items can be evaluated by the methods of this chapter:

	page
Center-of-gravity position	666
Component weights	663
Growth factor	643
Minimum tail rotor solidity	657
Optimum ratio of wing lift to rotor thrust	661
Ratio of useful load to gross weight	641

APPENDIX A

Characteristics of the Example Helicopter

Design gross weight, G.W.	20,000 lb
Minimum operating weight	10,700 lb
Power plants	two 2,000 h.p. turbines
Fuel tank capacity	3,000 lb
Parasite drag area, f	19.3 ft^2
Vertical drag ratio, $D_V/\text{G.W.}$	.04
Main Rotor	
Radius, R	30 ft
Disc area, A	2,827 ft^2
Tip speed, ΩR	650 ft/sec
Chord, c	2 ft
No. of blades, b	4
Solidity, σ	0.085
Blade area, A_b	240 ft^2
Airfoil	NACA 0012
Twist, θ_1	$-10°$
Blade cutout ratio, x_0	0.15
Hinge offset ratio, e/R	0.05
Blade-flapping inertia, I_b	2,870 slug ft^2
Lock no., γ	8.1
Polar moment of inertia, J	11,600 slug ft^2
Shaft incidence, i	0°
Height above C.G., h_M	7.5 ft

Tail Rotor

Radius, R_T	6.5 ft.
Disc area, A_T	133 ft^2
Tip speed, ΩR_T	650 ft/sec
Chord, c_T	1 ft
No. of blades, b_T	3
Solidity, σ_T	0.146
Blade area, A_{b_T}	19.4 ft^2
Airfoil	NACA 0012
Twist, θ_{1_T}	$-5°$
Lock no., γ_T	4
Polar moment of inertia, J_T	25 slug ft^2
Tail rotor moment arm, l_T	37 ft
Distance from fin, d/R	1.5 ft
Height above C.G., h_T	6 ft
Blocked area, S_B	31.5 ft
Delta-three angle, δ_3	$-30°$

Horizontal Stabilizer

Area, A_H	18 ft^2
Span, b_H	9 ft
Aspect ratio, A.R.$_H$	4.5
Taper ratio, c_T/c_R	.71
Sweep of mid-chord line, $\Lambda_{c/2_H}$	13°
Sweep of leading edge, $\Lambda_{L.E._H}$	15°
Airfoil	NACA 0012
Moment arm, l_H	33 ft
Height above C.G., h_T	-1.5 ft

Vertical Stabilizer

Area, A_V	33 ft^2
Span, b_V	7.7 ft
Aspect ratio, A.R.$_H$	1.8
Taper ratio, c_T/c_R	0.21
Sweep of mid-chord line, $\Lambda_{c/2_V}$	27°
Rudder deflection, δ_r	10°
Moment arm, l_V	35 ft
Height above C.G., h_V	3 ft

Fuselage

Length, L_F	57 ft
Width, W_F	8 ft
Height, H_F	10 ft
Wetted area, S_{W_F}	680 ft^2
Volume, V_F	4,600 ft^3
Fineness ratio, F.R.$_F$	5.2
Height above C.G., h_F	.5 ft

Aircraft inertias:

  Pitch $I_{yy} = 40,000$ slug ft^2
  Roll $I_{xx} = 5,000$ slug ft^2
  Yaw $I_{zz} = 35,000$ slug ft^2

Miscellaneous:

  Landing gear vertical design velocity = 15 ft/sec
  Effective transmission inertia (referenced to rotor rpm) = 20 slug ft^2

Power plant:

  Engine rpm   20,000
  Main rotor transmission rating   4000 h.p.
  Tail rotor transmission rating   800 h.p.
  Total engine nacelle wetted area   94 ft^2

MIL-STD-1374 PART I-TAB
NAME
DATE

GROUP WEIGHT STATEMENT
WEIGHT EMPTY

PAGE I-10
MODEL
REPORT

No.	Description	COMPONENT	GROUP
1	WING GROUP		0
2	BASIC STRUCTURE – CENTER SECTION		
3	– INTERMEDIATE PANEL		
4	– OUTER PANEL		
5	– GLOVE		
6	SECONDARY STRUCTURE – INCL. WING FOLD WEIGHT _____ LBS.		
7	AILERONS – INCL. BALANCE WEIGHT _____ LBS.		
8	FLAPS – TRAILING EDGE		
9	– LEADING EDGE		
10	SLATS		
11	SPOILERS		
12			
13			
14	ROTOR GROUP		1466
15	BLADE ASSEMBLY	830	
16	HUB & HINGE – INCL. BLADE FOLD WEIGHT _____ LBS.	636	
17			
18			
19	TAIL GROUP		281
20	STRUCT. – STABILIZER (INCL _____ LBS. SEC. STRUCT.)	37	
21	– FIN – INCL. DORSAL (INCL. _____ LBS. SEC. STRUCT.)	63	
22	VENTRAL		
23	ELEVATOR – INCL. BALANCE WEIGHT _____ LBS.		
24	RUDDERS – INCL. BALANCE WEIGHT _____ LBS.		
25	TAIL ROTOR – BLADES	181	
26	– HUB & HINGE		
27			
28	BODY GROUP		1801

#	Item	RUNNING	*STRUCT.	CONTROLS	
29	BASIC STRUCTURE – FUSELAGE OR HULL				
30	– BOOMS				
31	SECONDARY STRUCTURE – FUSELAGE OR HULL				
32	– BOOMS				
33	– SPEED BRAKES				
34	– DOORS, RAMPS, PANELS & MISC.				
35					
36					539
37	ALIGHTING GEAR GROUP – TYPE**				
38	LOCATION	RUNNING	*STRUCT.	CONTROLS	
39	MAIN				
40	NOSE/TAIL				
41	ARRESTING GEAR				
42	CATAPULTING GEAR				
43					
44					207
45	ENGINE SECTION OR NACELLE GROUP				
46	BODY – INTERNAL				
47	– EXTERNAL				
48	WING – INBOARD				
49	– OUTBOARD				
50					0
51	AIR INDUCTION GROUP				
52	– DUCTS				
53	– RAMPS, PLUGS, SPIKES				
54	– DOORS, PANELS & MISC.				
55					
56					
57	TOTAL STRUCTURE				4294

* CHANGE TO FLOATS AND STRUTS FOR WATER TYPE GEAR.

** LANDING GEAR "TYPE": INSERT "TRICYCLE", "TAIL WHEEL", "BICYCLE", "QUADRICYCLE", OR SIMILAR DESCRIPTIVE NOMENCLATURE.

MIL–STD–1374

2

673

MIL-STD-1374 PART I-TAB
NAME
DATE

GROUP WEIGHT STATEMENT
WEIGHT EMPTY

PAGE   I-11
MODEL
REPORT

No.	Item	COMPONENT	GROUP
58	PROPULSION GROUP		2773
59	ENGINE INSTALLATION	900	
60			
61			
62	ACCESSORY GEAR BOXES & DRIVE		
63	EXHAUST SYSTEM		
64	ENGINE COOLING		
65	WATER INJECTION	127	
66	ENGINE CONTROL		
67	STARTING SYSTEM		
68	PROPELLER INSTALLATION		
69	SMOKE ABATEMENT		
70	LUBRICATING SYSTEM		
71	FUEL SYSTEM	73	
72	TANKS – PROTECTED		
73	– UNPROTECTED		
74	PLUMBING, ETC.		
75			
76	DRIVE SYSTEM	1673	
77	GEAR BOXES, LUB SY & ROTOR BRK		
78	TRANSMISSION DRIVE		
79	ROTOR SHAFTS		
80			
81	FLIGHT CONTROLS GROUP		205
82	COCKPIT CTLS. (AUTOPILOT LBS.)	38	
83	SYSTEMS CONTROLS	167	
84			
85			

86	AUXILIARY POWER PLANT GROUP				150
87	INSTRUMENTS GROUP				172
88	HYDRAULIC & PNEUMATIC GROUP				88
89					
90	ELECTRICAL GROUP				548
91					
92	AVIONICS GROUP				400
93	EQUIPMENT				
94	INSTALLATION				
95					
96	ARMAMENT GROUP (INCL. PASSIVE PROT. LBS)				0
97	FURNISHINGS & EQUIPMENT GROUP				1130
98	ACCOMMODATION FOR PERSONNEL				
99	MISCELLANEOUS EQUIPMENT				
100	FURNISHINGS				
101	EMERGENCY EQUIPMENT				
102					
103	AIR CONDITIONING GROUP				160
104	ANTI-ICING GROUP				
105					
106	PHOTOGRAPHIC GROUP				0
107	LOAD & HANDLING GROUP				0
108	AIRCRAFT HANDLING				
109	LOADING HANDLING				
110	BALLAST				
111	MANUFACTURING VARIATION				
112	TOTAL CONTRACTOR CONTROLLED				80
113	TOTAL GFAE				
114	TOTAL WEIGHT EMPTY – PG 2-3				10000

3

MIL-STD-1374

675

MIL-STD-1374 PART I-TAB
NAME
DATE

GROUP WEIGHT STATEMENT
USEFUL LOAD AND GROSS WEIGHT

MODEL
REPORT

			NORMAL	FERRY		
115	LOAD CONDITION					
116						
117	CREW (NO. 2)		360	360		
118	PASSENGERS (NO. 30)		5100			
119	FUEL LOCATION TYPE	GALS.				
120	UNUSABLE FUSE JP-4	4.6 (6.5 lb/gal)	30	30		
121	INTERNAL FWD JP-4	228.5	1485	1485		
122	AFT JP-4	228.5	1485	1485		
123	CABIN JP-4	740		4809		
124						
125	EXTERNAL (FERRY)	1259		8182		
126						
127						
128	OIL					
129	TRAPPED					
130	ENGINE		40			
131						
132	FUEL TANKS (LOCATION)					
133	WATER INJECTION FLUID (GALS.)					
134						
135	BAGGAGE		1500			
136	CARGO					
137						
138	GUN INSTALLATIONS					
139	GUNS LOCAT. FIX. OR FLEX. QUANTITY CALIBER					
140						
141						
142	AMMO.					

676

No.	Item		
143			
144			
145	SUPP'TS*		
146	WEAPONS INSTALL.**		
147			
148			
149			
150			
151			
152			
153			
154			
155			
156			
157			
158			
159	CABIN TANK		481
160			
161	EXTERNAL TANKS		818
162	SURVIVAL KITS		100
163	LIFE RAFTS		50
164	OXYGEN		200
165	MISC.		
166			
167			
168			
169	TOTAL USEFUL LOAD	10000	18000
170	WEIGHT EMPTY	10000	10000
171	GROSS WEIGHT	20000	28000

* IF REMOVABLE AND SPECIFIED AS USEFUL LOAD.

**LIST STORES, MISSILES, SONOBUOYS, ETC., FOLLOWED BY RACKS, LAUNCHERS, CHUTES, ETC., THAT ARE NOT PART OF WEIGHT EMPTY. LIST IDENTIFICATION, LOCATION, AND QUANTITY FOR ALL ITEMS SHOWN INCLUDING INSTALLATION.

4

MIL–STD–1374

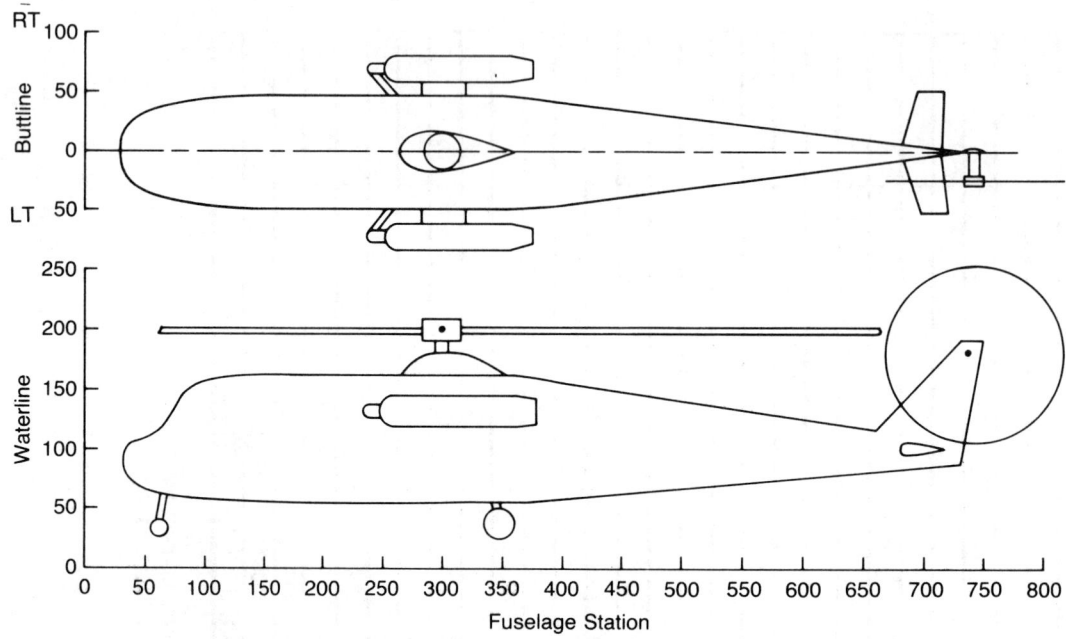

FIGURE A.1   Example Helicopter

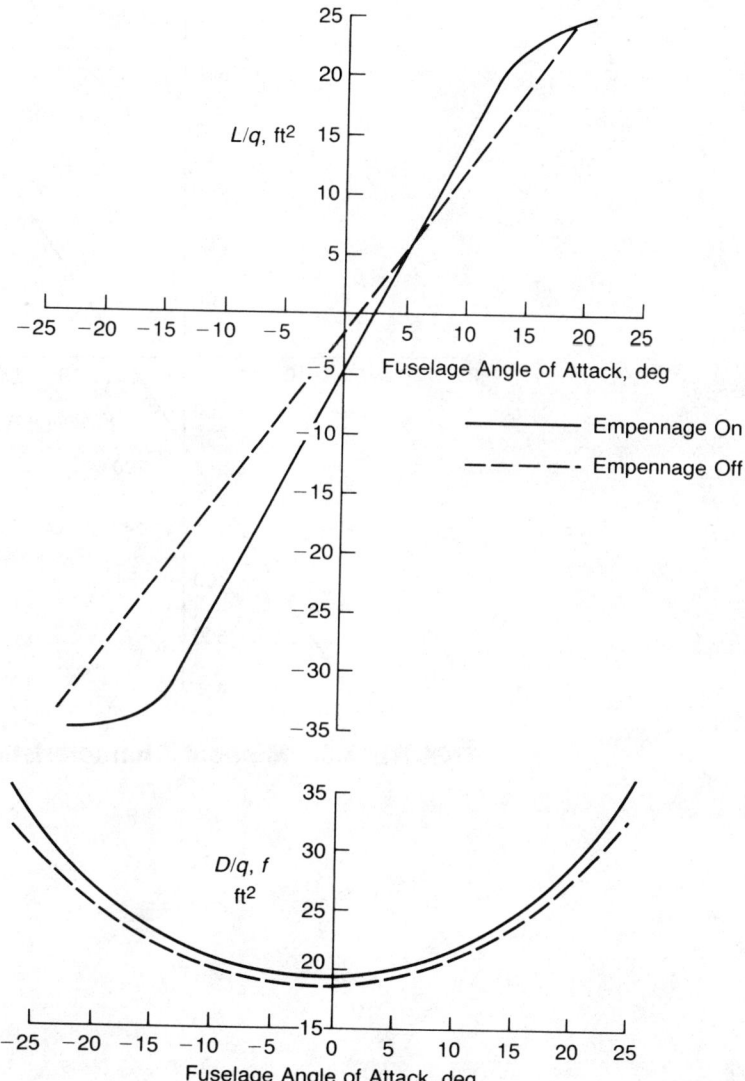

FIGURE A.2   Lift and Drag Characteristics of Airframe

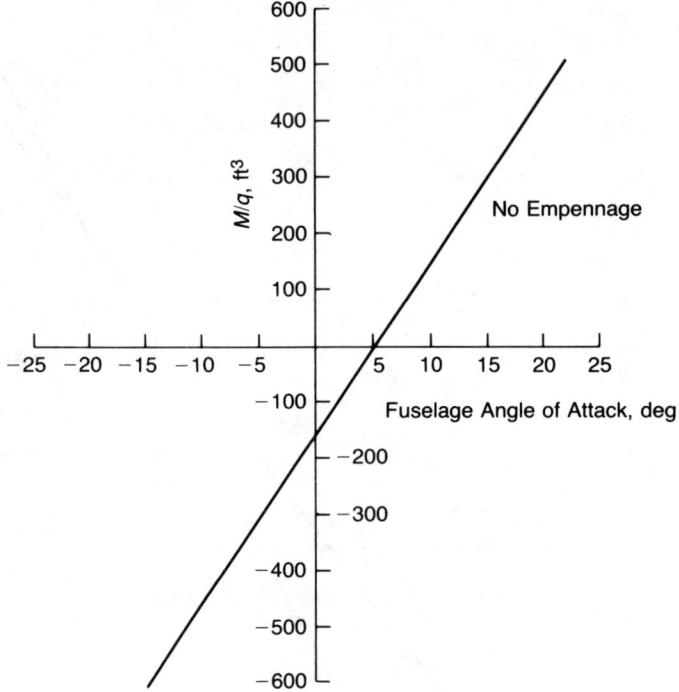

FIGURE A.3   Moment Characteristics of Airframe

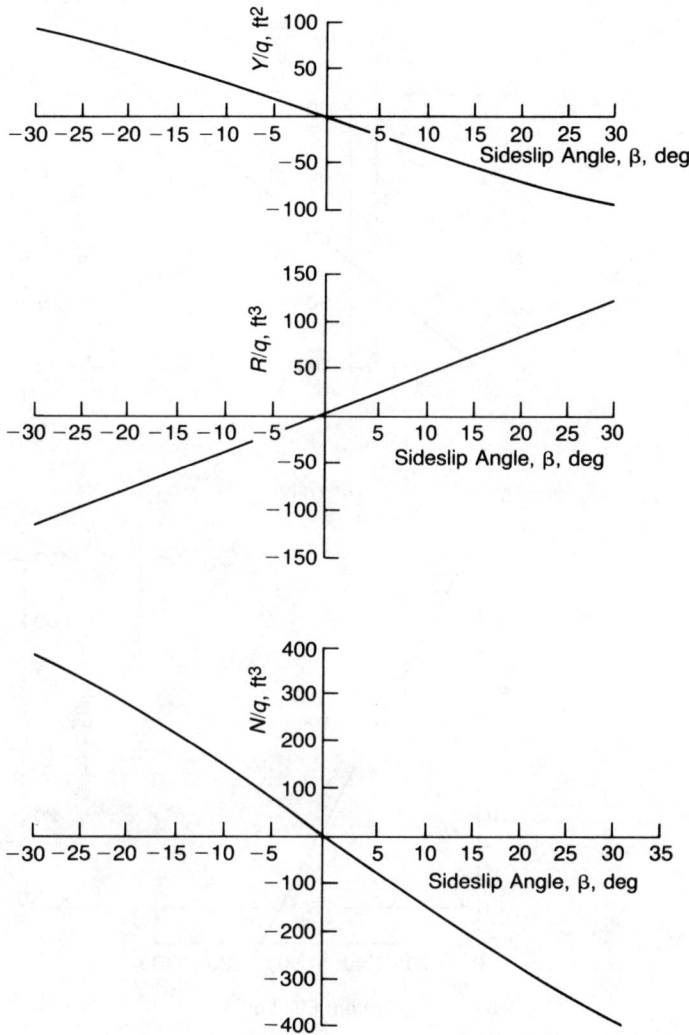

FIGURE A.4   Fuselage Lateral-Directional Aerodynamic Characteristics

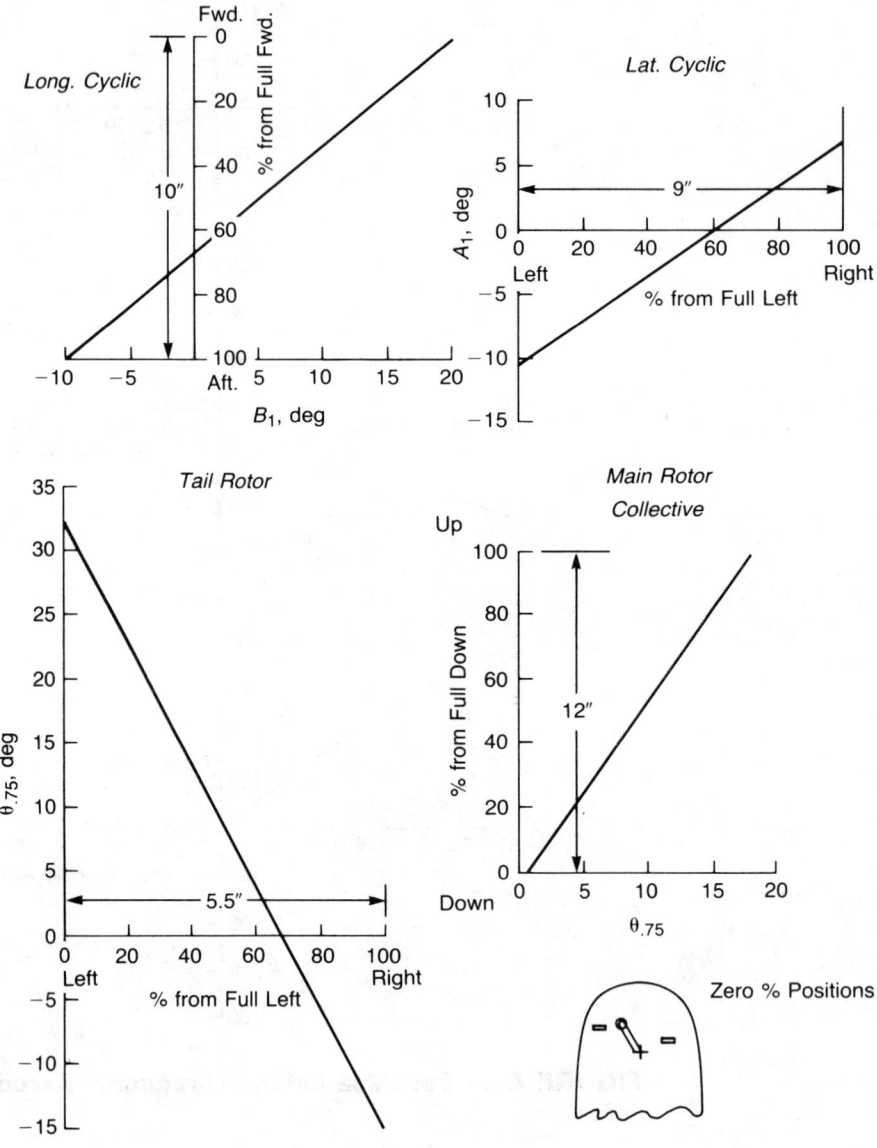

FIGURE A.5 Rigging Curves

APPENDIX B

Physical Parameters of Existing Helicopters

The following information has been graciously provided by the manufacturer of each helicopter.

AEROSPATIALE AS 332 L1

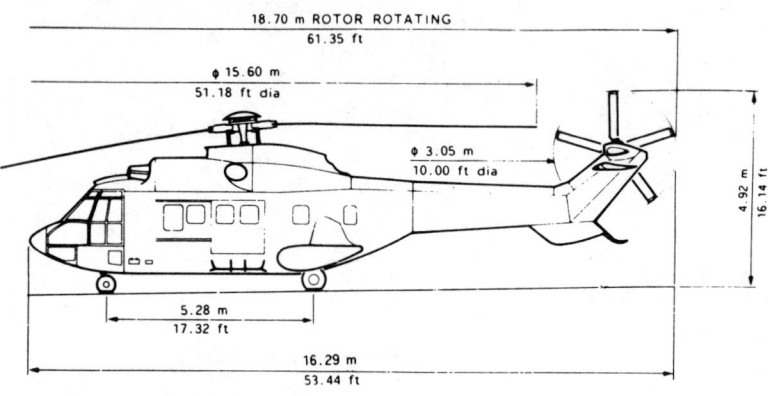

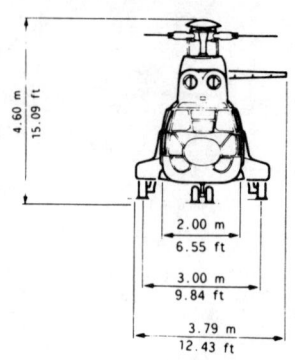

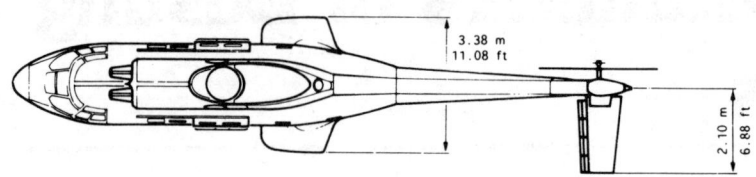

Weights (kg)		Engines	
Empty	4,420	Type	Turbomeca Makila 1A1
Maximum takeoff	8,600	Number	2
Fuel capacity	1,620	Maximum T.O. rating	2,712 kW
		Maximum usable power	2,133 kW

Rotor Parameters		Main Rotor	Tail Rotor
Radius (m)	R	7.79	1.525
Chord (m)	c	0.6	0.2
Solidity	σ	0.098	0.209
No. of blades	b	4	5
Tip speed (m/sec)	ΩR	217	204
Twist (deg)	θ_1	−12.06	−15.71
Hinge offset ratio	e/R	0.037	0.072
Airfoil		HAS 13112,13109,13106	NACA 23012–23010
Collective range (deg)		25	41
Longitudinal cyclic range (deg)		30	—
Lateral cyclic range (deg)		13.5	—
Polar moment of inertia (m²kg)	J	7,035	13.1

AEROSPATIALE SA 365N

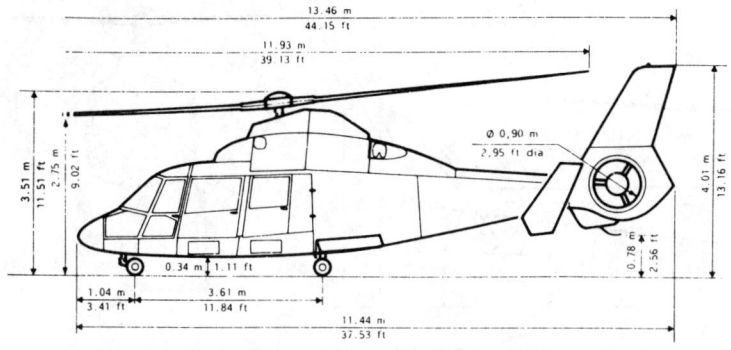

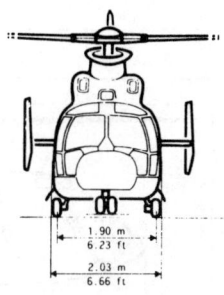

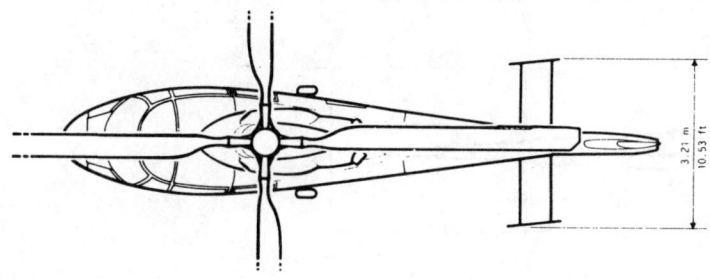

Weights (kg)		Engines	
Empty	1900	Type	Turbomeca Arriel 1C
Maximum takeoff	4000	Number	2
Fuel capacity	892	Maximum T.O. rating	984 kW
		Maximum usable power	899 kW

Rotor Parameters		Main Rotor	Fenestron
Radius (m)	R	5.965	.45
Chord (m)	c	.385	0.0435
Solidity	σ	0.082	—
No. of blades	b	4	13
Tip speed (m/sec)	ΩR	218	227
Twist (deg)	θ_1	−10.2	−8°
Hinge offset ratio	e/R	0.038	0
Airfoil		OA 212, OA 209	NACA 63A312
		OA 207	NACA 63A309
Collective range (deg)		13.75	67
Longitudinal cyclic range (deg)		26	—
Lateral cyclic range (deg)		13	—
Polar moment of inertia (m²kg)	J	2,090	0.35

AEROSPATIALE AS 350B

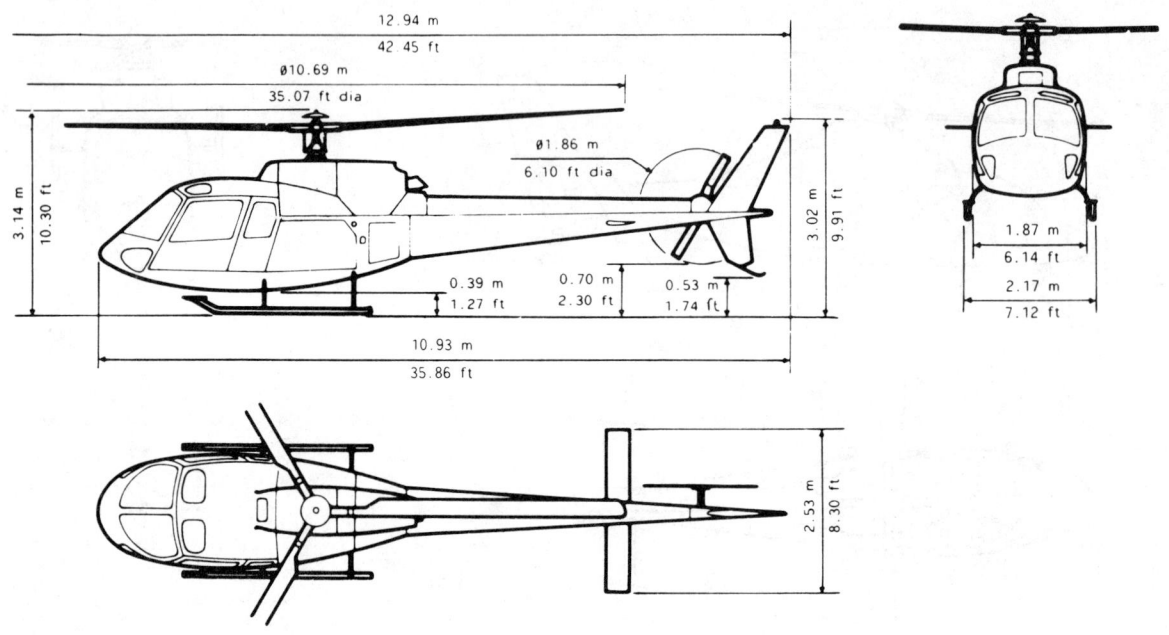

Weights (kg)		Engines	
Empty	1,051	Type	Turbomeca Arriel IB
Maximum takeoff	1,950	Number	1
Fuel capacity	405	Maximum T.O. rating	478 kW
		Maximum usable power	441 kW

Rotor Parameters		Main Rotor	Tail Rotor
Radius (m)	R	5.345	0.930
Chord (m)	c	0.3	0.185
Solidity	σ	0.0536	0.127
No. of blades	b	3	2
Tip speed (m/sec)	ΩR	213	199
Twist (deg)	θ_1	−12.275*	0
Hinge offset ratio	e/R	0.038	0
Airfoil		NACA 0012	NACA 0012
Collective range (deg)		15	27
Longitudinal cyclic range (deg)		26	—
Lateral cyclic range (deg)		12	—
Polar moment of inertia (m² kg)	J	995	1.06

*Linear twist from tip to rotor center.

686

AGUSTA A109

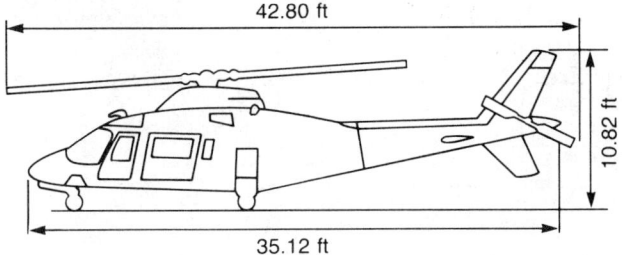

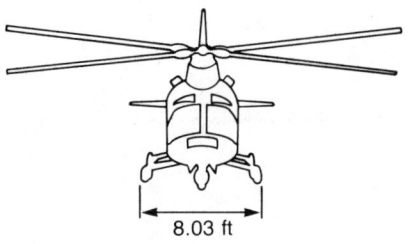

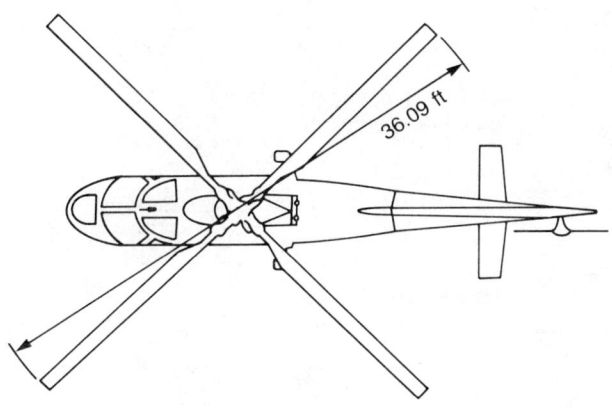

Weights (lb)			Engines	
Empty	3,300	Type		Allison 250-C20B
Maximum takeoff	5,730	Number		2
Fuel capacity	1,202	Maximum T.O. rating		840
		Maximum usable power		740

Rotor Parameters			Main Rotor	Tail Rotor
Radius (ft)	R		18.04	3.33
Chord (ft)	c		1.10	0.66
Solidity	σ		0.0775	0.1256
No. of blades	b		4	2
Tip speed (ft/sec)	ΩR		727	727
Twist (deg)	θ_1		−6	0
Hinge offset ratio	e/R		0.027	—
Airfoil			NACA 23011.3/13006	NACA 0016/0009
Collective range (deg)			0/+16	−7/+21
Longitudinal cyclic range (deg)			−10.5/+12.5	—
Lateral cyclic range (deg)			±6.25	—
Polar moment of inertia (slug ft²)	J		2,000	2

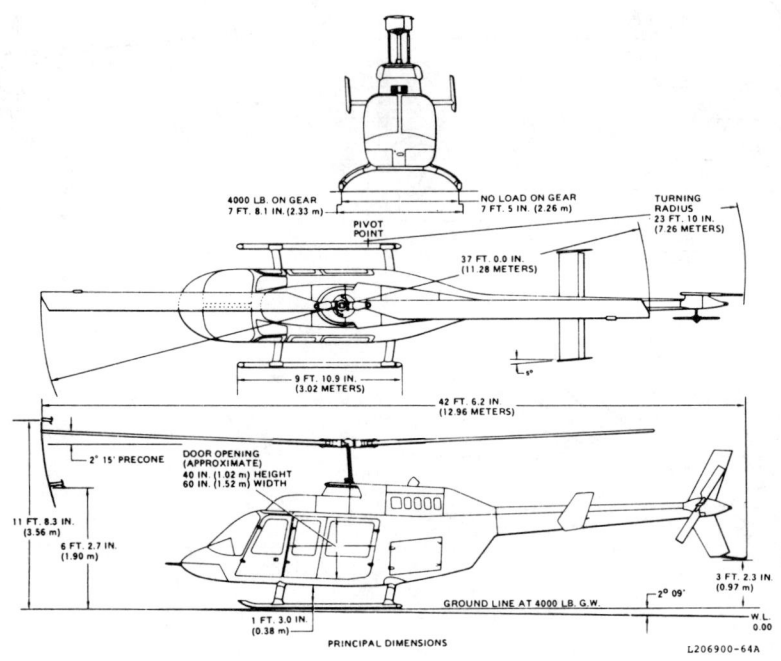

4000 LB. ON GEAR
7 FT. 8.1 IN. (2.33 m)

NO LOAD ON GEAR
7 FT. 5 IN. (2.26 m)

TURNING RADIUS
23 FT. 10 IN.
(7.26 METERS)

PIVOT POINT

37 FT. 0.0 IN.
(11.28 METERS)

9 FT. 10.9 IN.
(3.02 METERS)

42 FT. 6.2 IN.
(12.96 METERS)

2° 15' PRECONE

DOOR OPENING (APPROXIMATE)
40 IN. (1.02 m) HEIGHT
60 IN. (1.52 m) WIDTH

11 FT. 8.3 IN.
(3.56 m)

6 FT. 2.7 IN.
(1.90 m)

3 FT. 2.3 IN.
(0.97 m)

GROUND LINE AT 4000 LB. G.W.

2° 09'

W.L. 0.00

1 FT. 3.0 IN.
(0.38 m)

PRINCIPAL DIMENSIONS

L206900-64A

Weights (lb)			Engines	
Empty	2,223		Type	Allison 250-C30
Maximum takeoff	4,150		Number	1
Fuel Capacity	753		Maximum T.O. rating	692
			Maximum usable power	456

Rotor Parameters		Main Rotor	Tail Rotor
Radius (ft)	R	18.5	2.71
Chord (ft)	c	1.083	.439
Solidity	σ	.0373	.1031
No. of blades	b	2	2
Tip speed (ft/sec)	ΩR	763	722
Twist (deg)	θ_1	−11.1	0
Hinge offset ratio	e/R	—	—
Airfoil		11.3%MOD."DROOP SNOOT"	NACA 0012.5
Collective range (deg)		6 to 22	19.5 to −12.45
Longitudinal cyclic range (deg)		±10	—
Lateral cyclic range (deg)		±10	—
Polar moment of inertia (slug ft²)	J	724	5.9

688

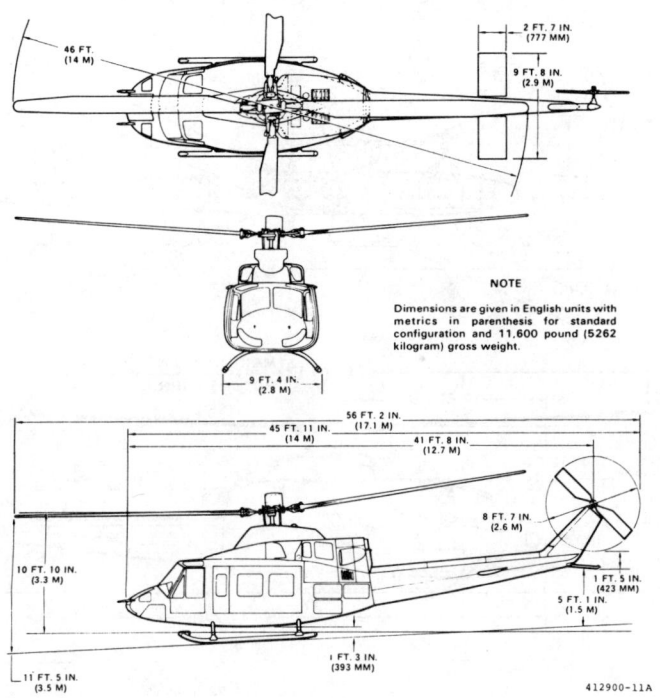

NOTE

Dimensions are given in English units with metrics in parenthesis for standard configuration and 11,600 pound (5262 kilogram) gross weight.

412900-11A

Weights (lbs)		Engines	
Empty	6,325	Type	Pratt & Whitney PT6-3B
Maximum takeoff	11,900	Number	2
Fuel capacity	2,248	Maximum T.O. rating	1,800
		Maximum usable power	1,400

Rotor Parameters		Main Rotor	Tail Rotor
Radius (ft)	R	23	4.25
Chord (ft)	c	1.25	.96
Solidity	σ	.0699	.1435
No. of blades	b	4	2
Tip speed (ft/sec)	ΩR	780	716
Twist (deg)	θ_1	−15.5	0
Hinge offset ratio	e/R	.025	—
Airfoil		FX71-H-080	NACA 0018 @ x = .25 Tapering to 0008.27@Tip
Collective range (deg)		0 to 16	19.85 to −10.15
Longitudinal cyclic range (deg)		± 11	—
Lateral cyclic range (deg)		− 12.6 to 6.7	—
Polar moment of inertia (slug ft²)	J	2,760	5.9

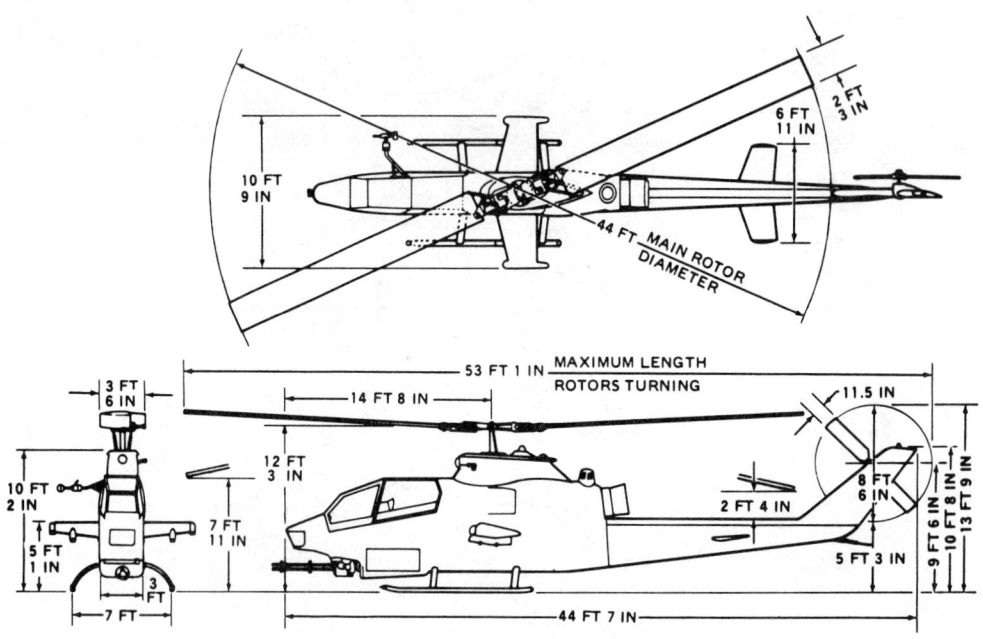

Weights (lb)		Engines	
Empty	6,598	Type	Lycoming 703
Maximum takeoff	10,000	Number	1
Fuel capacity	1,684	Maximum T.O. rating	1,800
		Maximum usable power	1,290

Rotor Parameters		Main Rotor	Tail Rotor
Radius (ft)	R	22	4.25
Chord (ft)	c	2.25	.96
Solidity	σ	.065	.1435
No. of blades	b	2	2
Tip speed (ft/sec)	ΩR	746	739
Twist (deg)	θ_1	−10.0	0
Hinge offset ratio	e/R	—	—
Airfoil		9.33% SYM. Sect. (Special)	NACA 0018 @ x = .25 Tapering to 0008.27 @ Tip
Collective range (deg)		8.5 to 24	19.85 to −10.15
Longitudinal cyclic range (deg)		±10	—
Lateral cyclic range (deg)		±10	—
Polar moment of inertia (slug ft²)	J	2,770	5.9

BELL 222B

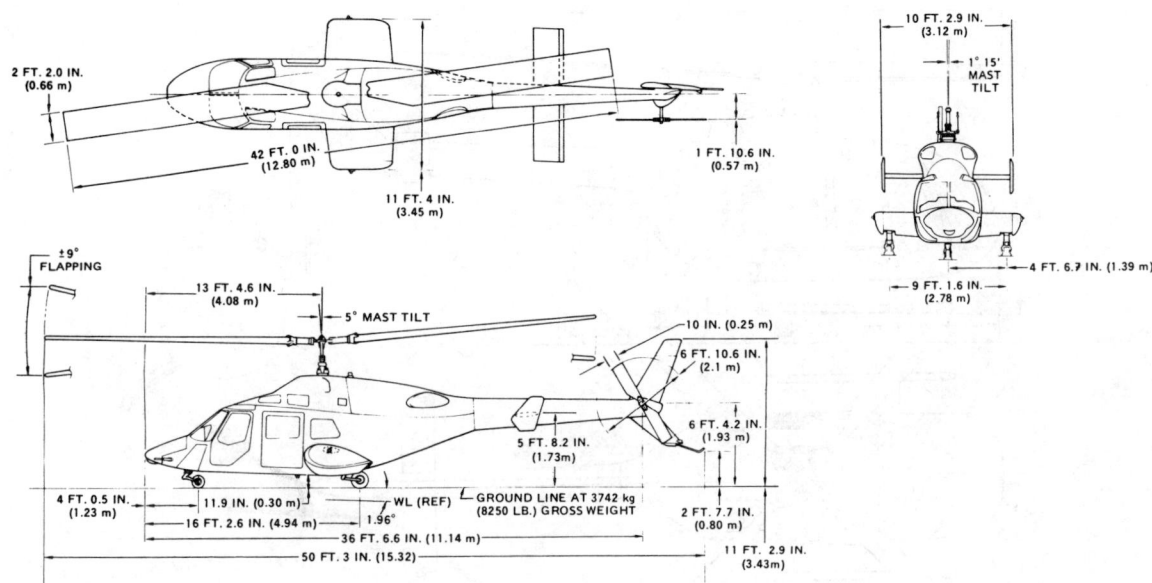

Weights (lb)		Engines	
Empty	4,929	Type	Lycoming 750C-2
Maximum takeoff	8,250	Number	2
Fuel capacity	1,275	Maximum T.O. rating	1,470
		Maximum usable power	1,088

Rotor Parameters		Main Rotor	Tail Rotor
Radius (ft)	R	21	3.45
Chord (ft)	c	2.167	.8
Solidity	σ	.0657	.1476
No. of blades	b	2	2
Tip speed (ft/sec)	ΩR	765	671
Twist (deg)	θ_1	−10.74	0
Hinge offset ratio	e/R	—	—
Airfoil		FX 71-H-080	BHT 10.9 FC
Collective range (deg)		11.7 ± 9	−1.7 to 25
Longitudinal cyclic range (deg)		−13 to +15	—
Lateral cyclic range (deg)		±9	—
Polar moment of inertia (slug ft²)	J	1,664	1.0

MBB BO 105 CB

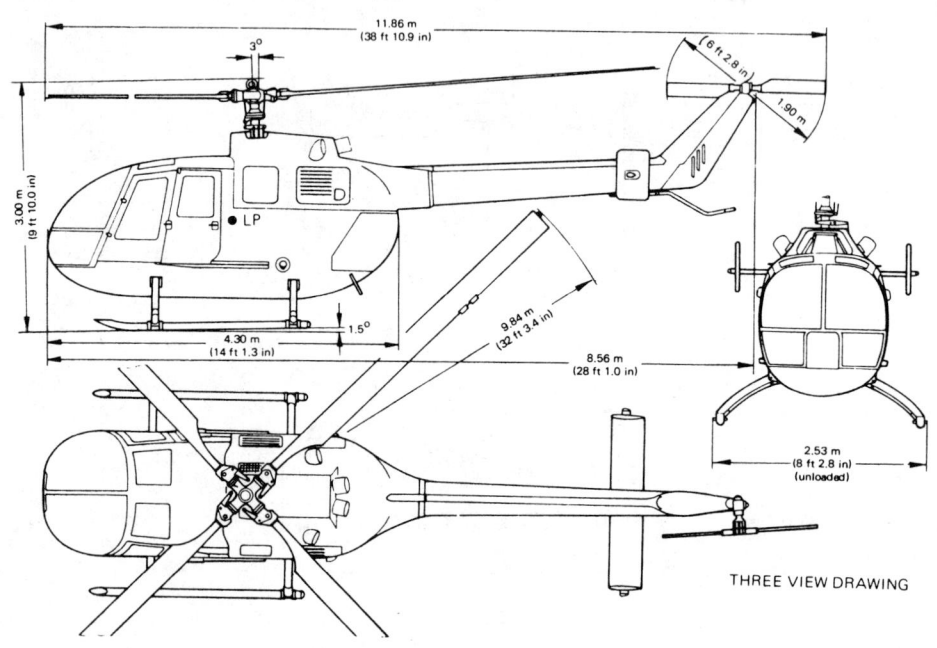

THREE VIEW DRAWING

Weights (kg)		Engines	
Empty	1276	Type	Allison 250 C20B
Maximum takeoff	2500	Number	2
Fuel capacity	456	Maximum T.O. rating (kW)	616
		Maximum usable power (kW)	588

Rotor Parameters		Main Rotor	Tail Rotor
Radius (m)	R	49.2	0.95
Chord (m)	c	0.27	0.18
Solidity	σ	0.07	0.12
No. of blades	b	4	2
Tip speed (m/sec)	ΩR	218	221
Twist (deg)	θ_1	$-8°$	0
Hinge offset ratio (effective)	e/R	0.14	0
Airfoil		NACA 23012	MBB-S102E
Collective range (deg)		-0.2 to 15.0	-8 to $+20$
Longitudinal cyclic range (deg)		-6 to $+11$	—
Lateral cyclic range (deg)		4.2(R) to 5.7(L)	—
Polar moment of inertia (kg-m²)	J	918	0.94

MBB/KAWASAKI BK117

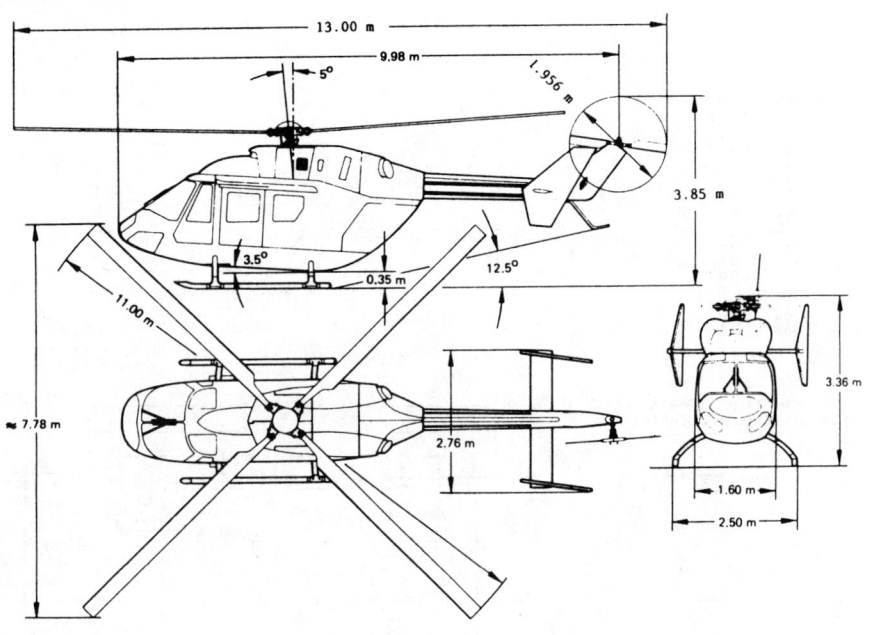

Weights (kg)			Engines	
Empty	1,700		Type	Lycoming LTS101-650B-1
Maximum takeoff	3,200		Number	2
Fuel capacity	478		Maximum T.O. rating (kW)	884
			Maximum usable power (kW)	632

Rotor Parameters			Main Rotor	Tail Rotor
Radius (m)	R		5.5	0.97
Chord (m)	c		0.32	0.2
Solidity	σ		0.074	0.13
No. of blades	b		4	2
Tip speed (m/sec)	ΩR		221	221
Twist (deg)	θ_1		−8	−6.5
Hinge offset ratio (effective)	e/R		0.12	0
Airfoil			NACA 23012/V23010	S102
Collective range (deg)			− 1.8 to 13.3	−10 to +20
Longitudinal cyclic range (deg)			−5 to +17.7	—
Lateral cyclic range (deg)			4.2(R) to 5.7(L)	—
Polar moment of inertia (kg-m²)	J		1,256	0.94

MCDONNELL DOUGLAS (HUGHES) 500E

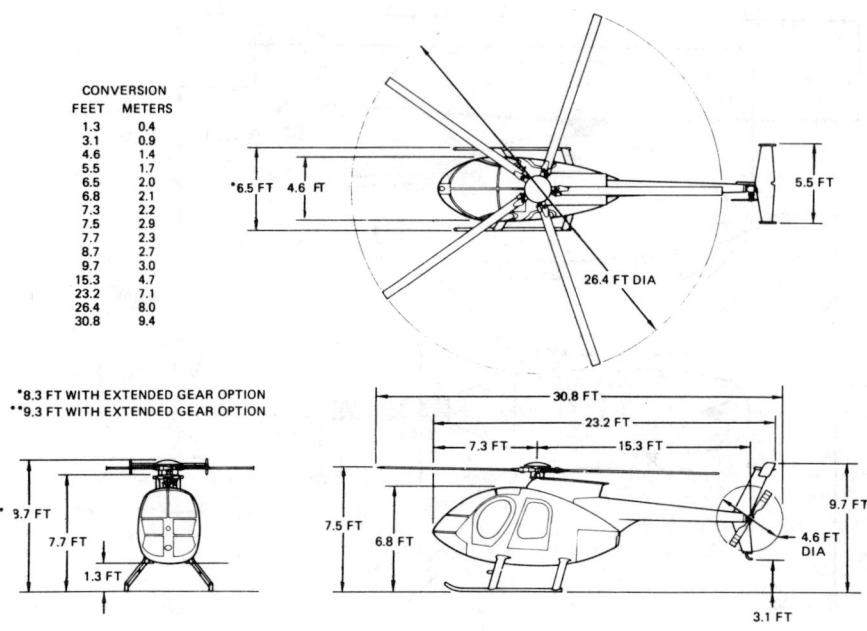

CONVERSION

FEET	METERS
1.3	0.4
3.1	0.9
4.6	1.4
5.5	1.7
6.5	2.0
6.8	2.1
7.3	2.2
7.5	2.9
7.7	2.3
8.7	2.7
9.7	3.0
15.3	4.7
23.2	7.1
26.4	8.0
30.8	9.4

*8.3 FT WITH EXTENDED GEAR OPTION
**9.3 FT WITH EXTENDED GEAR OPTION

Weights (lb)		Engines	
Empty	1,455	Type	Allison 250-C20B
Maximum takeoff	3,000	Number	1
Fuel capacity	404	Maximum T.O. rating	420
		Maximum usable power	375

Rotor Parameters		Main Rotor	Tail Rotor
Radius (ft)	R	13.2	2.3
Chord (ft)	c	.56	0.44
Solidity	σ	.068	.119
No. of blades	b	5	2
Tip speed (ft/sec)	ΩR	680	704
Twist (deg)	θ_1	−9	−8.6
Hinge offset ratio	e/R	.032	—
Airfoil		NACA 0015	NACA 63-415
Collective range (deg)		0 to 14.3	−13 to 27
Longitudinal cyclic range (deg)		7 to 17	—
Lateral cyclic range (deg)		−8.5 to 5.5	—
Polar moment of inertia (slug ft²)	J	263	.23

MCDONNELL DOUGLAS (HUGHES) AH-64

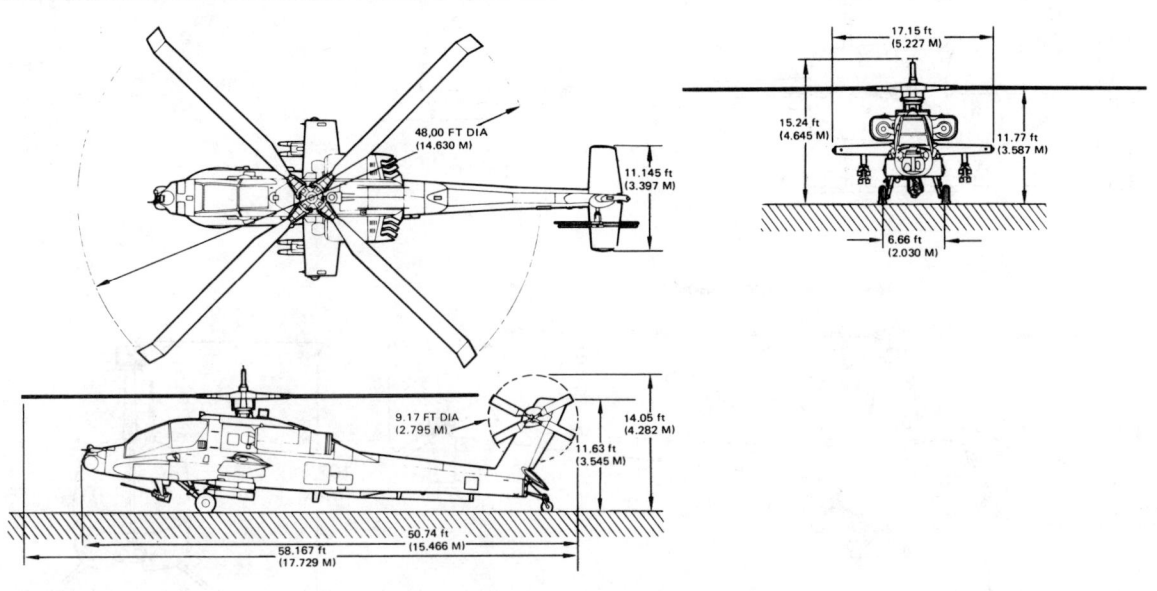

Weights (lb)			Engines	
Empty		10,268	Type	General Electric T700-GE-701
Maximum takeoff		17,650	Number	2
Fuel capacity		2,442	Maximum T.O. rating	3,392
			Maximum usable power	2,828

Rotor Parameters		Main Rotor	Tail Rotor
Radius (ft)	R	24	4.6
Chord (ft)	c	1.75	.83
Solidity	σ	.092	.231
No. of blades	b	4	4
Tip speed (ft/sec)	ΩR	726	677
Twist (deg)	θ_1	−9	−8
Hinge offset ratio	e/R	.038	—
Airfoil		HH-02	NACA 63-414
Collective range (deg)		+1 to +19	−15 to +27
Longitudinal cyclic range (deg)		−10 to +20	—
Lateral cyclic range (deg)		−10.5 to +7	—
Polar moment of inertia (slug ft²)	J	3,800	10

ROBINSON R22 BETA

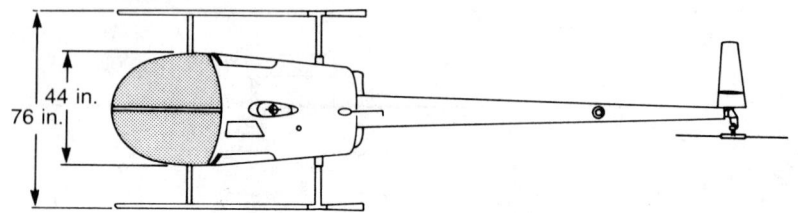

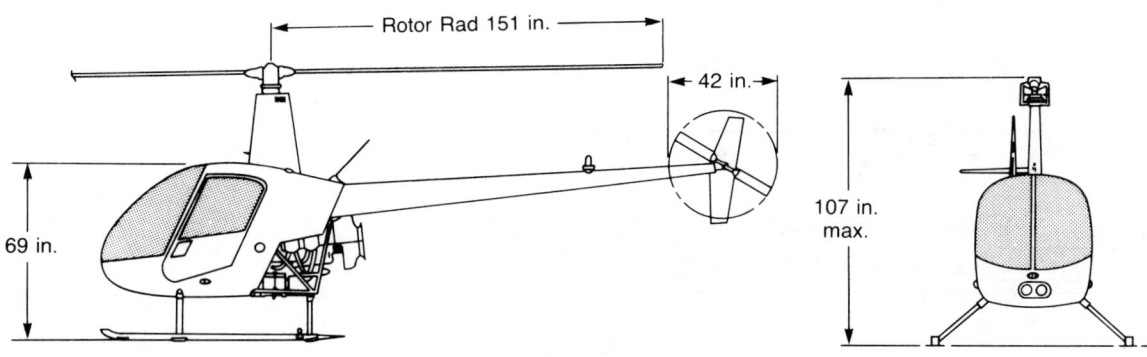

Weights (lb)		Engines	
Empty	830 lb	Type	Lycoming 0-320-B2C
Maximum takeoff	1,370 lb	Number	1
Fuel capacity: main	19.2 gal	Maximum T.O. rating	131 hp
auxiliary	10.6 gal	Maximum usable power	131 hp

Rotor Parameters		Main Rotor	Tail Rotor
Radius (ft)	R	12.58	1.75
Chord (ft)	c	.60	.33
Solidity	σ	.030	.12
No. of blades	b	2	2
Tip speed (ft/sec)	ΩR	699	623
Twist (deg)	θ_1	$-7°$	0
Hinge offset ratio	e/R	N/A	N/A
Airfoil		63-015	63-415
Collective range (deg)		+1.5 to +14.5	+19.5 left, -10.6 right
Longitudinal cyclic range (deg)		-9.0 to +11.0	—
Lateral cyclic range (deg)		-9.5 to +6.0	—
Polar moment of inertia (slug ft²)	J	86.3	.03

SCHWEIZER/HUGHES 300C

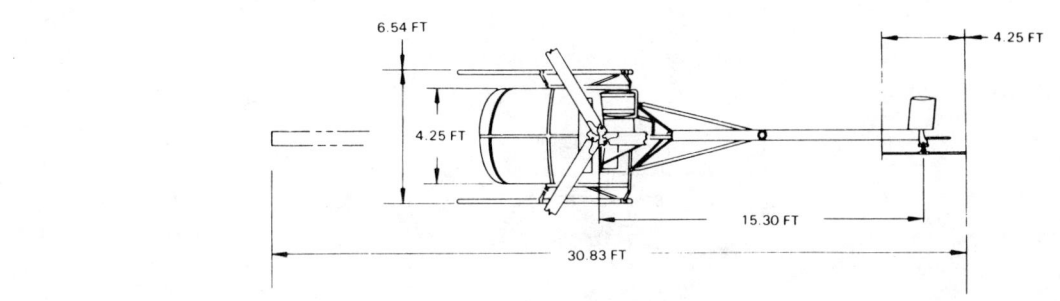

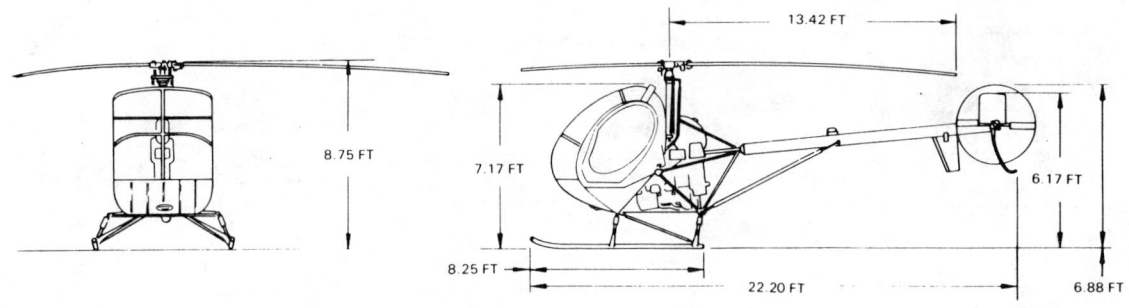

Weights (lb)		Engines	
Empty	1,120	Type	Lycoming HIO-360-D1A
Maximum takeoff	2,050	Number	1
Fuel capacity	180	Maximum T.O. rating	225
		Maximum usable power	190

Rotor Parameters		Main Rotor	Tail Rotor
Radius (ft)	R	13.42	2.12
Chord (ft)	c	.563	.39
Solidity	σ	.040	.116
No. of blades	b	3	2
Tip speed (ft/sec)	ΩR	662	690
Twist (deg)	θ_1	−8.7	−8
Hinge offset ratio	e/R	.014	—
Airfoil		NACA 0015	NACA 0014
Collective range (deg)		+3 to +15	−12 to +26
Longitudinal cyclic range (deg)		−7.5 to +9.5	—
Lateral cyclic range (deg)		−7 to 5.5	—
Polar moment of inertia (slug ft²)	J	161	.22

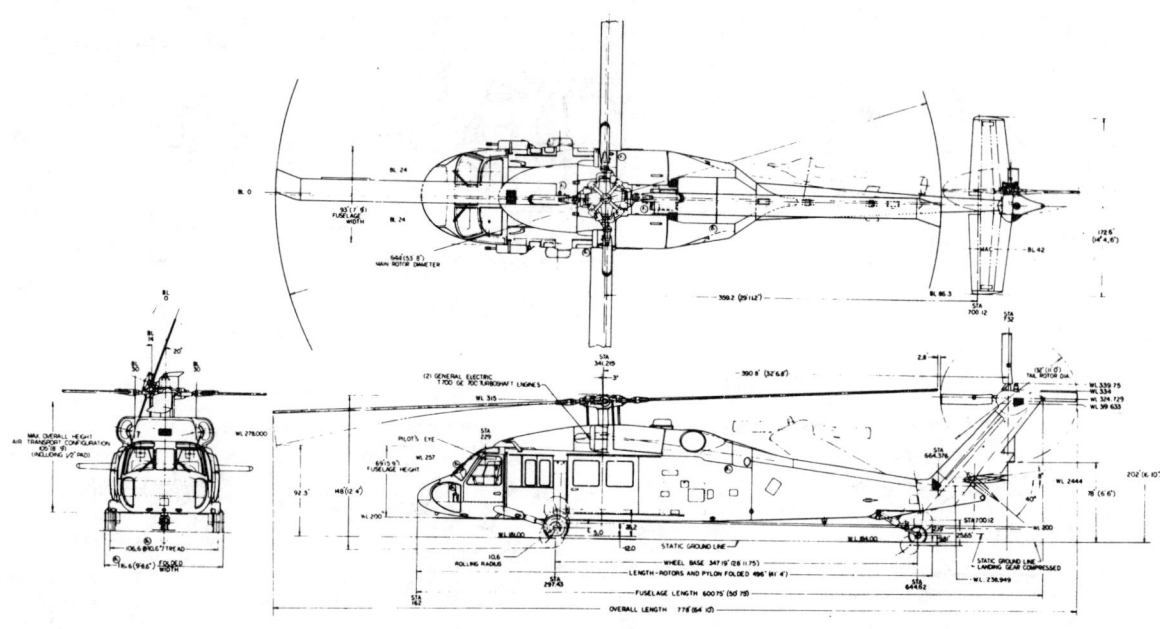

Weights (lb)		Engines	
Empty	10,901	Type	General Electric T700-GE-700
Maximum takeoff	22,000	Number	2
Fuel capacity	2,353	Maximum T.O. rating	3,086 SHP
		Maximum usable power	2,828 SHP

Rotor Parameters		Main Rotor	Tail Rotor
Radius (ft)	R	26.83	5.5
Chord (ft)	c	1.73	0.81
Solidity	σ	0.082	0.188
No. of blades	b	4	4
Tip speed (ft/sec)	ΩR	725	685
Equivalent linear twist (deg)	θ_1	−18	−17
Hinge offset ratio	e/R	.047	—
Airfoil		SC 1095	SC 1095
		SC 1095 R8	
Collective range (deg)		+9.9 to +25.9	−15.6 to +16.4
Longitudinal cyclic range (deg)		−12.3 to +16.5	—
Lateral cyclic range (deg)		−8.0 to +8.0	—
Polar moment of inertia (slug ft²)	J	7,060	19.5

SIKORSKY CH-53E

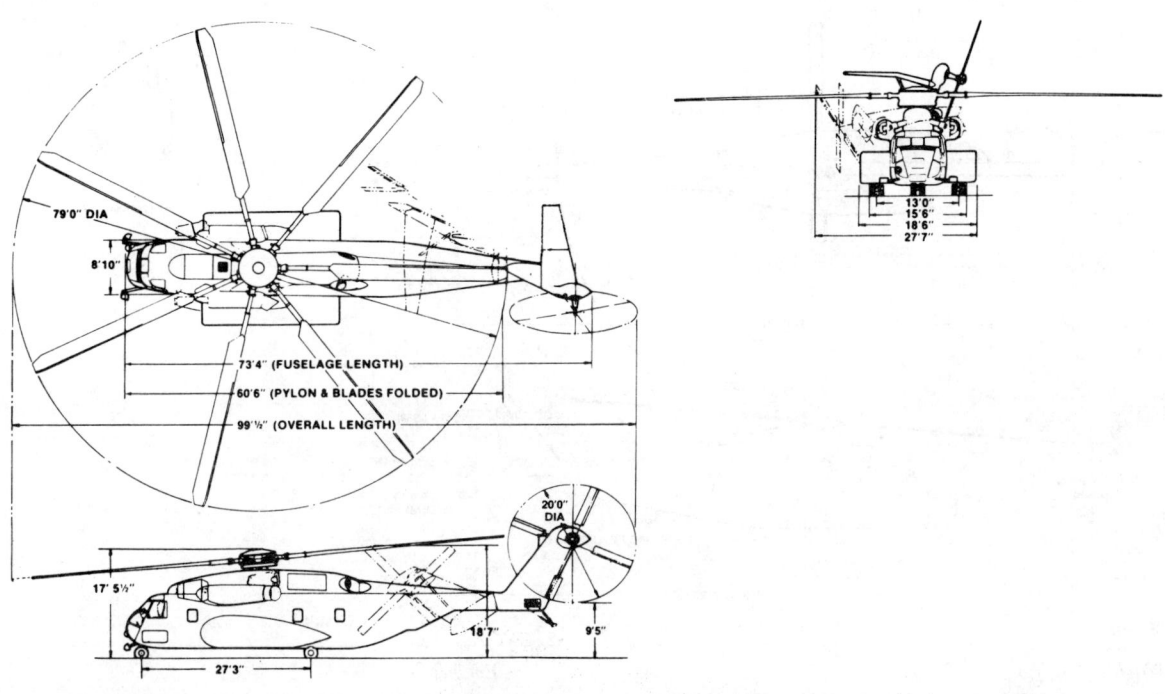

Weights (lb)		Engines	
Empty	33,009	Type	General Electric T64-GE-416
Maximum takeoff	69,750[1]	Number	3
Fuel capacity	6,682 norm	Maximum T.O. rating	13,140
	8,450 aux	Maximum usable power	11,570

Rotor Parameters		Main Rotor	Tail Rotor
Radius (ft)	R	39.5	10
Chord (ft)	c	2.44	1.28
Solidity	σ	0.138	0.163
No. of blades	b	7	4
Tip speed (ft/sec)	ΩR	732	733
Twist (deg)	θ_1	−20	−8
Equivalent linear hinge offset ratio	e/R	.063	.043
Airfoil		SC1095	NACA 0015
Collective range (deg)		−1.4 to 19.6	−10 to 24
Longitudinal cyclic range (deg)		−8.5 to 18.0	—
Lateral cyclic range (deg)		−9.8 to 6.1	—
Polar moment of inertia (slug ft²)	J	51,800	181

[1]Internal payload, 73,500 with external payload.

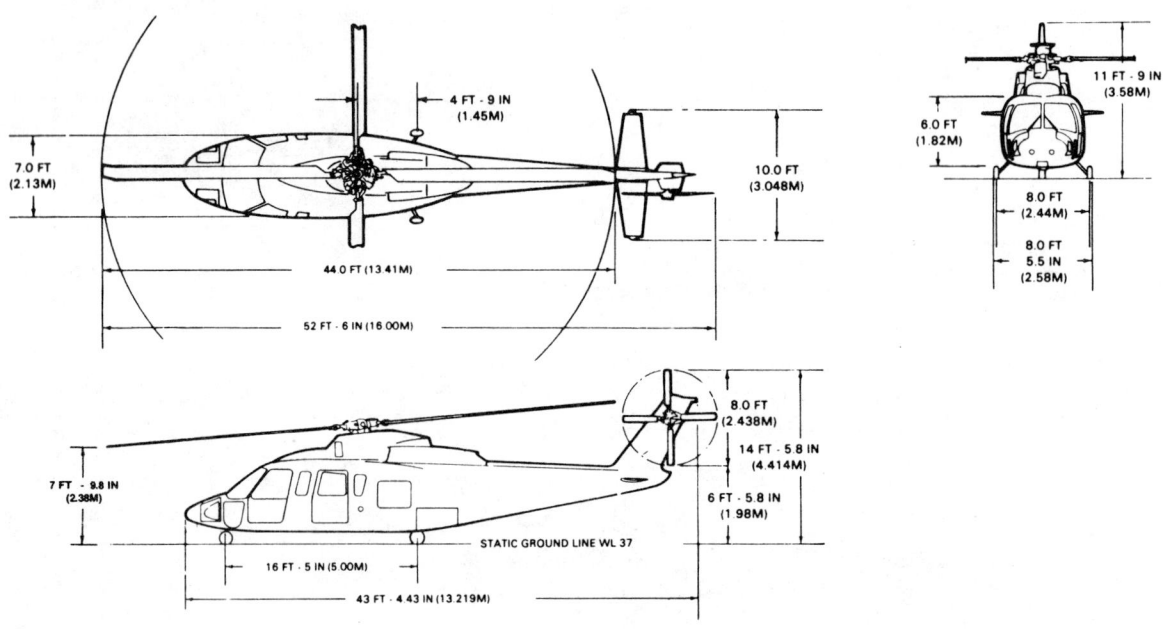

Weights (lb)		Engines	
Empty	5,546	Type	Allison 250-C20S
Maximum takeoff	10,300	Number	2
Fuel capacity	1,876	Maximum T.O. rating	1,300 SHP
		Maximum usable power	1,300 SHP

Rotor Parameters		Main Rotor	Tail Rotor
Radius (ft)	R	22	4
Chord (ft)	c	1.29	0.54
Solidity	σ	0.0747	0.1719
No. of blades	b	4	4
Tip speed (ft/sec)	ΩR	675	674
Equivalent linear twist (deg)	θ_1	−10	−8
Hinge offset ratio	e/R	.038	—
Airfoil		SC 1095	SC 1095
		SC 1095 R8	SC 1095 R8
Collective range (deg)		+4.0 to +22.7	−14.2 to 14.5
Longitudinal cyclic range (deg)		−14.75 to 19.25	—
Lateral cycle range (deg)		−7.1 to +9.7	—
Polar moment of inertia (slug ft²)	J	1,890	4.04

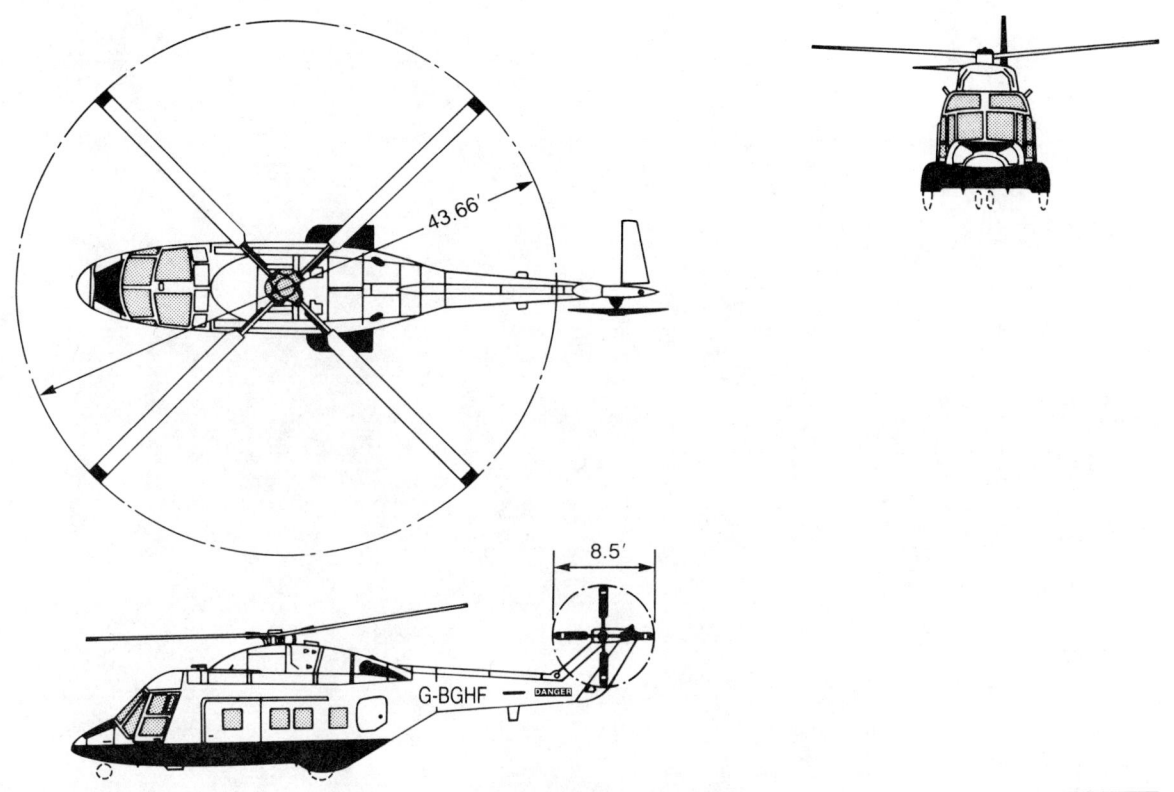

Weights (lb)			Engines	
Empty	7,700		Type	Rolls Royce Gem 41-1 MHS
Maximum takeoff	12,800		Number	2
Fuel capacity	2,346		Maximum T.O. rating	2,000 hp
			Maximum usable power	1,860 hp

Rotor Parameters			Main Rotor	Tail Rotor
Radius (ft)	R		21.83	4.25
Chord (ft)	c		1.293	0.62
Solidity	σ		0.0754	0.182
No. of blades	b		4	4
Tip speed (ft/sec)	ΩR		745	731
Twist (deg)	θ_1		8.32	0
Hinge offset ratio	e/R		.123 (effective)	.056
Airfoil			NPL 9615/9617	RAE 9670
Collective range (deg)			6.25 to 23.25	−9.2 to 31.4
Longitudinal cyclic range (deg)			−8.7 to +14	—
Lateral cyclic range (deg)			−7 to +8	—
Polar moment of inertia (slug ft²)	J		2,175	3.40

APPENDIX C

Atmospheric Charts

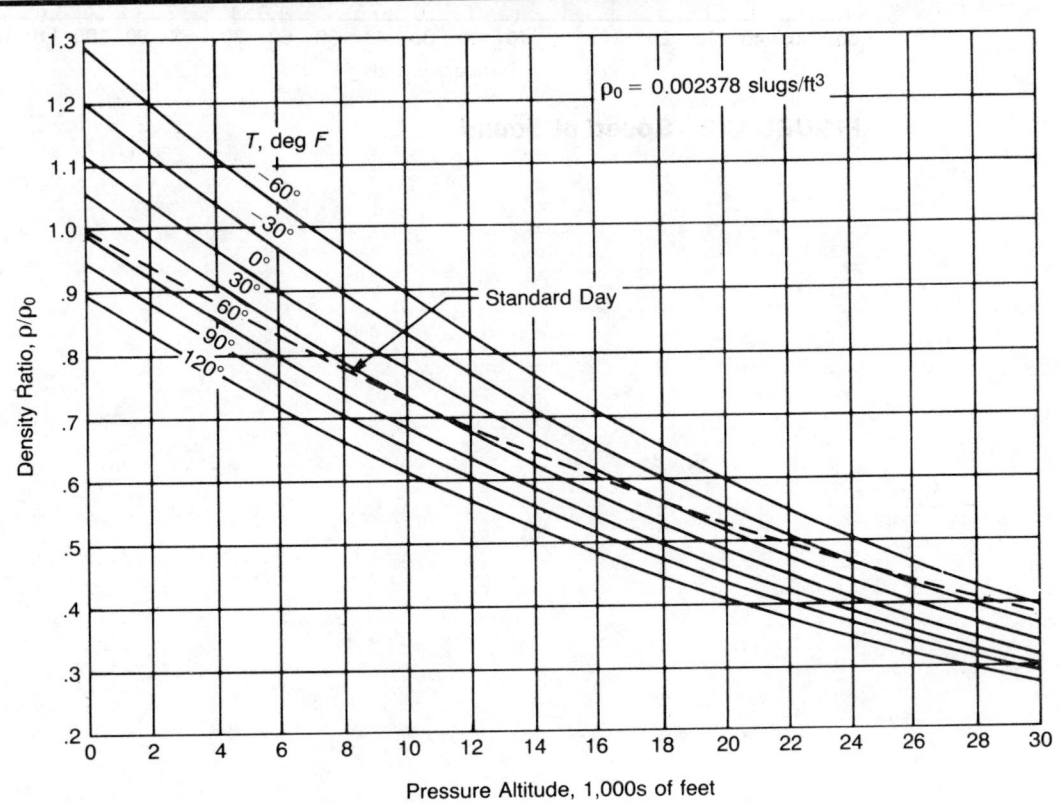

FIGURE C.1   Atmospheric Density Ratio

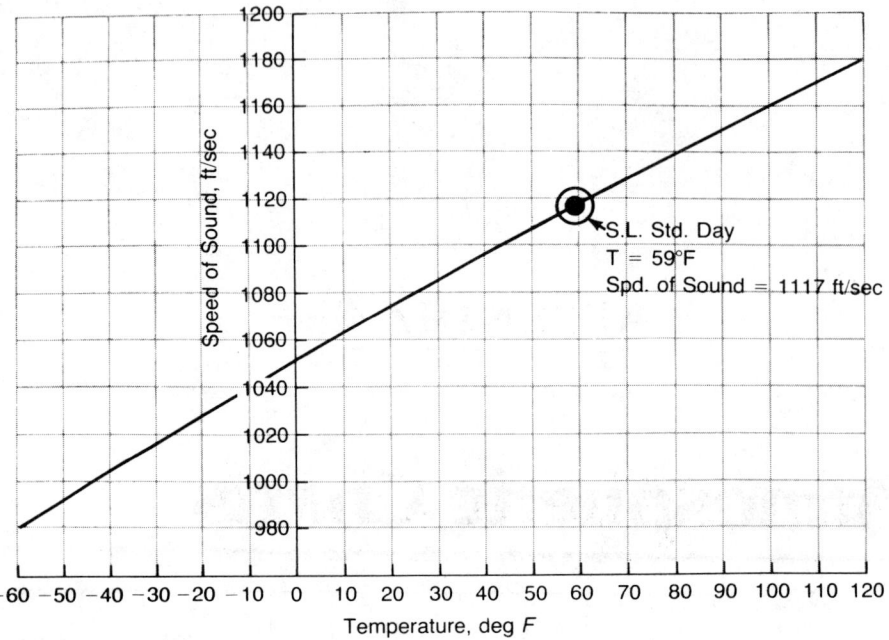

FIGURE C.2   Speed of Sound

APPENDIX D

List of Symbols

Symbol	Definition	Units
a	Slope of lift curve	1/rad, 1/deg
a_0	Coning angle	rad, deg
a_{1_s}	First harmonic coefficient of longitudinal blade flapping with respect to shaft	rad, deg
b	Number of blades	integers
b	Span of wing	ft
b_{1_s}	First harmonic coefficient of lateral blade flapping with respect to shaft	rad, deg
c	Chord of blade	ft
c	Damping	ft lb/rad/sec
c_d	Coefficient of drag; two-dimensional	
c_l	Coefficient of drag; two-dimensional	
c_m	Coefficient of moment; two-dimensional	
e	Hinge offset	ft
f	Equivalent flat plate area	ft^2

Symbol	Definition	Units
n	Load factor	g's
i	Incidence	rad, deg
k	Reduced frequency	
p	Roll rate	rad/sec
q	Pitch rate	rad/sec
q	Dynamic pressure	lb/ft^2
r	Radius of blade element	ft
r	Yaw rate	rad/sec
v	Induced velocity	ft/sec
x	Radius ratio of blade element	
A	Area of disc	ft^2
A_b	Area of blades	ft^2
A_1	First lateral harmonic of blade feathering	rad, deg
B	Tip loss factor	
B_1	First longitudinal coefficient of blade feathering	rad, deg
C_P	Coefficient of power	
C_Q	Coefficient of torque	
C_T	Coefficient of thrust	
D	Diameter of rotor	ft
D	Drag	lb
D.L.	Disc loading	lb/ft^2
F.M.	Figure of Merit	
G.W.	Gross weight	lb
H	Horizontal force on rotor	lb
I_b	Moment of inertia of blade about flapping hinge	slug ft^2
J	Polar moment of inertia	slug ft^2
L	Lift	lb

Symbol	Definition	Units
M	Mach no.	
M	Moment	ft lb
M_b	Static moment of blade about flapping hinge	ft lb
Q	Torque	ft lb
R	Rotor radius	ft
R.N.	Reynolds number	
R/C	Rate of climb	ft/min
T	Thrust	lb
U	Local velocity	ft/sec
V	Remote velocity	ft/sec, knots
Z	Height of rotor above ground	ft
α	Angle of attack	rad, deg
β	Blade flapping angle	rad, deg
β	Sideslip angle	rad, deg
γ	Blade Lock no.	
γ	Climb angle	rad, deg
δ	Thickness of boundary layer	ft, in
δ_3	Pitch-flap coupling angle	rad, deg
ε	Downwash angle	rad, deg
η	Sidewash angle	rad, deg
θ	Blade pitch	rad, deg
θ_1	Blade twist	rad, deg
λ	Inflow ratio with respect to swashplate	
λ'	Inflow ratio with respect to tip path plane	
μ	Tip speed ratio	
ρ	Density of air	slug/ft^3
σ	Solidity of rotor	
ϕ	Inflow angle	rad

Symbol	Definition	Units
χ	Wake skew angle	rad, deg
ψ	Azimuth position of blade	rad, deg
ω	Rotation of wake	rad/sec
ω_N	Natural frequency of blade flapping	rad/sec
Γ	Circulation	lb sec/ft³
Θ	Helicopter pitch angle	rad, deg
Λ	Sweep angle	rad, deg
Φ	Helicopter roll angle	rad, deg
Ψ	Helicopter yaw angle	rad, deg
Ω	Rotational speed of rotor	rad/sec
ΩR	Tip speed of rotor	ft/sec

Subscripts

b	Per blade	
i	Induced	
0	Profile	
F	Fuselage	
H	Horizontal stabilizer	
L	Local	
M	Main rotor	
Q	Torque	
T	Tail rotor	
T	Thrust	
TPP	Tip path plane	
V	Vertical stabilizer	

REFERENCES

CHAPTER 1

1.1   Carpenter, "Lift and Profile-Drag Characteristics of an NACA 0012 Airfoil Section as Derived from Measured Helicopter-Rotor Hovering Performance," NACA TN 4357, 1958.

1.2   Harris & McVeigh, "Uniform Downwash with Rotors Having a Finite Number of Blades," *JAHS*[a] 21-1, 1975.

1.3   Lock, "Tables for Use in an Improved Method of Airscrew Strip Theory Calculations," British R&M 1674.

1.4   Boatwright, "Measurements of Velocity Components in the Wake of a Full-Scale Helicopter Rotor in Hover," USAAMRDL TR 72-33, 1972.

1.5   Landgrebe, "An Analytical and Experimental Investigation of Helicopter Rotor Hover Performance and Wake Geometry Characteristics," USAAMRDL TR 71-24, 1971.

1.6   Clark, "Can Helicopter Rotors Be Designed for Low Noise and High Performance?" AHS 30th Forum, 1974.

1.7   Fradenburgh, "Aerodynamic Design of the Sikorsky S-76 Helicopter," *JAHS* 24-4, 1979.

1.8   Shivers, "Hovering Characteristics of a Rotor Having an Airfoil Section Designed for a Flying-Crane Type of Helicopter," NASA TND-742, 1961.

1.9   Stivers, "Effects of Subsonic Mach Numbers on the Forces and Pressure Distributions on Four NACA 64A-Series Airfoil Sections at Angles of Attack as high as 28°, NACA TN 3162, 1954.

---

[a]*JAHS—Journal of the American Helicopter Society*

1.10   Bellinger, "Experimental Investigation of Effects of Blade Section Camber and Planform Taper on Rotor Performance," USAAMRDL TR 72-4, 1972.

1.11   Durand & Glauert, *Aerodynamic Theory*, Division L, "Airplane Propellers." Berlin: Julius Springer, 1935.

1.12   Wu, Sigman, & Goorjian, "Optimum Performance of Hovering Rotors," NASA TMX 62138, 1972.

1.13   Heyson, "A Note on the Mean Value of Induced Velocity for a Helicopter Rotor," NASA TND-240, 1960.

1.14   Jenney, Olson, & Landgrebe, "A Reassessment of Rotor Hovering Performance Prediction Methods," *JAHS* 13-2, 1968.

1.15   Arcidiacono & Zincone, "Titanium UTTAS Main Rotor Blade," *JAHS* 21-2, 1976.

1.16   Kocurek & Tangler, "A Prescribed Wake Lifting Surface Hover Performance Analysis," *JAHS* 22-1, 1977.

1.17   Tanner & Yaggy, "Experimental Boundary Layer Study on Hovering Rotors," AHS 22nd Forum, 1966.

1.18   Knight & Hefner, "Analysis of Ground Effect on the Lifting Airscrew," NACA TN 835, 1941.

1.19   Hayden, "The Effect of the Ground on Helicopter Hovering Power Required," AHS 32nd Forum, 1976.

1.20   Zbrozek, "Ground Effect on the Lifting Rotor," British R&M 2347, 1950.

1.21   Fradenburgh, "The Helicopter and the Ground Effect Machine," *JAHS* 5-4, 1960.

1.22   Engelhardt & Teleki, "High Speed Flight Tests with the BO 105 Rigid Rotor Helicopter," Proceedings, 7th Annual Soc. Exp. Test Pilots, European Section Symposium, Munich, 1975.

1.23   Yeager, Young, & Mantay, "A Wind-Tunnel Investigation of Parameters Affecting Helicopter Directional Control at Low Speeds in Ground Effect," NASA TND 7694, 1974.

1.24   Johnston & Cook, "AH-56A Vehicle Development," AHS 27th Forum, 1971.

1.25   Landgrebe, Moffitt, & Clark, "Aerodynamic Theory for Advanced Rotorcraft," *JAHS* 22-2, 1977.

1.26   Knight & Hefner, "Static Thrust Analysis of the Lifting Airscrew," NACA TN 626, 1937.

1.27   Carpenter, "Effects of Compressibility on the Performance of Two Full-Scale Helicopter Rotors," NACA TR 1078, 1952.

1.28   Powell, "Hovering Performance of a Helicopter Rotor Using NACA 8-H-12 Airfoil Sections," NACA TN 3237, 1954.

1.29   Shivers & Carpenter, "Experimental Investigation on the Langley Helicopter Test Tower of Compressibility Effects on a Rotor Having NACA 63_2-015 Airfoil Sections," NACA TN 3850, 1956.

1.30   Rabbott, "Static Thrust Measurements of the Aerodynamic Loading on a Helicopter Rotor Blade," NASA TN 3688, 1956.

1.31   Powell & Carpenter, "Low Tip Mach Number Stall Characteristics and High Tip Mach Number Compressibility Effects on a Helicopter Rotor Having an NACA 0009 Tip Airfoil Section," NACA TN 4355, 1958.

1.32   Shivers & Carpenter, "Effects of Compressibility on Rotor Hovering Performance and Synthesized Blade-Section Characteristics Derived from Measured Rotor Performance of Blades Having NACA 0015 Airfoil Tip Sections," NACA TN 4356, 1958.

1.33   Jewel, "Compressibility Effects on the Hovering Performance of a Two-Blade 10-Foot-Diameter Helicopter Rotor Operating at Tip Mach Numbers Up to 0.98," NASA TND-245, 1960.

1.34   Shivers, "High-Tip-Speed Static-Thrust Tests of a Rotor Having NACA $63_{215}A018$ Airfoil Section with and without Vortex Generators Installed," NASA TND-376, 1960.

1.35   Rabbott, "Comparison of Theoretical and Experimental Model Helicopter Rotor Performance in Forward Flight," USATREC TR 61-103, 1961.

1.36   Shivers & Monahan, "Hovering Characteristics of a Rotor Having an Airfoil Section Designed for a Utility Type of Helicopter," NASA TND-1517, 1962.

1.37   Cassarino, "Effect of Root Cutout on Hover Performance," AFFDL-TR-70-70, 1970.

1.38   Benson, Dadone, Gormont, & Kohler, "Influence of Airfoils on Stall Flutter Boundaries of Articulated Helicopter Rotors," *JAHS* 18-1, 1973.

1.39   Landgrebe & Bellinger, "Experimental Investigation of Model Variable-Geometry and Ogee Tip Rotors," NASA CR 2275, 1974.

1.40   Rorke, "Hover Performance Tests of Full-Scale Variable Geometry Rotors," NASA CR 2713, 1976.

1.41   Gray, McMahon, Bird, Palfery, Samant, & Shivananda, "Helicopter Hovering Performance Studies," ARO 11630 1-E, 1976.

CHAPTER 2

2.1   Ferrell & Frederickson, "Flight Evaluation Compliance Test Techniques for Army Hot Day Hover Criteria," USAASTA Project 68-55, 1974.

2.2   Moffitt & Sheehy, "Prediction of Helicopter Rotor Performance in Vertical Climb and Sideward Flight," AHS 33rd Forum, 1977.

2.3   Yeates, "Flight Measurements of the Vibration Experienced by a Tandem Helicopter in Transition, Vortex-Ring State, Landing Approach and Yawed Flight," NACA TN 4409, 1958.

2.4   Jepson, "Some Considerations of the Landing and Take-Off Characteristics of Twin Engine Helicopters," *JAHS* 7-4, 1962.

2.5   Drees & Hendal, "Airflow Patterns in the Neighborhood of Helicopter Rotors," Aircraft Engineering, Vol. 23, April, 1951.

2.6   Azuma & Obata, "Induced Flow Variation of the Helicopter Rotor Operating in the Vortex Ring State," *Jour. of Aircraft*, July-August, 1968.

2.7   Wolkovitch, "Analytical Prediction of Vortex-Ring Boundaries for Helicopters in Steep Descents," *JAHS* 17-3, 1972.

2.8   Peters & Chen, "Momentum Theory, Dynamic Inflow, and the Vortex Ring State," *JAHS* 27-3, 1982.

2.9   Heyson, "A Momentum Analysis of Helicopters and Autogyros in Inclined Descent, with Comments on Operational Restrictions," NASA TND-7917, 1975.

2.10   Castles & Gray, "Empirical Relation between Induced Velocity, Thrust, and Rate of Descent of a Helicopter Rotor as Determined by Wind-Tunnel Tests of Four Model Rotors," NACA TN 2474, 1951.

2.11   Lehman, "Model Studies of Helicopter Tail Rotor Flow Patterns in and Out of Ground Effect," USAAVLABS TR 71-12, 1971.

2.12   Wiesner & Kohler, "Tail Rotor Design Guide," USAAMRDL TR 73-99, 1973.

2.13   Yeager, Young, & Mantay, "A Wind-Tunnel Investigation of Parameters Affecting Helicopter Directional Control at Low Speeds in Ground Effect," NASA TND-7694, 1974.

2.14   Prouty, "Development of the Empennage Configuration of the YAH-64 Advanced Attack Helicopter," USAAVRADCOM TR-82-D-22, 1983.

2.15   Carpenter & Fridovich, "Effect of a Rapid Blade-Pitch Increase on the Thrust and Induced-Velocity Response of a Full-Scale Helicopter Rotor," NACA TN 3044, 1953.

CHAPTER 3

3.1   Wiesner & Kohler, "Tail Rotor Design Guide," USAAMRDL TR 73-99, 1973.

3.2   Simons, Pacifico, & Jones, "The Movement, Structure and Breakdown of Trailing Vortices for a Rotor Blade," CAL/USAAVLABS Symposium, 1966.

3.3   Brotherhood & Stewart, "An Experimental Investigation of the Flow through a Helicopter Rotor in Forward Flight," British R&M 2734, 1949.

3.4   Ruddell, "Advancing Blade Concept (ABC) Development," *JAHS* 22-1, 1977.

3.5   Harris, "Articulated Rotor Blade Flapping Motion at Low Advance Ratio," *JAHS* 17-1, 1972.

3.6   Baskin, Vil'dgube, Vozhdayeu, & Maykapar, "Theory of the Lifting Airscrew," NASA TTF-823, 1976.

3.7   Prouty, "A Second Approximation to the Induced Drag of a Helicopter Rotor in Forward Flight," *JAHS* 21-3, 1976.

3.8   Piziali & DuWalt, "Computed Induced Velocity, Induced Drag, and Angle of Attack Distributions for a Two-Bladed Rotor," AHS 19th Forum, 1963.

3.9   Heyson & Katzoff, "Induced Velocities Near a Lifting Rotor with Non-uniform Disk Loading," NACA TR 1319, 1957.

3.10   Heyson, "Ground Effect for Lifting Rotors in Forward Flight," NASA TND-234, 1960.

3.11   Hoak, "USAF Stability and Control DATCOM," USAF DATCOM, 1960.

3.12   Cheesman & Bennett, "The Effect of the Ground on a Helicopter Rotor in Forward Flight," British R&M 3021, 1957.

3.13   Sheridan & Wiesner, "Aerodynamics of Helicopter Flight Near the Ground," AHS 33rd Forum, 1977.

3.14   Empey & Ormiston, "Tail-Rotor Thrust on a 5.5-Foot Helicopter Model in Ground Effect," AHS 30th Forum, 1974.

3.15   Wheatley, "An Aerodynamic Analysis of the Autogyro Rotor with a Comparison between Calculated and Experimental Results," NACA TR 487, 1934.

3.16   Bailey, "A Simplified Theoretical Method of Determining the Characteristics of a Lifting Rotor in Forward Flight," NACA Rep. 716, 1941.

3.17   Gessow & Myers, *Aerodynamics of the Helicopter*. Ungar Publishing Company, 1952.

3.18   Dooley, "Handling Qualities Considerations for NOE Flight," AHS 32nd Forum, 1976.

3.19   Johnson, "Comparison of Calculated and Measured Helicopter Rotor Lateral Flapping Angles," AVRADCOM TR 80-A-11, NASA TM 81213, 1980.

3.20   Hoerner, "Fluid Dynamic Drag," published by author, 1965.

3.21   LeNard & Boehler, "Inclusion of Tip Relief in the Prediction of Compressibility Effects on Helicopter Rotor Performance," USAAMRDL TR 73-71, 1973.

3.22   Anderson, "Aspect Ratio Influence at High Subsonic Speeds," *Jour. of Aero. Sci.* 23-9, 1956.

3.23   Abbott & Von Doenhoff, *Theory of Wing Sections*. New York: Dover, 1959.

3.24   Cresap, Duhon, Lynn, & Van Wyckhouse, "The 200+ Knot Flight Research Vehicle," AHS 21st Forum, 1965.

3.25   Blackburn, "High Speed Helicopter Research with Jet Thrust Augmentation," AHS 21st Forum, 1965.

3.26   Wilby & Grant, "Transonic Aerodynamics and the Helicopter Rotor (A 76-Z7864 AIAA)," Proceedings of Symposium Transonicum II, Gottingen, Germany. Berlin: Springer-Verlag, 1976.

3.27   McCloud, Biggers, & Stroub, "An Investigation of Full-Scale Helicopter Rotors at High Advance Ratios and Advancing Tip Mach Numbers," NASA TND4632, 1968.

3.28   Norman & Somsel, "Determination of Helicopter Rotor Blade Compressibility Effects—Prediction vs. Flight Test," AHS 23rd Forum, 1967.

3.29   Caradonna & Philippe, "The Flow Over a Helicopter Blade Tip in the Transonic Regime," Proceedings of 2nd European Rotorcraft and Powered Lift Aircraft Forum, Buckeburg, Germany (Paper 21), 1976.

3.30   Huber & Frommlett, "Development of a Bearingless Helicopter Tail Rotor," 6th European Rotorcraft Forum, 1980.

3.31   Bain & Landgrebe, "Investigation of Compound Helicopter Aerodynamic Interference Effects," USAAVLABS TR 67-44, 1967.

3.32   Critzos, Heyson, & Boswinkle, "Aerodynamic Characteristics of NACA 0012

Airfoil Section at Angles of Attack from 0° to 180°," NACA TN 3361, 1955.

3.33   Arcidiacono, "Aerodynamic Characteristics of a Model Helicopter Rotor Operating under Nominally Stalled Conditions in Forward Flight," *JAHS* 9-3, 1964.

3.34   Harris, Tarzanin, & Fisher, "Rotor High Speed Performance, Theory vs. Test," *JAHS* 15-3, 1970.

3.35   Harris, "Preliminary Study of Radial Flow Effects on Rotor Blades," *JAHS* 11-3, 1966.

3.36   Purser & Spearman, "Wind-Tunnel Tests at Low Speed of Swept and Yawed Wings Having Various Plan Forms," NACA TN 2445, 1951.

3.37   Shamroth & Kreskovsky, "A Weak Interaction Study of the Viscous Flow about Oscillating Airfoils," NASA CR 132425, 1974.

3.38   Crimi & Reeves, "A Method for Analyzing Dynamic Stall of Helicopter Rotor Blades," NASA CR 2009, 1972.

3.39   Ericsson & Reding, "Unsteady Airfoil Stall and Stall Flutter," NASA CR 111906, 1971.

3.40   Carta, "Investigation of Airfoil Dynamic Stall and Its Influence on Helicopter Control Loads," USAAMRDL TR 72-51, 1972.

3.41   Gormont, "A Mathematical Model of Unsteady Aerodynamic and Radial Flow for Application to Helicopter Rotors," USAAMRDL TR 72-67, 1973.

3.42   Liiva, Davenport, Gray, & Walton, "Two-Dimensional Tests of Airfoils Oscillating Near Stall," USAAVLABS TR 68-13, 1968.

3.43   Gray & Liiva, "Wind Tunnel Tests of Thin Airfoils Oscillating Near Stall," USAAVLABS TR 68-89, 1969.

3.44   Gross & Harris, "Prediction of Inflight Stalled Airloads from Oscillating Airfoil Data," AHS 25th Forum, 1969.

3.45   Theodorsen, "General Theory of Aerodynamic Instability and the Mechanism of Flutter," NACA TR 496, 1935.

3.46   Tanner, "Charts for Estimating Rotary Wing Performance in Hover and at High Forward Speeds," NASA CR 114, 1964.

3.47   Heyson, "Linearized Theory of Wind-Tunnel Jet-Boundary Corrections and Ground Effect for VTOL-STOL Aircraft," NASA TRR-124, 1962.

3.48   Bellinger, "Analytic Investigation of the Effects of Blade Flexibility, Unsteady Aerodynamics, and Variable Inflow on Helicopter Rotor Stall Characteristics," NASA CR 1769, 1971.

3.49   McCloud & McCullough, "Wind-Tunnel Tests of a Full-Scale Helicopter Rotor with Symmetrical and with Cambered Blade Sections at Advance Ratios from 0.3 to 0.4," NACA TN 4367, 1958.

3.50   Sweet, Jenkins, & Winston, "Wind-Tunnel Measurements on a Lifting Rotor at High Thrust Coefficients and High Tip-Speed Ratios," NASA TND-2462, 1964.

3.51   Jenkins, "Wind-Tunnel Investigation of a Lifting Rotor Operating at Tip-Speed Ratios from 0.65 to 1.45," NASA TND-2628, 1965.

3.52   Livingston, "Wind Tunnel Tests of a Full-Scale Rotor at High Speeds," USAAVLABS TR 65-42, 1965.

3.53   Niebanck, "Model Rotor Test Data for Verification of Blade Response and Rotor Performance Calculations," USAAMRDL TR 74-29, 1974.

3.54   Jepson, Moffitt, Hilzinger, & Bissell, "Analysis and Correlation of Test Data from an Advanced Technology Rotor System," NASA CR 3714, 1983.

CHAPTER 4

4.1   McKee & Naseth, "Experimental Investigation of the Drag of Flat Plates and Cylinders in the Slipstream of a Hovering Rotor," NACA TN 4239, 1958.

4.2   Hoerner, "Fluid Dynamic Drag," published by author, 1965.

4.3   McCroskey, Spalart, Laub, Maisel, & Maskew, "Airloads on Bluff Bodies, with Application to the Rotor-Induced Downloads on Tilt-Rotor Aircraft," 9th European Rotorcraft Forum, 1983.

4.4   Cassarino, "Effect of Rotor Blade Root Cutout on Vertical Drag," AAVLABS TR 70-59, 1970.

4.5   Fradenburgh, "Aerodynamic Factors Influencing Overall Hover Performance," AGARD CP 1111, 1972.

4.6   Fradenburgh, "Aerodynamic Design of the Sikorsky S-76 Helicopter," *JAHS* 24-4, 1979.

4.7   Lynn, Robinson, Batra, & Duhon, "Tail Rotor Design," Part I: "Aerodynamics," *JAHS* 15-4, 1970.

4.8   Morris, "A Wind Tunnel Investigation of Fin Force for Several Tail-Rotor and Fin Configurations," NASA LWP-995, 1971.

4.9   Harris et al., "High Performance Tandem Helicopter Study," USATREC TR 61-42, 1961.

4.10   Harned, "Development of the OH-6 for Maximum Performance and Efficiency," AHS 20th Forum, 1964.

4.11   Foster, "Tilt-Pylon and Wind Tunnel Tests," Bell R&D Conference, 1961.

4.12   Linville, "An Experimental Investigation of High-Speed Rotorcraft Drag," USAAMRDL TR 71-46, 1971.

4.13   Gillespie, "An Investigation of the Flow Field and Drag of Helicopter Fuselage Configurations," AHS 29th Forum, 1973.

4.14   Sweet & Jenkins, "Wind-Tunnel Investigation of the Drag and Static Stability Characteristics of Four Helicopter Fuselage Models," NASA TND 1363, 1962.

4.15   Biggers, McCloud, & Patterakis, "Wind-Tunnel Tests of Two Full-Scale Helicopter Fuselages," NASA TND 1548, 1962.

4.16   Keys & Wiesner, "Guidelines for Reducing Helicopter Parasite Drag," *JAHS* 20-1, 1975.

4.17   Skinner, "Army Preliminary Evaluation, YOH-58A Helicopter with a Flat-Plate Canopy," USAAEFA Proj. 75-20, 1975.

4.18   Moser, "Full Scale Wind Tunnel Investigation of Helicopter Drag," *JAHS* 6-1, 1961.

4.19   Michel, "Research and Design Progress toward High Performance Rotary Wing Aircraft," *JAHS* 5-2, 1960.

4.20   Roesch & Dequin, "Experimental Research on Helicopter Fuselage and Rotor Hub Wake Turbulence," AHS 39th Forum, 1983.

4.21   Keys & Wiesner, "Guidelines for Reducing Helicopter Parasite Drag," *JAHS* 20-1, 1975.

4.22   Sheehy & Clark, "A Method for Predicting Helicopter Hub Drag," USAAMRDL TR 75-48, 1976.

4.23   Pruyn & Miller, "Studies of Rotorcraft Aerodynamic Problems Aimed at Reducing Parasite Drag, Rotor-Airframe Interference Effects and Improving Airframe Static Stability," WADD TR 61-124, 1961.

4.24   Rosenstein & Stanzione, "Computer Aided Helicopter Design," AHS 37th Forum, 1981.

4.25   McHugh & Harris, "Have We Overlooked the Full Potential of the Conventional Rotor?" *JAHS* 21-3, 1976.

CHAPTER 5

5.1   Gustafson & Crim, "Flight Measurements and Analysis of Helicopter Normal Load Factors in Maneuvers," NACA TN 2990, 1953.

5.2   Wells & Wood, "Maneuverability—Theory and Application," *JAHS* 18-1, 1973.

5.3   Pegg, "An Investigation of the Height-Velocity Diagram Showing Effects of Density Altitude and Gross Weight," NASA TND-4536, 1968.

5.4   Arcidiacono, "Aerodynamic Characteristics of a Model Helicopter Rotor Operating under Nominally Stalled Conditions in Forward Flight," *JAHS* 9-3, 1964.

5.5   Harris, Tarzanin, & Fisher, "Rotor High Speed Performance, Theory vs. Test," *JAHS* 15-3, 1970.

5.6   Benson, Dadone, Gormont, & Kohler, "Influence of Airfoils on Stall Flutter Boundaries of Articulated Helicopter Rotors," *JAHS* 18-1, 1973.

5.7   Wells & Woods, "Maneuverability—Theory and Application," *JAHS* 18-1, 1973.

5.8   Gabel & Tarzanin, "Blade Torsional Tuning to Manage Large Amplitude Control Loads," *J. of Aircraft* 11-8, 1974.

5.9   McHugh & Harris, "Have We Overlooked the Full Potential of the Conventional Rotor?" *JAHS* 21-3, 1976.

5.10   McHugh, "What Are the Lift and Propulsive Force Limits at High Speed for the Conventional Rotor?" AHS 34th Forum, 1978.

5.11   Condon, Bailes, & Connor, "Height-Velocity Test, AH-1G Helicopter," USAASTA Project 69-13, 1971.

5.12   McIntyre, "A Simplified Study of High Speed Autorotation Entry Characteristics," AHS 26th Forum, 1970.

5.13   Ferrell, Frederickson, Shapley, & Kyker, "A Flight Research Investigation of Autorotational Performance and Height-Velocity Testing of a Single Main Rotor Single Engine Helicopter," USAAEFA Project 68-25, 1976.

5.14   Jepson, "Some Considerations of the Landing and Take-off Characteristics of Twin Engine Helicopters," *JAHS* 7-4, 1962.

5.15   Wood, "High Energy Rotor System," AHS 32nd Forum, 1976.

5.16   Fradenburgh, "A Simple Autorotative Flare Index," *JAHS* 29-3, 1984.

5.17   Schmitz & Vause, "Near-Optimal Takeoff Policy for Heavily Loaded Helicopters Exiting from Confined Areas," *J. of Aircraft*, 13-5, 1976.

5.18   Jenney & Decker, "Maneuverability Criteria for Weapons Helicopters," AHS 20th Forum, 1964.

5.19   McCutcheon, "S-67 Flight Test Program," AHS 28th Forum, 1972.

CHAPTER 6

6.1   Abbott & Von Doenhoff, *Theory of Wing Sections*. Dover, 1959.

6.2   McCullough & Gault, "Examples of Three Types of Stall," NACA TN 2502, 1951.

6.3   Loftin & Bursnall, "The Effects of Variations in Reynolds Number between 3.0×10^6 and 25.0×10^6 upon the Aerodynamic Characteristics of a Number of NACA 6-Series Airfoil Sections," NACA TR 964, 1950.

6.4   Hicks, Mendoza, & Bandettini, "Effects of Forward Contour Modification on the Aerodynamic Characteristics of the NACA 64_1-212 Airfoil Section," NASA TMX-3293, 1975.

6.5   Wortmann, "The Quest for High-Lift," AIAA Paper 74-1018, 1974.

6.6   Liebeck, "A Class of Airfoils Designed for High Lift in Incompressible Flow," *Jour. of Aircraft* 10-10, 1973.

6.7   Jacobs, Pinkerton, & Greenberg, "Tests of Related Forward-Camber Airfoils in the Variable-Density Wind Tunnel," NACA TR 610, 1937.

6.8   Maki & Hunton, "An Investigation at Subsonic Speeds of Several Modifications to the Leading-Edge Region of the NACA 64A410 Airfoil Section Designed to Increase Maximum Lift," NACA TN 3871, 1956.

6.9   Prouty, "A State-of-the-Art Survey of Two-Dimensional Airfoil Data," *JAHS* 20-4, 1975.

6.10   Lizak, "Two-Dimensional Wind Tunnel Tests of an H-34 Main Rotor Airfoil Section," USATRECOM TR 60-53, 1960.

6.11   Wootton, "The Effect of Compressibility on the Maximum Lift Coefficient of Aerofoils at Subsonic Airspeeds," *Jour. of Royal Aero. Soc.* Vol. 71, July, 1967.

6.12   Racisz, "Effects of Independent Variations of Mach Number and Reynolds Numbers on the Maximum Lift Coefficients of Four NACA 6-Series Airfoil Sections," NACA TN 2824, 1952.

6.13   Stivers, "Effects of Subsonic Mach Number on the Forces and Pressure

Distributions on Four NACA 64A-Series Airfoil Sections at Angles of Attack as High as 28°," NACA TN 3162, 1954.

6.14   Hoerner & Borst, "Fluid-Dynamic Lift," published by Mrs. Hoerner, 1975.

6.15   Critzos, Heyson, & Boswinkle, "Aerodynamic Characteristics of NACA 0012 Airfoil Section at Angles of Attack From 0° to 180°," NACA TN 3361, 1955.

6.16   Benson, Dadone, Gormont, & Kohler, "Influence of Airfoils on Stall Flutter Boundaries of Articulated Helicopter Rotors," *JAHS* 18-1, 1973.

6.17   Graham, Nitzberg, & Olson, "A Systematic Investigation of Pressure Distributions at High Speeds over Five Representative NACA Low-Drag and Conventional Airfoil Sections," NACA TR 832, 1945.

6.18   Sato, "On Peaky Airfoil Sections," NASA TTF 749, 1973.

6.19   Dadone, "Design and Analytical Study of a Rotor Airfoil," NASA CR 2988, 1978.

6.20   Blackwell & Hinson, "The Aerodynamic Design of an Advanced Rotor Airfoil," NASA CR 2961, 1977.

6.21   Davenport & Front, "Airfoil Sections for Helicopter Rotors—A Reconsideration," AHS 22nd Forum, 1966.

6.22   Gregory & Wilby, "NPL 9615 and NACA 0012—A Comparison of Aerodynamic Data," ARC CP 1261 (British), 1973.

6.23   Liiva, Davenport, Gray, & Walton, "Two-Dimensional Tests of Airfoils Oscillating Near Stall," USAAVLABS TR 68-13, 1968.

6.24   Jepson, "Two Dimensional Test of Four Airfoil Configurations with an Aspect Ratio of 7.5 and a 16-inch Chord Up to a Mach Number of 1.1," SER-50977 (Navy Contract N60921-73-C-0037), 1977.

6.25   Thibert & Gallot, "Advanced Research on Helicopter Blade Airfoils," 6th European Rotorcraft & Powered Lift Aircraft Forum, 1980.

6.26   Harris, "Two-Dimensional Aerodynamic Characteristics of the NACA 0012 Airfoil in the Langley 8-Foot Transonic Pressure Tunnel," NASA TM 81927, 1981.

6.27   McCroskey, "The Phenomenon of Dynamic Stall," NASA TM 81264, 1981.

6.28   Ham & Garelick, "Dynamic Stall Considerations in Helicopter Rotors," *JAHS* 13-2, 1968.

6.29   Johnson & Ham, "On the Mechanism of Dynamic Stall," *JAHS* 17-4, 1972.

6.30   Carta, "Effect of Unsteady Pressure Gradient Reduction on Dynamic Stall Delay," *Jour. of Aircraft* 8-10, 1971.

6.31   Ericsson & Reding, "'Spilled' Leading-Edge Vortex Effects on Dynamic Stall Characteristics—Engr. Note," *Jour. of Aircraft* 13-4, 1976.

6.32   Conner, "A Flight and Wind Tunnel Investigation of the Effect of Angle-of-Attack Rate on Maximum Lift Coefficient," NASA CR 321, 1965.

6.33   Gray & Liiva, "Wind Tunnel Tests of Thin Airfoils Oscillating Near Stall," USAAVLABS TR 68-89, 1969.

6.34   Carta, "Investigation of Airfoil Dynamic Stall and Its Influence on Helicopter Control Loads," USAAMRDL TR 72-51, 1972.

6.35   McCroskey, Carr, McAllister, Pucci, Lambert, & Indergrand, "Dynamic Stall on Advanced Airfoil Sections," *JAHS* 26-3, 1981.

6.36   Ericsson & Reding, "Unsteady Airfoil Stall," NASA CR 66787, 1969.

6.37   Shamroth & Kreskovsky, "A Weak Interaction Study of the Viscous Flow about Oscillating Airfoils," NASA CR 132425, 1974.

6.38   Crimi & Reeves, "A Method for Analyzing Dynamic Stall of Helicopter Rotor Blades," NASA CR 2009, 1972.

6.39   Gangwani, "Synthesized Airfoil Data Method for Prediction of Dynamic Stall and Unsteady Airloads," AHS 39th Forum, 1983.

6.40   Ericsson & Reding, "Unsteady Airfoil Stall and Stall Flutter," NASA CR 111906, 1971.

6.41   Ericsson & Reding, "Dynamic Stall Analysis in Light of Recent Numerical and Experimental Results," *Jour. of Aircraft* 13-4, 1976.

6.42   Bielawa, "Synthesized Unsteady Airfoil Data with Applications to Stall Flutter Calculations," AHS 31st Forum, 1975.

6.43   Gormont, "A Mathematical Model of Unsteady Aerodynamic and Radial Flow for Application to Helicopter Rotors," USAAMRDL TR 72-67, 1973.

6.44   Gross & Harris, "Prediction of Inflight Stalled Airloads from Oscillating Airfoil Data," AHS 25th Forum, 1969.

6.45   Fukushima & Dadone, "Comparison of Dynamic Stall Phenomena for Pitching and Vertical Translation Motions," NASA CR 2793, 1976.

6.46   Sipe & Gorenberg, "Effect of Mach Number, Reynolds Number, and Thickness Ratio on the Aerodynamic Characteristics of NACA 63A-Series Airfoil Sections," USATRECOM TR 65-28, 1965.

6.47   Van Dyke, "High-Speed Subsonic Characteristics of 16 NACA 6-Series Airfoil Sections," NACA TN 2670, 1952.

6.48   Wilson & Horton, "Aerodynamic Characteristics at High and Low Subsonic Mach Numbers of Four NACA 6-Series Airfoil Sections at Angles of Attack from −2° to 31°, NACA TM 876, 1953.

6.49   Wiesner & Kohler, "Tail Rotor Design Guide," USAAMRDL TR 73-99, 1973.

6.50   Gothert, "Airfoil Measurements in the DVL High-Speed Wind Tunnel," NACA TM 1240.

6.51   Tanner & Yaggy, "Experimental Boundary Layer Study on Hovering Rotors," AHS 22nd Forum, 1966.

6.52   Wortmann & Drees, "Design of Airfoils for Rotors," CAL/AVLABS Symposium, 1969.

6.53   Kemp, "An Analytical Study for the Design of Advanced Rotor Airfoils," NASA CR 112297, 1973.

6.54   Powell, "Hovering Performance of a Helicopter Rotor Using NACA 8-H-12 Airfoil Sections," NACA TN 3237, 1954.

6.55   Shivers & Monahan, "Hovering Characteristics of a Rotor Having an Airfoil Section Designed for a Utility Type of Helicopter," NASA TND-1517, 1962.

6.56   Philippe & Sagner, "Aerodynamic Forces Computation and Measurement on an Oscillating Aerofoil Profile with and without Stall," AGARD CP 111, 1972.

6.57   Dadone, "Helicopter Design DATCOM," Vol. I: "Airfoils," USAAMRDL CR 76-2, 1976.

6.58   Robinson, "Increasing Tail Rotor Thrust and Comments on Other Yaw Control Devices," *JAHS* 15-4, 1970.

6.59   Niebanck, "Model Rotor Test Data for Verification of Blade Response and Rotor Performance Calculations," USAAMRDL TR-74-29, 1974.

6.60   Reichert & Wagner, "Some Aspects of the Design of Rotor-Airfoil Shapes," AGARD CP 111, 1972.

6.61   Paul, "A Self Excited Rotor Blade Oscillation at High Subsonic Mach Numbers," AHS 24th Forum, 1968.

6.62   Prouty, "Aerodynamics," column in *Rotor & Wing International*, Vol. 18, No. 9 (August).

6.63   Carta, "An Analysis of the Stall Flutter Instability of Helicopter Rotor Blades," *JAHS* 12-4, 1967.

6.64   Carta, Commerford, & Carlson, "Determination of Airfoil and Rotor Blade Dynamic Stall Response," *JAHS* 18-2, 1973.

6.65   Liiva & Davenport, "Dynamic Stall of Airfoil Sections for High Speed Rotors," *JAHS* 14-2, 1969.

6.66   Carpenter, "Lift and Profile-Drag Characteristics of an NACA 0012 Airfoil Section as Derived from Measured Helicopter-Rotor Hovering Performance," NACA TN 4357, 1958.

6.67   Bellinger, "Experimental Investigation of Effects of Blade Section Camber and Planform Taper on Rotor Performance," USAAMRDL TR 72-4, 1972.

6.68   Rabbott, "Comparison of Theoretical and Experimental Model Helicopter Rotor Performance in Forward Flight," USATREC TR 61-103, 1961.

6.69   Livingston, "Wind Tunnel Tests of a Full-Scale Rotor at High Speeds," USAAVLABS TR 65-42, 1965.

6.70   Noonan & Bingham, "Aerodynamic Characteristics of Three Helicopter Rotor Airfoil Sections at Reynolds Numbers from Model Scale to Full Scale at Mach Numbers from 0.35 to 0.90," AVRADCOM TR 81-B-5, 1980.

6.71   Noonan & Bingham, "Two-Dimensional Aerodynamic Characteristics of Several Rotorcraft Airfoils at Mach Numbers from 0.35 to 0.90," NASA TMX-73920, 1977.

6.72   Shivers & Carpenter, "Effects of Compressibility on Rotor Hovering Performance and Synthesized Blade-Section Characteristics Derived from Measured Rotor Performance of Blades Having NACA 0015 Airfoil Tip Sections," NACA TN 4356, 1958.

6.73   Shivers & Carpenter, "Experimental Investigation on the Langley Helicopter Test Tower of Compressibility Effects on a Rotor Having NACA 63_2-015 Airfoil Sections," NACA TN 3850, 1956.

6.74   Wilby, Gregory, Quincey, "Aerodynamic Characteristics of NPL 9626 and NPL 9627, Further Aerofoils Designed for Helicopter Rotor Use," ARC CP 1262 (British), 1973.

6.75   Dadone, "Two-Dimensional Wind Tunnel Test of an Oscillating Rotor Airfoil," NASA CR 2914, 1977.

6.76   Noonan, "Experimental Investigation of a 10-Percent-Thick Helicopter Rotor Airfoil Section Designed with a Viscous Transonic Analysis Code," AVRADCOM TR 80-B-3, 1981.

6.77   Bingham, Noonan, & Sewall, "Two-Dimensional Aerodynamic Characteristics of an Airfoil Designed for Rotorcraft Application," AVRADCOM TR 81-B-6, 1981.

6.78   Bingham & Noonan, "Two-Dimensional Aerodynamic Characteristics of Three Rotorcraft Airfoils at Mach Numbers from 0.35 to 0.90," AVRADCOM TR 82-B-2, 1982.

6.79   Hicks & McCroskey, "An Experimental Evaluation of a Helicopter Rotor Section Designed by Numerical Optimization," NASA TM 78622, 1980.

6.80   Tanner, "Charts for Estimating Rotary Wing Performance in Hover and at High Forward Speeds," NASA CR 114, 1964.

CHAPTER 7

7.1   Crim, "Gust Experience of a Helicopter and an Airplane in Formation Flight," NACA TN 3354, 1954.

7.2   Amer, "Theory of Helicopter Damping in Pitch or Roll and a Comparison with Flight Measurements," NACA TN 2136, 1950.

CHAPTER 8

8.1   Hoak, "USAF Stability and Control Datcom," USAF DATCOM, 1960.

8.2   Harris, Kocurek, McLarty, Trept, "Helicopter Performance Methodology at Bell Helicopter Textron," AHS 35th Forum, 1979.

8.3   Hoerner & Borst, "Fluid-Dynamic Lift," published by Mrs. Hoerner, 1975.

8.4   Engelhardt & Teleki, "High-Speed Flight Tests with the BO 105 Rigid Rotor Helicopter," *Proceedings*, 7th Annual Soc. Exp. Test Pilots, European Section Symposium, Munich, 1975.

8.5   Roesch & Vuillet, "New Designs for Improved Aerodynamic Stability on Recent Aerospatiale Helicopters," AHS 37th Proceedings, 1981.

8.6   Sheridan, "Interactional Aerodynamics of the Single Rotor Helicopter Configuration," USARTL TR 78-23A, 1978.

8.7   Logan, Prouty, & Clark, "Wind Tunnel Tests of Large and Small Scale Rotor Hubs and Pylons," USAAVRADCOM TR-80-D-21, 1981.

8.8   Blake & Alanski, "Stability and Control of the YUH-61A," *JAHS* 22-1, 1977.

8.9   Heyson & Katzoff, "Induced Velocities Near a Lifting Rotor with Non-uniform Disc Loading," NACA TR 1319, 1957.

8.10   Cooper, "YUH-60A Stability and Control," *JAHS* 23-3, 1978.

8.11   Amer, Prouty, Walton, & Engle, "Handling Qualities of Army/Hughes YAH-64 Advanced Attack Helicopter," AHS 34th Forum, 1978.

8.12   Bain & Landgrebe, "Investigation of Compound Helicopter Aerodynamic Interference Effects," USAAVLABS TR 67-44, 1967.

8.13   Jepson, Moffitt, Hilzinger, & Bissell, "Analysis and Correlation of Test Data from an Advanced Technology Rotor System," NASA CR 3714, 1983.

8.14   Pope, *Basic Wing and Airfoil Theory*. New York: McGraw-Hill, 1951.

8.15   Abbott & Von Doenhoff, *Theory of Wing Sections*. New York: Dover, 1959.

8.16   Dadone, "Helicopter Design DATCOM" Vol. I. "Airfoils," USAAMRDL CR 76-2, 1976.

8.17   Von Mises, *Theory of Flight*. New York: McGraw-Hill, 1944.

8.18   Biggers, McCloud, & Patterakis, "Wind-Tunnel Tests of Two Full-Scale Helicopter Fuselages," NASA TND 1548, 1962.

8.19   Jenkins, Winston, & Sweet, "A Wind Tunnel Investigation of the Longitudinal Aerodynamic Characteristics of Two Full-Scale Helicopter Fuselage Model with Appendages," NASA TND-1364, 1962.

CHAPTER 9

9.1   Hansen, "Toward a Better Understanding of Helicopter Stability Derivatives," *JAHS* 29-1, 1984.

9.2   Harrison, "An Integrated Approach to Effective Analytical Support of Helicopter Design and Development," 6th European Rotorcraft Forum, 1980.

9.3   Amer, "Charts for Estimation of Longitudinal-Stability Derivatives in Forward Flight," NASA TN 2309, 1951.

9.4   Hohenemser, "Stability in Hovering of the Helicopter with Central Rotor Location," AMC Translation F-TS-687-RE, 1946.

9.5   Salmirs & Tapscott, "The Effects of Various Combinations of Damping and Control Power on Helicopter Handling Qualities During Both Instrument and Visual Flight," NASA TND-58, 1959.

9.6   Anonymous, "Helicopter Flying and Ground Handling Qualities: General Requirements for," MIL-H-8501A, 1962.

9.7   Tapscott & Sommer, "A Flight Study with a Large Helicopter Showing Trends of Lateral and Longitudinal Control Response with Size," NASA TND-3600, 1966.

9.8   Pegg & Connor, "Effects of Control-Response Characteristics on the Capability of a Helicopter for Use as a Gun Platform," NASA TND-464, 1960.

9.9   Corliss & Carico, "A Flight Investigation of Roll-Control Sensitivity, Damping, and Cross-Coupling in a Low-Altitude Lateral Maneuvering Task," NASA TM 84376, USAAVRADCOM TR-83-A-16, 1983.

9.10   Pegg & Connor, "Effects of Control-Response Characteristics on the Capability of a Helicopter for Use as a Gun Platform," NASA TND-464, 1960.

9.11   Sinclair & Kereliuk, "Evaluation of the Effects of Lateral and Longitudinal Aperiodic Modes on Helicopter Instrument Flight Handling Qualities," NAE-AN-15 (Canada), 1983.

9.12   Padfield & DuVal, "Applications of System Identification Methods to the Prediction of Helicopter Stability, Control and Handling Qualities," NASA Conference Publication 2219, 1982.

9.13   Amer, "Method for Studying Helicopter Longitudinal Maneuver Stability," NACA TN 3022, 1953.

9.14   Blake & Alansky, "Stability and Control of the YUH-61A," *JAHS* 22-1, 1977.

9.15   Seckel, *Stability and Control of Airplanes and Helicopters*. New York: Academic Press, 1964.

9.16   Faulkner & Kloster, "Lateral-Directional Stability: Theoretical Analysis and Flight Test Experience," 9th European Forum, 1983.

CHAPTER 10

10.1   Vega, "Advance Technology Impacts on Rotorcraft Weight," AHS 40th Forum, 1984.

10.2   Prouty, "A Second Approximation to the Induced Drag of a Helicopter Rotor in Forward Flight," *JAHS* 21-3, 1976.

10.3   Fradenburgh, "Aerodynamic Factors Influencing Overall Hover Performance," AGARD CP 111, 1972.

10.4   Arcidiacono & Zincone, "Titanium UTTAS Main Rotor Blade," *JAHS* 21-2, 1976.

10.5   Fradenburgh, "Aerodynamic Design of the Sikorsky S-76 Helicopter," *JAHS* 24-4, 1979.

10.6   Caradonna & Philippe, "The Flow over a Helicopter Blade Tip in the Transonic Regime," *Proceedings*, 2nd European Rotorcraft and Powered Lift Aircraft Forum, Buckeburg, Germany (Paper 21), 1976.

10.7   Balch, "Experimental Study of Main Rotor/Tail Rotor/Airframe Interaction in Hover," AHS 39th Forum, 1983.

10.8   Prouty, "Development of the Empennage Configuration of the YAH-64 Advanced Attack Helicopter," USAAVRADCOM TR-82-D-22, 1983.

10.9   Leverton, "Reduction of Helicopter Noise by Use of a Jet Tail Rotor," 6th European Rotorcraft Forum," 1980.

10.10   Wiesner & Kohler, "Tail Rotor Design Guide," USAAMRDL TR 73-99, 1973.

10.11   Amer, Prouty, Walton, & Engle, "Handling Qualities or Army/Hughes YAH-64 Advanced Attack Helicopter," AHS 34th Forum, 1978.

10.12   Castles & Gray, "Empirical Relation between Induced Velocity, Thrust, and Rate of Descent of a Helicopter Rotor as Determined by Wind-Tunnel Tests of Four Model Rotors," NACA TN 2474, 1951.

10.13   Gerdes, Jackson, & Beno, "Directional Control Development Experiences Associated with the UH-60A Utility Helicopter," USAAVRADCOM TR-82-D-26, 1983.

10.14   Prouty & Amer, "The YAH-64 Empennage and Tail Rotor—A Technical History," AHS 38th Forum, 1982.

10.15   Cooper, "YUH-60A Stability and Control," *JAHS* 23-3, 1978.

10.16   Shinn, "Impact of Emerging Technology on the Weight of Future Aircraft," AHS 40th Forum, 1984.

10.17   Schwartzberg, Smith, Means, Law, & Chappell, "Single Rotor Helicopter Design and Performance Estimation Programs," USAAMRDL, SR10, 77-1, 1977.

10.18   Swan & Schmidt, "Integrating New Technology into Weight Methodology," AHS 40th Forum, 1984.

Index